OXIDE SURFACES

SURFACTANT SCIENCE SERIES

1. Nonionic Surfactants, *edited by Martin J. Schick* (see also Volumes 19, 23, and 60)
2. Solvent Properties of Surfactant Solutions, *edited by Kozo Shinoda* (see Volume 55)
3. Surfactant Biodegradation, *R. D. Swisher* (see Volume 18)
4. Cationic Surfactants, *edited by Eric Jungermann* (see also Volumes 34, 37, and 53)
5. Detergency: Theory and Test Methods (in three parts), *edited by W. G. Cutler and R. C. Davis* (see also Volume 20)
6. Emulsions and Emulsion Technology (in three parts), *edited by Kenneth J. Lissant*
7. Anionic Surfactants (in two parts), *edited by Warner M. Linfield* (see Volume 56)
8. Anionic Surfactants: Chemical Analysis, *edited by John Cross*
9. Stabilization of Colloidal Dispersions by Polymer Adsorption, *Tatsuo Sato and Richard Ruch*
10. Anionic Surfactants: Biochemistry, Toxicology, Dermatology, *edited by Christian Gloxhuber* (see Volume 43)
11. Anionic Surfactants: Physical Chemistry of Surfactant Action, *edited by E. H. Lucassen-Reynders*
12. Amphoteric Surfactants, *edited by B. R. Bluestein and Clifford L. Hilton* (see Volume 59)
13. Demulsification: Industrial Applications, *Kenneth J. Lissant*
14. Surfactants in Textile Processing, *Arved Datyner*
15. Electrical Phenomena at Interfaces: Fundamentals, Measurements, and Applications, *edited by Ayao Kitahara and Akira Watanabe*
16. Surfactants in Cosmetics, *edited by Martin M. Rieger* (see Volume 68)
17. Interfacial Phenomena: Equilibrium and Dynamic Effects, *Clarence A. Miller and P. Neogi*
18. Surfactant Biodegradation: Second Edition, Revised and Expanded, *R. D. Swisher*
19. Nonionic Surfactants: Chemical Analysis, *edited by John Cross*
20. Detergency: Theory and Technology, *edited by W. Gale Cutler and Erik Kissa*
21. Interfacial Phenomena in Apolar Media, *edited by Hans-Friedrich Eicke and Geoffrey D. Parfitt*
22. Surfactant Solutions: New Methods of Investigation, *edited by Raoul Zana*
23. Nonionic Surfactants: Physical Chemistry, *edited by Martin J. Schick*
24. Microemulsion Systems, *edited by Henri L. Rosano and Marc Clausse*
25. Biosurfactants and Biotechnology, *edited by Naim Kosaric, W. L. Cairns, and Neil C. C. Gray*
26. Surfactants in Emerging Technologies, *edited by Milton J. Rosen*
27. Reagents in Mineral Technology, *edited by P. Somasundaran and Brij M. Moudgil*
28. Surfactants in Chemical/Process Engineering, *edited by Darsh T. Wasan, Martin E. Ginn, and Dinesh O. Shah*
29. Thin Liquid Films, *edited by I. B. Ivanov*
30. Microemulsions and Related Systems: Formulation, Solvency, and Physical Properties, *edited by Maurice Bourrel and Robert S. Schechter*
31. Crystallization and Polymorphism of Fats and Fatty Acids, *edited by Nissim Garti and Kiyotaka Sato*
32. Interfacial Phenomena in Coal Technology, *edited by Gregory D. Botsaris and Yuli M. Glazman*
33. Surfactant-Based Separation Processes, *edited by John F. Scamehorn and Jeffrey H. Harwell*
34. Cationic Surfactants: Organic Chemistry, *edited by James M. Richmond*
35. Alkylene Oxides and Their Polymers, *F. E. Bailey, Jr., and Joseph V. Koleske*
36. Interfacial Phenomena in Petroleum Recovery, *edited by Norman R. Morrow*
37. Cationic Surfactants: Physical Chemistry, *edited by Donn N. Rubingh and Paul M. Holland*
38. Kinetics and Catalysis in Microheterogeneous Systems, *edited by M. Grätzel and K. Kalyanasundaram*
39. Interfacial Phenomena in Biological Systems, *edited by Max Bender*
40. Analysis of Surfactants, *Thomas M. Schmitt* (see Volume 96)

41. Light Scattering by Liquid Surfaces and Complementary Techniques, *edited by Dominique Langevin*
42. Polymeric Surfactants, *Irja Piirma*
43. Anionic Surfactants: Biochemistry, Toxicology, Dermatology. Second Edition, Revised and Expanded, *edited by Christian Gloxhuber and Klaus Künstler*
44. Organized Solutions: Surfactants in Science and Technology, *edited by Stig E. Friberg and Björn Lindman*
45. Defoaming: Theory and Industrial Applications, *edited by P. R. Garrett*
46. Mixed Surfactant Systems, *edited by Keizo Ogino and Masahiko Abe*
47. Coagulation and Flocculation: Theory and Applications, *edited by Bohuslav Dobiáš*
48. Biosurfactants: Production • Properties • Applications, *edited by Naim Kosaric*
49. Wettability, *edited by John C. Berg*
50. Fluorinated Surfactants: Synthesis • Properties • Applications, *Erik Kissa*
51. Surface and Colloid Chemistry in Advanced Ceramics Processing, *edited by Robert J. Pugh and Lennart Bergström*
52. Technological Applications of Dispersions, *edited by Robert B. McKay*
53. Cationic Surfactants: Analytical and Biological Evaluation, *edited by John Cross and Edward J. Singer*
54. Surfactants in Agrochemicals, *Tharwat F. Tadros*
55. Solubilization in Surfactant Aggregates, *edited by Sherril D. Christian and John F. Scamehorn*
56. Anionic Surfactants: Organic Chemistry, *edited by Helmut W. Stache*
57. Foams: Theory, Measurements, and Applications, *edited by Robert K. Prud'homme and Saad A. Khan*
58. The Preparation of Dispersions in Liquids, *H. N. Stein*
59. Amphoteric Surfactants: Second Edition, *edited by Eric G. Lomax*
60. Nonionic Surfactants: Polyoxyalkylene Block Copolymers, *edited by Vaughn M. Nace*
61. Emulsions and Emulsion Stability, *edited by Johan Sjöblom*
62. Vesicles, *edited by Morton Rosoff*
63. Applied Surface Thermodynamics, *edited by A. W. Neumann and Jan K. Spelt*
64. Surfactants in Solution, *edited by Arun K. Chattopadhyay and K. L. Mittal*
65. Detergents in the Environment, *edited by Milan Johann Schwuger*
66. Industrial Applications of Microemulsions, *edited by Conxita Solans and Hironobu Kunieda*
67. Liquid Detergents, *edited by Kuo-Yann Lai*
68. Surfactants in Cosmetics: Second Edition, Revised and Expanded, *edited by Martin M. Rieger and Linda D. Rhein*
69. Enzymes in Detergency, *edited by Jan H. van Ee, Onno Misset, and Erik J. Baas*
70. Structure–Performance Relationships in Surfactants, *edited by Kunio Esumi and Minoru Ueno*
71. Powdered Detergents, *edited by Michael S. Showell*
72. Nonionic Surfactants: Organic Chemistry, *edited by Nico M. van Os*
73. Anionic Surfactants: Analytical Chemistry, Second Edition, Revised and Expanded, *edited by John Cross*
74. Novel Surfactants: Preparation, Applications, and Biodegradability, *edited by Krister Holmberg*
75. Biopolymers at Interfaces, *edited by Martin Malmsten*
76. Electrical Phenomena at Interfaces: Fundamentals, Measurements, and Applications, Second Edition, Revised and Expanded, *edited by Hiroyuki Ohshima and Kunio Furusawa*
77. Polymer-Surfactant Systems, *edited by Jan C. T. Kwak*
78. Surfaces of Nanoparticles and Porous Materials, *edited by James A. Schwarz and Cristian I. Contescu*
79. Surface Chemistry and Electrochemistry of Membranes, *edited by Torben Smith Sørensen*
80. Interfacial Phenomena in Chromatography, *edited by Emile Pefferkorn*

81. Solid–Liquid Dispersions, *Bohuslav Dobiáš, Xueping Qiu, and Wolfgang von Rybinski*
82. Handbook of Detergents, *editor in chief: Uri Zoller*
    Part A: Properties, *edited by Guy Broze*
83. Modern Characterization Methods of Surfactant Systems, *edited by Bernard P. Binks*
84. Dispersions: Characterization, Testing, and Measurement, *Erik Kissa*
85. Interfacial Forces and Fields: Theory and Applications, *edited by Jyh-Ping Hsu*
86. Silicone Surfactants, *edited by Randal M. Hill*
87. Surface Characterization Methods: Principles, Techniques, and Applications, *edited by Andrew J. Milling*
88. Interfacial Dynamics, *edited by Nikola Kallay*
89. Computational Methods in Surface and Colloid Science, *edited by Małgorzata Borówko*
90. Adsorption on Silica Surfaces, *edited by Eugène Papirer*
91. Nonionic Surfactants: Alkyl Polyglucosides, *edited by Dieter Balzer and Harald Lüders*
92. Fine Particles: Synthesis, Characterization, and Mechanisms of Growth, *edited by Tadao Sugimoto*
93. Thermal Behavior of Dispersed Systems, *edited by Nissim Garti*
94. Surface Characteristics of Fibers and Textiles, *edited by Christopher M. Pastore and Paul Kiekens*
95. Liquid Interfaces in Chemical, Biological, and Pharmaceutical Applications, *edited by Alexander G. Volkov*
96. Analysis of Surfactants: Second Edition, Revised and Expanded, *Thomas M. Schmitt*
97. Fluorinated Surfactants and Repellents: Second Edition, Revised and Expanded, *Erik Kissa*
98. Detergency of Specialty Surfactants, *edited by Floyd E. Friedli*
99. Physical Chemistry of Polyelectrolytes, *edited by Tsetska Radeva*
100. Reactions and Synthesis in Surfactant Systems, *edited by John Texter*
101. Protein-Based Surfactants: Synthesis, Physicochemical Properties, and Applications, *edited by Ifendu A. Nnanna and Jiding Xia*
102. Chemical Properties of Material Surfaces, *Marek Kosmulski*
103. Oxide Surfaces, *edited by James A. Wingrave*

ADDITIONAL VOLUMES IN PREPARATION

Interfacial Electrokinetics and Electrophoresis, *edited by Ángel V. Delgado*

Polymers in Particulate Systems, *edited by Vincent A. Hackley, Ponisseril Somasundaran, and Jennifer A. Lewis*

OXIDE SURFACES

edited by
James A. Wingrave
University of Delaware
Newark, Delaware

CRC Press
Taylor & Francis Group
Boca Raton London New York

CRC Press is an imprint of the
Taylor & Francis Group, an **informa** business

CRC Press
Taylor & Francis Group
6000 Broken Sound Parkway NW, Suite 300
Boca Raton, FL 33487-2742

First issued in paperback 2020

ISBN-13: 978-0-367-45515-6 (pbk)
ISBN-13: 978-0-8247-0000-3 (hbk)

Visit the Taylor & Francis Web site at
http://www.taylorandfrancis.com

and the CRC Press Web site at
http://www.crcpress.com

Preface

The scope of any discussion on oxide surfaces is virtually limitless when one considers that most of the surfaces in the universe are covered with oxides of some sort. Possibly as a result of the vastness of this subject, few if any books have been published that attempt to concentrate into a single volume all the knowledge associated with oxide surfaces. As a result of the magnitude of the subject, this book was undertaken with three goals in mind: (1) to collect a wide-ranging group of topics relating to oxide surfaces, (2) to meld the chapters so that they augment and amplify each other, and (3) to write rigorous, tutorial discussions rather than literature reviews.

In attempting to meet the first goal it was necessary to determine just which topics to include. As an organizational criterion it seemed reasonable to divide the topics of the book into two sections according to interfacial type: (1) solid–liquid interfaces and (2) solid–gas (or –vapor) interfaces. (Although the subject of solid–solid oxide interfaces has many important applications, such as electronics, adhesion, heat transfer, etc., these topics were excluded.) Within these two broad categories, the topics covered by each chapter concern specific areas of oxide surface technology.

The methods for quantifying adsorption and desorption of polymeric and nonpolymeric solutes at the solution–oxide interface are discussed in Chapters 1–4. Both kinetic and equilibrium adsorption behavior are considered as well as the activity coefficient effect on the adsorption process.

Oxide surfaces are influenced to a large degree by the molecular composition of the bulk solid oxide. The structure and properties of oxide surfaces as they are influenced by the composition of the contiguous phases of solution and solid oxide are discussed in Chapters 5 and 6.

Although the aqueous solution–metal oxide interface has been exhaustively studied, literature is beginning to emerge on the nonaqueous–metal oxide interface. Many of the current research findings are examined in Chapter 7.

The solution–oxide interface has been found to be a particularly suitable location for polymerization, with the result being novel polymer–solution interfaces.

Techniques presented in Chapter 8 describe very controlled coating methods for solution–oxide interfaces with molecularly thin, homogeneous polymer films.

As with all colloid behavior, surface properties become more important than bulk properties as specific surface areas increase. When oxide solids are formulated into high-surface-area membranes, unique surface properties result. In Chapter 9, the synthesis and novel properties of oxide membranes are examined.

The second section of the book addresses phenomena associated with the gas–oxide (and vapor–oxide) interface. The question of adsorption at the solid–gas interface has been explored using innumerable theoretical methods. In Chapters 10 and 11, two of the more successful methods based on statistical mechanics and thermodynamics are examined. These treatments represent the current state of gas–solid adsorption theory in the field of gas–solid adsorption, a field that has been actively evolving for many decades.

In summary, the chapters in this book provide a readable yet rigorous treatment of current topics related to the fluid–metal oxide interface. The chapter topics were chosen and organized in an attempt to provide a valuable research guide to the fascinating subject of metal oxide interfaces.

James A. Wingrave

Contents

Contributors

David Andelman School of Physics and Astronomy, Tel Aviv University, Tel Aviv, Israel

Nianjiong Bei Department of Chemical Engineering, University of California, Los Angeles, Los Angeles, California

Patrick V. Brady Geochemistry Department, Sandia National Laboratories, Albuquerque, New Mexico

José Ignacio Calvo Departamento de Termodinámica y Física Aplicada, Universidad de Valladolid, Valladolid, Spain

William H. Casey Department of Land, Air, and Water Resources, and Department of Geology, University of California, Davis, Davis, California

Yoram Cohen Department of Chemical Engineering, University of California, Los Angeles, Los Angeles, California

Martinus A. Cohen Stuart Physical Chemistry and Colloid Science, Wageningen University, Wageningen, The Netherlands

Arie de Keizer Physical Chemistry and Colloid Science, Wageningen University, Wageningen, The Netherlands

Geir Martin Førland Faculty of Engineering, Bergen College, Bergen, Norway

Antonio Hernández Departamento de Termodinámica y Física Aplicada, Universidad de Valladolid, Valladolid, Spain

Harald Høiland Faculty of Engineering, University of Bergen, Bergen, Norway

Jeng-Dung Jou Strategic Technology, Printronix, Inc., Irvine, California

James L. Krumhansl Geochemistry Department, Sandia National Laboratories, Albuquerque, New Mexico.

Roland R. Netz Department of Theory, Max-Planck-Institute for Colloids and Interfaces, Potsdam, Germany

Van Nguyen Department of Chemical Engineering, University of California, Los Angeles, Los Angeles, California

Jan Nordin Department of Land, Air, and Water Resources, University of California, Davis, Davis, California

Laura Palacio Departamento de Termodinámica y Física Aplicada, Universidad de Valladolid, Valladolid, Spain

Brian L. Phillips Chemical Engineering and Material Science, University of California, Davis, Davis, California

Pedro Prádanos Departamento de Termodinámica y Física Aplicada, Universidad de Valladolid, Valladolid, Spain

József Tóth Research Institute of Applied Chemistry, University of Miskolc, Miskolc-Egyetemváros, Hungary

Georgi N. Vayssilov Department of Chemistry, University of Sofia, Sofia, Bulgaria

James A. Wingrave Department of Chemistry and Biochemistry, University of Delaware, Newark, Delaware

Wayne Yoshida Department of Chemical Engineering, University of California, Los Angeles, Los Angeles, California

OXIDE SURFACES

1
Equilibrium Adsorption at the Solution–Metal Oxide Interface

JAMES A. WINGRAVE University of Delaware, Newark, Delaware

I. ADSORPTION AT THE SOLUTION–METAL OXIDE INTERFACE

A. Introduction

This chapter is devoted largely to the derivation of adsorption equations for the adsorption of solute at a solution–metal oxide (S–MO) interface. The derivations are based on rigorous Gibbs–Lewis thermodynamics [1], employing both mass action and mass balance equations. The thermodynamic mass action/mass balance (TMAB) approach is relatively new, appearing first in the literature in the mid-1990s [2–4]. Many of the subtle but important distinctions between TMAB and older adsorption models will be examined in Section VII.

The bulk of this chapter will be devoted to the derivation of adsorption equations using the TMAB model for different types of solute molecules, i.e., protons, monovalent cations, divalent anions, weak acid anions, etc. For each different solute type, the derivation will repeat many steps. Thus, while the reader may find repetition among different adsorption equation derivations, each derivation will be essentially complete and not require further reference.

The positive adsorption of a solute at the S–MO interface results whenever the chemical potential change (at constant temperature and pressure) is negative for a solute moved from a bulk solution and bonded to or adsorbed at a S–MO interface. The chemical potential imbalance diminishes as solute adsorption and concentration changes at the S–MO interface. The complete disappearance of a chemical potential difference (at constant temperature and pressure) for the solute in the bulk liquid and the S–MO interface serves as the definition of adsorption equilibrium for the solute. This singular definition of adsorption has produced a large number of adsorption models which strive to translate the chemical potential model of equilibrium adsorption into measurable variables and constants. Many of these adsorption equations will be examined in detail later in this chapter.

In the following, a chemical thermodynamic description of the S–MO interface will be carefully developed from rigorous Gibbs–Lewis thermodynamics. From this Gibbs-Lewis model of the S–MO interface, solute adsorption equations will be derived with the TMAB model and examined with respect to: (1) solution pH, (2) thermodynamic mass action parameters, (3) activity coefficients, (4) solute types (cations, anions, weak acid anions, polymers, etc.), and (5) various other adsorption equations.

B. Description of the Solvent–Metal Oxide Interface

Metal oxides are matrices of oxygen and metal atoms combined at various ratios into a three-dimensional solid. The three-dimensional structure of metal oxides is discussed in detail in Chapters 5, 6 and 9. For adsorption at the S–MO interface the bulk structure of a metal oxide even a few atomic diameters distant from the S–MO interface plays only a minor role in the adsorption process. Therefore, metal oxide bulk structure will only be discussed when it affects adsorption processes directly.

The surfaces of metal oxides are composed of metal atoms to which lone oxygen atoms are bonded. These lone oxygen atoms may or may not be bonded to other ions or molecules and represent what will henceforth be referred to as "surface sites." The surface site density of most metal oxide surfaces ranges from about 0.1 to 1.0 nm^2 per surface site, such that a square meter of surface would contain, nominally, 1×10^{18} individual surface sites.

When metal oxide surfaces are contacted with water, the surface sites behave amphoterically by exchanging protons with the water phase. Even "pure" water will contain a small amount of adsorbable solutes such as protons and hydroxide ions. Hence, from an adsorption perspective, there is no such thing as "pure" water. Evidence of the amphoteric behavior of metal oxide surface sites is seen in the pH dependence of zeta potentials for metal oxide particles ranging from positive to negative values: positive when an excess of protons adsorb at low water pH and negative when the same surface-site complexes are deficient in protons at high water pH. From the changing protonation state of surface sites in contact with a water phase one would anticipate that the adsorption process will involve the binding energy between the proton and surface site as well as the electrostatic influence dependent upon the amount of positively charged protons which are adsorbed.

When metal oxide surfaces are contacted with aqueous solutions, the surface-site complexes form among surface sites and all solutes in the solution, including protons and hydroxide ions. The solute–surface site complexes form just as complexes form among solute molecules and ions in a bulk solution. And as with solution complexes, the amphoteric nature of the solute–surface site complexes imparts a significant pH dependence to the relative abundance of each different type of solute–surface site complex.

Implicit in all adsorption equations derived in this chapter is the assumption that the total number of all surface sites, complexed plus uncomplexed, remains constant throughout any adsorption process. If significant dissolution of the metal oxide solid occurs, the validity of this assumption can be violated. However, when new surface sites are created in the same number as surface sites are removed by dissolution, the total surface-site number remains contant. Therefore, a significant amount of metal oxide dissolution may occur without violating the constant surface-site requirement for all the adsorption equations. Discussions on the mechanism and conditions under which metal oxide dissolution occurs is discussed in detail in Chapter 5.

Adsorption at the nonaqueous–metal oxide interface will not be examined in this chapter. However, based on discussions in this chapter and Chapter 7, adsorption equations for the nonwater solvent–metal oxide interfaces can be derived by using similar methods.

II. PROTON ADSORPTION EQUATIONS

A. Chemical Thermodynamics of Proton Adsorption

The exchange of protons between water and an amphoteric metal oxide interface will produce three different types of surface sites: (1) unprotonated anionic sites each with a unit negative valence, (2) monoprotonated nonionic sites each with a zero valence, and (3) diprotonated cationic sites each with a unit positive valence. Given sufficient time to reach equilibrium, the proton exchange will produce a S–MO interface in which the number of each type of site and the number of protons in solution becomes constant. To specify the stochiometry of this equilibrium proton distribution, the total number of surface sites will be designated as n_s^T with the number of each specific type of surface site designated as: (1) n_{SO}, for unprotonated anionic sites, (2) n_{SOH}, for monoprotonated nonionic sites, and (3) n_{SOH_2}, for diprotonated cationic sites.

The surface-site mole fraction for each protonated surface-site type can therefore be defined relative to n_s^T as

$$\theta_{SO} \equiv \frac{n_{SO}}{n_s^T} \tag{1}$$

$$\theta_{SOH} \equiv \frac{n_{SOH}}{n_s^T} \tag{2}$$

$$\theta_{SOH_2} \equiv \frac{n_{SOH_2}}{n_s^T} \tag{3}$$

The molarity of protons, χ_H, and hydroxide ions, χ_{OH}, in aqueous solution is defined in terms of the volume of the solution phase, V_w, and the number of moles of protons, n_H^w, and hydroxide ions, n_{OH}^w, in the solution phase as

$$\chi_H \equiv \frac{n_H^w}{V_w} \tag{4}$$

$$\chi_{OH} \equiv \frac{n_{OH}^w}{V_w} \tag{5}$$

with pH and pOH defined as the negative common logarithm of χ_H and χ_{OH}:

$$\text{pH} = -\log \chi_H \tag{6}$$

$$\text{pOH} \equiv -\log \chi_{OH} \tag{7}$$

and pH and pOH related through the ion product of water as

$$\text{pH} + \text{pOH} = 14 \tag{8}$$

Equation (8) makes it unnecessary to specify both χ_H and χ_{OH} in adsorption problems. Therefore, adsorption equations will be derived in terms of χ_H knowing that χ_{OH} can always be determined from Eqs (6) to (8) if needed.

For the sake of clarity, surface-site mole fractions will be designated with the lower case Greek symbol, θ, while molar solution concentrations will be designated with the lower case Greek symbol, χ. The exchange of protons between a metal oxide interface and a contiguous solution can be visualized as shown in Fig. 1.

Although the precise chemistry of proton and solute exchange may differ slightly from that shown in Fig. 1 for different S–MO interfaces, the amphoteric

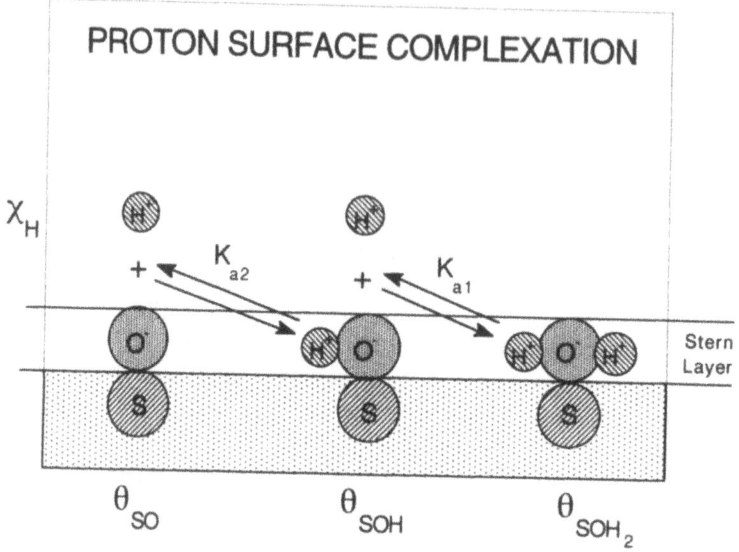

FIG. 1 Interfacial complexation reactions of protons at the solvent–metal oxide interface. Molarity of protons in aqueous solution, χ_H. Surface-site mole fraction of: unprotonated surface sites, θ_{SO}; monoprotonated surface sites, θ_{SOH}; diprotonated surface sites, θ_{SOH_2}.

and ion-exchange properties for the S–MO interface are experimentally established. The TMAB adsorption equations are derived by applying Gibbs–Lewis thermodynamics to the amphoteric and ion-exchange behavior, not to any specific solute exchange chemistry at the S–MO interface. The chemistry shown in Fig. 1 is an heuristic model for the amphoteric behavior of a S–MO interface.

Using the chemical information above, it will now be possible to derive explicit mass balance and mass action equations which describe proton adsorption quantitatively. The mole fractions in Eqs (1)–(3) and shown in Fig. 1 represent all the surface sites in an amphoteric S–MO system. Therefore, the sum or mass balance of the surface-site mole fractions must equal unity:

$$1 = \theta_{SO} + \theta_{SOH} + \theta_{SOH_2} \quad \text{(Surface site mass balance)} \tag{9}$$

Mass action equations are derived for an equilibrium chemical reaction by equating the chemical potential of the reactants and products since, by definition, the chemical potential of reactants and products of a chemical reaction must be equal at equilibrium. For the protonation reactions shown in Fig. 1, chemical potentials, μ_H, μ_{SO}, μ_{SOH}, and μ_{SOH_2}, for each of the four chemicals, H^+, SO^-, SOH, and ${SOH_2}^+$, respectively, can be written following Lewis *et al.* [1], in terms of their respective standard state chemical potentials (μ_H^*, μ_{SO}^*, μ_{SOH}^*, and μ_{SOH_2}), solution molarities, χ_H, S–MO surface mole fractions (θ_{SO}, θ_{SOH}, and θ_{SOH_2}), and activity coefficients (f_H,f_{SO},f_{SOH}, and f_{SOH_2}) to give

$$\mu_H = \mu_H^* + kT \ln \chi_H + kT \ln f_H \tag{10}$$

$$\mu_{SO} = \mu_{SO}^* + kT \ln \theta_{SO} + kT \ln f_{SO} \tag{11}$$

$$\mu_{SOH} = \mu^*_{SOH} + kT \ln \theta_{SOH} + kT \ln f_{SOH} \tag{12}$$

$$\mu_{SOH_2} = \mu^*_{SOH_2} + kT \ln \theta_{SOH_2} + kT \ln f_{SOH_2} \tag{13}$$

For the protonation reaction shown in Fig. 1, which is designated by the symbol, K_{a1}, the chemical reaction is

$$SOH_2^+ \leftrightarrow SOH + H^+ \quad \text{(Proton adsorption reaction)} \tag{14}$$

The chemical potentials of all the reactants must equal the chemical potentials of all the products in Eq. (14) at equilibrium, so

$$\mu_{SOH_2} = \mu_{SOH} + \mu_H \tag{15}$$

Substitution of Eqs (10), (12), and (13) into Eq. (15) gives, upon rearrangement,

$$-(\mu^*_{SOH} + \mu^*_H - \mu^*_{SOH_2}) = +kT \ln\left(\frac{\theta_{SOH}\chi_H}{\theta_{SOH_2}}\right) + kT \ln\left(\frac{f_{SOH}f_H}{f_{SOH_2}}\right) \tag{16}$$

The standard state chemical potential terms on the left-hand side (l.h.s.) of Eq. (16) vary only with temperature and solvent type so it is conventional to define the mole fraction mass action constant or adsorption constant, K_{a1}, as

$$kT \ln(K_{a1}) \equiv -(\mu^*_{SOH} + \mu^*_H - \mu^*_{SOH_2}) \tag{17}$$

Equations (16) and (17) can be combined to give a mass action equation for the reaction represented in Fig. 1 by the adsorption constant, K_{a1}, as

$$K_{a1} = \left(\frac{\theta_{SOH}\chi_H}{\theta_{SOH_2}}\right)\left(\frac{f_{SOH}f_H}{f_{SOH_2}}\right) \equiv \left(\frac{\theta_{SOH}\chi_H}{\theta_{SOH_2}}\right)Z^{\epsilon_{K_{a1}}} \quad \left(\begin{array}{l}\text{Proton mass} \\ \text{action equation I}\end{array}\right) \tag{18}$$

The other proton adsorption equation can be written:

$$SOH \leftrightarrow SO^- + H^+ \quad \text{(Proton adsorption equation)} \tag{19}$$

By an analogous procedure, using Eqs (10)–(12), the mass action equation for the reaction in Eq. (19), which is represented by the adsorption constant, K_{a2}, in Fig. 1, can be derived:

$$K_{a2} = \left(\frac{\theta_{SO}\chi_H}{\theta_{SOH}}\right)\left(\frac{f_{SO}f_H}{f_{SOH}}\right) \equiv \left(\frac{\theta_{SO}\chi_H}{\theta_{SOH}}\right)Z^{\epsilon_{K_{a2}}} \quad \left(\begin{array}{l}\text{Proton mass} \\ \text{action equation II}\end{array}\right) \tag{20}$$

In Eqs (18) and (20) the activity coefficient effects have been grouped into the variables, $Z^{\epsilon_{K_{a1}}}$ and $Z^{\epsilon_{K_{a2}}}$. Evaluation and in depth discussion of activity coefficients will be pursued in Chapter 4 and in Section II.E.

Also of interest are the units of the mass action constants and activity coefficients. For the latter, Lewis [1, p. 253] stated, "As the activity coefficient is the quantity that is invariably tabulated, it would seem to be preferable to make it . . . dimensionless . . . ." Given that the activity coefficient is dimensionless, the mass action constants in Eqs (18) and (20) must necessarily have units of molar solution concentration. However, Eq. (17) would argue that mass action constants must be dimensionless. The answer to this dilemma begins with the fact that mass action constants are actually defined in terms of mole fractions (strictly, in terms of relative activities, see [1]).

If we consider Eq. (20) as an example, the concentration quotient must first be written in terms of mole fractions or $\theta_{SO}/\theta_{SOH} \cdot \chi_H/1M$, since surface-site mole fractions are already mole fractions. When this rigorous concentration quotient is substituted into Eq. (20) and rearranged, we see that the mass action constant, K_{a2}, in Eq. (20) is actually a product of a *unitless* thermodynamic mass action constant, $K_{a2(thermo)}$, and a unit molar activity, 1 M, or $K_{a2} \equiv 1\,M \cdot K_{a2(thermo)}$. The unitless thermodynamic mass action constants, $K_{a1(thermo)}$, as defined in Eq. (17) are defined strictly in terms of mole fractions, not concentrations, and thus must be unitless. Practically then, mass action constants, K_{a1} and K_{a2}, are considered here and throughout the literature to have the same units as the concentration quotients while rigorously the mass action constants in Eqs (18) and (20), respectively, will be recognized as the products, $K_{a1} \equiv 1\,M \cdot K_{a1(thermo)}$ and $K_{a2} \equiv 1\,M \cdot K_{a2(thermo)}$.

Equations (18) and (20) can be rearranged to give expressions for the surface mole fractions, θ_{SOH} and θ_{SOH_2}, as

$$\theta_{SOH} = \left(\frac{\theta_{SO}\chi_H}{K_{a2}}\right)Z^{\epsilon_{K_{a2}}} \tag{21}$$

$$\theta_{SOH_2} = \left(\frac{\theta_{SO}\chi_H^2}{K_{a1}K_{a2}}\right)Z^{\epsilon_{K_{a12}}} \tag{22}$$

B. Proton Adsorption Equation

A sufficient number of equations have now been derived to make possible the derivation of explicit adsorption equations for proton adsorption. This can best be seen by collecting the relevant equations into a table in order to examine the number of equations and unknowns which have been introduced so far (see Table 1).

An equal number of equations and unknown variables allows the elimination of all the unknown variables in the proton adsorption equations to be derived. Substitution of Eqs (21) and (22) into Eq. (9) results in the elimination of the surface mole fraction variables, θ_{SOH} and θ_{SOH_2}:

TABLE 1 Tabulation of Equations and Unknowns Used for Derivation of the Proton Adsorption Equations

Unknowns		Equation	
No.	Variable	No.	Description
2	χ_{SOH}, χ_{SOH_2}	(18) or (21)	$K_{a1} = \left(\frac{\theta_{SOH}\chi_H}{\theta_{SOH_2}}\right)Z^{\epsilon_{K_{a1}}}$
1	χ_{SO}	(20) or (22)	$K_{a2} = \left(\frac{\theta_{SO}\chi_H}{\theta_{SOH}}\right)Z^{\epsilon_{K_{a2}}}$
0		(9)	$1 = \theta_{SO} + \theta_{SOH} + \theta_{SOH_2}$
3 unknowns			3 equations

$$1 = \theta_{\mathrm{SO}}\left(1 + \frac{\chi_{\mathrm{H}} Z^{\epsilon_{K_{a2}}}}{K_{a2}} + \frac{\chi_{\mathrm{H}}^2 Z^{\epsilon_{K_{a12}}}}{K_{a1} K_{a2}}\right) \tag{23}$$

Equation (23) can be solved for θ_{SO} to give an adsorption equation in terms of the molar solution concentration of protons, χ_{H}, and the proton adsorption mass action constants, K_{a1} and K_{a2}:

$$\theta_{\mathrm{SO}} = \frac{K_{a1} K_{a2}}{K_{a1} K_{a2} + \chi_{\mathrm{H}} K_{a1} Z^{\epsilon_{K_{a2}}} + \chi_{\mathrm{H}}^2 Z^{\epsilon_{K_{a12}}}} \quad \left(\begin{matrix}\text{Explicit proton} \\ \text{adsorption equation}\end{matrix}\right) \tag{24}$$

In a similar fashion, adsorption equations for the mole fraction of the other surface sites can be derived from Eqs (9), (18) and (20) to give

$$\theta_{\mathrm{SOH}} = \frac{\chi_{\mathrm{H}} K_{a1} Z^{\epsilon_{K_{a2}}}}{K_{a1} K_{a2} + \chi_{\mathrm{H}} K_{a1} Z^{\epsilon_{K_{a2}}} + \chi_{\mathrm{H}}^2 Z^{\epsilon_{K_{a12}}}} \quad \left(\begin{matrix}\text{Explicit proton} \\ \text{adsorption equation}\end{matrix}\right) \tag{25}$$

$$\theta_{\mathrm{SOH_2}} = \frac{\chi_{\mathrm{H}}^2 Z^{\epsilon_{K_{a12}}}}{K_{a1} K_{a2} + \chi_{\mathrm{H}} K_{a1} Z^{\epsilon_{K_{a2}}} + \chi_{\mathrm{H}}^2 Z^{\epsilon_{K_{a12}}}} \quad \left(\begin{matrix}\text{Explicit proton} \\ \text{adsorption equation}\end{matrix}\right) \tag{26}$$

Equations (24)–(26) are explicit equations for proton adsorption at the S–MO interface derived from rigorous Gibbs–Lewis thermodynamics. From a knowledge of K_{a1} and K_{a2} the mole fraction of amphoteric surface sites can be calculated as a function of solution pH with the exception of the activity coefficient quotient, Z. As an approximation, Z can be set to unity so that $Z^{\epsilon_{K_{a2}}} = Z^{\epsilon_{K_{a12}}} = 1$, which eliminates all nonideal adsorption effects from the adsorption equations. The resulting ideal proton adsorption equations are

$$\theta_{\mathrm{SO}}^{\mathrm{ID}} = \frac{K_{a1} K_{a2}}{K_{a1} K_{a2} + \chi_{\mathrm{H}} K_{a1} + \chi_{\mathrm{H}}^2} \quad \left(\begin{matrix}\text{Ideal proton} \\ \text{adsorption equation}\end{matrix}\right) \tag{27}$$

$$\theta_{\mathrm{SOH}}^{\mathrm{ID}} = \frac{\chi_{\mathrm{H}} K_{a1}}{K_{a1} K_{a2} + \chi_{\mathrm{H}} K_{a1} + \chi_{\mathrm{H}}^2} \quad \left(\begin{matrix}\text{Ideal proton} \\ \text{adsorption equation}\end{matrix}\right) \tag{28}$$

$$\theta_{\mathrm{SOH_2}}^{\mathrm{ID}} = \frac{\chi_{\mathrm{H}}^2}{K_{a1} K_{a2} + \chi_{\mathrm{H}} K_{a1} + \chi_{\mathrm{H}}^2} \quad \left(\begin{matrix}\text{Ideal proton} \\ \text{adsorption equation}\end{matrix}\right) \tag{29}$$

In Section II.E, activity coefficients effects on proton adsorption will be discussed.

Although the surface-site mole fractions could be calculated directly from Eqs (24)–(29) a more instructive method of evaluating these equations is graphically using a master variable plot [5] commonly used for acid–base and metal complexation reactions in solution.

C. Master Variable Plots for Proton Adsorption

With pH as the master variable and x-axis, the log of the surface-site concentrations can be represented on the y-axis. Each equation, (24)–(26), will be plotted on the master variable plot to show graphically the pH dependence of each type of surface site.

Equations (24)–(26) all share a common denominator which makes a significant contribution to the predicted pH dependence of each type of surface site. The three pH-dependent denominator terms arise because the pH dependence of each surface-site mole fraction is directly dependent upon the two mass action ionization constants, K_{a1} and K_{a2}: (1) at high pH where pH > pK_{a2}, (2) at intermediate pH where pK_{a1} < pH < pK_{a2}, and (3) at low pH where pH < pK_{a1}. Between each of the three regions are two system points defined by a single pH equal to K_{a1} and K_{a2}: (1) the second system point defined at pH = pK_{a2}, and (2) the first system point defined at pH = pK_{a1}. We will examine Eq. (24) over the three different pH ranges and at two different system points beginning at the highest pH (most basic) and decreasing the pH through both system points and all three straight lines. We will begin by calculating the pH dependence of the unprotonated anionic surface site, θ_{SO}, using Eq. (24).

Line at highest pH: pH > pK_{a2} $\left(K_{a1}K_{a2} \gg \chi_H K_{a1} Z^{\epsilon K_{a2}} \gg \chi_H^2 Z^{\epsilon K_{a12}}\right)$

$$\theta_{SO}(\text{pH} > \text{p}K_{a2}) = \frac{K_{a1}K_{a2}}{K_{a1}K_{a2} + \chi_H K_{a1} Z^{\epsilon K_{a2}} + \chi_H^2 Z^{\epsilon K_{a12}}} \cong \frac{K_{a1}K_{a2}}{K_{a1}K_{a2}} = 1 \tag{30}$$

and

$$\log \theta_{SO}(\text{pH} > \text{p}K_{a2}) = \log 1 = 0 \tag{31}$$

Second system point at high pH: Decreasing pH until pH = pK_{a2} gives $K_{a1}K_{a2} = \chi_H K_{a1} Z^{\epsilon K_{a2}} \gg \chi_H^2 Z^{\epsilon K_{a12}}$, so

$$\chi_H = K_{a2} Z^{-\epsilon K_{a2}} \cong K_{a2} \quad \text{(Ideal equations)} \tag{32}$$

$$\log \chi_H = -\text{pH} \cong \log K_{a2} = -\text{p}K_{a2} \tag{33}$$

also

$$\theta_{SO}(\text{pH} = \text{p}K_{a2}) = \frac{K_{a1}K_{a2}}{K_{a1}K_{a2} + \chi_H K_{a1} Z^{\epsilon K_{a2}} + \chi_H^2 Z^{\epsilon K_{a12}}} \cong \frac{K_{a1}K_{a2}}{2K_{a1}K_{a2}} = \frac{1}{2} \tag{34}$$

$$\log \theta_{SO}(\text{pH} = \text{p}K_{a2}) = \log 0.5 = -0.301 \tag{35}$$

Line at intermediate pH: At intermediate pH where pK_{a1} < pH < pK_{a2}, then, $\chi_H K_{a1} Z^{\epsilon K_{a2}} \gg K_{a1}K_{a2} \cong \chi_H^2 Z^{\epsilon K_{a12}}$, so

$$\theta_{SO}(\text{pH} < \text{p}K_{a2}) = \frac{K_{a1}K_{a2}}{K_{a1}K_{a2} + \chi_H K_{a1} Z^{\epsilon K_{a2}} + \chi_H^2 Z^{\epsilon K_{a12}}} \cong \frac{K_{a2}}{\chi_H Z^{\epsilon K_{a2}}} \cong \frac{K_{a2}}{\chi_H} \tag{36}$$

and

$$\log \theta_{SO}(\text{pH} < \text{p}K_{a2}) = -\log \chi_H + \log K_{a2} = \text{pH} - \text{p}K_{a2} \tag{37}$$

First system point at low pH: Decreased pH until pH = pK_{a1}, then, $K_{a1}K_{a2} \ll \chi_H K_{a1} Z^{\epsilon K_{a2}} = \chi_H^2 Z^{\epsilon K_{a12}}$, so

$$K_{a1} Z^{\epsilon K_{a2}} = \chi_H Z^{\epsilon K_{a12}} \cong \chi_H \cong K_{a1} \quad \text{(Ideal equations)} \tag{38}$$

$$\log \chi_H = -\text{pH} \cong \log K_{a1} = -\text{p}K_{a1} \tag{39}$$

also

$$\theta_{SO}(\text{pH} = \text{p}K_{a1}) = \frac{K_{a1}K_{a2}}{K_{a1}K_{a2} + \chi_H K_{a1} Z^{\epsilon K_{a2}} + \chi_H^2 Z^{\epsilon K_{a12}}} \cong \frac{K_{a2}}{2\chi_H Z^{\epsilon K_{a2}}} \tag{40}$$

$$\theta_{SO}(\chi_H = K_{a1}) \cong \frac{K_{a2}}{2\chi_H Z^{\epsilon K_{a2}}} \cong \frac{K_{a2}}{2K_{a1} Z^{\epsilon K_{a2}}} \cong \frac{K_{a2}}{2K_{a1}} \tag{41}$$

$$\log\theta_{SO}(\text{pH} = \text{p}K_{a2}) = \log 0.5 + \log\frac{K_{a2}}{K_{a1}} = -0.301 + \log\frac{K_{a2}}{K_{a1}} \tag{42}$$

Line at lowest pH: At lowest pH where pH $< \text{p}K_{a1}$, then, $K_{a1}K_{a2} \ll \chi_H K_{a1} Z^{\epsilon K_{a2}} \ll \chi_H^2 Z^{\epsilon K_{a12}}$, so

$$\theta_{SO}(\text{pH} < \text{p}K_{a1}) = \frac{K_{a1}K_{a2}}{K_{a1}K_{a2} + \chi_H K_{a1} Z^{\epsilon K_{a2}} + \chi_H^2 Z^{\epsilon K_{a12}}} \cong \frac{K_{a1}K_{a2}}{\chi_H^2 Z^{\epsilon K_{a12}}} \cong \frac{K_{a1}K_{a2}}{\chi_H^2} \tag{43}$$

and

$$\log\theta_{SO}(\text{pH} < \text{p}K_{a1}) = -\log\chi_H^2 + \log K_{a1}K_{a2} = 2\text{pH} - (\text{p}K_{a1} + \text{p}K_{a2}) \tag{44}$$

The master plot line for the unprotonated anionic surface-site mole fraction, θ_{SO}, can now be plotted if the two surface-site mass action constants are known. For demonstration, values of $\text{p}K_{a1} = 4.0$ and $\text{p}K_{a2} = 8.0$ will be chosen and used in the equations derived above to give:

Line at highest pH: pH $> \text{p}K_{a2}$

$$\log\theta_{SO}(\text{pH} > \text{p}K_{a2}) = \log 1 = 0 \tag{31}$$

Second system point – high pH: pH $= \text{p}K_{a2}$

$$\log\theta_{SO}(\text{pH} = \text{p}K_{a2}) = \log 0.5 = -0.301 \tag{35}$$

Line at intermediate pH: $\text{p}K_{a1} <$ pH $< \text{p}K_{a2}$

$$\log\theta_{SO}(\text{p}K_{a1} < \text{pH} < \text{p}K_{a2}) = \text{pH} - \text{p}K_{a2} = \text{pH} - 8.0 \tag{37a}$$

First system point – low pH: pH $= \text{p}K_{a1}$

$$\log\theta_{SO}(\text{pH} = \text{p}K_{a1}) = \log\frac{K_{a2}}{2K_{a1}} = \log\frac{(1.0\times 10^{-8})}{2(1.0\times 10^{-4})} = -4.301 \tag{42a}$$

Line at lowest pH: pH $< \text{p}K_{a1}$

$$\log\theta_{SO}(\text{pH} < \text{p}K_{a1}) = 2\text{pH} - (\text{p}K_{a1} + \text{p}K_{a2}) = 2\text{pH} - 12.0 \tag{44a}$$

The master plot line for θ_{SO} can now be plotted in Fig. 2 beginning with Eq. (31) at the highest pH, which is a horizontal line at $\log\theta_{SO} = 0$, until the pH drops to the second system point where $\log\theta_{SO} = -0.301$ (indicated with a circle in Fig. 2) as given by Eq. (35). In the intermediate pH region where $\text{p}K_{a1} <$ pH $< \text{p}K_{a2}$, Eq. (37) is the equation for a straight line of slope +1, with *y*-intercept (where pH = 0) of −8.0 (indicated with a square in Fig. 2). At the first-system point, Eq. (42) gives log $\theta_{SO} = -4.301$ (indicated with a circle in Fig. 2). In the lowest pH region where pH $< \text{p}K_{a1}$, Eq. (44) is the equation for a straight line of slope +2, with *y*-intercept (where pH = 0) of −12.0 (indicated with a square in Fig. 2).

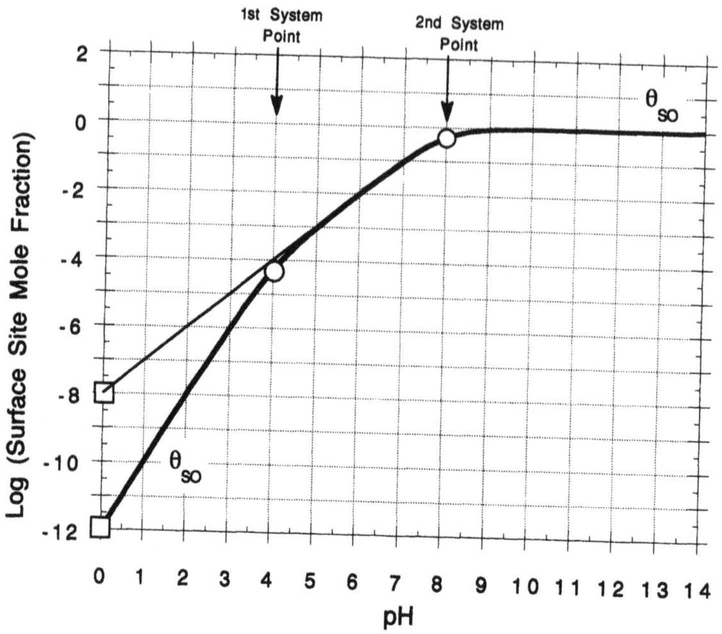

FIG. 2 Calculated pH dependence of unprotonated anionic surface site mole fraction, θ_{SO}, in the absence of solute molecules other than protons and hydroxide ions. Proton adsorption constants: $pK_{a1} = 4.0$, $pK_{a2} = 8.0$. Circles represent system points and squares represent *y*-intercepts.

In a similar fashion, equations for master variable plot lines can be derived for the nonionic, monoprotonated surface sites, θ_{SOH}, and the cationic, diprotonated surface sites, θ_{SOH_2}.

We will begin by deriving the master variable plot equations for the nonionic, monoprotonated surface sites, θ_{SOH}, from Eq. (25):

$$\theta_{SOH} = \frac{\chi_H K_{a1} Z^{\epsilon_{K_{a2}}}}{K_{a1}K_{a2} + \chi_H K_{a1} Z^{\epsilon_{K_{a2}}} + \chi_H^2 Z^{\epsilon_{K_{a12}}}} \tag{25}$$

Line at highest pH: $\mathrm{pH} > \mathrm{p}K_{a2}\ (K_{a1}K_{a2} \gg \chi_H K_{a1} Z^{\epsilon_{K_{a2}}} \gg \chi_H^2 Z^{\epsilon_{K_{a12}}})$

$$\theta_{SOH}(\mathrm{pH} > \mathrm{p}K_{a2}) \cong \frac{\chi_H K_{a1} Z^{\epsilon_{K_{a2}}}}{K_{a1}K_{a2}} \cong \frac{\chi_H Z^{\epsilon_{K_{a2}}}}{K_{a2}} \cong \frac{\chi_H}{K_{a2}} \tag{45}$$

$$\log \theta_{SOH}(\mathrm{pH} > \mathrm{p}K_{a2}) \cong \log \frac{\chi_H}{K_{a2}} \cong -\mathrm{pH} + \mathrm{p}K_{a2} \cong -\mathrm{pH} + 8.0 \tag{46}$$

$$\log \theta_{SOH}(\mathrm{pH} = 14.0) \cong \log \frac{\chi_H}{K_{a2}} \cong -\mathrm{pH} + 8.0 \cong -6.0 \tag{47}$$

Second system point – high pH: $\mathrm{pH} = \mathrm{p}K_{a2}\ (K_{a1}K_{a2} \cong \chi_H K_{a1} Z^{\epsilon_{K_{a2}}} \gg \chi_H^2 Z^{\epsilon_{K_{a12}}})$

$$\theta_{SOH}(\mathrm{pH} = \mathrm{p}K_{a2}) \cong \frac{\chi_H K_{a1} Z^{\epsilon_{K_{a2}}}}{2\chi_H K_{a1} Z^{\epsilon_{K_{a2}}}} = \frac{1}{2} \tag{48}$$

$$\log \theta_{SOH}(\mathrm{pH} = \mathrm{p}K_{a2}) = \log \tfrac{1}{2} = -0.301 \tag{49}$$

Line – intermediate pH: $pK_{a1} < pH < pK_{a2}(\chi_H K_{a1} Z^{\epsilon K_{a2}} \gg \chi_H^2 Z^{\epsilon K_{a12}} \cong K_{a1}K_{a2})$

$$\theta_{SOH}(K_{a1} < pH < pK_{a2}) \cong \frac{\chi_H K_{a1} Z^{\epsilon K_{a2}}}{\chi_H K_{a1} Z^{\epsilon K_{a2}}} = 1.0 \tag{50}$$

$$\log \theta_{SOH}(pK_{a1} < pH < pK_{a2}) = \log 1.0 = 0 \tag{51}$$

First system point – low pH: $pH = pK_{a1}\ (\chi_H^2 Z^{\epsilon K_{a12}} = \chi_H K_{a1} Z^{\epsilon K_{a2}} \gg K_{a1}K_{a2})$

$$\theta_{SOH}(pH = pK_{a1}) \cong \frac{\chi_H K_{a1} Z^{\epsilon K_{a2}}}{2\chi_H K_{a1} Z^{\epsilon K_{a2}}} \cong \frac{1}{2} \tag{52}$$

$$\log \theta_{SOH}(pH = pK_{a1}) = \log \tfrac{1}{2} = -0.301 \tag{53}$$

Line at lowest pH: $pH < pK_{a1}\ (\chi_H^2 Z^{\epsilon K_{a12}} \gg \chi_H K_{a1} Z Z^{\epsilon K_{a2}} \gg K_{a1}K_{a2})$

$$\theta_{SOH}(pH < pK_{a1}) \cong \frac{\chi_H K_{a1} Z^{\epsilon K_{a2}}}{\chi_H^2 Z^{\epsilon K_{a12}}} \cong \frac{K_{a1} Z^{\epsilon K_{a2}}}{\chi_H Z^{\epsilon K_{a12}}} \cong \frac{K_{a1}}{\chi_H} \tag{54}$$

$$\log \theta_{SOH}(pH < pK_{a1}) = +pH - pK_{a1} = +pH - 4.0 \tag{55}$$

$$\log \theta_{SOH}(pH = 0) = +pH - pK_{a1} = +0.0 - 4.0 = -4.0 \tag{56}$$

In a similar fashion, the master variable plot equations for the cationic, diprotonated surface sites, θ_{SOH_2}, can be derived from Eq. (26):

$$\theta_{SOH_2} = \frac{\chi_H^2 Z^{\epsilon K_{a12}}}{K_{a1}K_{a2} + \chi_H K_{a1} Z^{\epsilon K_{a2}} + \chi_H^2 Z^{\epsilon K_{a12}}} \tag{26}$$

Line at lowest pH: $pH < pK_{a1}\ (\chi_H^2 Z^{\epsilon K_{a12}} \gg \chi_H K_{a1} Z^{\epsilon K_{a2}} \gg K_{a1}K_{a2})$

$$\log \theta_{SOH_2}(pH < pK_{a1}) \cong \log \frac{\chi_H^2 Z^{\epsilon K_{a12}}}{\chi_H^2 Z^{\epsilon K_{a12}}} = \log 1 = 0 \tag{57}$$

First system point – low pH: $pH = pK_{a1}\ (\chi_H^2 Z^{\epsilon K_{a12}} = \chi_H K_{a1} Z^{\epsilon K_{a2}} \gg K_{a1}K_{a2})$

$$\log \theta_{SOH_2}(pH = pK_{a1}) = \log \frac{\chi_H^2 Z^{\epsilon K_{a12}}}{2\chi_H^2 Z^{\epsilon K_{a12}}} = \log \frac{1}{2} = -0.301 \tag{58}$$

Line – intermediate pH: $pK_{a1} < pH < pK_{a2}(\chi_H K_{a1} Z^{\epsilon K_{a2}} \gg \chi_H^2 Z^{\epsilon K_{a12}} \cong K_{a1}K_{a2})$

$$\log \theta_{SOH_2}(pK_{a1} < pH < pK_{a2}) \cong \log \frac{\chi_H^2 Z^{\epsilon K_{a12}}}{\chi_H K_{a1} Z^{\epsilon K_{a2}}} \cong -pH + pK_{a1} = -pH + 4.0 \tag{59}$$

Second system point – high pH: $pH = pK_{a2}\ (K_{a1}K_{a2} \cong \chi_H K_{a1} Z^{\epsilon K_{a2}} \gg \chi_H^2 Z^{\epsilon K_{a12}})$

$$\begin{aligned}\log \theta_{SOH_2}(pH = pK_{a2}) &\cong \log \frac{\chi_H^2 Z^{\epsilon K_{a12}}}{2K_{a1} Z^{\epsilon K_{a2}}} \cong \log \frac{\chi_H^2 Z^{\epsilon K_{a12}}}{2\chi_H K_{a1} Z^{\epsilon K_{a2}}} \cong \log \frac{K_{a2}^2 Z^{\epsilon K_{a12}}}{2K_{a1}K_{a2} Z^{\epsilon K_{a2}}} \\ &\cong \log \frac{K_{a2} Z^{\epsilon K_{a12}}}{2K_{a1} Z^{\epsilon K_{a2}}} \cong \log \frac{(1.0 \times 10^{-8})}{2(1.0 \times 10^{-4})} = -4.301\end{aligned} \tag{60}$$

Line at highest pH: $\mathrm{pH} > \mathrm{p}K_{a2}$ $(K_{a1}K_{a2} \gg \chi_H K_{a1} Z^{\epsilon K_{a2}} \gg \chi_H^2 Z^{\epsilon K_{a12}})$

$$\log \theta_{SOH_2}(\mathrm{pH} < \mathrm{p}K_{a1}) \cong \log \frac{\chi_H^2 Z^{\epsilon K_{a12}}}{K_{a1}K_{a2}} \cong -2\mathrm{pH} - \log(K_{a1}K_{a2}) \cong -2\mathrm{pH} + 12.0 \tag{61}$$

The equations derived above can now be used to construct master plot lines for all three surface site mole fractions, θ_{SO}, θ_{SOH}, and θ_{SOH_2}. The symmetry of these master plot lines can be seen from the equations in Table 2.

Several observations can be made from Fig. 3:

- The line for any particular surface species mole fraction changes slope at the pH equal to the $\mathrm{p}K_a$ of the surface sites. For example, the transition from horizontal to diagonal for the SOH_2 surface species line occurs at $\mathrm{pH} = \mathrm{p}K_{a1} = 4.0$. At $\mathrm{pH} = \mathrm{p}K_{a2} = 8.0$, the master plot line for SOH_2 changes from a slope of -1 to -2.
- A vertical line at a $\mathrm{pH} = 6.0$ indicates that $\theta_{SO} = \theta_{SOH_2}$. If Eqs (24) and (26) are substituted, the equations describing this intersection are

$$\chi_H = \sqrt{K_{a1}K_{a2}} \quad \mathrm{pH} = \frac{\mathrm{p}K_{a1} + \mathrm{p}K_{a2}}{2} \tag{62}$$

  From the equality of the two oppositely charged surface species, $\theta_{SO} = \theta_{SOH_2}$, one would conclude that the net electrostatic surface charge would be zero since the monoprotonated surface sites, θ_{SOH}, contribute no electrostatic charge. The point of electroneutrality for any S–MO interface is the isoelectric point (IEP) or zero point of charge (ZPC), which occurs at $\mathrm{pH} = 6.0$ for the data in Fig. 3.
- Any vertical line drawn on Fig. 3 will intersect the mole fraction lines for all three surface species. These points of intersection specify the precise equilibrium mole fraction of all three species at a particular pH, which can be determined graphically from Fig. 3. As an example, consider a vertical line on Fig. 3 at $\mathrm{pH} = 6.5$

TABLE 2 Ideal Master Variable Plot Equations

Line or point	$\log \theta_{SO}$	$\log \theta_{SOH}$	$\log \theta_{SOH_2}$
$\mathrm{pH} < \mathrm{p}K_{a1}$ Acidic/low pH	$= +2\mathrm{pH} -(\mathrm{p}K_{a1} + \mathrm{p}K_{a2})$	$= +\mathrm{pH} - \mathrm{p}K_{a1}$	$= 0.0$
$\mathrm{pH} = \mathrm{p}K_{a1}$ First system point	$= -4.301$	$= -0.301$	$= -0.301$
$\mathrm{p}K_{a1} < \mathrm{pH} < \mathrm{p}K_{a2}$ Intermediate pH	$= +\mathrm{pH} - \mathrm{p}K_{a2}$	$= 0.0$	$= -\mathrm{pH} + \mathrm{p}K_{a1}$
$\mathrm{pH} = \mathrm{p}K_{a2}$ Second system point	$= -0.301$	$= -0.301$	$= -4.301$
$\mathrm{pH} > \mathrm{p}K_{a2}$ Basic/high pH	$= 0.0$	$= -\mathrm{pH} + \mathrm{p}K_{a2}$	$= -2\mathrm{pH} +(\mathrm{p}K_{a1} + \mathrm{p}K_{a2})$

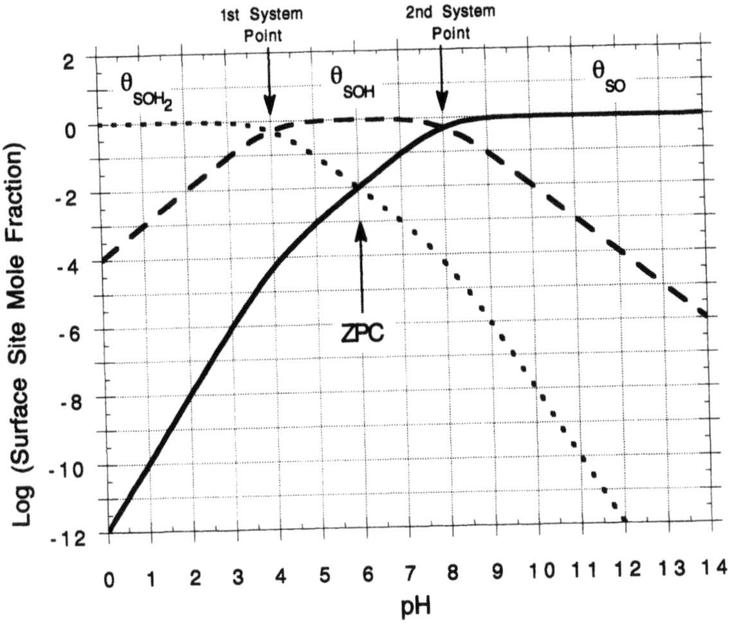

FIG. 3 Calculated surface-site mole fractions in the absence of solute molecules other than protons and hydroxide ions. Proton adsorption constants: p$K_{a1} = 4.0$, p$K_{a2} = 80$, ZPC = pK_{a1} + p$K_{a2}/2 = 6.0$.

$$\log\theta_{SO} = -1.5 : \theta_{SO} = 0.0316$$
$$\log\theta_{SOH} = 0.0 : \theta_{SOH} = 1.0$$
$$\log\theta_{SOH_2} = -2.5 : \theta_{SOH_2} = 0.00316$$

Calculated surface-site mole fractions can be obtained by using the appropriate equation for pH = 6.5, and assuming ideal adsorption ($Z = 1.0$) as

$$\theta_{SO} = \frac{K_{a1}K_{a2}}{K_{a1}K_{a2} + \chi_H K_{a1} Z^{\epsilon_{K_{a2}}} + \chi_H^2 Z^{\epsilon_{K_{a12}}}} \cong \frac{K_{a1}K_{a2}}{K_{a1}K_{a2} + \chi_H K_{a1} + \chi_H^2} \tag{24a}$$

$$\begin{aligned}\theta_{SO} &\cong \frac{K_{a1}K_{a2}}{K_{a1}K_{a2} + \chi_H K_{a1} + \chi_H^2} \\ &\cong \frac{(1.0 \times 10^{-4})(1.0 \times 10^{-8})}{(1.0 \times 10^{-4})(1.0 \times 10^{-8}) + (3.16 \times 10^{-7})(1.0 \times 10^{-4}) + (3.16 \times 10^{-7})^2} \\ &\cong \frac{1.0 \times 10^{-12}}{1.0 \times 10^{-12} + 3.16 \times 10^{-11} + 1.0 \times 10^{-13}} \cong \frac{1.0 \times 10^{-12}}{327 \times 10^{-13}} = 0.0306\end{aligned} \tag{63}$$

$$\theta_{SOH} = \frac{\chi_H K_{a1} Z^{\epsilon_{K_{a2}}}}{K_{a1}K_{a2} + \chi_H K_{a1} Z^{\epsilon_{K_{a2}}} + \chi_H^2 Z^{\epsilon_{K_{a12}}}} \cong \frac{\chi_H K_{a1}}{k_{a1}K_{a2} + \chi_H K_{q1} + \chi_H^2} \tag{25a}$$

$$\theta_{\mathrm{SOH}} \cong \frac{\chi_{\mathrm{H}}K_{\mathrm{a1}}}{K_{\mathrm{a1}}K_{\mathrm{a2}}+\chi_{\mathrm{H}}K_{\mathrm{a1}}+\chi_{\mathrm{H}}^2}$$

$$\cong \frac{(3.16\times10^{-7})(1.0\times10^{-4})}{(1.0\times10^{-4})(1.0\times10^{-8})+(3.16\times10^{-7})(1.0\times10^{-4})+(3.16\times10^{-7})^2}$$

$$\cong \frac{3.16\times10^{-11}}{1.0\times10^{-12}+3.16\times10^{-11}+1.0\times10^{-13}} \cong \frac{3.16\times10^{-11}}{327\times10^{-13}} = 0.966 \tag{64}$$

$$\theta_{\mathrm{SOH_2}} = \frac{\chi_{\mathrm{H}}^2 Z^{\epsilon K_{\mathrm{a12}}}}{K_{\mathrm{a1}}K_{\mathrm{a2}}+\chi_{\mathrm{H}}K_{\mathrm{a1}}Z^{\epsilon K_{\mathrm{a2}}}+\chi_{\mathrm{H}}^2 Z^{\epsilon K_{\mathrm{a12}}}} \cong \frac{\chi_{\mathrm{H}}^2}{K_{\mathrm{a1}}K_{\mathrm{a2}}+\chi_{\mathrm{H}}K_{\mathrm{a1}}+\chi_{\mathrm{H}}^2} \tag{26a}$$

$$\theta_{\mathrm{SOH_2}} \cong \frac{\chi_{\mathrm{H}}^2}{K_{\mathrm{a1}}K_{\mathrm{a2}}+\chi_{\mathrm{H}}K_{\mathrm{a1}}+\chi_{\mathrm{H}}^2}$$

$$\cong \frac{(3.16\times10^{-7})^2}{(1.0\times10^{-4})(1.0\times10^{-8})+(3.16\times10^{-7})(1.0\times10^{-4})+(3.16\times10^{-7})^2}$$

$$\cong \frac{1.0\times10^{-13}}{1.0\times10^{-12}+3.16\times10^{-11}+1.0\times10^{-13}} \cong \frac{1.0\times10^{-13}}{327\times10^{-13}} = 0.00306 \tag{65}$$

The surface-site mole fractions determined graphically and by calculation show good agreement. The master plot method is seen to be a reliable and convenient method for representing and determining the pH dependence of surface-site ionization behavior, which requires only a minimum number of data (K_{a1} and K_{a2}) to construct. Values of K_{a1} and K_{a2} are available from the literature [6].

D. Titration Plots for Proton Adsorption

Owing to the amphoteric nature of metal oxide surface sites, acid–base titrations are commonly performed. Since surface sites can accept up to two protons, titration plots of metal oxide powders are qualitatively similar to titration plots of diprotic weak acids (or weak bases) in solution. As with diprotic weak acids, the quantitative shape of metal oxide powder titration plots is determined by the two proton mass action constants, K_{a1} and K_{a2}. If we take the same proton mass action constant values used in Section II.C ($pK_{a1} = 4.0$ and $pK_{a2} = 8.0$) a titration plot similar to the one shown in Fig. 4 would result.

Based on the principles of titration plots, inflection points of minimum slope are half-equivalence points, and inflection points of maximum slope are equivalence points. In Fig. 4, before addition of any titrant at the lowest pH, the mole fraction of cationic diprotonated surface sites is virtually unity or $\theta_{SOH_2} \cong 1.0$ and $\theta_{SOH} \cong \theta_{SO} \cong 0.0$. As titrant is added, the SOH_2 surface sites are converted largely into SOH surface sites. At the half-equivalence point 1, where one unit of titrant has been added, the concentrations of the di- and mono-protonated surface sites are nearly equal, so $\theta_{SOH_2} \cong \theta_{SOH}$ and $\theta_{SO} \cong 0.0$. Based on the discussion in Section II.C it was concluded that, when $\theta_{SOH_2} \cong \theta_{SOH}$, then $pH = pK_{a1}$. It is clear then that the first half-equivalence point in the titration plot in Fig. 4 corresponds to the first system point in the master variable plot in Fig. 3.

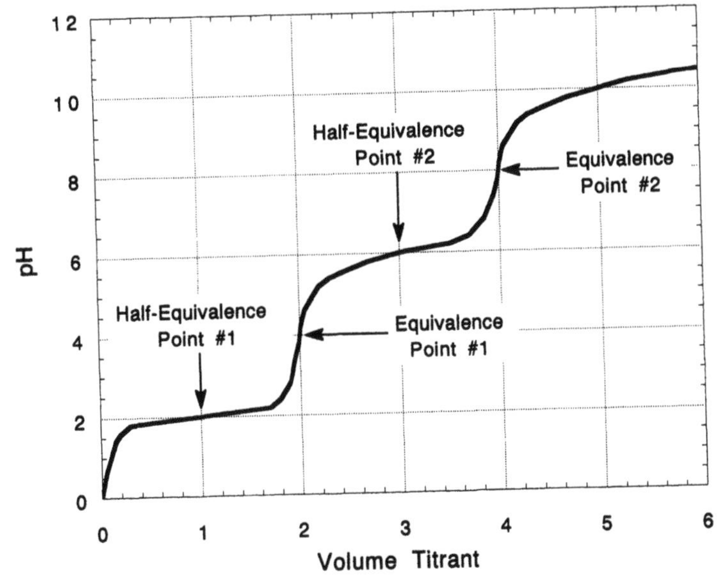

FIG. 4 Titration plot for a hypothetical metal oxide powder in the absence of solute molecules other than protons and hydroxide ions. Proton adsorption constants: $pK_{a1} = 4.0$, $pK_{a2} = 8.0$, ZPC $= pK_{a1} + pK_{a2}/2 = 6.0$, $Z^{\epsilon K_{a2}} = Z^{\epsilon K_{a12}} = 1$.

As additional titrant is added, the first equivalence point is eventually reached with two units of titrant added. As equivalence point 1 is approached, additional SOH_2 surface sites are converted into mostly SOH and a few SO surface sites. At equivalence point 1 in Fig. 4, the concentration of the monoprotonated surface sites predominates. However, the few di- and un-protonated surface sites which exist are equal in number or $\theta_{SOH} \gg \theta_{SOH_2} = \theta_{SO}$. From Fig. 3, this condition can be seen as the ZPC. Thus, equivalence point 1 on a titration plot of a metal oxide powder is the ZPC and its pH value can be obtained from the master variable plot (Fig. 3).

With continued addition of titrant, the second half-equivalence point is eventually reached. As half-equivalence point 2 is approached, SOH surface sites are converted largely into SO^- surface sites. At the half-equivalence point 2 in Fig. 4, the concentration of the mono- and un-protonated surface sites is nearly equal, so $\theta_{SO} \cong \theta_{SOH}$ and $\theta_{SOH_2} \cong 0.0$. Based on the discussion in Section II.C it was concluded that, when $\theta_{SO} \cong \theta_{SOH}$, then pH $= pK_{a2}$. It is clear then that the second half-equivalence point in the titration plot in Fig. 4 corresponds to the second system point in the master variable plot in Fig. 3. Results from Sections II.B, C, and D are tabulated in Table 3.

E. Activity Coefficients for Proton Adsorption

The purpose of this section will be to make a semiquantitative estimate of the magnitude of activity coefficient effects in the proton adsorption process at the S–MO interface. In Section II.B, proton adsorption equations were derived with and without activity coefficient effects: Eqs (24)–(26) and (27)–(29), respectively. The

TABLE 3 Comparison of Common Features for Titration and Master Variable Plots

Titration plot	Master variable plot	pH equation
Half-equivalence point 1	First system point	$\mathrm{pH} = \mathrm{p}K_{\mathrm{a1}}$
Equivalence point 1	Zero point of charge (ZPC)	$\mathrm{pH} = \dfrac{\mathrm{p}K_{\mathrm{a1}} + \mathrm{p}K_{\mathrm{a2}}}{2}$
Half-equivalence point 2	Second system point	$\mathrm{pH} = \mathrm{p}K_{\mathrm{a2}}$

former will be referred to as "AC" and the latter as "ID" proton adsorption equations. To demonstrate the role of activity coefficients, explicit activity coefficient equations derived in Chapter 4 will be used to calculate the activity coefficient quotients in Eqs (24)–(26), then plotted in a master variable plot. Comparisons of the master variable plot data for the AC and ID proton adsorption equations will allow an assessment of the relative effect activity coefficients have on proton adsorption.

Tables 2 and 3 and Figs 2–4 are compilations and plots of ideal adsorption behavior because all the information contained in these tables and plots is based on ideal proton adsorption equations, (27)–(29), which were derived from Eqs (24)–(26) by setting all activity coefficients to unity. To evaluate the effect activity coefficients have on adsorption, we will begin with explicit equations for the activity coefficient quotients derived for proton adsorption in Chapter 4:

$$Z^{\epsilon_{K_{\mathrm{a1}}}} = (\mathrm{e}^{-\kappa q_{\mathrm{B}}})^{\epsilon_{K_{\mathrm{a1}}}} \quad \text{where} \quad \epsilon_{K_{\mathrm{a1}}} = \frac{1}{1+\kappa a_{\mathrm{H}}} + \frac{2\bar{\nu}_{\mathrm{s}} - 1}{1+\kappa a_{\mathrm{s}}} \tag{66}$$

$$Z^{\epsilon_{K_{\mathrm{a2}}}} = (\mathrm{e}^{-\kappa q_{\mathrm{B}}})^{\epsilon_{K_{\mathrm{a2}}}} \quad \text{where} \quad \epsilon_{K_{\mathrm{a2}}} = \frac{1}{1+\kappa a_{\mathrm{H}}} + \frac{2\bar{\nu}_{\mathrm{s}} + 1}{1+\kappa a_{\mathrm{s}}} \tag{67}$$

$$Z^{\epsilon_{K_{\mathrm{a12}}}} \equiv Z^{\epsilon_{K_{\mathrm{a1}}}} Z^{\epsilon_{K_{\mathrm{a2}}}} = (\mathrm{e}^{-\kappa q_{\mathrm{B}}})^{\epsilon_{K_{\mathrm{a12}}}} \quad \text{where} \quad \epsilon_{K_{\mathrm{a12}}} = \frac{2}{1+\kappa a_{\mathrm{H}}} + \frac{4\bar{\nu}_{\mathrm{s}}}{1+\kappa a_{\mathrm{s}}} \tag{68}$$

where, κ^{-1} is the Debye–Hückel diffuse ion layer thickness [7–9], q_{B} is the Bjerrum association length, which is 3.44 Å in aqueous solution at 25°C (Chapter 4, this volume) [10, 11], a_{H} and a_{s} are the effective diameters of the proton and surface site in the aqueous S–MO interfacial system [9, 12], and $\bar{\nu}_{\mathrm{s}}$ is the mean surface-site valence (Chapter 4).

Equations (66)–(68) express the activity coefficient effects for the proton adsorption processes represented by mass action Eqs (20)–(22), respectively. Since general explicit equations for activity coefficients cannot be derived solely from thermodynamics [13, 14], Eqs (66)–(68) have been derived by combining electrostatic theory with Gibbs–Lewis thermodynamics. For solute ions, the procedure is commonly referred to as the Debye–Hückel model (Chapter 4) [8, 9, 15], while for surface site ions the activity coefficient derivations are based on the primitive interfacial model (chapter 4) [16].

Values chosen for the evaluation of the activity coefficient quotients are given in Table 4.

TABLE 4 Variables Used to Calculate Activity Coefficient Quotients

Variable	Value	Comments
κ (0.01 M)	$3.29 \times 10^{8}\ \mathrm{m}^{-1}$	Varies directly with solute ion concentration. For 1:1 electrolyte
q_B	$3.44 \times 10^{-10}\ \mathrm{m}$	Bjerrum association length (10, 11]
$Z \equiv e^{-\kappa q_B}$	0.893	Chapter 4
$a_H = 9\ \mathrm{\mathring{A}} \cong a_s$	$3.29 \times 10^{8}\ \mathrm{m}^{1}$	Proton effective diameter is 9 Å and reasonable value for surface site also (Chapter 4) [9, 12]
$\bar{\nu}_s$	$= \theta_{SOH_2} - \theta_{SO}$	Valence: $SOH_2^+ = +1$; $SO^- = -1$

Substituting values from Table 4 into Eqs (66)–(68):

$$Z = e^{-q_B} = e^{-[(3.29\times 10^{+8}\ \mathrm{m}^{-1})(3.44\times 10^{-10}\ \mathrm{m})]} = 0.893 \tag{69}$$

$$1 + \kappa a = 1 + (3.29 \times 10^{+8}\ \mathrm{m}^{-1})(9 \times 10^{-10}\ \mathrm{m}) = 1.2961 \tag{70}$$

$$\epsilon_{K_{a1}} = \frac{1}{1+\kappa a_H} + \frac{2\bar{\nu}_s - 1}{1+\kappa a_s} = \frac{2\bar{\nu}_s}{1+\kappa a} = \frac{2(\theta_{SOH_2} - \theta_{SO})}{1.2961} = 1.543(\theta_{SOH_2} - \theta_{SO}) \tag{71}$$

$$Z^{\epsilon_{K_{a1}}} = (e^{-\kappa q_B})^{\epsilon_{K_{a1}}} = (0.893)^{1.543^{(\theta_{SOH_2}-\theta_{SO})}} = (0.840)^{(\theta_{SOH_2}-\theta_{SO})} \tag{72}$$

$$\epsilon_{K_{a2}} = \frac{1}{1+\kappa a_H} + \frac{2\bar{\nu}_s + 1}{1+\kappa a_s} = \frac{2(1+\nu_s)}{1+\kappa a} = \frac{2(1+\theta_{SOH_2} - \theta_{SO})}{1.2961} = 1.543(1+\theta_{SOH_2} - \theta_{SO}) \tag{73}$$

$$Z^{\epsilon_{K_{a2}}} = (e^{-\kappa q_B})^{\epsilon_{K_{a2}}} = (0.893)^{1.543(1+\theta_{SOH_2}-\theta_{SO})} = (0.840)^{(1+\theta_{SOH_2}-\theta_{SO})} \tag{74}$$

$$\epsilon_{K_{a12}} = \epsilon_{K_{a1}} + \epsilon_{K_{a2}} = 3.086(0.5 + \theta_{SOH_2} - \theta_{SO}) \tag{75}$$

$$Z^{\epsilon_{K_{a12}}} = (e^{-\kappa q_B})^{\epsilon_{K_{a12}}} = (0.893)^{3.086(0.5+\theta_{SOH_2}-\theta_{SO})} = (0.705)^{(0.5+\theta_{SOH_2}-\theta_{SO})} \tag{76}$$

A master variable plot for proton adsorption can now be constructed, using activity coefficient quotient Eqs (72), (74), and (76), in proton adsorption Eqs (24)–(29). The results are plotted in Fig. 5.

Solid lines in Fig. 5 represent ideal (ID) proton adsorption. Dashed lines in Fig. 5 represent proton adsorption when activity effects (AC) are included. Several aspects of Fig 5 are noteworthy:

- The ideal surface-site mole fractions for ID proton adsorption were calculated for Fig. 5 with Eqs (27)–(29). The surface-site mole fractions for AC proton adsorption with activity coefficient effects included were calculated by substituting Eqs (72), (74), and (76) into adsorption equations (24)–(26).
- The small deviation between ID and AC proton adsorption can be seen to be in the order a few per cent variation in surface-site mole fraction and this only when surface-site mole fractions are in the part per million range, so the error is a few per cent of a few parts per million, which puts the variation between ID and AC in the parts per billion (10^9) range.

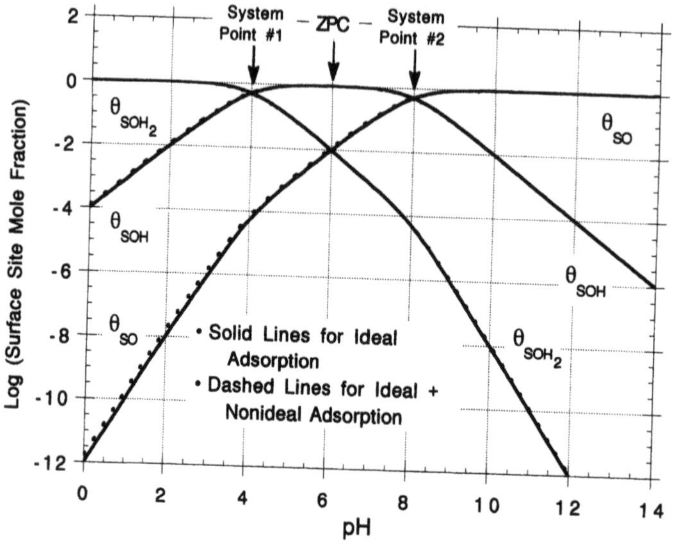

FIG. 5 Master variable plot for proton adsorption including non-ideal adsorption effects using Debye–Hückel and primitive interfacial activity coefficients.

- Activity coefficients vary inversely with electrolyte concentration. For ID proton adsorption, surface-site mole fractions are independent of electrolyte concentration, but electrolyte-concentration dependent for AC proton adsorption. The example calculations in Fig. 5 were for a 1:1 electrolyte at a 0.01 M concentration. It was assumed that the adsorbed amount of both ions in this 0.01 M electrolyte was insignificant compared to that of protons and hydroxide ions. Such ions are commonly referred to as nonpotential determining or indifferent ions becuse the amount of these ions which adsorbs is so small that the electrostatic potential of the interface is not significantly affected.
- The small deviation between ID and AC proton adsorption seen in Fig. 5 can be seen as asymmetric with pH. Specifically, the maximum deviation between AC and ID proton adsorption occurs at low pH. This is a result of the fact that protons are cations and, at low pH, S–MO interfaces tend to be positively charged. The AC proton adsorption equations correctly predict a greater difference in ID and AC proton adsorption at low pH. The source of this enhanced low pH deviation between ID and AC proton adsorption can be seen if the calculated activity coefficient quotients, $Z^{\epsilon K_{a1}}$, $Z^{\epsilon K_{a2}}$, and $Z^{\epsilon K_{a12}}$, are plotted as a function of pH.

The enhanced deviation between ID and AC proton adsorption at low pH can be seen to result from the pH dependence of the two activity coefficient quotients, $Z^{\epsilon K_{a2}}$ and $Z^{\epsilon K_{a12}}$ (Fig. 6), which appear in the AC proton adsorption equations (24)–(26). (Note that the activity coefficient quotient, $Z^{\epsilon K_{a1}}$, does not appear in proton adsorption equations.) The activity coefficient quotients, $Z^{\epsilon K_{a2}}$ and $Z^{\epsilon K_{a12}}$, drop significantly from unity at low pH, which is the source of the enhanced deviation between ID and AC proton adsorption at low pH.

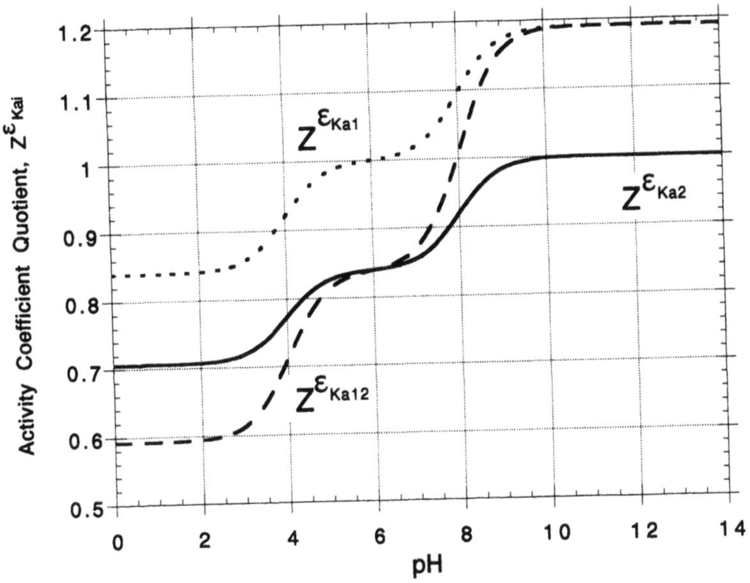

FIG. 6 Variation of activity coefficient quotients for proton adsorption with pH.

III. CATION ADSORPTION EQUATIONS

A. Monovalent Cation Adsorption

Since metal oxide surfaces are amphoteric, with different surface site types, different types of solutes will adsorb differently on each different type of surface site. In general, cation adsorption on all three types of surface site is possible. However, only the predominant adsorption reaction of a cation with an anionic surface site will be considered in the following derivation. Therefore, the principle reaction by which cation adsorption occurs will be written in terms of a cation solute M^{ν_m}, reacting with an anionic metal oxide surface site SO^-, to give an adsorbed surface complex, $SOM^{\nu_m - 1}$. If the cation is monovalent, M^+, the surface complex is uncharged with the structure, SOM, which is formed as

$$M^+ + SO^- \leftrightarrow SOM \quad \text{(Cation adsorption reaction)} \tag{77}$$

Proton adsorption and desorption from the surface will occur in addition to cation adsorption and desorption, as shown in Fig. 7.

From Fig. 7 one can see that cation and proton adsorption are competitive processes so that any cation adsorption equation must also include proton adsorption behavior. The adsorbed cation mole fraction, θ_{SM}, is defined just like surface-site mole fractions by division of the number of surface sites complexed with a cation, n_{SM}, by the total number of surface sites, n_s^T,

$$\theta_{SM} \equiv \frac{n_{SM}}{n_s^T} \tag{78}$$

The surface-site balance for the surface sites at a cation S–MO interface is written in analogy with the proton mass balance equation (9) as

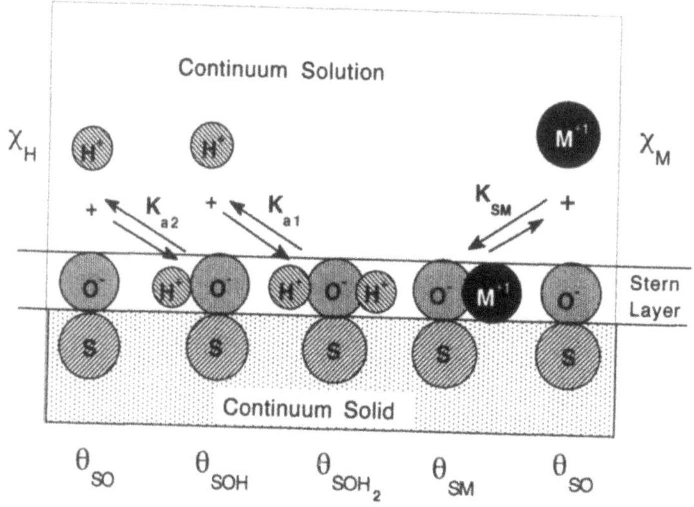

FIG. 7 Interfacial complexation reactions of protons and the cation, M^+, at a S–MO interface.

$$1 = \theta_{SO} + \theta_{SOH} + \theta_{SOH_2} + \theta_{SM} \quad \begin{pmatrix}\text{Surface site mass} \\ \text{balance equation}\end{pmatrix} \tag{79}$$

Before proceeding with an evaluation of the surface sites complexed with a cation, we will examine the fraction of surface sites *not* complexed with a cation. From Eq. (79) the mole fraction of surface sites not complexed with a cation is

$$\theta_{SO} + \theta_{SOH} + \theta_{SOH_2} = 1 - \theta_{SM} \tag{80}$$

If α_{SO} is taken as the mole fraction of anionic surface sites not complexed with a cation, then the definition for α_{SO} can be written by using Eqs (24)–(26):

$$\begin{aligned}\alpha_{SO} &\equiv \frac{n_{SO}}{n_{SO} + n_{SOH} + n_{SOH_2}} = \frac{\theta_{SO}}{\theta_{SO} + \theta_{SOH} + \theta_{SOH_2}} \\ &= \frac{K_{a1}K_{a2}}{K_{a1}K_{a2} + \chi_H K_{a1} Z^{\epsilon K_{a2}} + \chi_H^2 Z^{\epsilon K_{a12}}}\end{aligned} \tag{81}$$

Combining Eqs (80) and (81) gives the desired expression for θ_{SO}:

$$\theta_{SO} = \alpha_{SO}(\theta_{SO} + \theta_{SOH} + \theta_{SOH_2}) = \alpha_{SO}(1 - \theta_{SM}) \tag{82}$$

By analogy for θ_{SOH} and θ_{SOH_2}:

$$\alpha_{SOH} = \frac{\theta_{SOH}}{\theta_{SO} + \theta_{SOH} + \theta_{SOH_2}} = \frac{\chi_H K_{a1} Z^{\epsilon K_{a2}}}{K_{a1}K_{a2} + \chi_H K_{a1} Z^{\epsilon K_{a2}} + \chi_H^2 Z^{\epsilon K_{a12}}} \tag{83}$$

$$\theta_{SOH} = \alpha_{SOH}(\theta_{SO} + \theta_{SOH} + \theta_{SOH_2}) = \alpha_{SOH}(1 - \theta_{SM}) \tag{84}$$

$$\alpha_{SOH_2} = \frac{\theta_{SOH_2}}{\theta_{SO} + \theta_{SOH} + \theta_{SOH_2}} = \frac{\chi_H^2 Z^{\epsilon K_{a2}}}{K_{a1}K_{a2} + \chi_H K_{a1} Z^{\epsilon K_{a2}} + \chi_H^2 Z^{\epsilon K_{a12}}} \tag{85}$$

$$\theta_{SOH_2} = \alpha_{SOH_2}(\theta_{SO} + \theta_{SOH} + \theta_{SOH_2}) = \alpha_{SOH_2}(1 - \theta_{SM}) \tag{86}$$

Equations (81)–(86) will be used in the cation adsorption equation.

The surface mole fraction variables, θ_{SO}, θ_{SOH} and θ_{SOH_2}, in Eq. (79) can be replaced with Eqs (21) and (22) as was done before for proton adsorption. However, the cation-surface mole fraction term introduces an additional unknown, which in turn requires an additional equation in order to derive a cation adsorption equation. To this end, two additional chemical potential equations for the monovalent cation, M^+, in solution, μ_M, and the adsorbed cation, μ_{SM}, are written:

$$\mu_M = \mu_M^* + kT \ln \chi_M + kT \ln f_M \tag{87}$$

where an explicit activity coefficient equation can be written from Chapter 4:

$$f_M = Z^{\epsilon_m} \quad \text{where } \epsilon_m = \frac{\nu_m^2}{1 + \kappa a_m} \tag{88}$$

where κ^{-1} is the Debye–Hückel diffuse ion layer thickness [8, 9,15], a_m is the effective diameter of the metal ion, M^+, in the aqueous solution [9, 12], and $\bar{\nu}_s$ is the mean surface site valence (Chapter 4). In a like manner for the surface-site complex:

$$\mu_{SM} = \mu_{SM}^* + kT \ln \theta_{SM} + kT \ln f_{SM} \tag{89}$$

and

$$f_{SM} = Z^{\epsilon_{sm}} \quad \text{where} \quad \epsilon_{sm} = \frac{(\nu_m - 1 - \bar{\nu}_s)^2}{1 + \kappa a_s} \tag{90}$$

where a_s is the effective diameter of the surface site in the aqueous S–MO interface [9, 12]. For the cation adsorption reaction given in Eq. (77), the chemical potentials of the products and reactants must be equal at equilibrium, so

$$\mu_{SM} = \mu_{SO} + \mu_M \tag{91}$$

Substitution of Eqs (11), (87) and (89) into Eq. (91) gives

$$-(\mu_{SM}^* - \mu_M^* - \mu_{SO}^*) = +kT \ln\left(\frac{\theta_{SM}}{\theta_{SO}\chi_M}\right) + kT \ln\left(\frac{f_{SM}}{f_{SO} f_M}\right) \tag{92}$$

Defining the adsorption constant, K_{SM}, for the adsorption of cation, M^+, as

$$kT \ln(K_{SM}) \equiv -(\mu_{SM}^* - \mu_M^* - \mu_{SO}^*) \tag{93}$$

Combining Eqs (92) and (93) gives the mass action equation for adsorption of cation, M^+, as

$$K_{SM} = \left(\frac{\theta_{SM}}{\theta_{SO}\chi_M}\right)\left(\frac{f_{SM}}{f_{SO} f_M}\right) \equiv \left(\frac{\theta_{SM}}{\theta_{SO}\chi_M}\right) Z^{\epsilon_{K_{sm}}} \quad \left(\begin{array}{c}\text{Cation mass}\\ \text{action equation}\end{array}\right) \tag{94}$$

with

$$\left(\frac{f_{SM}}{f_{SO} f_M}\right) \equiv Z^{\epsilon_{K_{sm}}} \quad \text{where } \epsilon_{K_{sm}} = \epsilon_{sm} - \epsilon_{SO} - \epsilon_M = -\frac{\nu_m(\nu_m - 2 - \bar{\nu}_s)}{1 + \kappa a_s} - \frac{\nu_m^2}{1 + \kappa a_m} \tag{95}$$

A sufficient number of equations have now been derived to make possible the derivation of the adsorption equations for cation adsorption. This can best be seen by collecting the relevant equations into a table in order to examine the number of equations and unknowns which have been discussed so far.

From Table 5, one can see that four equations in six unknowns are available for the derivation of the cation adsorption equation. Since the adsorption of interest

TABLE 5 Tabulation of Equations and Unknowns Used for Derivation of the Monovalent Cation Adsorption Equations

Unknowns		Equation	
No.	Variable	No.	Description
3	χ_{SOH}, χ_{SOH_2}, $\bar{\nu}_s$	(18)	$K_{a1} = \left(\frac{\theta_{SOH}\chi_H}{\theta_{SOH_2}}\right)Z^{\epsilon_{K_{a1}}}$
1	χ_{SO}	(20)	$K_{a2} = \left(\frac{\theta_{SO}\chi_H}{\theta_{SOH}}\right)Z^{\epsilon_{K_{a2}}}$
2	χ_M, θ_{SM}	(94)	$K_{SM} = \left(\frac{\theta_{SM}}{\theta_{SO}\chi_M}\right)Z^{\epsilon_{K_{sm}}}$
		(79)	$(1 = \theta_{SO} + \theta_{SOH} + \theta_{SOH_2} + \theta_{SM})$
6 unknowns			4 equations

is for the cation, the dependent variable will be θ_{SM}. Algebraically, one can expect that an adsorption equation derived from four equations can be expressed in terms of two of the six independent variables. This is satisfactory because many adsorption equations have conventionally been expressed as $\theta_{SM} = \theta_{SM}(\chi_M, Z^{\epsilon_{K_{SM}}})$. To derive the cation adsorption equation, we begin by solving the cation adsorption mass action equation for θ_{SM}:

$$\theta_{SM} = \theta_{SO}(\chi_M K_{SM} Z^{-\epsilon_{K_{sm}}}) \tag{96}$$

Substituting the mass balance equation (82) into Eq. (96):

$$\theta_{SM} = \left(\frac{1-\theta_{SM}}{\alpha_{SO}^{-1}}\right)(\chi_M K_{SM} Z^{-\epsilon_{K_{sm}}}) \tag{97}$$

Removing fractions and multiplying out all products:

$$\alpha_{SO}^{-1}\theta_{SM} = \chi_M K_{SM} Z^{-\epsilon_{K_{sm}}} - \chi_M K_{SM} Z^{-\epsilon_{K_{sm}}}\theta_{SM} \tag{98}$$

Collecting common terms in θ_{SM}:

$$\theta_{SM}(\alpha_{SO}^{-1} + \chi_M K_{SM} Z^{-\epsilon_{K_{sm}}}) = \chi_M K_{SM} Z^{-\epsilon_{K_{sm}}} \tag{99}$$

Solving for θ_{SM} gives the desired monovalent cation adsorption equation:

$$\theta_{SM} = \frac{\chi_M K_{SM} Z^{-\epsilon_{K_{sm}}}}{\alpha_{SO}^{-1} + \chi_M K_{SM} Z^{-\epsilon_{K_{sm}}}} \qquad \left(\begin{array}{c}\text{Cation adsorption}\\ \text{equation, } \theta_{SM}\end{array}\right) \tag{100}$$

Equation (100) is a rigorous adsorption equation which allows calculation of θ_{SM}, the fraction of surface sites complexed with the cation, M^+, from a knowledge of the activity coefficients, the equilibrium molar solution concentration, χ_M, of cation M^+, and the mass action constants for proton and cation surface complexation, K_{a1}, K_{a2}, and K_{SM}, respectively. Several important observations about the cation adsorption equations (98) and (100) can be made:

- The pH dependence for cation adsorption is determined by the α_{SO}^{-1} term. This pH dependence does not require different interfacial adsorption planes of differing electrostatic potential for different solute molecules such as inner- and outer-sphere complexes.
- Terms which contain solute mole fractions are products of two variables other than the molar cation solution concentration: (1) the mass action complexation constant, and (2) the activity coefficient quotient. Hence, the amount of cation adsorbed, θ_{SM}, is not only dependent on χ_M, but also on the affinity of the cation for the surface sites, K_{SM}, and the electrostatic interaction of the cation with other molecules, $Z^{-\epsilon_{K_{sm}}}$. The three-variable product, $\chi_M K_{SM} Z^{-\epsilon_{K_{sm}}}$, can be understood to be an "effective" molar cation solution concentration.
- Two of the three denominator terms in α_{SO}^{-1} are pH dependent in Eqs (98) and (100) insuring that predicted cation adsorption will increase as pH increases. Since one of the terms in α_{SO}^{-1} is unity, one can see that, unless $\chi_M K_{SM} Z^{-\epsilon_{K_{sm}}} \gg 1$, nearly complete adsorption, i.e., $\theta_{SM} \rightarrow 1$, is impossible, in general. For example, if $\chi_M K_{SM} Z^{-\epsilon_{K_{sm}}} = 1$, then the maximum possible cation adsorption is, $\theta_{SM} = 0.5$. Note also that when $\chi_M K_{SM} Z^{-\epsilon_{K_{sm}}} > 1$, nearly complete adsorption will always be possible although it might occur at pH values which are experimentally inaccessible.

Adsorption equation (100) expresses cation adsorption, θ_{SM}, defined as a function of the moles of cation adsorbed, n_{SM}, per mole of surface sites, n_s^T, or n_{SM}/n_s^T. In the literature (see Sec. VII), other measures of adsorption can be found. For example, cation adsorption, Γ_{SM}, for the Gibbs adsorption equation is usually expressed in terms of the moles of cation adsorbed, n_{SM}, per total solution–solid interfacial area, A_s^T, or n_{SM}/A_s^T. The conversion factor between θ_{SM} and Γ_{SM} is clearly A_s^T/n_s^T, which is the surface area per mole of surface sites. If the area of a single surface site is A_s, then the surface area for a mole of surface sites is $A_s N_A$, where N_A is Avogadro's number. The relationship between θ_{SM} and Γ_{SM} can be seen to be

$$\Gamma_{SM} \equiv \frac{n_{SM}}{A_s^T} = \left(\frac{n_{SM}}{n_s^T}\right) \cdot \left(\frac{n_s^T}{A_s^T}\right) = \theta_{SM} \cdot \left(\frac{n_s^T}{A_s^T}\right) = \frac{\theta_{SM}}{A_s N_A} \tag{101}$$

By combining Eqs (100) and (101) the Gibbs adsorption isotherm equation for the moles of cation adsorbed per square meter of pigment surface area, Γ_{SM}, can be written as

$$\Gamma_{SM} = \frac{\chi_M K_{SM} Z^{-\epsilon_{K_{sm}}}}{A_s N_A (\alpha_{SO}^{-1} + \chi_M K_{SM} Z^{-\epsilon_{K_{sm}}})} [=] \frac{\text{Moles cation adsorbed, } n_{SM}}{\text{Metal oxide surface area, } A_s^T} \left(\begin{array}{c}\text{Cation}\\ \text{adsorption}\\ \text{equation}\\ \Gamma_{SM}\end{array}\right) \tag{102}$$

Another measure of adsorption, $\widehat{\theta}_{SM}$, can be defined as the number of cations, n_{SM}, removed from solution as a result of adsorption per the total number of cations in the system, n_M^T. Since θ_{SM} is defined as the moles of surface sites complexed with cation, n_{SM}, per mole of surface sites, n_S^T, then the relationship between $\widehat{\theta}_{SM}$ and θ_{SM} can be expressed as

$$\widehat{\theta}_{SM} \equiv \left(\frac{n_{SM}}{n_M^T}\right) = \left(\frac{n_{SM}}{n_s^T}\right)\left(\frac{n_s^T}{V_w^T}\right)\left(\frac{V_w^T}{n_M^T}\right) = (\theta_{SM})(c_s^T)\left(\frac{1}{c_M^T}\right) = \theta_{SM}\left(\frac{c_s^T}{c_M^T}\right) \tag{103}$$

Two new concentration variables are defined in Eq. (103) as

$$c_s^T = \frac{n_s^T}{V_w^T} \tag{104}$$

$$c_M^T = \frac{n_M^T}{V_w^T} \tag{105}$$

Note that four symbols for concentration have now been introduced, so that concentration variables with different units can be easily distinguished:

- Equilibrium surface mole fractions are designated with the lower case Greek symbol, θ.
- Equilibrium molar solution concentrations are designated with the lower case Greek symbol, χ.
- Initial or formal or pre-equilibrium solution mole fractions are designated with the symbol, c^T. The symbol c_M^T represents the molar cation solution concentration prior to adsorption. The symbol c_s^T is the "effective" molar concentration of surface sites that would be in solution if the surface sites were homogeneously distributed in the aqueous phase. This is a hypothetical solution concentration which does not imply or require that surface sites be distributed evenly throughout the solution phase like solute ions. Instead, c_s^T is simply a ratio of mole of molecules defined above.

Equations (104) and (105) allow the total or initial number of moles of surface sites and cations to be expressed as solution concentrations. Substituting Eq. (100) into (103) gives the adsorption equation:

$$\hat{\theta}_{SM} = \frac{c_s^T(\chi_M K_{SM} Z^{-\epsilon_{K_{sm}}})}{c_M^T(\alpha_{SO}^{-1} + \chi_M K_{SM} Z^{-\epsilon_{K_{sm}}})} [=] \frac{\text{Moles cation removed from solution, } n_{SM}}{\text{Total moles of cation, } n_M^T} \left(\begin{array}{c}\text{Cation} \\ \text{adsorption} \\ \text{equation} \\ \widehat{\theta}_{SM}\end{array}\right) \tag{106}$$

Due to the difficulty in measuring the equilibrium cation solution concentration, χ_M, one further set of equations will be derived in order to eliminate χ_M. Elimination of χ_M from the heretofore derived cation adsorption equations will require an additional equation in no new unknown variables, which can be obtained from a mass balance over the cation M^+. If a mass balance over all the S–MO interfacial system for the cation M^+ is performed, it is clear that all the molecules of cation M^+ in the system n_M^T must either be in solution, n_M, or adsorbed, n_{SM}, so a mass balance can be written:

$$n_M^T = n_M + n_{SM} \tag{107}$$

Equation (107) can be converted into solution molar concentrations by division by the total volume of solution in the system, V_w.

$$c_M^T \equiv \frac{n_M^T}{V_w} = \frac{n_M}{V_w} + \frac{n_{SM}}{n_s^T} \cdot \frac{n_s^T}{V_w} = \chi_M + \theta_{SM} c_s^T \quad \begin{pmatrix} \text{Cation mass} \\ \text{balance equation} \end{pmatrix} \tag{108}$$

Note that the volume of the metal oxide solid is excluded from V_w.

Equation (108) represents a new equation in no new unknowns. If an equation summary is now tabulated and Eq. (108) is included, one can see in Table 6 that the number of equations and unknowns now differ by just one.

An equal number of equations and variables allows the elimination of all the unknown variables in all the cation adsorption equations derived so far if activity coefficient effects are ignored, so that $\bar{\nu}_s$ does not require evaluation. Evaluation of $\bar{\nu}_s$ will be discussed at the end of this section.

Combining Eqs (100) and (108) allows the equilibrium mole fraction, χ_M, to be replaced in the adsorption equation (100) by the more convenient formal or initial molar concentrations c_s^T and c_M^T to give

$$\theta_{SM} = \frac{(c_M^T - c_s^T \theta_{SM}) K_{SM} Z^{-\epsilon K_{sm}}}{\alpha_{SO}^{-1} + (c_M^T - c_s^T \theta_{SM}) K_{SM} Z^{-\epsilon K_{sm}}} \tag{109}$$

Equation (109) is quadratic in θ_{SM}:

$$\begin{aligned} &\theta_{SM}^2 (c_s^T K_{SM} Z^{-\epsilon K_{SM}}) \\ &- \theta_{SM} \left[\alpha_{SO}^{-1} + (c_M^T + c_s^T) K_{SM} Z^{-\epsilon K_{SM}} \right] \\ &+ c_M^T K_{SM} Z^{-\epsilon K_{SM}} = 0 \end{aligned} \tag{110}$$

TABLE 6 Tabulation of Equations and Unknowns Used for Derivation of the Monovalent Cation Adsorption Equations

Unknowns		Equation	
No.	Variable	No.	Description
3	$\chi_{SOH}, \chi_{SOH_2}, \bar{\nu}_s$	(18)	$K_{a1} = \left(\frac{\theta_{SOH} \chi_H}{\theta_{SOH_2}}\right) Z^{\epsilon K_{a1}}$
1	χ_{SO}	(20)	$K_{a2} = \left(\frac{\theta_{SO} \chi_H}{\theta_{SOH}}\right) Z^{\epsilon K_{a2}}$
2	χ_M, θ_{SM}	(94)	$K_{SM} = \left(\frac{\theta_{SM}}{\theta_{SO} \chi_M}\right) Z^{\epsilon K_{sm}}$
		(79)	$(1 = \theta_{SO} + \theta_{SOH} + \theta_{SOH_2} + \theta_{SM})$
0		(108)	$c_M^T = \chi_M + c_s^T \theta_{SM}$
6 unknowns			5 equations

The exact solution of Eq. (110) is

$$\theta_{SM} = \left[\frac{\alpha_{SO}^{-1} + (c_s^T + c_M^T)K_{SM}Z^{-\epsilon\kappa_{sm}}}{2c_s^T K_{SM}Z^{-\epsilon\kappa_{sm}}}\right] - \sqrt{\left[\frac{\alpha_{SO}^{-1} + (c_s^T + c_M^T)K_{SM}Z^{-\epsilon\kappa_{sm}}}{2c_s^T K_{SM}Z^{-\epsilon\kappa_{sm}}}\right]^2 - \frac{c_M^T}{c_s^T}} \quad \begin{pmatrix}\text{Cation adsorption}\\ \text{equation – quadratic}\\ \text{solution for } \theta_{SM}\end{pmatrix} \tag{111}$$

Using Eq. (103) to convert Eq. (111) from θ_{SM} to $\widehat{\theta}_{SM}$:

$$\widehat{\theta}_{SM} = \left[\frac{\alpha_{SO}^{-1} + (c_s^T + c_M^T)K_{SM}Z^{-\epsilon\kappa_{sm}}}{2c_M^T K_{SM}Z^{-\epsilon\kappa_{sm}}}\right] - \sqrt{\left[\frac{\alpha_{SO}^{-1} + (c_s^T + c_M^T)K_{SM}Z^{-\epsilon\kappa_{sm}}}{2c_M^T K_{SM}Z^{-\epsilon\kappa_{sm}}}\right]^2 - \frac{c_s^T}{c_M^T}} \quad \begin{pmatrix}\text{Cation adsorption}\\ \text{equation – quadratic}\\ \text{solution for } \widehat{\theta}_{SM}\end{pmatrix} \tag{112}$$

An approximate linear solution of Eq. (109) for θ_{SM} yields

$$\theta_{SM} \cong \frac{c_M^T K_{SM}Z^{-\epsilon\kappa_{sm}}}{\alpha_{SO}^{-1} + (c_s^T + c_M^T)K_{SM}Z^{-\epsilon\kappa_{sm}}} [=] \frac{n_{SM}}{n_s^T} \quad \begin{pmatrix}\text{Cation adsorption}\\ \text{equation – linear}\\ \text{solution for } \theta_{SM}\end{pmatrix} \tag{113}$$

where the approximation sign denotes that Eq. (113) is the approximate linear solution which results from dropping the squared term in the original quadratic equation (109). Since θ_{sm} must always be a fraction less than one, Eq. (113) would be expected to be a very good approximation to Eq. (111), deviating significantly only as $\theta_{sm} \to 1$.

Approximate linear adsorption equations can also be derived for the other two adsorption variables $\widehat{\theta}_{sm}$ and Γ_{sm}. Substitution of Eq. (113) into Eq. (103)

$$\widehat{\theta}_{sm} = \frac{c_s^T K_{sm}Z^{-\epsilon\kappa_{sm}}}{\alpha_{SO}^{-1} + (c_s^T + c_M^T)K_{sm}Z^{-\epsilon\kappa_{sm}}} \tag{114}$$

Substitution of Eq. (113) into Eq. (101)

$$\Gamma_{sm} = \frac{c_M^T K_{sm}Z^{-\epsilon\kappa_{sm}}}{A_A\left[\alpha_{SO}^{-1} + (c_s^T + c_M^T)K_{sm}Z^{-\epsilon\kappa_{sm}}\right]} \tag{115}$$

Note that five symbols for concentration have now been introduced so that concentration variables with different units can be easily distinguished:

- Equilibrium surface mole fractions are designated with the lower case Greek symbol, θ.
- Equilibrium molar solution concentrations are designated with the lower case Greek symbol, χ.
- Initial or total or pre-equilibrium solution and surface site molar concentrations are designated with the symbol, c^T.
- Equilibrium mole fraction of solute ions removed from solution are designated with the "capped" lower case Greek symbol, $\widehat{\theta}$.

- Equilibrium molar surface concentrations per unit surface area (Gibbs adsorption equation surface concentration) are designated with the upper case Greek symbol, Γ.

To examine the monovalent cation adsorption equation behavior we will begin by comparing adsorption predicted by the quadratic adsorption equation (111) and approximate linear solution equation (113). In both cases we will assume ideal adsorption behavior by setting activity coefficient quotients to unity, $Z^{\epsilon K_{sm}} = Z^{\epsilon K_{a2}} = Z^{\epsilon K_{a12}} = 1$.

Two important aspects of the cation adsorption equations are demonstrated in Fig. 8. First, when adsorption is measured as the mole fraction of surface sites complexed with a cation, θ_{SM}, then, as expected, θ_{SM} can be seen to decrease as the number of surface sites, c_s^T, approaches the number of cations available to adsorb, c_M^T, in moving from line b to line d in Fig. 8. As c_s^T exceeds c_M^T, the number of surface sites exceeds the number of cations available for adsorption so that a value of unity is impossible under any conditions for θ_{SM}.

The other important point about Fig. 8 is that, as c_s^T approaches and then exceeds c_M^T, the linear adsorption equation becomes much less accurate than the quadratic for predicting cation adsorption. Of course, both equations accurately predict cation adsorption when $c_s^T \ll c_M^T$.

Having derived the cation adsorption equations, it will now be of interest to examine how adsorption varies with the variables and constants in those equations. As a first example, the role of three variables and constants on cation adsorption predicted by Eq. (114) can be examined for ideal adsorption where the activity coefficient quotients have been set to unity. The role of activity coefficient quotients on adsorption will be examined later and in Chapter 4.

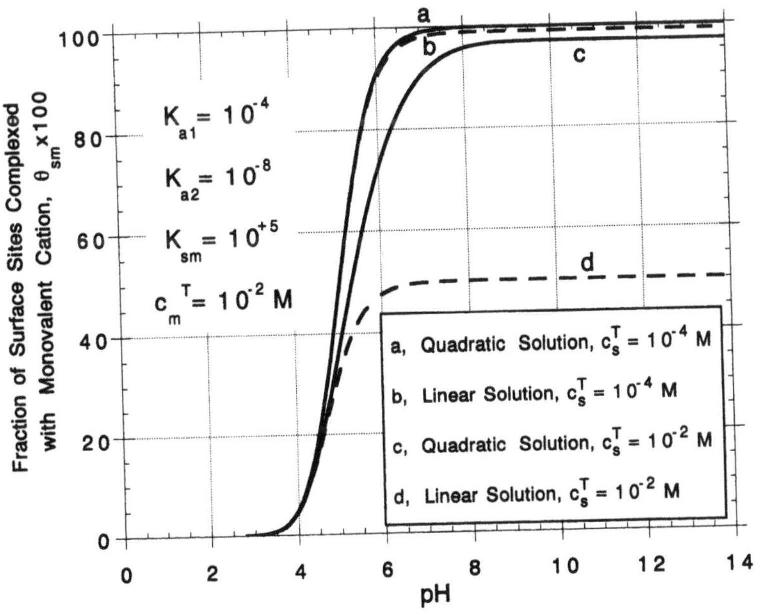

FIG. 8 Comparison of quadratic and approximate linear cation adsorption equations (111) and (113), respectively.

The lines in Fig. 9 give the cation adsorption behavior at a solution–solid interface as a function of constant pH with total cation mole fraction, c_M^T, and total surface-site mole fraction, c_s^T, held constant. The dominant trend in cation adsorption is that as pH increases so does cation adsorption because the fraction of anionic surface sites increases with increasing pH. The role of adsorption constants can be seen to be very significant. Adsorption occurs at a lower pH as the pK_{a1} and pK_{a2} increase and vice versa for the K_{SM}. Since the former are defined as dissociation constants [see Eqs (14) and (19)] while the latter is an association or formation constant [see Eq. (77)], the different relationship of these adsorption constants with adsorbed cation mole fraction is not surprising.

In Fig. 9, a plot of Eq. (114) results in predicted adsorption as a function of pH with total cation and surface-site mole fraction held constant. For the Gibbs plot in Fig. 10 of equation (115), adsorption is plotted as the areal surface concentration, Γ_{SM}, as a function of increasing total cation mole fraction [or equilibrium solution mole fraction with Eq. (102)] at constant pH and total surface-site concentration.

The pH dependence of cation adsorption at a S–MO interface is evident from lines a–c in Fig. 10. Also, the direct relationship of K_{SM} with adsorption and the inverse relationship of K_{a1} and K_{a2} with adsorption can be seen from lines c–e. The effect of c_s^T on adsorption can be determined by comparing lines b and f.

Before leaving the subject of monovalent cation adsorption at the S–MO interface, the role of activity coefficients on cation adsorption should be examined. Activity coefficient effects on adsorption of two hypothetical monovalent cations, both from 0.01 M solutions of 1:1 electrolyte will be examined. The cations will differ

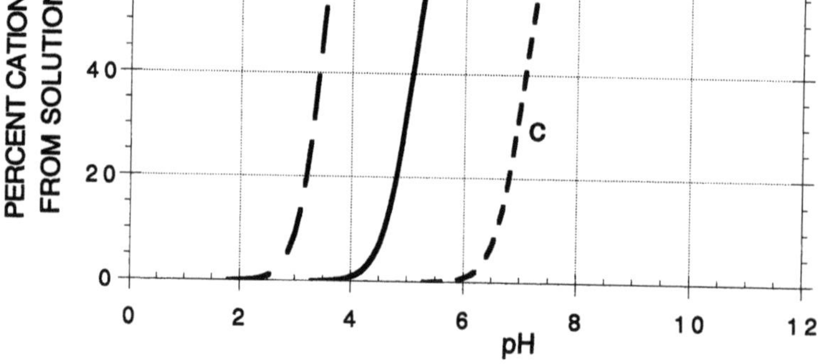

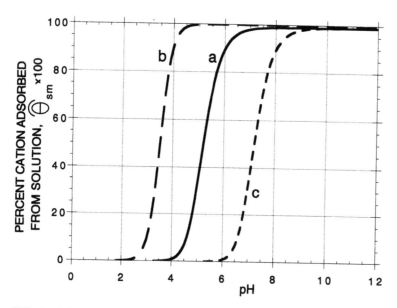

FIG. 9 Adsorption plot of Eq. (114) of fraction of cation adsorbed from solution, $\hat{\theta}_{SM}$, versus pH with $Z^{\epsilon K_{sm}} = Z^{\epsilon K_{a2}} = Z^{\epsilon K_{a12}} = 1$. Line a, $pK_{a1} = 5$, $pK_{a2} = 7$, $\log K_{SM} = 5$, $c_s^T = 1 \times 10^{-3}$, $c_M^T = 1 \times 10^{-6}$; line b, $pK_{a1} = 5$, $pK_{a2} = 7$, $\log K_{SM} = 8$, $c_s^T = 1 \times 10^{-3}$, $c_M^T = 1 \times 10^{-6}$; line c, $pK_{a1} = 7$, $pK_{a2} = 9$, $\log K_{SM} = 5$, $c_s^T = 1 \times 10^{-3}$, $c_M^T = 1 \times 10^{-6}$.

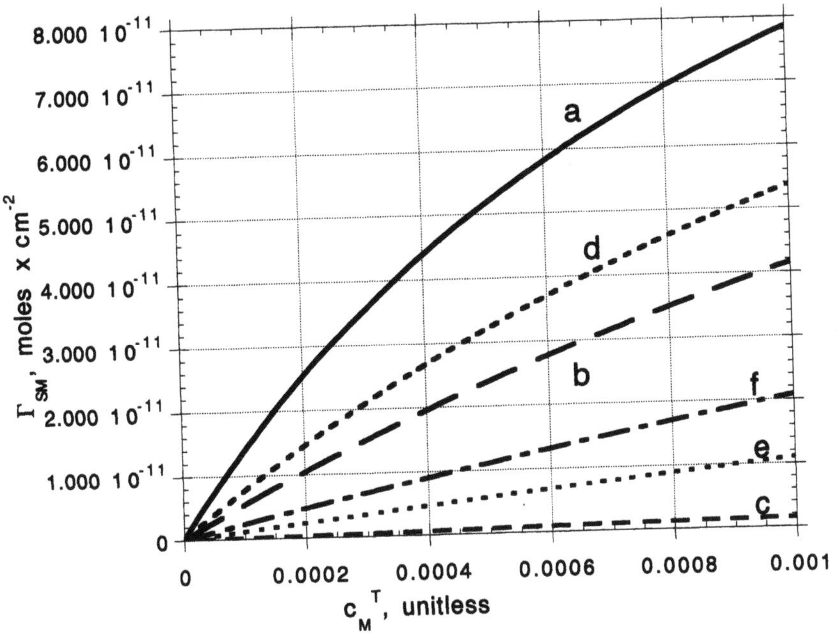

FIG. 10 Plot of Gibbs adsorption equation (115), Γ_{SM} versus c_M^T. Variables and constants used in calculation of lines, $Z^{\epsilon K_{sm}} = Z^{\epsilon K_{a2}} = Z^{\epsilon K_{a12}} = 1$. Line a, pH = 6, p$K_{a1}$ = 5, pK_{a2} = 7, log K_{SM} = 5, $c_s^T = 1 \times 10^{-3}$; line b, pH = 5, pK_{a1} = 5, pK_{a2} = 7, log K_{SM} = 5, $c_s^T = 1 \times 10^{-3}$; line c, pH = 4, pK_{a1} = 7, pK_{a2} = 9, log K_{SM} = 5, $c_s^T = 1 \times 10^{-3}$; line d, pH = 4, pK_{a1} = 5, pK_{a2} = 7, log K_{SM} = 7, $c_s^T = 1 \times 10^{-3}$; line e, pH = 4, pK_{a1} = 4.5, pK_{a2} = 6.5, log K_{SM} = 5, $c_s^T = 1 \times 10^{-3}$; line f, pH = 5, pK_{a1} = 5, pK_{a2} = 7, log K_{SM} = 5, $c_s^T = 5 \times 10^{-3}$.

in their bonding affinity. Cation 1 will adsorb strongly with a mass action adsorption constant of $K_{SM1} = 1.0 \times 10^5$, while the more weakly adsorbed cation 2 will have a mass action adsorption constant of $K_{SM1} = 1.0 \times 10^2$. The remainder of the data used in the calculations are given in Table 7 with the results plotted in Fig. 11.

All the data in Table 7 are self-explanatory with the exception of the mean surface-site electrostatic charge, $\bar{\nu}_s$, which is simply the fraction of positive less negative surface sites. For proton adsorption, $\bar{\nu}_s = \theta_{SOH_2} - \theta_{SO}$ (Table 3). When monovalent cations adsorb, the cation–surface site complex has a zero electrostatic charge. Thus, when monovalent cations adsorb, $\bar{\nu}_s$ becomes the fraction of positive less negative surface sites which are *not* bonded to a cation, or using Eqs (82) and (86):

$$\bar{\nu}_s = \theta_{SOH_2} - \theta_{SO}|_{\text{uncomplexed}} = (\alpha_{SOH_2} - \alpha_{SO})(1 - \theta_{SM}) \tag{116}$$

Combining Eqs (81), (85), and (116) gives the desired equation for $\bar{\nu}_s$:

$$\bar{\nu}_s = \left(\frac{\chi_H^2 Z^{\epsilon K_{a12}} - K_{a1} K_{a2}}{K_{a1} K_{a2} + \chi_H K_{a1} Z^{\epsilon K_{a1}} + \chi_H^2 Z^{\epsilon K_{a12}}} \right) (1 - \theta_{SM}) \tag{117}$$

Using adsorption equation (113) with the data in Table 7 and Eq. (117), the pH dependence of monovalent cation adsorption, θ_{sm}, at the S-MO interface can be

TABLE 7 Variables Used to Calculate Activity Coefficient Quotients

Variable	Value	Comments
q_B	3.44×10^{-10} m	Bjerrum association length [10, 11]
κ(0.01M)	3.29×10^{8} m^{-1}	Varies directly with solute ion concentration. For 1:1 electrolyte
$Z \equiv e^{-\kappa q_B}$	0.893	Chapter 4
$a_H = 9$ Å $\cong a_s$ $a_H = 9$ Å $\cong a_m$	3.29×10^{8} m^{-1}	Proton effective diameter is 9 Å and reasonable value for surface site and monovalent cation also (Chapter 4) [9,12]
$\bar{\nu}_s =$	$(\alpha_{SOH_2} - \alpha_{SO}) \times (1 - \theta_{SM})$	Equation (117)
K_{a1}	1.0×10^{-4}	Typical but arbitrarily chosen value
K_{a2}	1.0×10^{-8}	Typical but arbitrarily chosen value
$K_{sm1} =$	1.0×10^{5}	Cation 1, strongly adsorbed
$K_{sm2} =$	1.0×10^{2}	Cation 2, weakly adsorbed
c_s^T	1.0×10^{-4} M	Typical but arbitrarily chosen value
c_M^T	1.0×10^{-2} M	Typical but arbitrarily chosen value

calculated with the result shown in Fig. 11. Use of Eq. (117) to calculate activity coefficient quotients found in mass action equations is done iteratively by: (1) calculation of θ_{SM}^T under ideal adsorption conditions where $Z = 1$, using Eq. (113); (2) calculation of $\bar{\nu}_s$ using the ideal value of θ_{SM}^T in Eq. (117); (3) calculation of θ_{SM}^T using the calculated activity coefficient quotients in adsorption equation (113); and (4) repeat steps 2 and 3.

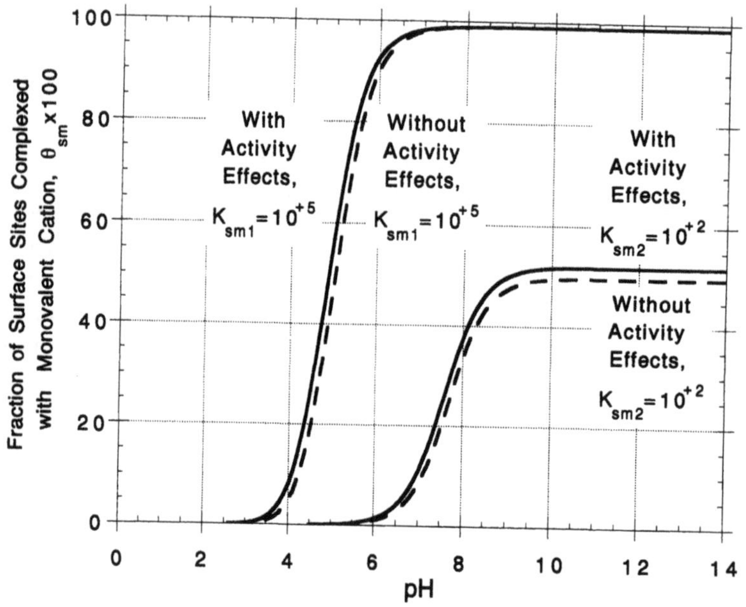

FIG. 11 Activity coefficient effects on monovalent cation adsorption. Adsorption data calculated using Eq. (113) and the raw data in Table 7.

Using the iterative calculation scheme, described for Eq. (117) to include activity effects in adsorption, required only one iteration. Differences in calculated divalent cation adsorption were $\ll 0.1\%$ between the first and second iterations.

At a pH of 5, the deviation between ID and AC adsorption curves for the case of strong adsorption ($K_{SM1} = 10^5$) is at a maximum and with a deviation of $\Delta\theta_{SM} = 0.0901$. In pH units the maximum deviation occurs at $\theta_{SM} = 0.50$, with a difference between ID and AC curves of $\Delta pH = 0.15$. The deviations between ID and AC for the weak adsorption curve ($K_{SM2} = 10^2$) is even smaller. It is also noteworthy that as K_{SM} decreases, the adsorbed cation mole fraction, even at high pH, will not reach unity. This is consistent with smaller values of K_{SM} being indicative of weaker adsorption energies between cations and surfaces sites.

Due to the difficulty of obtaining good adsorption data for S–MO interfacial systems, one concludes that differences between ID and AC cation adsorption equations are probably within the accuracy of experimental cation adsorption data. Not surprisingly, the same conclusion regarding differences between ID and AC adsorption equations and experimental adsorption data was reached for adsorption of protons, which are also monovalent cations.

Before leaving the subject of monovalent cation adsorption, it is instructive to replot the cation adsorption data in Fig. 10 in the $\log\theta_{SM}$ versus pH format so that cation adsorption and proton adsorption, shown in Fig. 3, can be compared. Adsorption data in Fig. 12 can be compared with proton adsorption data in Fig. 3. When cation adsorption is weak, $K_{SM2} = 10^2$, and the competition for surface sites between protons and cations is evidenced by the slope change in Fig. 12.

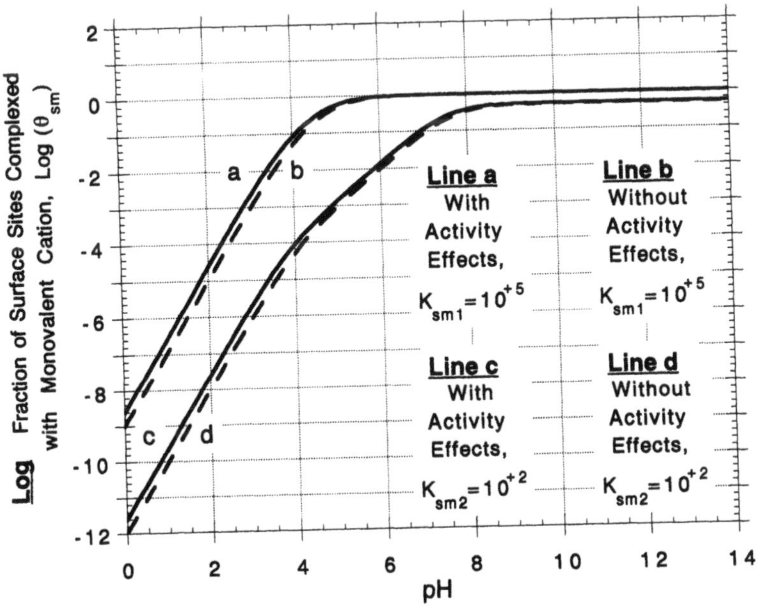

FIG. 12 Replot of Fig. 10 with $\log\theta_{sm}$ for activity coefficient effects on monovalent cation adsorption using Eq. (113) and the raw data in Table 7.

B. Divalent Cation Adsorption

When monovalent cations adsorb on metal oxide surfaces from aqueous solution, the most likely surface complex which forms results from a single monovalent cation adsorbing on a single surface site. When a divalent cation adsorbs at the S–MO interface, several surface complex stoichiometries are possible. Of the multiple stoichiometries which might form, the relative abundance of each will be determined by the mass action constant: larger indicating greater abundance and smaller indicating less. When mass action constants are determined experimentally, the relative significance of each divalent cation surface-site complex is immediately apparent. However, without experimentally determined mass action constants as a guide, one has no alternative but to choose the most "reasonable" divalent cation surface site complexes for inclusion in the adsorption equation. If many possible cation–surface site complexes are included in the adsorption equation derivation, the magnitude of each mass action constant (each mass action constant a cation–surface site complex stoichiometry) can be determined by fitting the resulting adsorption equation with experimental adsorption results.

For the following derivation, the three possible divalent cation–surface site complexes which will be considered are: (1) a single cation plus surface site, SOM^+, (2) a monohydroxylated cation plus surface site, SOMOH, and (3) a single cation plus double surface site, $^{SO}_{SO}M$. The disposition of these chemical interactions for divalent cation adsorption at the S–MO interface is shown in Fig. 13.

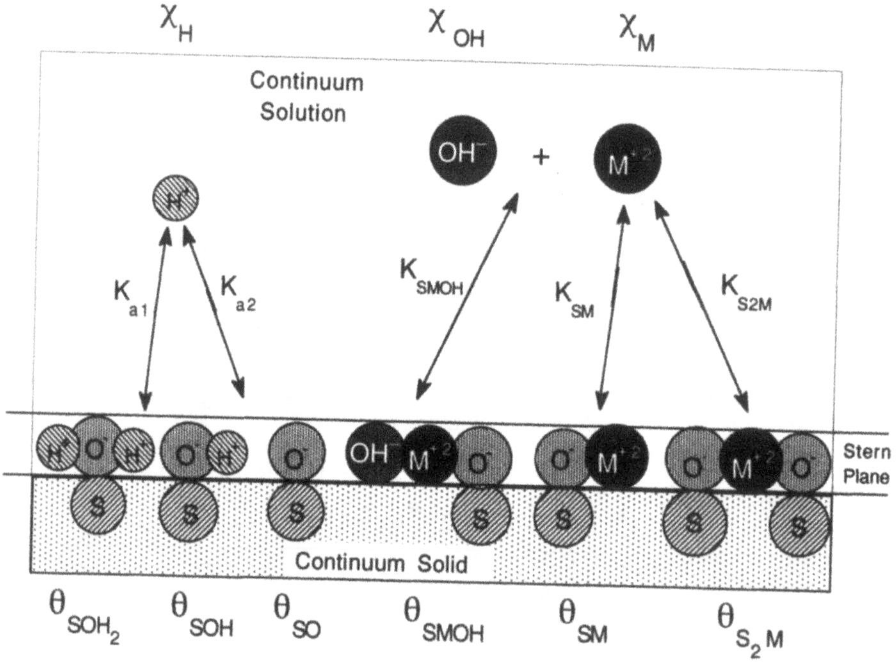

FIG. 13 Interfacial complexation reactions for divalent cation adsorption.

Mass action equations for the three surface complexes shown in Fig. 13 for divalent cation M^{2+} adsorption at the aqueous S–MO interface are:

1. A single cation plus surface site, SOM^+:

$$SO^- + M^{2+} = SOM^+ \quad \vdots \quad K_{SM} = \frac{\theta_{SM}}{\theta_{SO}\chi_M}\left(\frac{f_{SM}}{f_{SO}f_M}\right) = \frac{\theta_{SM}}{\theta_{SO}\chi_M}Z^{\epsilon_{K_{sm}}} \tag{118}$$

where

$$\epsilon_{K_{sm}} = \epsilon_{sm} - \epsilon_{SO} - \epsilon_m = -\frac{4\bar{\nu}_s}{1+\kappa a_s} - \frac{4}{1+\kappa a_m} \tag{119}$$

2. A monohydroxylated cation plus surface site, SOMOH:

$$SOM^+ + OH^- = SOMOH \quad \vdots \quad K_{SMOH} = \frac{\theta_{SMOH}}{\theta_{SM}\chi_{OH}}\left(\frac{f_{SMOH}}{f_{SM}f_{OH}}\right) = \frac{\theta_{SMOH}}{\theta_{SM}\chi_{OH}}Z^{\epsilon_{K_{SMOH}}} \tag{120}$$

where,

$$\epsilon_{K_{SMOH}} = \epsilon_{SMOH} - \epsilon_{SOM} - \epsilon_{OH} = -\frac{1+3\bar{\nu}-\bar{\nu}_s^2}{1+\kappa a_s} - \frac{1}{1+\kappa a_{OH}} \tag{121}$$

3. A single cation plus two surface sites, ${}^{SO}_{SO}M$:

$$2SO^- + M^{2+} = {}^{SO}_{SO}M \quad \vdots \quad K_{S_2M} = \frac{\theta_{S_2M}}{\theta_{SO}^2\chi_M}\left(\frac{f_{S_2M}}{f_{SO}^2 f_M}\right) = \frac{\theta_{S_2M}}{\theta_{SO}^2\chi_M}Z^{\epsilon_{K_{S_2M}}} \tag{122}$$

where,

$$\epsilon_{KS_2M} = \epsilon_{S_2M} - 2\epsilon_{SO} - \epsilon_M = -\frac{2+4\bar{\nu}_s}{1+\kappa a_s} - \frac{4}{1+\kappa a_M} \tag{123}$$

The mass action equations can be rearranged for expressions of the surface site and solute concentrations as

$$\theta_{SM} = \theta_{SO}\chi_M K_{SM} Z^{-\epsilon_{K_{sm}}} \tag{124}$$

$$\theta_{SMOH} = \theta_{SM}K_{SMOH}\chi_{OH}Z^{-\epsilon_{K_{SMOH}}} = \frac{\theta_{SM}K_{SMOH}K_w Z^{-\epsilon_{K_{SMOH}}}}{\chi_H} \tag{125}$$

or by substituting Eq. (124) into (123):

$$\theta_{SMOH} = \theta_{SO}\frac{\chi_M K_{SM}K_{SMOH}K_w Z^{-\epsilon_{K_{SMOH}}}Z^{-\epsilon_{K_{SMOH}}}}{\chi_H} \equiv \theta_{SO}\frac{\chi_M K''_{SMOH}Z^{-\epsilon''_{K_{SMOH}}}}{\chi_H} \tag{126}$$

where

$$\epsilon''_{K_{SMOH}} \equiv \epsilon_{K_{SMOH}} + \epsilon_{K_{SM}} = -\frac{1+7\bar{\nu}_s-\bar{\nu}_s^2}{1+\kappa a_s} - \frac{4}{1+\kappa a_M} - \frac{1}{1+\kappa a_{OH}} \tag{127}$$

$$\theta_{S_2M} = \theta_{SO}^2\chi_M K_{S_2M}Z^{-\epsilon_{K_{s_2m}}} \tag{128}$$

The surface-site mass balance equation is the sum of the moles of all surface sites shown in Fig. 13, which gives

$$1 = n_{SO} + n_{SOH} + n_{SOH_2} + n_{SM} + n_{SMOH} + 2n_{S_2M} \tag{129}$$

Note the coefficient of 2 in the last term, which arises for the surface-site complex consisting of two surface sites in a surface-site mass balance:

$$1 = \theta_{SO} + \theta_{SOH} + \theta_{SOH_2} + \theta_{SM} + \theta_{SMOH} + 2\theta_{S_2M} \equiv \theta_{SO} + \theta_{SOH} + \theta_{SOH_2} + \theta_{SM}^{T} \tag{130}$$

Solving Eq. (130) for the total mole fraction of surface sites complexed to a cation, θ_{SM}^{T}:

$$\theta_{SM}^{T} = 1 - (\theta_{SO} + \theta_{SOH} + \theta_{SOH_2}) \tag{131}$$

Rearranging Eq. (81) and substituting into Eq. (131):

$$\theta_{SO} + \theta_{SOH} + \theta_{SOH_2} = \theta_{SO}\alpha_{SO}^{-1} = 1 - \theta_{SM}^{T} \tag{132}$$

$$\theta_{SO} = \alpha_{SO}(1 - \theta_{SM}^{T}) \quad \vdots \quad \theta_{SM}^{T} = 1 - \alpha_{SO}^{-1}\theta_{SO} \tag{133}$$

From comparison of Eq. (81), derived for monovalent cation adsorption, and Eq. (133), derived for divalent cation adsorption, one can see that these mass balance equations have a general form.

In a similar fashion the mass balance equation for the divalent cation can be written as a mass balance over all cations in the system with the total moles of cation, n_M^T, distributed in solution as n_M, or among three types of surface complexes, n_{SM}, n_{SMOH}, or n_{S_2M}, which can be summed to give

$$n_M^T = n_M + n_{SM} + n_{SMOH} + n_{S_2M} \tag{134}$$

Division by V_w and substitution of mass action equations gives the divalent cation mass balance equation:

$$\begin{aligned} \frac{n_M^T}{V_w} \equiv c_M^T &= \frac{n_M}{V_w} + \frac{n_{SM}}{n_s^T}\cdot\frac{n_s^T}{V_w} + \frac{n_{SMOH}}{n_s^T}\cdot\frac{n_s^T}{V_w} + \frac{n_{S_2M}}{n_s^T}\cdot\frac{n_s^T}{V_w} \\ &= \chi_M + \theta_{SM}c_s^T + \theta_{SMOH}c_s^T + \theta_{S_2M}c_s^T \\ &= \chi_M + c_s^T(\theta_{SM} + \theta_{SMOH} + \theta_{S_2M}) \equiv \chi_M + \theta_M^T c_s^T \end{aligned} \tag{135}$$

Defining variables for the parenthetical expressions gives the divalent cation mass balance equation in the same form as found for the monovalent cation in Eq. (108):

$$c_M^T \equiv \chi_M + \theta_M^T c_s^T \tag{136}$$

Comparing Eqs (129) and (134) one notes the coefficients of 2 and 1 for the n_{S_2M} term, respectively. This coefficient difference for surface site and cation mass balances is due to the stoichiometry of the $(SO)_2M$ surface site complex. As a result, two different cation complexed surface-site mole fractions must be defined:

$$\theta_{SM}^{T} \equiv \theta_{SM} + \theta_{SMOH} + 2\theta_{S_2M} \tag{137}$$

$$\theta_{M}^{T} \equiv \theta_{SM} + \theta_{SMOH} + \theta_{S_2M} \tag{138}$$

In order to combine cation and surface-site mass balance equations it will be necessary to find the relationship between surface complex mole fractions. Substituting Eq. (138) into (137):

$$\theta_{SM}^{T} = 2\theta_{M}^{T} - \theta_{SM} - \theta_{SMOH} \tag{139}$$

substituting Eqs (122), (124), and (131) into Eq. (137):

$$\theta_{SM}^{T} = 2\theta_{M}^{T} - \left(\chi_M K_{SM} Z^{-\epsilon_{K_{sm}}} + \frac{\chi_M K''_{SMOH} Z^{-\epsilon''_{K_{SMOH}}}}{\chi_H}\right)(1-\theta_{SM}^{T})\alpha_{SO} \tag{140}$$

and solving for θ_{SM}^{T}:

$$\theta_{SM}^{T} = \frac{2\alpha_{SO}^{-1}\theta_{M}^{T} - \chi_M K_{SM} Z^{-\epsilon_{K_{sm}}} + \dfrac{\chi_M K''_{SMOH} Z^{-\epsilon''_{K_{SMOH}}}}{\chi_H}}{\alpha_{SO}^{-1} - \chi_M K_{SM} Z^{-\epsilon_{K_{sm}}} + \dfrac{\chi_M K''_{SMOH} Z^{-\epsilon''_{K_{SMOH}}}}{\chi_H}} \equiv \frac{2\alpha_{SO}^{-1}\theta_{M}^{T} - \chi_M X_{1SO}}{\alpha_{SO}^{-1} - \chi_M X_{1SO}} \tag{141}$$

where X_{1SO} represents the sum of the products of all mass action constants and activity coefficient quotients products involving only one SO^- surface site:

$$X_{1SO} \equiv K_{SM} Z^{-\epsilon_{K_{sm}}} + \frac{K''_{SMOH} Z^{-\epsilon''_{K_{SMOH}}}}{\chi_H} \tag{142}$$

The utility of the variable X_{1SO} will become apparent in later derivations.

Another useful equation can also be derived from Eqs (133) and (140):

$$1 - \alpha_{SO}^{-1}\theta_{SO} = 2\theta_{M}^{T} - \left(\chi_M K_{SM} Z^{-\epsilon_{K_{sm}}} + \frac{\chi_M K''_{SMOH} Z^{-\epsilon''_{K_{SMOH}}}}{\chi_H}\right)\theta_{SO} \tag{143}$$

and solving for θ_{SO}:

$$\theta_{SO} = \frac{2\theta_{M}^{T} - 1}{\chi_M K_{SM} Z^{-\epsilon_{K_{sm}}} + \dfrac{\chi_M K''_{SMOH} Z^{-\epsilon''_{K_{SMOH}}}}{\chi_H} - \alpha_{SO}^{-1}} = \frac{2\theta_{M}^{T} - 1}{\chi_M X_{1SO} - \alpha_{SO}^{-1}} \tag{144}$$

The last variable requiring an explicit equation is the mean surface-site electrostatic charge, $\bar{\nu}_s$. From Fig. 13 one can see that of the six surface site complexes shown only three have a finite electrostatic charge: (1) −1 electrostatic charge on the SO^- surface site complex, (2) +1 electrostatic charge on the SOH_2^+ surface site complex, and (3) +1 electrostatic charge on the SOM^+ surface site complex. The mean surface-site electrostatic charge, $\bar{\nu}_s$, is the sum of the surface site fractions for these electrostatically charged surface sites:

$$\bar{\nu}_s = \theta_{SOH_2} + \theta_{SM} - \theta_{SO} \tag{145}$$

The appearance of the surface mole fraction for a surface complex requires the derivation of an α_{SM} equation:

$$\alpha_{SM} \equiv \frac{\theta_{SM}}{\theta_{SM} + \theta_{SMOH} + 2\theta_{S_2M}} = \frac{\theta_{SM}}{\theta_{SM}^{T}} \tag{146}$$

Substitution of Eqs (124), (125), and (128) into Eq. (146):

$$\alpha_{SM} \equiv \frac{K_{SM} Z^{-\epsilon_{K_{sm}}}}{K_{SM} Z^{-\epsilon_{K_{sm}}} + \dfrac{K_w K_{SM} K_{SMOH} Z^{-\epsilon_{K_{SM}}} Z^{-\epsilon_{K_{SMOH}}}}{\chi_H} + \theta_{SO} K_{S_2M} Z^{-\epsilon_{K_{s_2m}}}} \tag{147}$$

Equation (145) can now be written in terms of variables which can all be readily evaluated as

$$\bar{\nu}_s = \theta_{SOH_2} + \theta_{SM} - \theta_{SO} = (\alpha_{SOH_2} - \alpha_{SO})(1 - \theta_{SM}^{T}) + \alpha_{SM}\theta_{SM}^{T} \tag{148}$$

As in Section III.A for monovalent cation adsorption, Eq. (148) allows calculation of the mean surface-site electrostatic charge, $\bar{\nu}_s$, once the fraction of surface sites complexed with a cation, θ_{SM}^T, has been determined. Use of Eq. (148) to calculate activity coefficient quotients found in mass action equations is found iteratively by calculation of: (1) θ_{SM}^T under ideal adsorption conditions where $Z = 1$, (2) calculation of $\bar{\nu}_s$ using the ideal value of θ_{SM}^T, (3) calculation of θ_{SM}^T using calculated activity coefficient quotients in the adsorption equation, and (4) repeat steps 3 and 4.

All the equations needed to derive the divalent cation adsorption equation have now been derived. A listing of the equations is given in Table 8.

Table 8 shows the eight independent equations which have been derived for the eight unknown variables necessary to specify the adsorption of divalent cations at the aqueous S–MO interface as depicted in Fig. 13. This suite of equations can be solved iteratively or solved simultaneously for θ_M^T. No iterative solution will be attempted here. Instead, simulaneous solutions of the equations in Table 7 will be explored.

The solution for θ_M^T begins with Eq. (138), which is a mass balance over all moles of adsorbed divalent cations. Substitution of mass action equations (118), (120), and (122) gives

$$\theta_M^T = \theta_{SO}\chi_M\left(K_{SM}Z^{-\epsilon_{K_{sm}}} + \frac{K''_{SMOH}Z^{-\epsilon''_{K_{SMOH}}}}{\chi_H} + \theta_{SO}K_{S_2M}Z^{-\epsilon_{K_{S_2m}}}\right) \tag{149}$$

TABLE 8 Tabulation of Equations and Unknowns Used for Derivation of the Divalent Cation Adsorption Equations

Unknowns		Equation	
No.	Variable	No.	Description
3	$\chi_{SOH}, \chi_{SOH_2}, \bar{\nu}_s$	(18)	$K_{a1} = \left(\frac{\theta_{SOH}\chi_H}{\theta_{SOH_2}}\right)Z^{\epsilon_{K_{a1}}}$
1	χ_{SO}	(20)	$K_{a2} = \left(\frac{\theta_{SO}\chi_H}{\theta_{SOH}}\right)Z^{\epsilon_{K_{a2}}}$
2	χ_M, θ_{SM}	(118)	$K_{SM} = \frac{\theta_{SM}}{\theta_{SO}\chi_M}Z^{\epsilon_{K_{sm}}}$
1	θ_{SMOH}	(120)	$K_{SMOH}\frac{\theta_{SMOH}}{\theta_{SM}\chi_{OH}}Z^{\epsilon_{K_{SMOH}}}$
1	θ_{S_2M}	(122)	$K_{S_2M} = \frac{\theta_{S_2M}}{\theta_{SO}^2\chi_M}Z^{\epsilon_{K_{S_2M}}}$
0		(133)	$\theta_{SO} = \alpha_{SO}(1 - \theta_{SM}^T)$
0		(148)	$\bar{\nu}_s = (\alpha_{SOH_2} - \alpha_{SO})(1 - \theta_{SM}^T) + \alpha_{SM}\theta_{SM}^T$
0		(135)	$c_M^T = \chi_M + c_s^T\theta_M^T$
8 unknowns			8 equations

Substitution of Eq. (144) into (149) to eliminate θ_{SO} gives

$$\begin{aligned}\theta_{\mathrm{M}}^{\mathrm{T}} &= \theta_{\mathrm{SO}}\chi_{\mathrm{M}}\left(X_{1\mathrm{SO}} + \theta_{\mathrm{SO}}K_{\mathrm{S_2M}}Z^{-\epsilon K_{\mathrm{s_2m}}}\right)\\ &= \chi_{\mathrm{M}}\left(\frac{2\theta_{\mathrm{M}}^{\mathrm{T}} - 1}{\chi_{\mathrm{M}}X_{1\mathrm{SO}} - \alpha_{\mathrm{SO}}^{-1}}\right)\left[X_{1\mathrm{SO}} + \left(\frac{2\theta_{\mathrm{M}}^{\mathrm{T}} - 1}{\chi_{\mathrm{M}}X_{1\mathrm{SO}} - \alpha_{\mathrm{SO}}^{-1}}\right)X_{2\mathrm{SO}}\right]\end{aligned} \tag{150}$$

where $X_{2\mathrm{SO}}$ represents the product sum of all mass action constant–activity coefficient quotients involving two SO^- surface sites:

$$X_{2\mathrm{SO}} \equiv K_{\mathrm{S_2M}}Z^{-\epsilon K_{\mathrm{s_2m}}} \tag{151}$$

Since Eq. (151) is quadratic in $\theta_{\mathrm{M}}^{\mathrm{T}}$ it will be solved by grouping terms in like powers of $\theta_{\mathrm{M}}^{\mathrm{T}}$:

$$\begin{aligned}(\theta_{\mathrm{M}}^{\mathrm{T}})^2(4\chi_{\mathrm{M}}X_{2\mathrm{SO}}) - \theta_{\mathrm{M}}^{\mathrm{T}}(\alpha_{\mathrm{SO}}^{-2} - \chi_{\mathrm{M}}^2X_{1\mathrm{SO}}^2 + 4\chi_{\mathrm{M}}X_{2\mathrm{SO}})&\\ + \chi_{\mathrm{M}}(\alpha_{\mathrm{SO}}^{-1}X_{1\mathrm{SO}} - \chi_{\mathrm{M}}X_{1\mathrm{SO}}^2 + X_{2\mathrm{SO}}) = 0&\end{aligned} \tag{152}$$

with solution:

$$\begin{aligned}\theta_{\mathrm{M}}^{\mathrm{T}} = {}&\frac{\alpha_{\mathrm{SO}}^{-2} - \chi_{\mathrm{M}}^2X_{1\mathrm{SO}}^2 + 4\chi_{\mathrm{m}}X_{2\mathrm{SO}}}{8\chi_{\mathrm{m}}X_{2\mathrm{SO}}}\\ &- \frac{\sqrt{(\alpha_{\mathrm{SO}}^{-2} - \chi_{\mathrm{M}}^2X_{1\mathrm{SO}}^2 + 4\chi_{\mathrm{M}}X_{2\mathrm{SO}})^2 - 4(\alpha_{\mathrm{SO}}^{-1}\chi_{\mathrm{M}}X_{2\mathrm{SO}} - \chi_{\mathrm{M}}^2X_{1\mathrm{SO}}^2 + 2\chi_{\mathrm{M}}X_{2\mathrm{SO}})}}{8\chi_{\mathrm{M}}X_{2\mathrm{SO}}}\end{aligned} \tag{153}$$

Equation (153) is the adsorption equation for divalent cation adsorption, $\theta_{\mathrm{M}}^{\mathrm{T}}$, at the aqueous S–MO interface as a function of cation solution molar concentration, χ_{M}.

The adsorption equation for divalent cation adsorption, $\theta_{\mathrm{M}}^{\mathrm{T}}$, can be expressed as a function of total or initial cation solution molar concentration, $c_{\mathrm{M}}^{\mathrm{T}}$, by first solving the cation mass balance equation (135) for χ_{M}:

$$\chi_{\mathrm{M}} \equiv c_{\mathrm{M}}^{\mathrm{T}} - \theta_{\mathrm{M}}^{\mathrm{T}}c_{\mathrm{s}}^{\mathrm{T}} \tag{154}$$

then substituting the result into Eq. (152) to give

$$\begin{aligned}&(\theta_{\mathrm{M}}^{\mathrm{T}})^2\left[4(c_{\mathrm{M}}^{\mathrm{T}} - \theta_{\mathrm{M}}^{\mathrm{T}}c_{\mathrm{s}}^{\mathrm{T}})X_{2\mathrm{SO}}\right]\\ &- \theta_{\mathrm{M}}^{\mathrm{T}}\left[\alpha_{\mathrm{SO}}^{-2} - (c_{\mathrm{M}}^{\mathrm{T}} - \theta_{\mathrm{M}}^{\mathrm{T}}c_{\mathrm{s}}^{\mathrm{T}})^2X_{1\mathrm{SO}}^2 + 4(c_{\mathrm{M}}^{\mathrm{T}} - \theta_{\mathrm{M}}^{\mathrm{T}}c_{\mathrm{s}}^{\mathrm{T}})X_{2\mathrm{SO}}\right]\\ &+ (c_{\mathrm{M}}^{\mathrm{T}} - \theta_{\mathrm{M}}^{\mathrm{T}}c_{\mathrm{s}}^{\mathrm{T}})\left[\alpha_{\mathrm{SO}}^{-1}X_{1\mathrm{SO}} - (c_{\mathrm{M}}^{\mathrm{T}} - \theta_{\mathrm{M}}^{\mathrm{T}}c_{\mathrm{s}}^{\mathrm{T}})X_{1\mathrm{SO}}^2 + X_{2\mathrm{SO}}\right] = 0\end{aligned} \tag{155}$$

Equation (155) is cubic in $\theta_{\mathrm{M}}^{\mathrm{T}}$ and can be multiplied out and the terms collected to give

$$\begin{aligned}&(\theta_{\mathrm{M}}^{\mathrm{T}})^3\left[c_{\mathrm{s}}^{\mathrm{T}}(c_{\mathrm{s}}^{\mathrm{T}}X_{1\mathrm{SO}}^2 - 4X_{2\mathrm{SO}})\right]\\ &- (\theta_{\mathrm{M}}^{\mathrm{T}})^2\left[c_{\mathrm{s}}^{\mathrm{T}}(2c_{\mathrm{M}}^{\mathrm{T}} + c_{\mathrm{s}}^{\mathrm{T}})X_{1\mathrm{SO}}^2 - 4(c_{\mathrm{M}}^{\mathrm{T}} + c_{\mathrm{s}}^{\mathrm{T}})X_{2\mathrm{SO}}\right]\\ &+ \theta_{\mathrm{M}}^{\mathrm{T}}\left\{\left[(c_{\mathrm{M}}^{\mathrm{T}})^2X_{1\mathrm{SO}} + 2c_{\mathrm{M}}^{\mathrm{T}}c_{\mathrm{s}}^{\mathrm{T}}X_{1\mathrm{SO}} - c_{\mathrm{s}}^{\mathrm{T}}\alpha_{\mathrm{SO}}^{-1}\right]X_{1\mathrm{SO}} - (4c_{\mathrm{M}}^{\mathrm{T}} + c_{\mathrm{s}}^{\mathrm{T}})X_{2\mathrm{SO}} - \alpha_{\mathrm{SO}}^{-2}\right\}\\ &+ c_{\mathrm{M}}^{\mathrm{T}}\left[(\alpha_{\mathrm{SO}}^{-1} - c_{\mathrm{M}}^{\mathrm{T}}X_{1\mathrm{SO}})X_{1\mathrm{SO}} + X_{2\mathrm{SO}}\right] = 0\end{aligned} \tag{156}$$

In most cases iterative solutions of Eq. (156) are more practical than analytical ones. However, intuitive analysis is not possible from iterative solutions. For this reason, approximate linear solutions of Eq. (156) will be derived and examined.

The approximate linear solution to Eq. (156), solving for θ_M^T, is

$$\theta_M^T = \frac{c_M^T[(\alpha_{SO}^{-1} - c_M^T X_{1SO})X_{1SO} + X_{2SO}]}{\alpha_{SO}^{-2} - [(c_M^T)^2 X_{1SO} + 2c_M^T c_s^T X_{1SO} - c_s^T \alpha_{SO}^{-1}]X_{1SO} + (4c_M^T + c_s^T)X_{2SO}} \tag{157}$$

The pH dependence of adsorption equation (157) can be seen in Fig. 14.

The adsorption behavior shown in Fig. 14 for divalent cation adsorption was determined by using the approximate linear solution equation (157). Based on the linear approximation in this equation one would expect good accuracy only for low adsorption, say, $\theta_M^T < 0.3$, which is consistent with the findings in Fig. 8 for monovalent cation adsorption. However, useful information can be gleaned from Fig. 14 in the region $\theta_M^T < 0.3$:

1. Divalent cation adsorption dramatically increases when the mass action constant K_{SM} increases, indicating stronger cation adsorption for the SOM^+ cation–surface site complex as seen from lines a and c.

2. Divalent cation adsorption is essentially unaffected when either mass action constant K_{SMOH} (line b) or K_{S_2M} (line d) increases. The explanation for this behavior requires further investigation. The similarity of behavior among adsorption lines a, b, and d suggests that enhanced adsorption through formation of the SOMOH and/or $(SO)_2M$ cation–surface site complexes (by increasing K_{SMOH} and K_{S_2M}, respectively) depletes adsorption of such complexes of other stoichiometries in such a way as to keep constant the total number of moles of cation adsorbed. The role of the SOMOH and $(SO)_2M$ cation–surface complexes in adsorption of divalent cations may become more pronounced as $\theta_M^T \to 1$, where the more accurate equation (156) should be used to calculate adsorption.

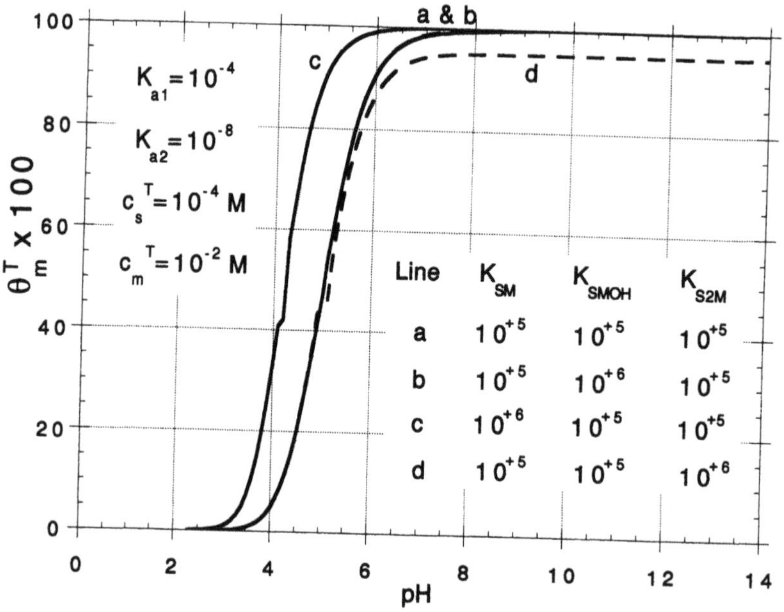

FIG. 14 Divalent cation adsorption at the aqueous S–MO interface from Eq. (157).

The fact that the surface complexes SOMOH and $(SO)_2M$ apparently contribute very little to cation adsorption (at least for $\theta_M^T < 0.3$) provides an opportunity to derive a much simpler adsorption equation. If the surface complexes SOMOH and $(SO)_2M$ are assumed to contribute little to divalent cation adsorption, the adsorption equation (150) when combined with Eq. (142) reduces to

$$\theta_M^T = \theta_{SO}\chi_M(K_{SM}Z^{-\epsilon_{Ksm}}) = \chi_M\left(\frac{2\theta_M^T - 1}{\chi_M K_{SM}Z^{-\epsilon_{Ksm}} - \alpha_{SO}^{-1}}\right)(K_{SM}Z^{-\epsilon_{Ksm}}) \tag{158}$$

Solving for θ_M^T:

$$\theta_M^T = \frac{\chi_M K_{SM}Z^{-\epsilon_{Ksm}}}{\alpha_{SO}^{-1} + \chi_M K_{SM}Z^{-\epsilon_{Ksm}}} \tag{159}$$

Equation (159) for divalent cation adsorption can be seen to be identical in form to the adsorption equation (100) for monovalent cation adsorption. Therefore, since formation of the surface complexes SOMOH and $(SO)_2M$ seem to have a minimal effect on the calculated adsorption of divalent cations, the simple adsorption equation (159) should accurately predict both monovalent and divalent cation adsorption at the aqueous S–MO interface. Following the monovalent cation adsorption derivation, the adsorption equation for divalent cations in terms of total cation and surface site concentration, c_M^T and c_s^T, respectively, would be

$$\theta_M^T \cong \frac{c_M^T K_{SM}Z^{-\epsilon_{Ksm}}}{\alpha_{SO}^{-1} + (c_s^T + c_M^T)K_{SM}Z^{-\epsilon_{Ksm}}} \tag{160}$$

If it is shown experimentally that significant formation of surface complexes SOMOH and $(SO)_2M$ occurs, then use of the more complicated adsorption equations (153) or (156) would be advisable.

The adsorption equations allow calculation of θ_M^T. The definition of θ_M^T is

$$\theta_M^T \equiv \frac{n_{SM} + n_{SMOH} + n_{S_2M}}{n_s^T} [=] \frac{\text{Moles of cation adsorbed}}{\text{Total moles of surface sites}} \tag{161}$$

Other measures of adsorption, which were defined for monovalent cation adsorption in Eqs (101) and (103), are based on θ_M^T and may be useful:

- The mole fraction of surface sites complexed with divalent cations, θ_{SM}^T, can be determined from Eq. (141).
- Moles of divalent cation adsorbed per total mole of cation in the system $\hat{\theta}_M^T$:

$$\hat{\theta}_M^T \equiv \frac{n_{SM} + n_{SMOH} + n_{S_2M}}{n_M^T} = \left(\frac{n_{SM} + n_{SMOH} + n_{S_2M}}{n_s^T}\right)\left(\frac{n_s^T}{V_w^T}\right)\left(\frac{V_w^T}{n_M^T}\right)$$

$$= (\theta_M^T)(c_s^T)\left(\frac{1}{c_M^T}\right) = (\theta_M^T)\left(\frac{c_s^T}{c_M^T}\right) \tag{162}$$

- Moles of divalent cation adsorbed per unit area of S–MO interface, Γ_M^T:

$$\Gamma_M^T \equiv \frac{n_{SM} + n_{SMOH} + n_{S_2M}}{A_s^T} = \left(\frac{n_{SM} + n_{SMOH} + n_{S_2M}}{n_s^T}\right)\left(\frac{n_s^T}{A_s^T}\right)$$

$$= \theta_M^T\left(\frac{n_s^T}{A_s^T}\right) = \frac{\theta_M^T}{A_s N_A} \qquad (163)$$

C. Trivalent Cation Adsorption

As with divalent cation adsorption at the aqueous S–MO interface, numerous cation–surface site stoichiometries are possible. Since there is no means to predict which stoichiometries are predominant and which are insignificant, the trivalent cation adsorption equation derivation will consider six of the more likely complexes. Modification of the resulting adsorption equation can then proceed as experimental evidence identifies which stoichiometries are in fact more prevalent. Since the derivation of mass action and mass balance equations will proceed as for other cations, the derivations will be given without much discussion.

For the following derivation, six possible trivalent cation–surface site complexes will be considered: (1) a single cation plus surface site, SOM^{2+}; (2) a monohydroxylated cation plus surface site, $SOMOH^+$; (3) a dihydroxylated cation plus surface site, $SOM(OH)_2^0$; (4) a single cation plus double surface site, $(SO)_2M^+$; (5) a monohydroxylated cation plus double surface site, $(SO)_2MOH^0$; and (6) a single cation plus triple surface site, $(SO)_3M^0$. The disposition of three of these cation–surface site stoichiometries for trivalent cation adsorption at the S–MO interface is shown in Fig. 15.

The mass action equations for all six cation–surface site complexes are derived as above:

1. A single cation plus surface site, SOM^{2+}:

$$SO^- + M^{3+} = SOM^{2+} \quad \vdots \quad K_{SM} = \frac{\theta_{SM}}{\theta_{SO}\chi_M}\left(\frac{f_{SM}}{f_{SO}f_M}\right) = \frac{\theta_{SM}}{\theta_{SO}\chi_M} Z^{\epsilon_{K_{sm}}} \qquad (164)$$

where

$$\epsilon_{K_{sm}} = \epsilon_{sm} - \epsilon_{SO} - \epsilon_m = \frac{3(1 - 2\bar{\nu}_s)}{1 + \kappa a_s} - \frac{9}{1 + \kappa a_m} \qquad (165)$$

$$\theta_{SM} = \theta_{SO}\chi_M K_{SM} Z^{-\epsilon_{K_{sm}}} \qquad (166)$$

2. A monohydroxylated cation plus surface site, $SOMOH^+$:

$$SOM^{2+} + OH^- = SOMOH^+ \quad \vdots \quad K_{SMOH} = \frac{\theta_{SMOH}}{\theta_{SM}\chi_{OH}}\left(\frac{f_{SMOH}}{f_{SM}f_{OH}}\right)$$

$$= \frac{\theta_{SMOH}}{\theta_{SM}\chi_{OH}} Z^{\epsilon_{K_{SMOH}}} \qquad (167)$$

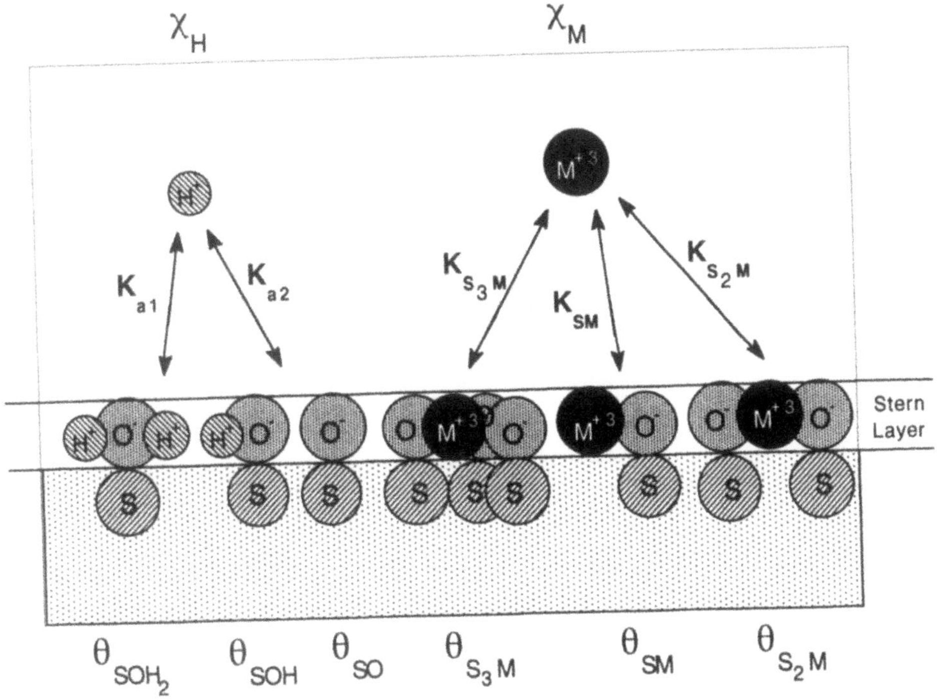

FIG. 15 Interfacial complexation reactions for trivalent cation adsorption at the aqueous S–MO interface.

where

$$\epsilon_{K_{\mathrm{SMOH}}} = \epsilon_{\mathrm{SMOH}} - \epsilon_{\mathrm{SOM}} - \epsilon_{\mathrm{OH}} = -\frac{3-2\bar{\nu}_{\mathrm{s}}}{1+\kappa a_{\mathrm{s}}} - \frac{1}{1+\kappa a_{\mathrm{OH}}} \tag{168}$$

$$\begin{aligned}\theta_{\mathrm{SMOH}} &= \chi_{\mathrm{OH}}\theta_{\mathrm{SM}}K_{\mathrm{SMOH}}Z^{-\epsilon_{K_{\mathrm{SMOH}}}} = \frac{\theta_{\mathrm{SM}}K_{\mathrm{w}}K_{\mathrm{SMOH}}Z^{-\epsilon_{K_{\mathrm{SMOH}}}}}{\chi_{\mathrm{H}}}\\ &= \theta_{\mathrm{SO}}\frac{\chi_{\mathrm{M}}K_{\mathrm{w}}K_{\mathrm{SM}}K_{\mathrm{SMOH}}Z^{-\epsilon_{K_{\mathrm{SMOH}}}}Z^{-\epsilon_{K_{\mathrm{SM}}}}}{\chi_{\mathrm{M}}} \equiv \theta_{\mathrm{SO}}\frac{\chi_{\mathrm{M}}K''_{\mathrm{SMOH}}Z^{-\epsilon''_{K_{\mathrm{SMOH}}}}}{\chi_{\mathrm{H}}}\end{aligned} \tag{169}$$

3. A dihydroxylated cation plus surface site, $\mathrm{SOM(OH)_2^0}$;

$$\mathrm{SOMOH^+ + OH^- = SOM(OH)_2} \tag{170}$$

$$K_{\mathrm{SM(OH)_2}} = \frac{\theta_{\mathrm{SM(OH)_2}}}{\theta_{\mathrm{SMOH}}\chi_{\mathrm{OH}}}\left(\frac{f_{\mathrm{SM(OH)_2}}}{f_{\mathrm{SMOH}}f_{\mathrm{OH}}}\right) = \frac{\theta_{\mathrm{SM(OH)_2}}}{\theta_{\mathrm{SMOH}}\chi_{\mathrm{OH}}}Z^{\epsilon_{K_{\mathrm{SM(OH)_2}}}} \tag{171}$$

where

$$\epsilon_{K_{\mathrm{SM(OH)_2}}} = \epsilon_{\mathrm{SM(OH)_2}} - \epsilon_{\mathrm{SMOH}} - \epsilon_{\mathrm{OH}} = -\frac{1-2\bar{\nu}_{\mathrm{s}}}{1+\kappa a_{\mathrm{s}}} - \frac{1}{1+\kappa a_{\mathrm{OH}}} \tag{172}$$

$$\theta_{SM(OH)_2} = \chi_{OH}\theta_{SMOH}K_{SM(OH)_2}Z^{-\epsilon_{K_{SM(OH)_2}}}$$
$$= \theta_{SO}\frac{\chi_M K_w^2 K_{SMOH}K_{SM(OH)_2}Z^{-\epsilon_{K_{SM}}}Z^{-\epsilon_{K_{SMOH}}}Z^{-\epsilon_{K_{SM(OH)_2}}}}{\chi_H^2}$$
$$\equiv \theta_{SO}\frac{\chi_M K''_{SMOH}Z^{-\epsilon''_{K_{SMOH}}}}{\chi_H^2} \tag{173}$$

4. A single cation plus double surface site, $(SO)_2M^+$:

$$2SO^- + M^{3+} = (SO)_2M^+ \quad \vdots \quad K_{S_2M} = \frac{\theta_{S_2M}}{\theta_{SO}^2\chi_M}\left(\frac{f_{S_2M}}{f_{SO}^2 f_M}\right) = \frac{\theta_{S_2M}}{\theta_{SO}^2\chi_M}Z^{-\epsilon_{K_{S_2M}}} \tag{174}$$

where

$$\epsilon_{K_{S_2M}} = \epsilon_{S_2M} - 2\epsilon_{SO} - \epsilon_M = -\frac{1+6\bar{\nu}_s+\nu_s^2}{1+\kappa a_s} - \frac{9}{1+\kappa a_M} \tag{175}$$

$$\theta_{S_2M} = \theta_{SO}^2\chi_M K_{S_2M}Z^{-\epsilon_{K_{S_2m}}} \tag{176}$$

5. A monohydroxylated cation plus double surface site, $(SO)_2MOH^0$, and

$$(SO)_2M^+ + OH^- = (SO)_2MOH^0 \tag{177}$$

$$K_{S_2MOH} = \frac{\theta_{S_2MOH}}{\theta_{S_2M}\chi_{OH}}\left(\frac{f_{S_2MOH}}{f_{S_2M}f_{OH}}\right) = \frac{\theta_{S_2MOH}}{\theta_{S_2M}\chi_{OH}}Z^{-\epsilon_{K_{S_2MOH}}} \tag{178}$$

where

$$\epsilon_{K_{S_2MOH}} = \epsilon_{S_2MOH} - \epsilon_{S_2M} - \epsilon_{OH} = -\frac{1-2\bar{\nu}_s}{1+\kappa a_s} - \frac{1}{1+\kappa a_{OH}} \tag{179}$$

$$\theta_{S_2MOH} = \chi_{OH}\theta_{S_2M}K_{S_2MOH}Z^{-\epsilon_{K_{S_2MOH}}} = \frac{\theta_{S_2M}K_w K_{S_2MOH}Z^{-\epsilon_{K_{S_2MOH}}}}{\chi_H}$$
$$= \theta_{SO}^2\frac{\chi_M K_w K_{S_2M}K_{S_2MOH}Z^{-\epsilon_{K_{S_2M}}}Z^{-\epsilon_{K_{S_2M}}}}{\chi_H} \equiv \theta_{SO}^2\frac{\chi_M K''_{S_2MOH}Z^{-\epsilon''_{K_{S_2M}}}}{\chi_H} \tag{180}$$

6. A single cation plus triple surface site, $(SO)_3M^0$:

$$3SO^- + M^{3+} = (SO)_3M \quad \vdots \quad K_{S_3M} = \frac{\theta_{S_3M}}{\theta_{SO}^3\chi_M}\left(\frac{f_{S_3M}}{f_{SO}^3 f_M}\right) = \frac{\theta_{S_3M}}{\theta_{SO}^3\chi_M}Z^{\epsilon_{K_{S_3M}}} \tag{181}$$

where

$$\epsilon_{K_{S_3M}} = \epsilon_{S_3M} - 3\epsilon_{SO} - \epsilon_M = -\frac{3+6\bar{\nu}_s+2\bar{\nu}_s^2}{1+\kappa a_s} - \frac{9}{1+\kappa a_M} \tag{182}$$

$$\theta_{S_3M} = \theta_{SO}^3\chi_M K_{S_3M}Z^{-\epsilon_{K_{S_3m}}} \tag{183}$$

The surface-site mass balance equation equates the total number of surface sites, n_s^T, to the three surface sites containing no cations and the six stoichiometric trivalent cation–surface site complexes:

$$1 = n_{SO} + n_{SOH} + n_{SOH_2} + n_{SM} + n_{SMOH} + n_{SM(OH)_2} + 2n_{S_2M} + 2n_{S_2MOH} + 3n_{S_3M} \tag{184}$$

Since Eq. (184) is a mass balance over surface sites and the stoichiometry for the S_2M, S_2MOH, and S_3M complexes contains 2, 2, and 3 surface sites, respectively, for each complex, the necessity for the factors is clear.

Division by n_s^T converts Eq. (184) to surface-site mole fractions:

$$1 = \theta_{SO} + \theta_{SOH} + \theta_{SOH_2} + \theta_{SM} + \theta_{SMOH} + \theta_{SM(OH)_2} + 2\theta_{S_2M} + 2\theta_{S_2MOH} + 3\theta_{S_3M} \tag{185}$$

The surface-site mole fraction of surface sites complexed to a trivalent cation, θ_{MS}^T, is defined as

$$\theta_{SM}^T \equiv \theta_{SM} + \theta_{SMOH} + \theta_{SM(OH)_2} + 2\theta_{S_2M} + 2\theta_{S_2MOH} + 3\theta_{S_3M} \tag{186}$$

Combining Eqs (185) and (186):

$$\theta_{SM}^T = (1 - (\theta_{SO} + \theta_{SOH} + \theta_{SOH_2}) \tag{187}$$

Rearranging Eq. (81) and substituting into Eq. (186):

$$\theta_{SO} + \theta_{SOH} + \theta_{SOH_2} = \theta_{SO}\alpha_{SO}^{-1} = 1 - \theta_{SM}^T \tag{188}$$

$$\theta_{SO} = \alpha_{SO}(1 - \theta_{SM}^T) \quad \vdots \; \theta_{SM}^T = 1 - \alpha_{SO}^{-1}\theta_{SO} \tag{189}$$

From a comparison of Eq. (81) derived for monovalent cation adsorption and Eqs (133) and (189) derived for divalent and trivalent cation adsorption, respectively, one can see that these mass balance equations have the same general form.

In a similar fashion the mass balance equation for the trivalent cation can be written as a mass balance over all cations in the system with the total moles of cation, n_M^T, distributed in solution as n_M or among six types of surface complexes, n_{SM}, n_{SMOH}, $n_{SM(OH)_2}$, n_{S_2M}, n_{S_2MOH}, or n_{S_3M}, which can be summed to give

$$n_M^T = n_M + n_{SM} + n_{SMOH} + n_{SM(OH)_2} + n_{S_2M} + n_{S_2MOH} + n_{S_3M} \tag{190}$$

division by V_w and substitution of mass action equations gives the trivalent cation mass balance equation:

$$\begin{aligned} c_M^T \equiv \frac{n_M^T}{V_w} = \frac{n_M}{V_w} + \frac{n_{SM}}{n_s^T}\cdot\frac{n_s^T}{V_w} + \frac{n_{SMOH}}{n_s^T}\cdot\frac{n_s^T}{V_w} + \frac{n_{SM(OH)_2}}{n_s^T}\cdot\frac{n_s^T}{V_w} \\ + \frac{n_{S_2M}}{n_s^T}\cdot\frac{n_s^T}{V_w} + \frac{n_{S_2MOH}}{n_s^T}\cdot\frac{n_s^T}{V_w} + \frac{n_{S_3M}}{n_s^T}\cdot\frac{n_s^T}{V_w} \end{aligned} \tag{191}$$

$$c_M^T = \chi_M + c_s^T(\theta_{SM} + \theta_{SMOH} + \theta_{SM(OH)_2} + \theta_{S_2M} + \theta_{S_2MOH} + \theta_{S_3M}) \tag{192}$$

Defining the variable θ_M^T for the parenthetical expression in Eq. (192) gives the trivalent cation mass balance equation in the same form as found for the mono- and di-valent cation in Eqs (108) and (136), respectively:

$$c_M^T \equiv \chi_M + \theta_M^T c_s^T \tag{193}$$

In Eqs (184) and (190) different coefficients are seen to arise due to the stoichiometry of the surface site complexes. These coefficients result in different definitions of adsorbed cation. The two different cation complexed surface-site mole fractions which arise are

$$\theta^{T}_{SM} \equiv \theta_{SM} + \theta_{SMOH} + \theta_{SM(OH)_2} + 2\theta_{S_2M} + 2\theta_{S_2MOH} + 3\theta_{S_3M} \tag{194}$$

$$\theta^{T}_{M} \equiv \theta_{SM} + \theta_{SMOH} + \theta_{SM(OH)_2} + \theta_{S_2M} + \theta_{S_2MOH} + \theta_{S_3M} \tag{195}$$

In order to combine cation and surface-site mass balance equations it will be necessary to find the relationship between surface complex mole fractions. Substituting Eq. (194) into (195):

$$\begin{aligned}\theta^{T}_{SM} &= 2\theta^{T}_{M} - \theta_{SM} - \theta_{SMOH} - \theta_{SM(OH)_2} + \theta_{S_3M}\\ &\cong 2\theta^{T}_{M} - \theta_{SM} - \theta_{SMOH} - \theta_{SM(OH)_2}\end{aligned} \tag{196}$$

Substituting mass action equations (164), (167), (171), and (189) into Eq. (196):

$$\theta^{T}_{M} = 2\theta^{T}_{M} - \chi_{M}X_{1SO}(1 - \theta^{T}_{SM})\alpha_{SO} \tag{197}$$

where X_{1SO} represents the sum of the products of all mass action constants and activity coefficient quotients products involving only one SO^{-} surface site:

$$X_{1SO} \equiv K_{SM}Z^{-\epsilon_{K_{sm}}} + \frac{K''_{SMOH}Z^{-\epsilon_{K''_{SMOH}}}}{\chi_{H}} + \frac{K''_{SM(OH)_2}Z^{-\epsilon_{K''_{SM(OH)_2}}}}{\chi^{2}_{H}} \tag{198}$$

solving for θ^{T}_{SM}:

$$\theta^{T}_{SM} = \frac{2\alpha^{-1}_{SO}\theta^{T}_{M} - \chi_{M}X_{1SO}}{\alpha^{-1}_{SO} - \chi_{M}X_{1SO}} \tag{199}$$

Another useful equation can also be derived from Eqs (189) and (196):

$$1 - \alpha^{-1}_{SO}\theta_{SO} = 2\theta^{T}_{M} - \chi_{M}X_{1SO}\theta_{SO} \tag{200}$$

solving for θ_{SO}:

$$\theta_{SO} = \frac{2\theta^{T}_{M} - 1}{\chi_{M}X_{1SO} - \alpha^{-1}_{SO}} \tag{201}$$

The last variable requiring an explicit equation is the mean surface-site elctrostatic charge, $\bar{\nu}_s$. From the surface-site mass balance equation (182) one can see that of the nine surface site complexes shown, five have a finite electrostatic charge: (1) -1 electrostatic charge on the SO^{-} surface site complex, (2) $+1$ electrostatic charge on the $SOH_2{}^{+}$ surface site complex, (3) $+2$ electrostatic charge on the SOM^{2+} surface site complex, (4) $+1$ electrostatic charge on the $SOMOH^{+}$ surface site complex, and (5) $+1$ electrostatic charge on the $(SO)_2M^{+}$ surface site complex. The mean surface-site electrostatic charge, $\bar{\nu}_s$, is the sum of the surface site fractions for these electrostatically charged surface sites:

$$\bar{\nu}_s = \theta_{SOH_2} + 2\theta_{SM} + \theta_{SMOH} + \theta_{S_2M} - \theta_{SO} \tag{202}$$

The appearance of the surface mole fraction for a surface complex requires the derivation of an α_{SM} equation:

$$\alpha_{SM} \equiv \frac{\theta_{SM}}{\theta_{SM} + \theta_{SMOH} + \theta_{SM(OH)} + 2\theta_{S_2M} + 2\theta_{S_2MOH} + 3\theta_{S_3M}} = \frac{\theta_{SM}}{\theta^{T}_{SM}} \tag{203}$$

Substitution of Eqs (166), (169), (173), (176), (180), (183), and (198) into Eq. (203):

$$\alpha_{SM} \equiv \frac{K_{SM} Z^{-\epsilon K_{sm}}}{X_{1SO} + 2\theta_{SO} X_{2SO} + 3\theta_{SO}^2 X_{3SO}} \tag{204}$$

where

$$X_{2SO} \equiv K_{s_2m} Z^{-\epsilon K_{s_2m}} + \frac{K_w K_{S_2M} K_{S_2MOH} Z^{-\epsilon K_{S_2M}} Z^{-\epsilon K_{S_2MOH}}}{\chi_H} \tag{205}$$

$$X_{3SO} \equiv K_{S_3M} Z^{-\epsilon K_{S_3M}} \tag{206}$$

Similar equations can be written for α_{SMOH} and α_{S_2M}:

$$\alpha_{SMOH} = \frac{\dfrac{K_w K_{SM} K_{SMOH} Z^{-\epsilon K_{SM}} Z^{-\epsilon K_{SMOH}}}{\chi_H}}{X_{1SO} + 2\theta_{SO} X_{2SO} + 3\theta_{SO}^2 X_{3SO}} \tag{207}$$

$$\alpha_{S_2M} = \frac{K_{s_2m} Z^{-\epsilon K_{s_2m}}}{X_{1SO} + 2\theta_{SO} X_{2SO} + 3\theta_{SO}^2 X_{3SO}} \tag{208}$$

Equation (204) can now be written in terms of variables which can all be readily evaluated as

$$\begin{aligned} \bar{\nu}_s &= \theta_{SOH_2} + 2\theta_{SM} + \theta_{SMOH} + \theta_{S_2M} - \theta_{SO} \\ &= (\alpha_{SOH_2} - \alpha_{SO})(1 - \theta_{SM}^T) + (2\alpha_{SM} + \alpha_{SMOH} + \alpha_{S_2M})\theta_{SM}^T \end{aligned} \tag{209}$$

As in Sections III.A and III.B, Eq. (207) allows iterative calculation of the mean surface-site electrostatic charge, $\bar{\nu}_s$, once the fraction of surface sites complexed with a cation, θ_{SM}^T, has been determined.

All the equations needed to obtain the trivalent cation adsorption equation have now been derived. A listing of the equations is given in Table 9.

When mass action equations from Table 9 are substituted into the surface-site mass balance equations (195), (198), (205), and (206) for all adsorbed cations:

$$\theta_M^T = \theta_{SO} \chi_M (X_{1SO} + \theta_{SO} X_{2SO} + \theta_{SO}^2 X_{3SO}) \tag{210}$$

Substituting Eq. (201) into (210):

$$\begin{aligned} \theta_M^T = \left(\frac{2\chi_M \theta_M^T - \chi_M}{\chi_M X_{1SO} - \alpha_{SO}^{-1}}\right) \Bigg[X_{1SO} &+ \left(\frac{2\theta_M^T - 1}{\chi_M X_{1SO} - \alpha_{SO}^{-1}}\right) X_{2SO} \\ &+ \left(\frac{2\theta_M^T - 1}{\chi_M X_{1SO} - \alpha_{SO}^{-1}}\right)^2 X_{3SO} \Bigg] \end{aligned} \tag{211}$$

Equation (211) is cubic and can be solved algebraically or iteratively for θ_M^T. For purposes of discussion, we will make an approximate solution based on the conclusions reached for divalent cation adsorption, which is to assume that the surface complexes, $SOMOH^+$, $SOM(OH)_2$, $(SO)_2M^+$, $(SO)_2MOH$, and $(SO)_3M$, make only a small contribution to the adsorption process. When experimental evidence suggests that any or all of these surface site complexes play a significant role in the adsorption process, the adsorption calculations for trivalent cations should be performed with Eq. (211).

TABLE 9 Tabulation of Equations and Unknowns Used for Derivation of an Adsorption Equation for a Trivalent Cation Adsorbing on Multiple Surface Sites

Unknowns		Equation	
No.	Variable	No.	Description
3	χ_{SOH}, χ_{SOH_2}, $\bar{\nu}_s$	(18)	$K_{a1} = \left(\frac{\theta_{SOH}\chi_H}{\theta_{SOH_2}}\right) Z_{SH}^{-1}$
1	χ_{SO}	(20)	$K_{a2} = \left(\frac{\theta_{SO}\chi_H}{\theta_{SOH}}\right) Z_{SH}^{-1}$
2	χ_M, θ_{SM}	(164)	$K_{SM} = \left(\frac{\theta_{SM}}{\theta_{SO}\chi_M}\right) Z^{\epsilon K_{SM}}$
1	θ_{SMOH}	(167)	$K_{SMOH} = \frac{\theta_{SMOH}}{\theta_{SM}\chi_{OH}} Z^{\epsilon K_{SMOH}}$
1	θ_{SMOH}	(171)	$K_{SM(OH)_2} = \frac{\theta_{SM(OH)_2}}{\theta_{SMOH}\chi_{OH}} Z^{\epsilon K_{SM(OH)_2}}$
1	θ_{S_2M}	(174)	$K_{S_2M} = \left(\frac{\theta_{S_2M}}{\theta_{SO}^2\chi_M}\right) Z^{\epsilon K_{S_2M}}$
1	θ_{S_2MOH}	(178)	$K_{S_2MOH} = \frac{\theta_{S_2MOH}}{\theta_{S_2M}\chi_{OH}} Z^{\epsilon K_{S_2MOH}}$
1	θ_{S_3M}	(181)	$K_{S_3M} = \left(\frac{\theta_{S_3M}}{\theta_{SO}^3\chi_M}\right) Z^{\epsilon K_{S_3M}}$
0		(185)	$1 = \theta_{SO} + \theta_{SOH} + \theta_{SOH_2} + \theta_{SM}^T$
0		(209)	$\bar{\nu}_s = (\alpha_{SOH_2} - \alpha_{SO})(1 - \theta_{SM}^T) + (2\alpha_{SM} + \alpha_{SMOH} + \alpha_{S_2M})\theta_{SM}^T$
0		(192)	$c_M^T = \chi_M + c_S^T\theta_M^T$
11 unknowns			11 equations

Assuming that SOM^{2+} is the only significant surface complex for trivalent cation adsorption, adsorption equation (211) reduces to

$$\theta_M^T = \frac{\chi_M K_{SM} Z^{-\epsilon K_{sm}}}{\alpha_{SO}^{-1} + \chi_M K_{SM} Z^{-\epsilon K_{sm}}} \tag{212}$$

Elimination of the equilibrium trivalent cation molar solution concentration, χ_M, for the total trivalent cation molar solution concentration, c_M^T, results when Eq. (192) is substituted into Eq. (212):

$$\theta_M^T \cong \frac{c_M^T K_{SM} Z^{-\epsilon K_{sm}}}{\alpha_{SO}^{-1} + (c_s^T + C_M^T) K_{SM} Z^{-\epsilon K_{sm}}} \tag{213}$$

where the equal sign signifies an approximate linear solution to the quadratic adsorption equation.

D. Predicted and Experimental Cation Adsorption

The next step is to examine how well the cation adsorption equations predict experimental cation adsorption. Cation adsorption data for different cations on different metal oxide particles measured by different researchers are shown in Fig. 16.

In Fig. 16, the adsorption curves were calculated using Eq. (213). Values for the constants and variables, K_{a1}, K_{a2}, c_s^T, and c_M^T, were taken from the respective study or other literature sources if no value was specified. Only the mass action adsorption constant, K_{SM}, was determined independently of those given by the original authors. The adsorption constant, K_{SM}, in Eq. (213) was chosen to give the best fit between the experimental and theoretical data. The values of K_{SM} obtained in this way are reasonable since one would expect trivalent Cr to adsorb more strongly than divalent Ca. The strength of adsorption is evidenced by the fact that Cr(III) adsorbs at a pH lower than that of Ca(II), suggesting that the former will displace protons in order to adsorb while Ca(II) adsorption does not begin until a higher pH is reached where protons have already left the surface and created anionic surface sites. Since the adsorption data for the cation Ag(I) fall between those of Cr(III) and Ca(II), one can assume that hematite is a stronger cation adsorber than silica.

The experimental adsorption data points in Fig. 16 can be seen to agree well with the adsorption curves predicted by Eq. (213). Improvement in the fit of the predicted and experimental data in Fig. 16 would be expected if two additional considerations are made: activity coefficient effects and different adsorbed structures

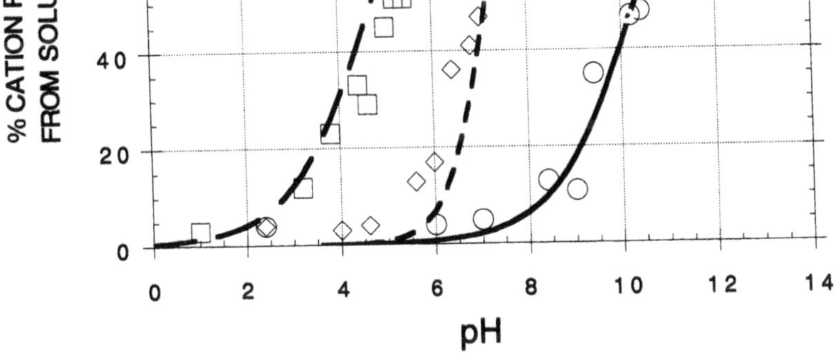

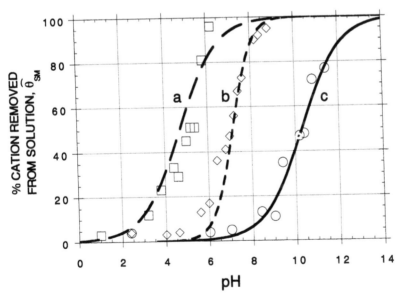

FIG. 16 Experimental adsorption data compared with Eq. (213) for ideal adsorption: $K^{\epsilon K_{a2}} = K^{\epsilon K_{a12}} = K^{\epsilon K_{SM}} = 1$. Adsorption data for curves: (a) James and Healy [17] for Cr(III) adsorption on silica, $K_{Cr(III)/SiO_2} = 2 \times 10^8$; (b) Davis and Leckie [18] for Ag(I) adsorbed on hematite, $K_{Ag(I)/Fe_2O_3} = 1 \times 10^6$; (c) James and Healy [17] for Ca(II) adsorption on silica, $K_{Ca(II)/SiO_2} = 2 \times 10^2$. Proton adsorption constants for: silica, p$K_{a1} = 2.8$, p$K_{a2} = 6.8$; hematite, p$K_{a1} = 5.0$, p$K_{a2} = 10.7$.

and stoichiometries such as multisurface site adsorption, adsorption of metal hydroxy solutes, etc. In short, the different stoichiometries of the different surface sites add terms to the adsorption equation containing independent variables, such as pH and cation solution mole fraction to different (mathematical) powers. The different chemical surface structures manifest themselves as terms with different independent variable orders and mass action constants, which give better fit to the equations. Some examples of the effects that activity coefficients and other adsorbed structures have on adsorption and how they can be included in the adsorption equations have been demonstrated in Sections III.A and III.B.

Hence, the proper mass action treatment can be expected to provide an excellent fit between experimental and predicted adsorption data. In addition, adsorption equations expanded to include more than one adsorbed complex are based on mass action thermodynamics in which the only unknown variables introduced are mass action constants; no parameters are introduced or needed to specify the fraction of each type of surface complex. Instead, once the mass action constants are determined for the respective surface complexes, the fraction of each surface complex can be calculated directly from a knowledge of the interfacial system, such as pH, initial mole fraction of each solute added, mole fraction of solute added, etc.

IV. ANION ADSORPTION EQUATIONS

A. Monovalent Anion Adsorption Equations

Monovalent anion adsorption on metal oxide surfaces from aqueous solution is similar to monovalent cation adsorption in that it can be assumed that a single anion–surface site complex is the predominant surface complex formed. The obvious difference is that anions adsorb on cationic surface sites, unlike cations, which adsorb on anionic surface sites. The chemical reactions creating the monovalent anion–surface site complex are depicted in Fig. 17.

The chemical reaction and mass action equation for the surface complex shown in Fig. 17 for the monovalent anion, A^-, adsorption at the aqueous S–MO interface is

$$SOH_2{}^+ + A^- = SOH_2{}^+A^0 \tag{214}$$

$$K_{SA} = \frac{\theta_{SA}}{\theta_{SOH_2}\chi_A}\left(\frac{f_{SA}}{f_{SOH_2}f_A}\right) = \frac{\theta_{SA}}{\theta_{SOH_2}\chi_A}Z^{\epsilon_{K_{SA}}} \tag{215}$$

where

$$\epsilon_{K_{SA}} = \epsilon_{SA} - \epsilon_{SOH_2} - \epsilon_A = +\frac{2\bar{\nu}_s - 1}{1+\kappa a_s} - \frac{1}{1+\kappa a_A} \tag{216}$$

The mass action equation can be solved for the surface site mole fraction as

$$\theta_{SA} = \theta_{SOH_2}\chi_A K_{SA} Z^{-\epsilon_{K_{SA}}} \tag{217}$$

The surface-site mass balance equation is the sum of the moles of all surface sites shown in Fig. 17 to give

$$1 = n_{SO} + n_{SOH} + n_{SOH_2} + n_{SOH_2A} \equiv n_{SO} + n_{SOH} + n_{SOH_2} + n_{SA} \tag{218}$$

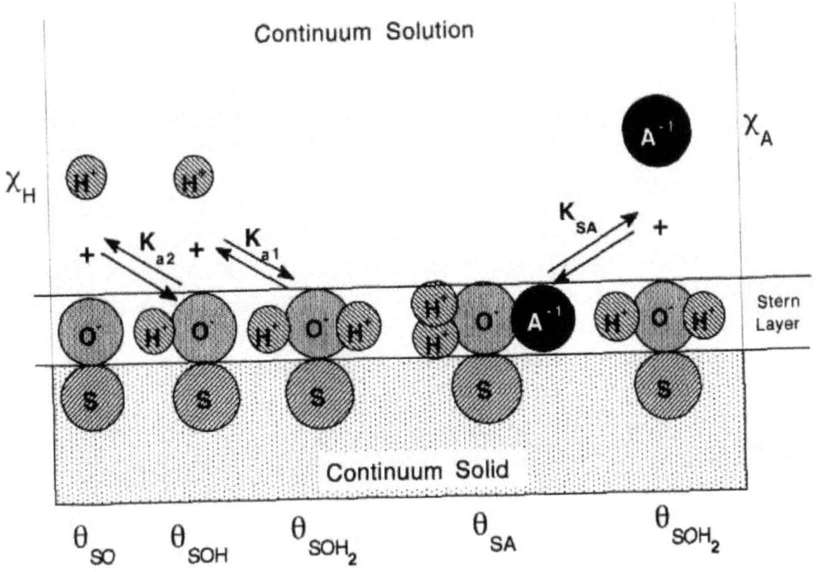

FIG. 17 Monovalent anion adsorption at the aqueous S—MO interface.

or division by the total moles of surface sites, n_s^I, converts Eq. (218) into mole fractions as

$$1 = \theta_{SO} + \theta_{SOH} + \theta_{SOH_2} + \theta_{SA} \tag{219}$$

Solving Eq. (219) for the total mole fraction of surface sites complexed to an anion,

$$\theta_{SA} = 1 - (\theta_{SO} + \theta_{SOH} + \theta_{SOH_2}) \tag{220}$$

Since anion adsorption involves cationic surface sites, $SOH_2{}^+$, Eqs (85) and (86) [not Eqs (81) and (82) as for cation adsorption] are rearranged and substituting into Eq. (220):

$$\theta_{SO} + \theta_{SOH} + \theta_{SOH_2} = \theta_{SOH_2}\alpha_{SOH_2}^{-1} = 1 - \theta_{SA} \tag{221}$$

$$\theta_{SOH_2} = \alpha_{SOH_2}(1 - \theta_{SA}) \quad \vdots \quad \theta_{SA} = 1 - \alpha_{SOH_2}^{-1}\theta_{SOH_2} \tag{222}$$

Comparison of Eq. (86) derived for monovalent cation adsorption and Eq. (220) derived for monovalent anion adsorption, one can see that these mass balance equations still have the same general form.

In a similar fashion the mass balance equation for the monovalent anion can be written as a mass balance over the anion, A^-, everywhere in the system with the total moles of anion, n_A^T, distributed in solution as n_A or in a surface complex, n_{SA}, summed to give

$$n_A^T = n_A + n_{SA} \tag{223}$$

division by V_w and substitution of mass action equations gives the monovalent anion mass balance equation:

$$\frac{n_A^T}{V_w} \equiv c_A^T = \frac{n_A}{V_w} + \frac{n_{SA}}{n_s^T} \cdot \frac{n_s^T}{V_w} = \chi_A + \theta_{SA} c_s^T \tag{224}$$

The last variable requiring an explicit equation is the mean surface-site electrostatic charge, $\bar{\nu}_s$. From Fig. 17 one can see that of the four surface site complexes shown only two have a finite electrostatic charge: (1) −1 electrostatic charge on the SO^- surface site complex, and (2) +1 electrostatic charge on the $SOH_2{}^+$ surface site complex. The mean surface-site electrostatic charge, $\bar{\nu}_s$, is the sum of the surface site fractions for these electrostatically charged surface sites:

$$\bar{\nu}_s = \theta_{SOH_2} - \theta_{SO} \tag{225}$$

Equation (225) can now be written in terms of variables which can all be readily evaluated as

$$\nu_s = \theta_{SOH_2} - \theta_{SO} = (\alpha_{SOH_2} - \alpha_{SO})(1 - \theta_{SA}) \tag{226}$$

Reviewing Section III.A one will note that the equation for the mean surface site electrostatic charge, $\bar{\nu}_s$, is the same for monovalent anion [Eq. (226)] and monovalent cation [Eq. (116)] adsorption. Moreover, evaluation of Eqs (226) and (116) can be accomplished with the same iterative procedure: (1) calculate θ_{SA} under ideal adsorption conditions where $Z = 1$, (2) calculate $\bar{\nu}_s$ using the ideal value of θ_{SA}, (3) calculate θ_{SA} using calculated activity coefficient quotients in the adsorption equation, and (4) repeat. All the equations needed to obtain the monovalent anion adsorption equation have now been derived. A listing of the equations is given in Table 10.

TABLE 10 Tabulation of Equations and Unknowns Used for Derivation of the Monovalent Anion Adsorption Equations

Unknowns		Equation	
No.	Variable	No.	Description
3	χ_{SOH}, χ_{SOH_2}, $\bar{\nu}_s$	(18)	$K_{a1} = \left(\frac{\theta_{SOH}\chi_H}{\theta_{SOH_2}}\right) Z^{\epsilon_{K_{a1}}}$
1	χ_{SO}	(20)	$K_{a2} = \left(\frac{\theta_{SO}\chi_H}{\theta_{SOH}}\right) Z^{\epsilon_{K_{a2}}}$
2	χ_A, θ_{SA}	(215)	$K_{SA} = \left(\frac{\theta_{SA}}{\theta_{SOH_2}\chi_A}\right) Z^{\epsilon_{K_{SA}}}$
0		(219)	$1 = \theta_{SO} + \theta_{SOH} + \theta_{SOH_2} + \theta_{SA}$
0		(226)	$\bar{\nu}_s = (\alpha_{SOH_2} - \alpha_{SO})(1 - \theta_{SA})$
0		(224)	$c_A^T = \chi_A + c_s^T\theta_{SA}$
6 unknowns			6 equations

An equal number of equations and variables allows the derivation of the monovalent anion adsorption equation with all unknown variables eliminated. To derive the anion adsorption equation, we begin by substituting Eq. (222) into (217):

$$\theta_{SA} = \theta_{SO}(\chi_A K_{SA} Z^{-\epsilon\kappa_{SA}}) = \left(\frac{1-\theta_{SA}}{\alpha_{SO}^{-1}}\right)(\chi_A K_{SA} Z^{-\epsilon\kappa_{SA}}) \tag{227}$$

Solving for θ_{SA} gives the desired monovalent anion adsorption equation:

$$\theta_{SA} = \frac{\chi_A K_{SA} Z^{-\epsilon\kappa_{SA}}}{\alpha_{SO}^{-1} + \chi_A K_{SA} Z^{-\epsilon\kappa_{SA}}} \quad \left(\begin{matrix}\text{Anion adsorption}\\ \text{equation, } \theta_{SA}\end{matrix}\right) \tag{228}$$

Substituting Eq. (224) replaces the equilibrium mole fraction, χ_A, by the more convenient total or formal molar solution concentrations, c_s^T and c_M^T, to give

$$\theta_{SA} = \frac{(c_A^T - c_s^T\theta_{SA})K_{SA}Z^{-\epsilon\kappa_{SA}}}{\alpha_{SOH_2}^{-1} + (c_A^T - c_s^T\theta_{SA})K_{SA}Z^{-\epsilon\kappa_{SA}}} \tag{229}$$

Equation (229) is quadratic in θ_{SA}:

$$\begin{aligned}&\theta_{SA}^2(c_s^T K_{SA} Z^{-\epsilon\kappa_{SA}})\\ &- \theta_{SA}\left[\alpha_{SOH_2}^{-1} + (C_A^T + C_s^T)K_{SA}Z^{-\epsilon\kappa_{SA}}\right]\\ &+ c_A^T K_{SA} Z^{-\epsilon\kappa_{SA}} = 0\end{aligned} \tag{230}$$

The exact solution of Eq. (230) is

$$\theta_{SA} = \left[\frac{\alpha_{SOH_2}^{-1} + (c_s^T + c_A^T)K_{SA}Z^{-\epsilon\kappa_{SA}}}{2c_s^T K_{SA} Z^{-\epsilon\kappa_{SA}}}\right] - \sqrt{\frac{\left[\alpha_{SOH_2}^{-1} + (c_s^T + c_A^T)K_{SA}Z^{-\epsilon\kappa_{SA}}\right]^2}{4c_s^T C_A^T (K_{SA}Z^{-\epsilon\kappa_{SA}})^2} - \frac{c_A^T}{c_s^T}} \quad \left(\begin{matrix}\text{Anion adsorption}\\ \text{equation -- quadratic}\\ \text{solution for } \theta_{SA}\end{matrix}\right) \tag{231}$$

Using Eq. (103) to convert Eq. (231) from θ_{SA} to $\hat{\theta}_{SA}$:

$$\hat{\theta}_{SA} = \left[\frac{\alpha_{SOH_2}^{-1} + (c_s^T + c_A^T)K_{SA}Z^{-\epsilon\kappa_{SA}}}{2c_A^T K_{SA} Z^{-\epsilon\kappa_{SA}}}\right] - \sqrt{\frac{c_s^T\left[\alpha_{SOH_2}^{-1} + (c_s^T + c_A^T)K_{SA}Z^{-\epsilon\kappa_{SA}}\right]^2}{c_A^T (2c_A^T K_{SA}Z^{-\epsilon\kappa_{SA}})^2} - \frac{c_s^T}{c_A^T}} \quad \left(\begin{matrix}\text{Anion adsorption}\\ \text{equation -- quadratic}\\ \text{solution for } \hat{\theta}_{SA}\end{matrix}\right) \tag{232}$$

Approximate linear anion adsorption equations can be derived in analogy with the cation adsorption equations (102), (113), and (114), respectively, as

$$\Gamma_{SA} = \frac{\chi_A K_{SA} Z^{-\epsilon_{K_{SA}}}}{A_s N_A(\alpha_{SOH_2}^{-1} + \chi_{SA} K_{SA} Z^{-\epsilon_{K_{SA}}})} \tag{233}$$

$$\theta_{SA} \cong \frac{c_A^T K_{SA} Z^{-\epsilon_{K_{SA}}}}{\alpha_{SOH_2}^{-1} + (c_s^T + c_A^T) K_{SA} Z^{-\epsilon_{K_{SA}}}} \tag{234}$$

$$\widehat{\theta}_{SA} \cong \frac{c_s^T K_{SA} Z^{-\epsilon_{K_{SA}}}}{\alpha_{SOH_2}^{-1} + (c_s^T + c_A^T) K_{SA} Z^{-\epsilon_{K_{SA}}}} \tag{235}$$

Anion adsorption can be plotted in many forms including, but not limited to, Eqs (228)–(235). Monovalent anion adsorption, $\widehat{\theta}_{SA}$, as a function of pH is plotted in Fig. 18.

When comparing Figs 9 and 18, one can see that cation and anion adsorption is symmetrically inverted through lines perpendicular to the pH axis. Qualitatively, the adsorption behavior can be seen to decrease with increasing pH for anions in Fig. 18 while the opposite pH dependence is seen in Fig. 9 for cation adsorption. Quantitatively, differences in mass action adsorption constants have opposite effects on adsorption of cations and anions. The role of solute concentration on anion and cation adsorption is the same.

B. Divalent Anion Adsorption Equations

As discussed in Section III.B for divalent cation adsorption, when a divalent anion adsorbs at the aqueous S–MO interface several surface complex stoichiometries are possible. The derivation methods and equations will bear many similarities to those of divalent cation adsorption.

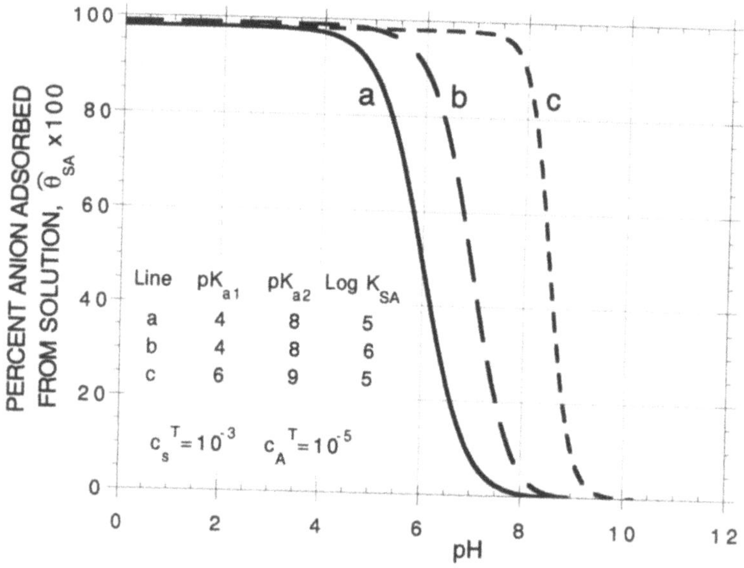

FIG. 18 Monovalent anion adsorption plot of Eq. (235) of percentage anion adsorbed from solution, $\widehat{\theta}_{SA}$, versus pH with $K^{\epsilon_{K_{a2}}} = K^{\epsilon_{K_{a12}}} = K^{\epsilon_{K_{SA}}} = 1$.

Of the multiple stoichiometries which might form, the relative abundance of each will be determined by the mass action constant: larger indicating greater abundance and smaller indicating less. Lacking experimental evidence as to which divalent anion–surface site complexes are predominant, adsorption equations will be derived for what are arguably the three most likely anion complexes. The three possible divalent anion–surface site complexes for the divalent anion, A^{2-}, which will be considered are: (1) a single anion plus surface site, SOH_2A^-; (2) a monohydrogen anion plus surface site, SOH_2HA^0; and (3) a single anion plus double surface site, $(SOH_2)_2A$. The disposition of these chemical interactions for divalent anion adsorption at the S–MO interface is shown in Fig. 19.

Mass action equations for the three surface complexes shown in Fig. 19 for divalent anion A^{2-} adsorption at the S–MO interface are:

1. A single anion plus surface site, SOH_2A^-:

$$SOH_2{}^+ + A^{2-} = SOH_2A^- \tag{236}$$

$$K_{SA} = \frac{\theta_{SA}}{\theta_{SOH_2}\chi_A}\left(\frac{f_{SA}}{f_{SOH_2}f_A}\right) = \frac{\theta_{SA}}{\theta_{SOH_2}\chi_A}Z^{\epsilon_{K_{SA}}} \tag{237}$$

where

$$\epsilon_{K_{SA}} = \epsilon_{SA} - \epsilon_{SOH_2} - \epsilon_A = +\frac{4\bar{v}_s}{1+\kappa a_s} - \frac{4}{1+\kappa a_A} \tag{238}$$

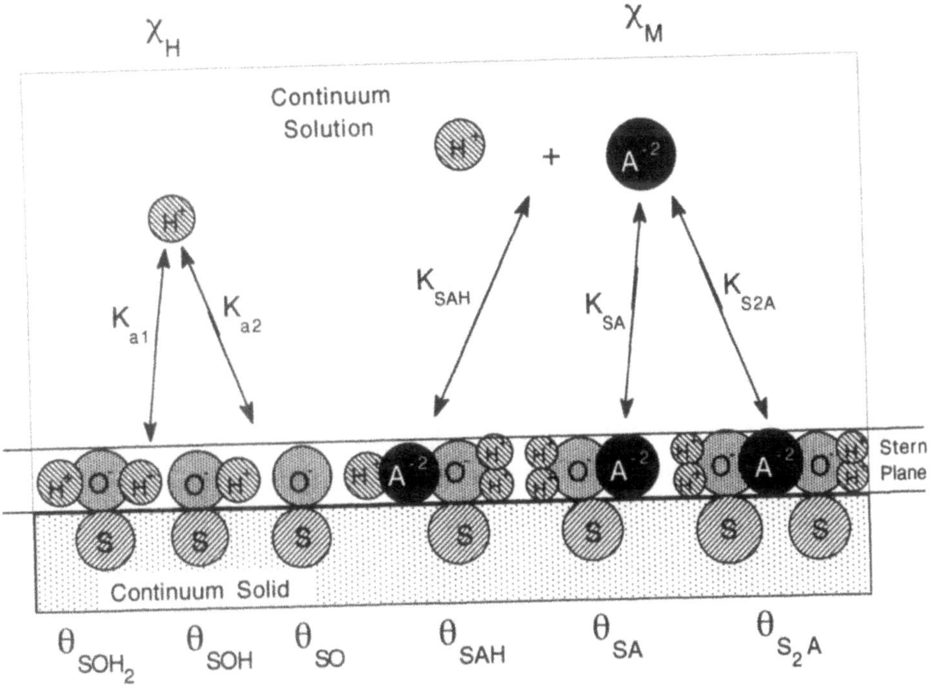

FIG. 19 Interfacial complexation reactions for divalent anion adsorption.

2. A monohydrogen anion plus surface site, SOH_2AH:

$$SOH_2A^- + H^+ = SOH_2AH \tag{239}$$

$$K_{SAH} = \frac{\theta_{SAH}}{\theta_{SA}\chi_H}\left(\frac{f_{SAH}}{f_{SA}f_H}\right) = \frac{\theta_{SAH}}{\theta_{SA}\chi_H}Z^{\epsilon_{K_{SAH}}} \tag{240}$$

where

$$\epsilon_{K_{SAH}} = \epsilon_{SAH} - \epsilon_{SA} - \epsilon_H = -\frac{2\bar{\nu}_s + 1}{1 + \kappa a_s} - \frac{1}{1 + \kappa a_H} \tag{241}$$

3. A single anion plus two surface sites, $(SOH_2)_2A$:

$$2SOH_2{}^+ + A^{2-} = (SOH_2)_2A \tag{242}$$

$$K_{S_2A} = \frac{\theta_{S_2A}}{\theta^2_{SOH_2}\chi_A}\left(\frac{f_{S_2A}}{f^2_{SOH_2}f_A}\right) = \frac{\theta_{S_2A}}{\theta^2_{SOH_2}\chi_A}Z^{\epsilon_{K_{S_2A}}} \tag{243}$$

where

$$\epsilon_{K_{S_2A}} = \epsilon_{S_2A} - 2\epsilon_{SOH_2} - \epsilon_A = -\frac{2 - 4\bar{\nu}_s + \bar{\nu}_s^2}{1 + \kappa a_s} - \frac{4}{1 + \kappa a_A} \tag{244}$$

The mass action equations can be rearranged for expressions of the surface site and solute concentrations as

$$\theta_{SA} = \theta_{SOH_2}\chi_A K_{SA} Z^{-\epsilon_{K_{SA}}} \tag{245}$$

$$\theta_{S_2A} = \theta^2_{SOH_2}\chi_A K_{S_2A} Z^{-\epsilon_{K_{S_2A}}} \tag{246}$$

$$\begin{aligned}\theta_{SAH} &= \theta_{SA}\chi_H K_{SAH} Z^{-\epsilon_{K_{SAH}}} = \theta_{SOH_2}\chi_A\chi_H K_{SA}K_{SAH}Z^{-\epsilon_{K_{SA}}}Z^{-\epsilon_{K_{SAH}}}\\ &\equiv \theta_{SOH_2}\chi_A\chi_H K''_{SAH} Z^{-\epsilon''_{K_{SAH}}}\end{aligned} \tag{247}$$

where

$$\epsilon''_{K_{SAH}} \equiv \epsilon_{K_{SAH}} + \epsilon_{K_{SA}} = +\frac{2\bar{\nu}_s - 1}{1 + \kappa a_s} - \frac{1}{1 + \kappa a_H} - \frac{4}{1 + \kappa a_A} \tag{248}$$

The surface-site mass balance equation is the sum of the moles of all surface sites shown in Fig. 19 to give

$$1 = n_{SO} + n_{SOH} + n_{SOH_2} + n_{SA} + n_{SAH} + 2n_{S_2A} \tag{249}$$

Note the coefficient of 2 in the last term, which arises for the surface consisting of two surface sites in a surface-site mass balance.

$$\begin{aligned}1 &= \theta_{SO} + \theta_{SOH} + \theta_{SOH_2} + \theta_{SA} + \theta_{SAH} + 2\theta_{S_2A}\\ &\equiv \theta_{SO} + \theta_{SOH} + \theta_{SOH_2} + \theta^T_{SA}\end{aligned} \tag{250}$$

Solving Eq. (250) for the total mole fraction of surface sites complexed to a cation, θ^T_{SA}:

$$\theta^T_{SA} = 1 - (\theta_{SO} + \theta_{SOH} + \theta_{SOH_2}) \tag{251}$$

Rearranging Eq. (86) and substituting into Eq. (248):

$$\theta_{SO} + \theta_{SOH} + \theta_{SOH_2} = \theta_{SOH_2}\alpha_{SOH_2}^{-1} = 1 - \theta_{SA}^{T} \tag{252}$$

$$\theta_{SOH_2} = \alpha_{SOH_2}(1 - \theta_{SA}^{T}) \quad \vdots \quad \theta_{SA}^{T} = 1 - \alpha_{SOH_2}^{-1}\theta_{SOH_2} \tag{253}$$

Comparison of Eq. (133) derived for divalent cation adsorption and Eq. (253) derived for divalent anion adsorption, one can see that these mass balance equations differ by their use of cationic surface sites in the latter. This change will make derivation of anion adsorption equations from mass balance equations easier since cationic surface-site mole fractions, θ_{SOH_2}, are found in the anion mass action equations.

In a similar fashion the mass balance equation for the divalent anion can be written as a mass balance over the anions everywhere in the system with the total moles of anion, n_A^T, distributed in solution as n_A or among three types of surface complexes, n_{SA}, n_{SAH}, or n_{S_2A}, which can be summed to give

$$n_A^T = n_A + n_{SA} + n_{SAH} + n_{S_2A} \tag{254}$$

division by V_w and substitution of mass action equations gives the divalent anion mass balance equation:

$$\begin{aligned} \frac{n_A^T}{V_w} \equiv c_A^T &= \frac{n_A}{V_w} + \frac{n_{SA}}{n_s^T}\cdot\frac{n_s^T}{V_w} + \frac{n_{SAH}}{n_s^T}\cdot\frac{n_s^T}{V_w} + \frac{n_{S_2A}}{n_s^T}\cdot\frac{n_s^T}{V_w} \\ &= \chi_A + c_s^T(\theta_{SA} + \theta_{SAH} + \theta_{S_2A}) \equiv \chi_A + \theta_A^T c_s^T \end{aligned} \tag{255}$$

Defining variables for the parenthetical expressions gives the divalent anion mass balance equation in the same form as found for the monovalent anion in Eq. (224):

$$c_A^T \equiv \chi_A + \theta_A^T c_s^T \tag{256}$$

Comparing Eqs (250) and (254) one notes the coefficients of 2 and 1 for the n_{S_2A} term, respectively. This coefficient difference for surface site and anion mass balances is due to the stoichiometry of the $(SOH_2)_2A$ surface site complex. As a result, two different anion complexed surface-site mole fractions must be defined:

$$\theta_{SA}^T \equiv \theta_{SA} + \theta_{SAH} + 2\theta_{S_2A} \tag{257}$$

$$\theta_A^T \equiv \theta_{SA} + \theta_{SAH} + \theta_{S_2A} \tag{258}$$

In order to combine anion and surface-site mass balance equations it will be necessary to find the relationship between surface complex mole fractions. Substituting Eq. (258) into (257):

$$\theta_{SA}^T = 2\theta_A^T - \theta_{SA} - \theta_{SAH} \tag{259}$$

Substituting Eqs (245) and (247) into Eq. (259):

$$\theta_{SA}^T = 2\theta_A^T - \left(\chi_A K_{SA} Z^{-\epsilon_{K_{SA}}} + \chi_H \chi_A K''_{SAH} Z^{-\epsilon''_{K_{SAH}}}\right)(1 - \theta_{SA}^T)\alpha_{SOH_2} \tag{260}$$

and solving for θ_{SA}^T:

$$\theta_{SA}^T = \frac{2\alpha_{SOH_2}^{-1}\theta_A^T - \chi_A K_{SA} Z^{-\epsilon_{K_{SA}}} - \chi_H \chi_A K''_{SAH} Z^{-\epsilon''_{K_{SAH}}}}{\alpha_{SOH_2}^{-1} - \chi_A K_{SA} Z^{-\epsilon_{K_{SA}}} - \chi_H \chi_A K''_{SAH} Z^{-\epsilon''_{K_{SAH}}}} \equiv \frac{2\alpha_{SOH_2}^{-1}\theta_A^T - \chi_A X_{1SOH_2}}{\alpha_{SOH_2}^{-1} - \chi_A X_{1SOH_2}} \tag{261}$$

where X_{1SOH_2} represents the sum of the products of all mass action constants and activity coefficient quotient products involving only one $SOH_2{}^+$ surface site:

$$X_{1SOH_2} \equiv K_{SA} Z^{-\epsilon K_{SA}} + \chi_H K''_{SAH} Z^{-\epsilon'' K_{SAH}} \tag{262}$$

Another useful equation can also be derived from Eqs (253) and (260):

$$1 - \alpha_{SOH_2}^{-1}\theta_{SOH_2} = 2\theta_A^T - \left(\chi_A K_{SA} Z^{-\epsilon_{K_{SA}}} + \chi_H \chi_A K''_{SAH} Z^{-\epsilon''_{K_{SAH}}}\right)\theta_{SOH_2} \tag{263}$$

solving for θ_{SOH_2}:

$$\theta_{SOH_2} = \frac{2\theta_A^T - 1}{\chi_A K_{SA} Z^{-\epsilon_{K_{SA}}} + \chi_H \chi_A K''_{SAH} Z^{-\epsilon''_{K_{SAH}}} - \alpha_{SOH_2}^{-1}} = \frac{2\theta_A^T - 1}{\chi_A X_{1SOH_2} - \alpha_{SOH_2}^{-1}} \tag{264}$$

The last variable requiring an explicit equation is the mean surface-site electrostatic charge, $\bar{\nu}_s$. From Fig. 19 one can see that of the six surface site complexes shown only three have a finite electrostatic charge: (1) −1 electrostatic charge on the SO^{-1} surface site complex, (2) +1 electrostatic charge on the $SOH_2{}^+$ surface site complex, and (3) −1 electrostatic charge on the SOH_2A^- surface site complex. The mean surface-site electrostatic charge, $\bar{\nu}_s$, is the sum of the surface site fractions for these electrostatically charged surface sites:

$$\bar{\nu}_s = \theta_{SOH_2} - \theta_{SO} - \theta_{SA} \tag{265}$$

The appearance of the surface mole fraction for a surface complex requires the derivation of an α_{SA} equation:

$$\alpha_{SA} \equiv \frac{\theta_{SA}}{\theta_{SA} + \theta_{SAH} + 2\theta_{S_2A}} = \frac{\theta_{SA}}{\theta_{SA}^T} \tag{266}$$

Substitution of Eqs (245)–(247) into Eq. (266):

$$\alpha_{SA} \equiv \frac{K_{SA} Z^{-\epsilon_{K_{SA}}}}{K_{SA} Z^{-K_{SA}} + \chi_H K''_{SAH} Z^{-\epsilon''_{SAH}} + \theta_{SOH_2} K_{S_2A} Z^{-\epsilon_{K_{S_2A}}}} \tag{267}$$

Equation (265) can now be written in terms of variables which can all be readily evaluated as

$$\bar{\nu}_s = \theta_{SOH_2} - \theta_{SO} - \theta_{SA} = (\alpha_{SOH_2} - \alpha_{SO})(1 - \theta_{SA}^T) + \alpha_{SA}\theta_{SA}^T \tag{268}$$

As in Section III.B for divalent cation adsorption, Eq. (268) allows calculation of the mean surface site electrostatic charge, $\bar{\nu}_s$, once the fraction of surface sites complexed with a cation, θ_{SA}^T, has been determined. Also, as for divalent cations, Eq. (268) can be used to calculate activity coefficient quotients found in mass action equations iteratively by: (1) calculation of θ_{SA}^T under ideal adsorption conditions where $Z = 1$,

(2) calculation of $\bar{\nu}_s$ using the ideal value of θ_{SA}^T, (3) calculation of θ_{SA}^T using calculated activity coefficient quotients in the adsorption equation, and (4) repeat steps 2 and 3.

All the equations needed to derive the divalent cation adsorption equation have now been derived. A listing of the equations is given in Table 11.

Table 11 shows the eight independent equations which have been derived for the eight unknown variables necessary to specify the adsorption of divalent anions at the aqueous S–MO interface as depicted in Fig. 19. This suite of equations can be solved iteratively or solved simultaneously for θ_A^T. No iterative solution will be attempted here. Instead, simultaneous solutions of the equations in Table 11 will be explored.

The solution for θ_A^T begins with Eq. (258), which is a mass balance over all moles of adsorbed divalent anion. Substitution of mass action equations (245)–(247):

$$\theta_A^T = \theta_{SOH_2}\chi_A\left(K_{SA}Z^{-\epsilon_{K_{SA}}} + \chi_H K''_{SAH}Z^{-\epsilon_{K''_{SAH}}} + \theta_{SOH_2}K_{S_2A}Z^{-\epsilon_{K_{S_2A}}}\right) \tag{269}$$

Substitution of Eq. (253) into (269) to eliminate θ_{SOH_2} gives

$$\begin{aligned}\theta_A^T &= \theta_{SOH_2}\chi_A\left(X_{1SOH_2} + \theta_{SOH_2}K_{S_2A}Z^{-\epsilon_{K_{S_2A}}}\right)\\ &= \chi_A\left(\frac{2\theta_A^T - 1}{\chi_A X_{1SOH_2} - \alpha_{SOH_2}^{-1}}\right)\left[X_{1SOH_2} + \left(\frac{2\theta_A^T - 1}{\chi_A X_{1SOH_2} - \alpha_{SOH_2}^{-1}}\right)X_{2SOH_2}\right]\end{aligned} \tag{270}$$

TABLE 11 Tabulation of Equations and Unknowns Used for Derivation of the Divalent Anion Adsorption Equations

Unknowns		Equation	
No.	Variable	No.	Description
3	χ_{SOH}, χ_{SOH_2}, $\bar{\nu}_s$	(18)	$K_{a1} = \left(\frac{\theta_{SOH}\chi_H}{\theta_{SOH_2}}\right)Z^{\epsilon_{K_{a1}}}$
1	χ_{SO}	(20)	$K_{a2} = \left(\frac{\theta_{SO}\chi_H}{\theta_{SOH}}\right)Z^{\epsilon_{K_{a2}}}$
2	χ_A, θ_{SA}	(237)	$K_{SA} = \frac{\theta_{SA}}{\theta_{SOH_2}\chi_A} = Z^{\epsilon_{K_{SA}}}$
1	θ_{SAH}	(240)	$K_{SAH} = \frac{\theta_{SAH}}{\theta_{SA}\chi_H}Z^{\epsilon_{K_{SAH}}}$
1	θ_{S_2A}	(243)	$K_{S_2A} = \frac{\theta_{S_2A}}{\theta_{SOH_2}^2\chi_A}Z^{\epsilon_{K_{S_2A}}}$
0		(253)	$\theta_{SOH_2} = \alpha_{SOH_2}(1 - \theta_{SA}^T)$
0		(268)	$\bar{\nu}_s = (\alpha_{SOH_2} - \alpha_{SO})(1 - \theta_{SA}^T)(1 - \theta_{SA}^T) + \alpha_{SA}\theta_{SA}^T$
0		(256)	$c_A^T = \chi_A + c_s^T\theta_A^T$
8 unknowns			8 equations

where X_{2SOH_2} represents the sum of all mass action constant–activity coefficient quotient products involving two SOH_2^+ surface sites:

$$X_{2SOH_2} \equiv K_{S_2A} Z^{-\epsilon_{K_{S_2A}}} \tag{271}$$

Since Eq. (271) is quadratic in θ_A^T, it will be solved by grouping terms in like powers of θ_A^T:

$$(\theta_A^T)^2(4\chi_A X_{2SOH_2}) - \theta_A^T(\alpha_{SOH_2}^{-2} - \chi_A^2 X_{1SOH_2}^2 + 4\chi_A X_{2SOH_2}) + \chi_A(\alpha_{SOH_2}^{-1} X_{1SOH_2} - \chi_A X_{1SOH_2}^2 + X_{2SOH_2}) = 0 \tag{272}$$

with the solution:

$$\theta_A^T = \frac{\alpha_{SOH_2}^{-2} - \chi_A^2 X_{1SOH_2}^2 + 4\chi_A X_{2SOH_2}}{8\chi_A X_{2SOH_2}} - \sqrt{\frac{(\alpha_{SOH_2}^{-2} - \chi_A^2 X_{1SOH_2}^2 + 4\chi_A X_{2SOH_2})^2 - (16\chi_A^2 X_{2SOH_2})(\alpha_{SOH_2}^{-1} X_{1SOH_2} - \chi_A X_{1SOH_2}^2 + X_{2SOH_2})}{8\chi_A X_{2SOH_2}}} \tag{273}$$

Equation (273) is the adsorption equation for divalent anion adsorption, θ_A^T, at the aqueous S–MO interface as a function of anion solution molar concentration, χ_A.

The adsorption equation for divalent anion adsorption, θ_A^T, can be expressed as a function of total or initial anion solution molar concentration, c_A^T, by first solving the anion mass balance equation (256) for χ_A:

$$\chi_A = c_A^T - \theta_A^T c_s^T \tag{274}$$

then substituting the result into Eq. (272) to give

$$(\theta_A^T)^2[4(c_A^T - \theta_A^T c_s^T)X_{2SOH_2}] - \theta_A^T[\alpha_{SOH_2}^{-2} - (c_A^T - \theta_A^T c_s^T)^2 X_{1SOH_2}^2 + 4(c_A^T - \theta_A^T c_s^T)X_{2SOH_2}] + (c_A^T - \theta_A^T c_s^T)[\alpha_{SOH_2}^{-1} X_{1SOH_2} - (c_A^T - \theta_A^T c_s^T)X_{1SOH_2}^2 + X_{2SOH_2}] = 0 \tag{275}$$

Equation (275) is cubic in θ_A^T and can be multiplied out and the terms collected to give

$$(\theta_A^T)^3[c_s^T(c_s^T X_{1SOH_2}^2 - 4X_{2SOH_2})] - (\theta_A^T)^2[c_s^T(2c_A^T + c_s^T)X_{1SOH_2}^2 - 4(c_A^T + c_s^T)X_{2SOH_2}] + \theta_A^T\{[(c_A^T)^2 X_{1SOH_2} + 2c_A^T c_s^T X_{1SOH_2} - c_s^T \alpha_{SOH_2}^{-1}]X_{1SOH_2} - (4c_A^T + c_s^T)X_{2SOH_2} - \alpha_{SOH_2}^{-2}\} + c_A^T[(\alpha_{SOH_2}^{-1} - c_A^T X_{1SOH_2})X_{1SOH_2} + X_{2SOH_2}] = 0 \tag{276}$$

In most cases iterative solutions of Eq. (276) are more practical than analytical solutions. However, intuitive analysis is not possible from iterative solutions. For this reason, approximate linear solutions of Eq. (276) will be derived and examined.

The approximate linear solution to Eq. (276), solving for θ_A^T:

$$\theta_A^T \cong \frac{c_A^T\left[(\alpha_{SOH_2}^{-1} - c_A^T X_{1SOH_2})X_{1SOH_2} + X_{2SOH_2}\right]}{\begin{array}{c}\alpha_{SOH_2}^{-2} - \left[(c_A^T)^2 X_{1SOH_2} + 2c_A^T c_s^T X_{1SOH_2} - c_s^T \alpha_{SOH_2}^{-1}\right] \\ X_{1SOH_2} + (4c_A^T + c_s^T)X_{2SOH_2}\end{array}} \tag{277}$$

The pH dependence of adsorption equation (277) can be seen in Fig. 20.

The adsorption behavior shown in Fig. 20 for divalen cation adsorption was determined by using the approximate linear solution equation (277). Based on the linear approximation in this equation one would expect good accuracy only for low adsorption, say $\theta_M^T < 0.3$, which is consistent with the findings in Fig. 14 for divalent cation adsorption. However, useful information can be gleaned from Fig. 20 in the region $\theta_M^T < 0.3$:

- Divalent anion adsorption dramatically increases when the mass action constant K_{SA} increases, indicating stronger anion adsorption for the SOH_2A^- anion–surface site complex, as seen for lines a and c.
- Divalent anion adsorption is unaffected when either mass action constant, K_{SAH} (line b) or K_{S_2A} (line d), increases. The explanation for this behavior requires further investigation. The similarity of behavior among adsorption lines a, b, and d suggests that enhanced adsorption through formation of the SOH_2AH and/or $(SOH_2)_2A$ anion–surface site complexes (by increasing K_{SAH} and K_{S_2A}, respectively) depletes adsorption of complexes of other stoichiometries in such a way as to keep constant the total number of moles of anion adsorbed.

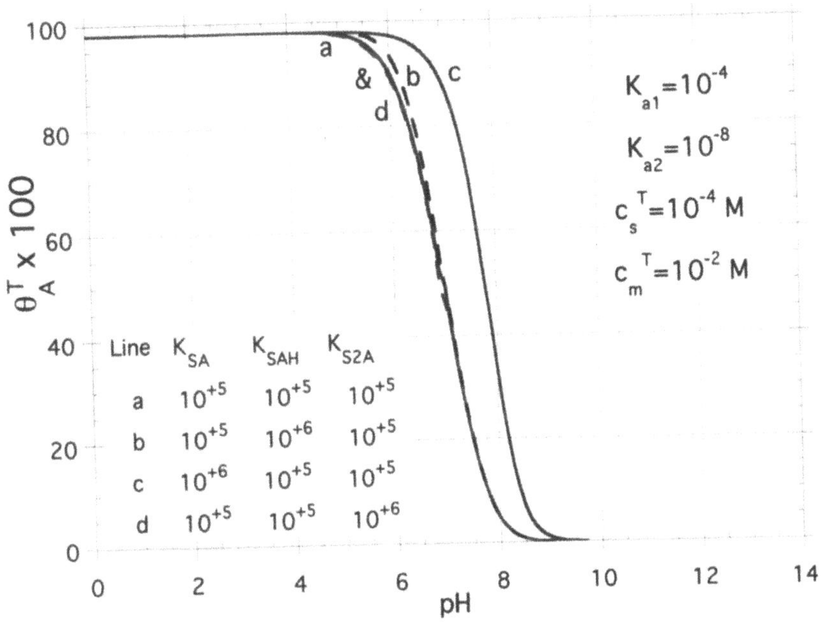

FIG. 20 Divalent anion adsorption at the aqueous S–MO interface from Eq. (277).

The fact that the surface complexes SOH_2AH and $(SOH_2)_2A$ apparently contribute very little to anion adsorption provides an opportunity to derive a much simpler adsorption equation. If the surface complexes SOH_2AH and $(SOH_2)_2A$ are assumed to contribute little to divalent anion adsorption, adsorption equation (277), when combined with Eq. (253), reduces to

$$\theta_A^T = \theta_{SOH_2}\chi_A(K_{SA}Z^{-\epsilon\kappa_{SA}}) = \chi_A\left(\frac{2\theta_A^T - 1}{\chi_A K_{SA}Z^{-\epsilon\kappa_{SA}} - \alpha_{SOH_2}^{-1}}\right)(K_{SA}Z^{-\epsilon\kappa_{SA}}) \quad (278)$$

Solving for θ_A^T:

$$\theta_A^T = \frac{\chi_A K_{SA}Z^{-\epsilon\kappa_{SA}}}{\alpha_{SOH_2}^{-1} + \chi_A K_{SA}Z^{-\epsilon\kappa_{SA}}} \quad (279)$$

Equation (279) for divalent anion adsorption can be seen to be identical in form to the adsorption equation (228) for monovalent anion adsorption. Therefore, since formation of the surface complexes SOH_2AH and/$(SOH_2)_2A$ seems to have a minimal effect on the calculated adsorption of divalent anions, the simple adsorption equation (228) should accurately predict both monovalent and divalent anion adsorption at the aqueous S–MO interface. Following the monovalent anion adsorption derivation, the adsorption equation for divalent anions in terms of total anion and surface site concentration, c_A^T and c_s^T, respectively, would be

$$\theta_A^T \cong \frac{c_A^T K_{SA}Z^{-\epsilon\kappa_{SA}}}{\alpha_{SOH_2}^{-1} + (c_s^T + C_A^T)K_{SA}Z^{-\epsilon\kappa_{SA}}} \quad (280)$$

If it is shown experimentally that significant formation of surface complexes SOH_2AH and $(SOH_2)_2A$ occurs, then use of the more complicated adsorption equations (276) or (277) would be advisable.

The adsorption equations allow calculation of θ_A^T. The definition of θ_A^T is

$$\theta_A^T \equiv \frac{n_{SA} + n_{SAH} + n_{S_2A}}{n_s^T} [=] \frac{\text{Moles of cation adsorbed}}{\text{Total moles of surface sites}} \quad (281)$$

Other measures of adsorption, which were defined for monovalent anion adsorption in Eqs (233) and (235), are based on θ_A^T and may be useful:

- The mole fraction of surface sites complexed with divalent anions, θ_A^T,can be determined from Eq. (276), (277), or (280).
- Moles of divalent cation adsorbed per total mole of anion in the system, $\hat{\theta}_A^T$:

$$\hat{\theta}_A^T \equiv \frac{n_{SA} + n_{SAH} + n_{S_2A}}{n_A^T} = \left(\frac{n_{SA} + n_{SAH} + n_{S_2A}}{n_s^T}\right)\left(\frac{n_s^T}{V_w}\right)\left(\frac{V_w}{n_A^T}\right)$$
$$= (\theta_A^T)(c_s^T)\left(\frac{1}{c_A^T}\right) = (\theta_A^T)\left(\frac{c_s^T}{c_A^T}\right) \quad (282)$$

- Moles of divalent anion adsorbed per unit area of S–MO interface, Γ_A^T:

$$\Gamma_A^T \equiv \frac{n_{SA} + n_{SAH} + n_{S_2A}}{A_s^T} = \left(\frac{n_{SA} + n_{SAH} + n_{S_2A}}{n_s^T}\right)\left(\frac{n_s^T}{A_s^T}\right)$$

$$= \theta_A^T\left(\frac{n_S^T}{A_S^T}\right) = \frac{\theta_A^T}{A_s^T N_A} \tag{283}$$

C. Predicted and Experimental Anion Adsorption

The adsorption data of Davis and Leckie [76] was fit with the single adsorption site equation (282) under ideal conditions by using the mass action constant as a fitting parameter (Fig. 21). The fit can be seen as excellent for a value of the mass action parameter of $K_{SA} = 6.0 \times 10^3$. Moreover, the excellent fit is achieved with the simplest adsorption equation (280), which assumes only a single adsorption site in spite of the fact that sulfate is divalent. This result is strong evidence, but not proof, that the predominant mechanism by which sulfate adsorbs on geothite is single anion–single surface complex formation. While this is true for the data plotted, it is not necessarily true in general. More complicated sulfate–surface site complexes may form under other conditions and/or for other anions and metal oxides.

V. WEAK ACID ANION ADSORPTION EQUATIONS

A. Adsorption Equation for Anion of a Monoprotic Weak Acid

The adsorption equation for the anion of a monoprotic weak acid, HA, is derived just like the monoprotic anion adsorption equation, but with one additional ioniza-

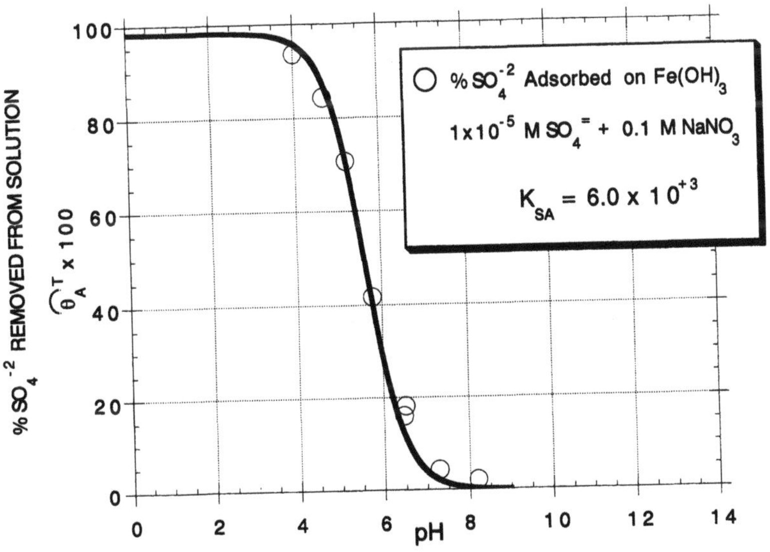

FIG. 21 Adsorption data from Davis and Leckie [76] fitted with Eq. (280) for sulfate adsorption on Geothite, FeOOH or Fe(OH)$_3$, at the aqueous S–MO interface. $K^{\epsilon_{a2}} = K^{\epsilon_{a12}} = K^{\epsilon_{K_{SSO_4^{-2}}}} = 1$.

tion reaction. In the water phase, a weak acid will exist mainly in the protonated form, HA. This solution protonation dramatically reduces the available anion, A^-, available for adsorption. It will be assumed that the protonated form of the acid, although in much greater aqueous concentration than the anion, adsorbs so weakly that its surface mole fraction can be neglected. The reactions from which the anion adsorption equation will be derived are shown in Fig. 22.

For weak acid anion adsorption the same mass balance equations will be used as were derived for the monovalent anion adsorption in Section IV.A, with an additional mass action equation for the solution acid ionization:

$$HA = H^+ + A^- \quad : \quad K_{HA} = \frac{\chi_H \chi_A}{\chi_{HA}} \left(\frac{f_H f_A}{f_{HA}} \right) = \frac{\chi_H \chi_A}{\chi_{HA}} Z^{\epsilon_{K_{HA}}} \tag{284}$$

where

$$\begin{aligned} \epsilon_{K_{HA}} &= \epsilon_H + \epsilon_A - \epsilon_{HA} = +\frac{1}{1+\kappa a_H} + \frac{1}{1+\kappa a_A} - \frac{0}{1+\kappa a_{HA}} \\ &= +\frac{1}{1+\kappa a_H} + \frac{1}{1+\kappa a_A} \end{aligned} \tag{285}$$

and

$$\chi_{HA} = \frac{\chi_H \chi_A}{K_{HA}} Z^{+\epsilon_{K_{HA}}} \tag{286}$$

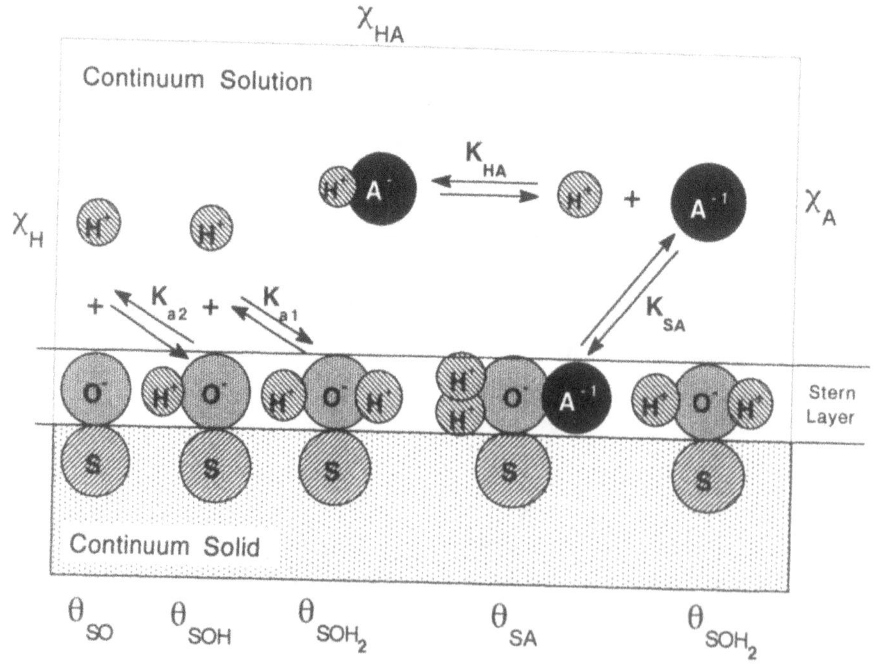

FIG. 22 Adsorption of anion of a monoprotic weak acid at the aqueous S–MO interface.

The surface-site mass balance equation is the sum over the same surface sites shown in Fig. 17 for the monovalent anion, so

$$1 = \theta_{SO} + \theta_{SOH} + \theta_{SOH_2} + \theta_{SA} \tag{287}$$

and

$$\theta_{SOH_2} = \alpha_{SOH_2}(1 - \theta_{SA}) \quad \vdots \quad \theta_{SA} = 1 - \alpha_{SOH_2}^{-1}\theta_{SOH_2} \tag{288}$$

In a similar fashion the mass balance equation for the monovalent anion can be written as a mass balance over the anion, A^-, everywhere in the system with the total moles of anion, n_A^T, distributed in solution as n_A or n_{HA} or in a surface complex, n_{SA}, summed to give

$$n_A^T = n_A + n_{HA} + n_{SA} \tag{289}$$

Division by V_w and substitution of mass action equations gives the monovalent anion mass balance equation:

$$\frac{n_A^T}{V_w} \equiv c_A^T = \frac{n_A}{V_w} + \frac{n_{HA}}{V_w} + \frac{n_{SA}}{n_s^T} \cdot \frac{n_s^T}{V_w} = \chi_A + \chi_{HA} + \theta_{SA} c_s^T \tag{290}$$

Substituting Eq. (286) into Eq. (290) eliminates χ_{HA}:

$$c_A^T = \chi_A \left(\frac{K_{HA} + \chi_H Z^{+\epsilon\kappa_{HA}}}{K_{HA}} \right) + \theta_{SA} c_s^T \equiv \alpha_{HA}^{-1} \chi_A + \theta_{SA} c_s^T \tag{291}$$

Note the similarity between the fractional acid ionization variable, α_{HA}, and the fractional surface-site ionization variables defined in Eqs (81)–(86).

From Fig. 22 one can see that the surface site species are the same as for the monovalent anion surface sites so $\bar{\nu}_s$ should be the same. Of the four surface site complexes shown only two have a finite electrostatic charge: (1) −1 electrostatic charge on the SO^- surface site complex, and (2) +1 electrostatic charge on the $SOH_2{}^+$ surface site complex. The mean surface-site electrostatic charge, $\bar{\nu}_s$, is the sum of the surface site fractions for these electrostatically charged surface sites:

$$\bar{\nu}_s = \theta_{SOH_2} - \theta_{SO} = (\alpha_{SOH_2} - \alpha_{SO})(1 - \theta_{SA}) \tag{292}$$

Reviewing Section IV.A one will note that the equation for the mean surface-site electrostatic charge, $\bar{\nu}_s$, is the same for a monovalent anion, independent of solution protonation of the anion as can be seen from Eqs (226) and (292). Clearly then, evaluation of Eqs (292) and (226) can be accomplished with the same iterative procedure: (1) calculate θ_{SA} under ideal adsorption conditions where $Z = 1$, (2) calculate $\bar{\nu}_s$ using the ideal value of θ_{SA}, (3) calculate θ_{SA} using calculated activity coefficient quotients in the adsorption equation, and (4) repeat. All the equations needed to obtain the monovalent anion adsorption equation have now been derived. A listing of the equations is given in Table 12.

An equal number of equations and variables allows the derivation of the monovalent anion adsorption equation with all unknown variables eliminated. To derive this equation, we substitute Eq. (288) into (215):

$$\theta_{SA} = \theta_{SO}(\chi_A K_{SA} Z^{-\epsilon\kappa_{SA}}) = \left(\frac{1 - \theta_{SA}}{\alpha_{SO}^{-1}} \right) (\chi_A K_{SA} Z^{-\epsilon\kappa_{SA}}) \tag{293}$$

TABLE 12 Tabulation of Equations and Unknowns Used for Derivation of the Monoprotic Weak Acid Anion Adsorption Equations

Unknowns		Equation	
No.	Variable	No.	Description
3	χ_{SOH}, χ_{SOH_2}, $\bar{\nu}_s$	(18)	$K_{a1} = \left(\frac{\theta_{SOH}\chi_H}{\theta_{SOH_2}}\right) Z^{\epsilon_{K_{a1}}}$
1	χ_{SO}	(20)	$K_{a2} = \left(\frac{\theta_{SO}\chi_H}{\theta_{SOH}}\right) Z^{\epsilon_{K_{a2}}}$
2	χ_A, θ_{SA}	(215)	$K_{SA} = \left(\frac{\theta_{SA}}{\theta_{SOH_2}\chi_A}\right) Z^{\epsilon_{K_{SA}}}$
1	χ_{HA}	(284)	$K_{HA} = \frac{\chi_H \chi_A}{\chi_{HA}} Z^{\epsilon_{K_{HA}}}$
0		(287)	$1 = \theta_{SO} + \theta_{SOH} + \theta_{SOH_2} + \theta_{SA}$
0		(292)	$\bar{\nu}_s = (\alpha_{SOH_2} - \alpha_{SO})(1 - \theta_{SA})$
0		(291)	$c_A^T = \chi_A \left(\frac{K_{HA} + \chi_H Z^{+\epsilon_{K_{HA}}}}{K_{HA}}\right) + \theta_{SA} c_S^T$
7 unknowns			7 equations

Solving for θ_{SA} gives the desired weak acid anion adsorption equation:

$$\theta_{SA} = \frac{\chi_A K_{SA} Z^{-\epsilon_{K_{SA}}}}{\alpha_{SO}^{-1} + \chi_A K_{SA} Z^{-\epsilon_{K_{SA}}}} \quad \left(\begin{array}{c}\text{Anion adsorption} \\ \text{equation, } \theta_{SA}\end{array}\right) \tag{294}$$

Substituting Eq. (291) replaces the equilibrium mole fraction χ_A by the more convenient total or inital molar solution concentrations c_s^T and c_A^T to give

$$\theta_{SA} = \frac{\alpha_{HA}(c_A^T - c_s^T \theta_{SA}) K_{SA} Z^{-\epsilon_{K_{SA}}}}{\alpha_{SOH_2}^{-1} + \alpha_{HA}(c_A^T - c_s^T \theta_{SA}) K_{SA} Z^{-\epsilon_{K_{SA}}}} \tag{295}$$

Note that, only when the solute mass balance equation is substituted, does the adsorption equation for the weak acid anion differ from that of the simple monovalent anion.

Equation (291) is quadratic in θ_{SA}:

$$\theta_{SA}^2(\alpha_{HA} c_s^T K_{SA} Z^{-\epsilon_{K_{SA}}}) - \theta_{SA}\left[\alpha_{SOH_2}^{-1} + \alpha_{HA}(c_A^T + c_s^T) K_{SA} Z^{-\epsilon_{K_{SA}}}\right] + \alpha_{HA} c_A^T K_{SA} Z^{-\epsilon_{K_{SA}}} = 0 \tag{296}$$

The exact solution of Eq. (296) is

$$\theta_{SA} = \left[\frac{\alpha_{SOH_2}^{-1} + \alpha_{HA}(c_s^T + c_A^T)K_{SA}Z^{-\epsilon\kappa_{SA}}}{2\alpha_{HA}c_s^T K_{SA}Z^{-\epsilon\kappa_{SA}}}\right] - \sqrt{\left[\frac{\alpha_{SOH_2}^{-1} + \alpha_{HA}(c_s^T + c_A^T)K_{SA}Z^{-\epsilon\kappa_{SA}}}{2\alpha_{HA}c_s^T K_{SA}Z^{-\epsilon\kappa_{SA}}}\right]^2 - \frac{c_A^T}{c_s^T}} \quad \left(\begin{array}{c}\text{Anion adsorption}\\ \text{equation – quadratic}\\ \text{solution for } \theta_{SA}\end{array}\right) \tag{297}$$

Using Eq. (103) to convert Eq. (297) from θ_{SA} to $\widehat{\theta}_{SA}$:

$$\widehat{\theta}_{SA} = \left[\frac{\alpha_{SOH_2}^{-1} + \alpha_{HA}(c_s^T + c_A^T)K_{SA}Z^{-\epsilon\kappa_{SA}}}{2\alpha_{HA}c_s^T K_{SA}Z^{-\epsilon\kappa_{SA}}}\right] - \sqrt{\left[\frac{\alpha_{SOH_2}^{-1} + \alpha_{HA}(c_s^T + c_A^T)K_{SA}Z^{-\epsilon\kappa_{SA}}}{2\alpha_{HA}c_s^T K_{SA}Z^{-\epsilon\kappa_{SA}}}\right]^2 - \frac{c_s^T}{c_A^T}} \quad \left(\begin{array}{c}\text{Cation adsorption}\\ \text{equation – quadratic}\\ \text{solution for } \widehat{\theta}_{SA}\end{array}\right) \tag{298}$$

Approximate linear anion adsorption equations can be derived in analogy with the cation adsorption equations (102), (111), and (112), respectively, as

$$\Gamma_{SA} = \frac{\chi_A K_{SA}Z^{-\epsilon\kappa_{SA}}}{A_s N_A(\alpha_{SOH_2}^{-1} + \chi_A K_{SA}Z^{-\epsilon\kappa_{SA}})} \tag{299}$$

$$\theta_{SA} \cong \frac{\alpha_{HA}c_A^T K_{SA}Z^{-\epsilon\kappa_{SA}}}{\alpha_{SOH_2}^{-1} + \alpha_{HA}(c_s^T + c_A^T)K_{SA}Z^{-\epsilon\kappa_{SA}}} \tag{300}$$

$$\widehat{\theta}_{SA} \cong \frac{\alpha_{HA}c_s^T K_{SA}Z^{-\epsilon\kappa_{SA}}}{\alpha_{SOH_2}^{-1} + \alpha_{HA}(c_s^T + c_A^T)K_{SA}Z^{-\epsilon\kappa_{SA}}} \tag{301}$$

Anion adsorption can be plotted in many forms including, but not limited to, Eqs (294)–(301). Monovalent anion adsorption, $\widehat{\theta}_{SA}$, as a function of pH is plotted in Fig. 23.

Adsorption of the anion of a weak acid is qualitatively similar to that of a simple anion (anion of a strong acid) at high pH as can be seen by comparing Figs 18 and 23. Quantitatively, two effects are noteworthy. First, as with simple anions in Fig. 18, increases in the adsorption mass action constant K_{SA} (indicating stronger anion–surface site bonds) increases anion adsorption as can be seen for lines a and b. The second effect at high pH is increased adsorption for the weaker acid, line c, than for the stronger acid, line a. The greater proton affinity by the weaker acid sites (line c) produces an increased number of cationic surface sites, which is more favorable for anion–surface site complex formation.

At low pH, adsorption decreases with pH for lines a and c, especially as the acid strength decreases ($K_{HA}\downarrow$ or $pK_{HA\uparrow}$). This behavior is a result of protonation of the weak acid ion, causing a decrease in the free anion available for adsorption.

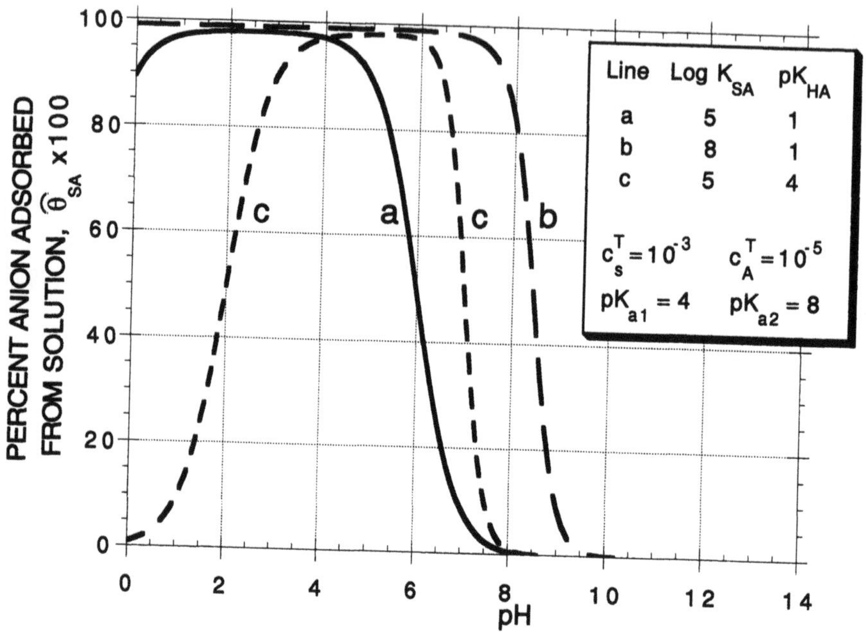

FIG. 23 Plot of Eq. (301) for monovalent weak acid anion adsorption. Percentage adsorbed from solution, $\widehat{\theta}_{SA}$, versus pH with $K^{\epsilon\kappa_{a2}} = K^{\epsilon\kappa_{a12}} = K^{\epsilon\kappa_{SA}} = 1$.

B. Adsorption Equation for Anion of a Diprotic Weak Acid

For weak diprotic acids two anions are capable of adsorbing. Deriving equations for adsorption of the anions of weak diprotic acids requires similar mass action and balance equations as were used in Sections IV.A and IV.B. In aqueous solution a weak diprotic weak acid, H_2A, will exist in three forms, H_2A, HA^-, and A^{2-}. Of these three solution species, two, HA^- and A^{2-}, will adsorb significantly at the aqueous S–MO interface under certain pH conditions. The mass action and mass balance equations for these solution ionization and adsorption reactions can be written as shown below.

1. Mass Action Equations for Solution Ionization

$$HA^- = H^+ + A^{2-} \quad : \quad K_{aHA} = \frac{\chi_H \chi_A Z^{\epsilon\kappa_{aHA}}}{\chi_{HA}} \quad : \quad \chi_{HA} = \frac{\chi_H \chi_A Z^{\epsilon\kappa_{aHA}}}{K_{aHA}} \tag{302}$$

$$H_2A = H^+ + HA^- \quad : \quad K_{aH_2A} = \frac{\chi_H \chi_{HA} Z^{\epsilon\kappa_{aH_2A}}}{\chi_{H_2A}} \tag{303}$$

$$\chi_{H_2A} = \frac{\chi_H \chi_{HA} Z^{\epsilon\kappa_{aH_2A}}}{K_{aH_2A}} = \frac{\chi_H^2 \chi_A Z^{\epsilon\kappa_{aHA}} Z^{\epsilon\kappa_{aH_2A}}}{K_{aHA} K_{aH_2A}} \tag{304}$$

where

$$\epsilon_{K_{aHA}} = \epsilon_H + \epsilon_A - \epsilon_{HA} = +\frac{1}{1+\kappa a_H} + \frac{4}{1+\kappa a_A} - \frac{1}{1+\kappa a_{HA}} \tag{305}$$

$$\begin{aligned}\epsilon_{K_{aH_2A}} &= \epsilon_H + \epsilon_{HA} - \epsilon_{H_2A} = +\frac{1}{1+\kappa a_H} + \frac{1}{1+\kappa a_{HA}} - \frac{0}{1+\kappa a_{H_2A}} \\ &= +\frac{1}{1+\kappa a_H} + \frac{1}{1+\kappa a_{HA}}\end{aligned} \tag{306}$$

2. Mass Action Equations for Adsorption of Anions

$$SOH_2{}^+ + A^{2-} = SOH_2A^- \quad : \quad K_{SA} = \frac{\theta_{SA}}{\theta_{SOH_2}\chi_A} Z^{\epsilon_{K_{SA}}} \tag{307}$$

$$\theta_{SA} = \theta_{SOH_2}\chi_A K_{SA} Z^{-\epsilon_{K_{SA}}} \tag{308}$$

$$SOH_2{}^+ + HA^- = SOH_2AH \quad : \quad K_{SAH} = \frac{\theta_{SAH}}{\theta_{SOH_2}\chi_{HA}} \tag{309}$$

$$\theta_{SAH} = \theta_{SOH_2}\chi_{HA} K_{SAH} Z^{-\epsilon_{K_{SAH}}} = \theta_{SOH_2}\chi_H\chi_A K_{aHA}^{-1} K_{SAH} Z^{\epsilon_{K_{aAH}}} Z^{-\epsilon_{K_{SAH}}} \tag{310}$$

where

$$\epsilon_{K_{SA}} = \epsilon_{SA} - \epsilon_{SOH_2} - \epsilon_A = +\frac{4\bar{v}_s}{1+\kappa a_s} - \frac{4}{1+\kappa a_A} \tag{311}$$

$$\epsilon_{K_{SAH}} = \epsilon_{SAH} - \epsilon_{SOH_2} - \epsilon_{HA} = +\frac{2\bar{v}_s - 1}{1+\kappa a_s} - \frac{1}{1+\kappa a_{HA}} \tag{312}$$

3. Surface-site Mass Balance Equation

$$1 = \theta_{SO} + \theta_{SOH} + \theta_{SOH_2} + \theta_{SA} + \theta_{SAH} \equiv \theta_{SO} + \theta_{SOH} + \theta_{SOH_2} + \theta_{SA}^T \tag{313}$$

and,

$$\theta_{SOH_2} = \alpha_{SOH_2}(1 - \theta_{SA}^T) \quad : \quad \theta_{SA}^T = 1 - \alpha_{SOH_2}^{-1}\theta_{SOH_2} \tag{314}$$

4. Anion Mass Balance Equation

$$\begin{aligned} n_A^T &= n_A + n_{HA} + n_{H_2A} + n_{SA} + n_{SAH} \\ \frac{n_A^T}{V_w} \equiv c_A^T &= \frac{n_A}{V_w} + \frac{n_{HA}}{V_w} + \frac{n_{H_2A}}{V_w} + \frac{n_{SA}}{n_s^T}\cdot\frac{n_s^T}{V_w} + \frac{n_{SAH}}{n_s^T}\cdot\frac{n_s^T}{V_w} \\ &= \chi_A + \chi_{HA} + \chi_{H_2A} + \theta_{SA}c_s^T + \theta_{SAH}c_s^T \equiv \chi_A + \chi_{HA} + \chi_{H_2A} + \theta_A^T c_s^T \end{aligned} \tag{315}$$

Substitution of Eqs (302) and (304) into Eq. (315):

$$c_A^T = \chi_A\left(1 + \frac{\chi_H Z^{\epsilon_{K_{aHa}}}}{K_{aHA}} + \frac{\chi_H^2 Z^{\epsilon_{K_{aHA}}} Z^{\epsilon_{K_{aH_2A}}}}{K_{aHA}K_{aH_2A}}\right) + \theta_A^T c_s^T \equiv \alpha_{H_2A}^{-1}\chi_A + \theta_A^T c_s^T \tag{316}$$

Note the similarity between the fractional acid ionization variable α_{H_2A} and α_{HA}, defined in Eq. (291). Also note that for this model, $\theta_A^T = \theta_{SA}^T$.

5. Mean Surface-site Electrostatic Charge

Of the five surface site complexes shown only three have a finite electrostatic charge: (1) −1 electrostatic charge on the SO^- surface site complex, (2) +1 electrostatic charge on the SOH_2^+ surface site complex, and (3) the anion surface site complex SOH_2A^-. The mean surface-site electrostatic charge, $\bar{\nu}_s$, is the sum of the surface site fractions for these electrostatically charged surface sites:

$$\bar{\nu}_s = \theta_{SOH_2} - \theta_{SO} - \theta_{SA} = (\alpha_{SOH_2} - \alpha_{SO} - \alpha_{SA})(1 - \theta_A^T) \tag{317}$$

where

$$\alpha_{SA} \equiv \frac{\theta_{SA}}{\theta_{SA} + \theta_{SAH}} = \frac{K_{SA} Z^{-\epsilon_{K_{SA}}}}{K_{SA} Z^{-\epsilon_{K_{SA}}} + \chi_H K_{aHA}^{-1} K_{SHA} Z^{-\epsilon_{K_{aHA}}} Z^{\epsilon_{SHA}}} \tag{318}$$

Reviewing Section IV.A one will note that the equation for the mean surface-site electrostatic charge, $\bar{\nu}_s$, is the same for a monovalent anion, independent of solution protonation of the anion as can be seen from Eqs (226) and (292). Clearly then, evaluation of Eqs (292) and (226) can be accomplished with the same iterative procedure: (1) calculate θ_A^T under ideal adsorption conditions where $Z = 1$, (2) calculate $\bar{\nu}_s$ using the ideal value of θ_A^T, (3) calculate θ_A^T using calculated activity coefficient quotients in the adsorption equation, and (4) repeat steps 2 and 3. All the equations needed to obtain the monovalent anion adsorption equation have now been derived.

To derive the anion adsorption equation, we substitute Eqs (308) and (310) into (315):

$$\begin{aligned}\theta_A^T &\equiv \theta_{SA} + \theta_{SAH} = \theta_{SOH_2} \chi_A \left(K_{SA} Z^{-\epsilon_{K_{SA}}} + \chi_H K_{aHA}^{-1} K_{SHA} Z^{-\epsilon_{K_{aHA}}} Z^{\epsilon_{SHA}}\right) \\ &= \left(\frac{1 - \theta_A^T}{\alpha_{SOH_2}^{-1}}\right) \chi_A \left(K_{SA} Z^{-\epsilon_{K_{SA}}} + \chi_H K_{aHA}^{-1} K_{SHA} Z^{-\epsilon_{K_{aHA}}} Z^{\epsilon_{SHA}}\right)\end{aligned} \tag{319}$$

Solving for θ_A^T gives the desired diprotic weak acid anion adsorption equation:

$$\theta_A^T = \frac{\chi_A (K_{SA} Z^{-\epsilon_{K_{SA}}} + \chi_H K_{aHA}^{-1} K_{SHA} Z^{-\epsilon_{K_{aHA}}} Z^{\epsilon_{SHA}})}{\alpha_{SOH_2}^{-1} + \chi_A \left(K_{SA} Z^{-\epsilon_{K_{SA}}} + \chi_H K_{aHA}^{-1} K_{SHA} Z^{-\epsilon_{K_{aHA}}} Z^{\epsilon_{SHA}}\right)} \tag{320}$$

Substituting Eqs (316) and (318) replaces the equilibrium mole fraction, χ_A, by the more convenient total or initial molar solution concentrations c_s^T and c_A^T to give

$$\theta_A^T = \frac{\alpha_{H_2A}(c_A^T - c_s^T \theta_{SA}^T) \alpha_{SA}^{-1} K_{SA} Z^{-\epsilon_{K_{SA}}}}{\alpha_{SOH_2}^{-1} + \alpha_{H_2A}(c_A^T - c_s^T \theta_{SA}^T) \alpha_{SA}^{-1} K_{SA} Z^{-\epsilon_{K_{SA}}}} \tag{321}$$

where, from Eq. (318):

$$\alpha_{SA}^{-1} K_{SA} Z^{-\epsilon_{K_{SA}}} = K_{SA} Z^{-\epsilon_{K_{SA}}} + \chi_H K_{aHA}^{-1} K_{SHA} Z^{-\epsilon_{K_{aHA}}} Z^{\epsilon_{SHA}} \tag{322}$$

Equation (321) is quadratic in θ_A^T, where $\theta_A^T = \theta_{SA}^T$:

$$\begin{aligned}&(\theta_A^T)^2 (\alpha_{H_2A} c_s^T \alpha_{SA}^{-1} K_{SA} Z^{-\epsilon_{K_{SA}}}) \\ &- \theta_A^T \left[\alpha_{SOH_2}^{-1} + \alpha_{H_2A}(c_A^T + c_s^T) \alpha_{SA}^{-1} K_{SA} Z^{-\epsilon_{K_{SA}}}\right] \\ &+ \alpha_{H_2A} c_A^T \alpha_{SA}^{-1} K_{SA} Z^{-\epsilon_{K_{SA}}} = 0\end{aligned} \tag{323}$$

The exact solution of Eq. (323) is

$$\theta_{\mathrm{A}}^{\mathrm{T}} = \left[\frac{\alpha_{\mathrm{SOH_2}}^{-1} + \alpha_{\mathrm{H_2A}}(c_{\mathrm{s}}^{\mathrm{T}} + c_{\mathrm{A}}^{\mathrm{T}})\alpha_{\mathrm{SA}}^{-1}K_{\mathrm{SA}}Z^{-\epsilon\kappa_{\mathrm{SA}}}}{2\alpha_{\mathrm{H_2A}}c_{\mathrm{s}}^{\mathrm{T}}\alpha_{\mathrm{SA}}^{-1}K_{\mathrm{SA}}Z^{-\epsilon\kappa_{\mathrm{SA}}}}\right] - \sqrt{\left[\frac{\alpha_{\mathrm{SOH_2}}^{-1} + \alpha_{\mathrm{H_2A}}(c_{\mathrm{s}}^{\mathrm{T}} + c_{\mathrm{A}}^{\mathrm{T}})\alpha_{\mathrm{SA}}^{-1}K_{\mathrm{SA}}Z^{-\epsilon\kappa_{\mathrm{SA}}}}{2\alpha_{\mathrm{H_2A}}c_{\mathrm{s}}^{\mathrm{T}}\alpha_{\mathrm{SA}}^{-1}K_{\mathrm{SA}}Z^{-\epsilon\kappa_{\mathrm{SA}}}}\right]^2 - \frac{c_{\mathrm{A}}^{\mathrm{T}}}{c_{\mathrm{s}}^{\mathrm{T}}}} \tag{324}$$

(Anion adsorption equation – quadratic solution for $\theta_{\mathrm{A}}^{\mathrm{T}}$)

Using equation (103) to convert equation (320) from $\theta_{\mathrm{A}}^{\mathrm{T}}$ to $\widehat{\theta}_{\mathrm{A}}^{\mathrm{T}}$,

$$\widehat{\theta}^{\mathrm{T}} = \left[\frac{\alpha_{\mathrm{SOH_2}}^{-1} + \alpha_{\mathrm{H_2A}}(c_{\mathrm{s}}^{\mathrm{T}} + c_{\mathrm{A}}^{\mathrm{T}})\alpha_{\mathrm{SA}}^{-1}K_{\mathrm{SA}}Z^{-\epsilon\kappa_{\mathrm{SA}}}}{2\alpha_{\mathrm{H_2A}}c_{\mathrm{A}}^{\mathrm{T}}\alpha_{\mathrm{SA}}^{-1}K_{\mathrm{SA}}Z^{-\epsilon\kappa_{\mathrm{SA}}}}\right] - \sqrt{\left[\frac{\alpha_{\mathrm{SOH_2}}^{-1} + \alpha_{\mathrm{H_2A}}(c_{\mathrm{s}}^{\mathrm{T}} + c_{\mathrm{A}}^{\mathrm{T}})\alpha_{\mathrm{SA}}^{-1}K_{\mathrm{SA}}Z^{-\epsilon\kappa_{\mathrm{SA}}}}{2\alpha_{\mathrm{H_2A}}c_{\mathrm{A}}^{\mathrm{T}}\alpha_{\mathrm{SA}}^{-1}K_{\mathrm{SA}}Z^{-\epsilon\kappa_{\mathrm{SA}}}}\right]^2 - \frac{c_{\mathrm{s}}^{\mathrm{T}}}{c_{\mathrm{A}}^{\mathrm{T}}}} \tag{325}$$

(Anion adsorption equation – quadratic solution for $\widehat{\theta}_{\mathrm{A}}^{\mathrm{T}}$)

Approximate linear anion adsorption equations can be derived from Eqs (320) and (323) in analogy with the cation adsorption equations (102), (111), and (112), respectively, as

$$\Gamma_{\mathrm{A}}^{\mathrm{T}} = \frac{\chi_{\mathrm{A}}\alpha_{\mathrm{SA}}^{-1}K_{\mathrm{SA}}Z^{-\epsilon\kappa_{\mathrm{SA}}}}{A_{\mathrm{s}}N_{\mathrm{A}}(\alpha_{\mathrm{SOH_2}}^{-1} + \chi_{\mathrm{A}}\alpha_{\mathrm{SA}}^{-1}K_{\mathrm{SA}}Z^{-\epsilon\kappa_{\mathrm{SA}}})} \tag{326}$$

$$\theta_{\mathrm{A}}^{\mathrm{T}} \cong \frac{\alpha_{\mathrm{H_2A}}c_{\mathrm{A}}^{\mathrm{T}}\alpha_{\mathrm{SA}}^{-1}K_{\mathrm{SA}}Z^{-\epsilon\kappa_{\mathrm{SA}}}}{\alpha_{\mathrm{SOH_2}}^{-1} + \alpha_{\mathrm{H_2A}}(c_{\mathrm{s}}^{\mathrm{T}} + c_{\mathrm{A}}^{\mathrm{T}})\alpha_{\mathrm{SA}}^{-1}K_{\mathrm{SA}}Z^{-\epsilon\kappa_{\mathrm{SA}}}} \tag{327}$$

$$\widehat{\theta}_{\mathrm{A}}^{\mathrm{T}} \cong \frac{\alpha_{\mathrm{H_2A}}c_{\mathrm{s}}^{\mathrm{T}}\alpha_{\mathrm{SA}}^{-1}K_{\mathrm{SA}}Z^{-\epsilon\kappa_{\mathrm{SA}}}}{\alpha_{\mathrm{SOH_2}}^{-1} + \alpha_{\mathrm{H_2A}}(c_{\mathrm{s}}^{\mathrm{T}} + c_{\mathrm{A}}^{\mathrm{T}})\alpha_{\mathrm{SA}}^{-1}K_{\mathrm{SA}}Z^{-\epsilon\kappa_{\mathrm{SA}}}} \tag{328}$$

Anion adsorption can be plotted in many forms including, but not limited to, Eqs (323)–(328). Diprotic weak acid anion adsorption, $\widehat{\theta}_{\mathrm{SA}}$, as a function of pH is plotted in Fig. 24.

The same dependence of anion adsorption at high and low pH is seen for the divalent anion of a weak acid in Fig. 24 as was seen for the monovalent anion adsorption of a weak acid in Fig. 23. The difference is that, with the diprotic acid, adsorption is diminished even more than for the monoprotic acid because with two solution ionization reactions for the former, even less anion is available for adsorption. The role of K_{SHA} was not explored.

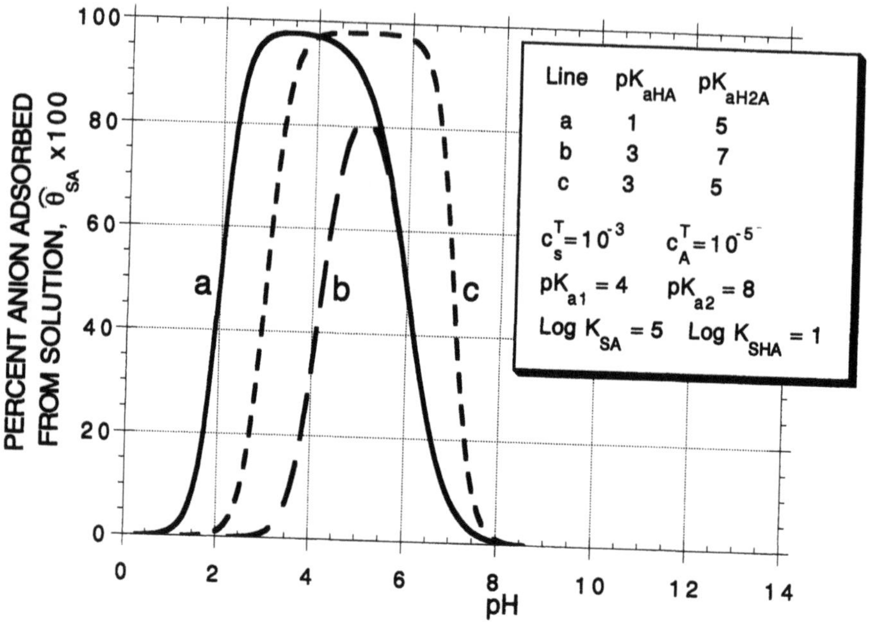

FIG. 24 Plot of Eq. (328) for divalent weak acid anion adsorption. Percentage anion adsorbed from Solution, $\widehat{\theta}_{SA}$, versus pH with $K^{\epsilon K_{a2}} = K^{\epsilon K_{a12}} = K^{\epsilon K_{SA}} = 1$.

C. Predicted and Experimental Weak Acid Anion Adsorption

The experimental benzoic acid adsorption data of Kummert and Stumm [19] will be used to compare with weak acid anion adsorption at the S–MO interface predicted by Eq. (299). The plot will be done for a constant pH = 5.0. Evaluation of Eq. (299) requires values for the four mass action constants: K_{a1}, K_{a2}, K_{aB}, and K_{SB}. The values for these constants are listed in Table 13 and were taken directly from the paper of Kummert and Stumm [19].

For fitting purposes the mass action constant values in Table 13 are substituted into Eq. (299). At a constant pH of 5.0, Eq. (299) reduces to the simple form:

$$\hat{\Gamma}_{SB}(\text{Ideal}) \cong \frac{\hat{n}_s^T(\chi_B^T K_{SB})}{1.2 + \chi_B^T K_{SB}} = \frac{\text{mmoles benzoic acid}}{\text{kg alumina}} \tag{329}$$

TABLE 13 Mass Action Constants from Kummert and Stumm [19]

Mass action constant	Value
Proton adsorption constant for alumina (K_{a1})	3.9×10^{-8}
Proton adsorption constant for alumina (K_{a2})	1.0×10^{-10}
Acid ionization constant for benzoic acid (K_{aB})	7.9×10^{-5}
Surface adsorption constant for benzoic acid on alumina (K_{SB})	2.0×10^{5}
Constant pH = 5.0 (χ_H)	1.0×10^{-5}

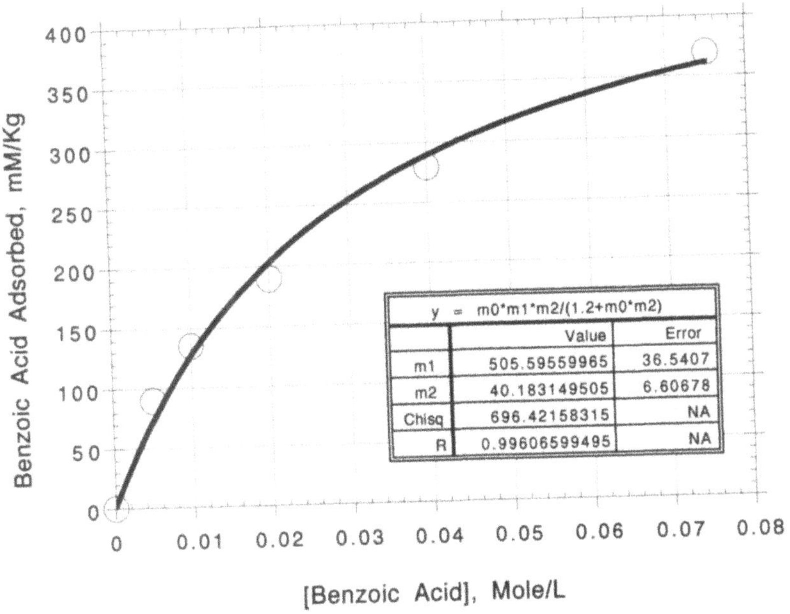

FIG. 25 Adsorption isotherm for benzoic acid adsorption on alumina at pH = 5.0. Data points from Kummert and Stumm [19]. Regression line fit for Eq. (329) with tabulated regression parameters.

In Eq. (329), the unknowns K_{SB} and $\hat{n}_s^T$ will be used as fitting parameters with χ_B^T as the indpendent variable. The regression analysis results of Eq. (329) are plotted in Fig. 25.

Equation (329) gives a statistically significant fit (i.e., $R = 0.996$) to the experimental adsorption data in Fig. 25. The parameters derived from the regression fit correspond to the two variables K_{SB} and $\hat{n}_s^T$, which can be compared to the values obtained by Kummert and Stumm [19], using a Langmuir adsoption equation. These data are summarized in Table 14.

In Table 14, the parameters determined for benzoic acid adsorption on alumina are contrasted. The values of Kummert and Stumm were determined from a simultaneous solution of multiple adsorption equations, so critical comparison of the adsorption parameters in Table 14 based on the two methods is difficult. It will

TABLE 14 Adsorption Parameters Obtained from Fitting Adsorption Equation (329) to Adsorption Data [19] in Fig. 25

Parameter	This study, Eq. (329)	Kummert and Stumm [19]
$\hat{n}_s^T$ (mol kg^{-1}), specific surface site concentration	0.506	0.62
K_{SB} (moles), molar mass action adsorption constant for benzoic acid on alumina at 22°C	40.2	3700

simply be noted that both methods predict a similar concentration of alumina surface while the TMAB Eq. (329) predicts weaker adsorption based on the smaller value for the adsorption constant. In both models, ideal adsorption has been assumed, i.e., $Z^{\epsilon_{K_{a2}}} = Z^{\epsilon_{K_{a12}}} = Z^{\epsilon_{K_{SA}}} = 1$.

VI. NONIONIC AND MULTIPLE ION ADSORPTION EQUATIONS

A. Nonionic Adsorption Equations

When a nonionic solute, U^0, adsorbs on to a S–MO interface, the type of surface site involved in the surface site complex is difficult to predict. In order to demonstrate nonionic solute adsorption at the three different types of surface site, a nonionic solute adsorption equation will be derived for adsorption of a nonionic solute, U^0, at all three surface sites, SO^-, SOH, and SOH_2^+. The adsorption processes can be visualized as shown in Fig. 26.

1. Mass Action Equations for Adsorption of Nonionics

$$SO^- + U^0 = SOU^- \quad \vdots \quad K_{SU} = \frac{\theta_{SU}}{\theta_{SO}\chi_U} Z^{\epsilon_{K_{SU}}} \tag{330}$$

$$\theta_{SU} = \theta_{SO}\chi_U K_{SU} Z^{-\epsilon_{K_{SU}}} \tag{331}$$

$$SOH + U^0 = SOHU \quad \vdots \quad K_{SHU} = \frac{\theta_{SHU}}{\theta_{SOH}\chi_U} Z^{\epsilon_{K_{SHU}}} \tag{332}$$

$$\theta_{SHU} = \theta_{SOH}\chi_U K_{SHU} Z^{-\epsilon_{K_{SHU}}} \tag{333}$$

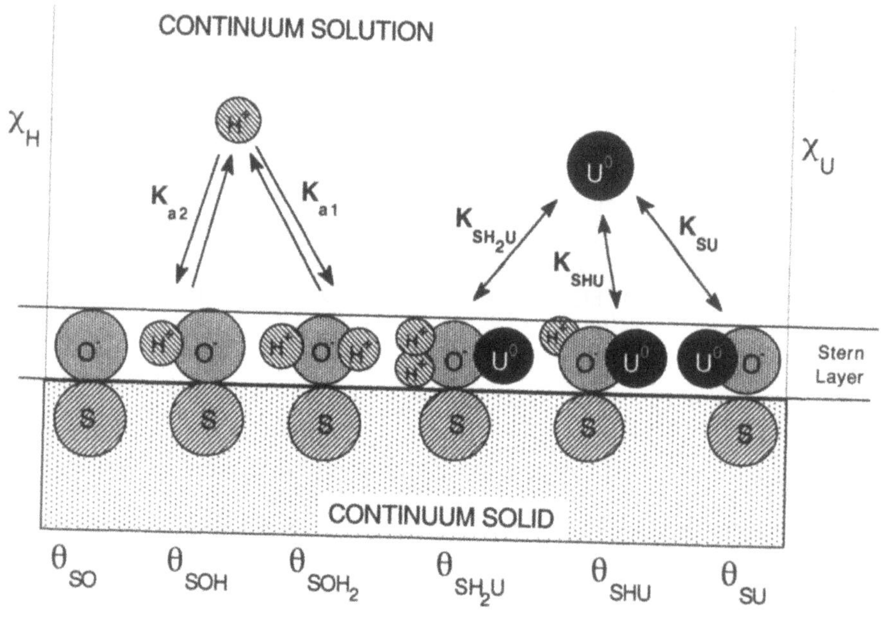

FIG. 26 Interfacial complexation reactions of protons and nonionic U^0 at the S–MO interface.

$$SOH_2^+ + U^0 = SOH_2U^+ \quad \vdots \quad K_{SH_2U} = \frac{\theta_{SH_2U}}{\theta_{SOH_2}\chi_U} Z^{\epsilon_{K_{SH_2U}}} \tag{334}$$

$$\theta_{SH_2U} = \theta_{SOH_2}\chi_U K_{SH_2U} Z^{-\epsilon_{K_{SH_2U}}} \tag{335}$$

where

$$\epsilon_{K_{SU}} = \epsilon_{SU} - \epsilon_{SO} - \epsilon_U = +\frac{(-1-\bar{\nu}_s)^2}{1+\kappa a_s} - \frac{(-1-\bar{\nu}_s)^2}{1+\kappa a_s} - \frac{0}{1+\kappa a_U} = 0 \tag{336}$$

$$\epsilon_{K_{SHU}} = \epsilon_{SHU} - \epsilon_{SOH} - \epsilon_U = +\frac{(-0-\bar{\nu}_s)^2}{1+\kappa a_s} - \frac{(-0-\bar{\nu}_s)^2}{1+\kappa a_s} - \frac{0}{1+\kappa a_U} = 0 \tag{337}$$

$$\epsilon_{K_{SH_2U}} = \epsilon_{SH_2U} - \epsilon_{SOH_2} - \epsilon_U = +\frac{(1-\bar{\nu}_s)^2}{1+\kappa a_s} - \frac{(1-\bar{\nu}_s)^2}{1+\kappa a_s} - \frac{0}{1+\kappa a_U} = 0 \tag{338}$$

2. Surface-site Mass Balance Equation

$$1 = \theta_{SO} + \theta_{SOH} + \theta_{SOH_2} + \theta_{SU} + \theta_{SHU} + \theta_{SH_2U} \equiv \theta_{SO} + \theta_{SOH} + \theta_{SOH_2} + \theta_U^T \tag{339}$$

and

$$\theta_{SOH} = \alpha_{SOH}(1-\theta_U^T) \quad \vdots \quad \theta_U^T = 1 - \alpha_{SOH}^{-1}\theta_{SOH} \tag{340}$$

3. Nonionic Mass Balance Equation

$$n_U^T = n_U + n_{SU} + n_{SHU} + n_{SH_2U} \tag{341}$$

$$\frac{n_U^T}{V_w} \equiv c_U^T = \frac{n_U}{V_w} + \frac{n_{SU}}{n_s^T}\cdot\frac{n_s^T}{V_w} + \frac{n_{SHU}}{n_s^T}\cdot\frac{n_s^T}{V_w} + \frac{n_{SH_2U}}{n_s^T}\cdot\frac{n_s^T}{V_w} \tag{341a}$$

$$c_U^T = \chi_U + \theta_{SU}c_s^T + \theta_{SHU}c_s^T + \theta_{SH_2U}c_s^T \equiv \chi_U + \theta_U^T c_s^T \tag{342}$$

4. Mean Surface-site Electrostatic Charge

Of the six surface site complexes shown only two have a finite electrostatic charge: (1) −1 electrostatic charge on the SO^- surface site complex, and (2) +1 electrostatic on the SOH_2^+ surface site complex. The mean surface-site electrostatic charge, $\bar{\nu}_s$, is the sum of the surface site fractions for these electrostatically charged surface sites:

$$\bar{\nu}_s = \theta_{SOH_2} - \theta_{SO} = (\alpha_{SOH_2} - \alpha_{SO})(1-\theta_U^T) \tag{343}$$

The fraction, α_{SHU}, of monoprotic nonionic surface sites, SOH, complexed with a nonionic solute molecule, U^0:

$$\alpha_{SHU} \equiv \frac{\theta_{SHU}}{\theta_{SU} + \theta_{SHU} + \theta_{SH_2U}} = \frac{\theta_{SOH}K_{SHU}Z^{-\epsilon_{K_{SHU}}}}{\theta_{SO}K_{SU}Z^{-\epsilon_{K_{SU}}} + \theta_{SOH}K_{SHU}Z^{-\epsilon_{K_{SHU}}} + \theta_{SOH_2}K_{SH_2U}Z^{-\epsilon_{K_{SH_2U}}}} \tag{344}$$

The equations required to derive the nonionic solute adsorption equation have been derived and are tabulated in Table 15.

Substituting proton mass action equations (18) and (20) into Eq. (344) and simplifying:

$$\alpha_{SHU} = \frac{\chi_H K_{a1} K_{SHU} Z^{-\epsilon_{K_{SHU}}}}{K_{a1}K_{a2}K_{SU}Z^{-\epsilon_{K_{a2}}}Z^{-\epsilon_{K_{SU}}} + \chi_H K_{a1} K_{SHU} Z^{-\epsilon_{K_{SHU}}} + \chi_H^2 K_{SH_2U} Z^{+\epsilon+K_{a1}} Z^{-\epsilon_{K_{SH_2U}}}} \tag{345}$$

Evaluation of Eq. (343) can be accomplished with the same iterative procedure: (1) calculate θ_U^T under ideal adsorption conditions where $Z = 1$, (2) calculate $\bar{\nu}_s$ using the ideal value of θ_U^T, (3) calculate θ_U^T using calculated activity coefficient quotients in the adsorption equation, and (4) repeat. All the equations needed to obtain the monovalent anion adsorption equation have now been derived.

To derive the nonionic adsorption equation, we substitute Eqs (331), (333), and (335) into (339):

$$\begin{aligned}\theta_U^T &\equiv \theta_{SU} + \theta_{SHU} + \theta_{SH_2U} \\ &= \chi_U(\theta_{SO}K_{SU}Z^{-\epsilon_{K_{SU}}} + \theta_{SOH}K_{SHU}Z^{-\epsilon_{K_{SHU}}} + \theta_{SOH_2}K_{SH_2U}Z^{-\epsilon_{K_{SH_2U}}})\end{aligned} \tag{346}$$

Substituting proton mass action equations (18) and (20) into Eq. (346) and simplifying:

TABLE 15 Tabulation of Equations and Unknowns Used for Derivation of the Nonionic Solute Adsorption Equations

Unknowns		Equation	
No.	Variable	No.	Description
3	χ_{SOH}, χ_{SOH_2}, $\bar{\nu}_s$	(18)	$K_{a1} = \left(\frac{\theta_{SOH}\chi_H}{\theta_{SOH_2}}\right) Z^{\epsilon_{K_{a1}}}$
1	χ_{SO}	(20)	$K_{a2} = \left(\frac{\theta_{SO}\chi_H}{\theta_{SOH}}\right) Z^{\epsilon_{K_{a2}}}$
2	χ_U, θ_{SU}	(331)	$K_{SU} = \frac{\theta_{SU}}{\theta_{SO}\chi_U} Z^{\epsilon_{K_{SU}}}$
1	θ_{SHU}	(333)	$K_{SHU} = \frac{\theta_{SHU}}{\theta_{SOH}\chi_U} Z^{\epsilon_{K_{SHU}}}$
1	θ_{SH_2U}	(335)	$K_{SH_2U} = \frac{\theta_{SH_2U}}{\theta_{SOH_2}\chi_U} Z^{\epsilon_{K_{SH_2U}}}$
0		(340)	$\theta_{SOH} = \alpha_{SOH}(1 - \theta_U^T)$
0		(343)	$\bar{\nu}_s = (\alpha_{SOH_2} - \alpha_{SO})(1 - \theta_U^T)$
0		(342)	$c_U^T = \chi_U + c_s^T\theta_U^T$
8 unknowns			8 equations

$$\theta_{U}^{T} = \theta_{SOH}\chi_{U}\left(\frac{K_{a1}K_{a2}K_{SU}Z^{-\epsilon\kappa_{a2}}Z^{-\epsilon\kappa_{SU}} + \chi_{H}K_{a1}K_{SHU}Z^{-\epsilon\kappa_{SHU}} + \chi_{H}^{2}K_{SH_2U}Z^{+\epsilon\kappa_{a1}}Z^{-\epsilon\kappa_{SH_2U}}}{\chi_{H}K_{a1}}\right)$$
$$= \theta_{SOH}\chi_{U}\alpha_{SHU}^{-1}K_{SHU}Z^{-\epsilon\kappa_{SHU}} \qquad (347)$$

Substituting Eq. (340) into Eq. (347):

$$\theta_{U}^{T} = \left(\frac{1-\theta_{U}^{T}}{\alpha_{SOH}^{-1}}\right)\chi_{U}\alpha_{SHU}^{-1}K_{SHU}Z^{-\epsilon\kappa_{SHU}} \qquad (348)$$

Solving for θ_{U}^{T}, gives the desired nonionic solute adsorption equation:

$$\theta_{U}^{T} = \frac{\chi_{U}\alpha_{SHU}^{-1}K_{SHU}Z^{-\epsilon\kappa_{SHU}}}{\alpha_{SOH}^{-1} + \chi_{U}\alpha_{SHU}^{-1}K_{SHU}Z^{-\epsilon\kappa_{SHU}}} \qquad (349)$$

Substituting Eq. (342) replaces the equilibrium mole fraction, χ_{U}, by the more convenient total or initial molar solution concentrations, c_{s}^{T} and c_{M}^{T}, to give

$$\theta_{U}^{T} = \frac{(c_{U}^{T} - c_{s}^{T}\theta_{U}^{T})\alpha_{SHU}^{-1}K_{SHU}Z^{-\epsilon\kappa_{SHU}}}{\alpha_{SOH}^{-1} + (c_{U}^{T} - c_{s}^{T}\theta_{U}^{T})\alpha_{SHU}^{-1}K_{SHU}Z^{-\epsilon\kappa_{SHU}}} \qquad (350)$$

Equation (350) is quadratic in θ_{U}^{T}:

$$\begin{aligned}&(\theta_{U}^{T})^{2}(c_{s}^{T}\alpha_{SHU}^{-1}K_{SHU}Z^{-\epsilon\kappa_{SHU}})\\ &- \theta_{U}^{T}\left[\alpha_{SOH}^{-1} + (c_{U}^{T} + c_{s}^{T})\alpha_{SHU}^{-1}K_{SHU}Z^{-\epsilon\kappa_{SHU}}\right]\\ &+ c_{U}^{T}\alpha_{SHU}^{-1}K_{SHU}Z^{-\epsilon\kappa_{SHU}} = 0\end{aligned} \qquad (351)$$

The exact solution of Eq. (351) is

$$\theta_{U}^{T} = \left[\frac{\alpha_{SOH}^{-1} + (c_{s}^{T} + c_{U}^{T})\alpha_{SHU}^{-1}K_{SHU}Z^{-\epsilon\kappa_{SHU}}}{2c_{s}^{T}\alpha_{SHU}^{-1}K_{SHU}Z^{-\epsilon\kappa_{SHU}}}\right] - \sqrt{\left[\frac{\alpha_{SOH}^{-1} + (c_{s}^{T} + c_{U}^{T})\alpha_{SHU}^{-1}K_{SHU}Z^{-\epsilon\kappa_{SHU}}}{2c_{s}^{T}\alpha_{SHU}^{-1}K_{SHU}Z^{-\epsilon\kappa_{SHU}}}\right]^{2} - \frac{c_{U}^{T}}{c_{s}^{T}}}$$
$$\left(\begin{array}{c}\text{Nonionic adsorption}\\ \text{equation – quadratic}\\ \text{solution for } \theta_{U}^{T}\end{array}\right) \qquad (352)$$

Using Eq. (103) to convert Eq. (352) from θ_{U}^{T} to $\widehat{\theta}_{U}^{T}$:

$$\widehat{\theta}_{U}^{T} = \left[\frac{\alpha_{SOH}^{-1} + (c_{s}^{T} + c_{U}^{T})\alpha_{SHU}^{-1}K_{SHU}Z^{-\epsilon\kappa_{SHU}}}{2c_{U}^{T}\alpha_{SHU}^{-1}K_{SHU}Z^{-\epsilon\kappa_{SHU}}}\right] - \sqrt{\left[\frac{\alpha_{SOH}^{-1} + (c_{s}^{T} + c_{U}^{T})\alpha_{SHU}^{-1}K_{SHU}Z^{-\epsilon\kappa_{SHU}}}{2c_{U}^{T}\alpha_{SHU}^{-1}K_{SHU}Z^{-\epsilon\kappa_{SHU}}}\right]^{2} - \frac{c_{s}^{T}}{c_{U}^{T}}}$$
$$\left(\begin{array}{c}\text{Nonionic adsorption}\\ \text{equation – quadratic}\\ \text{solution for } \widehat{\theta}_{U}^{T}\end{array}\right) \qquad (353)$$

Approximate linear anion adsorption equations can be derived from Eq. (351) in analogy with the cation adsorption equations (101) and (103), respectively, as

$$\Gamma_{\text{U}}^{\text{T}} = \frac{\chi_{\text{U}}\alpha_{\text{SHU}}^{-1}K_{\text{SHU}}Z^{-\epsilon_{K_{\text{SHU}}}}}{A_{\text{s}}N_{\text{A}}(\alpha_{\text{SOH}}^{-1} + \chi_{\text{U}}\alpha_{\text{SHU}}^{-1}K_{\text{SHU}}Z^{-\epsilon_{K_{\text{SHU}}}})} \tag{354}$$

$$\theta_{\text{U}}^{\text{T}} \cong \frac{c_{\text{U}}^{\text{T}}\alpha_{\text{SHU}}^{-1}K_{\text{SHU}}Z^{-\epsilon_{K_{\text{SHU}}}}}{\alpha_{\text{SOH}}^{-1} + (c_{\text{s}}^{\text{T}} + c_{\text{U}}^{\text{T}})\alpha_{\text{SHU}}^{-1}K_{\text{SHU}}Z^{-\epsilon_{K_{\text{SHU}}}}} \tag{355}$$

$$\widehat{\theta}_{\text{U}}^{\text{T}} \cong \frac{c_{\text{s}}^{\text{T}}\alpha_{\text{SHU}}^{-1}K_{\text{SHU}}Z^{-\epsilon_{K_{\text{SHU}}}}}{\alpha_{\text{SOH}}^{-1} + (c_{\text{s}}^{\text{T}} + c_{\text{U}}^{\text{T}})\alpha_{\text{SHU}}^{-1}K_{\text{SHU}}Z^{-\epsilon_{K_{\text{SHU}}}}} \tag{356}$$

It should be noted that the activity coefficient quotients, Z, are present even in nonionic solute adsorption equations because such quotients can never be zero in a solution or at a S–MO interface, which can have a net electrostatic surface site charge, $\bar{\nu}_{\text{s}}$. However, as seen in Eqs (336)–(338), two of the three terms in the exponent of Z are nonzero, but when combined into the exponent for the activity coefficient quotient for the adsorption complexation reaction, cancellation causes the exponents for all three adsorption reactions in Eqs (330), (332), and (334) to go to zero. Therefore, the activity coefficient quotients based on Coulombic interactions in adsorption equations (349)–(356) for nonionics are unity and can be ignored.

It should also be noted that Eqs (336)–(338) for activity coefficient quotient exponents have been derived for Coulombic interactions only. If electrostatic interactions other than Coulombic were considered, these exponents in Eqs (336)–(338) would not reduce to zero, but would have equations that would give small finite values. These effects are discussed in Chapter 4.

As one would expect, the adsorption behavior of nonionic solute molecules at the S–MO interface is qualitatively and quantitatively different than that observed for cations and anions. The nonionic solute adsorption behavior at the aqueous S–MO interface is shown in Fig. 27.

The varied mass action constants in Fig. 27 represent the relative strength of the nonionic–surface site complexation bond. The varied combinations of mass action constants for a single nonionic solute adsorbing on all three types of surface site produces many different shaped adsorption curves. As noted previously for different solutes, greater adsorption occurs when the mass action constant for a particular type of solute–surface site complex is largest.

B. Multiple Ion Adsorption Equations

Experimental adsorption studies for a particular ion are usually done with a salt having a counter ion which adsorbs very slightly. Ions which adsorb weakly are sometimes referred to as nonpotential determining ions because such ions have very little effect on the electrostatic potential of an interface.

In the case of a salt of two strongly adsorbing or potential determining ions, the adsorption of one ion can, under the right circumstances, affect the adsorption of the other. The fact that cations adsorb more readily at high pH and vice versa for anions, insures that in most cases adsorption of an ion of an electrolyte does not affect the adsorption of the counter ion. However, counter ion interference with adsorption is possible so an adsorption equation for a cation and anion will be

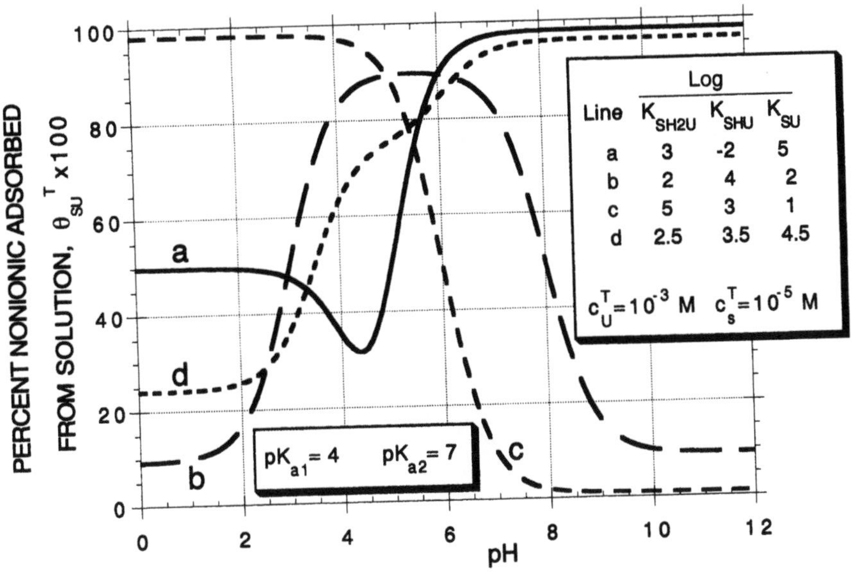

FIG. 27 Nonionic solute adsorption calculated with Eq. (355) for the aqueous S–MO interface.

derived to address quantitatively the question of counter ion adsorption interference. The adsorption reactions that must be considered in the adsorption equation derivation are given in Fig. 28.

1. Mass Action Equations for Adsorption of Monovalent Cation and Anion

$$SO^- + M^+ = SOM \quad : \quad K_{SM} = \frac{\theta_{SM}}{\theta_{SO}\chi_M} Z^{\epsilon_{K_{SM}}} \tag{357}$$

$$\theta_{SM} = (\theta_{SO})\chi_M K_{SM} Z^{-\epsilon_{K_{SM}}} \tag{358}$$

$$SOH_2{}^+ + A^- = SOH_2A \quad : \quad K_{SA} = \frac{\theta_{SA}}{\theta_{SOH_2}\chi_A} Z^{\epsilon_{K_{SA}}} \tag{359}$$

$$\begin{aligned}\theta_{SA} &= (\theta_{SOH_2})\chi_A K_{SA} Z^{-\epsilon_{K_{SA}}} + \left(\frac{\theta_{SO}\chi_H^2 Z^{\epsilon_{K_{a12}}}}{K_{a1}K_{a2}}\right)\chi_A K_{SA} Z^{-\epsilon_{K_{SA}}} \\ &= \theta_{SOH_2}\chi_A K_{SA} Z^{-\epsilon_{K_{SA}}} \equiv (\alpha_H\theta_{SO})\chi_A K_{SA} Z^{-\epsilon_{K_{SA}}}\end{aligned} \tag{360}$$

where

$$\epsilon_{K_{SM}} = \epsilon_{SM} - \epsilon_{SO} - \epsilon_M = -\frac{2\bar{v}_s + 1}{1 + \kappa a_s} - \frac{1}{1 + \kappa a_M} \tag{361}$$

$$\epsilon_{K_{SA}} = \epsilon_{SA} - \epsilon_{SOH_2} - \epsilon_A = +\frac{2\bar{v}_s - 1}{1 + \kappa a_s} - \frac{1}{1 + \kappa a_A} \tag{362}$$

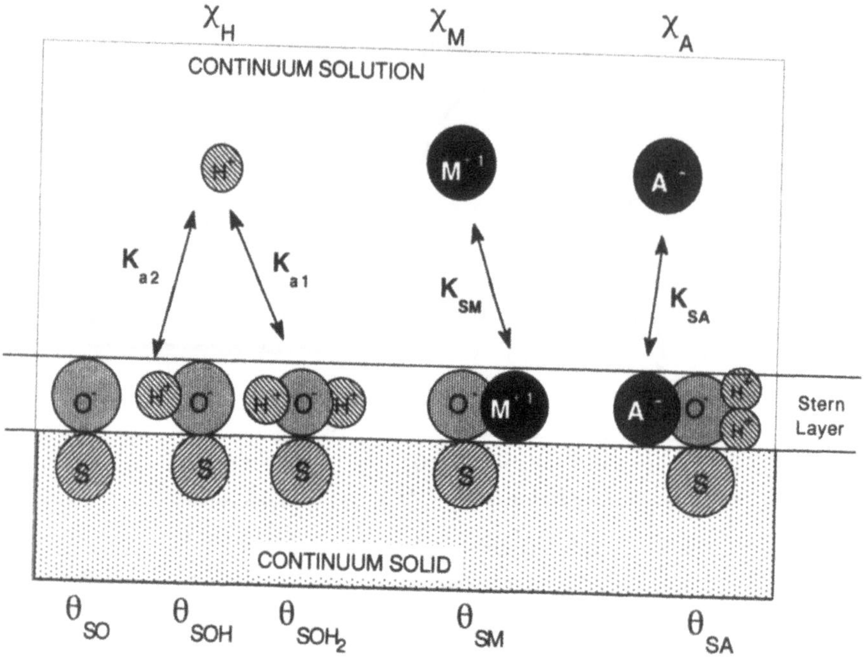

FIG. 28 Adsorption of monovalent cation and anion at the aqueous S–MO interface.

2. Surface-site Mass Balance Equation

$$1 = \theta_{SO} + \theta_{SOH} + \theta_{SOH_2} + \theta_{SM} + \theta_{SA} \tag{363}$$

and from Eqs (81) and (85):

$$\alpha_{SO} \equiv \frac{\theta_{SO}}{\theta_{SO} + \theta_{SOH} + \theta_{SOH_2}} = \frac{K_{a1}K_{a2}}{K_{a1}K_{a2} + \chi_H K_{a1} Z^{\epsilon K_{a1}} + \chi_H^2 Z^{\epsilon K_{a12}}} \tag{81a}$$

$$\alpha_{SOH_2} \equiv \frac{\theta_{SOH_2}}{\theta_{SO} + \theta_{SOH} + \theta_{SOH_2}} = \frac{\chi_H^2 Z^{\epsilon K_{a12}}}{K_{a1}K_{a2} + \chi_H K_{a1} Z^{\epsilon K_{a1}} + \chi_H^2 Z^{\epsilon K_{a12}}} \tag{85}$$

3. Cation and Anion Mass Balance Equation

$$c_M^T = \chi_M + \theta_{SM} c_s^T \tag{364}$$

$$c_A^T = \chi_A + \theta_{SA} c_s^T \tag{365}$$

4. Mean Surface Site Electrostatic Charge

Of the five surface site complexes shown only two have a finite electrostatic charge: (1) -1 electrostatic charge on the SO^- surface site complex, and (2) $+1$ electrostatic charge on the $SOH_2{}^+$ surface site complex, so $\bar{\nu}_s$:

$$\bar{\nu}_s = \theta_{SOH_2} - \theta_{SO} = (\alpha_{SOH_2} - \alpha_{SO})(1 - \theta_{SM} - \theta_{SA}) \tag{366}$$

where $\bar{\nu}_s$, θ_{SM}, and θ_{SA} are evaluated iteratively as described in Sections III and IV. The equations to be used to derive the adsorption equations are listed in Table 16.

To derive the cation adsorption equation, we begin with the fractions of surface sites complexed with a cation, α_{SM}, and an anion, α_{SA}, defined as,

$$\alpha_{SM} \equiv \frac{\theta_{SM}}{\theta_{SM} + \theta_{SA}} \quad : \quad \alpha_{SA} \equiv \frac{\theta_{SA}}{\theta_{SM} + \theta_{SA}} \tag{367}$$

Substituting Eq. (367) into Eq. (363),

$$1 = \theta_{SO} + \theta_{SOH} + \theta_{SOH_2} + \alpha_{SM}^{-1}\theta_{SM} \tag{368}$$

Substituting Eqs. (18), (20), (81), and (357) into Eq. (368) yields,

$$1 = \left(\frac{\theta_{SM}}{\chi_M K_{SM} Z^{-\epsilon\kappa_{SM}}}\right)\left(\alpha_{SO}^{-1} + \alpha_{SM}^{-1}\chi_M K_{SM} Z^{-\epsilon\kappa_{SM}}\right) \tag{369}$$

which upon solution for, θ_{SM}, gives the cation adsorption equation in terms of equilibrium cation solution molar concentration, χ_M,

$$\theta_{SM} = \frac{\chi_M K_{SM} Z^{-\epsilon\kappa_{SM}}}{\alpha_{SO}^{-1} + \alpha_{SM}^{-1}\chi_M K_{SM} Z^{-\epsilon\kappa_{SM}}} \tag{370}$$

TABLE 16 Tabulation of Equations and Unknowns Used for Derivation of the Nonionic Solute Adsorption Equations

Unknowns		Equation	
No.	Variable	No.	Description
3	χ_{SOH}, χ_{SOH_2}, $\bar{\nu}_s$	(18)	$K_{a1} = \left(\frac{\theta_{SOH}\chi_H}{\theta_{SOH_2}}\right)Z^{\epsilon\kappa_{a1}}$
1	χ_{SO}	(20)	$K_{a2} = \left(\frac{\theta_{SO}\chi_H}{\theta_{SOH}}\right)Z^{\epsilon\kappa_{a2}}$
2	χ_M, θ_{SM}	(357)	$K_{SM} = \frac{\theta_{SM}}{\theta_{SO}\chi_M}Z^{\epsilon\kappa_{SM}}$
2	χ_A, θ_{SA}	(359)	$K_{SA} = \frac{\theta_{SA}}{\theta_{SOH_2}\chi_A}Z^{\epsilon\kappa_A} = \frac{\theta_{SA}}{\alpha_H\theta_{SO}\chi_A}Z^{\epsilon\kappa_A}$
0		(363)	$1 = \theta_{SO} + \theta_{SOH} + \theta_{SOH_2} + \theta_{SM} + \theta_{SA}$
0		(366)	$\bar{\nu}_s = (\alpha_{SOH_2} - \alpha_{SO})(1 - \theta_M - \theta_A)$
0		(364)	$c_M^T = \chi_M + \theta_{SM}c_s^T$
0		(365)	$c_A^T = \chi_A + c_s^T\theta_{SA}$
8 unknowns			8 equations

The cation adsorption equation in terms of cation surface mole fraction, θ_{SM}, can be derived by substitution of cation mass balance equation (364) into Eq. (370):

$$\theta_{SM} = \frac{(c_M^T - c_s^T\theta_{SM})K_{SM}Z^{-\epsilon\kappa_{SM}}}{\alpha_{SO}^{-1} + \alpha_{SM}^{-1}(c_M^T - c_s^T\theta_{SM})K_{SM}Z^{-\epsilon\kappa_{SM}}} \tag{371}$$

which is quadratic in θ_{SM}:

$$\begin{aligned}&(\theta_{SM})^2(c_s^T\alpha_{SM}^{-1}K_{SM}Z^{-\epsilon\kappa_{SM}})\\ &-\theta_{SM}\left[\alpha_{SO}^{-1} + (\alpha_{SM}^{-1}c_M^T + c_s^T)K_{SM}Z^{-\epsilon\kappa_{SM}}\right]\\ &+c_M^TK_{SM}Z^{-\epsilon\kappa_{SM}} = 0\end{aligned} \tag{372}$$

The exact solution of Eq. (372) is

$$\theta_{SM} = \left[\frac{\alpha_{SO}^{-1} + (c_s^T + \alpha_{SM}^{-1}c_M^T)K_{SM}Z^{-\epsilon\kappa_{SM}}}{2c_s^T\alpha_{SM}^{-1}K_{SM}Z^{-\epsilon\kappa_{SM}}}\right] - \sqrt{\left[\frac{\alpha_{SO}^{-1} + (c_s^T + \alpha_{SM}^{-1}c_M^T)K_{SM}Z^{-\epsilon\kappa_{SM}}}{2c_s^T\alpha_{SM}^{-1}K_{SM}Z^{-\epsilon\kappa_{SM}}}\right]^2 - \frac{c_M^T}{\alpha_{SM}^{-1}c_s^T}}$$

$$\left(\begin{array}{c}\text{Cation adsorption}\\ \text{equation – quadratic}\\ \text{solution for } \theta_{SM}\end{array}\right) \tag{373}$$

The approximate linear solution of Eq. (372) is

$$\theta_{SM} \cong \frac{c_M^TK_{SM}Z^{-\epsilon\kappa_{SM}}}{\alpha_{SO}^{-1} + (c_s^T + \alpha_{SM}^{-1}c_M^T)K_{SM}Z^{-\epsilon\kappa_{SM}}} \tag{374}$$

Before proceeding to the anion adsorption equation derivation, evaluation of the reciprocal cation surface complex fraction, α_{SM}^{-1}, must be discussed. Substitution of mass action equations into the defining equation for α_{SM}^{-1}:

$$\alpha_{SM}^{-1} = \frac{\theta_{SM} + \theta_{SA}}{\theta_{SM}} = \frac{\chi_MK_{SM}Z^{-\epsilon\kappa_{SM}} + \chi_A\alpha_AK_{SA}Z^{-\epsilon\kappa_{SA}}}{\chi_MK_{SM}Z^{-\epsilon\kappa_{SM}}} \tag{375}$$

Comparing Eq. (375) with other fractional equations such as Eq. (81), (83), and (85) a subtle but important difference can be seen. All terms in fractional equations derived so far have included a common equilibrium molar solution concentration which subsequently cancelled. For example, in Eqs (81), (83), and (85) the molar proton concentration, χ_H, was present in all terms and as a consequence cancelled out of the equation. In Eq. (375) different equilibrium molar solution concentrations, χ_M and χ_A, appear in different terms making complete cancellation impossible. The solution is tedious so it will be given later in this section. A good approximation can be made by replacing the equilibrium molar solution concentrations, χ_M and χ_A, with the total molar solution concentrations, c_M^T and c_A^T, respectively:

$$\alpha_{SM}^{-1} = \frac{\chi_MK_{SM}Z^{-\epsilon\kappa_{SM}} + \chi_A\alpha_HK_{SA}Z^{-\epsilon\kappa_{SA}}}{\chi_MK_{SM}Z^{-\epsilon\kappa_{SM}}} \cong 1 + \frac{c_A^T\alpha_HK_{SA}Z^{-\epsilon\kappa_{SA}}}{c_M^TK_{SM}Z^{-\epsilon\kappa_{SM}}} \tag{376}$$

Equation (376) allows evaluation of α_{SM}^{-1} from constants and formal or initial molar solution concentrations with the approximation sign denoting the approximation,

$\chi_M \cong c_M^T$ and $\chi_A \cong c_A^T$. The approximation should give no significant error until the solution pH begins to approach zero. Since cation adsorption decreases in low pH solutions, any error incurred in the approximate α_{SM}^{-1} equation (376) should introduce only a small error in a small adsorption, hence negligible. At low pH the magnitude of error introduced by the approximation in Eq. (376) will depend on the relative values of K_{SM} and K_{SA}. When $K_{SM} \gg K_{SA}$, the approximation given by Eq. (376) will be negligible at all solution pH values, even in the high pH range.

Substituting Eq. (376) into adsorption equation (374) produces the desired, albeit approximate, cation adsorption equation:

$$\theta_{SM} \cong \frac{c_M^T K_{SM} Z^{-\epsilon\kappa_{SM}}}{\alpha_{SO}^{-1} + \left[c_s^T + \left(\dfrac{c_M^T K_{SM} Z^{-\epsilon\kappa_{SM}} + c_A^T \alpha_H K_{SA} Z^{-\epsilon\kappa_{SA}}}{c_M^T K_{SM} Z^{-\epsilon\kappa_{SM}}}\right) c_M^T\right] K_{SM} Z^{-\epsilon\kappa_{SM}}}$$

$$\cong \frac{c_M^T K_{SM} Z^{-\epsilon\kappa_{SM}}}{\alpha_{SO}^{-1} + \left[(c_s^T + c_M^T) K_{SM} Z^{-\epsilon\kappa_{SM}} + c_A^T \alpha_H K_{SA} Z^{-\epsilon\kappa_{SA}}\right]} \tag{377}$$

To derive the anion adsorption equation, we begin by substituting Eq. (367) into Eq. (363),

$$1 = \theta_{SO} + \theta_{SOH} + \theta_{SOH_2} + \alpha_{SA}^{-1}\theta_{SA} \tag{378}$$

Substituting Eqs. (18), (20) (81), and (359) into Eq. (378) yields,

$$1 = \left(\frac{\theta_{SA}}{\chi_A K_{SA} Z^{-\epsilon\kappa_{SA}}}\right)\left(\alpha_{SOH_2}^{-1} + \alpha_{SA}^{-1}\chi_A K_{SA} Z^{-\epsilon\kappa_{SA}}\right) \tag{379}$$

which upon solution for, θ_{SA}, gives the anion adsorption equation in terms of equilibrium anion solution molar concentration, χ_A,

$$\theta_{SA} = \frac{\chi_A K_{SA} Z^{-\epsilon\kappa_{SA}}}{\alpha_{SOH_2}^{1} + \alpha_{SA}^{-1}\chi_A K_{SA} Z^{-\epsilon\kappa_{SA}}} \tag{380}$$

The anion adsorption equation in terms of the anion surface mole fraction, θ_{SA}, can be derived by substitution of the cation mass balance equation (360) into (380):

$$\theta_{SA} = \frac{(c_A^T - c_s^T\theta_{SA}) K_{SA} Z^{-\epsilon\kappa_{SA}}}{\alpha_{SO}^{-1} + \alpha_{SA}^{-1}(c_A^T - c_s^T\theta_{SA}) K_{SA} Z^{-\epsilon\kappa_{SA}}} \tag{381}$$

Equation (381) is quadratic in θ_{SA}:

$$\begin{aligned}&(\theta_{SA})^2 (c_s^T \alpha_{SA}^{-1} K_{SA} Z^{-\epsilon\kappa_{SA}})\\&- \theta_{SA}\left[\alpha_{SOH_2}^{-1} + (\alpha_{SA}^{-1} c_A^T + c_s^T) K_{SA} Z^{-\epsilon\kappa_{SA}}\right]\\&+ c_A^T K_{SA} Z^{-\epsilon\kappa_{SA}} = 0\end{aligned} \tag{382}$$

The exact solution of Eq. (382) is

$$\theta_{SA} = \left[\frac{\alpha_{SOH_2}^{-1} + (c_s^T + \alpha_{SA}^{-1} c_A^T) K_{SA} Z^{-\epsilon\kappa_{SA}}}{2 c_s^T \alpha_{SA}^{-1} K_{SA} Z^{-\epsilon\kappa_{SA}}}\right] - \sqrt{\left[\frac{\alpha_{SOH_2}^{-1} + (c_s^T + \alpha_{SA}^{-1} c_A^T) K_{SA} Z^{-\epsilon\kappa_{SA}}}{2 c_s^T \alpha_{SA}^{-1} K_{SA} Z^{-\epsilon\kappa_{SA}}}\right]^2 - \frac{c_A^T}{\alpha_{SA}^{-1} c_s^T}}$$

$$\begin{pmatrix}\text{Anion adsorption} \\ \text{equation – quadratic} \\ \text{solution for } \theta_{SA}\end{pmatrix} \qquad (383)$$

The approximate linear solution of Eq. (382) is

$$\theta_{SA} \cong \frac{c_A^T K_{SA} Z^{-\epsilon\kappa_{SA}}}{\alpha_{SOH_2}^{-1} + (c_s^T + \alpha_{SA}^{-1} c_A^T) K_{SA} Z^{-\epsilon\kappa_{SA}}} \qquad (384)$$

Substitution of mass action equations into the defining equation for α_{SA}^{-1}:

$$\alpha_{SA}^{-1} = \frac{\theta_{SA} + \theta_{SM}}{\theta_{SA}} = \frac{\chi_A K_{SA} Z^{-\epsilon\kappa_{SA}} + \alpha_H^{-1} \chi_M K_{SM} Z^{-\epsilon\kappa_{SM}}}{\chi_A K_{SA} Z^{-\epsilon\kappa_{SA}}} \qquad (385)$$

The approximation to Eq. (385) is made by replacing the equilibrium molar solution concentrations, χ_M and χ_A, with the total molar solution concentrations, c_M^T and c_A^T, respectively:

$$\alpha_{SA}^{-1} = \frac{\chi_A K_{SA} Z^{-\epsilon\kappa_{SA}} + \chi_M \alpha_H^{-1} K_{SM} Z^{-\epsilon\kappa_{SM}}}{\chi_A K_{SA} Z^{-\epsilon\kappa_{SA}}} \cong 1 + \frac{c_M^T \alpha_H^{-1} K_{SM} Z^{-\epsilon\kappa_{SM}}}{c_A^T K_{SA} Z^{-\epsilon\kappa_{SA}}} \qquad (386)$$

Equation (386) allows evaluation of α_{SA}^{-1} from constants and total or initial molar solution concentrations with the approximation sign denoting the approximation, $\chi_M \cong c_M^T$ and $\chi_A \cong c_A^T$. The approximation should give no significant error until the solution pH becomes very high. Since anion adsorption decreases in high pH solutions, any error incurred in the approximate α_{SM}^{-1} equation (386) should introduce only a small error in a small adsorption, hence negligible. At high pH the magnitude of error introduced by the approximation in Eq. (386) will depend on the relative values of K_{SM} and K_{SA}. When $K_{SA} \gg K_{SM}$, the approximation given by Eq. (386) will be negligible at all solution pH values, even in the low pH range.

Substituting Eq. (386) into adsorption equation (384) produces the desired, albeit approximate, anion adsorption equation:

$$\theta_{SA} \cong \frac{c_A^T K_{SA} Z^{-\epsilon\kappa_{SA}}}{\alpha_{SOH_2}^{-1} + \left[c_s^T + \left(\frac{c_A^T K_{SA} Z^{-\epsilon\kappa_{SA}} + c_M^T \alpha_H^{-1} K_{SM} Z^{-\epsilon\kappa_{SM}}}{c_A^T K_{SA} Z^{-\epsilon\kappa_{SA}}}\right) c_A^T\right] K_{SA} Z^{-\epsilon\kappa_{SA}}}$$

$$\cong \frac{c_A^T K_{SA} Z^{-\epsilon\kappa_{SA}}}{\alpha_{SOH_2}^{-1} + [(c_s^T + c_A^T) K_{SA} Z^{-\epsilon\kappa_{SA}} + c_M^T \alpha_H^{-1} K_{SM} Z^{-\epsilon\kappa_{SM}}]} \qquad (387)$$

A rigorous solution for α_{SM}^{-1} and α_{SA}^{-1} begins by eliminating θ_{SO} between the mass action equations (358) and (360):

$$\theta_{SO} = \frac{\theta_{SM}}{\chi_M K_{SM} Z^{-\epsilon\kappa_{SM}}} \quad \vdots \quad \theta_{SOH_2} = \alpha_H \theta_{SO} = \frac{\theta_{SA}}{\chi_A K_{SA} Z^{-\epsilon\kappa_{SA}}} \tag{388}$$

where,

$$\alpha_H \equiv \frac{\chi_H^2 Z^{\epsilon\kappa_{a12}}}{K_{a1} K_{a2}} \tag{389}$$

The two expressions in Eq. (388) can be equated to eliminate θ_{SO}:

$$\frac{\theta_{SM}}{\chi_M K_{SM} Z^{-\epsilon\kappa_{SM}}} = \frac{\theta_{SA}}{\chi_A \alpha_H K_{SA} Z^{-\epsilon\kappa_{SA}}} \tag{390}$$

Substitution of mass balance equations (357) and (359):

$$\frac{c_M^T - \chi_M \theta_{SM}}{\chi_M K_{SM} Z^{-\epsilon\kappa_{SM}}} = \frac{c_A^T - \chi_A \theta_{SA}}{\chi_A \alpha_H K_{SA} Z^{-\epsilon\kappa_{SA}}} \tag{391}$$

Solving Eq. (391) for χ_A:

$$\chi_A = \frac{c_A^T \chi_M K_{SM} Z^{-\epsilon\kappa_{SM}}}{\chi_M K_{SM} Z^{-\epsilon\kappa_{SM}} + (c_M^T - \chi_M)\alpha_H K_{SA} Z^{-\epsilon\kappa_{SA}}} \tag{392}$$

Equation (392) can now be substituted into the cation mole fraction equation (375):

$$\alpha_{SM} = \frac{\chi_M K_{SM} Z^{-\epsilon\kappa_{SM}}}{\chi_M K_{SM} Z^{-\epsilon\kappa_{SM}} + \left[\dfrac{c_A^T \chi_M K_{SM} Z^{-\epsilon\kappa_{SM}}}{\chi_M K_{SM} Z^{-\epsilon\kappa_{SM}} + (c_M^T - \chi_M)\alpha_H K_{SA} Z^{-\epsilon\kappa_{SA}}}\right] \alpha_H K_{SA} Z^{-\epsilon\kappa_{SA}}} \tag{393}$$

which can be rearranged to collect terms in χ_M:

$$\begin{aligned}\alpha_{SM} &= \frac{\chi_M (K_{SM} Z^{-\epsilon\kappa_{SM}} - \alpha_H K_{SA} Z^{-\epsilon\kappa_{SA}}) + c_M^T \alpha_H K_{SA} Z^{-\epsilon\kappa_{SA}}}{\chi_M (K_{SM} Z^{-\epsilon\kappa_{SM}} - \alpha_H K_{SA} Z^{-\epsilon\kappa_{SA}}) + (c_M^T + c_A^T)\alpha_H K_{SA} Z^{-\epsilon\kappa_{SA}}} \\ &\equiv \frac{\chi_M \Delta K_M + c_M^T \alpha_H K_{SA} Z^{-\epsilon\kappa_{SA}}}{\chi_M \Delta K_M + (c_M^T + c_A^T)\alpha_H K_{SA} Z^{-\epsilon\kappa_{SA}}}\end{aligned} \tag{394}$$

Equation (394) is now ready to substitute into Eq. (370) so that the resulting adsorption equation will express the surface mole fraction of surface sites complexed with a cation, θ_{SM}, as a function of the equilibrium molar solution concentration, χ_M, of cation:

$$\begin{aligned}\theta_{SM} &= \frac{\chi_M K_{SM} Z^{-\epsilon\kappa_{SM}}}{\alpha_{SO}^{-1} + \left(\dfrac{\chi_M \Delta K_M + c_M^T \alpha_H K_{SA} Z^{-\epsilon\kappa_{SA}}}{\chi_M \Delta K_M + (c_M^T + c_A^T)\alpha_H K_{SA} Z^{-\epsilon\kappa_{SA}}}\right) \chi_M K_{SM} Z^{-\epsilon\kappa_{SM}}} \\ &= \frac{\chi_M K_{SM} Z^{-\epsilon\kappa_{SM}} (\chi_M \Delta K_M + c_M^T \alpha_H K_{SA} Z^{-\epsilon\kappa_{SA}})}{\begin{array}{c}\alpha_{SO}^{-1}(\chi_M \Delta K_M + c_M^T \alpha_H K_{SA} Z^{-\epsilon\kappa_{SA}}) + \\ \chi_M^2 K_{SM} Z^{-\epsilon\kappa_{SM}} \Delta K_M + \chi_M (c_M^T + c_A^T)\alpha_H K_{SM} K_{SA} Z^{-\epsilon\kappa_{SM}} Z^{-\epsilon\kappa_{SA}}\end{array}}\end{aligned} \tag{395}$$

As a final step, the equilibrium molar cation solution concentration, χ_{M}, can be eliminated by substituting the cation mass balance equation (364) into Eq. (395). When the combined equations are multiplied out and collected into like powers of θ_{SM} the result is a cubic equation of the form:

$$\begin{aligned}
&\theta_{\mathrm{SM}}^{3}[(c_{\mathrm{s}}^{\mathrm{T}})^{2}K_{\mathrm{SM}}Z^{-\epsilon\kappa_{\mathrm{SM}}}\Delta K_{\mathrm{M}}] \\
&-\theta_{\mathrm{SM}}^{2}[c_{\mathrm{s}}^{\mathrm{T}}(\alpha_{\mathrm{SO}}^{-1}+c_{\mathrm{M}}^{\mathrm{T}}K_{\mathrm{SM}}Z^{-\epsilon\kappa_{\mathrm{SM}}}+c_{\mathrm{s}}^{\mathrm{T}}K_{\mathrm{SM}}Z^{-\epsilon\kappa_{\mathrm{SM}}})\Delta K_{\mathrm{M}} \\
&+c_{\mathrm{s}}^{\mathrm{T}}(c_{\mathrm{M}}^{\mathrm{T}}+c_{\mathrm{s}}^{\mathrm{T}})K_{\mathrm{SM}}\alpha_{\mathrm{H}}K_{\mathrm{SA}}Z^{-\epsilon\kappa_{\mathrm{SA}}}Z^{-\epsilon\kappa_{\mathrm{SM}}}] \\
&+\theta_{\mathrm{SM}}[\alpha_{\mathrm{SO}}^{-1}c_{\mathrm{M}}^{\mathrm{T}}(\Delta K_{\mathrm{M}}+\alpha_{\mathrm{H}}K_{\mathrm{SA}}Z^{-\epsilon\kappa_{\mathrm{SA}}}) \\
&+c_{\mathrm{M}}^{\mathrm{T}}(c_{\mathrm{M}}^{\mathrm{T}}\Delta K_{\mathrm{M}}+2c_{\mathrm{s}}^{\mathrm{T}}\Delta K_{\mathrm{M}}+\alpha_{\mathrm{H}}K_{\mathrm{SA}}Z^{-\epsilon\kappa_{\mathrm{SA}}})K_{\mathrm{SM}}Z^{-\epsilon\kappa_{\mathrm{SM}}} \\
&+c_{\mathrm{M}}^{\mathrm{T}}(c_{\mathrm{M}}^{\mathrm{T}}+c_{\mathrm{A}}^{\mathrm{T}})K_{\mathrm{SM}}\alpha_{\mathrm{H}}K_{\mathrm{SA}}Z^{-\epsilon\kappa_{\mathrm{SM}}}Z^{\epsilon\kappa_{\mathrm{SA}}})] \\
&-[(c_{\mathrm{M}}^{\mathrm{T}})^{2}(\Delta K_{\mathrm{M}}+\alpha_{\mathrm{H}}K_{\mathrm{SA}}Z^{-\epsilon\kappa_{\mathrm{SA}}})K_{\mathrm{SM}}Z^{-\epsilon\kappa_{\mathrm{SM}}}]=0
\end{aligned}\tag{396}$$

By a similar procedure, rigorous equations for α_{SA} and θ_{SA} are derived to be

$$\begin{aligned}
\alpha_{\mathrm{SA}}&=\frac{\chi_{\mathrm{A}}(K_{\mathrm{SA}}Z^{-\epsilon\kappa_{\mathrm{SA}}}-\alpha_{\mathrm{H}}^{-1}K_{\mathrm{SM}}Z^{-\epsilon\kappa_{\mathrm{SM}}})+c_{\mathrm{A}}^{\mathrm{T}}\alpha_{\mathrm{H}}^{-1}K_{\mathrm{SM}}Z^{-\epsilon\kappa_{\mathrm{SM}}}}{\chi_{\mathrm{A}}(K_{\mathrm{SA}}Z^{-\epsilon\kappa_{\mathrm{SA}}}-\alpha_{\mathrm{H}}^{-1}K_{\mathrm{SM}}Z^{-\epsilon\kappa_{\mathrm{SM}}})+(c_{\mathrm{A}}^{\mathrm{T}}c_{\mathrm{M}}^{\mathrm{T}})\alpha_{\mathrm{H}}^{-1}K_{\mathrm{SM}}Z^{-\epsilon\kappa_{\mathrm{SM}}}} \\
&\equiv\frac{\chi_{\mathrm{A}}\Delta K_{\mathrm{A}}+c_{\mathrm{A}}^{\mathrm{T}}\alpha_{\mathrm{H}}^{-1}K_{\mathrm{SM}}Z^{-\epsilon\kappa_{\mathrm{SM}}}}{\chi_{\mathrm{A}}\Delta K_{\mathrm{A}}+(c_{\mathrm{A}}^{\mathrm{T}}+c_{\mathrm{M}}^{\mathrm{T}})\alpha_{\mathrm{H}}^{-1}K_{\mathrm{SM}}Z^{-\epsilon\kappa_{\mathrm{SM}}}}
\end{aligned}\tag{397}$$

$$\begin{aligned}
\theta_{\mathrm{SA}}&=\frac{\chi_{\mathrm{A}}K_{\mathrm{SA}}Z^{-\epsilon\kappa_{\mathrm{SA}}}}{\alpha_{\mathrm{SOH_2}}^{1}+\left(\dfrac{\chi_{\mathrm{A}}\Delta K_{\mathrm{A}}+c_{\mathrm{A}}^{\mathrm{T}}\alpha_{\mathrm{H}}^{-1}K_{\mathrm{SM}}Z^{-\epsilon\kappa_{\mathrm{SM}}}}{\chi_{\mathrm{A}}\Delta K_{\mathrm{A}}+(c_{\mathrm{A}}^{\mathrm{T}}+c_{\mathrm{M}}^{\mathrm{T}})\alpha_{\mathrm{H}}^{-1}K_{\mathrm{SM}}Z^{-\epsilon\kappa_{\mathrm{SM}}}}\right)\chi_{\mathrm{M}}K_{\mathrm{SM}}Z^{-\epsilon\kappa_{\mathrm{SM}}}} \\
&=\frac{\chi_{\mathrm{M}}K_{\mathrm{SM}}Z^{-\epsilon\kappa_{\mathrm{SM}}}(\chi_{\mathrm{M}}\Delta K_{\mathrm{A}}+c_{\mathrm{M}}^{\mathrm{T}}\alpha_{\mathrm{H}}K_{\mathrm{SA}}Z^{-\epsilon\kappa_{\mathrm{SA}}})}{\begin{array}{c}\alpha_{\mathrm{SO}}^{-1}(\chi_{\mathrm{M}}\Delta K_{\mathrm{A}}+c_{\mathrm{M}}^{\mathrm{T}}\alpha_{\mathrm{H}}K_{\mathrm{SA}}Z^{-\epsilon\kappa_{\mathrm{SA}}}+\chi_{\mathrm{M}}^{2}K_{\mathrm{SM}}Z^{-\epsilon\kappa_{\mathrm{SM}}}\Delta K_{\mathrm{A}}+\\ \chi_{\mathrm{M}}(c_{\mathrm{M}}^{\mathrm{T}}+c_{\mathrm{A}}^{\mathrm{T}})\alpha_{\mathrm{H}}K_{\mathrm{SM}}K_{\mathrm{SA}}Z^{-\epsilon\kappa_{\mathrm{SM}}}Z^{-\epsilon\kappa_{\mathrm{SA}}}\end{array}}
\end{aligned}\tag{398}$$

Equation (398) multiplied out and terms in like powers of θ_{SA} collected:

$$\begin{aligned}
&\theta_{\mathrm{SA}}^{3}[(c_{\mathrm{s}}^{\mathrm{T}})^{2}K_{\mathrm{SA}}Z^{-\epsilon\kappa_{\mathrm{SA}}}\Delta K_{\mathrm{A}}] \\
&-\theta_{\mathrm{SA}}^{2}[c_{\mathrm{s}}^{\mathrm{T}}(\alpha_{\mathrm{SOH_2}}^{-1}+c_{\mathrm{A}}^{\mathrm{T}}K_{\mathrm{SA}}Z^{-\epsilon\kappa_{\mathrm{SA}}}+c_{\mathrm{s}}^{\mathrm{T}}K_{\mathrm{SA}}Z^{-\epsilon\kappa_{\mathrm{SA}}})\Delta K_{\mathrm{A}} \\
&+c_{\mathrm{s}}^{\mathrm{T}}(c_{\mathrm{A}}^{\mathrm{T}}+c_{\mathrm{s}}^{\mathrm{T}})K_{\mathrm{SA}}\alpha_{\mathrm{H}}K_{\mathrm{SM}}Z^{-\epsilon\kappa_{\mathrm{SA}}}Z^{-\epsilon\kappa_{\mathrm{SM}}}] \\
&+\theta_{\mathrm{SA}}[\alpha_{\mathrm{SOH_2}}^{-1}c_{\mathrm{A}}^{\mathrm{T}}(\Delta K_{\mathrm{A}}+\alpha_{\mathrm{H}}K_{\mathrm{SM}}Z^{-\epsilon\kappa_{\mathrm{SM}}}) \\
&+c_{\mathrm{A}}^{\mathrm{T}}(c_{\mathrm{A}}^{\mathrm{T}}\Delta K_{\mathrm{A}}+2c_{\mathrm{s}}^{\mathrm{T}}\Delta K_{\mathrm{A}}+\alpha_{\mathrm{H}}^{-1}K_{\mathrm{SM}}Z^{-\epsilon\kappa_{\mathrm{SM}}})K_{\mathrm{SA}}Z^{-\epsilon\kappa_{\mathrm{SA}}} \\
&+c_{\mathrm{A}}^{\mathrm{T}}(c_{\mathrm{A}}^{\mathrm{T}}+c_{\mathrm{M}}^{\mathrm{T}})K_{\mathrm{SA}}\alpha_{\mathrm{H}}^{-1}K_{\mathrm{SM}}Z^{-\epsilon\kappa_{\mathrm{SA}}}Z^{-\epsilon\kappa_{\mathrm{SM}}})] \\
&-[(c_{\mathrm{A}}^{\mathrm{T}})^{2}(\Delta K_{\mathrm{A}}+\alpha_{\mathrm{H}}^{-1}K_{\mathrm{SM}}Z^{-\epsilon\kappa_{\mathrm{SM}}})K_{\mathrm{SA}}Z^{-\epsilon\kappa_{\mathrm{SA}}}]=0
\end{aligned}\tag{399}$$

Equations (396) and (399) are the rigorous cation and anion adsorption equations, respectively, for adsorption from a solution of an electrolyte, MA, at the aqueous S–MO interface. It can be compared to the approximate equations (374), (377), (384), and (387), which are far simpler and just as accurate at predicting cation and anion adsorption except at extremely low or high pH, respectively, where ion adsorption is so low as to be of little interest.

The competitive adsorption behavior of two ions in the simple electrolyte solute, MA, can be examined in Figs 29 and 30.

In Fig. 29, lines a and b show how the amount of cation adsorbed varies directly with the magnitude of cation adsorption mass action constant. Line c shows how when the mass action adsorption constant for the cation is invariant, competitive adsorption from the anion can occur. Cation adsorption for line c is delayed to even higher pH due to the anion mass adsorption constant being of equal magnitude. This same behavior can be seen for anion adsorption in Fig. 30.

VII. SURVEY OF ADSORPTION MODELS

A. Surface Ionization Reaction Conventions

The literature describing surface ionization as a mass action process datas from the early 1970s [4, 8, 20–27]. For about the next decade, ionization of amphoteric surface sites was described with the chemical equations (14) and (19). These equations are written in the classic form of acid ionization reactions so that the respective mass action equations (18) and (20) are written in terms of mass action constants, with a small subscript "a" indicating a mass action equation for an acid ionization reaction. In virtually all these early publications the "K acid ionization constants" for the ionization reactions, Eqs (14) and (19), and their respective mass action equations (18) and (20) are written with the symbols K_{a1} and K_{a2}, respectively.

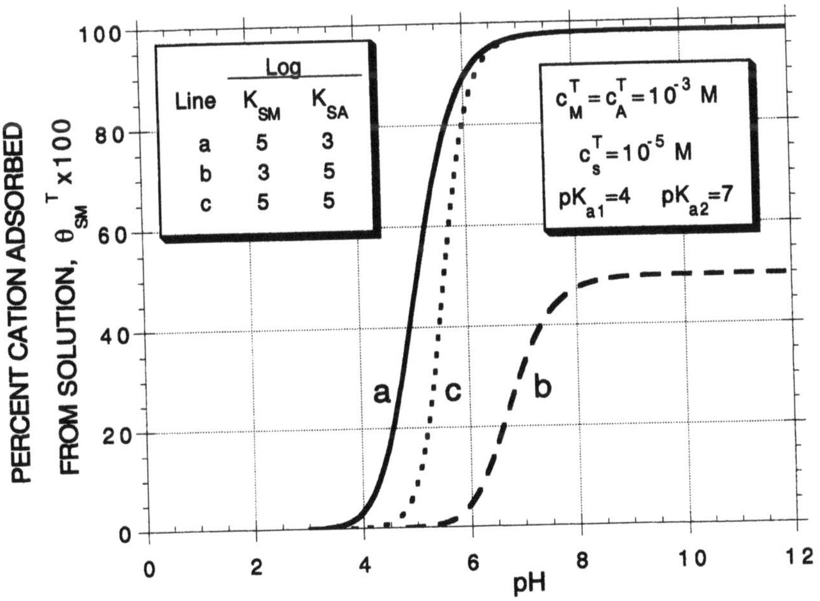

FIG. 29 Cation adsorption for an electrolyte MA at the S–MO interface calculated from cation adsorption Eq. (374) for ideal adsorption, $K^{\epsilon K_{a1}} = K^{\epsilon K_{a2}} = K^{\epsilon K_{a12}} = K^{\epsilon K_{SM}} = K^{\epsilon K_{SA}} = 1$.

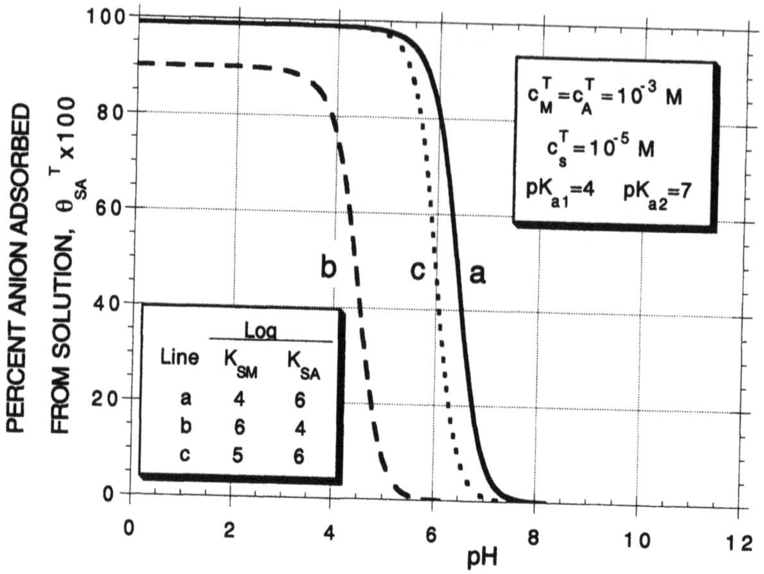

FIG. 30 Anion adsorption for an electrolyte MA at the S–MO interface calculated from anion adsorption Equation (384) for ideal adsorption, $K^{\epsilon K_{a1}} = K^{\epsilon K_{a2}} = K^{\epsilon K_{a12}} = K^{\epsilon K_{SM}} = K^{\epsilon K_{SA}} = 1$.

In the early 1980s, a change in the way the chemical equations for surface site reactions with protons began to appear in the literature [28–38]. The amphoteric surface site reactions were written as

$$SOH + H^+ \leftrightarrow SOH_2^+ \tag{400}$$

$$SO^- + H^+ \leftrightarrow SOH \tag{401}$$

Equations (400) and (401) are written as complexation or formation reactions not as acid ionization reactions like Eqs (14) and (19). The unfortunate aspect of these different conventions is that the same symbol was used in the literature to represent both the mass action constant for the acid ionization chemical reactions, Eqs (14) and (19), and the formation reactions, Eqs (400) and (401), as depicted in Table 17.

As a result of this confusing nomenclature the reader must ascertain the precise definition of the mass action constants, K_{a1} and K_{a2}, when reading the literature since both definitions are currently in use. Examples of the use of the two different definitions for K_{a1} and K_{a2} can be found in two chapters of this book: the "proton complexation formation" reactions are used in Chapter 5, while the "acid ionization" reactions are used in Chapter 6. To be consistent with the conventional use of the small subscript "a" in mass action constants to represent, "acid ionization" reactions, the former set of equations, (14) and (19), have been used throughout this chapter.

B. Overview of Adsorption Models

Adsorption at the solution–solid interface has been described in the literature with many different adsorption equations based on a bewildering array of differently

TABLE 17 Amphoteric Reactions for Surface Sites

Chemical reaction	Mass action equation	Equation type
$SOH_2^+ \leftrightarrow SOH + H^+$	$K_{a1} = \left(\frac{\theta_{SOH}\chi_H}{\theta_{SOH_2}}\right)\left(\frac{f_{SOH}f_H}{f_{SOH_2}}\right)$	Acid ionization (1970–present)
$SOH \leftrightarrow SO^- + H^+$	$K_{a2} = \left(\frac{\theta_{SO}\chi_H}{\theta_{SOH}}\right)\left(\frac{f_{SO}f_H}{f_{SOH}}\right)$	Acid ionization (1970–present)
$SOH + H^+ \leftrightarrow SOH_2^+$	$K_{a1} = \left(\frac{\theta_{SOH_2}}{\theta_{SOH}\chi_H}\right)\left(\frac{f_{SOH_2}}{f_{SOH}f_H}\right)$	Complexation formation (1980–present)
$SO^- + H^+ \leftrightarrow SOH$	$K_{a2} = \left(\frac{\theta_{SOH}}{\theta_{SO_2}\chi_H}\right)\left(\frac{f_{SOH}}{f_{SO}f_H}\right)$	Complexation formation (1980–present)

named adsorption theories. However, all these different adsorption equations are actually based on a limited number of different fundamental principles. For the following discussion, five of the most familiar theoretical approaches will be discussed: (1) partition coefficient (PC) model, (2) self-consistent lattice field (SCLF) model, (3) gas phase (GP) or Langmuir model, (4) Boltzmann potential distribution (BPD) model, and (5) thermodynamic mass action/mass balance (TMAB) model.

For all five adsorption models to be discussed, the surface of the solid contiguous to the solution is viewed as a plane of constant thermodynamic chemical (including electrical) potential. Adsorption is regarded as those solute molecules which partition into a monomolecular surface layer. However, these five models have significant distinguishing characteristics in the way they approach adsorption, as shown heuristically in Fig. 31.

The models depicted in Fig. 31 are generalizations and omit many details and variations of each model, which will be left for the following discussions.

- For the PC model, adsorption can be seen as purely partitioning behavior. This model ascribes a thermodynamic partitioning coefficient ("adsorption separation factor") to each solute to describe adsorption quantitatively. No explicit thermodynamic contribution of the surface site to the adsorption process nor competitive adsorption effects are considered.
- For the SCLF model adsorption can be seen as purely partitioning behavior. Adsorption is quantified in this model by assigning a statistical mechanical partition function to each solute molecule. The partition function recognizes three major features of adsorption: (1) entropy of adsorption, (2) energy of adsorption, and (3) solute–solvent interaction energy.
- The GP and BPD models treat adsorption as a thermodynamic mass action process so that the energetics of the important chemical species are incorporated into the adsorption equations: solute plus surface site = solute–surface

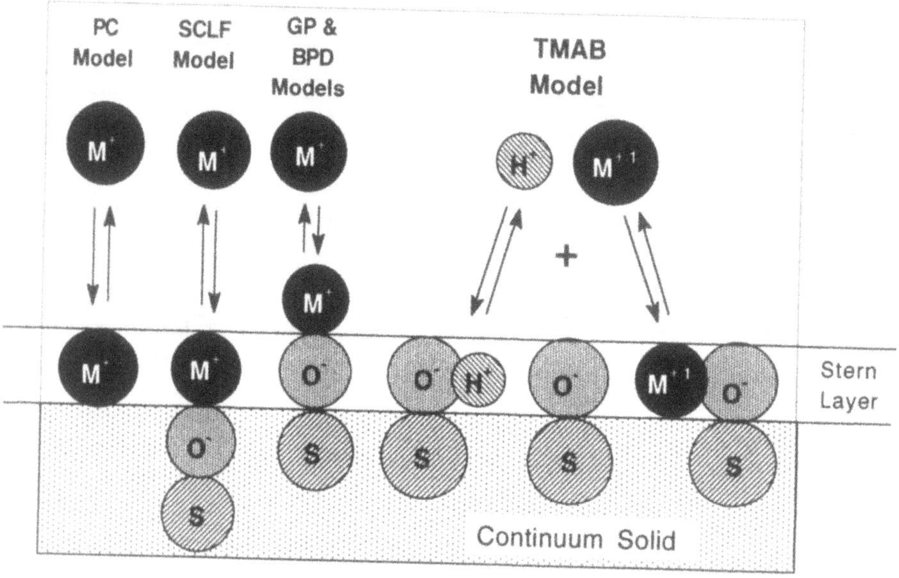

FIG. 31 Contrast of adsorption mechanisms underlying adsorption models for cation adsorption.

site complex. However, the GP model does not account for adsorption competition between protons and other solute ions because no mass balance equations are incorporated into the adsorption equations. With mass balance missing, adsorption equations lack pH dependence, which has led to many different schemes to reintroduce pH dependence back into these equations.

- The TMAB model combines the thermodynamic mass action and mass balance equations to produce adsorption equations which correctly predict adsorption at the S–MO interface and give correct pH dependence of adsorption. This model was used exclusively for all adsorption equations derived so far in this chapter. The TMAB model will be contrasted with the other adsorption models throughout the remainder of this section.

C. Partition Coefficient Models

The earliest solution–solid adsorption study using the PC model was published by Mair *et al.* [39] in 1950. The application of these and other early practitioners [40, 41] was to describe fractionation in liquid chromatography applications of a solution of two liquids, "1" and "2". The reaction for the adsorption process occurs among four species: (1) n_1, moles of solute 1 in the solution phase; (2) n_1^s, moles of solute 1 adsorbed in the surface phase; (3) n_2, moles of solute 2 in the solution phase; and (4) n_2^s, moles of solute 2 adsorbed in the surface phase, as

$$n_1 + n_2^s \Leftrightarrow n_2 + n_1^s \tag{402}$$

Relative adsorbability constants [41] for the adsorbed phase, K_a, and the liquid phase, K_ℓ, for each solute can be defined, respectively:

$$K_a \equiv \frac{\theta_1^s}{\theta_2^s} \tag{403}$$

$$K_\ell \equiv \frac{\chi_1}{\chi_2} \tag{404}$$

If Eqs (403) and (404) are divided and rearranged:

$$\frac{\theta_1^s}{\chi_1} = \frac{K_a}{K_\ell}\left(\frac{\theta_2^s}{\chi_2}\right) \equiv K_{1,2}\left(\frac{\theta_2^s}{\chi_2}\right) \tag{405}$$

Equation (405) [Eq. (1) in Refs 39 and 41] defines the adsorption separation factor [40], $K_{1,2}$.

In 1964, Everett [42–44] derived Eq. (405) with a thermodynamic treatment following the work of Butler [45] for vapor–solid adsorption. The distribution of two solutes, "1" and "2" (usually chosen as a solute, 1, and a solvent, 2, molecule) between two phases (solution and surface adsorbed), described by Eq. (402), is determined by requiring that at equilibrium the chemical potential for each solute is the same in both the solution, μ_i, and the surface, phase, μ_i^s, so

$$\mu_1 = \mu_1^s \quad : \quad \mu_2 = \mu_2^s \tag{406}$$

If the Gibbs–Lewis (Chapter 4, this volume) [1] expressions for chemical potential are substituted into Eqs (406), one can write, respectively:

$$\mu_1^* + kT \ln \chi_1 f_1 = \mu_1^{s*} + kT \ln \theta_1^s f_1^s \tag{407}$$

$$\mu_2^* + kT \ln \chi_2 f_2 = \mu_2^{s*} + kT \ln \theta_2^s f_2^s \tag{408}$$

If Eqs (407) and (408) are added and rearranged, the resulting equation gives the adsorption separation factor, $K_{1,2}$, for solute partitioning, which can be seen to be the quotient of the two individual relative adsorbabilities:

$$\left(\frac{\theta_1^S \chi_2}{\chi_1 \theta_2^s}\right)\left(\frac{f_1^s f_2}{f_1 f_2^s}\right) = \exp\left[-\frac{(\mu_1^{s*} - \mu_1^*) - (\mu_2^{s*} - \mu_2^*)}{kT}\right] = \frac{K_a}{K_\ell} \equiv K_{1,2} \tag{409}$$

Equation (409) derived by Everett [42, 43, 44] from rigorous Gibbs–Lewis thermodynamics provides three important pieces of information not available from the derivation of Mair *et al.* and others [39–41, 45];

- The adsorption separation factor, $K_{1,2}$, can be seen as a true constant (at constant temperature) since it equates to standard state chemical potentials (and temperature).
- Activity coefficients which appear in Eq. (409) make it applicable to non-ideal adsorption unlike the ideal equation (405).
- The relative adsorbabilities are not constants. In order for K_a and K_ℓ to be constant one can see that Eqs (403) and (404) can be derived from Eqs (407) and (408) only if $\mu_1 = \mu_2$ and $\mu_1^s = \mu_2^s$. Since the chemical potential of different solutes even in the same phase cannot be equal in the general

case, it is clear that K_a and K_ℓ cannot be constants because, in general, $\mu_1 \neq \mu_2$ and $\mu_1^s \neq \mu_2^s$.

Equation (409) is in a form which clearly defines the adsorption separation factor. However, Eq. (409) is not a very convenient adsorption equation since it contains four unknown variables: χ_1, θ_1^s, χ_2, and θ_2^s ($K_{1,2}$ is an unknown constant). Elimination of two unknowns is possible using solution and surface adsorbed mass balance equations. Two mass balance equations over both species, 1 and 2, in the solution and surface adsorbed phases can be written, respectively, as

$$1 = \chi_1 + \chi_2 \tag{410}$$

$$1 = \theta_1^s + \theta_2^s \tag{411}$$

Substitution of the mass balance equations (410) and (411) into Eq. (409), followed by rearrangement, gives the PC adsorption equation in its final form:

$$\begin{aligned}\theta_1^s &= \frac{\chi_1 K_{1,2}\left(\dfrac{f_1 f_2^s}{f_1^s f_2}\right)}{\chi_2 + \chi_1 K_{1,2}\left(\dfrac{f_1 f_2^s}{f_1^s f_2}\right)} \\ &= \frac{\chi_1 K_{1,2}(f_1 f_2^s)}{\chi_2(f_1^s f_2) + \chi_1\left[K_{1,2}(f_1 f_2^s) - (f_1^s f_2)\right]} \\ &= \frac{\chi_1 K_1(f_1 f_2^s)}{\chi_2 K_2(f_1^s f_2) + \chi_1\left[K_1(f_1 f_2^s) - K_2(f_1^s f_2)\right]}\end{aligned} \tag{412}$$

Equation (412) is an adsorption equation derived by applying rigorous thermodynamics to an adsorption model which depicts the adsorption process as solute partitioning between the solution and a hypothetical monomolecular surface phase. As a result of the rigorous thermodynamic basis, activity coefficients appear in the equation, making it applicable to nonideal adsorption.

Everett [43] discusses experimental evaluation methods for the activity coefficient, but offers no explicit activity coefficient equations in closed analytical form. Hence, in applications of Eq. (412) in the literature, the surface activity coefficients (and frequently the solution activity coefficients) are set to unity. While Eq. (412) is an adsorption equation based on rigorous thermodynamics, its application has always been as an ideal adsorption equation. Substitution of activity coefficient equations, described in Section II.E and Chapter 4, into Eq. (412) would extend its applicability to nonideal adsorption.

The other major limitation of Eq. (412) in accurately describing adsorption is the monomolecular surface phase inherent in the PC model. By ignoring surface sites which are amphoteric and with which solute molecules can form surface complexes, the PC adsorption equation (412) has the following shortcomings:

- No variable representing surface site concentration(s) appears explicitly.
- No explicit pH dependence appears.
- Only adsorption for a single solute and solvent species can be considered because the adsorbed surface mass balance equations (410) and (411) must sum to unity.

Of these three limitations, the first two are a direct result of the monomolecular surface phase model and cannot be changed without changing the PC model, such as recognizing surface site complexation. However, the third limitation is not a result of the model but is instead a result of the method of derivation of Eq. (412). This third limitation can be circumvented by: (1) expand surface-site balance equation (411) to include other types of adsorbed solutes, and (2) substitute mass action equations into the expanded surface phase balance equations, instead of the inverse which was done in the derivation of Eq. (412).

Before leaving the PC method, a comparison with the TMAB method for adsorption at the solution–solid interface will be instructive. Specifically, adsorption occurs by complex formation between solute and surface site molecules in the TMAB model while adsorption occurs by solute partitioning into a surface phase, without forming any specific bond with solid surface molecules, in the PC model. From a mechanistic point of view one can see that the TMAB and PC methods differ significantly. However, a more quantitative approach can be used to elicit the conditions under which these two models and their respective mass action constants can be compared.

Equation (409) is the general PC mass action equation for the solute–solvent exchange reaction equation (402). If solute "1" is a cation, M^m, and solute "2" is the solvent, w, then Eq. (411) becomes

$$K_{\mathrm{M,w}} = \left(\frac{\theta^{\mathrm{s}}_{\mathrm{M}}\chi_{\mathrm{w}}}{\chi_{\mathrm{M}}\theta^{\mathrm{s}}_{\mathrm{w}}}\right)\left(\frac{f^{\mathrm{s}}_{\mathrm{M}}f_{\mathrm{w}}}{f_{\mathrm{M}}f^{\mathrm{s}}_{\mathrm{w}}}\right) = \frac{K_{\mathrm{M}}}{K_{\mathrm{w}}} \tag{413}$$

The corresponding TMAB mass action equation for cation adsorption can be written from inspection of Eq. (94) as

$$K_{\mathrm{SM}} = \left(\frac{\theta_{\mathrm{SM}}}{\chi_{\mathrm{M}}\theta_{\mathrm{SO}}}\right)\left(\frac{f_{\mathrm{SM}}}{f_{\mathrm{M}}f_{\mathrm{SO}}}\right) \tag{414}$$

In Eqs (413) and (414) the solute activity in solution, $\chi_{\mathrm{M}}f_{\mathrm{M}}$, is defined identically in both models so that elimination of $\chi_{\mathrm{M}}f_{\mathrm{M}}$ between Eqs (413) and (414) gives upon rearrangement:

$$\frac{K_{\mathrm{M,w}}}{K_{\mathrm{SM}}} = \left[\left(\frac{\theta^{\mathrm{s}}_{\mathrm{M}}}{\theta_{\mathrm{SM}}}\right)\left(\frac{f^{\mathrm{s}}_{\mathrm{M}}}{f_{\mathrm{SM}}}\right)\right]\left[\left(\frac{\theta_{\mathrm{SO}}}{\theta^{\mathrm{s}}_{\mathrm{w}}}\right)\left(\frac{f_{\mathrm{SO}}}{f^{\mathrm{s}}_{\mathrm{w}}}\right)\right](\chi_{\mathrm{w}}f_{\mathrm{w}}) \tag{415}$$

Written in this form, the right-hand side (r.h.s.) of Eq. (415) becomes the product of four quotients enclosed in parentheses, with each quotient representing analogous quantities in each model. As defined by each model, these quotients will not, in general, reduce to unity, as one can see from Table 18.

The four variable quotients in Table 18 can be seen to differ significantly in definition. However, in most applications these eight variables will be evaluated experimentally from thermodynamic methods for activity coefficients and from experimental measurements of solution mole fractions of the solutes. When experimentally determined values for the eight variables in this table are used in mass action equations (413) and (414), the four quotients in Eq. (415) and Table 18 will reduce to unity. Under conditions where the four quotients become unity, then one

TABLE 18 Comparison of Analogous Variables in PC and TMAB Adsorption Models from Eq. (415)

Quotient		PC model	TMB model
Surface site quotients			
Without solute	$\left(\frac{\theta_{SO}}{\theta_{w}^{s}}\right)$	θ_{w}^{s}, Mole fraction of solvent in lattice surface sites	θ_{SO}, Mole fraction of solid surface molecules
With solute	$\left(\frac{\theta_{SM}}{\theta_{M}^{s}}\right)$	θ_{M}^{s}, Mole fraction of solvent in lattice surface sites	θ_{SM}, Mole fraction of solute–solid surface complexes
Activity coefficient quotients			
Without solute	$\left(\frac{f_{SO}}{f_{w}^{s}}\right)$	f_{w}^{s}, Activity coefficient of solvent molecule in lattice surface sites	f_{SO}, Activity coefficient of solid surface molecules
With solute	$\left(\frac{f_{M}^{s}}{f_{M}^{S}}\right)$	f_{M}^{s}, Activity coefficient of solvent molecule in lattice surface sites	f_{SM}, Activity coefficent of solute–solid surface complexes

can see that the mass action constants in the PC [$K_{M,w}$, in Eq. (413)] and the TMAB [K_{SM}, in Eq. (414)] models differ only by the activity of the solvent, $\chi_w f_w$.

D. Self-consistent Lattice Field Model

The SCLF method was developed by Koopal and coworkers [46–50] to describe adsorption of surfactant molecules at the solution–solid interface. The method derives from two earlier statistical thermodynamic lattice theories: (1) the Flory and Huggins [51] model describing properties of polymers in solution, and (2) the methods of Scheutjens and coworkers [52–55] developed to describe the properties of polymer molecules adsorbed at the solution–solid interface and in associated mesomorphic solution structures such as micelles and vesicles.

In the SCLF model, adsorption is depicted as the result of nonCoulombic interactions between the solute and solvent molecules and the electrostatic surface potential. The solution is divided into lattice sites which are arrayed in layers parallel to the surface. The equilibrium distribution of solute molecules between the solution and surface is determined from the partition function, Q, for a solution mixture [50, 52, 55] constituted of an entropic, Ω/Ω^{+}, and an energetic part, U, of general form [e.g., see Eq. (8) in Ref. 53]:

$$Q = \frac{\Omega}{\Omega^{+}} \exp\left(-\frac{U}{kT}\right) \tag{416}$$

The entropic part of the partition function is derived statistically, based not on distribution of individual solute and solvent segments but on the probability distribution of conformations of the entire solute molecule. The solvent and solute molecular conformations are allowed to distribute within each layer, using the Bragg–Williams approximation for random mixing.

The energetic part of the partition function contains two contributions: (1) the energy of adsorption, and (2) the energy of polymer-solvent interaction. Both energy contributions are represented by Flory–Huggins interaction parameters, χ, for the interaction energetics between a solute segment and a surface site and a solute segment and a solvent segment, respectively. In more recent publications [56], the energy of adsorption is segmented into two adsorption energies: (1) the solute–solid "contact interaction" energy, and (2) the solute–solid "electrostatic interaction" energy. The former is represented with the Flory–Huggins parameter and the latter with classical electrostatic theory for a sphere interacting with a semi-infinite continuum solid.

The SCLF model has several important features: (1) adsorption of polyfunctional, multisegmented solute molecules, such as surfactants, can be modeled; (2) the effects of flexibility of multisegmented solute and solvent molecules can be accounted for; (2) the volume fraction profiles of the solute molecular segments in the adsorbed layer (lattice layers near the solid surface) can be determined; (3) effects of counter ions and multiple solute molecules can be modeled; (4) electrostatic interactions are considered; (5) lateral interactions can be modeled; (6) since the model is based on the probability distribution of conformations of the entire solute molecule, the probability of conformations such as trains, loops, tails, and bridges can be determined specifically; and (7) the mean field approximation applied to solute distribution within each lattic layer results in computation time being linearly proportional to the number of segments in each solute or solvent molecule.

Limitations of the SCLF method include: (1) electrostatic Coulombic interactions between the solute and surface moleucles are ignored, (2) the dependence of calculated adsorption results on the model parameters (such as the solute–solute, solute–surface and solute–solvent Flory–Huggins interaction parameters, the lattice site size, etc.) is difficult to determine, (3) adsorption must be determined by simultaneous solution of probability density equations derived from partition functions so that no single analytical adsorption equation is possible, (4) effects of pH on surface sites can only be considered implicitly through the Flory–Huggins interaction parameters, and (5) the Flory–Huggins interaction parameters do not allow explicit consideration of the molecular or chemical characteristics of the surface site molecules.

E. Gas Phase or Langmuir Models

Adsorption of a single solute molecule by reaction with a surface site to form a surface complex is depicted in Fig. 32.

The adsorption complexation reaction shown in Fig. 32 and its corresponding mass action equation would be, respectively:

$$\mathrm{S} + \mathrm{Y} = \mathrm{SY} \quad : \quad K_{\mathrm{SY}} = \frac{\theta_{\mathrm{SY}}}{\theta_{\mathrm{S}}\chi_{\mathrm{Y}}}\frac{f_{\mathrm{SY}}}{f_{\mathrm{S}}f_{\mathrm{Y}}} \equiv \frac{\theta_{\mathrm{SY}}}{\theta_{\mathrm{S}}\chi_{\mathrm{Y}}}Z_{\mathrm{SY}} \tag{417}$$

The surface-site mass balance equation for the adsorption reaction in Fig. 32 would be

$$1 = \theta_{\mathrm{s}} + \theta_{\mathrm{Y}} \tag{418}$$

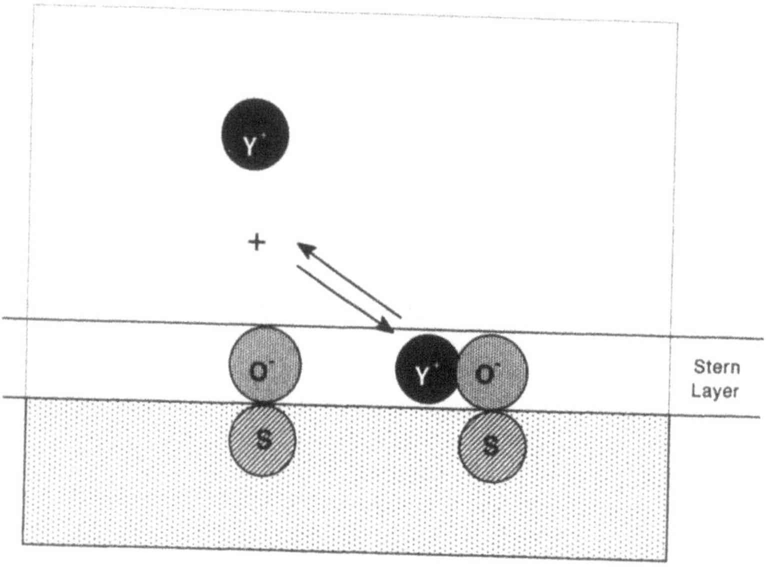

FIG. 32 Surface complex formation between a single solute molecule and a molecule of the solid at the solution–solid interface.

The adsorption equation can be derived directly by combining Eqs (417) and (418) to eliminate θ_s:

$$\theta_{SY} = \frac{\chi_Y K_{SY} Z_{SY}^{-1}}{1 + \chi_Y K_{SY} Z_{SY}^{-1}} \tag{419}$$

Equation (419) is the adsorption equation for nonideal single solute adsorption. If the activity coefficients are set to unity, Eq. (419) reduces to the GP adsorption equation for a single solute:

$$\theta_{SY} \cong \frac{\chi_Y K_{SY}}{1 + \chi_Y K_{SY}} \tag{420}$$

The mathematical form of Eq. (420) is the same as that of the Langmuir equation. Since this latter equation was derived for vapor adsorption at the vapor–solid interface [57–61], its success and popularity for vapor adsorption, where pH and nonideality can be ignored, is not surprising. At the same time one can see that an adsorption equation for solute adsorption at the solution–solid interface, such as the Langmuir equation, which makes no explicit allowance for solution pH and nonideality, is of questionable applicability to this latter type of adsorption.

While ideal adsorption is frequently a bad assumption for real systems, the difficulty of evaluating the activity coefficients (particularly surface activity coefficients) makes such an assumption popular with many authors. In addition to ignoring nonideality, many GP treatments do not have explicit pH dependence. However, as shown in previous sections, explicit pH dependence in surface-site complexation adsorption equations depends on proper surface-site mass balance equations, not activity coefficients.

A noteworthy difference between the GP and Langmuir adsorption equations is the definition of the constants in each respective equation. The GP equation is derived from mass action thermodynamics so the equation constants are mass action constants. The Langmuir equation is most commonly derived from either a kinetic [62, 63] or statistical thermodynamic basis [62, 64]. As a result the constants in the Langmuir equation are functions of either rate constants or partition functions, respectively.

Using the mass action adsorption constant, K_{SY}, as a fitting parameter, one can see that Fig. 33 appears to give satisfactory agreement between the mole fraction of surface sites complexed with solute, θ_{SY}, and equilibrium solute solution molar concentration, χ_Y, for the data of James and Healy [23] for Co(II) adsorption on silica.

The goodness of fit in Fig. 33 is deceptive because it requires significantly different values of the mass action adsorption constant, $K_{Co(II)/SiCO_2}$, to give a fit to adsorption data as pH changes. However, a pH-dependent mass action complexation constant violates the definition of a mass action constant as being a function of only standard state chemical potentials and temperature. The BPD adsorption model was an apparent attempt to circumvent this dilemma.

F. Boltzmann Potential Distribution Models

The lack of explicit pH dependence in adsorption equations derived from the GP model was an obvious deficiency. Among other things that changing solution pH does to a S–MO interface is change the electrostatic surface charge. The apparent conclusion drawn from these facts was that the pH dependence in adsorption was due to electrostatic interaction between solute and interface since the electrostatic

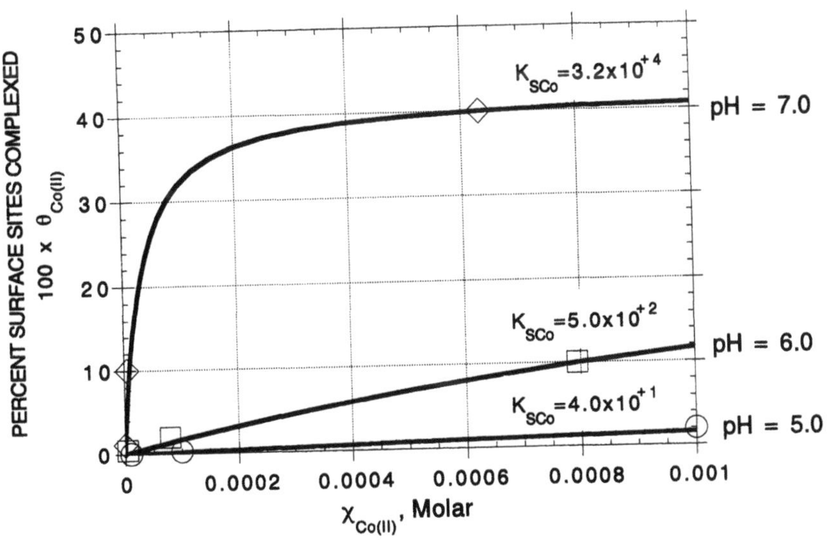

FIG. 33 GP or Langmuir adsorption equation (420) fitted to data for Co(II) adsorbed on silica by James and Healy [23].

surface charge was pH dependent. To quantify the electrostatic solute–interfacial interaction, the Boltzmann equation was the logical choice. As pH changed the electrostatic surface potential, the number of solute ions near the interface would change as predicted by the Boltzmann equation. The Boltzmann equation could, therefore, be used to predict the role of pH in solute adsorption at the S–MO interface.

The incorporation of the Boltzmann equation into surface complexation mass action equations is a result of hypothesizing that surface complexation can be described as a two-step process: (1) a solute molecule moves from a solution interior to the interface without bonding to a surface site, and (2) the unbound surface solute molecule then bonds to an adsorption site. In this way, the BPD models pictured adsorption as surface complex formation between a surface site and an unbound solute molecule already at the interface.

For the PBD model, two types of surface complexation reactions (shown for cation adsorption in Fig. 34) were proposed: (1) adsorption is depicted as the unbound surface solute molecule complexing with a surface site of opposite valence (r.h.s. of Fig. 34); and (2) the unbound surface solute molecule complexes with a neutrally charged, monoprotonaed surface site (l.h.s. of Fig. 34) displacing either a proton or a hydroxide ion from the monoprotonated surface site (for cation and anion adsorption, respectively).

The reaction and mass action equations for the two surface complexation reactions in Fig. 34, corresponding to the mass action constants K_{SM}^{INT} and $^*K_{SM}^{INT}$ can be written, respectively, as

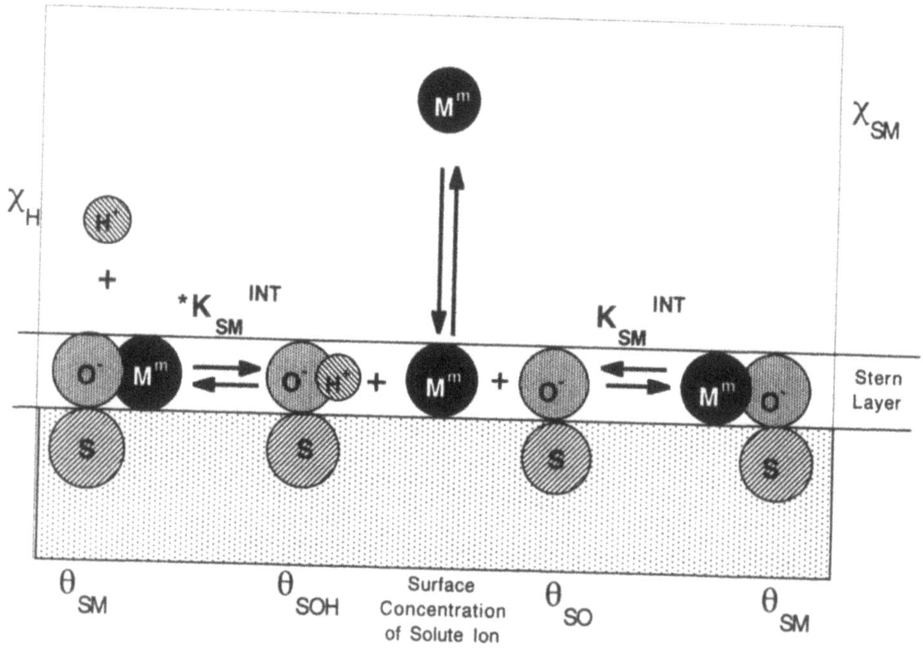

FIG. 34 Cation adsorption reactions for the Boltzmann potential distribution (BPD) mass action adsorption model.

BPD surface-site deprotonation reaction:

$$\mathrm{SOH} + \mathrm{M}_s^{+\nu_m} \Leftrightarrow \mathrm{SOM}^{\nu_m-1} + \mathrm{H}_s^+ : \mathrm{K}_{\mathrm{SM}}^{\mathrm{INT}} = \frac{\{\mathrm{SOM}^{\nu_m-1}\}\{\mathrm{H}_s^+\}}{\{\mathrm{SOH}\}\{\mathrm{M}_s^{+\nu_m}\}} \frac{f_{\mathrm{SOM}} f_{\mathrm{Hs}}}{f_{\mathrm{SOH}} f_{\mathrm{Ms}}} \tag{421}$$

BPD surface-site complexation reaction:

$$\mathrm{SO}^- + \mathrm{M}_s^{+\nu_m} \Leftrightarrow \mathrm{SOM}^{\nu_m-1} : K_{\mathrm{SM}}^{\mathrm{INT}} = \frac{\{\mathrm{SOM}^{\nu_m-1}\}}{\{\mathrm{SOH}\}\{\mathrm{M}_s^{+\nu_m}\}} \frac{f_{\mathrm{SOM}}}{f_{\mathrm{SOH}} f_{\mathrm{Ms}}} \tag{422}$$

where the mass action equation is written in terms of surface concentrations enclosed with the symbols { } and usually measured in units of mol/kg or mol/m^2, and the subscript "ss" signifies an unbound or uncomplexed solute in the interface. For the two surface complexation reactions represented in Fig. 34, the unbound interfacial concentration of protons $\{\mathrm{H}_s^+\}$ and cations $\{\mathrm{M}_s^+\}$ are rather ill-defined operationally and as such difficult to determine experimentally. To resolve this problem, the BPD model, as the name implies, invokes the Boltzmann equation to convert the unbound interfacial mole fraction of solute ions into the solution mole fraction of protons, $\chi_{\mathrm{H}} = [\mathrm{H}^+]$, and cations, $\chi_{\mathrm{M}} = [\mathrm{M}^{\nu_m}]$, as

$$\{\mathrm{H}_s^+\} = [\mathrm{H}^+]\exp\left(-\frac{e\psi_0}{kT}\right) \tag{423}$$

$$\{\mathrm{M}_s^{\nu_m}\} = [\mathrm{M}^{\nu_m}]\exp\left(\frac{-e\psi_1}{kT}\right) \tag{424}$$

where the braces and parentheses are used to be consistent with the literature. The symbol ψ_i represents the electrostatic potential that solute ions experience in an interface, with the subscript i signifying that each different type of interfacial solute molecule may be at a different perpendicular distance to the plane of the interface when the surface complexation or adsorption reaction occurs. Allowing solute molecules to adsorb at different perpendicular distances from the interface will be seen as a key feature of the BPD models.

It should be noted that no single convention for symbols exists to designate different types of solution and/or surface concentrations or even activities and activity coefficients in the open literature. Therefore, care must be taken to determine how concentration units are defined in each article. The brackets [ ] and { } used in this section are consistent with those used by Huang [6], Schindler [32], Sonnefeld [66], and Hohl and Stumm [22] for molar concentrations and mol/m^2, respectively. When Eqs (423) and (424) are substituted into Eqs (421) and (422), respectively, the result is the BPD mass action equations in its final form:

BPD mass action adsorption equation:

$$^*K_{\mathrm{SM}}^{\mathrm{INT}} = \frac{\{\mathrm{SOM}^{\nu_m-1}\}\{\mathrm{H}^+\}}{\{\mathrm{SOH}\}\{\mathrm{M}^{\nu_m}\}} \frac{f_{\mathrm{SOM}} f_{\mathrm{Hs}}}{f_{\mathrm{SOH}} f_{\mathrm{Ms}}} = \frac{\{\mathrm{SOM}^{\nu_m-1}\}[\mathrm{H}^+]}{\{\mathrm{SOH}\}[\mathrm{M}^{\nu_m}]} \exp\left[-\frac{\nu_m e(\psi_0 - \psi_1)}{kT}\right] \tag{425}$$

BPD mass action adsorption equation:

$$K_{\mathrm{SM}}^{\mathrm{INT}} = \frac{\{\mathrm{SOM}^{\nu_m-1}\}}{\{\mathrm{SO}\}\{\mathrm{M}^{\nu_m}\}} \frac{f_{\mathrm{SOM}}}{f_{\mathrm{SO}} f_{\mathrm{Ms}}} \cong \frac{\{\mathrm{SOM}^{\nu_m-1}\}}{\{\mathrm{SO}\}[\mathrm{M}^{\nu_m}]} \cong \frac{\{\mathrm{SOM}^{\nu_m-1}\}}{\{\mathrm{SO}\}[\mathrm{M}^{\nu_m}]} \exp\left(+\frac{e\psi_1}{kT}\right) \tag{426}$$

Equations (425) and (426) are the basis upon which all the different BPD adsorption models are based [31]. The rudiments of the BPD mass action equations (425) and (426) can best be understood by contrast with the corresponding rigorous mass action equation for cation adsorption based on Gibbs–Lewis thermodynamics described in Chapter 4, which can be written:

Gibbs–Lewis mass action equation for cation adsorption:

$$K_{SM} = \left(\frac{\theta_{SM}}{\theta_{SO}\chi_M}\right)\left(\frac{f_{SM}}{f_{SO}f_M}\right) = \left(\frac{\theta_{SM}}{\theta_{SO}\chi_M}\right)\left\{\frac{e^{\left[\frac{\nu_{SM}e\left(\psi_{SM}-\psi_{SM}^{ID}\right)}{2kT}\right]}}{e^{\left[\frac{\nu_{SO}e\left(\psi_{SO}-\psi_{SO}^{ID}\right)}{2kT}\right]}e^{\left[\frac{\nu_{M}e\left(\psi_{M}-\psi_{M}^{ID}\right)}{2kT}\right]}}\right\} \tag{427}$$

The BPD mass action equations (425) and (426) look deceptively similar to the rigorous Gibbs–Lewis mass action equation (427). However, several very significant differences exist:

- Activity coefficients for surface sites, SO^- and $SOM^{\nu_m^{-1}}$, are treated ideally, i.e., set to unity in the BPD equations.
- Taken together, Eqs (423)–(427) define activity coefficients as being equal to the Boltzmann equation. In rigorous thermodynamic equations one can see that activity coefficients depend on the difference in actual, ψ_i, and ideal, ψ_i^{ID}, electrostatic potentials, not just the ψ_i of the Boltzmann equation. The BPD model will incorrectly overestimate activity coefficient effects.
- In the BPD mass action equation (425), the electrostatic contribution to the mass action equation vanishes unless protons and hydroxide ions are arbitrarily required to adsorb into interfacial planes different from those of all other adsorbing solutes (i.e., inner and outer surface complexes, respectively). However, protons, hydroxide ions and all other solutes form surface complexes with the same surface sites, so that the assumption of different adsorption planes for different solutes seems unjustified.

When the BPD mass action equations are substituted into the GP adsorption equation (419) the result is the BPD adsorption equation:

$$\theta_{SM} = \frac{\chi_M {}^*K_{SM}^{INT} Z_{SM}^{-1}}{1 + \chi_M {}^*K_{SM}^{INT} Z_{SM}^{-1}} = \frac{\chi_M {}^*K_{SM}^{INT} \exp\left[-\frac{\nu_m e(\psi_0 - \psi_1)}{kT}\right]}{1 + \chi_M {}^*K_{SM}^{INT} \exp\left[-\frac{\nu_m e(\psi_0 - \psi_1)}{kT}\right]} \tag{428}$$

or

$$\theta_{SM} = \frac{\chi_M K_{SM}^{INT} Z_{SM}^{-1}}{1 + \chi_M K_{SM}^{INT} Z_{SM}^{-1}} = \frac{\chi_M K_{SM}^{INT} \exp\left(+\frac{e\psi_1}{kT}\right)}{1 + \chi_M K_{SM}^{INT} \exp\left(+\frac{e\psi_1}{kT}\right)} \tag{429}$$

Equations (428) and (429) represent the BPD equations for cation adsorption. Adsorption equations for other solute types can be derived and have a similar form. The pH dependence of adsorption rests entirely in ψ_i. As discussed in the bulk of this chapter, the pH dependence for adsorption at the S–MO interface will arise properly as a reciprocal surface site fraction, α_{SO}^{-1}, α_{SOH}^{-1}, or $\alpha_{SOH_2}^{-1}$, in place of the "1" in the denominator of Eqs (428) and (429) when surface-site mass balance equations are included in the derivation of the adsorption equation.

Due to the uncertainty in defining ψ_i, a number of adsorption models were developed over the last two decades. The main difference in these methods was what distance from the interfacial plane was each electrostatic potential. As a result, most of the models are named for how the surface plane location for the absorbing solutes was defined. The surface model used to define adsorption surface planes is shown in Fig. 35.

Based on the adsorption plane choice, the models will be broken into categories for discussed as follows:

1. Double plane models
   a. triple layer model;
   b. pseudo-Stern model.
2. Single plane models
   a. pseudo-standard state model;
   b. constant capacitance model and diffuse double layer model;
   c. Stern model.

(a) Triple Layer Model. The triple layer model [18, 21, 26, 32, 36] is a double plane model based on Eq. (428) in which protons and hydroxide ions are assumed to adsorb at a plane designated with the subscript 0, while all other solutes are

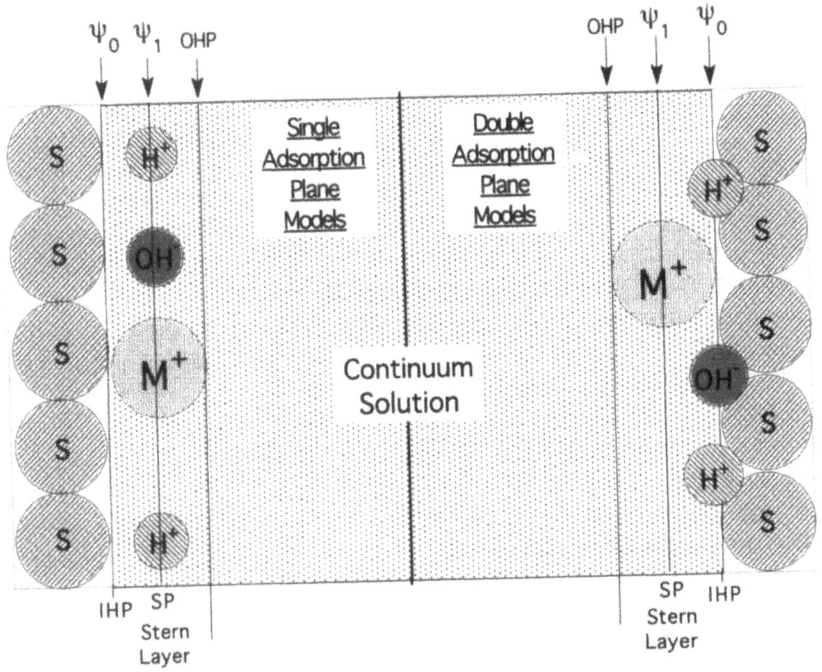

FIG. 35 Double and single adsorption plane models for adsorption at the solution–solid interface.

deemed to adsorb in a plane parallel to, but further away from, the solid designated with the subscript 1 (a third plane separates the solution and the two adsorbed layers of solute, hence the name, triple layer model). In alternative nomenclature, the 0 plane can be identified as the inner Helmholtz plane (IHP) and the 1 plane as the Stern plane (SP).

The three planes of the triple layer model divide the interface into three regions each with their own electrostatic potential gradients [31]: (1) between the IHP and the SP the electrostatic potential decreases linearly, (2) between the SP and the beginning of the solution the electrostatic potential decreases linearly, and (3) in the solution the electrostatic potential decays asymptotically as described by the Boltzmann equation.

The surface complexes formed between surface sites and protons or hydroxide ions formed at the IHP are referred to as inner surface complexes. Surface complexes formed between surface sites and solutes other than protons and hydroxide ions at the SP are referred to as outer surface complexes. While these adsorption layers might at first suggest that adsorbed layer structure might be determined from adsorption experiments, upon careful reflection, the assumption that proton and hydroxide ions complex with the same surface sites as do other solutes, but at different locations relative to the interface, seems curious at best. As a result of the arbitrary assignment of two adsorption planes it would seem that any information from the triple layer model about adsorbed layer structure must be equally arbitrary.

(b) Pseudo-Stern Model. When Stern proposed a model for the solution–solid interface [65], he envisioned only a single adsorption surface, at the IHP, parallel to the Stern plane, on which the centers of all the adsorbed solutes lie. Later workers [66] hypothesized that two adsorption planes must exist as shown for the double adsorption plane model in Fig. 35. Since double adsorption planes were not in the original Stern model, the designation of pseudo-Stern model [66] will be used to distinguish it from the true Stern models with a single adsorption plane.

The pseudo-Stern model (and Stern model) differs from the triple layer model in that: (1) the electrostatic potential remains constant between the OHP and the Stern plane in the former but decays linearly in this region in the latter [31], and (2) the pseudo-Stern model uses the BPD mass action equation (429). In the pseudo-Stern and triple layer models separate adsorption equations are used for adsorption of protons and hydroxide ions at the IHP and for adsorption of all other solute molecules at the OHP. The amount of adsorption at both planes is then adjusted in order to be self-consistent with an electrostatic balance over ψ_0, ψ_1, and the diffuse layer.

(c) Pseudo-Standard State Model. In an apparent effort to build pH dependence into the Langmuir equation for adsorption at the solution–solid interface, several researchers have proposed [17, 67–69] that electrostatic saturation, solvation and other contributions to the adsorption process are implicitly contained within the mass action constant. Pseudo-standard state methods seek to extract explicit expressions for these adsorption effects from the mass action adsorption constant. The equation defining the mass action constant, K, as a function of the difference in standard state chemical potentials between reactants and products is written as

$$K \equiv \exp\left(-\frac{\Delta\mu^*}{kT}\right) \tag{430}$$

where

$$\Delta\mu^* = \sum_i^{\text{Products}} \mu_i^* - \sum_j^{\text{Reactants}} \mu_j^* \tag{431}$$

For the pseudo-standard state model it is then hypothesized that the standard state chemical potential change, $\Delta\mu^*$, can be written as a sum of standard state chemical potential changes arising from different sources such as electrostatic, $\Delta\mu^*_{\text{coul}}$, solvation, $\Delta\mu^*_{\text{solv}}$, chemical, $\Delta\mu^*_{\text{chem}}$ (surface complex bond formation), etc.,

$$K^{\text{PSS}} \equiv \exp\left(-\frac{\Delta\mu^*_{\text{ele}} + \Delta\mu^*_{\text{solv}} + \Delta\mu^*_{\text{chem}} + \cdots}{kT}\right) \tag{432}$$

The form of Eq. (432) now suggests that explicit equations from electrostatics and mechanics can be substituted into Eq. (432). As an example, the electrostatic potential expression for

$$\Delta\mu^*_{\text{coul}} = e\psi_0 \tag{433}$$

is commonly substituted into Eq. (432) with the result:

$$K^{\text{PSS}} \equiv \exp\left(-\frac{\Delta\mu^*_{\text{solv}} + \Delta\mu^*_{\text{chem}} + \cdots}{kT}\right)\exp\left(-\frac{e\psi_0}{kT}\right) \tag{434}$$

In Eq. (434) the interfacial electrostatic potential appears explicitly. The pseudo-standard state model would thus appear to be consistent with the BPD mass action equations 428 and 429. However, Eq. (433) has been derived under some very peculiar circumstances. Specifically, the chemical potentials in Eqs (432) and (434) are all *standard state* chemical potentials while the electrostatic Eq (433) is for *variable, nonstandard states*. Hence, it would seem that the derivation of adsorption equations by the pseudo-standard state method either violates thermodynamic principles or defines some very peculiar standard states.

(d) Diffuse Layer and Constant Capacitance Models. These two models have many similar features [22, 24]. They both consider that only protons and hydroxide ions bond to the surface sites, where the surface potential is ψ_1, as shown for the single adsorption plane models in Fig. 35. The remaining solute molecules balance the electrostatic charge up to the OHP with an equal but opposite electrostatic charge in the diffuse ion solution layer. Under these conditions the protons and hydroxide ions were regarded as undergoing specific adsorption at the surface while all other solutes could only approach, but not specifically bond to, surface sites as part of the diffuse layer. In both methods the electrostatic charge of the interface resulted from the specifically adsorbed protons and hydroxide ions. The electrostatic interfacial potential was then determined to be a linear function of the interfacial electrostatic charge. Both methods utilize the BPD mass action equation (429) to describe proton and hydroxide ion adsorption.

In both models, the concentration of surface species was calculated from acid–base titration data to determine the speciation of the solution–solid interface. From a knowledge of the concentration of solution and interfacial species, the mass action constants were calculated directly. Mass action constants determined in this way were referred to as "microscopic" mass action constants to signify that their values were dependent upon the electrostatic potential of the interface. Intrinsic mass action constants are determined to be the microscopic mass action constant where the interfacial electrostatic charge is zero or the ZPC. The two methods relate the microscopic, K(micro), and intrinsic, K(int), mass action constants as

$$pK(\text{micro}) = pK(\text{int}) - b_{\pm}\frac{\sigma_0}{C} = pK(\text{int}) - b_{\pm}\psi_0 \tag{435}$$

For the constant capacitance model [22], the surface charge, σ_0, and K(micro) were experimentally determined, then the constant, $b_{\pm}/C$, was used as a constant fitting parameter to determine K(int). In the diffuse layer model [24], the electrostatic surface potential, ψ_0, was calculated directly from the linearized Gouy–Chapman potential, then along with the experimentally determined K(micro) was used to calculate K(int) directly. It can be seen that both the constant capacitance and diffuse layer models are based on the BPD mass equation when the logarithm is taken of Eq. (429).

(e) Stern Model. The Stern model is the interfacial model on which the BPD adsorption equations are based. As discussed above, the Stern model of a solution–solid interface [4, 57, 65, 70] depicts adsorption of all solutes at a single plane so that the centers of all the adsorbed solute molecules lie along the Stern plane where the electrostatic potential is ψ_1. In general, the centers of adsorbed solute molecules will undoubtedly lie slightly above or below the Stern plane due to differences in solute radii and adsorption bond length even on the smoothest of crystal surfaces. Except for these small and potentially determinate errors, the Stern model provides a model of the solution–solid interface which seems reasonable for the formation of a complex between a solute and surface site.

The BPD adsorption equations based on the Stern interfacial model also lack explicit pH dependence. Although the Stern model explicitly defines electrostatic potentials of the different planes parallel to the interface, no attempt is made to incorporate solute electrostastic effects or activity coefficients into the adsorption equations.

In all probability, the fact that the BPD methods lacked the correct pH dependence led previous researchers to solve a series of simultaneous equations rather than derive a single adsorption equation. In a review of adsorption literature, Westall and Hohl [31] describe some of the calculation strategies. In many of these simultaneous solutions, mass balance equations are commonly incorporated. However, in virtually all previous treatments authors chose to solve a series of simultaneous equations rather than derive a single adsorption equation. When mass balance and mass action equations are used to derive adsorption equations, TMAB adsorption equations result with the fractional surface site variables, $\alpha^{1-}_{\overline{\text{SO}}}$, $\alpha^{-1}_{\overline{\text{SOH}}}$, and $\alpha^{-1}_{\overline{\text{SOH}_2}}$, which give the proper pH dependence.

Before leaving the discussion of the BPD models, it will be worthwhile to examine the two different types of mass action constant used in the literature for these models as shown in Figs 34 and 35. According to the BPD models, adsorption of a cation, M^{ν_m}, was postulated to occur on either nonionic or anionic surface sites as shown in Figs 34 and 35.

TABLE 19 Relationship Among Proton and Cation Adsorption Mass Action Constants Found in the Literature

Surface site	Adsorption reaction	Mass action constant
Nonionic	$SOH + M^{\nu_m} \leftrightarrow SOM^{\nu_m - 1} + H^+$	$: {}^*K_{SM}^{INT}$ (421)
Protonation	$SO^- + H^+ \leftrightarrow SOH$	$: 1/K_{a2}^{INT}$ (20)
Anionic	$SO^- + M^{\nu_m} \leftrightarrow SOM^{\nu_m - 1}$	$: K_{SM}^{INT}$ (422)

By Hess's law, a relationship among the chemical equations in Eqs (20), (421), and (422), shown in Table 19, indicates that an algebraic relationship among the mass action equations must also exist. The Hess's law relationship for the mass action Eqs (20), (421), and (422) [6,29] can be seen from the following:

$$ {}^*K_{SM}^{INT} = \frac{\{SOM^{\nu_m - 1}\}\{H^+\}}{\{SOH\}\{M^{\nu_m}\}} \cong \frac{\{SOM^{\nu_m - 1}\}H^+]}{\{SOH\}[M^{\nu_m}]} \exp\left[-\frac{e(\psi_0 - \psi_\beta)}{kT}\right] \tag{436} $$

$$ \frac{1}{K_{a2}^{INT}} = \frac{\{SOH\}}{\{SO^-\}\{H^+\}} = \frac{\{SOH\}}{\{SO^-\}[H^+]} \exp\left(+\frac{e\psi_0}{kT}\right) \tag{437} $$

$$ K_{SM}^{INT} = \frac{\{SOM^{\nu_m - 1}\}}{\{SO^-\}\{M^{\nu_m}\}} = \frac{\{SOM^{\nu_m - 1}\}}{\{SO^-\}[M^{\nu_m}]} \exp\left(+\frac{e\psi_\beta}{kT}\right) \tag{438} $$

The subscripts on the electrostatic potentials are a result of models like the triple layer and pseudo-Stern models, which presume that protons and hydroxide ions adsorb at a surface plane with electrostatic potential, ψ_0, while all other solutes adsorb at a surface plane, further into the solution, with electrostatic surface potential, ψ_β. For other models, electrostatic surface potentials would be handled in a similar fashion. Comparing Eqs (20), (421), and (422), one can write:

$$ \begin{aligned} \frac{{}^*K_{SM}^{INT}}{K_{a2}^{INT}} &= \frac{\{SOM^{\nu_m - 1}\}[H^+]}{\{SOH\}[M^{\nu_m}]} \cdot \frac{\{SOH\}}{\{SO^-\}[H^+]} \exp\left[-\frac{e(\psi_0 - \psi_\beta)}{kT} + \frac{e\psi_0}{kT}\right] \\ &= \frac{\{SOM^{\nu_m - 1}\}}{[M^{\nu_m}]\{SO^-\}} \exp\left(+\frac{e\psi_\beta}{kT}\right) = K_{SM}^{INT} \end{aligned} \tag{439} $$

From Eq. (439), one can see that cation adsorption on an anionic or nonionic surface site (and on a cationic surface site) do not have to be treated as independent chemical reactions. Instead, these reactions are related through mass action constants for the proton and cation adsorption reactions as

$$ {}^*K_M^{INT} = K_{a2}^{INT} K_M^{INT} \tag{440} $$

Although the adsorption constants, K_M^{INT} and ${}^*K_M^{INT}$, are related quantities, they are generally quite different in value. For example, Kinniburgh *et al.* [71] have found adsorption constants for Ca(II) adsorbing on alumina gel of $K_{a2}^{INT} = 3.2 \times 10^{-11}$ and $K_M^{INT} = 1.6 \times 10^4$. From Eq. (440) a value of 5.1×10^{-7} is calculated for ${}^*K_M^{INT}$. Since adsorption constant values are reported in the literature in terms of both K_M^{INT} and ${}^*K_M^{INT}$, Eq. (440) provides a convenient conversion.

In a similar fashion, analogous mass action constants for anion, A^{-a}, adsorption are related as,

$$^{*}K_{A}^{INT} = K_{a1}^{INT}/K_{A}^{INT} \tag{441}$$

G. Thermodynamic Mass Action/Mass Balance Model

The earliest derivations of solute–metal oxide adsorption equations using the TMAB adsorption model appeared in the literature in the mid-1990s [2–4]. The details of the derivation of the TMAB adsorption equations have been explored in detail in previous sections of this chapter so no further review of the model is necessary. However, it is useful to compare the subtle but important differences between the TMAB adsorption models and all the other adsorption models discussed in this section as shown in Table 20.

In Table 20 the models discussed above are listed in the first column. In the second, third, and fourth columns the characteristics of the adsorption equations derived from these models are recorded. From these three columns, one can see that only the TMAB adsorption model provides a single, nonideal adsorption equation with explicit pH dependence. In addition, from the last two columns in the table, it can be seen that only with the TMAB model, are both surface sites and activity coefficients an integral part of the model on which the adsorption equations are based. These two properties are important if the role of the solid surface in adsorption is to be examined.

VIII. MISCELLANEOUS ADSORPTION TOPICS

A. Determination of Mass Action Constants and Solvation Effects

In adsorption equations, two types of mass action constants are required: (1) proton ionization constants, K_{a1} and K_{a2}; and (2) solute–surface site formation constants, K_{SY}. In the former case, probably the most reliable mass action constants can be determined from acid–base titration of the surface sites as discussed in Sections II.C and II.D. Once K_{a1} and K_{a2} are known, values for K_{SY} can be determined by fitting

TABLE 20 Comparison of Adsorption Models

Model	Theory	Practice	Explicit pH dependence	Surface sites	Electrostatic
PC	Multiple non-ID equations	Multiple ID equations	No	No	Activity coefficient
SCLF	Multiple non-ID equations	Multiple non-ID equations	No	No	Statistical mechanics
GP	Single ID equations	Single ID equations	No	Yes	None
BPD	Multiple ID equations	Multiple ID equations	No	Yes	Boltzmann equation
TMAB	Single non-ID equations	Single non-ID equations	Yes	Yes	Activity coefficient

adsorption equations to experimental data using K_{SY} as a fitting parameter. To correct for activity coefficient effects, ideal adsorption equations can be used to determine these mass action constants at different ionic strengths, then from a plot of mass action constant against ionic strength, the mass action constants, corrected for activity coefficient effects, can be determined by extrapolation to zero ionic strength. In addition to activity effects, solvation can modify mass action constants.

In many processes involving solutes, solution–solid interfaces, and adsorption, solvation energetics are an important consideration. Since, the magnitude of aqueous solvation energies is of the order of hydrogen bond energy, or nominally 10 kcal/mol, it is a subject worthy of examination.

When the adsorbing solute is a hydrated ion, some solvent water molecules must dissassociate from a solute ion and a surface site back into the bulk of the solution in order for the adsorption complexation reaction to occur. In other words, solvent–solute and solvent–surface site complexes must disassociate in order for the solute–surface site adsorption complex to form. Since mass action constants are determined experimentally, usually by fitting an adsorption equation to experimental data, measured mass action adsorption constants include hydration effects. However, all mass action constants used to this point have been anhydrous, including no hydration effects.

If we write the hydrated cation adsorption reaction showing j and k waters of hydration, W, initially on the hydrated surface site, $SO^- \cdot W_j$, and the initially hydrated cation, $M^{\nu_m} \cdot W_k$, respectively, which are lost when the cation–surface site complex, $SOM^{\nu_m - 1}$, forms, the chemical reaction and mass action equations in terms of activities of each reactant and product are

$$SO^- \cdot W_j + M^{\nu_m} \cdot W_k = SOM^{\nu_m-1} + (j+k)W \tag{442}$$

$$K_{SM,hyd} = \frac{a_{SM} a_W^{(j+k)}}{a_{SOW \cdot W_j} a_{M \cdot W_k}} \tag{443}$$

Note that the cation and surface site may have more than $(j+k)$ waters of hydration initially. The $(j+k)$ hydration waters are only those lost in the formation of the cation–surface site complex.

If we ignore waters of hydrations, as in earlier continuum solution reactions, the same reaction and mass action equations for anhydrous cation adsorption, analogous to Eqs (442) and (443), could be written as if hydration waters were absent:

$$SO^- + M^{\nu_m} = SOM^{\nu_m-1} \quad \vdots \quad K_{SM} = \frac{a_{SM}}{a_{SO} a_M} \tag{444}$$

Comparing the reactants in Eqs (442) and (444) one can see that the difference is hydration for which hydration and mass action equations can be written:

$$SO^- + jW = SO^- \cdot W_j \quad \vdots \quad K_{SO,hyd} = \frac{a_{SO \cdot W_j}}{a_{SO} a_W^j} \tag{445}$$

$$M^{\nu_m} + kW = M^{\nu_m} \cdot W_k \quad \vdots \quad K_{M,hyd} = \frac{a_M \cdot W_k}{a_M a_W^k} \tag{446}$$

If one now multiplies Eqs (443), (445), and (446) together, the result is

$$K_{\mathrm{SM,hyd}} \cdot K_{\mathrm{SO,hyd}} \cdot K_{\mathrm{M,hyd}} = \frac{a_{\mathrm{SM}} a_{\mathrm{W}}^{(j+k)}}{a_{\mathrm{SO\cdot W}_j} a_{\mathrm{M\cdot W}_k}} \cdot \frac{a_{\mathrm{SO\cdot W}_j}}{a_{\mathrm{SO}} a_{\mathrm{W}}^{j}} \cdot \frac{a_{\mathrm{M\cdot W}_k}}{a_{\mathrm{M}} a_{\mathrm{W}}^{k}} = \frac{a_{\mathrm{SM}}}{a_{\mathrm{SO}} a_{\mathrm{M}}} = K_{\mathrm{SM}} \tag{447}$$

Equation (447) clearly shows that the experimentally determined hydrous mass action constant, $K_{\mathrm{SM,hyd}}$, is related to the anhydrous constant, K_{SM}, through the hydrated mass action constants, $K_{\mathrm{SO,hyd}} \cdot K_{\mathrm{M,hyd}}$, for the two reactants: cation and surface site. The anhydrous mass action constant represents the (standard state) chemical potential for surface complex formation exclusive of any hydration effects. If the solvent changes, the experimentally determined $K_{\mathrm{SM,hyd}}$ will of course change but not K_{SM}, since it is independent of effects of solvent. Solvent changes will change all three mass action constants on the l.h.s. of Eq. (447) but in such a way that their product, $K_{\mathrm{SM,hyd}} \cdot K_{\mathrm{SO,hyd}} \cdot K_{\mathrm{M,hyd}}$, is a constant since the product equals K_{SM}, which is independent of solvent effects. The suite of chemical reactions for cation–surface site complex formation with and without waters of hydration can be written:

$$\left.\begin{array}{l} \mathrm{SO}^- \cdot W_j + \mathrm{M}^{\nu_m} \cdot \mathrm{W}_k \overset{K_{\mathrm{SM,hyd}}}{\longleftrightarrow} \\ \quad \updownarrow K_{\mathrm{SO,hyd}} \qquad\quad \updownarrow K_{\mathrm{M,hyd}} \\ \mathrm{SO}^- + j\mathrm{W} + M^{\nu_m} + k\mathrm{W} \overset{K_{\mathrm{SM}}}{\longleftrightarrow} \end{array}\right\} \mathrm{SOM}^{\nu_m - 1} + (j+k)\mathrm{W} \tag{448}$$

It can be seen that solute adsorption at the aqueous S–MO interface actually occurs as a hydrous process involving waters of hydration corresponding to the mass action constant, $K_{\mathrm{SM,hyd}}$. The adsorption equations derived in previous sections were based on anhydrous adsorption reactions from which Eqs (447) and (448) can be seen as simple shorthand and notation for the hydrous adsorption process. In other words, if waters of hydration were shown in all the adsorption equations derived above, the mathematics of the adsorption equations and their derivations would not change. Therefore, mass action constants appearing in all adsorption equations derived in previous sections implicitly include hydration effects.

One final consideration with regard to solvent effects on adsorption at the S–MO interface concerns the approximation of the continuum solution. Properties of continuum solutions are considered to be the same as bulk properties right up to the surface of an ion or surface site. For electrostatic properties in solutions, the continuum approximation assumes that the dielectric strength of the solution is that of bulk solvent right up to the surface of solutes and surface sites. Strictly speaking, the finite diameters of solvent molecules makes this impossible. Since solutes and solvents can only approach to within a molecular diameter, the question becomes one of whether the dielectric strength remains constant at distances of a molecular diameter.

In a previous study [20] it was argued that the dielectric strength of a solvent will change or saturate under the influence of the electric field surrounding an ion. If dielectric saturation of solvent near an interfacial or solution ion did occur, then waters of hydration would also affect the complexation bonding process through changes in the electrostatic variable, Z. However, calculations reported in more recent literature [72] have shown that in order for dielectric saturation to occur, it would have to occur at " . . . a distance that is smaller than even the smallest solvent molecule." Hence, it would be expected that electrostatic effects of ions on the dielectric strength of waters of hydration continguous to those ions would have a negligible effect on the adsorption process.

B. Lateral Interactions

In general, the term lateral interaction can be very broadly applied to many different interactions among molecules and ions in a S–MO interfacial system. Thermodynamics helps bring clarity to this issue if we use as an example the re-arranged mass action equation (94) for the adsorption of a cation, Y^{ν_y}:

$$\frac{\theta_{SY}}{\theta_{SO}\chi_Y} = K_{SY}\frac{f_{SO}f_Y}{f_{SY}} \equiv K_{SY}Z^{-\epsilon\kappa_{SY}} \tag{449}$$

From a thermodynamic point of view, predicting adsorption is simply being able to predict the concentration quotient on the l.h.s. of Eq. (449). Phenomena which increase adsorption, increase the concentration quotient and vice versa. Thermodynamic equation (449) can be seen to have two contributions: (1) a mass action constant, K_{SY}, representing the net standard state chemical potential energies [Eq. (93)] of the adsorption process; and (2) the activity coefficient quotient, $Z^{-\epsilon\kappa_{SY}}$ [Eq. (95)] representing any net chemical potential energy of the adsorption process which is not experienced when adsorption occurs under standard state conditions. Thermodynamically then, any aspect of adsorption must be attributable to either K_{SY} or $Z^{-\epsilon\kappa_{SY}}$.

Returning to lateral interactions, we will assume from the term "lateral" that only interactions among molecules and ions in the S–MO interfacial plane are to be considered. While many types of lateral interactions might be envisioned, we will consider two of the most important: (1) Coulombic-type long-range electrostatic interactions among neighboring ions and molecules; and (2) steric crowding, which is actually short-range electrostatic interactions. Since these effects are electrostatic interactions, they could be treated as activity coefficient effects as discussed in Chapter 4 [73], but with a little thought, a more insightful interpretation results.

Lateral interactions resulting from Coulombic long-range interaction energies are best included in the activity coefficient term, $Z^{-\epsilon\kappa_{SY}}$ and not the mass action constant, K_{SY}, because long-range Coulombic interactions vary with solute concentration as does $Z^{-\epsilon\kappa_{SY}}$, but not K_{SY}. For long-range lateral interactions in a real interface, an ion interacts electrostatically with neighboring ions and molecules. Assuming additivity of electrostatic potentials, long-range lateral interactions would be determined by summing each individual pair interaction. However, this approach is virtually impossible unless some molecular structure for the interface is known. A simpler approach, described in Chapter 4, was to determine lateral interactions in terms of the primitive interface model [74]. The primitive interfacial model treats the electrostatic potential of neighboring interfacial ions as a mean interfacial electrostatic charge located at the Stern plane. Thus, the explicit equations derived for the activity coefficient quotients, $Z^{-\epsilon\kappa_{SY}}$, in Chapter 4 and used throughout this chapter actually are the long-range Coulombic lateral interactions for the adsorption process or reaction based on the primitive interfacial model.

Short-range or steric interactions of an adsorbing ion or molecule with neighboring interfacial ions and molecules are best included in the mass action constant,

K_{SY}, and not the activity coefficient term, $Z^{-\epsilon K_{SY}}$, because short-range steric interactions are independent of solute concentration as is K_{SY}, but not $Z^{-\epsilon K_{SY}}$. Hence, when steric crowding occurs for adsorbing ions or molecules, the ratio of concentrations on the l.h.s. of Eq. (449) will change because sterically hindered and unhindered adsorption must have different mass action constants. If steric adsorption effects occur, multiple or surface concentration dependent mass action constants would replace the single mass action constants in the adsorption equations derived earlier.

In practice, the occurrence of steric effects in adsorption of simple ions is probably not common. Evidence for this lies in the fact that adsorption equations using single value/concentration independent mass action constants are frequently found to fit experimental adsorption data quite well. Also, since typical surface sites occupy an area of the order of 50–100 Å^2 based on surface electrostatic charge measurements [75] and solute ions occupy surface areas of about 10 Å^2 based on ionic radii of 2–3 Å, significant steric lateral interactions would seem unlikely for all but the very largest solute ions and molecules.

C. Polymer Adsorption

A thermodynamic treatment of polymer adsorption begins by presuming that the chemical potential for a polymer molecule is the sum of the chemical potential for each segment in the molecule. Following this model, the chemical potential for a polymer molecule of N identical segments in solution can be written as the sum of the chemical potentials of each of the i segments:

$$\mu_{\mathrm{Poly}} = \sum_{i=1}^{N} (\mu_i^* + kT \ln \chi_i + kT \ln f_i) \tag{450}$$

For the same polymer adsorbed at a S–MO interface, the segments complexed with a surface site will be designated with a subscript "sk" and those not complexed but in loops or tails near the interface with a subscript "sj":

$$\mu_{\mathrm{sPoly}} = \sum_{j\neq k=1}^{N} \sum_{k\neq j=1}^{N} [(\mu_{\mathrm{sj}}^* + kT \ln \chi_{sj} + kT \ln f_{\mathrm{sj}}) + (\mu_{\mathrm{sk}}^* + kT \ln \chi_{\mathrm{sk}} + kT \ln f_{\mathrm{sk}})] \tag{451}$$

For adsorption at an amphoteric S–MO interface we must define the valence on the solute which in turn defines the type of surface site. For this example we will choose a polyanionic polymer so that the surface sites participating in the bonding will be cationic with a chemical potential defined by Eq. (13) as

$$\mu_{\mathrm{SOH_2}} = \mu_{\mathrm{SOH_2}}^* + kT \ln \chi_{\mathrm{SOH_2}} + kT \ln f_{\mathrm{SOH_2}} \tag{13}$$

For the polymer molecule at equilibrium in a S–MO interface:

$$0 = \mu_{\mathrm{sPoly}} - \mu_{\mathrm{Poly}} - N_{\mathrm{s}} \mu_{\mathrm{SOH_2}} \tag{452}$$

where N_s is the number of polymer segments complexed with a surface site. Substituting Eqs (13), (450), and (451) into Eq. (452) and rearranging:

$$K_{\mathrm{sPoly}} = \left[\frac{(\chi_{\mathrm{sj}})^{N_w} (\theta_{\mathrm{sk}})^{N_s}}{(\chi_i)^N (\chi_{\mathrm{SOH_2}})^{N_s}}\right] \left[\frac{(f_{\mathrm{sj}})^{N_w} (f_{\mathrm{sk}})^{N_s}}{(f_i)^N (f_{\mathrm{SOH_2}})^{N_s}}\right] \tag{453}$$

where $N = N_w + N_s$. Equation (453) is the mass action expression derived for an anionic homopolymer adsorbing at the S–MO interface. Equation (453) is useful to examine polymer adsorption qualitatively. However, the form of Eq. (453) is deceptively simple for a number of reasons: (1) the exponents N, N_w, and N_s must be determined from a free-energy minimization calculation; and (2) for each segment the activity coefficient, f_m, will vary as a function of the perpendicular distance of a segment from the Stern plane (or some other interfacial definition). Each of these aspects requires a free-energy minimization calculation in order to evaluate the free-energy dependent parameters, which is beyond the scope of this chapter so only a cursory analysis will be given.

A free-energy minimization of the adsorption process would begin by writing a free-energy expression for the adsorption process:

$$\Delta\mu_{ads}(N_s, N_w, x) = \mu_{sPoly} - \mu_{Poly} - N_s\mu_{SOH_2} \tag{454}$$

followed by substitution of Eqs (13), (450), and (451), where x is the distance between the center of a segment and the Stern plane (or some other interfacial definition) measured perpendicularly to the Stern plane. An example of a free-energy minimization procedure can be found in Chapter 3.

D. Adsorption at an Heterogeneous Interface

An heterogeneous interface is characterized by an interface containing surface sites with different proton ionization and/or solute adsorption constants. To demonstrate how heterogeneous interfaces influence solute adsorption we will consider a monovalent cation adsorbing at an heterogeneous S–MO interface with two types of surface sites: type "1" and type "2". The total number of surface sites, n_s^T, will then be the sum of type 1 sites, $n_{s(1)}^T$, and type 2 sites, $n_{s(2)}^T$:

$$n_s^T = n_{s(1)}^T + n_{s(2)}^T \quad : \quad \frac{n_s^T}{n_s^T} = 1 = \theta_{s(1)}^T + \theta_{s(2)}^T \tag{455}$$

and the relative fraction of the two different types of sites, α_{S12}, given by

$$\alpha_{S12} \equiv \frac{n_{S(1)}^T}{n_{S(2)}^T} = \frac{\theta_{S(1)}^T}{\theta_{S(2)}^T} \quad : \quad \frac{\theta_{S(2)}^T}{\theta_{S(1)}^T + \theta_{S(2)}^T} = \theta_{S(2)}^T = \frac{1}{1+\alpha_{S12}} \tag{456}$$

One will note that α_{S12} can be considered as a known variable since it can be readily obtained from an acid–base titration of the metal oxide surface sites as described in Section II.D. If two different types of surface site are present on a metal oxide surface, instead of a single break at each of the two equivalence points as shown in Fig. 4, four equivalence points would be observed.

In some cases, four distinct equivalence points may not occur in which case a first or second derivative plot of Fig. 4 might delineate the four points. Another possibility is that the deprotonation of the surface sites does not occur at a distinct pH, but instead deprotonation occurs continuously over a pH range. Such behavior is observed in solution titrations of citric acid. In the case of citric acid, which is known to be triprotic, three pK_a values are "assigned" at equal increments of 1.6 pK_a units across the pH range where all three protons are extracted, viz., for citric acid, $pK_{a1} = 3.13$, $pK_{a2} = 4.76$, and $pK_{a1} = 6.40$. Of course, if there appears to be only two equivalence points, it is just possible that there *are* only two equivalence points, signifying that only one type of surface is present!

When adsorption of a cation, M^{ν_m}, reaches equilibrium, the total number of surface sites complexed with a cation, n_{SM}^{T}, can be written as the sum of type 1 surface sites complexed with the cation, $n_{SM(1)}^{T}$, and type 2 surface sites complexed with the cation, $n_{SM(2)}^{T}$:

$$n_{SM}^{T} = n_{SM(1)}^{T} + n_{SM(2)}^{T} \quad : \quad \frac{n_{SM}^{T}}{n_{s}^{T}} = \theta_{M}^{T} = \theta_{SM(1)}^{T} + \theta_{SM(2)}^{T} \tag{457}$$

A different mass action equation will be required for each surface site type so from Eq. (96) for monovalent cation adsorption:

$$\theta_{SM(1)} = \theta_{SO(1)}\left(\chi_M K_{SM(1)} Z^{-\epsilon\kappa_{SM(1)}}\right) \tag{458}$$

$$\theta_{SM(2)} = \theta_{SO(2)}\left(\chi_M K_{SM(2)} Z^{-\epsilon\kappa_{SM(2)}}\right) \tag{459}$$

Substituting Eqs (458) and (459) into Eq. (457):

$$\theta_{M}^{T} = \theta_{SM(1)}^{T} + \theta_{SM(2)}^{T} = \chi_M\left[\theta_{SO(1)} K_{SM(1)} Z^{-\epsilon\kappa_{SM(1)}} + \theta_{SO(2)} K_{SM(2)} Z^{-\epsilon\kappa_{SM(2)}}\right] \tag{460}$$

Solving Eq. (460) requires that $\theta_{SO(1)}$ and $\theta_{SO(2)}$ be expressed in explicit, known variables. We begin with the surface-site fraction equations analogous to Eq. (81):

$$\alpha_{SO(1)} = \frac{\theta_{SO(1)}}{\theta_{SO(1)} + \theta_{SOH(1)} + \theta_{SOH_2(1)}} = \frac{K_{a1(1)}K_{a2(1)}}{K_{a1(1)}K_{a2(1)} + \chi_H K_{a1(1)} Z^{\epsilon\kappa_{a1(1)}} + \chi_H^2 Z^{\epsilon\kappa_{a12(1)}}} \tag{461}$$

$$\alpha_{SO(2)} = \frac{\theta_{SO(2)}}{\theta_{SO(2)} + \theta_{SOH(2)} + \theta_{SOH_2(2)}} = \frac{K_{a1(2)}K_{a2(2)}}{K_{a1(2)}K_{a2(2)} + \chi_H K_{a1(2)} Z^{\epsilon\kappa_{a1(2)}} + \chi_H^2 Z^{\epsilon\kappa_{a12(2)}}} \tag{462}$$

We will now define two new surface-site fraction variables, $\alpha_{SO(1)}^{T}$ and $\alpha_{SO(2)}^{T}$, as

$$\alpha_{SO(1)}^{T} \equiv \frac{\theta_{SO(1)}}{\theta_{S(1)} + \theta_{SM(1)}} = \frac{1}{\dfrac{\theta_{S(1)}}{\theta_{SO(1)}} + \dfrac{\theta_{SM(1)}}{\theta_{SO(1)}}} = \frac{1}{\alpha_{SO(1)}^{-1} - \chi_M K_{SM(1)} Z^{-\epsilon\kappa_{SM(1)}}} \tag{463}$$

where Eq. (458) was substituted into the r.h.s. of Eq. (463). Combining Eqs (456) and (463):

$$\alpha_{SO(1)}^{T} \equiv \frac{\theta_{SO(1)}}{\theta_{S(1)} + \theta_{SM(1)}} = \theta_{SO(1)}(1 + \alpha_{S12}^{-1}) \tag{464}$$

Equations (463) and (464) gives the desired equation for $\theta_{SO(1)}$:

$$\theta_{SO(1)} = \frac{1}{(1 + \alpha_{S12}^{-1})(\alpha_{SO(1)}^{-1} + \chi_M K_{SM(1)} Z^{-\epsilon\kappa_{SM(1)}})} \tag{465}$$

A similar equation for $\theta_{SO(2)}$ can be derived in an analogous manner:

$$\alpha_{SO(2)}^{T} \equiv \frac{\theta_{SO(2)}}{\theta_{S(2)} + \theta_{SM(2)}} = \frac{1}{\dfrac{\theta_{S(2)}}{\theta_{SO(2)}} + \dfrac{\theta_{SM(2)}}{\theta_{SO(2)}}} = \frac{1}{\alpha_{SO(2)}^{-1} - \chi_M K_{SM(2)} Z^{-\epsilon\kappa_{SM(2)}}} \tag{466}$$

where Eq. (459) was substituted into the r.h.s. of Eq. (465). Combining Eqs (456) and (466):

$$\alpha^{\mathrm{T}}_{\mathrm{SO(2)}} \equiv \frac{\theta_{\mathrm{SO(2)}}}{\theta_{\mathrm{S(2)}} + \theta_{\mathrm{SM(2)}}} = \theta_{\mathrm{SO(2)}}(1 + \alpha_{\mathrm{S12}}) \tag{467}$$

Equations (463) and (464) and gives the desired equation for $\theta_{\mathrm{SO(2)}}$:

$$\theta_{\mathrm{SO(2)}} = \frac{1}{(1 + \alpha_{\mathrm{S12}})(\alpha^{-1}_{\mathrm{SO(2)}} + \chi_{\mathrm{M}} K_{\mathrm{SM(2)}} Z^{-\epsilon \kappa_{\mathrm{SM(2)}}})} \tag{468}$$

Substituting Eqs (465) and (468) into Eq. (460) gives the adsorption equation for cation adsorption on a heterogeneous S–MO interface with two different types of surface sites:

$$\theta^{\mathrm{T}}_{\mathrm{M}} = \chi_{\mathrm{M}} \left[\frac{K_{\mathrm{SM(1)}} Z^{-\epsilon \kappa_{\mathrm{SM(1)}}}}{(1 + \alpha^{-1}_{\mathrm{S12}})(\alpha^{-1}_{\mathrm{SO(1)}} + \chi_{\mathrm{M}} K_{\mathrm{SM(1)}} Z^{-\epsilon \kappa_{\mathrm{SM(1)}}})} + \frac{K_{\mathrm{SM(2)}} Z^{-\epsilon \kappa_{\mathrm{SM(2)}}}}{(1 + \alpha^{-1}_{\mathrm{S12}})(\alpha^{-1}_{\mathrm{SO(2)}} + \chi_{\mathrm{M}} K_{\mathrm{SM(2)}} Z^{-\epsilon \kappa_{\mathrm{SM(2)}}})} \right] \tag{469}$$

IX. SUMMARY

Based on rigorous Gibbs–Lewis thermodynamic mass action and mass balance equations, adsorption equations (TMAB model) for solute adsorption at the S–MO interface have been derived and contrasted with older adsorption equations in the literature. The most signfiicant advantage of the TMAB adsorption equations is the explicit pH dependence which arises in the derivation. The TMAB method also includes activity effects of solute and surface site ions and molecules in the form of activity coefficient quotients. Explicit equations for the activity coefficient quotients, derived from the primitive interfacial model in Chapter 4, are given for each adsorption equation. However, activity effects are shown to be minimal for solute concentrations of ≤ 0.01 M.

TMAB adsorption equations are shown to give excellent agreement with the pH and solute concentration dependence of solute adsorption for different solute–metal oxide adsorption data. The effects of different solute–surface site complex stoichiometries are examined for TMAB adsorption equations derived to include many different stoichiometries. In most cases these additional stoichiometries have no significant effect on total solute adsorption, but no general statement is possible without a more in-depth study. Of particular importance is the potential interference of adsorption of a particular ion by the counter ion in the salt. No experimental data were found to confirm this effect, but TMAB adsorption equations derived for adsorption of a single ion from a solution of a salt of that ion predicted signifciant adsorption interference between the counter ion and ion when both ion and counter ion have mass action adsorption constants of similar magnitude.

REFERENCES

1. GN Lewis, M Randall, KS Pitzer, L Brewer. Thermodynamics. 2nd ed. New York: McGraw-Hill, 1961, pp 153–154, 242–243, 249–253.
2. JA Wingrave. "Cation Adsorption Equation for the Solution-Metal Oxide Interface", paper 15, Proceedings of the Fifteenth International Conference of the Paint Research Association, Brussels, Belgium, 1995.
3. JA Wingrave. J Colloid Interface Sci 183:579, 1996.
4. VA Tertykh, V V Yanishpolskii. In: A Dabrowski, VA Tertykh, eds. Adsorption on New and Modified Sorbents. Amsterdam: Elsevier Science, 1996, Ch 3.3.
5. AJ Bard. Chemical Equilibrium. New York: Harper & Row, 1996, Ch 8.
6. CP Huang. In: MA Anderson, AJ Rubins, eds. Adsorption of Inorganics at Solid–Liquid Interfaces. Ann Arbor, MI: Ann Arbor Science Publishers, 1981, pp 183–216.
7. HS Harn, BO Owen. The Physical Chemistry of Electrolyte Solutions. 2nd ed. New York: Reinhold, 1960, pp 34–37.
8. MC Gupta. Statistical Thermodynamics. New York: Wiley, 1990, pp 345–349.
9. IM Klotz, RM Rosenberg. Chemical Thermodynamcis. 4th ed. Malabar, FL: Krieger, 1991, Ch 20.
10. N Bjerrum. Kgl Danske Vidensk Selskab 7:9, 1926.
11. HS Harn, BO Owen. Kgl Dansk Vidensk Selshkab 7:42–45, 1926.
12. J Keilland. J Am Chem Soc 59:1675, 1937.
13. K Denbigh. The Principles of Chemical Equilibrium. 4th ed. London:Cambridge University Press, 1981, pp 215–216.
14. HA Laitinen. Chemical Analysis. New York: McGraw-Hill, 1960, p 24.
15. L Onsager. Chem Rev 13:73–89, 1933. See p 76.
16. P Attard. In: I Prigogine, SA Rice, eds. Advances in Chemical Physics. Vol XCII. New York: John Wiley, 1996, p 1.
17. RO James, TW Healy. J Colloid Interface Sci 40:65–81, 1972.
18. JA Davis, JO Leckie. J Colloid Interface Sci 67:90, 1978.
19. R Kummert, W Stumm. J Colloid Interface Sci 75:373, 1980.
20. RO James, TW Healy. J Colloid Interface Sci 40:65–81, 1972.
21. JA Davis, JO Leckie. J Colloid Interface Sci 67:90–107, 1978.
22. H Hohl, WJ Stumm. J Colloid Interface Sci 55:281–288, 1976.
23. RO James, TW Healy. J Colloid Interface Sci 40:42–52, 1972.
24. CP Huang, WJ Stumm. J Colloid Interface Sci 43:409–420, 1973.
25. DE Yates, S Levine, TW Healy. Trans Faraday Soc 70:1807–1818, 1974.
26. JA Davis, RO James, JO Leckie. J Colloid Interface Sci 63:480–499, 1978.
27. S Levine, AL Smith. Disc Faraday Soc 52:290–301, 1971.
28. R Kummert, W Stumm. J Colloid Interface Sci 75:373–385, 1980.
29. G Sposito. The Surface Chemistry of Soils. New York: Oxford University Press, 1984, Ch 5.
30. DL Sparks. Soil Physical Chemistry. Boca Raton, FL: CRC Press, 1986, Ch 4, Sect III; Ch 6, Sect III.
31. J Westall, H Hohl. Adv Colloid Interface Sci 12:265–294, 1980.
32. PW Schindler. In: MA Anderson, AJ Rubin, eds. Adsorption of Inorganics at Solid–Liquid Interfaces. Ann Arbor, MI: Ann Arbor Science Publishers, 1981, pp 1–49.
33. RM McKenzie. Aust J Soil Res 18:61–73, 1980.
34. FMM Morel, JC Westall, JG Yeasted. In: MA Anderson, AJ Rubin, eds. Adsorption of Inorganics at Solid–Liquid Interfaces. Ann Arbor, MI: Ann Arbor Science Publishers, 1981, Ch 7.
35. JC Westall. In: W Stumm, ed. Aquatic Surface Chemistry. New York: Wiley, 1987, Ch 1.
36. RO James, PJ Stiglich, TW Healy. In: PH Tewari, ed. Adsorption from Aqueous Solutions. New York: Plenum Press, 1981, pp 19–40.

37. IS Balistrieri, JW Murray. Am J Sci 281:788–806, 1981.
38. PW Schindler, W Stumm. In: W Stumm, ed. Aquatic Surface Chemistry. New York: Wiley, 1987, pp 83–110.
39. BJ Mair, JW Westhaver, FD Rossini. Ind Eng Chem 42:1279–1286, 1950.
40. RW Schiessler, CN Rowe. J Am Chem Soc 75:4611–4612, 1953.
41. A Klinkenber. Rec Trav Chim 78:83–90, 1959.
42. DH Everett. Trans Faraday Soc 60:1803–1813, 1964.
43. DH Everett. Trans Faraday Soc 61:2478–1495, 1965.
44. DH Everett. In: J Rouquerol, KWS Sing, eds. Adsorption at the Gas–Solid Interface. Amsterdam: Elsevier, 1982, pp 1–20.
45. JAV Butler. Proc Roy Soc A 135:348–375, 1932.
46. LK Koopal, EM Lee, MR Böhmer. J Colloid Interface Sci 170:85–97, 1995.
47. EM Lee, LK Koopal. J Colloid Interface Sci 177:478–489, 1996.
48. MR Böhmer, LK Koopal. Langmuir 8:2649–2659, 1992.
49. MR Böhmer, LK Koopal. Langmuir 8:1478–1484, 1990.
50. MR Böhmer, LK Koopal. Langmuir 8:1594–1602, 1992.
51. PJ Flory. Principles of Polymer Chemistry. Ithaca, NY: Cornell University Press, 1953.
52. JMHM Scheutjens, GJ Fleer. J Phys Chem 83:1619–1635, 1979.
53. JMHM Scheutjens, GJ Fleer. J Phys Chem 84:178–190, 1980.
54. FAM Leermakers, JMHM Scheutjens, J Lyklema. Biochim Biophys Acta 1024:139–151, 1990.
55. OA Evers, JMHM Scheutjens, GJ Fleer. Macromolecules 23:5221–5233, 1990.
56. MR Böhmer, OA Evers, JMHM Scheutjens. Macromolecules 23:2288–2301, 1990.
57. WH van Riemsdijk, JCM de Wit, LK Koopal, GH Bolt. J Colloid Interface Sci 116:511–522, 1987.
58. LK Koopal, WH van Riemsdijk, JCM de Wit, MF Benedetti. J Colloid Interface Sci 166:51–60, 1994.
59. BB Johnson. Environ Sci Technol 24:112–118, 1990.
60. H Seki, A Suzuki. J Colloid Interface Sci 190:206–211, 1997.
61. DP Rodda, JD Wells, BB Johnson. J Colloid Interface Sci 184:564–569, 1996.
62. AW Adamson. Physical Chemistry of Surfaces. 3rd ed. New York: Wiley, 1976, pp 553–557.
63. I Langmuir. J Am Chem Soc 40:1361–1402, 1918.
64. RH Fowler, EA Guggenheim. Statistical Thermodynamics. Cambridge, UK: Cambridge University Press, 1956.
65. O Stern. Z Elektrochem 30:508–516, 1924.
66. JW Bowden, MDA Bolland, AM Posner, JP Quirk. Nature Phys Sci 245:81–83, 1973.
67. RJ Crawford, IH Harding, DE Mainwaring. Langmuir 9:3057–3062, 1993.
68. N Spanos, A Lycourghiotis. Langmuir 10:2351–2362, 1994.
69. KB Agashe, JR Regalbuto. J Colloid Interface Sci 185:174–189, 1997.
70. J Sonnefeld. Colloid Polym Sci 273:932–938, 1995.
71. DG Kinniburgh, ML Jackson, JK Syers. Soil Sci Soc Am J 40:796–799, 1976.
72. JN Israelachvili. Intermolecular and Surface Forces. 2nd ed. London: Academic Press, 1992, p 43.
73. JN Israelachvili. Intermolecular and Surface Forces. 2nd ed. London: Academic Press, 1992, p 11.
74. P Attard. In: I Prigogine, SA Rice, eds. Advances in Chemical Physics. Vol XCII. New York: John Wiley, 1987, pp 1–159.
75. JN Israelachvili. Intermolecular and Science Forces. 2nd ed. London: Academic Press, 1992, Ch 12.
76. JA Davis, JO Leckie. J Colloid Interface Sci 74:32–43, 1980.

2
Adsorbed and Grafted Polymers at Equilibrium

ROLAND R. NETZ Max-Planck-Institute for Colloids and Interfaces, Potsdam, Germany

DAVID ANDELMAN Tel Aviv University, Tel Aviv, Israel

I. INTRODUCTION

In this chapter, we review the basic mechanisms underlying adsorption of long-chain molecules on solid surfaces such as oxides. We concentrate on the physical aspects of adsorption and summarize the main theories which have been proposed. This chapter should be viewed as a general introduction to the problem of polymer adsorption at thermodynamical equilibrium. For a selection of previous review articles see Refs 1–4, while more detailed treatments are presented in two books on this subject [5, 6]. We do not attempt to explain any specific polymer/oxide system and do not emphasize experimental results and techniques. Rather, we detail how concepts taken from statistical thermodynamics and interfacial science can explain general and *universal* features of polymer adsorption. The present chapter deals with equilibrium properties whereas Chapter 3 by Cohen Stuart and de Keizer is about kinetics.

A. Types of Polymers

The polymers considered here are taken as linear and long chains, such as schematically depicted in Fig. 1a. We do not address the more complicated case of branched polymers at interfaces, although a considerable amount of work has been done on such systems [7]. In Fig. 1b we schematically present an example of a branched polymer. Moreover, we examine mainly homopolymers where the polymers are composed of the same repeated unit (monomer). We discuss separately, in Section VIII, extensions to adsorption of block copolymers and to polymers that are terminally grafted to the surface on one side ("polymer brushes"). In most of this review we shall assume that the chains are neutral. The charged case, i.e., where each or a certain fraction of monomers carries an electric charge, as depicted in Fig. 1c, is still not very well understood and depends on additional parameters such as the surface charge density, the polymer charge, and the ionic strength of the solution. We address shortly adsorption of polyelectrolytes in Section V. Furthermore, the chains are considered to be flexible. The statistical thermodynamics of flexible chains is rather well developed and the theoretical concepts can be applied with a considerable degree of confidence. Their large number of conformations play a crucial role in the

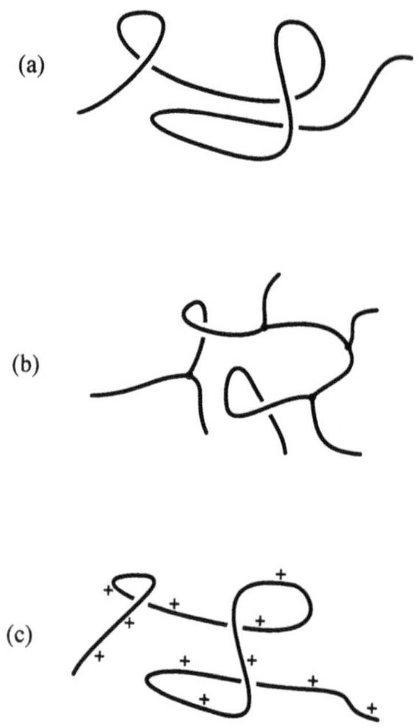

FIG. 1 Schematic view of different polymers. (a) Linear homopolymers, which are the main subject of this chapter; (b) branched polymers; (c) charged polymers or polyelectrolytes, with a certain fraction of charged groups.

adsorption, causing a rather *diffusive* layer extending away from the surface into the solution. This is in contrast to rigid chains, which usually form dense adsorption layers on surfaces.

B. Solvent Conditions

Polymers in solution can experience three types of solvent conditions. The solvent is called "good" when the monomer–solvent interaction is more favorable than the monomer–monomer one. Single polymer chains in good solvents have "swollen" spatial configurations, reflecting the effective repulsion between monomers. In the opposite case of "bad" (sometimes called "poor") solvent conditions, the effective interaction between monomers is attractive, leading to collapse of the chains and to their precipitation from the solution (phase separation between the polymer and the solvent). In the third and intermediate solvent condition, called "theta" solvent, the monomer–solvent and monomer–monomer interactions are equal in strength. The chains are still soluble, but their spatial configurations and solution properties differ from the good-solvent case.

The theoretical concepts and methods leading to these three classes make up a large and central part of polymer physics and are summarized in textbooks [7–12]. In general, the solvent quality depends mainly on the specific chemistry determining the interaction between the solvent molecules and monomers. It also can be changed by varying the temperature.

C. Adsorption and Depletion

Polymers can adsorb spontaneously from solution on to surfaces if the interaction between the polymer and the surface is more favorable than that of the solvent with the surface. For example, a polymer like poly(ethylene oxide) (PEO) is soluble in water but will adsorb on various hydrophobic surfaces and on the water/air interface. This is the case of equilibrium adsorption where the concentration of the polymer monomers increases close to the surface with respect to their concentration in the bulk solution. We discuss this phenomenon at length both on the level of a single polymer chain (valid only for extremely dilute polymer solutions), see Section II, and for polymers adsorbing from (semidilute) solutions, see Section III. In Fig. 2a we schematically show the volume fraction profile $\phi(z)$ of monomers as a function of the distance z from the adsorbing substrate. In the bulk, i.e., far away from the substrate surface, the volume fraction of the monomers is ϕ_b, whereas, at the surface, the corresponding value is $\phi_s > \phi_b$. The theoretical models address questions in relation to the polymer conformations at the interface, the local concentration of polymer in the vicinity of the surface, and the total amount of adsorbing polymer chains. In turn, the knowledge of the polymer interfacial behavior is used to calculate

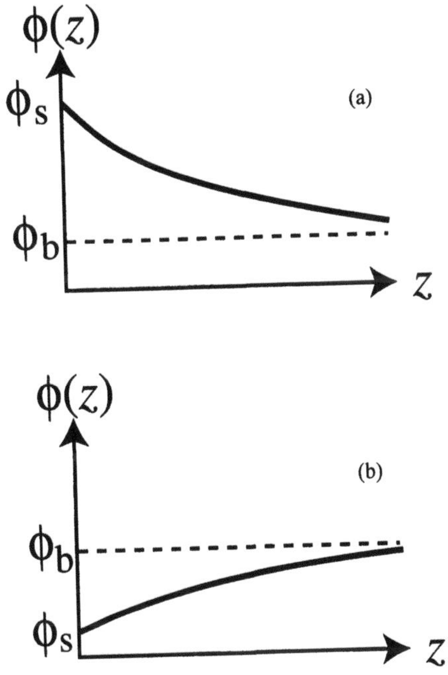

FIG. 2 Schematic profile of the monomer volume fraction $\phi(z)$ as a function of the distance z from a flat substrate as appropriate: (a) for the case of adsorption, where the substrate attracts monomers, leading to an increase in the polymer concentration close to the wall; and (b) for the case of depletion, where the substrate repels the monomers, leading to a decrease of the polymer concentration close to the wall. The symbol ϕ_b denotes the bulk volume fraction, i.e., the monomer volume fraction infinitely far away from the wall, and ϕ_s denotes the surface volume fraction right at the substrate surface.

thermodynamical properties like the surface tension in the presence of polymer adsorption.

The opposite case of *depletion* occurs when the monomer–surface interaction is less favorable than the solvent–surface interaction. This is, e.g., the case for polystyrene in toluene, which is depleted from a mica substrate. The depletion layer is defined as the layer adjacent to the surface from which the polymers are depleted. Their concentration in the vicinity of the surface is lower than the bulk value, as shown schematically in Fig. 2b.

D. Surface–Polymer Interactions

Equilibrium adsorption of polymers is only one of the methods used to create a change in the polymer concentration close to a surface. For an adsorbed polymer, it is interesting to look at the detailed conformation of a single polymer chain at the substrate. One distinguishes sections of the polymer which are bound to the surface, so-called trains, sections which form loops, and the end sections of the polymer chain, which can form dangling tails. This is schematically depicted in Fig. 3a. Two other methods to produce polymer layers at surfaces are commonly used for polymers which do not spontaneously adsorb on a given surface.

1. In the first method, the polymer is chemically attached (grafted) to the surface by one of the chain ends, as shown in Fig. 3b. In good solvent conditions the polymer chains look like "mushrooms" on the surface when the distance between grafting points is larger than the typical size of the chains. In some cases, it is possible to induce a much higher density of the grafting, resulting in a polymer "brush" extending in the perpendicular direction from the surface, as is discussed in detail in Section VIII.

2. A variant on the grafting method is to use a diblock copolymer made out of two distinct blocks, as shown in Fig. 3c. The first block is insoluble and is attracted to the substrate, and thus acts as an "anchor" fixing the chain to the surface; it is drawn as a thick line in Fig. 3c. It should be long enough to cause irreversible fixation on the surface. The other block is a soluble one (the "buoy"), forming the brush layer. For example, fixation on hydrophobic surfaces from a water solution can be made using a polystyrene–poly(ethylene oxide) (PS–PEO) diblock copolymer. The PS block is insoluble in water and is attracted towards the substrate, whereas the PEO forms the

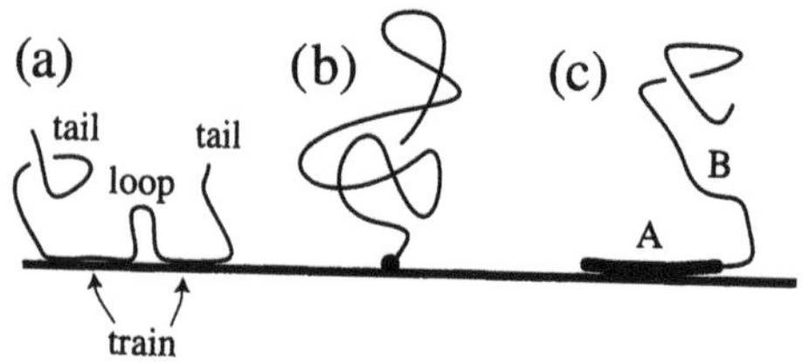

FIG. 3 The different adsorption mechanisms discussed in this chapter. (a) Adsorption of a homopolymer, where each monomer has the same interaction with the substrate; the "tail," "train," and "loop" sections of the adsorbing chain are shown. (b) Grafting of an end-functionalized polymer via a chemical or a physical bond. (c) Adsorption of a diblock copolymer where one of the two blocks is attached to the substrate surface, while the other is not.

brush layer. The process of diblock copolymer fixation has a complex dynamics during the formation stage, but is very useful in applications [10].

E. Surface Characteristics

Up to now we have outlined the polymer properties. What about the surface itself? Clearly, any adsorption process will be sensitive to the type of surface and its internal structure. As a starting point we assume that the solid surface is atomically smooth, flat, and homogeneous, as shown in Fig. 4a. This ideal solid surface is impenetrable to the chains and imposes on them a surface interaction. The surface potential can be short ranged, affecting only the monomers which are in direct contact with the substrate or in close vicinity of the surface. In other cases, the surface can have a longer range effect, like van der Waals, or electrostatic interactions, if it is charged.

Interesting extensions beyond ideal surface conditions are expected in several situations: (1) rough or corrugated surfaces, such as depicted in Fig. 4b; (2) surfaces that are curved, e.g., adsorption on spherical colloidal particles, see Fig. 4c; (3) substrates that are chemically inhomogeneous, i.e., which show some lateral organization, as shown schematically in Fig. 4d; (4) surfaces that have internal degrees of freedom like surfactant monolayers; and (5) polymers adsorbing on "soft" and

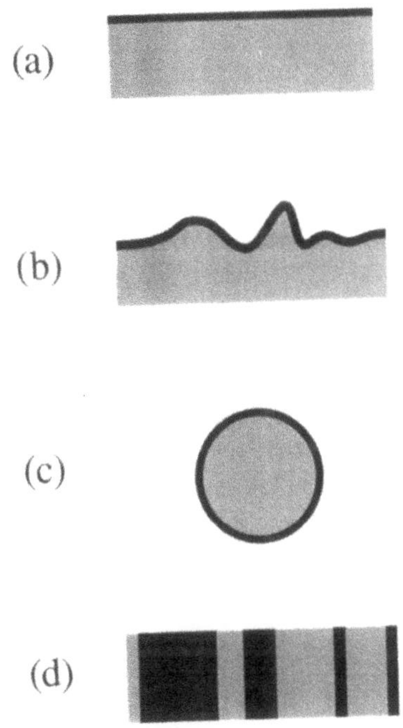

FIG. 4 Different possibilities of substrates. (a) The prototype, a flat, homogeneous substrate. (b) A corrugated, rough substrate; note that experimentally, every substrate exhibits some degree of roughness on some length scale. (c) A spherical adsorption substrate, such as a colloidal particle; if the colloidal radius is much larger than the polymer size, curvature effects (which means the deviation from the planar geometry) can be neglected. (d) A flat but chemically heterogeneous substrate.

"flexible" interfaces between two immiscible fluids or at the liquid/air surface. We briefly mention those situations in Sections V–VII.

F. Polymer Physics

Before turning to the problem of polymer adsorption let us briefly mention some basic principles of polymer theory. For a more detailed exposure the reader should consult the books by Flory, de Gennes, or Des Cloizeaux and Jannink [8, 11, 12]. The main parameters needed to describe a flexible polymer chain are the polymerization index N, which counts the number of repeat units or monomers, and the Kuhn length a, which corresponds to the spatial size of one monomer or the distance between two neighboring monomers. The monomer size ranges from 1.5 Å, as for example for polyethylene, to a few nanometers for biopolymers [8]. In contrast to other molecules or particles, a polymer chain contains not only translational and rotational degrees of freedom, but also a vast number of conformational degrees of freedom. For typical polymers, different conformations are produced by torsional rotations of the polymer backbone bonds, as shown schematically in Fig. 5a for a polymer consisting of three bonds of length a each. A satisfactory description of flexible chain conformations is achieved with the (bare) statistical weight for a polymer consisting of $N+1$ monomers:

$$\mathscr{P}_N = \exp\left\{-\frac{3}{2a^2}\sum_{i=1}^{N}(\mathbf{r}_{i+1}-\mathbf{r}_i)^2\right\} \tag{1}$$

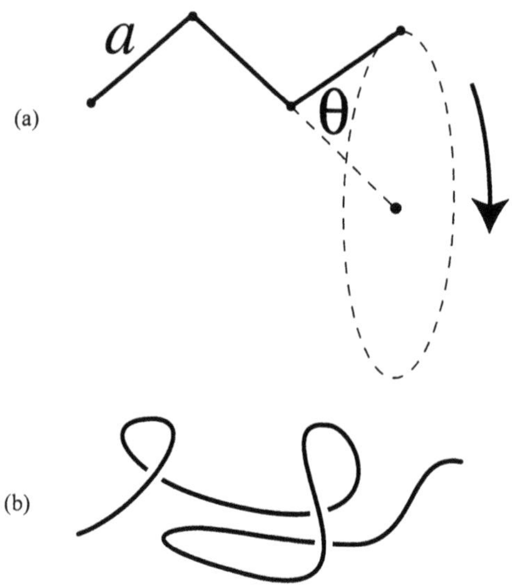

FIG. 5 (a) A polymer chain can be described as a chain of bonds of length a, with fixed torsional angles θ, reflecting the chemical bond structure, but otherwise freely rotating joints. (b) The simplified model, appropriate for theoretical calculations, consists of a structureless line, governed by some bending rigidity or line tension; this model chain is used when the relevant length scales are much larger than the monomer size, a.

which assures that each bond vector, given by $\mathbf{r}_{i+1} - \mathbf{r}_i$ with $i = 1, \ldots, N$, treated for convenience as a fluctuating Gaussian variable, has a mean length given by the Kuhn length, i.e.,

$$\langle(\mathbf{r}_i - \mathbf{r}_{i+1})^2\rangle = a^2$$

In most theoretical approaches, it is useful to take the simplification one step further and represent the polymer as a continuous line, as shown in Fig. 5b, with the statistical weight for each conformation given by Eq. (1) in the continuum limit. The Kuhn length a in this limit loses its geometric interpretation as the monomer size, and simply becomes an elastic parameter which is tuned such as to ensure the proper behavior of the large-scale properties of this continuous line, as is detailed below. Additional effects, a local bending rigidity, preferred bending angles (as relevant for *trans-gauche* isomery encountered for saturated carbon backbones), and hindered rotations can be taken into account by defining an effective polymerization index and an effective Kuhn length. In that sense, we always talk about effective parameters N and a, without saying so explicitly. Clearly, the total polymer length in the completely extended configuration is $L = aN$. However, the average spatial extent of a polymer chain in solution is typically much smaller. An important quantity characterizing the size of a polymer coil is the average end-to-end radius R_e. For the simple Gaussian polymer model defined above, we obtain:

$$R_e^2 = \langle(\mathbf{r}_{N+1} - \mathbf{r}_1)^2\rangle = a^2 N \tag{2}$$

In a more general way, one describes the scaling behavior of the end-to-end radius for large values of N as $R_e \sim aN^{\nu}$. For an ideal polymer chain, i.e., for a polymer whose individual monomers do not interact with each other, the above result implies $\nu = 1/2$. This result holds only for polymers where the attraction between monomers (as compared with the monomer–solvent interaction) cancels the steric repulsion due to the impenetrability of monomers. This situation can be achieved in special solvent conditions called "theta" solvent as was mentioned above. In a theta solvent, the polymer chain is not as swollen as in good solvents but is not collapsed on itself either, as it is under bad solvent conditions.

For good solvents, the steric repulsion dominates and the polymer coil takes a much more open structure, characterized by an exponent $\nu \simeq 3/5$ [8]. The general picture that emerges is that the typical spatial size of a polymer coil is much smaller than the extended length $L = aN$, but larger than the size of the ideal chain $aN^{1/2}$. The reason for this peculiar behavior is entropy combined with the favorable interaction between monomers and solvent molecules in good solvents. The number of polymer configurations having a small end-to-end radius is large, and these configurations are entropically favored over configurations characterized by a large end-to-end radius, for which the number of possible polymer conformations is drastically reduced. It is this conformational freedom of polymer coils which leads to salient differences between polymer adsorption and that of simple liquids.

Finally, in bad solvent conditions, the polymer and the solvent are not compatible. A single polymer chain collapses on itself in order to minimize the monomer–solvent interaction. It is clear that in this case, the polymer size, like any space filling object, scales as $N \sim R_e^3$, yielding $\nu = 1/3$.

II. SINGLE-CHAIN ADSORPTION

Let us consider now the interaction of a single polymer chain with a solid substrate. The main effects particular to the adsorption of polymers (as opposed to the adsorption of simple molecules) are due to the reduction of conformational states of the polymer at the substrate, which is due to the impenetrability of the substrate for monomers [13–18]. The second factor determining the adsorption behavior is the substrate–monomer interaction. Typically, for the case of an adsorbing substrate, the interaction potential $V(z)$ between the substrate and a single monomer has a form similar to the one shown in Fig. 6, where z measures the distance of the monomer from the substrate surface:

$$V(z) \simeq \begin{cases} \infty & \text{for } z < 0 \\ -U & \text{for } 0 < z < B \\ -bz^{-\tau} & \text{for } z > B \end{cases} \tag{3}$$

The separation of $V(z)$ into three parts is done for convenience. It consists of a hard wall at $z = 0$, which embodies the impenetrability of the substrate, i.e., $V(z) = \infty$ for $z < 0$. For positive z we assume the potential to be given by an attractive well of depth U and width B. At large distances, $z > B$, the potential can be modeled by a long-ranged attractive tail decaying as $V(z) \sim -bz^{-\tau}$.

For the important case of (unscreened and nonretarded) van der Waals interactions between the substrate and the polymer monomers, the potential shows a decay governed by the exponent $\tau = 3$ and can be attractive or repulsive, depending on the solvent, the chemical nature of the monomers, and the substrate material. The decay power $\tau = 3$ follows from the van der Waals pair interaction, which decays as the inverse sixth power with distance, by integrating over the three spatial dimensions of the substrate, which is supposed to be a semi-infinite half-space [19].

The strength of the potential well is measured by $U/(k_B T)$, i.e., by comparing the potential depth U with the thermal energy $k_B T$. For strongly attractive poten-

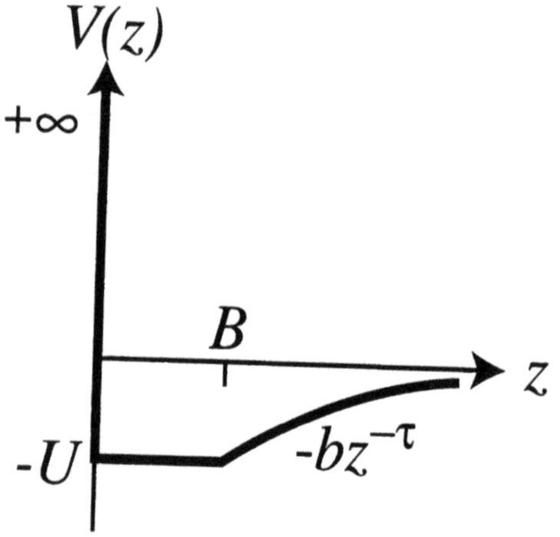

FIG. 6 A typical surface potential felt by a monomer as a function of the distance z from an adsorbing wall. First, the wall is impenetrable; then, the attraction is of strength U and range B. For separations larger than B, typically a long-ranged tail exists and is modeled by $-bz^{-\tau}$.

tials, i.e., for large U or, equivalently, for low temperatures, the polymer is strongly adsorbed and the thickness of the adsorbed layer, D, approximately equals the potential range B. The resulting polymer structure is shown in Fig. 7a, where the width of the potential well, B, is denoted by a broken line.

For weakly attractive potentials, or for high temperatures, we anticipate a weakly adsorbed polymer layer, with a diffuse layer thickness D much larger than the potential range B. This structure is depicted in Fig. 7b. For both cases shown in Fig. 7, the polymer conformations are unperturbed on a spatial scale of the order of D; on larger length scales, the polymer is broken up into decorrelated *polymer blobs* [11, 12], which are denoted by dotted circles in Fig. 7. The idea of introducing polymer blobs is related to the fact that very long and flexible chains have different spatial arrangements at small and large length scales. Within each blob the short-range interaction is irrelevant, and the polymer structure inside the blob is similar to the structure of an unperturbed polymer far from the surface. Since all monomers are connected, the blobs themselves are linearly connected and their spatial arrangement represents the behavior on large length scales. In the adsorbed state, the formation of each blob leads to an entropy loss of the order of one k_BT (with a numerical prefactor of order unity, which is neglected in this scaling argument), so the total entropy loss of a chain of N monomers is $\mathcal{F}_{\text{rep}} \sim k_BT(N/g)$, where g denotes the number of monomers inside each blob.

Using the scaling relation $D \simeq ag^{\nu}$ for the blob size dependence on the number of monomers g, the entropy penalty for the confinement of a polymer chain to a width D above the surface can be written as [20]

$$\frac{\mathcal{F}_{\text{rep}}}{k_BT} \simeq N\left(\frac{a}{D}\right)^{1/\nu} \tag{4}$$

The adsorption behavor of a polymer chain results from a competition between the attractive $V(z)$, which tries to bind the monomers to the substrate, and the entropic

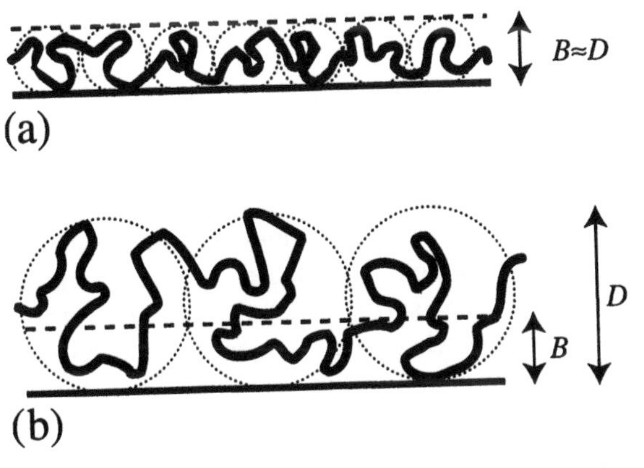

FIG. 7 Schematic drawing of single-chain adsorption: (a) in the limit of strong coupling, the polymer decorrelates into a whole number of blobs (shown as dotted circles) and the chain is confined to a layer thickness D, of the same order of magnitude as the potential range B; (b) in the case of weak coupling, the width of the polymer layer D is much larger than the interaction range B and the polymer forms large blobs, within which the polymer is not perturbed by the surface.

repulsion $\mathscr{F}_{\mathrm{rep}}$, which tries to maximize entropy and, therefore, favors a delocalized state where a large fraction of the monomers are located farther away from the surface.

It is of interest to compare the adsorption of long-chain polymers with the adsorption of small molecular solutes. Small molecules adsorb on to a surface only if there is a bulk reservoir with nonzero concentration in equilibrium with the surface. An infinite polymer chain $N \to \infty$ behaves differently as it remains adsorbed also in the limit of zero bulk concentration. This corresponds to a true thermodynamic phase transition in the limit $N \to \infty$ [21]. For a finite polymer length, however, the equilibrium behavior is, in some sense, similar to the adsorption of small molecules, and a nonzero bulk polymer concentration is needed for the adsorption of polymer chains on the substrate. For fairly long polymers, the desorption of a single polymer is almost a "true" phase transition, and corrections due to finite (but long) polymer length are often below experimental resolution.

A. Mean-Field Regime

Fluctuations of the local monomer concentration are of importance to the description of polymers at surfaces owing to the many possible chain conformations. These fluctuations are treated theoretically using field-theoretical or transfer-matrix techniques. In a field-theoretical formalism, the problem of accounting for different polymer conformations is converted into a functional integral over different monomer–concentration profiles [12]. Within transfer-matrix techniques, the Markov-chain property of ideal polymers is exploited to re-express the conformational polymer fluctuations as a product of matrices [22].

However, there are cases where fluctuations in the local monomer concentration become unimportant. The adsorption behavior of a single polymer chain is then obtained using simple *mean-field theory* arguments. Mean-field theory is a very useful approximation applicable in many branches of physics, including polymer physics. In a nutshell, each monomer is placed in a "field", generated by the averaged interaction with all the other monomers.

The mean-field theory can be justified for two cases: (1) a *strongly adsorbed polymer chain*, i.e., a polymer chain which is entirely confined inside the potential well; and (2) the case of *long-ranged attractive surface potentials*. To proceed, we assume that the adsorbed polymer layer is confined with an average thickness D, as depicted in Fig. 7a or 7b. Within mean-field theory, the polymer chain experiences an average of the surface potential, $\langle V(z)\rangle$, which is replaced by the potential evaluated at the average distance from the surface, $\langle z\rangle \simeq D/2$. Therefore, $\langle V(z)\rangle \simeq V(D/2)$. Further stringent conditions when such a mean-field theory is valid are detailed below. The full free energy of one chain, $\mathscr{F}$, of polymerization index N, can be expressed as the sum of the repulsive entropic term, Eq. (4), and the average potential:

$$\frac{\mathscr{F}}{k_{\mathrm{B}}T} \simeq N\left(\frac{a}{D}\right)^{1/\nu} + N\frac{V(D/2)}{k_{\mathrm{B}}T} \tag{5}$$

Let us consider first the case of a strongly adsorbed polymer, confined to a potential well of depth $-U$. In this case the potential energy per monomer becomes

$V(D/2) \simeq -U$. Comparing the repulsive entropic term with the potential term, we find the two terms to be of equal strength for a well depth $U^* \simeq k_B T(a/D)^{1/\nu}$. Hence, the strongly adsorbed state, which is depicted in Fig. 7a, should be realized for a high attraction strength $U > U^*$. For smaller attraction strength, $U < U^*$, the adsorbed chain will actually be adsorbed in a layer of width D much larger than the potential width B, as shown in Fig. 7b. Since the threshold energy U^* is proportional to the temperature, it follows that at high temperatures it becomes increasingly difficult to confine the chain. In fact, for an ideal chain, with $\nu = 1/2$, the resulting scaling relation for the critical well depth, $U^* \sim k_B T(a/D)^2$, agrees with exact transfer-matrix predictions for the adsorption threshold in a square-well potential [23].

We turn now to the case of a weakly adsorbed polymer layer. The potential depth is smaller than the threshold, i.e., $U < U^*$, and the stability of the weakly adsorbed polymer chains (depicted in Fig. 7b) has to be examined. The thickness D of this polymer layer follows from the minimization of the free energy, Eq. (5), with respect to D, where we use the asymptotic form of the surface potential, Eq. (3), for large separations. The result is

$$D \simeq \left(\frac{a^{1/\nu} k_B T}{b}\right)^{\nu/(1-\nu\tau)} \tag{6}$$

Under which circumstances is the prediction of Eq. (6) correct, at least on a qualitative level? It turns out that the prediction for D, Eq. (6), obtained within the simple mean-field theory, is correct if the attractive tail of the substrate potential in Eq. (3) decays for large values of z slower than the entropic repulsion in Eq. (4) [24]. In other words, the mean-field theory is valid for weakly adsorbed polymers only for $\tau < 1/\nu$. This can already be guessed from the functional form of the layer thickness, Eq. (6), because for $\tau > 1/\nu$ the layer thickness D goes to zero as b diminishes. Clearly an unphysical result. For ideal polymers (theta solvent, $\nu = 1/2$), the validity condition is $\tau < 2$, whereas for swollen polymers (good solvent conditions, $\nu = 3/5$), it is $\tau < 5/3$. For most interactions (including van der Waals interactions with $\tau = 3$) this condition on τ is not satisfied, and fluctuations are in fact important, as is discussed in the next section.

There are two notable exceptions. The first is for charged polymers close to an oppositely charged surface, in the *absence* of salt ions. Since the attraction of the polymer to an infinite, planar, and charged surface is linear in z, the interaction is described by Eq. (3) with an exponent $\tau = -1$, and the inequality $\tau < 1/\nu$ is satisfied. For charged surfaces, Eq. (6) predicts the thickness D to increase to infinity as the temperature increases or as the attraction strength b (proportional to the surface charge density) decreases. The resulting exponents for the scaling of D follow from Eq. (6) and are $D \sim (T/b)^{1/3}$ for ideal chains, and $D \sim (T/b)^{3/8}$ for swollen chains [25–27].

A second example where the mean-field theory can be used is the adsorption of polyampholytes on charged surfaces [28]. Polyampholytes are polymers consisting of negatively and positively charged monomers. In cases where the total charge on such a polymer adds up to zero, it might seem that the interaction with a charged surface should vanish. However, it turns out that local charge fluctuations (i.e., local spontaneous dipole moments) lead to a strong attraction of polyampholytes to charged substrates. In the absence of salt this attractive interaction has an algebraic decay with an exponent $\tau = 2$ [28]. On the other hand, in the presence of salt, the effective interaction is exponentially screened, yielding a decay faster than the fluctuation repulsion, Eq. (4). Nevertheless, the mean-field theory, embodied in the free energy expression, Eq. (5), can be used to predict the adsorption phase behavior within the strongly adsorbed case (i.e., far from any desorption transition) [29].

B. Fluctuation Dominated Regime

Here we consider the weakly adsorbed case for substrate potentials which decay (for large separations from the surface) faster than the entropic repulsion, Eq. (4), i.e., $\tau > 1/\nu$. This applies, e.g., to van der Waals attractive interaction between the substrate and monomers, screened electrostatic interactions, or any other short-ranged potential. In this case, fluctuations play a decisive role. In fact, for *ideal chains*, it can be rigorously proven (using transfer-matrix techniques) that all potentials decaying faster than z^{-2} for large z have a continuous adsorption transition at a finite critical temperature T^* [24]. This means that the thickness of the adsorbed polymer layer diverges for $T \to T^*$ as

$$D \sim (T^* - T)^{-1} \tag{7}$$

The power law divergence of D is universal. Namely, it does not depend on the specific functional form and strength of the potential as long as they satisfy the above condition.

The case of *nonideal chains* is much more complicated. First, progress has been made by de Gennes who recognized the analogy between the partition function of a self-avoiding chain and the correlation function of an n-component spin model in the zero-component ($n \to 0$) limit [30]. The adsorption behavior of nonideal chains has been treated by field-theroetical methods using the analogy to surface critical behavior of magnets (again in the $n \to 0$ limit) [6, 31]. The resulting behavior is similar to the ideal-chain case and shows an adsorption transition at a finite temperature, and a continuous increase towards infinite layer thickness characterized by a power law divergence as a function of $T - T^*$ [31].

The complete behavior for ideal and swollen chains can be described using scaling ideas in the following way. The entropic loss due to the confinement of the chain to a region of thickness D close to the surface is again given by Eq. (4). Assuming that the adsorption layer is much thicker than the range of the attractive potential $V(z)$, the attractive potential can be assumed to be localized at the substrate surface $V(z) \simeq V(0)$. The attractive free energy of the chain to the substrate surface can then be written as [32]

$$\mathcal{F}_{\mathrm{att}} \simeq -\tilde{\gamma} k_{\mathrm{B}}(T^* - T) N f_1 = -\gamma_1 a^2 N f_1 \tag{8}$$

where f_1 is the probability of finding a monomer at the substrate surface, and $\tilde{\gamma}$ is a dimensionless interaction parameter. Two surface excess energies are typically being used: $\gamma_1 = \tilde{\gamma} k_{\mathrm{B}}(T^* - T)/a^2$ is the excess of energy per unit area, while $\gamma_1 a^2$ is the excess of energy per monomer at the surface. Both are positive for the attractive case (adsorption) and negative for the depletion case. The dependence of γ_1 on T in Eq. (8) causes the attraction to vanish at a critical temperature, $T = T^*$, in accord with our expectations.

The contact probability for a swollen chain with the surface, f_1, can be calculated as follows [33]. In order to force the chain of polymerization index N to be in contact with the wall, one of the chain ends is pinned to the substrate. The number of monomers which are in contact with the surface can be calculated using field-theoretical methods and is given by N^φ, where φ is called the *surface crossover exponent* [6, 31]. The fraction of bound monomers follows to be $f_1 \sim N^{\varphi-1}$, and thus goes to zero as the polymer length increases, for $\varphi < 1$. Now instead of speaking of the entire chain, we refer to a "chain of blobs" (see Fig. 7) adsorbing on the surface, each blob consisting of g monomers. We proceed by assuming that the size of an adsorbed blob D scales with the number of monomers per blob g similarly as in the bulk, $D \sim ag^\nu$, as is indeed confirmed by field-theoretical calculations. The fraction of bound monomers can be expressed in terms of D and is given by

$$f_1 \sim \left(\frac{D}{a}\right)^{(\varphi-1)/\nu} \tag{9}$$

Combining the entropic repulsion, Eq. (4), and the substrate attraction, Eqs (8) and (9), the total free energy is given by

$$\frac{\mathcal{F}}{k_B T} \simeq N\left(\frac{a}{D}\right)^{1/\nu} - N\frac{\tilde{\gamma}(T^* - T)}{T}\left(\frac{D}{a}\right)^{(\varphi-1)/\nu} \tag{10}$$

Minimization with respect to D leads to the final result:

$$D \simeq a\left[\frac{\tilde{\gamma}(T^* - T)}{T}\right]^{-\nu/\varphi} \simeq a\left(\frac{\gamma_1 a^2}{k_B T}\right)^{-\nu/\varphi} \tag{11}$$

For ideal chains, one has $\varphi = \nu = 1/2$, and thus we recover the prediction from the transfer-matrix calculations, Eq. (7). For nonideal chains, the crossover exponent φ is, in general, different from the swelling exponent ν. However, extensive Monte Carlo computer simulations point to a value for φ very close to ν, such that the adsorption exponent ν/φ appearing in Eq. (11) is very close to unity for polymers embedded in three-dimensional space [31].

A further point which has been calculated using field theory is the behavior of the monomer volume fraction $\phi(z)$ close to the substrate. From rather general arguments borrowed from the theory of critical phenomena, one expects a power-law behavior for $\phi(z)$ at sufficiently small distances from the substrate [31, 33, 34]:

$$\phi(z) \simeq \phi_s(z/a)^m \tag{12}$$

recalling that the monomer density is related to $\phi(z)$ by $c(z) = \phi(z)/a^3$.

In the following, we relate the so-called *proximal exponent m* with the two other exponents introduced above, ν and φ. First note that the surface value of the monomer volume fraction, $\phi_s = \phi(z \approx a)$, for one adsorbed blob follows from the number of monomers at the surface per blob, which is given by $f_1 g$, and the cross-sectional area of a blob, which is of the order of D^2. The surface volume fraction is given by

$$\phi_s \sim \frac{f_1 g a^2}{D^2} \sim g^{\varphi - 2\nu} \tag{13}$$

Using the scaling prediction, Eq. (12), we see that the monomer volume fraction at the blob center, $z \simeq D/2$, is given by $\phi(D/2) \sim g^{\varphi-2\nu}(D/a)^m$, which (again using $D \sim ag^{\nu}$) can be rewritten as $\phi(D/2) \simeq g^{\varphi-2\nu+m\nu}$.

On the other hand, at a distance $D/2$ from the surface, the monomer volume fraction should have decayed to the average monomer volume fraction $a^3 g/D^3 \simeq g^{1-3\nu}$ inside the blob since the statistics of the chain inside the blob is the same for a chain in the bulk. By direct comparison of the two volume fractions, we see that the exponents $\varphi - 2\nu + m\nu$ and $1 - 3\nu$ have to match in order to have a consistent result, yielding

$$m = \frac{1 - \varphi - \nu}{\nu} \tag{14}$$

For an ideal chain (theta solvents), one has $\varphi = \nu = 1/2$. Hence, the proximal exponent vanishes, $m = 0$. This means that the proximal exponent has no mean-field analog, explaining why it was discovered only within field-theoretical calculations [6, 31]. In the presence of correlations (good solvent conditions) one has $\varphi \simeq \nu \simeq 3/5$ and thus $m \simeq 1/3$.

Using $D \simeq ag^{\nu}$ and Eq. (11), the surface volume fraction, Eq. (13), can be rewritten as

$$\phi_s \sim \left(\frac{D}{a}\right)^{(\varphi-2\nu)/\nu} \sim \left(\frac{\gamma_1}{k_B T}\right)^{(2\nu-\varphi)/\varphi} \simeq \frac{\gamma_1}{k_B T} \tag{15}$$

where in the last approximation appearing in Eq. (15) we used the fact that $\varphi \simeq \nu$. The last result shows that the surface volume fraction within one blob can become large if the adsorption energy γ_1 is large enough as compared with $k_B T$. Experimentally, this is very often the case, and additional interactions (such as multibody interactions) between monomers at the surface have, in principle, to be taken into account.

After having discussed the adsorption behavior of a single chain, a word of caution is in order. Experimentally, one never looks at single chains adsorbed to a surface. First, this is because one always works with polymer solutions, where there is a large number of polymer chains contained in the bulk reservoir, even when the bulk monomer (or polymer) concentration is quite low. Second, even if the bulk polymer concentration is very low, and in fact so low that polymers in solution rarely interact with each other, the surface concentration of polymer is enhanced relative to that in the bulk. Therefore, adsorbed polymers at the surface usually do interact with neighboring chains, owing to the higher polymer concentration at the surface [34].

Nevertheless, the adsorption behavior of a single chain serves as a basis and guideline for the more complicated adsorption scenarios involving many-chain effects. It will turn out that the scaling of the adsorption layer thickness D and the proximal volume fraction profile, Eqs (11) and (12), are not affected by the presence of other chains. This finding as well as other many-chain effects on polymer adsorption is the subject of the next section.

III. POLYMER ADSORPTION FROM SOLUTION

A. The Mean-Field Approach: Ground State Dominance

In this section we look at the equilibrium behavior of many chains adsorbing on (or equivalently depleting from) a surface in contact with a bulk reservoir of chains at equilibrium. The polymer chains in the reservoir are assumed to be in a semidilute concentration regime. The semi-dilute regime is defined by $c > c^*$, where c denotes the monomer concentration (per unit volume) and c^* is the concentration where individual chains start to overlap. Clearly, the overlap concentration is reached when the average bulk monomer concentration exceeds the monomer concentration inside a polymer coil. To estimate the overlap concentration c^*, we simply note that the average monomer concentration inside a coil with dimension $R_e \sim aN^\nu$ is given by $c^* \sim N/R_e^3 \sim N^{1-3\nu}/a^3$.

As in the previous section, the adsorbing surface is taken as an ideal and smooth plane. Neglecting lateral concentration fluctuations, one can reduce the problem to an effective one-dimensional problem, where the monomer concentration depends only on the distance z from the surface, $c = c(z)$. The two boundary conditions are: $c_b = c(z \to \infty)$ in the bulk, while $c_s = c(z = 0)$ on the surface.

In addition to the monomer concentration c, it is more convenient to work with the monomer volume fraction: $\phi(z) = a^3 c(z)$ where a is the monomer size. While the bulk value (far away from the surface) is fixed by the concentration in the reservoir, the value on the surface at $z = 0$ is self-adjusting in response to a given surface interaction. The simplest phenomenological surface interaction is linear in the surface polymer concentration. The resulting contribution to the surface free energy (per unit area) is

$$F_s = -\gamma_1 \phi_s \tag{16}$$

where $\phi_s = a^3 c_s$ and a positive (negative) value of $\gamma_1 = \tilde{\gamma} k_B (T - T^*)/a^2$, defined in the previous section, enhances adsorption (depletion) of the chains on (from) the surface. However, F_s represents only the local reduction in the interfacial free energy due to the adsorption. In order to calculate the full interfacial free energy, it is important to note that monomers adsorbing on the surface are connected to other monomers belonging to the same polymer chain. The latter accumulate in the vicinity of the surface. Hence, the interfacial free energy does not only depend on the surface concentration of the monomers but also on their concentration in the *vicinity* of the surface. Due to the polymer flexibility and connectivity, the entire adsorbing layer can have a considerable width. The *total* interfacial free energy of the polymer chains will depend on this width and is quite different from the interfacial free energy for simple molecular liquids.

There are several theoretical approaches for treating this polymer adsorption. One of the simplest approaches which yet gives reasonable qualitative results is that of Cahn–de Gennes [35, 36]. In this approach, it is possible to write down a continuum functional which describes the contribution to the free energy of the polymer chains in the solution. This procedure was introduced by Edwards in the 1960s [15] and was applied to polymers at interfaces by de Gennes [36]. Below, we present such a continuum version which can be studied analytically. Another approach is a discrete one, where the monomers and solvent molecules are put on a lattice. The latter approach is quite useful in computer simulations and numerical self-consistent field (SCF) studies, and is reviewed elsewhere [5].

In the continuum approach and using a mean-field theory, the bulk contribution to the adsorption free energy is written in terms of the local monomer volume fraction $\phi(z)$, neglecting all kinds of monomer–monomer correlations. The total reduction in the surface tension (interfacial free energy per unit area) is then

$$\gamma - \gamma_0 = -\gamma_1 \phi_s + \int_0^\infty \mathrm{d}z \left[L(\phi) \left(\frac{\mathrm{d}\phi}{\mathrm{d}z} \right)^2 + F(\phi) - F(\phi_b) + \mu(\phi - \phi_b) \right] \tag{17}$$

where γ_0 is the bare surface tension of the surface in contact with the solvent, but without the presence of the monomers in solution, and γ_1 was defined in Eq. (16). The stiffness function $L(\phi)$ represents the energy cost of local concentration fluctuations and its form is specific to long polymer chains. For low polymer concentration it can be written as [11]

$$L(\phi) = \frac{k_B T}{a^3} \left(\frac{a^2}{24\phi} \right) \tag{18}$$

where $k_B T$ is the thermal energy. The other terms in Eq. (17) come from the Cahn–Hilliard free energy of mixing of the polymer solution, μ being the chemical potential, and [8]

$$F(\phi) = \frac{k_B T}{a^3} \left(\frac{\phi}{N} \log \phi + \frac{1}{2} v \phi^2 + \frac{1}{6} w \phi^3 + \cdots \right) \tag{19}$$

where N is the polymerization index. In the following, we neglect the first term in Eq. (19) (translational entropy), as can be justified in the long-chain limit, $N \gg 1$. The second and third dimensionless virial coefficients are v and w, respectively. Good, bad, and theta solvent conditions are achieved, respectively, for positive, negative, or zero v. We concentrate hereafter only on good solvent conditions, $v > 0$, in which case the higher order w term can be safely neglected. In addition, the local monomer density is assumed to be small enough to justify the omission of higher virial coefficients. Note that for small molecules the translational entropy always acts in favor of desorbing from the surface. As was discussed in Section I, the vanishingly small translational entropy for polymers results in a stronger adsorption (as compared with low molecular weight solutes) and makes the polymer adsorption much more of an irreversible process.

The key feature in obtaining Eq. (17) is the so-called *ground state dominance*, where, for long enough chains $N \gg 1$, only the lowest energy eigenstate (ground state) of a diffusion-like equation is taken into account. This approximation gives us the leading behavior in the $N \to \infty$ limit [21]. It is based on the fact that the weight of the first excited eigenstate is smaller than that of the ground state by an exponential factor: $\exp(-N\Delta E)$ where $\Delta E = E_1 - E_0 > 0$ is the difference in the eigenvalues between the two eigenstates. Clearly, close to the surface more details on the polymer conformations can be important. The adsorbing chains have tails (end sections of the chains that are connected to the surface by only one end), loops (mid-sections of the chains that are connected to the surface by both ends), and trains (sections of the chains that are adsorbed on the surface), as depicted in Fig. 3a. To some extent it is possible to obtain profiles of the various chain segments even within mean-field theory, if the ground state dominance condition is relaxed as is discussed below.

Taking into account all those simplifying assumptions and conditions, the mean-field theory for the interfacial free energy can be written as

$$\gamma - \gamma_0 = -\gamma_1 \phi_s + \frac{k_B T}{a^3} \int_0^\infty \mathrm{d}z \left[\frac{a^2}{24\phi} \left(\frac{\mathrm{d}\phi}{\mathrm{d}z} \right)^2 + \frac{1}{2} v [\phi(z) - \phi_b]^2 \right] \tag{20}$$

where the monomer bulk chemical potential μ is given by $\mu = \partial F / \partial \phi|_b = v\phi_b$.

It is also useful to define the total amount of monomers per unit area which take part in the adsorption layer. This is the so-called surface excess Γ; it is measured experimentally using, e.g., ellipsometry, and is defined as

$$\Gamma = \frac{1}{a^3} \int_0^\infty \mathrm{d}z [\phi(z) - \phi_b] \tag{21}$$

The next step is to minimize the free energy functional, Eq. (20), with respect to both $\phi(z)$ and $\phi_s = \phi(0)$. It is more convenient to re-express Eq. (20) in terms of $\psi(z) = \phi^{1/2}(z)$ and $\psi_s = \phi_s^{1/2}$:

$$\gamma - \gamma_0 = -\gamma_1 \psi_s^2 + \frac{k_B T}{a^3} \int_0^\infty \mathrm{d}z \left[\frac{a^2}{6} \left(\frac{\mathrm{d}\psi}{\mathrm{d}z} \right)^2 + \frac{1}{2} v [\psi^2(z) - \psi_b^2]^2 \right] \tag{22}$$

Minimization of Eq. (22) with respect to $\psi(z)$ and ψ_s leads to the following profile equation and boundary condition:

$$\begin{aligned} \frac{a^2}{6} \frac{\mathrm{d}^2 \psi}{\mathrm{d}z^2} &= v\psi(\psi^2 - \psi_b^2) \\ \frac{1}{\psi_s} \frac{\mathrm{d}\psi}{\mathrm{d}z}\bigg|_s &= -\frac{6a}{k_B T} \gamma_1 = -\frac{1}{2D} \end{aligned} \tag{23}$$

The second equation sets a boundary condition on the logarithmic derivative of the monomer volume fraction, $\mathrm{d}\log\phi/\mathrm{d}z|_s = 2\psi^{-1}\mathrm{d}\psi/\mathrm{d}z|_s = -1/D$, where the strength of the surface interaction γ_1 can be expressed in terms of a length $D \equiv k_B T/(12a\gamma_1)$. Note that exactly the same scaling of D on γ_1/T is obtained in Eq. (11) for the single-chain behavior if one sets $\nu = \varphi = 1/2$ (ideal chain exponents). This is strictly valid at the upper critical dimension ($d = 4$) and is a very good approximation in three dimensions.

The profile equation (23) can be integrated once, yielding

$$\frac{a^2}{6} \left(\frac{\mathrm{d}\psi}{\mathrm{d}z} \right)^2 = \frac{1}{2} v [\psi^2 - \psi_b^2]^2 \tag{24}$$

The above differential equation can now be solved analytically. We first present the results in more detail for polymer adsorption ($\gamma_1 > 0$) and then repeat the main findings for polymer depletion ($\gamma_1 < 0$).

1. Polymer Adsorption

Setting $\gamma_1 > 0$ as is applicable for the adsorption case, the first-order differential equation (24) can be integrated and, together with the boundary condition, Eq. (23), yields:

$$\phi(z) = \phi_b \coth^2\left(\frac{z + z_0}{\xi_b}\right) \tag{25}$$

where the length $\xi_b = a/\sqrt{3v\phi_b}$ is the Edwards correlation length characterizing the exponential decay of concentration fluctuations in the bulk [11, 15]. The length z_0 is not an independent length since it depends on D and ξ_b, as can be seen from the boundary condition, Eq. (23):

$$z_0 = \frac{\xi_b}{2} \operatorname{arcsinh}\left(\frac{4D}{\xi_b}\right) = \xi_b \operatorname{arccoth}\left(\sqrt{\phi_s/\phi_b}\right) \tag{26}$$

Furthermore, ϕ_s can be directly related to the surface interaction γ_1 and the bulk value ϕ_b:

$$\frac{\xi_b}{2D} = \frac{6a^2\gamma_1}{k_B T\sqrt{3v\phi_b}} = \sqrt{\frac{\phi_b}{\phi_s}}\left(\frac{\phi_s}{\phi_b} - 1\right) \tag{27}$$

In order to be consistent with the semidilute concentration regime, the correlation length ξ_b should be smaller than the size of a single chain, $R_e = aN^\nu$, where $\nu \simeq 3/5$ is the Flory exponent in good solvent conditions. This sets a lower bound on the polymer concentration in the bulk, $c > c^*$.

So far, three length scales have been introduced: the Kuhn length or monomer size a, the adsorbed-layer width D, and the bulk correlation length ξ_b. It is more convenient for the discussion to consider the case where those three length scales are quite separate: $a \ll D \ll \xi_b$. Two conditions must be satisfied. On the one hand, the adsorption parameter is not large, $12a^2\gamma_1 \ll k_B T$ in order to have $D \gg a$; on the other, the adsorption energy is large enough to satisfy $12a^2\gamma_1 \gg k_B T\sqrt{3v\phi_b}$ in order to have $D \ll \xi_b$. The latter inequality can also be regarded as a condition for the polymer bulk concentration. The bulk correlation length is large enough if indeed the bulk concentration (assumed to be in the semidilute concentration range) is not too large. Roughly, let us assume in a typical case that the three length scales are well separated: a is of the order of a few angstroms, D of the order of a few dozen angstroms, and ξ_b of the order of a few hundred angstroms.

When the above two inequalities are satisfied, three spatial regions of adsorption can be differentiated: the proximal, central, and distal regions, as is outlined below. In addition, as soon as $\xi_b \gg D$, $z_0 \simeq 2D$, as follows from Eq. (26).

- Close enough to the surface, $z \sim a$, the adsorption profile depends on the details of the short-range interactions between the surface and monomers. Hence, this region is not universal. In the proximal region, for $a \gg z \gg D$, corrections to the mean-field theory analysis (which assumes the concentration to be constant) are presented below similarly to the treatment of the single-chain section. These corrections reveal a new scaling exponent characterizing the concentration profile. They are of particular importance close to the adsorption/desorption transition.

- In the distal region, $z \gg \xi_b$, the excess polymer concentration decays exponentially to its bulk value:

$$\phi(z) - \phi_b \simeq 4\phi_b e^{-2z/\xi_b} \tag{28}$$

as follows from Eq. (25). This behavior is very similar to the decay of fluctuations in the bulk, with ξ_b being the correlation length.

- Finally, in the central region (and with the assumption that ξ_b is the largest length scale in the problem), $D \ll z \ll \xi_b$, the profile is universal and from Eq. (25) it can be shown to decay with a powere law:

$$\phi(z) = \frac{1}{3v}\left(\frac{a}{z+2D}\right)^2 \tag{29}$$

A sketch of the different scaling regions in the adsorption profile is given in Fig. 8a. Included in this figure are corrections in the proximal region, which is discussed further below.

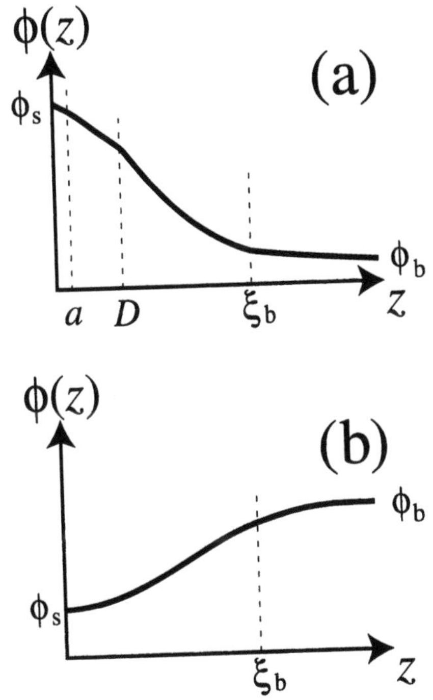

FIG. 8 (a) The schematic density profile for the case of adsorption from a semidilute solution; we distinguish a layer of molecular thickness $z \sim a$ where the polymer density depends on details of the interaction with the substrate and the monomer size, the proximal region $a < z < D$ where the decay of the density is governed by a universal power law (which cannot be obtained within mean-field theory), the central region for $D < z < \xi_b$ with a self-similar profile, and the distal region for $\xi_b < z$, where the polymer concentration relaxes exponentially to the bulk volume fraction ϕ_b. (b) The density profile for the case of depletion, where the concentration decrease close to the wall ϕ_s relaxes to its bulk value ϕ_b at a distance of the order of the bulk correlation length ξ_b.

A special consideration should be given to the formal limit of setting the bulk concentration to zero, $\phi_b \to 0$ (and equivalently $\xi_b \to \infty$), which denotes the limit of an adsorbing layer in contact with a polymer reservoir of vanishing concentration. It should be emphasized that this limit is not consistent with the assumption of a semidilute polymer solution in the bulk. Still, some information on the polymer density profile close to the adsorbing surface, where the polymer solution is locally semidilute [34] can be obtained. Formally, we take the limit $\xi_b \to \infty$ in Eq. (25), and obtain the limiting expression given by Eq. (29), which does not depend on ξ_b. The profile in the central region decays algebraically. In the case of zero polymer concentration in the bulk, the natural cut-off is not ξ_b but rather R_e, the coil size of a single polymer in solution. Hence, the distal region loses its meaning and is replaced by a more complicated scaling regime [37]. The length D can be regarded as the layer thickness in the $\xi_b \to \infty$ limit in the sense that a finite fraction of all the monomers are located in this layer of thickness D from the surface. Another observation is that $\phi(z) \sim 1/z^2$ for $z \gg D$. This power law is a result of the mean-field theory, and its modification is discussed below.

It is now possible to calculate within the mean-field theory the two physical quantities that are measured in many experiments: the surface tension reduction $\gamma - \gamma_0$ and the surface excess Γ.

The surface excess, defined in Eq. (21), can be calculated in a close form by inserting Eq. (25) into Eq. (21),

$$\Gamma = \frac{1}{\sqrt{3v}a^2}(\phi_s^{1/2} - \phi_b^{1/2}) = \frac{\xi_b \phi_b}{a^3}\left(\sqrt{\frac{\phi_s}{\phi_b}} - 1\right) \tag{30}$$

For strong adsorption, we obtain from Eq. (27) that $\phi_s \simeq (a/2D)^2/3v \gg \phi_b$, and Eq. (30) reduces to

$$\Gamma = \frac{1}{3va^2}\left(\frac{a}{D}\right) \sim \gamma_1 \tag{31}$$

while the surface volume fraction scales as $\phi_s \sim \gamma_1^2$. As can be seen from Eqs (31) and (29), the surface excess as well as the entire profile does not depend (to leading order) on the bulk concentration ϕ_b. We note again that the strong adsorption condition is always satisfied in the $\phi_b \to 0$ limit. Hence, Eq. (31) can be obtained directly by integrating the profile in the central region, Eq. (29).

Finally, let us calculate the reduction in surface tension for the adsorbing case. Inserting the variational equations (23) in Eq. (20) yields:

$$\gamma - \gamma_0 = -\gamma_1 \phi_s + \frac{k_B T \sqrt{3v}}{9a^2}\phi_s^{3/2}\left[1 - 3\left(\frac{\phi_b}{\phi_s}\right) + 2\left(\frac{\phi_b}{\phi_s}\right)^{3/2}\right] \tag{32}$$

The surface term in Eq. (32) is negative while the second term is positive. For strong adsorption this reduction of γ does not depend on ϕ_b and reduces to

$$\gamma - \gamma_0 \sim -\left(\frac{\gamma_1 a^2}{k_B T}\right)^3 \frac{k_B T}{a^2} + \mathcal{O}(\gamma_1^{4/3}) \tag{33}$$

where the leading term is just the contribution of the surface monomers.

2. Polymer Depletion

We highlight the main differences between the polymer adsorption and polymer depletion. Keeping in mind that $\gamma_1 < 0$ for depletion, the solution of the same profile equation (24), with the appropriate boundary condition, results in

$$\phi(z) = \phi_b \tanh^2\left(\frac{z + z_0}{\xi_b}\right) \tag{34}$$

which is schematically plotted in Fig. 8b. The limit $\phi_b \to 0$ cannot be taken in the depletion case since depletion with respect to a null reservoir has no meaning. However, we can, alternatively, look at the strong depletion limit, defined by the condition $\phi_s \ll \phi_b$. Here we find

$$\phi(z) = 3v\phi_b^2\left(\frac{z + 2D}{a}\right)^2 \tag{35}$$

In the same limit, we find for the surface volume fraction $\phi_s \sim \phi_b^2\gamma_1^{-2}$, and the exact expression for the surface excess, Eq. (30), reduces to

$$\Gamma = -\frac{1}{a^2}\sqrt{\frac{\phi_b}{3v}} \simeq -\frac{\phi_b\xi_b}{a^3} \tag{36}$$

The negative surface excess can be directly estimated from a profile varying from ϕ_b to zero over a length scale of order ξ_b.

The dominating behavior for the surface tension can be calculated from Eq. (20) where both terms are now positive. For the strong depletion case we obtain:

$$\gamma - \gamma_0 \simeq \frac{k_B T}{a^2}\left(\frac{a}{\xi_b}\right)^3 \sim \phi_b^{3/2} \tag{37}$$

B. Beyond Mean-Field Theory: Scaling Arguments for Good Solvents

One of the mean-field theory results that should be corrected is the scaling of the correlation length with ϕ_b. In the semidilute regime, the correlation length can be regarded as the average mesh size created by the overlapping chains. It can be estimated using very simple scaling arguments [11]: The volume fraction of monomers inside a coil formed by a subchain consisting of g monomers is $\phi \sim g^{1-3\nu}$ where ν is the Flory exponent. The spatial scale of this subchain is given by $\xi_b \sim ag^{\nu}$. Combining these two relations, and setting $\nu \simeq 3/5$, as appropriate for good solvent conditions, we obtain the known scaling of the correlation length:

$$\xi_b \simeq a\phi_b^{3/4} \tag{38}$$

This relation corrects the mean-field theory result $\xi_b \sim \phi_b^{-1/2}$, which can be obtained from, e.g., Eq. (20).

1. Scaling for Polymer Adsorption

We repeat here an argument due to de Gennes [36]. The main idea is to assume that the relation Eq. (38) holds locally: $\phi(z) = [\xi(z)/a]^{-4/3}$, where $\xi(z)$ is the local "mesh size" of the semidilute polymer solution. Since there is no other length scale in the problem beside the distance from the surface, z, the correlation length $\xi(z)$ should scale as the distance z itself, $\xi(z) \simeq z$, leading to the profile:

$$\phi(z) \simeq \left(\frac{a}{z}\right)^{4/3} \tag{39}$$

We note that this argument holds only in the central region $D \ll z \ll \xi_b$. It has been confirmed experimentally using neutron scattering [38] and neutron reflectivity [39]. Equation (39) satisfies the distal boundary condition: $z \to \xi_b$, $\phi(z) \to \phi_b$, but for $z > \xi_b$ we expect the regular exponential decay behavior of the distal region, Eq. (28). De Gennes also proposed (without a rigorous proof) a convenient expression for $\phi(z)$, which has the correct crossover from the central to the mean-field proximal region [36]:

$$\phi(z) = \phi_s\left(\frac{\frac{4}{3}D}{z+\frac{4}{3}D}\right)^{4/3} \simeq \left(\frac{a}{z+\frac{4}{3}D}\right)^{4/3} \tag{40}$$

Note that the above equation reduces to Eq. (39) for $z \gg D$. The extrapolation of Eq. (40) also gives the correct definition of D: $D^{-1} = -\mathrm{d}\log\phi/\mathrm{d}z|_s$. In addition, ϕ_s is obtained from the extrapolation to $z = 0$ and scales as

$$\phi_s = \phi(z=0) = \left(\frac{a}{D}\right)^{4/3} \tag{41}$$

For strong adsorption ($\phi_s \gg \phi_b$), we have

$$\begin{aligned}
\phi_s &\simeq \left(\frac{a}{D}\right)^{4/3} \sim \gamma_1^2 \\
D &\simeq a\left(\frac{k_B T}{a^2\gamma_1}\right)^{3/2} \sim \gamma_1^{-3/2} \\
\Gamma &\simeq a^2\left(\frac{a^2\gamma_1}{k_B T}\right)^{1/2} \sim \gamma_1^{1/2} \\
\gamma - \gamma_0 &\simeq -\frac{k_B T}{a^2}\phi_s^{3/2} \sim -\gamma_1^3
\end{aligned} \tag{42}$$

It is interesting to note that although D and Γ have different scaling with the surface interaction γ_1 in the mean-field theory and scaling approaches, ϕ_s and $\gamma - \gamma_0$ have the same scaling using both approaches. This is a result of the same scaling $\phi_s \sim \gamma_1^2$, which, in turn, leads to $\gamma - \gamma_0 \simeq \gamma_1\phi_s \sim \gamma_1^3$.

2. Scaling for Polymer Depletion

For polymer depletion, similar arguments [36] suggest the following scaling form for the central and mean-field proximal regions, $a < z < \xi_b$:

$$\phi(z) = \phi_b\left(\frac{z+\frac{5}{3}D}{\xi_b}\right)^{5/3} \tag{43}$$

where the depletion thickness is $\xi_b - D$, whereas in the strong depletion regime ($\phi_s \ll \phi_b$):

$$\phi_s \simeq \phi_b\left(\frac{D}{\xi_b}\right)^{5/3} \sim \phi_b^{9/4}\gamma_1^{-5/2}$$
$$D = a\left(\frac{a^2\gamma_1}{k_B T}\right)^{-3/2}$$
$$\Gamma \simeq -\phi_b a^{-3}(\xi_b - D) \sim \phi_b^{1/4}$$
$$\gamma - \gamma_0 \simeq -\frac{k_B T}{a^2}\phi_b^{3/2} \tag{44}$$

Note that the scaling of the surface excess and surface tension with the bulk concentration ϕ_b is similar to that obtained by the mean-field theory approach in Section III.A.2.

C. Proximal Region Corrections

So far we have not addressed any corrections in the proximal region: $a < z < D$ for the many-chain adsorption. In the mean-field theory picture the profile in the proximal region is featureless and saturates smoothly to its extrapolated surface value, $\phi_s > 0$. However, in relation to surface critical phenomena, which is in particular relevant close to the adsorption–desorption phase transition (the so-called "special" transition), the polymer profile in the proximal region has a scaling form with another exponent m:

$$\phi(z) \simeq \phi_s\left(\frac{a}{z}\right)^m \tag{45}$$

where $m = (1 - \varphi - \nu)/\nu$ is the proximal exponent, Eq. (14). This is similar to the single-chain treatment in Section II.

For good solvents, one has $m \simeq 1/3$, as was derived using analogies with surface critical phenomena, exact enumeration of polymer configurations, and Monte Carlo simulations [31]. It is different from the exponent 4/3 of the central region.

With the proximal region correction, the polymer profile can be written as [33]

$$\phi(z) \simeq \begin{cases} \phi_s & \text{for } 0 < z < a \\ \phi_s\left(\frac{a}{z}\right)^{1/3} & \text{for } a < z < D \\ \phi_s\left(\frac{a}{z}\right)^{1/3}\left(\frac{D}{z+D}\right) & \text{for } D < z < \xi_b \end{cases} \tag{46}$$

where

$$\phi_s = \frac{a}{D} \tag{47}$$

The complete adsorption profile is shown in Fig. 8a. By minimization of the free energy with respect to the layer thickness D it is possible to show that D is proportional to $1/\gamma_1$:

$$D \sim \gamma_1^{-1} \tag{48}$$

in accord with the exact field-theoretical results for a single chain as discussed in Section II.

The surface concentration, surface excess, and surface tension have the following scaling [33]:

$$\phi_s \simeq \frac{a}{D} \sim \gamma_1$$
$$\Gamma \simeq a^{-3} D\left(\frac{a}{D}\right)^{4/3} \sim \gamma_1^{1/3}$$
$$\gamma - \gamma_0 \simeq -\frac{\gamma_1 a^2}{k_B T}\gamma_1 \sim \gamma_1^2 \tag{49}$$

Note the differences in the scaling of the surface tension and surface excess in Eq. (49) as compared with the results obtained with no proximity exponent ($m = 0$) in the previous section, Eq. (42).

At the end of our discussion of polymer adsorption from solutions, we would like to add that for the case of adsorption from dilute solutions, there is an intricate crossover from the single-chain adsorption behavior, as discussed in Section II, to the adsorption from semidilute polymer solutions, as discussed in this section [34]. Since the two-dimensional adsorbed layer has a local polymer concentration higher than that of the bulk, it is possible that the adsorbed layer forms a two-dimensional semidilute state, while the bulk is a truly dilute polymer solution. Only for extremely low bulk concentrations or for very weak adsorption energies has the adsorbed layer a single-chain structure with no chain crossings between different polymer chains.

D. Loops and Tails

It was realized quite some time ago that the so-called central region of an adsorbed polymer layer is characterized by a rather broad distribution of loop and tail sizes [5, 40, 41]. A loop is defined as a chain region located between two points of contact with the adsorbing surface, and a tail is defined as the chain region between the free end and the closest contact point to the surface, while a train denotes a chain section which is tightly bound to the substrate (see Fig. 3a). The relative statistical weight of loops and tails in the adsorbed layer is clearly of importance to applications. For example, it is expected that polymer loops which are bound at both ends to the substrate are more prone than tails to entanglements with free polymers and, thus, lead to enhanced friction effects. It was found in detailed numerical mean-field theory calculations that the external part of the adsorbed layer is dominated by dangling tails, while the inner part is mostly built up by loops [5, 40].

Recently, an analytical theory was formulated which correctly takes into account the separate contributions of loops and tails and which thus goes beyond the *ground state dominance* assumption made in ordinary mean-field theories. The theory predicts that a crossover between tail-dominated and loop-dominated regions occurs at some distance $z^* \simeq aN^{1/(d-1)}$ [42] from the surface, where d is the dimension of the embedding space. It is well known that mean-field theory behavior can be formally obtained by setting the embedding dimensionality equal to the upper critical dimension, which is for self-avoiding polymers given by $d = 4$ [12]. Hence, the above expression predicts a crossover in the adsorption behavior at a distance $z^* \simeq aN^{1/3}$. For good-solvent conditions in three dimensions ($d = 3$), $z^* \simeq aN^{1/2}$.

In both cases, the crossover occurs at a separation much smaller than the size of a free polymer $R_e \sim aN^{\nu}$ where, according to the classical Flory argument [8], $\nu = 3/(d+2)$.

A further rather subtle result of these improved mean-field theories is the occurrence of a depletion hole, i.e., a region at a certain separation from the adsorbing surface where the monomer concentration is smaller than the bulk concentration [42]. This depletion hole results from an interplay between the depletion of free polymers from the adsorbed layer and the slowly decaying density profile resulting from dangling tails. It occurs at a distance from the surface comparable with the radius of gyration of a free polymer, but also shows some dependence on the bulk polymer concentration. These and other effects, related to the occurrence of loops and tails in the adsorbed layer, have been recently reviewed [43].

IV. INTERACTION BETWEEN TWO ADSORBED LAYERS

One of the many applications of polymers lies in their influence on the behavior of colloidal particles suspended in a solvent [10]. If the polymers do not adsorb on the surface of the colloidal particles but are repelled from it, a strong attraction between the particles results from this polymer–particle depletion, and can lead to polymer-induced flocculation [44]. If the polymers adsorb uniformly on the colloidal surface (and under good-solvent conditions), they show the experimentally well-known tendency to stabilize colloids against flocculation, i.e., to hinder the colloidal particles from coming so close that van der Waals attractions will induce binding. We should also mention that, in other applications, adsorbing high molecular weight polymers are used in the opposite sense as flocculants to induce binding between unwanted submicrometer particles, thereby, removing them from the solution. It follows that adsorbing polymers can have different effects on the stability of colloidal particles, depending on the detailed parameters.

Hereafter, we assume the polymers to form an adsorbed layer around the colloidal particles, with a typical thickness much smaller than the particle radius, such that curvature effects can be neglected. In that case, the effective interaction between the colloidal particles with adsorbed polymer layers can be traced back to the interaction energy between two planar substrates covered with polymer adsorption layers. In the case when the approach of the two particles is slow and the adsorbed polymers are in *full equilibrium* with the polymers in solution, the interaction between two opposing adsorbed layers is predominantly attractive [45, 46], mainly because polymers form bridges between the two surfaces. Recently, it has been shown that there is a small repulsive component to the interaction at large separations [47].

The typical equilibration times of polymers are extremely long. This holds in particular for adsorption and desorption processes, and is due to the slow diffusion of polymers and their rather high adsorption energies. Note that the adsorption energy of a polymer can be much higher than $k_B T$ even if the adsorption energy of a single monomer is small since there are typically many monomers of a single chain attached to the surface. Therefore, for the typical time scales of colloid contacts, the adsorbed polymers are not in equilibrium with the polymer solution. This is also true for most of the experiments done with a surface-force apparatus, where two polymer layers adsorbed on crossed mica cylinders are brought into contact.

In all these cases one has a *constrained equilibrium* situation, where the polymer configurations and thus the density profile can adjust only with the constraint that the total adsorbed polymer excess stays constant. This case has been first considered by de Gennes [45] and he found that two fully saturated adsorbed layers will strongly repel each other if the total adsorbed amount of polymer is not allowed to decrease. The repulsion is mostly due to osmotic pressure and originates from the steric interaction between the two opposing adsorption layers. It was experimentally verified in a series of force-microscope experiments on PEO layers in water (which is a good solvent for PEO)[48].

In other experiments, the formation of the adsorbtion layer is stopped before the layer is fully saturated. The resulting adsorption layer is called *undersaturated*. If two of these undersaturated adsorption layers approach each other, a strong attraction develops, which only at smaller separation changes to an osmotic repulsion [49]. The theory developed for such nonequilibrium conditions predicts that any surface excess lower than the one corresponding to full equilibrium will lead to attraction at large separations [50]. Similar mechanisms are also at work in colloidal suspensions, if the total surface available for polymer adsorption is large compared to the total polymer added to the solution. In this case, the adsorption layers are also undersaturated, and the resulting attraction is utilized in the application of polymers as flocculation agents [10].

A distinct mechanism, which also leads to attractive forces between adsorption layers, was investigated in experiments with dilute polymer solutions in bad solvents. An example is given by PS in cyclohexane below the theta temperature [51]. The subsequently developed theory [52] showed that the adsorption layers attract each other since the local concentration in the outer part of the adsorption layers is enhanced over the dilute solution and lies in the unstable two-phase region of the bulk phase diagram. Similar experiments were repeated at the theta temperature [53].

The force apparatus was also used to measure the interaction between depletion layers [54], as realized with PS in toluene, which is a good solvent for PS but does not favor the adsorption of PS on mica. Surprisingly, the resultant depletion force is too weak to be detected.

The various regimes and effects obtained for the interaction of polymer solutions between two surfaces have recently been reviewed [55]. It transpires that force-microscope experiments done on adsorbed polymer layers form an ideal tool for investigating the basic mechanisms of polymer adsorption, colloidal stabilization, and flocculation.

V. ADSORPTION OF POLYELECTROLYTES

Adsorption of charged chains (polyelectrolytes) on to charged surfaces is a difficult problem, which is only partially understood from a fundamental point of view. This is the case in spite of the prime importance of polyelectrolyte adsorption in many applications [5]. We comment here briefly on the additional features that are characteristic for the adsorption of charged polymers on surfaces.

A polyelectrolyte is a polymer where a fraction f of its monomers are charged. When the fraction is small, $f \ll 1$, the polyelectrolyte is weakly charged, whereas when f is close to unity, the polyelectrolyte is strongly charged. There are two common ways to control f [56]. One way is to polymerize a heteropolymer using charged and neutral monomers as building blocks. The charge distribution along the chain is quenched ("frozen") during the polymerization stage, and it is characterized by the fraction of charged monomers on the chain, f. In the second way, the polyelectrolyte is a weak polyacid or polybase. The effective charge of each monomer is controlled by the pH of the solution. Moreover, this annealed fraction depends on the local electric potential. This is in particular important to adsorption processes since the local electric field close to a strongly charged surface can be very different from its value in the bulk solution.

Electrostatic interactions play a crucial role in the adsorption of polyelectrolytes [5, 57, 58]. Besides the fraction f of charged monomers, the important parameters are the surface charge density (or surface potential in the case of conducting surfaces), the amount of salt (ionic strength of low molecular weight electrolyte) in solution, and, in some cases, the solution pH. For polyelectrolytes the electrostatic interactions between the monomers themselves (same charges) are always repulsive, leading to an effective stiffening of the chain [59, 60]. Hence, this interaction will *favor* the adsorption of single polymer chains, since their configurations are already rather extended [61], but it will *oppose* the formation of dense adsorption layers close to the surface [62]. A special case is that of *polyampholytes*, where the charge groups on the chain can be positive as well as negative, resulting in a complicated interplay of attraction and repulsion between the monomers [28, 29]. If the polyelectrolyte chains and the surface are oppositely charged, the electrostatic interactions between them will *enhance* the adsorption.

The role of the salt can be conveniently expressed in terms of the Debye–Hückel screening length, defined as

$$\lambda_{\mathrm{DH}} = \left(\frac{8\pi c_{\mathrm{salt}} e^2}{\varepsilon k_{\mathrm{B}} T} \right)^{-1/2} \tag{50}$$

where c_{salt} is the concentration of monovalent salt ions, e is the electronic charge, and $\varepsilon \simeq 80$ is the dielectric constant of the water. Qualitatively, the presence of small positive and negative ions at thermodynamical equilibrium screens the r^{-1} electrostatic potential at distances $r > \lambda_{\mathrm{DH}}$, and roughly changes its form to $r^{-1} \exp(-r/\lambda_{\mathrm{DH}})$. For polyelectrolyte adsorption, the presence of salt has a complex effect. It simultaneously screens the monomer–monomer repulsive interactions as well as the attractive interactions between the oppositely charged surface and polymer.

Two limiting adsorbing cases can be discussed separately:

1. A noncharged surface on which the chains tend to adsorb. Here, the interaction between the surface and the chain does not have an electrostatic component. However, as the salt screens the monomer–monomer electrostatic repulsion, it leads to enhancement of the adsorption.
2. The surface is charged but does not interact with the polymer besides the electrostatic interaction. This is called the pure electrosorption case. At low salt concentration, the polymer charge completely compensates the surface charge. At high salt concentration some of the compensation is done by the salt, leading to a decrease in the amount of adsorbed polymer.

In practice, electrostatic and other types of interactions with the surface can occur in parallel, making the analysis more complex. An interesting phenomenon of *overcompensation* of surface charges by the polyelectrolyte chains is observed, where the chains form a condensed layer and reverse the sign of the total surface charge. This is used, e.g., to build a multilayered structure of cationic and anionic polyelectrolytes – a process that can be continued for a few dozen or even a few hundred times [63–65]. The phenomenon of overcompensation is discussed in Refs 62 and 66, but is still not very well understood.

Adsorption of polyelectrolytes from semidilute solutions is treated either in terms of a discrete multi-Stern layer model [5, 67, 68] or in a continuum approach [62, 69, 70]. In the latter approach, the concentration of polyelectrolytes as well as the electric potential close to the substrate are considered as continuous functions. Both the polymer chains and the electrostatic degrees of freedom are treated on a mean-field theory level. In some cases the salt concentration is considered explicitly [62, 70], while in other works (e.g., [69]) it induces a screened Coulombic interaction between the monomers and the substrate.

In a recent work [62], a simple theory has been proposed to treat polyelectrolyte adsorption from a semidilute bulk. The surface was treated as a surface with constant electric potential. (Note that, in other works, the surface is considered to have a constant charge density.) In addition, the substrate is assumed to be impenetrable by the requirement that the polymer concentration at the wall is zero.

Within a mean-field theory it is possible to write down the coupled profile equations of the polyelectrolyte concentration and electric field, close to the surface, assuming that the small counterions (and salt) concentration obeys a Boltzmann distribution. From numerical solutions of the profile equations as well as scaling arguments the following picture emerges. For very low salt concentration, the surface excess of the polymers Γ and the adsorbed layer thickness D are decreasing functions of f : $\Gamma \sim D \sim f^{-1/2}$. This effect arises from a delicate competition between an enhanced attraction to the substrate, on one hand, and an enhanced electrostatic repulsion between monomers, on the other.

Added salt will screen both the electrostatic repulsion between monomers and the attraction to the surface. In the presence of salt, for low f, Γ scales as $f/c_{\text{salt}}^{1/2}$ until it reaches a maximum value at $f^* \sim (c_{\text{salt}}v)^{1/2}$, v being the excluded volume parameter of the monomers. At this special value, $f = f^*$, the electrostatic contribution to the monomer–monomer excluded volume $v_{\text{el}} \sim f^2\lambda_{\text{DH}}^2$ is exactly equal to the non electrostatic v. For $f > f^*$, $v_{\text{el}} > v$ and the surface excess is a descending function of f, because of the dominance of monomer–monomer electrostatic repulsion. It scales as $c_{\text{salt}}^{1/2}/f$. Chapter 7 of Ref. 5 contains a fair amount of experimental results on polyelectrolyte adsorption.

VI. POLYMER ADSORPTION ON HETEROGENEOUS SURFACES

Polymer adsorption can be coupled in a subtle way with lateral changes in the chemical composition or density of the surface. Such a surface undergoing lateral rearrangements at thermodynamical equilibrium is called an *annealed* surface [71, 72]. A Langmuir monolayer of insoluble surfactant monolayers at the air/water interface is an example of such an annealed surface. As a function of the temperature change, a Langmuir monolayer can undergo a phase transition from a high-temperature homogeneous state to a low-temperature demixed state, where dilute and dense phases coexist. Alternatively, the transition from a dilute phase to a dense one may be induced by compressing the monolayer at constant temperature, in which case the adsorbed polymer layer contributes to the pressure [73]. The domain boundary between the dilute and dense phases can act as a nucleation site for adsorption of bulky molecules [74].

The case where the insoluble surfactant monolayer interacts with a semidilute polymer solution solubilized in the water subphase was considered in some detail. The phase diagrams of the mixed surfactant/polymer system were investigated within the framework of mean-field theory [75]. The polymer enhances the fluctuations of the monolayer and induces an upward shift of the critical temperature. The critical concentration is increased if the monomers are more attracted (or at least less repelled) by the surfactant molecules than by the bare water/air interface. In the case where the monomers are repelled by the bare interface but attracted by the surfactant molecules (or vice versa), the phase diagram may have a triple point. The location of the polymer desorption transition line (i.e., where the substrate–polymer interaction changes from being repulsive to being attractive) appears to have a great effect on the phase diagram of the surfactant monolayer.

VII. POLYMER ADSORPTION ON CURVED INTERFACES AND FLUCTUATING MEMBRANES

The adsorption of polymers on rough substrates is of much interest in applications. One example is the reinforcement of rubbers by filler particles such as carbon black or silica particles [76]. Theoretical models considered sinusoidal surfaces [77] and rough and corrugated substrates [78, 79]. In all cases, enhanced adsorption was found and rationalized in terms of the excess surface available for adsorption.

The adsorption on macroscopically curved bodies leads to slightly modified adsorption profiles, and also to contribution to the elastic bending moduli of the adsorbing surfaces. The elastic energy of liquid-like membrane can be expressed in terms of two bending moduli, κ and κ_G. The elastic energy (per unit area) is

$$\frac{\kappa}{2}(c_1 + c_2)^2 + \kappa_G c_1 c_2 \tag{51}$$

where κ and κ_G are the elastic bending modulus and the Gaussian bending modulus, respectively. The reciprocals of the principal radii of curvature of the surface are given by c_1 and c_2. Quite generally, the effective κ_G turns out to be positive and thus favors the formation of surfaces with negative Gaussian curvature, as, for example, an "egg-carton" structure consisting of many saddles. On the other hand, the effective κ is reduced, leading to a more deformable and flexible surface due to the adsorbed polymer layer [71, 80].

Of particular interest is the adsorption of strongly charged polymers on oppositely charged spheres, because this is a geometry encountered in many collodial science applications and in molecular biology as well [81–85].

In other works, the effects of a modified architecture of the polymers on the adsorption behavior was considered. For example, the adsorption of star polymers [86] and random copolymers [87] was considered.

Note that some polymers exhibit a transition into a glassy state in concentrated adsorbed layers. This glassy state depends on the details of the molecular interaction, which are not considered here. It should be kept in mind that such high-concentration effects can slow down the dynamics of adsorption considerably and will prolong the reach of equilibrium.

VIII. TERMINALLY ATTACHED CHAINS

The discussion has so far assumed that all monomers of a polymer are alike and therefore show the same tendency to adsorb to the substrate surface. For industrial and technological applications, one is often interested in *end-functionalized polymers.* These are polymers which attach with one end only to the substrate, as is depicted in Fig. 3b, while the rest of the polymer is not particularly attracted to (or even repelled from) the grafting surface. Hence, it attains a random-coil structure in the vicinity of the surface. Another possibility of block copolymer grafting (Fig. 3c) will be also briefly discussed.

The motivation to study such terminally attached polymers lies in their enhanced power to stabilize particles and surfaces against flocculation. Attaching a polymer by one end to the surface opens up a much more effective route to stable surfaces. Bridging and creation of polymer loops on the same surface, as encountered in the case of homopolymer adsorption, do not occur if the main-polymer section is chosen such that it does not adsorb to the surface.

Experimentally, the end-adsorbed polymer layer can be built in several different ways, depending on the application in mind. First, one of the polymer ends can be *chemically* bound to the grafting surface, leading to a tight and irreversible attachment [88], shown schematically in Fig. 3b. The second possibility consists of *physical* adsorption of a specialized end-group which favors interaction with the substrate. For example, PS chains have been used which contain a zwitterionic end group that adsorbs strongly on mica sheets [89].

Physical grafting is also possible with a suitably chosen diblock copolymer (Fig. 3c), e.g., a PS–PVP diblock in the solvent toluene at a quartz substrate [90]. Toluene is a *selective solvent* for this diblock, i.e., the PVP [poly(vinylpyridine)] block is strongly adsorbed to the quartz substrate and forms a collapsed anchor, while the PS block is under good-solvent conditions, not adsorbing to the substrate and thus extending into the solvent. General adsorption scenarios for diblock copolymers have been theoretically discussed, both for selective and nonselective solvents [91]. Special consideration has been given to the case when the asymmetry of the diblock copolymer, i.e., the length difference between the two blocks, decreases [91].

Yet another experimental realization of grafted polymer layers is possible with diblock copolymers which are anchored at the liquid–air [92] or at a liquid–liquid interface between two immiscible liquids [93]; this scenario offers the advantage that the surface pressure can be directly measured. A well studied example is a diblock copolymer of PS–PEO. The PS block is shorter and functions as the anchor at the air–water interface as it is not miscible in water. The PEO block is miscible in water, but because of attractive interaction with the air–water interface it forms a quasi-two dimensional layer at very low surface coverage. As the pressure increases and the area per polymer decreases, the PEO block is expelled from the surface and forms a quasi-polymer brush.

In the following we simplify the discussion by assuming the polymers to be irreversibly grafted at one end to the substrate. Let us consider the good-solvent case in the absence of any polymer attraction to the surface. The important new parameter that enters the discussion is the grafting density (or area per chain) σ, which is the inverse of the average area that is available for each polymer at the grafting surface. For small grafting densities, $\sigma < \sigma^*$, the polymers will be far apart from each other and hardly interact, as schematically shown in Fig. 9a. The overlap grafting density is $\sigma^* \sim a^{-2}N^{-6/5}$ for swollen chains, where N is the polymerization index [94].

For large grafting densities, $\sigma > \sigma^*$, the chains overlap. Since we assume the solvent to be good, monomers repel each other. The lateral separation between the polymer coils is fixed by the grafting density, so that the polymers stretch away from the grafting surface in order to avoid each other, as depicted in Fig. 9b. The resulting structure is called a polymer "brush", with a vertical height h which greatly exceeds the unperturbed coil radius [94, 95]. Similar stretched structures occur in many other situations, such as diblock copolymer melts in the strong segregation regime, or polymer stars under good-solvent conditions [96]. The universal occurrence of stretched polymer configurations in many seemingly disconnected situations warrants a detailed discussion of the effects obtained with such systems.

A. Grafted Polymer Layer: A Mean-Field Theory Description

The scaling behavior of the polymer height can be analyzed by using a Flory-like mean-field theory, which is a simplified version of the original Alexander theory [95].

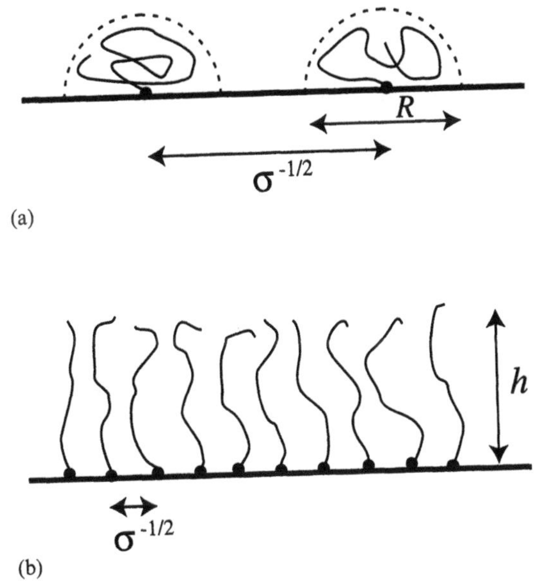

FIG. 9 For grafted chains, one distinguishes (a) the mushroom regime, where the distance between chains, $\sigma^{-1/2}$, is larger than the size of a polymer coil, and (b) the brush regime, where the distance between chains is smaller than the unperturbed coil size. Here, the chains are stretched away from the wall due to repulsive interactions between monomers. The brush height h scales linearly with the polymerization index, $h \sim N$, and is thus larger than the unperturbed coil radius $R_e \sim aN^{\nu}$.

The stretching of the chain leads to an entropic free energy loss of $h^2/(a^2N)$ per chain, and the repulsive energy density resulting from unfavorable monomer–monomer contacts is proportional to the squared monomer density times the dimensionless excluded-volume parameter v (introduced in Section III). The free energy per chain is then

$$\frac{\mathcal{F}}{k_BT} = \frac{3h^2}{2a^2N} + 2a^3v\left(\frac{\sigma N}{h}\right)^2\frac{h}{\sigma} \tag{52}$$

where the numerical prefactors were chosen for convenience. The equilibrium height is obtained by minimizing Eq. (52) with respect to h, and the result is [95]

$$h = N(2va^5\sigma/3)^{1/3} \tag{53}$$

The vertical size of the brush scales linearly with the polymerization index N, a clear signature of the strong stretching of the polymer chains. At the overlap threshold, $\sigma^* \sim N^{-6/5}$, the height scales as $h \sim N^{3/5}$, and thus agrees with the scaling of an unperturbed chain radius in a good solvent, as it should. The simple scaling calculation predicts the brush height h correctly in the asymptotic limit of long chains and strong overlap. It has been confirmed by experiments [88–90] and computer simulations [97, 98].

The above scaling result assumes that all chains are stretched to exactly the same height, leading to a step-like shape of the density profile. Monte Carlo and numerical mean-field calculations confirm the general scaling of the brush height, but exhibit a more rounded monomer density profile which goes continuously to zero at the outer perimeter [97]. A large step towards a better understanding of stretched polymer systems was made by Semenov [99], who recognized the importance of *classical paths* for such systems.

The classical polymer path is defined as the path which minimizes the free energy, for a given start and end position, and thus corresponds to the most likely path a polymer takes. The name follows from the analogy with quantum mechanics, where the classical motion of a particle is given by the quantum path with maximal probability. Since for strongly stretched polymers the fluctuations around the classical path are weak, it is expected that a theory which takes into account only classical paths is a good approximation in the strong-stretching limit. To quantify the stretching of the brush, let us introduce the (dimensionless) stretching parameter β, defined as

$$\beta \equiv N\left(\frac{3v^2\sigma^2a^4}{2}\right)^{1/3} = \frac{3}{2}\left(\frac{h}{aN^{1/2}}\right)^2 \tag{54}$$

where $h \equiv N(2v\sigma a^5/3)^{1/3}$ is the brush height according to Alexander's theory [cf. Eq. (53)]. The parameter β is proportional to the square of the ratio of the Alexander prediction for the brush height h and the unperturbed chain radius $R_0 \sim aN^{1/2}$, and, therefore, is a measure of the stretching of the brush. Constructing a classical theory in the infinite-stretching limit, defined as the limit $\beta \to \infty$, it was shown independently by Milner et al. [100] and Skvortsov et al. [101] that the resulting normalized monomer volume-fraction profile only depends on the vertical distance from the grafting surface. It has in fact a *parabolic* profile given by

$$\phi(z) = \left(\frac{3\pi}{4}\right)^{2/3} - \left(\frac{\pi z}{2h}\right)^2 \tag{55}$$

The brush height, i.e., the value of z for which the monomer density becomes zero, is given by $z^* = (6/\pi^2)^{1/3}h$. The parabolic brush profile has subsequently been confirmed in computer simulations [97, 98] and experiments [88] as the limiting density profile in the strong-stretching limit, and constitutes one of the cornerstones in this field. Intimately connected with the density profile is the distribution of *polymer end-points*, which is nonzero everywhere inside the brush, in contrast with the original scaling description leading to Eq. (53).

However, deviations from the parablic profile become progressively important as the length of the polymers N or the grafting density σ decreases. In a systematic derivation of the mean-field theory for Gaussian brushes [102] it was shown that the theory is characterized by a single parameter, namely, the stretching parameter β. In the limit $\beta \to \infty$, the difference between the classical approximation and the mean-field theory vanishes, and one obtains the parabolic density profile. For finite β the full mean-field theory and the classical approximation lead to different results and both show deviations from the parabolic profile.

In Fig. 10 we show the density profiles for four different values of β, obtained with the full mean-field theory [102]. The parameter values used are $\beta = 100$ (solid line), $\beta = 10$ (broken line), $\beta = 1$ (dotted–dashed line), and $\beta = 0.1$ (dotted line). For comparison, we also show the asymptotic result according to Eq. (55) as a thick dashed line. In contrast to earlier numerical implementations [5], the self-consistent mean-field equations were solved in the continuum limit, in which case the results only depend on the single parameter β, and direct comparison with other continuum theories becomes possible. Already for $\beta = 100$ the density profile obtained within

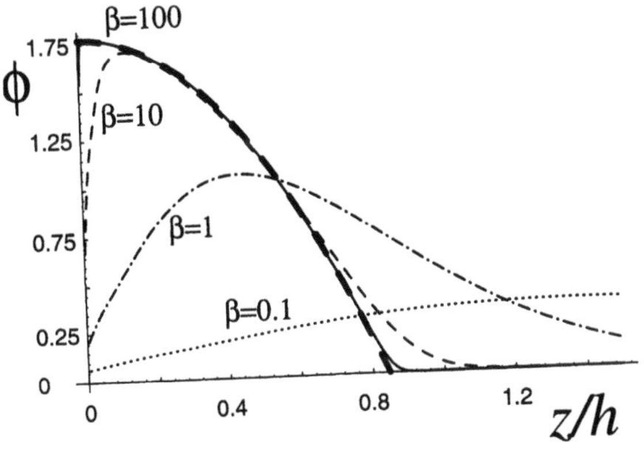

FIG. 10 Results for the density profile of a strongly compressed brush, as obtained within a mean-field theory calculation. As the compression increases, described by the stretching parameter β, which varies from 0.1 (dots) to 1 (dash–dots), 10 (dashes), and 100 (solid line), the density profile approaches the parabolic profile (shown as a thick, dashed line) obtained within a classical-path analysis. (Adapted from Ref. 102.)

mean-field theory is almost indistinguishable from the parabolic profile denoted by a thick dashed line.

Experimentally, the highest values of β achievable are in the range of $\beta \simeq 20$, and therefore deviations from the asymptotic parabolic profile are important. For moderately large values of $\beta > 10$, the classical approximation (not shown here), derived from the mean-field theory by taking into account only one polymer path per end-point position, is still a good approximation, as judged by comparing density profiles obtained from both theories [102], except very close to the surface. The classical theory misses completely the depletion effects at the substrate, which mean-field theory correctly takes into account. Depletion effects at the substrate lead to a pronounced density depression close to the grafting surface, as is clearly visible in Fig. 10.

A further interesting question concerns the behavior of individual polymer paths. As we already discussed for the infinite-stretching theories ($\beta \to \infty$), there are polymer paths ending at any distance from the surface. Analyzing the paths of polymers which end at a common distance from the wall, two rather unexpected features are obtained: (1) free polymer ends are in general stretched, and (2) the end-points lying close to the substrate are pointing towards the surface (such that the polymer paths first move away from the grafting surface before moving towards the substrate), and end-points lying beyond a certain distance from the substrate point away from the surface (such that the paths move monotonically towards the surface). We should point out that these two features have recently been confirmed in molecular-dynamics simulations [103]. They are neither an artifact of the continuous self-consistent theory used in Ref. 102 nor are they due to the neglect of fluctuations. These are interesting results, especially since it has been long assumed that free polymer ends are unstretched, based on the assumption that no forces act on free polymer ends.

Let us now turn to the thermodynamic behavior of a polymer brush. Using the Alexander description, we can calculate the free energy per chain by putting the result for the optimal brush height, Eq. (53), into the free-energy expression, Eq. (52). The result is

$$\mathcal{F}/k_BT \sim N(v\sigma a^2)^{2/3} \tag{56}$$

In the presence of excluded-volume correlations, i.e., when the chain overlap is rather moderate, the brush height h is still correctly predicted by the Alexander calculation, but the prediction for the free energy is in error. Including correlations [95], the free energy is predicted to scale as $\mathcal{F}/k_BT \sim N\sigma^{5/6}$. The osmotic surface pressure Π is related to the free energy per chain by

$$\Pi = \sigma^2 \frac{\partial \mathcal{F}}{\partial \sigma} \tag{57}$$

and should thus scale as $\Pi \sim \sigma^{5/3}$ in the absence of correlations and as $\Pi \sim \sigma^{11/6}$ in the presence of correlations. However, these theoretical predictions do not compare well with experimental results for the surface pressure of a compressed brush [92]. At present, there is no explanation for this discrepancy. An alternative theoretical method for studying tethered chains is the so-called single-chain mean-field method [104], where the statistical mechanics of a single chain is treated exactly, and the interactions with the other chains are taken into account on a mean-field level. This method is especially useful for short chains, where fluctuation effects are important, and dense systems, where excluded-volume interactions play a role. The calculated profiles and brush heights agree very well with those of experiments and computer simulations, and moreover explain the pressure isotherms measured experimentally [92] and in molecular-dynamics simulations [105].

As we described earlier, the main interest in end-adsorbed or grafted polymer layers stems from their ability to stabilize surfaces against van der Waals' attraction. The force between colloids with grafted polymers is repulsive if the polymers do not adsorb on the grafting substrates [106]. This is in accord with our discussion of the interaction between adsorption layers, where attraction was found to be mainly caused by bridging and creation of polymer loops, which of course is absent for nonadsorbing brushes. A stringent test of brush theories was possible with accurate experimental measurements of the repulsive interaction between two opposing grafted polymer layers by using a surface-force apparatus [89]. The resultant force could be fitted very nicely by the infinite-stretching theory of Milner [107]. It was also shown that polydispersity effects, although rather small experimentally, have to be taken into account theoretically in order to obtain a good fit of the data [108].

B. Solvent, Substrate, and Charge Effects on Polymer Grafting

So far we have assumed that the polymer grafted layer is in contact with a good solvent. In this case, the grafted polymers try to minimize their contacts by stretching out into the solvent. If the solvent is bad, the monomers try to avoid the solvent by forming a collapsed brush, the height of which is considerably reduced with respect to the good-solvent case. It turns out that the collapse transition, which leads to phase separation in the bulk, is smeared out for the grafted layer and does not correspond to a true phase transition [109]. The height of the collapsed layer scales linearly in σN, which reflects the constant density within the brush, in agreement with experiments [110]. Some interesting effects have been described theoretically [111] and experimentally [110] for brushes in mixtures of good and bad solvents, and can be rationalized in terms of a partial solvent demixing.

For a theta solvent ($T = T_\theta$) the relevant interaction is described by the third-virial coefficient; using a simple Alexander approach similar to the one leading to Eq. (53), the brush height is predicted to vary with the grafting density as $h \sim \sigma^{1/2}$, in agreement with computer simulations [112].

Up to now we have discussed planar grafting layers. Typically, however, polymers are grafted to curved surfaces. The first study taking into account curvature effects of stretched and tethered polymers was done in the context of star polymers [113]. It was found that chain tethering in the spherical geometry leads to a universal density profile, showing a densely packed core, an intermediate region where correlation effects are negligible and the density decays as $\phi(r) \sim 1/r$, and an outside region where correlations are important and the density decays as $\phi \sim r^{-4/3}$. These considerations were extended using the infinite-stretching theory by Ball et al. [114], self-conistent mean-field theories [115], and molecular-dynamics simulations [116]. Of particular interest is the behavior of the bending rigidity of a polymer brush, which can be calculated from the free energy of a cylindrical and a spherical brush and forms a conceptually simple model for the bending rigidity of a lipid bilayer [117].

A different scenario is obtained with special functionalized lipids with attached water-soluble polymers. If such lipids are incorporated into lipid vesicles, the water-soluble polymers [typically one uses PEG (polyethylene glycol) for its nontoxic properties] form well-separated mushrooms, or, at higher concentration of PEG lipid, a dense brush. These modified vesicles are very interesting in the context of drug delivery, because they show prolonged circulation times in vivo [118]. This is probably the result of steric serum protein-binding inhibition due to the hydrophilic brush coat provided by the PEG lipids. Since the lipid bilayer is rather flexible and undergoes thermal bending fluctuations, there is an interesting coupling between the polymer density distribution and the membrane shape [119]. For nonadsorbing, anchored polymers, the membrane will bend away from the polymer due to steric repulsion, but for adsorbing anchored polymers the membrane will bend towards the polymer [119].

The behavior of a polymer brush in contact with a solvent, which is by itself also a polymer, consisting of chemically identical but somewhat shorter chains than the brush, had been first considered by de Gennes [94]. A complete scaling description has been given only recently [120]. One distinguishes different regimes where the polymer solvent is expelled to various degrees from the brush. A somewhat related question concerns the behavior of two opposing brushes in a solvent which consists of a polymer solution [121]. Here one distinguishes a regime where the polymer solution leads to a strong attraction between the surfaces via the ordinary depletion interaction (cf. [44]), but also a high polymer concentration regime where the attraction is not strong enough to induce colloidal flocculation. This phenomenon is called colloidal restabilization [121].

Another important extension of the brush theory is obtained with charged polymers [122], showing an interesting interplay of electrostatic interactions, polymer elasticity, and monomer–monomer repulsion. Considering a mixed brush made of mutually incompatible grafted chains, a novel transition to a brush characterized by a lateral composition modulation was found [123]. Even more complicated spatial structures are obtained with grafted diblock copolymers [124]. Finally, we would like to mention in passing that these static brush phenomena have interesting consequences on the dynamic properties of polymer brushes [125].

IX. CONCLUDING REMARKS

We have reviewed simple physical concepts underlying the main theories which deal with equilibrium and static properties of polymers adsorbed or grafted to substrates. Most of the review dealt with somewhat ideal situations: smooth and flat surfaces which are chemically homogeneous; long and linear homopolymer chains on which chemical properties can be averaged; and simple phenomenological types of interactions between the monomers and the substrate as well as between the monomers and the solvent.

Even with all the simplifying assumptions, the emerging physical picture is quite rich and robust. Adsorption of polymers from dilute solutions can be understood in terms of a single-chain adsorption on the substrate. Mean-field theory is quite successful, but in some cases fluctations in the local monomer concentration play an important role. Adsorption from more concentrated solutions offers rather complex and rich density profiles, with several regimes (proximal, central, distal).

Each regime is characterized by a different physical behavior. We have reviewed the principal theories used to model the polymer behavior. We also mentioned briefly more recent ideas about the statistics of polymer loops and tails.

The second part of this review was about polymers which are terminally grafted on one end to the surface and are called polymer brushes. The theories here are quite different since the statistics of the grafted layer depend crucially on the fact that the chain is not attracted to the surface but is forced to be in contact with the surface since one of its ends is chemically or physically bonded to the surface. Here as well we reviewed the classical mean-field theory and more advanced theories giving the concentration profiles of the entire polymer layer as well as that of the polymer free ends.

We also discussed additional factors that have an effect on the polymer adsorption and grafted layers: the quality of the solvent, undulating and flexible substrates such as fluid/fluid interfaces or lipid membranes; adsorption and grafted layer of charged polymers (polyelectrolytes); and adsorption and grafting on curved surfaces such as spherical colloidal particles.

Although our main aim was to review the theoretical progress in this field, we mentioned many relevant experiments. In this active field several advanced experimental techniques are used to probe adsorbed or grafted polymer layers: neutron scattering, small-angle high-resolution X-ray scattering, light scattering using fluorescent probes, ellipsometry, and surface isotherms as well as use of the surface-force apparatus to measure forces between two surfaces.

The aim of this chapter was to review the wealth of knowledge on how flexible macromolecules such as linear polymer chains behave as they are adsorbed or grafted to a surface (like an oxide). This chapter should be viewed as a general introduction to these phenomena. Although the chapter does not offer any details about specific oxide/polymer systems, it can serve as a starting point to understanding more complex systems such as encountered in applications and real-life experiments.

ACKNOWLEDGMENTS

We would like to thank I. Borukhov and H. Diamant for discussions and comments. One of us (DA) would like to acknowledge partial support from the Israel Science Foundation founded by the Israel Academy of Sciences and Humanities – Centers of Excellence Program, the Israel–US Binational Science Foundation (BSF) under grant no. 98-00429, and the Tel Aviv University Basic Research Fund. This work was completed during a visit by both of us to the Institute for Theoretical Physics, University of California, Santa Barbara (UCSB).

REFERENCES

1. MA Cohen Stuart, T Cosgrove, B Vincent. Adv Colloid Interface Sci 24:143–239, 1986.
2. PG de Gennes. Adv Colloid Interface Sci 27:189–209, 1987.
3. I Szleifer. Curr Opin Colloid Interface Sci 1:416–423, 1996.
4. GJ Fleer, FAM Leermakers. Curr Opin Colloid Interface Sci 2:308–314, 1997.
5. GJ Fleer, MA Cohen Stuart, JMHM Scheutjens, T Cosgrove, B Vincent. Polymers at Interfaces. London: Chapman & Hall, 1993.
6. E Eisenriegler. Polymers near Surfaces. Singapore: World Scientific, 1993.

7. AYu Grosberg, AR Khokhlov. Statistical Physics of Macromolecules. New York: AIP Press, 1994.
8. PJ Flory. Principles of Polymer Chemistry. Ithaca, NY: Cornell University, 1953.
9. H Yamakawa. Modern Theory of Polymer Solutions. New York: Harper & Row, 1971.
10. DH Napper. Polymeric Stabilization of Colloidal Dispersions. London: Academic Press, 1983.
11. PG de Gennes. Scaling Concepts in Polymer Physics. Ithaca, NY: Cornell University, 1979.
12. J des Cloizeaux, J Jannink. Polymers in Solution. Oxford: Oxford University, 1990.
13. HL Frish, R Simha, FR Eirish. J Chem Phys 21:365, 1953; R Simha, HL Frish, FR Eirish. J Phys Chem 57:584, 1953.
14. A Silberberg. J Phys Chem 66:1872, 1962; 66:1884, 1962.
15. SF Edwards. Proc Phys Soc (London) 85:613, 1965; 88:255, 1966.
16. EA DiMarzio. J Chem Phys 42:2101, 1965; EA DiMarzio, FL McCrackin. J Chem Phys 43:539, 1965; C Hoeve, EA DiMarzio, P Peyser. J Chem Phys 42:2558, 1965.
17. RJ Rubin. J Chem Phys 43:2392, 1965.
18. IS Jones, P Richmond. J Chem Soc, Faraday Trans 2:73, 1977.
19. JN Israelachvili. Intermolecular and Surface Forces. London: Academic Press, 1992.
20. M Daoud, PG de Gennes. J Phys (France) 38:85, 1977.
21. PG de Gennes. Rep Prog Phys 32:187, 1969.
22. PJ Flory. Statistical Mechanics of Chain Molecules. Munich: Hanser Press, 1988.
23. JMJ van Leeuwen, HJ Hilhorst. Physica A 107:319, 1981; TW Burkhardt. J Phys A 14:L63, 1981; DM Kroll. Z Phys B 41:345, 1981.
24. R Lipowsky, A Baumgärtner. Phys Rev A 40:2078, 1989; R Lipowsky. Physica Scripta T29:259, 1989.
25. RR Netz. Phys Rev E 51:2286, 1995.
26. OV Borisov, EB Zhulina, TM Birshtein. J Phys II (France) 4:913, 1994.
27. X Châtellier, TJ Senden, JF Joanny, JM di Meglio. Europhys Lett 41:303, 1998.
28. JF Joanny. J Phys II France 4:1281, 1994; AV Dobrynin, M Rubinstein, JF Joanny. Macromolecules 30:4332, 1997.
29. RR Netz, JF Joanny. Macromolecules 31:5123, 1998.
30. PG de Gennes. Phys Lett A 38:339, 1972.
31. E Eisenriegler, K Kremer, K Kinder. J Chem Phys 77:6296–6320, 1982; E Eisenriegler. J Chem Phys 79:1052–1064, 1983.
32. PG de Gennes. J Phys (Paris) 37:1445–1452, 1976.
33. PG de Gennes, P Pincus. J Phys Lett (Paris) 44:L241–L246, 1983.
34. E Bouchaud, M Daoud. J Phys (Paris) 48:1991–2000, 1987.
35. JW Cahn, JE Hilliard. J Chem Phys 28:258–267, 1958.
36. PG de Gennes. Macromolecules 14:1637–1644, 1981.
37. O Guiselin. Europhys Lett 17:225–230, 1992.
38. L Auvray, JP Cotton. Macromolecules 20:202, 1987.
39. LT Lee, O Guiselin, B Farnoux, A Lapp. Macromolecules 24:2518, 1991; O Guiselin, LT Lee, B Farnoux, A Lapp. J Chem Phys 95:4632, 1991; O Guiselin. Europhys Lett 1:57, 1992.
40. JMHM Scheutjens, GJ Fleer, MA Cohen Stuart. Colloids Surfaces 21:285, 1986; GJ Fleer, JMHM Scheutjens, MA Cohen Stuart. Colloids Surfaces 31:1, 1988.
41. A Johner, JF Joanny, M Rubinstein. Europhys Lett 22:591, 1993.
42. AN Semenov, JF Joanny. Europhys Lett 29:279, 1995; AN Semenov, J Bonet-Avalos, A Johner, JF Joanny. Macromolecules 29:2179, 1996; A Johner, J Bonet-Avalos, CC van der Linden, AN Semenov, JF Joanny. Macromolecules 29:3629, 1996.
43. AN Semenov, JF Joanny, A Johner. In: A Grosberg, ed. Theoretical and Mathematical Models in Polymer Research. Boston: Academic Press, 1998.
44. JF Joanny, L Leibler, PG de Gennes. J Polym Sci: Polym Phys Ed 17:1073–1084, 1979.

45. PG de Gennes. Macromolcules 15:492–500, 1982.
46. JMHM Scheutjens, GJ Fleer. Macromolecules 18:1882, 1985; GJ Fleer, JMHM Scheutjens. J Coll Interface Sci 111:504, 1986.
47. J Bonet Avalos, JF Joanny, A Johner, AN Semenov. Europhys Lett 35:97, 1996; J Bonet Avalos, A Johner, JF Joanny. J Chem Phys 101:9181, 1994.
48. J Klein, PF Luckham. Nature 300:429, 1982; Macromolecules 17:1041–1048, 1984.
49. J Klein, PF Luckham. Nature 308:836, 1984; Y Almog, J Klein. J Colloid Interface Sci 106:33, 1985.
50. G Rossi, PA Pincus. Europhys Lett 5:641, 1988; Macromolecules 22:276–283, 1989.
51. J Klein. Nature 288:248, 1980; J Klein, PF Luckham. Macromolecules 19:2007–2010, 1986.
52. J Klein, P Pincus. Macromolecules 15:1129, 1982; K Ingersent, J Klein, P Pincus. Macromolecules 19:1374–1381, 1986.
53. K Ingersent, J Klein, P Pincus. Macromolecules 23:548–560, 1990.
54. PF Luckham, J Klein. Macromolecules 18:721–728, 1985.
55. J Klein, G Rossi. Macromolecules 31:1979–1988, 1998.
56. I Borukhov, D Andelman, H Orland. Eur Phys J B 5:869–880, 1998.
57. MA Cohen Stuart. J Phys (France) 49:1001–1008, 1988.
58. MA Cohen Stuart, GJ Fleer, J Lyklema, W Norde, JMHM Scheutjens. Adv Colloid Interface Sci 34:477–535, 1991.
59. JL Barrat, JF Joanny. Europhys Lett 24:333, 1993.
60. RR Netz, H Orland. Eur Phys J B 8:81, 1999.
61. RR Netz, JF Joanny. Macromolecules 32:9013, 1999.
62. I Borukhov, D Andelman, H Orland. Europhys Lett 32:499, 1995; Macromolecules 31:1665–1671, 1998; J Phys Chem B 103:5042–5057, 1999.
63. G Decher. Science 277:1232, 1997; M Lösche, J Schmitt, G Decher, WG Bouwman, K Kjaer. Macromolecules 31:8893, 1998.
64. E Donath, GB Sukhorukov, F Caruso, SA Davis, H Möhwald. Angew Chem Int Ed 16:37, 1998; GB Sukhorukov, E Donath, SA Davis, H Lichtenfeld, F Caruso, VI Popov, H Möhwald. Polym Adv Technol 9:759, 1998.
65. F Caruso, RA Caruso, H Möhwald. Science 282:1111, 1998; F Caruso, K Niikura, DN Furlong, Y Okahata. Langmuir 13:3422, 1997.
66. JF Joanny. Eur Phys J B 9:117, 1999.
67. HA van der Schee, J Lyklema. J Phys Chem 88:6661, 1984; J Papenhuijzen, HA van der Schee, GJ Fleer. J Colloid Interface Sci 104:540, 1985; OA Evers, GJ Fleer, JHMH Scheutjens, J Lyklema. J Colloid Interface Sci 111:446, 1985; HGM van de Steeg, MA Cohen Stuart, A de Keizer, BH Bijsterbosch. Langmuir 8:8, 1992.
68. P Linse. Macromolecules 29:326, 1996.
69. M Muthukumar. J Chem Phys 86:7230, 1987.
70. R Varoqui, A Johner, A Elaissari. J Chem Phys 94:6873, 1991; R Varoqui. J Phys II (France) 3:1097, 1993.
71. PG de Gennes. J Phys Chem 94:8407, 1990.
72. D Andelman, JF Joanny. Macromolecules 24:6040–6041, 1991; JF Joanny, D Andelman. Makromol Chem Macromol Symp 62:35–41, 1992; D Andelman, JF Joanny. J Phys II (France) 3:121–138, 1993.
73. V Aharonson, D Andelman, A Zilman, PA Pincus, E Raphaël. Physica A 204:1–16, 1994; 227:158–160, 1996.
74. RR Netz, D Andelman, H Orland. J Phys II (France) 6:1023–1047, 1996.
75. X Châtellier, D Andelman. Europhys Lett 32:567–572, 1995; X Châtellier, D Andelman. J Phys Chem 22:9444–9455, 1996.
76. TA Vilgis, G Heinrich. Macromolcules 27:7846, 1994; G Huber, TA Vilgis. Eur Phys J B 3:217, 1998.

77. D Hone, H Ji, PA Pincus. Macromolecules 20:2543, 1987; H Ji, D Hone. Macromolecules 21:2600, 1988.
78. M Blunt, W Barford, R Ball. Macromolcules 22:1458, 1989.
79. CM Marques, JF Joanny. J Phys (France) 49:1103, 1988.
80. JT Brooks, CM Marques, ME Cates. Europhys Lett 14:713, 1991; J Phys II (France) 1:673, 1991; F Clement, JF Joanny. J Phys II (France) 7:973, 1997.
81. F von Goeler, M Muthukumar. J Chem Phys 100:7796, 1994.
82. T Wallin, P Linse. Langmuir 12:305, 1996; J Phys Chem 100:17873, 1996; J Phys Chem B 101:5506, 1997.
83. E Gurovitch, P Sens. Phys Rev Lett 82:339, 1999.
84. EM Mateescu, C Jeppesen, P Pincus. Europhys Lett 46:493, 1999.
85. RR Netz, JF Joanny. Macromolecules 32:9026, 1999.
86. A Halperin, JF Joanny. J Phys II (France) 1:623, 1991.
87. CM Marques, JF Joanny. Macromolecules 23:268, 1990; B van Lent, JMHM Scheutjens. J Phys Chem 94:5033, 1990; JP Donley, GH Fredrickson. Macromolecules 27:458, 1994.
88. P Auroy, L Auvray, L Leger. Phys Rev Lett 66:719, 1991; Macromolecules 24:2523, 1991; Macromolecules 24:5158, 1991.
89. HJ Taunton, C Toprakcioglu, LJ Fetters, J Klein. Nature 332:712, 1988; Macromolecules 23:571, 1990.
90. JB Field, C Toprakcioglu, L Dai, G Hadziioannou, G Smith, W Hamilton. J Phys II (France) 2:2221, 1992.
91. CM Marques, JF Joanny, L Leibler. Macromolecules 21:1051, 1988. CM Marques, JF Joanny. Macromolecules 22:1454, 1989.
92. MS Kent, LT Lee, B Farnoux, F Rondelez. Macromolecules 25:6240, 1992; MS Kent, LT Lee, BJ Factor, F Rondelez, GS Smith. J Chem Phys 103:2320, 1995; HD Bijsterbosch, VO de Haan, AW de Graaf, M Mellema, FAM Leermakers, MA Cohen Stuart, AA van Well. Langmuir 11: 4467, 1995; MC Fauré, P Bassereau, MA Carignano, I Szleifer, Y Gallot, D Andelman. Eur Phys J B 3:365, 1998.
93. R Teppner, M Harke, H Motschmann. Rev Sci Instrum 68:4177, 1997; R Teppner, H Motschmann. Macromolecules 31:7467, 1998.
94. PG de Gennes. Macromolecules 13:1069, 1980.
95. S Alexander. J Phys (France) 38:983, 1977.
96. A Halperin, M Tirell, TP Lodge. Adv Polym Sci 100:31, 1992.
97. T Cosgrove, T Heath, B van Lent, F Leermakers, J Scheutjens. Macromolecules 20:1692, 1987.
98. M Murat, GS Grest. Macromolecules 22:4054, 1989; A Chakrabarti, R Toral. Macromolecules 23:2016, 1990; PY Lai, K Binder. J Chem Phys 95:9288, 1991.
99. AN Semenov. Sov Phys JETP 61:733, 1985.
100. ST Milner, TA Witten, ME Cates. Europhys Lett 5:413, 1988; Macromolecules 21:21610, 1988; ST Milner. Science 251:905, 1991.
101. AM Skvortsov, IV Pavlushkov, AA Gorbunov, YB Zhulina, OV Borisov, VA Pryamitsyn. Polym Sci 30:1706, 1988.
102. RR Netz, M Schick. Europhys Lett 38:37, 1997; Macromolecules 31:5105, 1998.
103. C Seidel, RR Netz. Macromolecules 33:634, 2000.
104. MA Carignano, I Szleifer. J Chem Phys 98:5006, 1993; J Chem Phys 100:3210, 1994; Macromolecules 28:3197, 1995; J Chem Phys 102:8662, 1995. A detailed summary of tethered layers is given in I Szleifer, MA Carignano. Adv Chem Phys XCIV:165, 1996.
105. GS Grest. Macromolecules 27:418, 1994.
106. TA Witten, PA Pincus. Macromolecules 19:2509, 1986; EB Zhulina, OV Borisov, VA Priamitsyn. J Colloid Surface Sci 137:495, 1990.
107. ST Milner. Europhys Lett 7:695, 1988.

108. ST Milner, TA Witten, ME Cates. Macromolecules 22:853, 1989.
109. A Halperin. J Phys (France) 49:547, 1988; YB Zhulina, VA Pryamitsyn, OV Borisov. Polym Sci 31:205, 1989; EB Zhulina, OV Borisov, VA Pryamitsyn, TM Birshtein. Macromolecules 24:140, 1991; DRM Williams. J Phys II (France) 3:1313, 1993.
110. P Auroy, L Auvray. Macromolecules 25:4134, 1992.
111. JF Marko. Macromolecules 26:313, 1993.
112. PY Lai, K Binder. J Chem Phys 97:586, 1992; GS Grest, M Murat. Macromolecules 26:3108, 1993.
113. M Daoud, JP Cotton. J Phys (France) 43:531, 1982.
114. RC Ball, JF Marko, ST Milner, AT Witten. Macromolecules 24:693, 1991; H Li, TA Witten. Macromolecules 27:449, 1994.
115. N Dan, M Tirrell. Macromolecules 25:2890, 1992.
116. M Murat, GS Grest. Macromolecules 24:704, 1991.
117. ST Milner, TA Witten. J Phys (France) 49:1951, 1988.
118. TM Allen, C Hansen, F Martin, C Redemann, A Yau-Young. Biochim Biophys Acta 1006:29, 1991; MJ Parr, SM Ansell, LS Choi, PR Cullis. Biochim Biophys Acta 1195:21, 1994.
119. R Lipowsky. Europhys Lett 30:197, 1995; C Hiergeist, R Lipowsky. J Phys II (France) 6:1465, 1996; C Hiergeist, VA Indrani, R Lipowsky. Europhys Lett 36:491, 1996; M Breidenich, RR Netz, R Lipowsky. Europhys Lett 49:431, 2000.
120. M Aubouy, GH Fredrickson, P Pincus, E Raphael. Macromolecules 28:2979, 1995.
121. AP Gast, L Leibler. Macromolecules 19:686, 1986.
122. P Pincus. Macromolecules 24:2912, 1991; OV Porisov, TM Birshtein, EB Zhulina. J Phys II (France) 1:521, 1991; F Csaijka, RR Netz, C Seidel. Eur Phys J E 4:505, 2001.
123. JF Marko, TA Witten. Phys Rev Lett 66:1541, 1991.
124. G Brown, A Chakrabarti, JF Marko. Macromolecules 28:7817, 1995; EB Zhulina, C Singhm, AC Balazs. Macromolecules 29:8254, 1996.
125. A Halperin, S Alexander. Europhys Lett 6:329, 1988; A Johner, JF Joanny. Macromolecules 23:5299, 1990; C Ligoure, L Leibler. J Phys (France) 51:1313, 1990; ST Milner. Macromolecules 25:5487, 1992; A Johner, JF Joanny. J Chem Phys 98:1647, 1993.

3
Adsorption Kinetics of Polymeric Molecules

MARTINUS A. COHEN STUART and ARIE DE KEIZER Wageningen University, Wageningen, The Netherlands

I. INTRODUCTION

From a molecular dynamics perspective, the solid/liquid interface is a peculiar world. On one side of the interface there is the liquid with typical relaxation times in the nano- or even pico-second range and on the other side there is the solid phase where molecular translation is virtually excluded. Without more information, all one can say about a molecule at the interface is that its dynamic state will be intermediate between these two extremes, which does not really provide much insight. One anticipates, perhaps, that the polymer–surface interaction potentials will play a most important role.

During the last decades the theory of thermodynamic equilibrium of polymers at interfaces has been elaborated in detail [1–4], but a well-developed theory of adsorption kinetics, including a description of relaxation processes in the adsorbed layer, does not exist except for the part played by transport in solution. For this reason we approach the subject of the present chapter, the kinetics of polymer adsorption on oxide surfaces, primarily from an empirical direction. We have carefully selected literature data from which, in our opinion, we can gain insight into the rate of the various processes by which adsorbed layers are built up or broken down. In order to discuss these data from a proper perspective, we first review some basic theoretical elements as applicable to adsorption kinetics, and we discuss some published theoretical work (e.g., Monte Carlo simulations). Experimental techniques suited for measurements of adsorption kinetics are also briefly reviewed. The remainder of the chapter is devoted to discussing measurements of adsorption from simple solutions, desorption by flushing, and several variations of polymer/polymer exchange. Some special effects of charges on the polymer, or of confined geometries, on the kinetics of polymer adsorption or exchange are treated in separate sections.

II. THEORETICAL CONSIDERATIONS

A. Polymer Adsorption Kinetics: Outline

Adsorption is the process whereby matter dispersed in a solution accumulates at an interface. The adsorption kinetics of any substance, be it a dissolved polymer, an ion, or a colloidal particle can, therefore, be described in very similar terms, and there is a good deal of similarity between the literature dealing with electrochemical reactions at electrodes [5], deposition of colloidal particles on surfaces [6–8], and adsorption of, say, surfactants or polymers. There are always two essential steps, namely, (1) *transport* towards the surface, and (2) *attachment*. Once the molecules are attached to the surface, they may undergo further changes specific for the system at hand.

In the present discussion we deal with polymers. These are flexible, deformable objects. One therefore has to consider one additional aspect, namely, *changes in shape* upon adsorption (unfolding, spreading). For the reverse pathway (desorption) analogous steps ("coiling up", detachment) must be considered. In exchange processes, where one species adsorbs and another desorbs, one has all these processes occurring; in addition the surface processes (unfolding and coiling up) may be coupled in some way. In this section, we first review theory (as far as available) that is relevant to each of the subprocesses mentioned.

B. Transport

A consideration of systems out of equilibrium, in which material fluxes occur, has to start with the equation of continuity. For each component i there is such an equation [9]:

$$\frac{\mathrm{d}c_i}{\mathrm{d}t} = -\bar{\nabla}\cdot\bar{J}_i \tag{1}$$

where c_i and t denote local concentration of species i and time, respectively, and $\bar{J}_i$ is the flux of that species. We will consider three contributions to the flux, namely, those due to convection, those due to diffusion, and those coming from "chemical" processes (association, dissociation) in the solution:

$$\bar{J}_i = \bar{J}_{i,\mathrm{conv}} + \bar{J}_{i,\mathrm{diff}} + \bar{J}_{i,\mathrm{source}} \tag{2}$$

The convective contribution to $\bar{J}_i$ is included in the form of a term $c_i\bar{v}$, where $\bar{v}$ is the local velocity. The diffusive term comes from Fick's law:

$$\bar{J}_{i,\mathrm{diff}} = -D_i\bar{\nabla}c_i \tag{3}$$

where D_i is the diffusion coefficient of component i and $\bar{\nabla}$ denotes the gradient operator, in Cartesian coordinates

$$\bar{i}\frac{\partial}{\partial x} + \bar{j}\frac{\partial}{\partial y} + \bar{k}\frac{\partial}{\partial z}.$$

Sometimes, an adsorbing substance is present in the form of various species, e.g., as single molecules and in an associated form, e.g., in aggregates or micelles. If that is the case, the various transport equations are coupled through the association–dissociation equilibria by several "source" terms in them. These are generally written as chemical reactions, i.e., as $J_{\mathrm{source}} = \sum_i k_i c_i^n$, where k_i is a rate constant for the process (association, dissociation, etc.) and n is the order of that process.

Solving the appropriate set of equations for the system under consideration will eventually yield $\bar{J}_i$ as a function of time and position. For the remainder of our discussion we shall consider simple plane interfaces, where $\bar{J}_i$ is a function of the distance z normal to the interface only. The rate of adsorption $d\Gamma_i/dt$ can then be found simply as

$$\frac{d\Gamma_i}{dt} = J_{i,z=0} = D_i\left(\frac{\partial c_i}{\partial z}\right)_{z=0} \tag{4}$$

Polymers diffuse relatively slowly, but adsorb often very strongly, so that saturation is already possible from a very low concentration. Because, at low concentration, transport is very slow and often rate limiting, convection can be very effective in enhancing the rate.

An important question is: what boundary conditions hold? Of course, far from the interface we have the imposed bulk concentration c^b. Another boundary condition can be imposed at $z = 0$; the usual approach is to define a *subsurface concentration* c^s of free molecules near the surface. This is essentially the concentration arrived at upon extrapolation to the surface of the concentration profile $c(z)$: $c^s = \lim_{z\to 0} c(z)$. Of course, $z = 0$ coincides with the surface itself so that, strictly speaking, one cannot speak of free molecules at $z = 0$. Actually, one should consider the concentration at $z = R$, where R is the size of a polymer coil. In practical situations, however, R is so small compared to the thickness of the stagnant layer that this difference is of no consequence for c^s.

For adsorption from a dilute solution, the characteristic time to saturate the surface usually becomes long compared to the time needed to establish a steady-state diffusion layer. This implies that local concentrations become time independent, and $\sum_i \bar{\nabla} \cdot \bar{J}_i = 0$, hence $\sum_i \bar{J}_i$ is independent of z. If we have a single adsorbing species we can then write:

$$\frac{d\Gamma}{dt} = k^{tr}(c^b - c^s) = \frac{J_0}{c^b}(c^b - c^s) \tag{5}$$

where k^{tr} is the transport coefficient, and the *limiting flux* $J_0 = k^{tr}c^b$ is the flux obtained when c^s is zero, i.e., usually at the beginning of the adsorption process when the surface, to a large extent, is still empty. We assume the absence of source terms. The transport coefficient $k^{tr} = J_0/c^b$ depends on the diffusion coefficient D and the geometry of the flow field; it is a constant for a given experimental set-up and chosen adsorbate/solvent system. Some useful flow geometries are discussed in some detail in Section III.B, and expressions for the transport coefficient k^{tr} are listed.

C. Attachment

At the surface, molecules can attach or detach. This gives rise to two fluxes, one forward, $(d\Gamma/dt)|_+$, and one backward, $(d\Gamma/dt)|_-$; the net flux (adsorption rate) equals:

$$\frac{d\Gamma}{dt} = \left.\frac{d\Gamma}{dt}\right|_+ - \left.\frac{d\Gamma}{dt}\right|_- \tag{6}$$

The forward flux will generally respond to the subsurface concentration c^s and the fraction of unoccupied surface area $1 - \theta$. If the adsorbed molecules do not spread after initial adsorption, i.e., the effective molar area is constant, we can write as a first approximation:

$$\left.\frac{d\Gamma}{dt}\right|_{+} = k^{a}(1-\theta)c^{s} \tag{7}$$

where k^a (in m s^{-1}) is an adsorption rate coefficient, and $\theta = \Gamma/\Gamma_m$, Γ_m being the amount of adsorbed polymer per unit area at saturation. We note in passing that for polymers k^a varies, as a rule, with coverage Γ. As a first approximation the backward flux will be proportional to Γ. At equilibrium, the net rate must be zero, and the subsurface concentration c^s can be identified with the equilibrium concentration (c^{eq}) corresponding to the actual value of Γ. It follows that

$$\left.\frac{d\Gamma}{dt}\right|_{+} = \left.\frac{d\Gamma}{dt}\right|_{-} = k^{d}\theta = k^{a}(1-\theta)c^{eq} \tag{8}$$

with k^d (in mol m^{-2} s^{-1}) the desorption rate constant. The ratio between the adsorption and desorption rate constants is equal to the equilibrium constant for adsorption, $K = k^a/k^d$ (in mol^{-1} m^3). Off equilibrium, the value of $(d\Gamma/dt)|_{-}$ is still given by $k^a(1-\theta)c^{eq}$ provided that $(d\Gamma/dt)|_{-}$ is uniquely determined by θ. Hence, we can write for the net adsorption rate:

$$\frac{d\Gamma}{dt} = k^{a}(1-\theta)(c^{s} - c^{eq}) \tag{9}$$

For small molecules it is often assumed that both attachment and detachment are so rapid (i.e., k^a and k^d are large) that there is almost equilibrium between the adsorbed layer and the subsurface concentration throughout the adsorption process ("local equilibrium") [10]. This means that the equilibrium adsorption isotherm determines the value of c^s, namely, $c^s \approx c^{eq}(\Gamma)$, and that other kinetic factors can be entirely ignored. For polymers, however, this cannot be generally supposed.

Combining Eqs (5) and (9) we arrive at a simple, yet fairly general adsorption rate equation:

$$\frac{d\Gamma}{dt} = \frac{c^{b} - c^{eq}}{\dfrac{1}{k^{tr}} + \dfrac{1}{k^{a}(1-\theta)}} \tag{10}$$

We have to be aware that for complicated deformable molecules such as polymers, k^a may depend on θ. As a rule of thumb, we can say that polymer adsorption isotherms are characterized by a very high affinity, represented by a high value of K. The equilibrium concentration is extremely (undetactably) small up to values very close to saturation, then it suddenly shoots up very steeply [1] and c^{eq} becomes equal to c^b. This implies that the rate of adsorption drops strongly near saturation, so that the kinetic curve has a sharp break (see Fig. 1). As long as we are not very close to saturation we can set c^{eq} equal to zero and rewrite Eq. (10) in a linearized form:

$$\frac{c_b/\Gamma_m}{d\theta/dt} = \frac{1}{k^{tr}} + \frac{1}{k^{a}(1-\theta)} \tag{11}$$

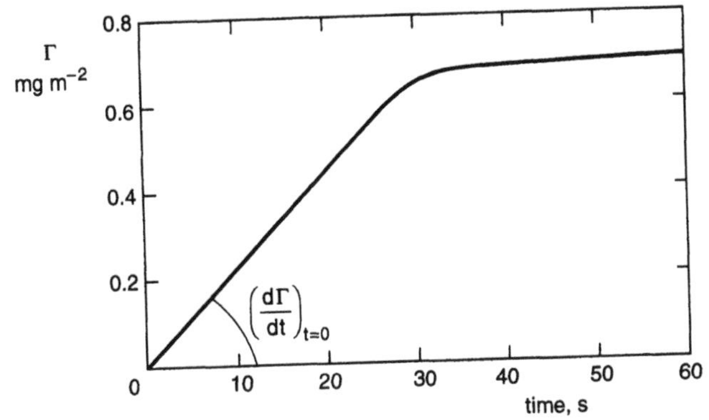

FIG. 1 Typical example of the adsorption kinetics of a polymer from an aqueous solution on to a flat surface using stagnation-point flow.

For the case where k^a is strictly constant, the adsorption rate equation, Eq. (11), can be easily integrated:

$$\frac{\theta}{k^{tr}} - \frac{\ln(1-\theta)}{k^a} = \frac{c_b}{\Gamma_m} t \tag{12}$$

It is perhaps instructive to explore some consequences of Eqs (10) and (11). If $k^{tr} \ll k^a(1-\theta)$ we can neglect the second term on the right-hand side of Eq. (11). The transport to the subsurface fully determines the net adsorption rate independent of θ. For polymer adsorption the initial adsorption rate is often fully determined by transport, and k^{tr} can be obtained directly from this initial rate. If $k^{tr} \gg k^a(1-\theta)$ the adsorption process is determined by the rate of attachment and thus depends on the degree of coverage θ. The value of k^a can be obtained as a function of Γ using Eq. (11). If the adsorption is not of an extremely high affinity (e.g., in the case of short polymers) the equilibrium concentration is relevant and Eq. (10) applies.

What can we conclude about the rate of desorption? Polymer adsorption is characterized by an extremely high value of the equilibrium constant K. As for physical reasons the attachment rate k^a cannot be extremely high, it follows that k^d must be extremely low. An interesting consequence is that the desorption rate is very small, even for flushing with pure solvent ($c^b = 0$). Also, the equilibrium concentration will be almost zero for most of the isotherm.

Close to saturation the value for the equilibrium concentration will be significant and will approach c^b. One would expect the rate of desorption to be negligible because the probability of detachment of *all* adsorbed segments is extremely low. However, in this case a different mechanism for desorption based on self-exchange of the polymer molecules will occur. Detached segments of the desorbing molecules are replaced by segments of the adsorbing molecule. Accordingly, the rate of exchange between like molecules increases with the concentration of molecules in the subsurface. The constant k^d will be higher, resulting in finite (and equal) rates of adsorption and desorption. A *net* flux will be absent according to Eq. (10), realizing that $(c^b - c^{eq}) \to 0$. Self-exchange will be treated in more detail in Section II.E.

If adsorption occurs, it means that there is some kind of attraction between monomer units and the surface. If no other interactions are present, one would expect the attachment to be rapid [$k^a(1-\theta)$ is large], in particular as long as the surface is (largely) empty. As an adsorbed layer becomes more dense ($1-\theta$ approaching zero), however, it may itself form a barrier to incoming molecules and slow down their attachment rate.

A rather general way to calculate the effect of interactions on the rate of a process is by the theory of Kramers [11]. According to this theory, the rate at which barrier crossing occurs depends on the equilibrium shape of the barrier $U(z)$:

$$k^a = \frac{D}{(1-\theta)\int_{\infty}^{0} dz \exp(-U(z)/kT)} \tag{13}$$

(In contrast to previous treatments [12, 13], Eq. (13) emphasizes the effect of the degree of coverage on the overall "resistance" by a separate factor $1-\theta$.) For a rigid object (such as, e.g., a colloidal particle), $U(z)$ can simply be taken to be the (Helmholtz) energy of interaction at separation z. However, a polymer is flexible and can adjust its shape to the local field. We should, therefore, average over all allowed conformations of a polymer molecule that has an arbitrary segment at the distance of closest approach z. In other words, one has to evaluate the partition function $Q(z)$ of a chain at a distance z from the surface, in the imposed field of the surface:

$$k^a = \frac{D}{(1-\theta)\int_{\infty}^{0} dz \frac{Q(z)}{Q(\infty)}} \tag{14}$$

Theoretical considerations of k^a suggest that adsorbed layers of homopolymers are usually sufficiently "transparent" so that attachment is not likely to be the rate-limiting step [note, however, that we need to compare $k^a(1-\theta)c^b$ to J_0] [14]. For dense "brushes" formed by block copolymers this may be different [11, 15, 16].

A rather special case is that of polyelectrolytes. Here, we have electrostatic interactions acting both between pairs of molecules, and between one molecule and the surface. In aqueous solutions of low ionic strength, these interactions may extend over tens of nanometers. An electrostatic barrier may, therefore, impede adsorption considerably, just like the charge on colloidal particles prohibits their aggregation. We should also realize that such an electrostatic barrier may well build up as a consequence of the adsorption of the polyelectrolyte itself. Kinetic *limitation* is, therefore, very likely to occur in these systems. An attempt to estimate the height of the electrostatic barrier as it builds up during polyelectrolyte adsorption, using Kramers' theory, has been recently made for the case of a simple constant-charge surface and a constant-charge polymer [13]. This treatment is approximate as it neglects complications that may occur in real systems and that may strongly influence the electrostatic effect, such as charge regulation of the surface (typical for many inorganic oxides), charge regulation of the polymer, inhomogeneity of the surface due to partial coverage with polymer, specific ion binding, etc.

D. Conformational Changes

The rate at which molecules change their conformation (shape) after attachment cannot easily be studied experimentally. Yet, it seems rather obvious that such a process will occur. Given enough free area, the situation of lowest free energy of an adsorbed flexible chain will usually be that of a relatively flat ("pancake") structure, so that there is a driving force towards spreading. However, in the case of strong polymer/surface interactions, friction may also occur because the attached monomer units sit in deep potential wells with a low probability of detachment: monomers cannot "slide" over the surface but have to "hop" very slowly. One obvious consequence of the spreading process is that the available free area decreases. If each molecular "pancake" protects the area it covers against attachment by others, one anticipates a coupling between rate of attachment and rate of spreading. Furthermore, crowding on the surface may lead to competition between chains attempting to spread, which slows down these processes even more.

One may take a phenomenological approach and postulate that molecules, once attached, spread according to a suitably chosen rate equation, and that the area thus covered becomes inaccessible for adsorbing molecules. For example, Pefferkorn and Elaissari [17], and later van Eijk and coworkers [18, 19], explored the consequences of a "growing disk" model, leading to an exponential spreading law:

$$a(t;t') = a_0 + \alpha(1 - \mathrm{e}^{-(t-t')/\tau}) \tag{15}$$

where $a(t;t')$ is the area occupied by a molecule at time t and adsorbed at time t', a_o its value at first contact, and α its increment at residence time $t = \infty$; τ is the characteristic spreading time. From Eqs (11) and (15) we can derive the contribution of the molecules to the covered area fraction θ at a given time t that arrived during a time interval $\mathrm{d}t'$. $\Gamma_\mathrm{m} = a(t;t')^{-1}$ becomes a function of the spreading time.

$$\mathrm{d}\theta = \frac{k^\mathrm{a}(1-\theta(t'))c^\mathrm{b}a(t;t')\mathrm{d}t'}{1 + k^\mathrm{a}/k^\mathrm{tr}(1-\theta(t'))} \tag{16}$$

If $k^\mathrm{tr} \gg k^\mathrm{a}$ or $\theta \cong 1$ the denominator becomes unity. Integration over all molecules (i.e., over all arrival times t') gives the total occupied area fraction:

$$\theta(t) = k^\mathrm{a}c^\mathrm{b}\int_0^t \mathrm{d}t'(1-\theta(t'))\left[a_0 + \alpha(1 - \mathrm{e}^{-(t-t')/\tau})\right] \tag{17}$$

Equation (17) was also derived by van Eijk and Cohen Stuart [18] except that for the adsorption rate constant the transport rate constant was used. However, if adsorption is transport limited, no $1-\theta$ term should be introduced into Eq. (16) because the diffusion layer thickness is much larger than the distances between the empty surface patches. Differentiating Eq. (17) twice with respect to t, and introducing dimensionless variables ($1-\theta \to \beta$, $k^\mathrm{a}c^\mathrm{b}a_0\tau \to J$, $1+\alpha/a_0 \to \sigma$, $t/\tau \to t$), one can rewrite this equation as

$$\frac{\mathrm{d}^2\beta}{\mathrm{d}t^2} + (1+J)\frac{\mathrm{d}\beta}{\mathrm{d}t} + J\sigma\beta = 0 \tag{18}$$

which is equivalent to the equation which describes the motion of a strongly damped oscillator. Solving it, one finds that for a low attachment rate (low J, e.g., low c^b) the total adsorption rises gradually and a final adsorbed amount $(a_0 + \alpha)^{-1}$ is reached. At higher J the adsorption rises faster, and high adsorptions up to a_0^{-1} are reached. Should the molecules continue to spread even after the surface has reached full coverage, one expects that part of the molecules have to desorb. In other words, an *overshoot* is expected in the $\Gamma(t)$ curve.

Typical values for the spreading time range from tenths of seconds to 1 or 2 h [18]. From self-exchange experiments [20, 21] there is clear evidence that the relaxation process in the saturated layer proceeds for much longer times, typically from about 10 h to days. The main origin of this effect may be due to changes in the number of segment–surface contacts, heterogeneity of the surface, and/or entanglements. Some surface fractions may even adsorb apparently irreversibly. In Section IV some experimental results will be presented.

Recently, several other groups have tried to model the kinetics of polymer adsorption (and exchange) by taking into account diffusion and some aspects of reconformation of the adsorbed polymer at the surface [22–24].

In order to achieve some insight into rates of attachment and spreading, simulations would seem helpful. Indeed some groups have carried out Monte Carlo studies of attachment and reconformation processes [16, 25–31]. The obvious advantage of such simulations is that the rather complex influence of topological restrictions such as entanglements can (in principle) be realistically included in the model.

An extensive Monte Carlo study was carried out by Zajac and Chakrabarti [25]. In this study the surface friction was assumed to be small. As a result, the chains adsorbing first on the bare surface rapidly unfolded to flat structures with many surface contacts and few dangling loops or ends. Late arrivals, however, were forced to adopt looser, more weakly bound conformations. Hence, one can consider the layer as consisting of two populations, one with the high bound fraction and one rather loosely attached.

Recently, Hasegawa and Doi [32] extended the self-consistent field theory to dynamic problems. The evolution of the density profiles and the polymer conformation were calculated numerically; the results were qualitatively consistent with experimental observations.

E. Exchange Processes and Trapping

As we have seen above, rates of adsorption and desorption are for many polymers almost entirely controlled by transport in the solution adjacent to the adsorbent surface. Therefore, we do not learn much from simple adsorption and desorption studies about rates of processes occurring *in* the surface layer. A better way to study such processes is by maintaining saturation, i.e., by supplying molecules that can take the place of the ones that desorb. In other words, we should do exchange experiments. In such experiments, c^{eq} remains high and transport is fast so that we may hope that surface process(es) become(s) rate determining and, hence, measurable. Two cases may be considered: (1) the adsorbing and desorbing species are identical (apart from a label that allows one to distinguish them experimentally), "self-exchange", and (2) the adsorbing and desorbing polymers are different, either in molar mass or in chemical composition. In case (1), there is no adsorption preference; the surface excess free energy does not change during the process. Therefore, this method probes the dynamics of an adsorbed layer *as it would be in full equilibrium*. [In experiments, the criterion that the surface excess free energy remains strictly constant may be hard to meet. Very minor differences between the labeled and unlabeled species (e.g., deuterium labeling, or a minor difference in molar mass) may be enough to disturb the equilibrium and create an adsorption preference.] A review on the effects of competition and displacements in sequential polymer adsorption has been published by Kawaguchi [33].

Fu and Santore [21] proposed a theoretical description of the exchange rate for two identical polymers A and B (e.g., a labeled and unlabeled polymer) similar to the treatment of adsorption of a single polymer presented above. The fluxes are described by

$$J_A = k_A^{tr}(c_A^b - c_A^s) \qquad (19)$$
$$J_B = k_B^{tr}(c_B^b - c_B^s)$$

For two components A and B the kinetics of the exchange reaction at the interface can be described by

$$\frac{d\Gamma_B}{dt} = -\frac{d\Gamma_A}{dt} = k^{ex}\theta_A c_B^s - k^{ex'}\theta_B c_A^s \qquad (20)$$

The exchange rate constants are related by the equilibrium constant for the exchange reaction of A against B:

$$K_{B-A} = \frac{k^{ex}}{k^{ex'}} = \left[\frac{\Gamma_B c_A}{\Gamma_A c_B}\right]_{\text{equilibrium}} \qquad (21)$$

The fluxes must be equal to the exchange rates: $J_B = -J_A = d\Gamma_B/dt = -d\Gamma_A/dt$. Furthermore, if we assume that $c_A^b = 0$, $c_B^b = \text{constant}$, $k_A^{tr} = k_B^{tr} = k^{tr}$, and $\Gamma_A + \Gamma_B = \text{constant} = \Gamma_m$, it is possible to eliminate the subsurface concentrations of components A and B. The exchange rate can be written in an equation similar to Eq. (11) for the adsorption rate of a single adsorbing polymer:

$$\frac{c_B^b/\Gamma_m}{d\theta_B/dt} = \frac{1}{k^{tr}} + \frac{1}{k^{ex}(1-\theta_B)} + \frac{\theta_B}{(1-\theta_B)k^{tr}K_{B-A}} \qquad (22)$$

Upon integration of the resulting $d\theta_B/dt$, the following relation is obtained:

$$\frac{\theta_B}{k^{tr}} - \frac{\ln(1-\theta_B)}{k^{ex}} - \frac{\theta_B + \ln(1-\theta_B)}{k^{tr}K_{B-A}} = \frac{c_b t}{\Gamma_m} \qquad (23)$$

If the rate of transport to the subsurface (k^{tr}) is high, the exchange follows an exponential law:

$$1 - \theta_B = \exp(-k^{ex}c^b t/\Gamma_m)$$

Self-exchange of homopolymers has also been studied with Monte Carlo simulations [25]. One interesting result emerging from these simulations is that adsorbed layers of homopolymers, when exposed to a solution of the same polymers, can readily exchange attached molecules against free ones. However, not all molecules participate equally in this exchange process. There appears to be a relation between the conformation and the residence time, namely, the more flattened the molecules, the longer their residence times. This is logical, as desorption has to be preceded by a certain "coiling up" process, which is the more extensive and time consuming the further the molecule is away from the coiled state. The distribution of residence times was found to be bimodal with one peak at short residence times and one at long residence times.

In most simulations, the interaction between the monomer unit and the surface is modeled by a low potential near an ideally flat impenetrable surface. In other words, the friction is taken to be very low. This is plausible for polymers near a liquid surface where attached monomers will be mobile in in-plane directions; however, it is not very good for polymers on solids such as oxides, because here attachment is mainly to strongly localized sites on the surface. It would seem that this enhances the differences in dynamics between flat conformations and coiled conformations, but no studies of this effect are known.

In case (2) (exchange between unlike molecules) the lowering of the free energy is an extra driving force for the exchange process. As a first approximation, Eqs (22) and (23) can easily be extended for dissimilar polymeric molecules. This situation has also been considered by Zajac and Chakrabarti [25]. One observation in this study is that the adsorption rate of the displacer is initially higher than the desorption rate of weakly adsorbed chains, thus producing a small overshoot in the total adsorption. Once a certain amount of displacing chains is firmly attached, they spread out thereby producing loosely bound polymers, which subsequently desorb. Hence, the rate of desorption (displacement) depends strongly on the spreading rate of the displacing polymers. Yet many questions remain. Do incoming chains make one single first contact which grows out, thus expelling surrounding molecules, or do they make multiple initial contacts, thereby creating loops that also "trap" molecules that happen to be under such a loop? How stable are such loops? An attempt to take topological constraints ("surface reptation") into account in a theory of adsorption kinetics was made by Semenov and Joanny [14]. However, the polymer/surface friction was neglected. The question of surface trapping has been addressed experimentally; we discuss this in Section IV.

III. EXPERIMENTAL ASPECTS OF THE DYNAMICS OF POLYMER ADSORPTION

A. General

In principle, dynamic aspects of polymer adsorption can be determined with the same methods as one uses to characterize static properties of the adsorbed polymer layer. Fleer et al. [1] have presented an overview of experimental methods for the determination of adsorption isotherms, the adsorbed layer thickness, the bound fraction, and the volume fraction profile. However, in order to determine the dynamics of some property of the adsorbed polymer layer, the characteristic time of the experimental method should be shorter than that of the process investigated. Moreover, the geometry of the experimental system is often of crucial importance. These factors severely limit the applicability of some experimental methods. In this section we will particularly review those methods which have been successfully applied for characterizing the kinetics of polymer adsorption.

As described above, adsorption equilibrium is obtained if the rate of adsorption equals the rate of desorption. For a basic understanding of the dynamics of the adsorption process it is important to determine experimentally the rate of adsorption as well as that of desorption. For polymer adsorption the characterization of the kinetics differs in several aspects from the adsorption of low molar mass substances.

1. Polymers adsorb if some critical segmental interaction energy has been surpassed; if so, the adsorption has a high-affinity character. As a consequence, desorption by flushing the surface with the solvent typically results in an extremely slow desorption.
2. After attachment to a surface a polymer will unfold. The attachment is a relatively fast process, but unfolding requires several segments to detach simultaneously, and therefore it is a rather slow process.

The time scale for the kinetics of polymer adsorption, as found in experiments, ranges from seconds to days or even weeks. As it is not possible to establish the rate of desorption by simply lowering the solution concentration, an alternative, and fruitful, method to study the dynamics of the adsorbed polymer is to measure the *exchange* against other polymeric or monomeric molecules.

In the literature, several methods are presented for determining rates of adsorption, desorption, or exchange at oxidic solid–liquid interfaces (see Section D). The most powerful methods measure the adsorbed amount, or a quantity related to the adsorbed amount, "in situ." Before reviewing the different experimental methods, we will discuss the influence of the hydrodynamics and the properties of the solid–liquid interface.

B. Controlling Hydrodynamic Conditions

An important aspect of the kinetics of polymer adsorption is the transfer of the polymer from solution to the interface by convection and/or diffusion. At a bare surface, and in the absence of a barrier, the rate of adsorption is limited by mass transfer (see Section II.B). In this kind of experiment hydrodynamic conditions must be well defined in order to allow quantitative conclusions with respect to the kinetics of adsorption. However, for slow processes, like rearrangements in the polymer layer or slow polymer exchange experiments, the rate of mass transfer to the surface does not significantly contribute to the overall rate. In this case the design of the experimental cell is not critical and it suffices that the solution is stirred.

According to Eq. (5) the rate of mass transfer in the stationary state is proportional to the concentration gradient. The mass transfer coefficient k^{tr} depends on the flow-cell geometry and the hydrodynamic conditions. Adamczyk and coworkers [6–8] have analyzed a few well-defined situations and have given the corresponding expressions for k^{tr}. In Fig. 2 the flow patterns are given for three flow-cell geometries. The following expressions for k^{tr} have been derived:

$$k^{tr} = 0.776\bar{v}^{1/3} D^{2/3} \left(\frac{\alpha_{spf}}{R^2}\right)^{1/3} \quad \text{(stagnation-point flow)} \tag{24}$$

$$k^{tr} = 0.776\bar{v}^{1/3} D^{2/3} \left(\frac{1}{xb}\right)^{1/3} \quad \text{(flow through a narrow slab)} \tag{25}$$

$$k^{tr} = 0.855\bar{v}^{1/3} D^{2/3} \left(\frac{1}{xR}\right)^{1/3} \quad \text{(flow through a capillary)} \tag{26}$$

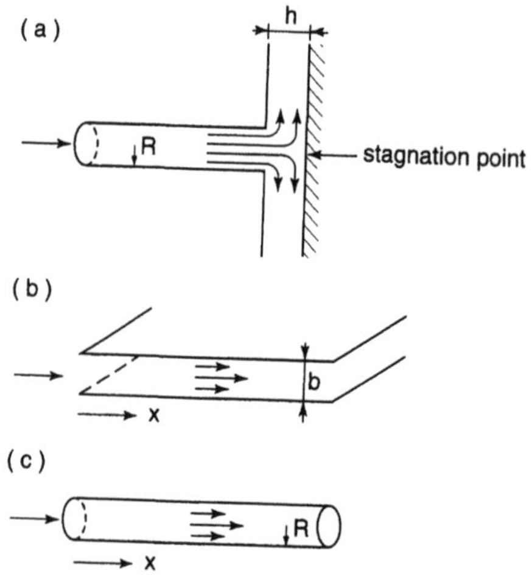

FIG. 2 Geometry of some flow cells with simple well-defined flow patterns: (a) impinging jet with a stagnation-point flow; (b) a slab; (c) a cylindrical capillary.

where $\bar{v}$ is the flow rate (for stagnation-point flow in the inlet tube), D is the diffusion constant of the polymer, R is the radius of the inlet tube [Eq. (24)] or the cylindrical capillary [Eq. (26)], b is the distance between the parallel plates, and x is the lateral position along the slab or the capillary. The dimensionless parameter α_{spf} reflects the intensity of the flow near the stagnation point. It depends on h/R and is roughly proportional to the Reynolds number, $Re = \bar{v}R/\nu$ (ν is the kinematic viscosity). It is interesting to see that the equations for the three geometries have a striking similarity.

In order to study the kinetics of adsorption or desorption with flow cells as described above, an in situ method is needed for the determination of the adsorbed mass or another property of the adsorbed molecules. A survey of the different methods is given in Section III.D.

In many cases, polymer adsorption is studied at the surface of dispersed particles of, e.g., insoluble oxides, or particles packed in a column. In this case it is very difficult to control convection and diffusion properly and to obtain reproducible results if mass transfer contributes significantly to the overall rate of adsorption. To obtain qualitatively significant trends it is necessary to choose carefully stirring or liquid flow conditions, to use particles with a well-defined geometry and a narrow size distribution, and to avoid experiments at very low concentrations. The adsorbed amount on dispersed particles is usually obtained by determining the amount taken up from solution at different time intervals. In principle, any technique for quantitative analysis of the polymer concentration in the supernatant can be applied in that case. Problems with separation of the supernatant and the dispersion are inherent for indirect methods.

C. Properties of the Solid–Liquid Interface

Kinetic studies of polymer adsorption on different substrates can be performed with dispersed materials or with ideally smooth surfaces. For systematic studies, well-defined flat surfaces like silicon wafers with an appropriate surface layer are preferred. If transport through the solution is rate limiting, hydrodynamic conditions must also be well defined.

In most model studies macroscopic *flat* homogeneous surfaces are used for studying the kinetics of polymer adsorption. Silicon wafers, glass, metal surfaces, single crystal surfaces, mica, etc., can serve as the substrate itself, or as a base for deposition of another kind of material. Roughness or heterogeneity may strongly influence the kinetics of reconformation or spreading on the surface and should, therefore, be controlled carefully. With flat surfaces only methods which directly measure a property of the adsorbed layer can be used, because the total surface area is too small to determine depletion from solution. For fast processes a direct method is always a prerequisite.

Although the macroscopic curvature of the surface can have a small effect on the equilibrium properties of the adsorbed polymer layer, it will have a strong effect if the kinetics of polymer adsorption are dominated by diffusion. Examples of frequently applied well-defined geometries were presented in the previous section. Wang et al. [34] have studied exchange kinetics in a spherical geometry. It was shown that the influence of bulk diffusion is reduced with decreasing radius of the adsorbent particle. For slow processes the macroscopic curvature will not be very important, and powders with poorly defined macroscopic curvatures can still be used.

The kinetics of polymer adsorption on porous substrates is much more difficult to tackle. Besides adsorption, desorption, and exchange, size exclusion has to be taken into account. Also, most in situ methods are not applicable to porous substrates. A major difficulty is that with all available methods smeared-out properties are measured while it is likely that strong gradients in the axial direction of the cylindrical pore are present. The process of "axial equilibration" is poorly understood and in many cases extremely slow. Most studies were performed with porous substrates with broad pore size and shape distributions. Controlled-pore glasses, zeolites, or porous membranes could be used as model systems with pores of molecular size. Application of glass capillaries is interesting for controlling the hydrodynamics in a curved system.

D. "In Situ" Methods for Studying Dynamics of Polymer Adsorption

In principle every method suited for measuring some property of an adsorbed polymer as a function of time can be used for studying one or more aspects of the dynamics of polymer adsorption. Some methods can only be used for polymer adsorption on flat (oxidic) surfaces, others require higher amounts of adsorbed polymers and therefore can only be applied in dispersions with sufficient surface area. Recently some reviews on experimental methods have been published [1, 35]. In this section we will discuss some frequently applied methods applicable for oxidic surfaces.

1. Optical Reflection Methods

Optical reflection methods are powerful tools for determining thickness and/or adsorbed amount of a polymer at an interface. Different reflection techniques are applied in practice. Well known is ellipsometry which measures the change in ellipticity of the electric field vectors [36]. In principle, ellipsometry can measure both the adsorbed amount and the layer thickness on a flat surface in situ, but for very thin layers this is not entirely unambiguous.

The method requires a highly reflective substrate for sufficient optical contrast. On the other hand, reflectometry measures the change in *intensity* of the parallel (p) and perpendicular (s) components of the reflected light. The reflectivity depends on the angle of incidence. Whereas for the s-polarized incident light the reflectivity gradually increases with increasing angle of incidence, for p-polarized light the reflectivity for a "perfect" (Fresnel) interface vanishes at the so-called Brewster angle. It is this property which is used in fixed-angle reflectometry, as will be described in more detail below. In scanning angle reflectometry the reflectivity of the p-signal is measured at different angles around the Brewster angle, and from changes of the reflectivity due to adsorption both the thickness and the adsorbed mass of the layer can be calculated. Different types of scanning Brewster angle reflectometers have been developed [37–40] for detecting polymer adsorption on pure silica, high refractive index glass substrates, and etched silicate glass.

Fixed-angle reflectometry measures the reflectivity at the Brewster angle on a flat surface [41–43]. It is usually combined with a stagnation-point flow cell. A linearly polarized laser beam is reflected from the adsorbing surface and split into its parallel (p) and perpendicular (s) components of which the intensities (I_p and I_s) are measured continuously. The output signal is equal to the ratio between the intensity of the parallel (I_p) and perpendicular (I_s) components $S = I_p/I_s$. In order to obtain a detectable change in S upon adsorption, the reflecting substrate must be covered with a dielectric film of appropriate thickness. This has the additional advantage that the signal increment ΔS is linear in Γ so that, after calibration, the adsorbed mass can be easily calculated from the expression:

$$\Gamma = \frac{\Delta S}{S_0} \frac{1}{A_s} \tag{27}$$

where S_0 is the output signal in the absence of adsorbate, ΔS is the effect of an adsorbed layer, and A_s is the sensitivity factor. In a typical reflectometer (Fig. 3), monochromatic laser light (1) is linearly polarized and passes a 45° glass prism (2). This beam arrives at the oxide–solution interface (3) with an angle of incidence close to the Brewster angle. When a silicon wafer is used as a substrate and water as solvent the angle of incidence will be around 71°. After reflection at the interface and refraction at the prism the beam is split into its p- and s-polarized components by means of a beam splitter (4). Both components are detected separately by two photodiodes (5), and the ratio between the intensities of the parallel and perpendicular components is the output signal S (6). The solution is introduced into the cell by a switching valve (7). A typical example of a reflectometer experiment with stagnation-point flow at a flat surface is presented in Fig. 4. Roughly, we can subdivide the curve into three regions. After polymer injection the process is fully transport controlled and S changes linearly with time. Near saturation the adsorption rate slows down due to the presence of a barrier caused by molecules that are already adsorbed. At saturation the adsorbed amount is constant. Usually, effects due to unfolding and reconformation of the polymer are too slow and/or the change in adsorbed amount is too small to be detected by reflectometry. Often, solvent injection does not change the output signal, which indicates that the polymer is

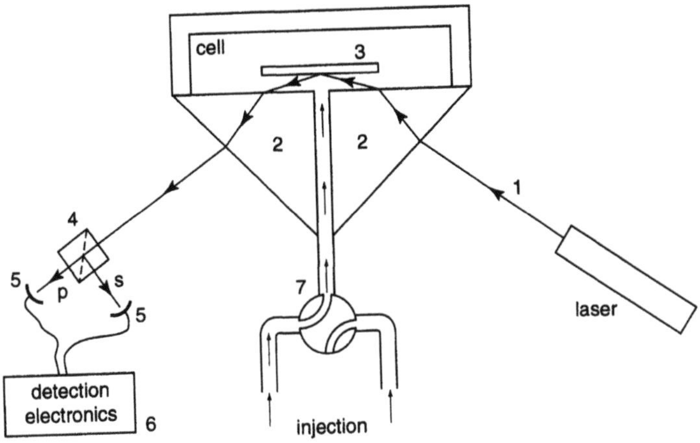

FIG. 3 Schematic representation of a reflectometer with a stagnation-point flow cell. (Adapted from Ref. 41.)

not noticeably desorbed by flushing with solvent. This behavior is characteristic for high-affinity polymer adsorption.

2. FTIR–ATR and TIRF

Fourier transform infrared spectroscopy (FTIR) in the mode of attenuated total reflection (ATR) has been successfully applied for studying polymer adsorption or exchange on the surface of an infrared prism consisting of, e.g., germanium or silicon [44, 45]. In an ATR cell a beam of light is totally reflected at the boundary of the interface between the prism and the solution. The prism has a refractive index higher than that of the solution. An evanescent wave penetrates the medium of lower refractive index (polymer solution side) with a penetration depth of the order of

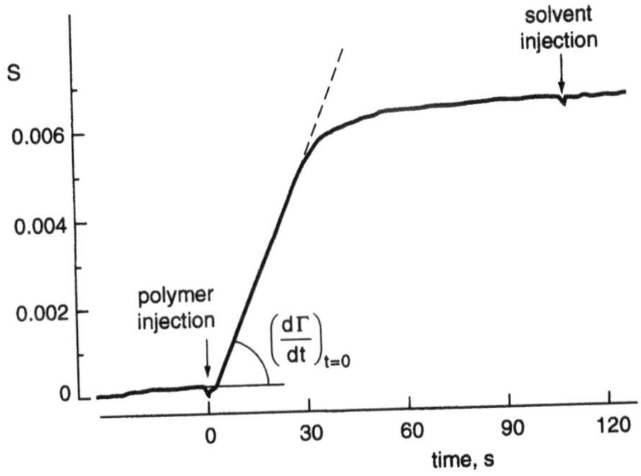

FIG. 4 Typical example of a reflectometer experiment with stagnation-point flow to a flat surface. Adsorption of PEO onto silica; $M = 400\,\mathrm{kg\,mol^{-1}}$, $c = 10\,\mathrm{mg\,dm^{-3}}$, Re = 12.2, and $dS/d\Gamma = -8.9 \times 10^{-3}\,\mathrm{m^2\,mg^{-1}}$. (Adapted from Ref. 41.)

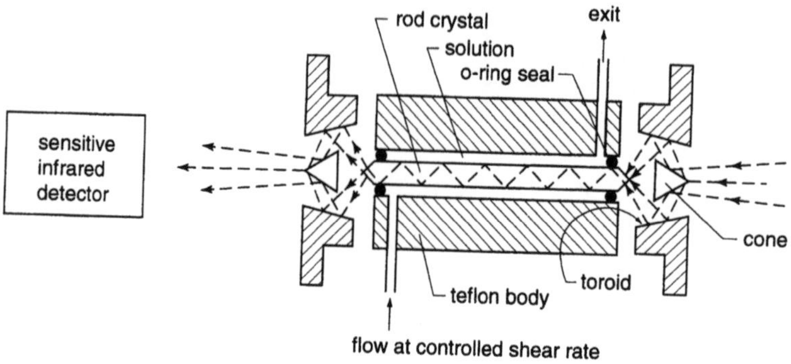

FIG. 5 Schematic diagram of an ATR infrared flow cell. (From Ref. 44.)

the wavelength. As a consequence the measured absorbance peaks contain only information about the surface excess of the polymer and not about the segmental distribution. Johnson and Granick [45] applied deuterated and nondeuterated polymer samples in order to distinguish the adsorption of the exchanging polymers. For the internal reflection element they used a silicon rod with conical ends (Fig. 5). The angle of incidence was 45°. The infrared beam sustained about seven internal reflections as it traveled the length of the cylindrical prism.

Total internal reflection fluorescence (TIRF) has also been used for characterization of the kinetics of polymer adsorption [46–55]. A typical set-up [56] is shown in Fig. 6. A light beam is totally reflected at the solid side (2) of a solid–solution interface resulting, on the solution side (1), in an evanescent field with the wavelength of the incident light similar to that of FTIR–ATR. The solution contains fluorescent molecules which can adsorb at the solid–solution interface. Owing to the limited

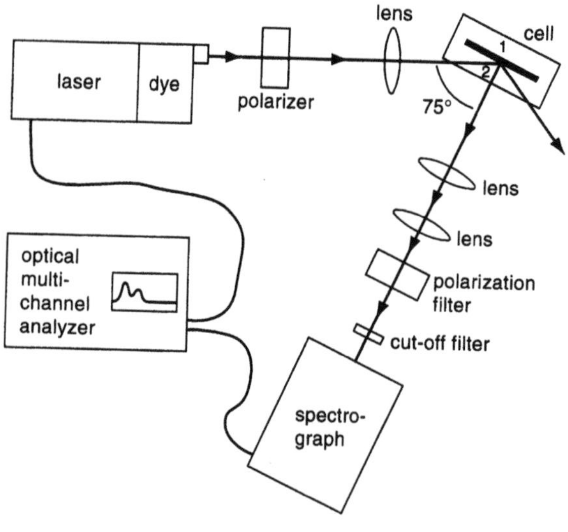

FIG. 6 Schematic diagram of a TIRF set-up. (From Ref. 56.)

depth of the evanescent field (~ 1 μm) the fluorescence comes mainly from the adsorbed molecules. In addition to the adsorbed mass, TIRF (with polarized light) allows determination of the orientation of the adsorbed molecules and the concentration profile. As the signal is recorded directly, the method is highly suitable for studying kinetic aspects of polymer adsorption. A disadvantage is that a fluorescent label must be attached to the polymer molecule, which possibly affects the equilibrium state and the dynamics of the adsorbed polymer. However, Fu and Santore [21] have shown recently that, with a proper choice of the fluorescent label (they used a terminal coumarin dye as fluorescent label on a PEO molecule), the effect was less than that of isotopic labeling of the entire chain with deuterium as applied by FTIR–ATR.

3. Other Experimental Methods

By means of *neutron reflectivity* [57] and *neutron scattering* [58] one can determine density profiles of adsorbed polymer layers, but the data acquisition time is usually of the order of hours and therefore not very suitable for studying the dynamics of polymer layers.

The kinetics of polymer adsorption has also been studied in capillary flow by measuring the electrokinetic *streaming potential* [10]. In this method one measures the ζ-potential which can be related to the position of the surface of shear, from which the thickness of the adsorbed polymer layer may be inferred. The method is particularly sensitive for detecting tails protruding into the solution.

Nuclear magnetic resonance (NMR) allows the determination of the fraction of adsorbed segments of a polymer on a solid interface [59, 60]. The method has been applied by Cosgrove and Fergie-Woods [61] to polystyrene adsorbed on a carbon substrate from d-chloroform. Two NMR techniques were used. With conventional high-resolution NMR the observed signal is due to protons on polymer segments with sufficient mobility and therefore only contains contributions from residual polymers in solution and from segments in loops and tails with sufficient mobility. With the solid-state NMR technique, signals from segments which are totally immobilized are separated from segments with sufficient mobility. Segments in the second or third layer may be undetectable with high-resolution NMR, but can be observed with solid-state NMR. It is also possible to measure the exchange of nondeuterated and deuterated polystyrene molecules. A typical time constant for the exchange rate of polystyrenes with a molar mass of 9600 amounts to 4.5 h (compare the corresponding measurements for oxides in Section III.E). Carbon-13 NMR relaxation measurements have been applied to measure the local mobility in adsorbed poly(-dimethylsiloxane) layers [62].

Electron spin resonance (ESR) has been used to measure the rates of polymer adsorption and exchange [63]. Similarly to NMR, the method is based on a mobility criterion to measure the bound polymer fraction. However, it has the advantage that it can be readily carried out in systems rich in ^{1}H such as water and that it has a much higher sensitivity. A major drawback of ESR is that the polymer must be spin labeled and again the label can influence the conformation and the adsorbed amount at the surface. For this method, only sufficiently concentrated dispersions can be used. For that reason it is only useful for relatively slow processes in which geometrical factors determining transport are of less importance.

IV. EXPERIMENTAL STUDIES OF THE KINETICS OF POLYMER ADSORPTION ON OXIDES

A. General

Researchers studying polymer adsorption are always faced with the typical dynamic behavior of such systems: slow equilibration, hysteresis, and apparent irreversibility. During the last two decades several experimental investigations have been performed in order to obtain a better understanding of the dynamics of the process of polymer adsorption [64–66]. These studies have also led to a better insight into the mechanism of polymer adsorption in general.

The study of the dynamics of polymer adsorption is hampered by the different time regimes in which the various subprocesses take place. In principle, three kinds of experiments can be distinguished: adsorption on a bare or partially covered surface, desorption, and exchange experiments. Typically, the time scale for adsorption (on a bare surface) is of the order of seconds to minutes, desorption against pure solvent is (almost) completely inhibited, whereas the rate of polymer/polymer exchange processes may have widely varying values, depending on the molecular mass of the polymers and the nature of their interaction with the substrate. A special case is the exchange of identical polymeric molecules between bulk and surface in the equilibrium situation (self-exchange). The time constant for the latter process is typically of the order of hours, depending on molar mass.

Different time-dependent properties of the adsorbed polymer layer can be used to study the dynamic behavior of polymer adsorption. In most investigations only the increase or decrease of the adsorbed mass with time is determined. However, these experiments do not give sufficient information with respect to rearrangement processes taking place within the adsorbed layer. For characterizing these rearrangements, measurements of the bound fraction, train densities, or the layer thickness, are indispensable.

In the following sections some systematic investigations of the kinetics of homo- and co-polymer adsorption are discussed. For the adsorption of neutral polymers the transport from the bulk to the surface usually dominates the process. Exchange experiments with neutral polymers will help to unravel the mechanisms of attachment, detachment, and rearrangement. The kinetics of polyelectrolyte adsorption, and the kinetics of polymer adsorption on porous substrates are discussed in separate sections.

B. Kinetics of Adsorption and Desorption of Random Coil Homopolymers

In the adsorption process one can distinguish three major contributions : (1) transport of polymeric molecules from the bulk phase to the subsurface, (2) attachment to the surface, and (3) rearrangement of the polymer molecule at the surface. In the absence of long-range repulsion (e.g., charged systems) attachment to the surface is (extremely) fast. The transport rate depends on the diffusion coefficient of the polymer and the geometry of the measuring cell.

Dijt et al. [41] have studied the kinetics of polymer adsorption on a flat silica surface in a stagnation-point flow cell with a fixed-angle reflectometer. The effect of solution concentration, flow rate, and molar mass on the rate of adsorption of poly(ethylene oxide) (PEO) has been studied; PEO is a neutral, flexible polymer adsorbing with a high affinity on to silica. A typical experimental result is shown in Fig. 4. The linear section up to 80% of saturation indicates that the adsorption rate is mainly determined by transport from the bulk phase to the subsurface. According to Eq. (11), $k^{a} \gg k^{tr}$, indicating the absence of any barrier. The initial adsorption rate appears to be proportional to the solution concentration over a broad concentration range (1–100 mg dm^{-3}) in agreement with Eq. (11) (Fig. 7). Also, the flow rate (or the Reynolds number) follows the theoretically predicted trend $(\mathrm{d}\Gamma/\mathrm{d}t)_{t=0} \sim \mathrm{Re}^{2/3}$, where the Reynolds number Re is defined as

$$\mathrm{Re} = \frac{UR}{\nu} \tag{28}$$

with R being the radius of the inlet tube, U the mean fluid velocity at the end of the inlet tube, and ν the kinematic viscosity. From this result it follows that the initial adsorption rate is equal to the limiting flux.

In Fig. 8 the effect of molar mass on the initial adsorption rate is presented for three flow rates (expressed as Re). The rate of adsorption decreases with increasing molar mass, because of a lowering of the diffusion constant. From the scaling theory of polymer solutions [4] it follows that in a good solvent $D \sim M^{-\gamma}$ with $\gamma = 0.5$ for low M and 0.588 for high M. According to Eq. (24) a limiting slope of $2/3\gamma$ is predicted, which is in perfect agreement with the experimental results, indicating again that the initial adsorption rate is entirely determined by the mass transfer rate.

Near saturation the net polymer flux will gradually decrease to zero. In Fig. 9 the adsorption is plotted against $t(\mathrm{d}\Gamma/\mathrm{d}t)_{t=0}$. For small molecules the process is mass transfer rate limited up to 95% of the plateau value, and for large molecules up to at least 70% of the plateau. This behavior is in agreement with the high-affinity character of the adsorption isotherm.

The evolution of the layer density and the layer thickness of PEO during adsorption onto silica glass has been investigated by Fu and Santore [43]. These authors have compared the change of the TIRF and reflectivity signals with time for two different molar masses of PEO. The reflectivity of the p-polarized laser light was

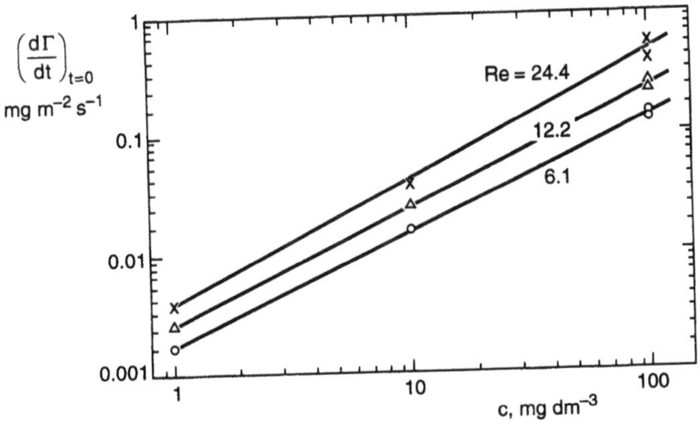

FIG. 7 Effect of PEO concentration on the initial adsorption rate for three values of the Reynolds number Re; $M = 246\,\mathrm{kg\,mol^{-1}}$. (Adapted from Ref. 41.)

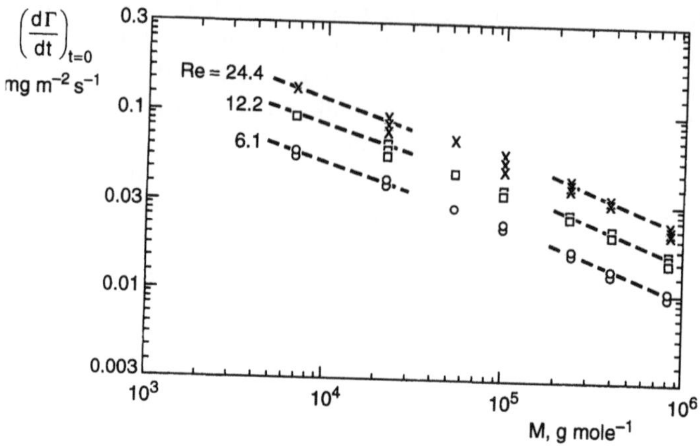

FIG. 8 Effect of molar mass on the initial adsorption rate for three values of the Reynolds number; polymer concentration: $10\,\mathrm{mg\,dm^{-3}}$. The dashed lines are drawn with a slope according to the theory (see text). (Adapted from Ref. 41.)

detected for a pure homogeneous substrate, i.e., without a spacer. Under these conditions the reflectivity depended quadratically (rather than linearly) on Γ. The TIRF curves increased linearly with time, but the reflectivity curves were concave (Fig. 10). The TIRF signal is proportional to the adsorbed amount whereas the square root of the reflectivity is proportional to the product of the adsorbed amount and the PEO layer density. Hence, from the two data sets the layer thickness and the adsorbed amount can be derived separately. The result is shown in Fig. 11. With increasing degree of coverage the layer becomes less dense due to the formation of loops and tails.

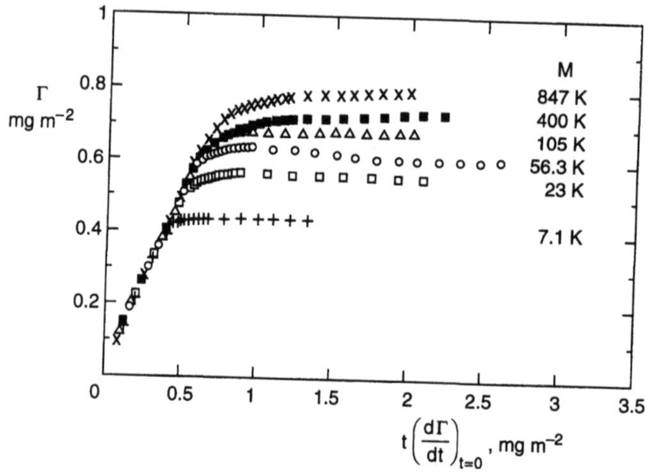

FIG. 9 Comparison of the adsorption rate near saturation for different molar masses; polymer concentration: $10\,\mathrm{mg\,dm^{-3}}$, Re = 12.2. (Adapted from Ref. 41.)

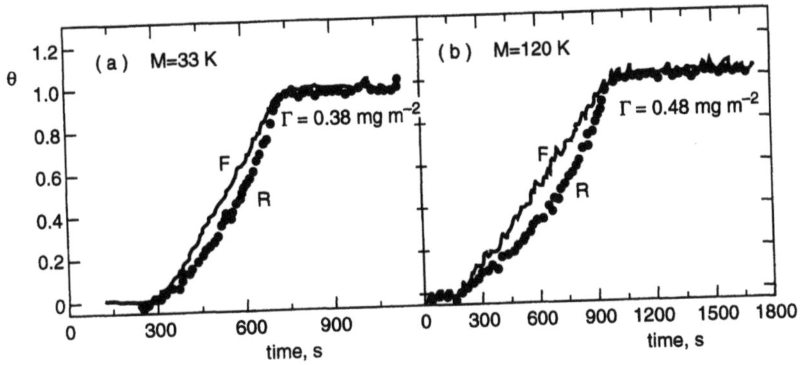

FIG. 10 Normalized adsorption signals obtained from Brewster angle reflectivity (R-curves) and TIRF (F-curves) for coumarin dye-labeled PEO at two molar masses. Adsorption in the plateau is indicated. (Adapted from Ref. 43.)

Dijt et al. [41] also studied the desorption kinetics with reflectometry for PEO molecules adsorbed on a silica surface by replacing the polymer solution with solvent. For molar masses above $10^5\,g\,mol^{-1}$ no detectable decrease in the adsorbed amount takes place on the time scale of the experiment (hours to days). For lower molar mass some desorption is observed; the decrease approaches about 15% for $M = 7100\,g\,mol^{-1}$. Owing to the high-affinity character of PEO on silica, complete detachment of the polymeric molecules is highly suppressed. With reflectometry one detects the change in the total mass present in the surface layer. In order to check if changes in the volume fraction profile take place after replacing the polymer solution by solvent, measurements of the layer thickness are more relevant.

Dijt et al. [10] determined the hydrodynamic thickness of a PEO layer on a glass capillary by streaming potential measurements. The capillary was first saturated with polymers at a concentration of $100\,g\,m^{-3}$ after which pure solvent was flushed through the capillary. In Fig. 12a the hydrodynamic layer thickness δ_h is plotted as a function of the flushing time for different molar masses. A substantial

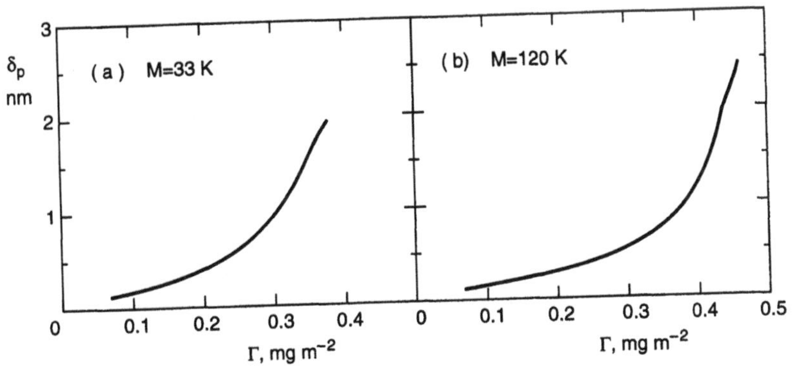

FIG. 11 Dependence of the layer thickness δ_p on the adsorbed amount of PEO on silica glass. (Adapted from Ref. 43.)

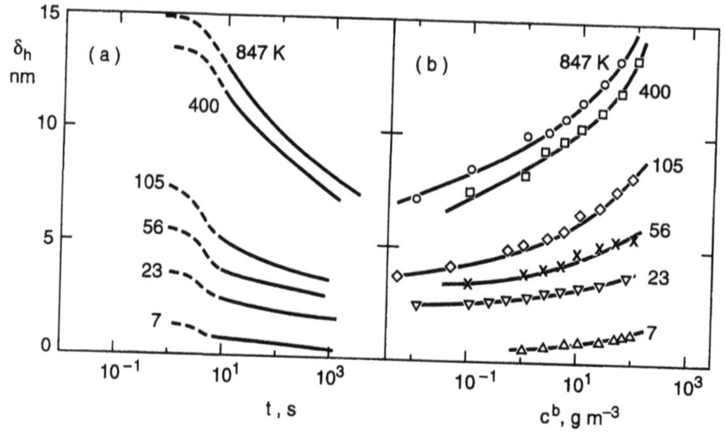

FIG. 12 (a) Desorption kinetics of PEO during injection of solvent obtained from streaming potential measurements in a cylindrical capillary; (b) equilibrium hydrodynamic layer thickness as a function of solution concentration. Experiments were performed for six molar masses, as indicated in the figure. (Adapted from Ref. 10.)

decrease in layer thickness is observed. In Fig. 12b the equilibrium hydrodynamic layer thickness is presented as a function of the solution concentration. It is striking that flushing with solvent hardly affects the maximum adsorbed mass, whereas the hydrodynamic layer thickness decreases by at least a factor of two. The layer thicknesses observed after a long desorption period correspond quite well with the thicknesses obtained by adsorption at low solution concentrations. Now, the question arises if the change in hydrodynamic layer thickness can be attributed to a rearrangement effect of the adsorbed molecules or to real desorption of some adsorbed molecules. In keeping with the results of Fu and Santore [43], discussed above, Dijt et al. [10] conclude that the thickness relaxation must be attributed to a real desorption of a tiny amount of polymer. The result indicates that at relatively high polymer solution concentrations a few molecules are very loosely attached to the surface. This has also been found for the adsorption of polystyrene (PS) on silica [67]. These loosely bound molecules are the ones that mostly influence the hydrodynamic layer thickness. This effect will be discussed in more detail below.

Several studies have been performed on the adsorption and desorption kinetics of PS on silica [67–72]. In comparison with PEO, PS is a less flexible molecule owing to its bulky styrene side-groups. Initial adsorption rates of PS from decalin on silica as a function of the molar mass have been measured by fixed-angle reflectometry [72]. The results are presented in Fig. 13 in a double logarithmic plot. The adsorption rates scale with molar mass as $D \sim M^{-\gamma}$ with $\gamma = 0.5$ for low as well as high molar masses independent of the solution concentration. The value of 0.5 is consistent with the finding that at $\approx 22°C$ decalin is nearly a Θ-solvent for PS. As for PEO, the rate of adsorption of PS at low coverages is completely determined by the rate of mass transfer from the bulk solution. The adsorption rate decreases at higher coverages. The effect of molar mass on the adsorption–time curves is presented in Fig. 14. The rate of adsorption is again constant up to about 70% of the plateau value. Plateau values are reached in about 1–3 min, depending on the molar mass. Desorption is

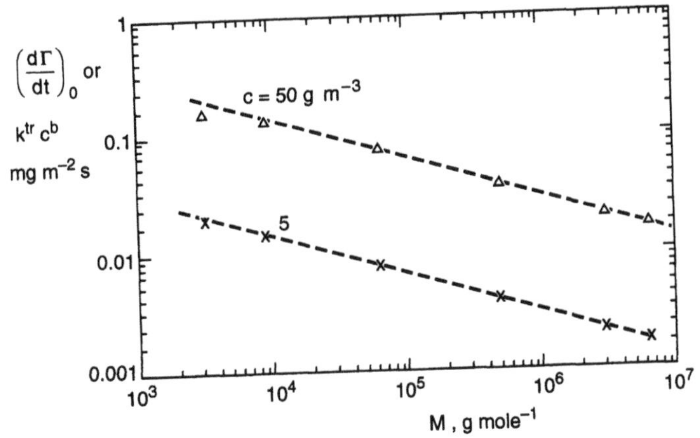

FIG. 13 Effect of molar mass on the initial adsorption rate of polystyrene in decalin solution on to silica for two solution concentrations. (Adapted from Ref. 72.)

extremely slow for $M > 100{,}000$. For a molar mass of 9200 about 20% of the plateau value of the adsorbed mass is desorbed after injection of pure solvent. As for PEO, at moderate molar mass, a fraction of the adsorbed polymer molecules are relatively weakly bound. These results agree well with the simulations of Zajac and Chakrabarti [25]. Also, Schneider et al. [67] concluded that, for the adsorption of PS on silica, a bimodal distribution of bound fractions was present, with some molecules adsorbed in a relatively flat position and others in a more extended conformation.

The rate of adsorption of PS on to a silica surface was also studied by Pefferkorn and coworkers [68, 69]. These authors used a radioactive tracer method for determining the rate of adsorption from CCl_4 on nonporous silica glass beads. They found that the rate constant decreased proportionally to the fraction of area

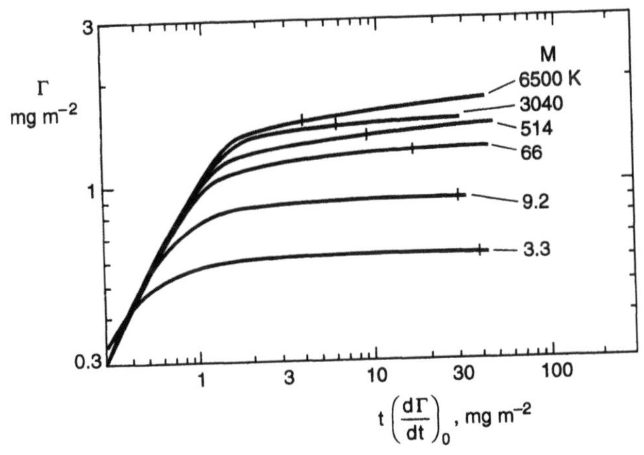

FIG. 14 Comparison of the kinetics of adsorption of polystyrene onto silica from decalin solutions for different molar masses. (Adapted from Ref. 72.)

covered with polymer. From Eq. (11) it must be concluded that in this experiment $k^a \gg k^d$. At higher solution concentrations the rate of adsorption becomes very low and molecules adsorb with large loops and tails.

Dynamic aspects of adsorbed layers of PS on oxidic surfaces have also been studied by surface-force measurements [73]. The effect of compression was measured for PS adsorbed on mica from cyclopentane near Θ conditions. If compression was performed slowly the layers seemed to become "irreversibly" compressed as was concluded from the absence of long-range bridging attraction. Even after several days the polymer layers did not relax.

For the adsorption of PS from decalin, Dijt et al. [72] calculated k^{tr} and $k^a(\Gamma) \times (1-\theta)$ from the corresponding kinetic curves of $\Gamma(t)$ by applying Eq. (11). The results are plotted in Fig. 15. For a simple Langmuir adsorption model one would expect k^a to be a constant. However, from Fig. 15 it can be derived that the adsorption rate constant is approximately proportional to $(1-\theta)^{1.7}$. It follows that for these polymers the net rate of adsorption is more strongly suppressed near saturation. This is again an indication that the first molecules adsorb in a flat conformation whereas the last molecules can adsorb only in a more extended conformation after some coiling up of the already adsorbed polymeric molecules.

We concluded that the process of desorption of adsorbed polymers by "quietly" flushing with solvent is in general slow and only a small fraction of adsorbed molecules is able to detach. However, strong flow fields may affect adsorbed layers in ways which are poorly understood. Several studied were performed on porous bodies or capillaries [74, 75]. Here, we discuss two examples of polymer adsorption and desorption on flat surfaces in the presence of different kinds of shear forces [76–79]. Using ellipsometry, Lee and Fuller [77] studied flow-enhanced desorption of PS at a cyclohexane–chromium interface. It was concluded that hydrodynamic forces can detach adsorbed molecules. The rate of detachment increases with the shear rate, but, interestingly, also with molar mass. Apparently, the total hydrodynamic drag on an adsorbed chain increases with the chain length.

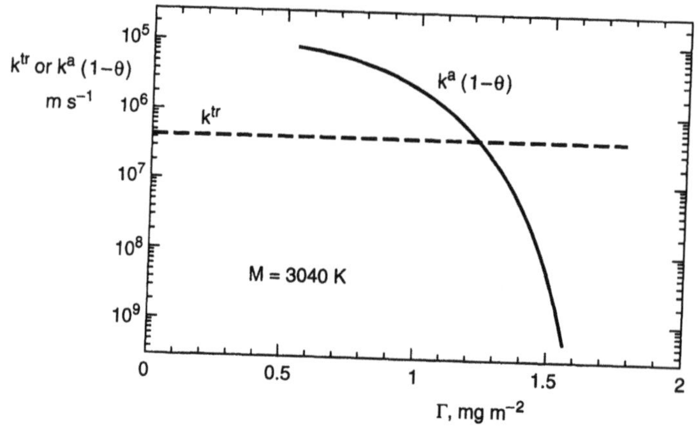

FIG. 15 Transport rate coefficient k^{tr} and attachment rate coefficient $k^a \times (1-\theta)$ as a function of the adsorbed amount Γ for the adsorption of polystyrene from decalin on silica using a stagnation-point flow cell. Polymer concentration: $50\,g\,m^{-3}$. (Adapted from Ref. 72.)

Moreover, reattachment of the segments probably becomes more difficult for large molecules. The effect of shear rate nicely illustrates that a polymer can adsorb with high affinity, even if the interaction energy of a single segment with the surface is relatively low. In Fig. 16 the effect of molar mass on the rate of detachment is shown for the case of high flow rate. For the lowest molar mass, complete desorption occurs after about 2 h. For low flow rates (not shown) the polymer is only partially desorbed.

Chang and Chung [80] have studied the desorption of PEO from PS latex particles under high shear with photon correlation spectroscopy. It was concluded that the segmental interaction energy has a larger effect than the molar mass. Besio et al. [79] studied the adsorption and desorption of PS from cyclohexane at the Θ-temperature on a chromium mirror by ellipsometry. An elongational velocity field (parallel to the surface) at the stagnation point was obtained by directing a jet of polymer solution or solvent perpendicular to the surface (cf. Fig. 2a). Under strong elongational flow a high molar mass PS ($M = 2 \times 10^7$) formed a layer with a thickness that was 10–20 times smaller than under weak flow, whereas the adsorbed mass was about three times higher. It was concluded that due to elongation the polymer chains are attached in stretched configurations, having many more bound segments than under weak flow conditions. A strong impinging solvent jet causes a complicated deposition pattern around the stagnation point with a radially varying deposition density. In order of increasing distance from the stagnation point one finds (1) a thick polymer layer at the stagnation point, (2) a bare ring, (3) a ring with a thick polymer layer, and (4) far away from the stagnation point an adsorbed layer under equilibrium conditions.

It must be concluded that adsorption and desorption can be dramatically affected by external forces such as (strong) shear forces exerted on adsorbed polymer molecules.

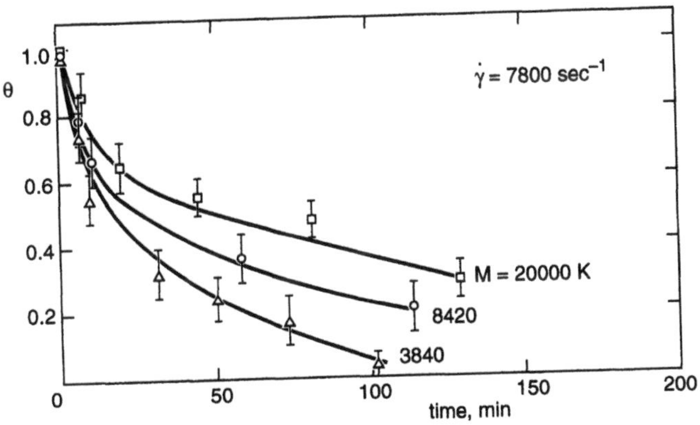

FIG. 16 Detachment of polystyrene from the chromium–cyclohexane interface due to shear forces. Molar mass dependence: $M = 3.84 \times 10^6$, 8.42×10^6, and 20×10^6. (Adapted from Ref. 78.)

C. Kinetics of Adsorption and Desorption of Block Copolymers

Several studies have been performed with respect to the kinetics of diblock copolymer adsorption onto oxidic surfaces [12, 81–89]. Two classes of systems must be distinguished: nonselective solvents (both blocks soluble) and selective solvents (only one block soluble). In nonselective solvents the transport is not different from that of homopolymers. However, the adsorption is usually selective in that one block adsorbs more strongly than the other. This may lead to some kind of rearrangement. Tripp and Hair [85] studied the kinetics of adsorption of PS–PEO block copolymers from chloroform (nonselective) on silica particles by measuring segment–surface interactions with infrared spectroscopy. Three different kinetic regimes could be distinguished, which were interpreted as follows. In the first regime the polymeric molecules adsorb without any rearrangement. In the second regime, surface sites are fully covered, and the incoming polymer adsorbs presumably by displacement of the weakly attached PS segments of the already adsorbed polymer by PEO segments of the arriving polymer. In the third step a slow rearrangement of the adsorbed polymer layer takes place because the polymers that arrive later are attached with fewer PEO segments.

Diblock copolymers in a selective solvent form micelles in solution. The insoluble blocks, being prone to avoid contact with the solvent, have the stronger tendency to attach to the surface so that the soluble tails extend into the solution, forming a thick layer. If the hydrophilic block has no affinity for the surface, intact micelles cannot adsorb without undergoing large and unfavorable deformations. Still, they can act as a source which supplies free molecules to the solution near the interface. Thus, the kinetics of adsorption of a diblock copolymer not only depend on the diffusion coefficient of the single polymer, but also on the diffusion coefficient of the micelles, and on the exchange rate between unimers and micelles. For this complex transport an explicit general expression is not available. However, some special cases can be considered: (1) if the exchange rate is slow as compared to the rate of diffusion through the stagnant layer, diffusion of both micelles and unimers can be treated separately; and (2) if exchange is fast, at every position z from the surface the unimer concentration will be equal to the critical micelle concentration (cmc) if the total concentration of diblock copolymers at a given position is above the cmc. Bijsterbosch et al. [90] have solved the transport equations for both limiting cases. This leads to three situations, for each of which the flux is plotted as a function of polymer concentration in Fig. 17. In (a) if there is no exchange and micelles are repelled by the surface, the flux increases linearly up to the cmc and stays constant afterwards; (b) if the "static" micelles also adsorb, the flux continues to increase linearly with concentration after the cmc, but with a smaller slope because of a smaller diffusion constant of the micelles; (c) an interesting situation occurs for fast exchange with nonadsorbing micelles. After the cmc the flux continues to increase. There is no discontinuity at the cmc, but at higher concentration the slope decreases, approaching the same value as that for adsorbing micelles. However, the absolute values are always higher than in case (b).

Bijsterbosch et al. [90] studied the adsorption on silica and titania from aqueous micellar solutions of diblock copolymers of hydrophobic poly(dimethylsiloxane) (PDMS) and hydrophilic poly(2-ethyl-2-oxazoline) (PEtOx) of different molar mass but constant PEtOx/PDMS ratio (5:6). Both blocks of the copolymers had

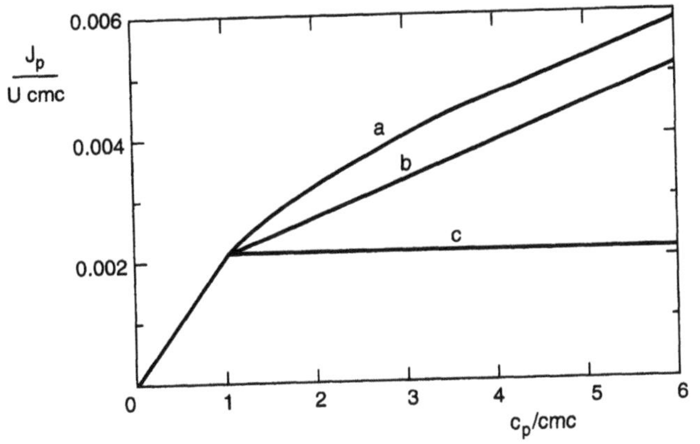

FIG. 17 The flux towards the surface in a stagnation-point flow as a function of the total diblock polymer concentrations. The flux J_p is scaled by the mean velocity U and the cmc, and the polymer concentration c_p is scaled by the cmc. Calculations were made by Bijsterbosch et al. [90]. Curve a assumes fast exchange between unimers and micelles; in curve b the exchange rate is zero, but micelles adsorb; and in curve c there is no exchange and micelles do not adsorb.

affinity for silica from water, but only the hydrophobic block had affinity for the titania surface. Unfortunately, the cmc values of all the polymers were too low to allow measurements at $c <$ cmc. When the initial rates of adsorption on titania were plotted versus concentration, a linear relation was found with a slope that scaled with the micellar diffusion rate. Apparently, exchange was sufficiently rapid to achieve substantial transport. On silica, the rates were very similar, which corroborates the conclusion. The fast exchange may have been due to the high mobility of the polymer in the PDMS core: PDMS has a very low glass temperature.

Layers of the diblock copolymer PS–PVP [polystyrene/poly-(2-vinylpyridine)] formed from toluene solutions on mica have been investigated by atomic-force microscopy (AFM), surface-force measurements, and ellipsometry [91–93]. These polymers form very stable micelles with (presumable) glassy cores. This leads to layers consisting of distinct small blobs of polymer in which one recognizes the original micelles. Clearly, the formation of a homogeneous, uniform layer (expected on thermodynamic grounds) is prohibited because micelles cannot coalesce. Pelletier et al. [88] studied the formation of ultrathin layers of PS–PVP from toluene on mica by using a surface-force apparatus. The absolute value of the surface coverage was determined by comparison with AFM measurements. It was shown that the thickness of the anchoring PVP layer increased with increasing surface coverage, whereas the volume fraction of PVP in this layer did not change. Recently, Maas et al. [94] studied layers of PS–PVP diblock copolymers adsorbed from chloroform on to silica. Two routes were used: adsorption from dilute solution, and spin coating of a thick layer followed by annealing at 60°C and dissolution of nonattached polymer in excess of chloroform. The adsorbed amounts determined by ellipsometry are given in Fig. 18. Surprisingly, the two methods give very different absolute values, but the dependence on block length is strikingly similar. Clearly, at least one of the routes does not lead to equilibrium; perhaps equilibrium is not even reached at all. The spin-coating method gives the highest coverages.

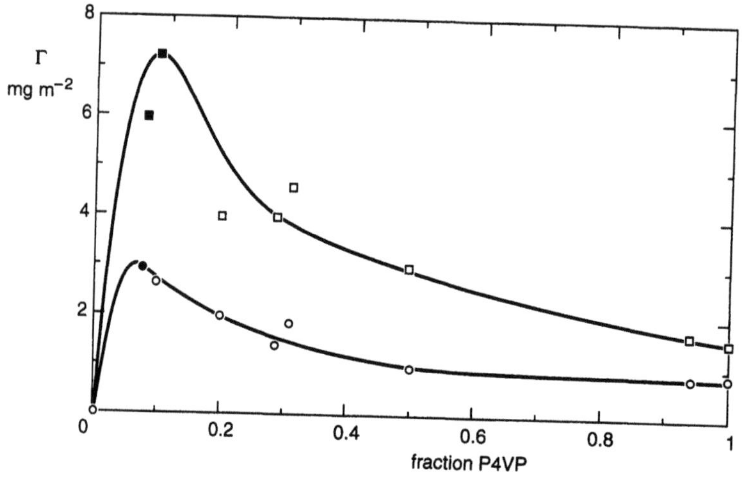

FIG. 18 Surface coverage (Γ) as a function of the fraction of P4VP: (□) spin coating (10,000 ppm, $CHCl_3$, 3000 rpm, annealed for 24 h at 145°C); (○) dip coating (100 ppm, $CHCl_3$, 30 min). Open symbols are used for polymers in the anchor-dominated regime and filled symbols for polymers which are in the crossover or buoy-dominated regime. (From Ref. 94.)

D. Exchange of Adsorbed Polymers by Polymers in Solution

Desorption of an adsorbed polymer by flushing with solvent proceeds particularly slowly and, in practice, it is impossible to desorb a significant amount in this way. However, we have seen that desorption of a polymer can be accomplished under the influence of an external shear force. In addition, desorption may be caused by competition with a polymeric or a monomeric displacer. In this section we will discuss exchange by a chemically identical polymer or by a polymer differing in size or chemical composition.

In the exchange of two polymers at a surface (1 against 2) four processes can be distinguished: (1) transport of polymer 2 from the bulk phase to the subsurface, (2) simultaneous attachment and detachment, (3) rearrangement of the mixed polymer layer at the surface, and (4) transport of polymer 1 from the subsurface to the bulk. If the adsorbed polymer layer is fully equilibrated for exchange against an identical competing polymer, step (3) is fast and, hence has no effect on the overall rate. In studying exchange, one should take care that the polymer samples are highly monodisperse, because results can be strongly disturbed by small fractions of higher molar masses [21].

The group of Granick [20, 95, 96] has studied the exchange of protonated polystyrene (p-PS) with deuterated polystyrene (d-PS) on oxidized silicon. The temperature was chosen close to the Θ-temperature of both samples (T_Θ equals 34.5°C and 30°C for p-PS and d-PS in cyclohexane, respectively) in order to minimize a kinetic barrier for adsorption due to an osmotic repulsion. The surface coverage of each polymer was monitored separately by ATR–FTIR making use of the difference in absorption frequencies for the carbon–hydrogen and carbon–deuterium vibrations. The experiments were carried out as follows: first, p-PS was adsorbed and after a waiting period p-PS in solution was replaced by d-PS with an almost identical degree of polymerization. The desorption (by displacement) of p-PS as a function of time is plotted semilogarithmically in Fig. 19. The data were fitted empirically with a so-called stretched exponential function:

$$\Gamma(t)/\Gamma_0 = \exp[(t/\tau_{\text{off}})^{\beta}] \tag{29}$$

For displacement of p-PS by d-PS a pure exponential decay is found with β equal to 1, in accordance with Eq. (23) if k^{tr} is large. This means that the time constant for exchange will be much lower than that for rearrangement. The effective time constant for desorption, τ_{off}, follows directly from the slope of the curves and can be interpreted as the time in which the adsorbed amount reduces to $1/e$ (where $\ln e = 1$). The value of τ_{off} depends strikingly on the aging time τ_{aging} as is shown in Fig. 20. Apparently, polymer molecules are more tightly bound to the surface after longer aging times. This effect can be attributed to an increase in segment–surface contacts and entanglements of the molecules in the polymer layer. The time constant τ_{off} becomes constant at large values of τ_{aging}. The value of τ_{off} appears to depend exponentially on the chain length as shown in Fig. 21. In another experiment, PS was displaced by poly(methyl methacrylate) (PMMA) (Fig. 22). At 40°C exchange was almost complete after about 15 h, whereas at 25°C the process was much slower. The data could be fitted well to a stretched exponential function. The value of τ_{off} decreased strongly with temperature, but the value of β was almost constant at $\beta \approx 0.5$ (Fig. 23). It was concluded that the weakly adsorbing p-PS was sterically pinned to the surface by the more strongly adsorbing d-PMMA. The desorption times τ_{off} tended to diverge near 25°C. The effect of molar mass on the desorption time constant τ_{off} could be described by a power law ($\sim M^{2.3}$). It was concluded that topological constraints which impede diffusion of the desorbing polymer away from the surface play a dominant role. The process has a strong similarity to entanglements in bulk systems. More direct evidence of such "trapping" effects will be discussed below.

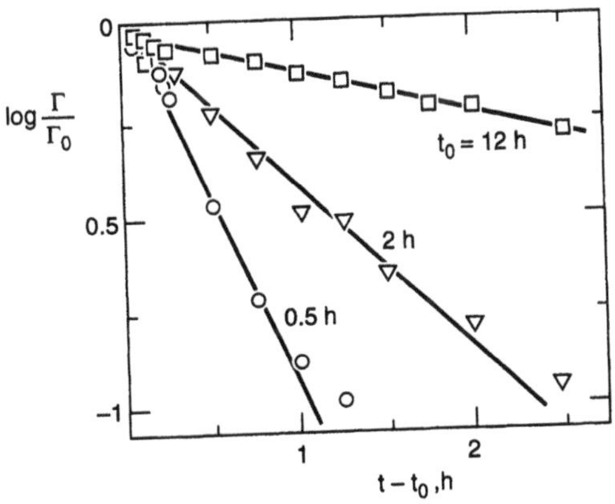

FIG. 19 Kinetics of exchange of p-PS by d-PS at a silicon oxide surface in cyclohexane for three aging periods. $T = 35°C$, solution concentration $1\,\text{mg}\,\text{mL}^{-1}$, and $N_w = 5500$. (Adapted from Ref. 20.)

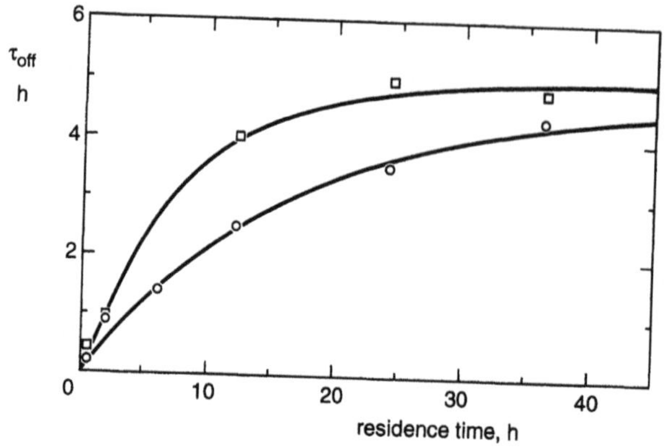

FIG. 20 Displacement time constants plotted against aging times of initially adsorbed p-PS layer ($N_w = 5750$). Circles: 0.1 mg mL^{-1}; squares: 1.0 mg mL^{-1}. (Adapted from Ref. 20.)

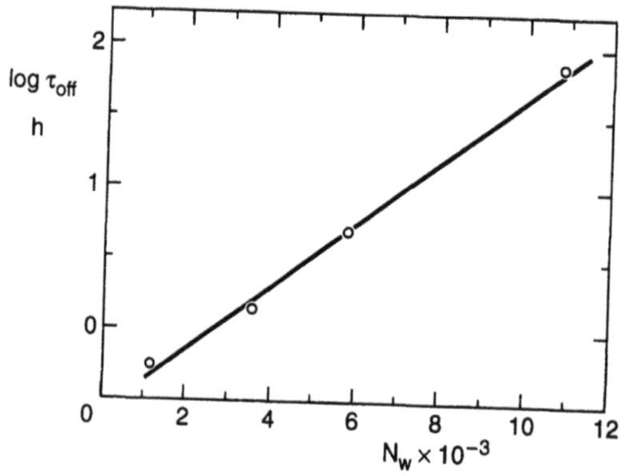

FIG. 21 Dependence of the equilibrated displacement time constant as a function of the degree of polymerization of adsorbed p-PS; N_w of d-PS amounts of 5500. (Adapted from Ref. 20.)

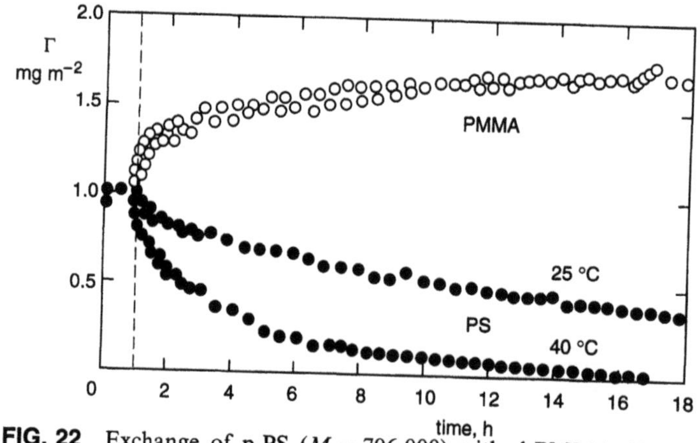

FIG. 22 Exchange of p-PS ($M = 706{,}000$) with d-PMMA ($M = 146{,}000$) on an oxidized silicon surface at 25° and 40°C. Polymer concentrations 1.0 mg mL^{-1}; aging time 1 h. (Adapted from Ref. 100.)

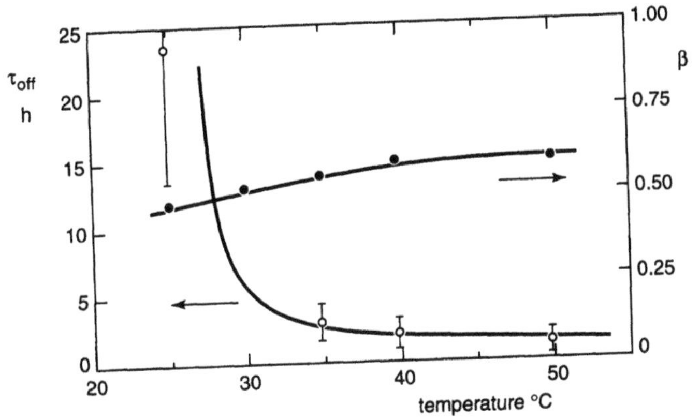

FIG. 23 Effect of temperature on the characteristic desorption time constant τ_{off} and the exponent β. Experiments according to Johnson and Granick [100].

Diffusion of branched molecules in bulk solution is much slower than that of linear polymers. Accordingly, one expects also a large difference in the adsorption behavior between linear and branched molecules. Thus, Johnson et al. [97] compared the exchange of linear and star-branched polystyrenes of matched molar mass by PMMA. It appeared that the star-branched molecule was hardly exchanged against the relatively high molar mass PMMA whereas the linear polymer of almost the same molar mass readily desorbed (Fig. 24). This observation strongly supports the conclusion that topological effects are important.

Schneider et al. [67] studied the kinetics of adsorption of atactic PMMA polymers of different molar mass from dilute CCl_4 solution on to oxidized silicon. They measured the number of hydrogen bonds to the surface (as calculated from the intensity of the infrared carbonyl peak) as a function of the total mass adsorbed (Fig. 25). A unique relation was obtained, independent of molar mass, solution concentration, solvent, and temperature. It was concluded that the first chains

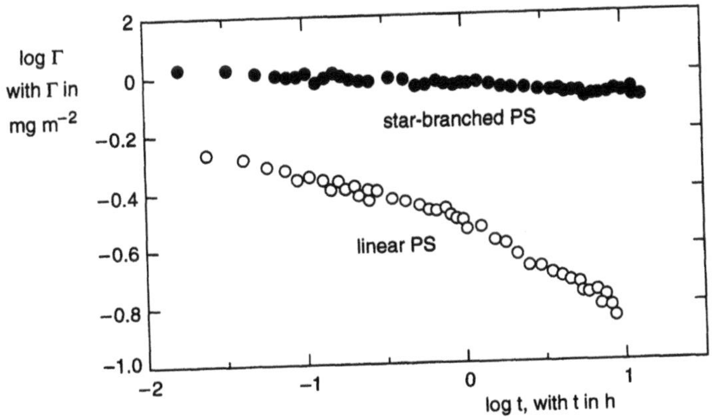

FIG. 24 Exchange of linear ($M = 220{,}000$) and star-branched polystyrene (3 arms with $M = 55{,}000$) against PMMA ($M = 820{,}000$). (From Ref. 97.)

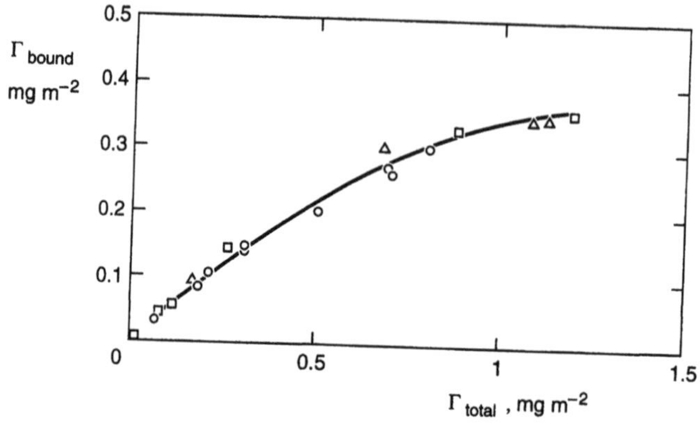

FIG. 25 Bound mass of atactic PMMA polymers of molar masses ranging from 7700 to 130,000 plotted against total mass adsorbed. (From Ref. 67.)

could spread on the surface to a relatively flat position, whereas later arrived molecules were bound more loosely. Clearly, the latter fraction largely determines the hydrodynamic layer thickness; also, these are the molecules which will first desorb.

Dijt et al. [98] have studied the exchange kinetics of PEO on oxidized silica in water by fixed-angle reflectometry, using stagnation-point flow. In Fig. 26 the adsorption of PEO with $M = 7000$ (1) and $M = 400{,}000$ (2), and a mixture of the two polymers $(1 + 2)$ is plotted as a function of time. The difference $(1 + 2) - 2$ indicates the exchange of the low molar mass by the high molar mass PEO. Four sections in the curve $(1 + 2)$ can be distinguished. The first part is completely determined by the transport of (1) from the bulk to the surface. In the second part, long PEO molecules replace the short ones, but the long molecules spread out on the surface to form a flat layer with about the same thickness as the layer of the small PEO molecules, leading to a constant adsorbed mass. After replacement of all small PEO molecules, the high molar mass molecules start to absorb in a more extended conformation. After about 150 s the adsorbed mass hardly changes any more.

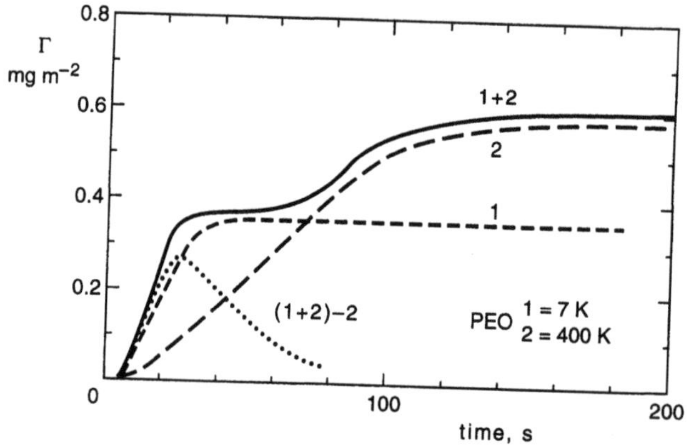

FIG. 26 Exchange kinetics of PEO for adsorption from aqueous solution on silica. Bulk concentrations are $5\,\text{mg}\,\text{mL}^{-1}$ (1) and $10\,\text{mg}\,\text{mL}^{-1}$ (2). (From Ref. 98.)

Recently, Fu and Santore [21] studied the effect of aging and interfacial relaxations on the kinetics of self-exchange of native PEO and PEO with a terminal fluorescent label (C-PEO; C = coumarin dye) adsorbed on silica. The total mass was obtained by reflectivity measurements while the evolution of the adsorbed C-PEO was detected by TIRF. The experiments were performed in a slit cell. A typical result for relatively young adsorbed layers is shown in Fig. 27. From A to B, C-PEO was preadsorbed at 5 ppm bulk solution, at B the surface was flushed with blank solution, and at time zero C-PEO was exchanged for PEO by injecting a 5 ppm PEO solution. In the second sequence (A′B′) PEO was preadsorbed and the adsorbed layer was exchanged for C-PEO. The value of the equilibrium constant K_{A-B} can be determined experimentally and amounts to 2.5 for the exchange of C-PEO against PEO ($M = 35{,}000$), which means that C-PEO has a slightly higher affinity for the silica surfaces. This gives rise to an asymmetry of the two experimental sequences in Fig. 27. The ratio k^{tr}/k^{ex} is treated as a fitting parameter. If k^{tr}/k^{ex} approaches zero, the surface exchange rate exceeds the rate of diffusion and the process becomes transport limited. The case of pure transport limitation is represented in Fig. 27 by the dashed curves. Equation (23), which includes attachment kinetics, fits the experimental results much better. Deviation at a high degree of coverage is attributed to the presence of a fraction which is extremely tightly bound. For a higher molar mass, deviations from Eq. (23) are larger, presumably due to a higher degree of entanglement and a higher number of segment–surface contacts.

In the experiments shown in Fig. 27 the adsorbed layer was allowed only a short period of relaxation. In order to determine if this layer was already fully in equilibrium, experiments were performed at different aging periods of the initially adsorbed polymer layer. A typical result is shown in Fig. 28. It can be concluded that even this relative short and flexible PEO polymer needs at least about 10 h to relax entirely. With increasing age of the adsorbed polymer layer a substantial fraction of

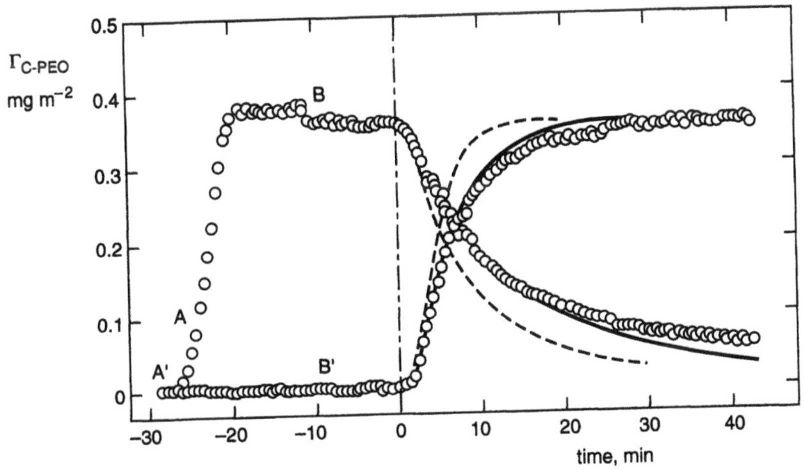

FIG. 27 Self-exchange of native PEO and C-PEO ($M = 33{,}000$). Evolution of the adsorption of C-PEO as measured by TIRF. Dashed lines represent exchange kinetics assuming transport limitation; the full curves are a fit according to Eq. (23) with surface exchange rate constants $k_{ex} = 1500\,\mathrm{cm^3\,g^{-1}\,s^{-1}}$ and $k'_{ex} = 600\,\mathrm{cm^3\,g^{-1}\,s^{-1}}$. (Adapted from Ref. 21.)

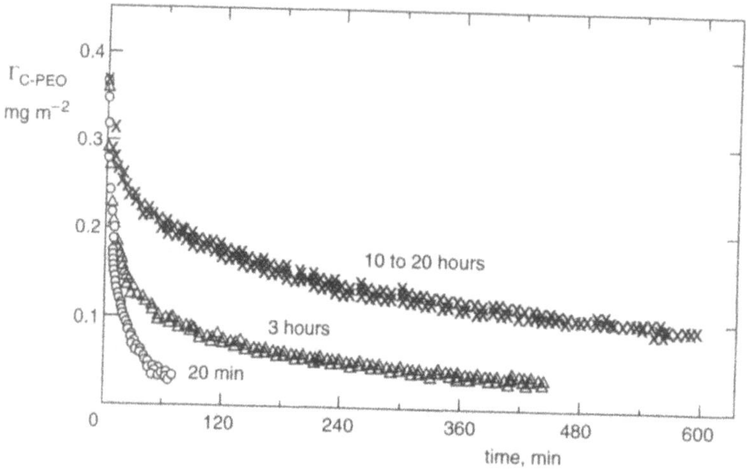

FIG. 28 Displacement kinetics of C-PEO ($M = 33{,}000$) by native PEO of the same molar mass and a constant bulk concentration of $5\,\mathrm{mg\,L^{-1}}$. Layer ages are indicated in the figure. (Adapted from Ref. 21.)

the adsorbed molecules seems to become "trapped" on the surface so that they appear irreversibly attached.

Several studies were performed in which an adsorbed polymer was replaced by a polymer with a chemically different chain [95, 99, 100]. If PS is replaced by a polymer with a higher affinity, the adsorbed amount does not decay exponentially, but follows a stretched exponential ($\beta < 1$). Johnson and Granick [100] exchanged PS with PMMA from CCl_4 on oxidized silicon at two temperatures (Fig. 22). The net surface interaction energies are $\approx 1\,\mathrm{kT}$ (kT being the thermal energy) for PS and $\approx 4\,\mathrm{kT}$ for PMMA. Contrary to the exchange of d- and p-PS, no effect of the aging time of the previous adsorbed PS layer could be detected. Self-exchange of PMMA appeared to be extremely slow [45, 101]. Dijt et al. [99] studied the exchange of pairs of the following polymers from decalin solutions: PS, poly(tetrahydrofuran), and poly(butyl methacrylate) on to silica by reflectometry in stagnation-point flow. The authors conclude that not only the importance of the segmental interaction energy, but also the stiffness of the polymer chain is an important factor.

From the results presented above it follows that (self-) exchange experiments contribute significantly to a better insight into the relaxation behavior of an adsorbed polymer layer. There is strong evidence that in most cases the adsorbed polymers continue to relax over many hours or even days. It is questionable whether full relaxation is always obtained: the polymers may be trapped in a metastable state. The irreversible behavior must be attributed to an increasing number and/or strength of segment–surface contacts ("more friction") and entanglements of the adsorbed polymeric molecules. Also, the stiffness of the polymer chain can play a role.

E. Polyelectrolytes

As discussed in Section II, the rate of polyelectrolyte attachment may be considerably influenced by the presence of a repulsive electric field at the interface. This influence can be taken into account in the attachment rate constant k^a, as expressed in Eqs (9) and (14). Even when the polyelectrolyte and the surface initially carry opposite charges, so that attraction prevails during most of the adsorption process, one must expect that eventually the charge on the adsorbing molecule compensates that of the surface so that the attraction vanishes. In fact, theoretical considerations have shown that in the equilibrium situation there must be almost always some overcompensation, even when the polyelectrolyte/surface interactions are entirely electrostatic in nature [14, 102]. This implies that newly approaching molecules must often overcome an electrostatic barrier, the height of which depends on the effective charge of the surface plus polyelectrolyte, i.e., on the degree of coverage.

When polyelectrolyte adsorption is mainly driven by electrostatic attraction, the overcompensation is expected to depend on ionic strength, being small at low salt concentration, but increasing when salt is added. Therefore, very pronounced attachment barriers are not expected. In many cases, however, short-range interactions contribute to the adsorption affinity or even determine it completely. In such cases, the equilibrium situation is characterized by a strong accumulation of charge on the surface, the electrostatic barrier can become very high, and one may expect that the equilibrium adsorbed amount is often not reached at all.

Systematic studies of these effects are very scarce, however. This may be due to the difficulty that there is no unambiguous way to ascertain that a given experimental system is equilibrated. It has been noted by several authors that adsorbed amounts of polyelectrolytes may depend on the way the samples are prepared, in particular with respect to variations in pH and in salt concentration. However, the question whether this is due to a blocking of *adsorption* or *desorption* is left open.

A systematic study of adsorption/desorption hysteresis effects should provide insight. One such study was done with respect to the adsorption of carboxymethylcellulose (CMC) on metal oxides like rutile (TiO_2) and hematite (Fe_2O_3) [103]. These oxides have point-of-zero-charge (pzc) values of 6 and 8.5, respectively, whereas CMC has a negative charge arising from its carboxyl groups.

Figure 29 shows results obtained when adsorption was measured along a pH cycle. Fixing the pH and then measuring the adsorbed amount leads to the lower curve. Clearly, the adsorbed amount is lower the higher the pH, i.e., the more negative the polymer charge and the less positive/more negative the substrate charge. When the pH is increased *after* adsorption has taken place, one finds the upper curve. These adsorbed amounts are considerably higher. A decrease in adsorbed amount is not observed until the pH becomes very high.

A quantitative interpretation of this very large hysteresis (a complete loop cannot even be obtained in the experimentally accessible pH range) is as yet not possible. However, it can hardly be ignored that CMC adheres very strongly to rutile and hematite. Several authors [104] have pointed out that polysaccharides tend to form a co-ordination complex with metal ions. Such a complex can be very strong. Further confirmation of this mechanism comes from experiments of CMC adsorption on to silica. Pure silica does not bind polysaccharides, but modified silica with a small number of 3d-metal ions in the surface does [104]. We therefore tend to conclude that in equilibrium a large adsorbed amount can be maintained on account of the strong anchoring of monomers on the surface and that, owing to kinetic limitation, the adsorbed amount measured at fixed pH is much lower than the equilibrium adsorption.

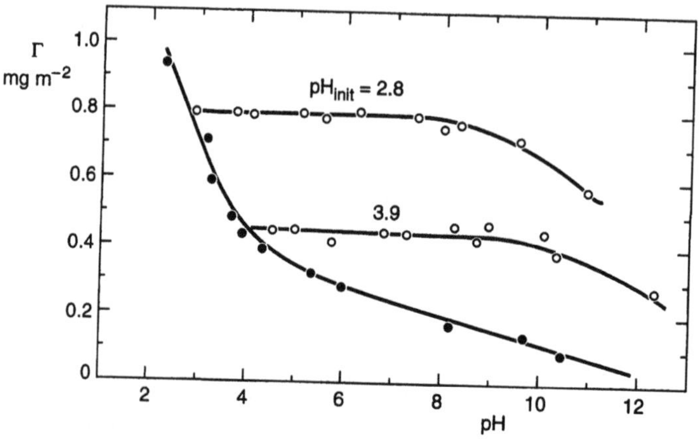

FIG. 29 Effect of successive pH increases after adsorbing CMC on hematite at two low initial pH values and 0.01 M NaCl. (Adapted from Ref. 103.)

To some extent, this is supported by the finding that the adsorbed amount increases substantially when the *ionic strength* is increased. Unfortunately, the pH cycles have not been fully determined for a range of salt concentrations, so that it cannot be proven that there is a salt concentration large enough to suppress all hysteresis. Recently, a more complete study on the salt dependence of adsorption/desorption pH cycles was carried out which clearly showed the complete disappearance of hysteresis at high ionic strength. However, this study did not consider oxide surfaces so that we shall not discuss it further here [105].

Since the kinetic barrier results from the electrostatic repulsion between polyelectrolyte and surface, one expects that its height depends not only on the effective surface charge and the salt concentration, but also on the effective charge of the polyelectrolyte. The latter charge depends, of course, on pH, but also on molar mass: a long chain experiences more repulsion and will thus adsorb more slowly than a short one. If the chains are long enough, their adsorption may even become entirely blocked. From a thermodynamic point of view, one expects an opposite trend, namely, that longer chains adsorb preferentially over shorter ones; for neutral polymers, there is ample evidence that this occurs (see, e.g., Ref. 1).

Direct measurements of polyelectrolyte adsorption rates as a function of molar mass have, to our knowledge, not been reported. However, there is indirect evidence of *kinetic blocking effects* in adsorption experiments with polydisperse samples [106, 107]. De Laat et al. [106] have analyzed the fractionation occurring during adsorption of polyacrylic acids on to barium titanate powder particles. As would be expected on the basis of transport rates, the initially formed layer consisted mainly of the smallest molecules in the sample. Over times of up to 24 days, these were gradually replaced by a fraction in the center of the molar mass distribution. The longest molecules in the sample were totally excluded. As a result the molar mass distribution in the solution became bimodal, with one peak at low M and one at high M (Fig. 30). In a similar experiment, Geffroy et al. [107] found that the molar masses of adsorbed molecules did indeed correspond to the center portion of the original distribution (Fig. 31). Hence, these two complementary studies support the view that

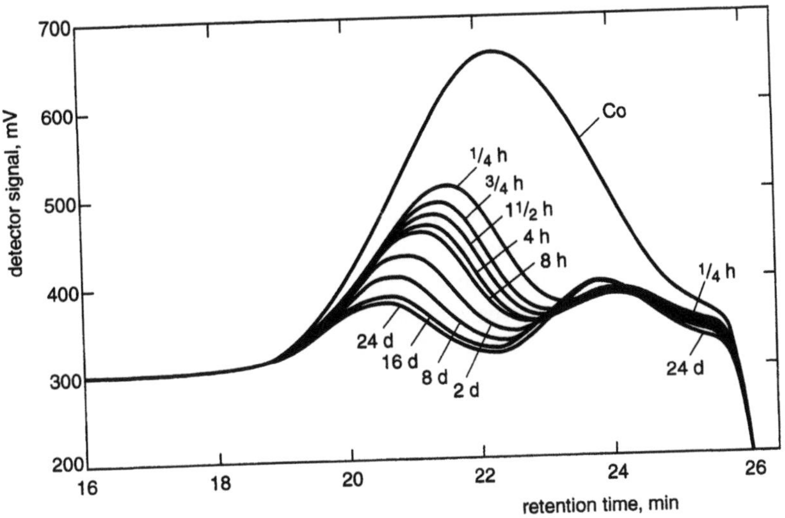

FIG. 30 Change in molar mass fractionation with time in the adsorption of poly(acrylic acid) salt on to $BaTiO_3$ in the absence of salt. (Adapted from Ref. 106.)

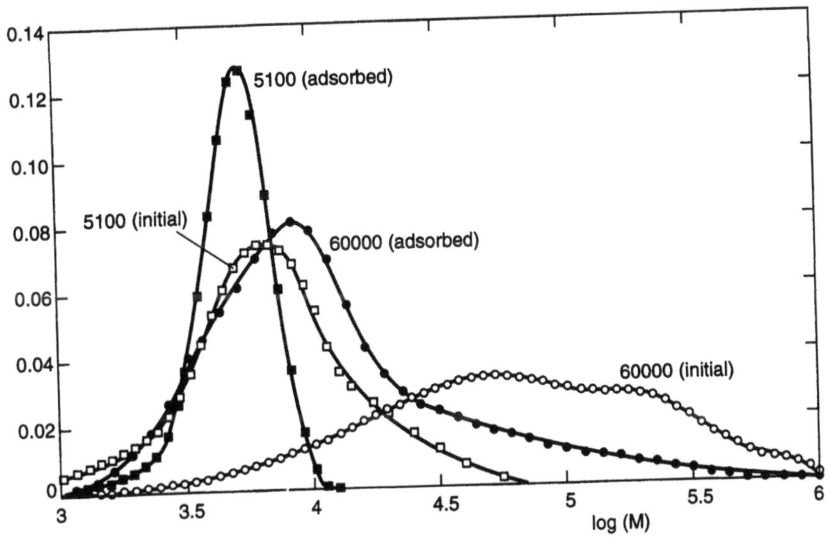

FIG. 31 Molar mass distribution of sodium polyacrylate before and after adsorption on $CaCO_3$. Open symbols represent the initial distribution; filled symbols represent the distribution of the adsorbed molecules after adsorption during 5 days. Initial molar masses are indicated in the figure; pH values are between 8 and 9. (Adapted from Ref. 107.)

the situation of lowest free energy (i.e., with the largest molecules on the surface) is not reached in these polyelectrolyte systems, but that the composition of the adsorbed layers is determined by kinetics, i.e., by the height of the electrostatic barrier.

We have seen that polymer/substrate interactions play an important role in surface processes such as unfolding and spreading. If these interactions are of an electrostatic nature, one expects that they can be influenced by means of, e.g., salt concentration. Experiments by Hoogeveen et al. [108] on exchange between short and long polyelectrolytes at moderate ionic strengths (where attachment barriers are no longer relevant) indicate that an increase in salt concentration has an accelerating effect on spreading and exchange. In Fig. 32 we present the adsorbed mass as a function of time for a mixture of two *poly(vinylpyridinium)* samples ($M = 12{,}000$ and 120,000 respectively). At a salt concentration of 0.1 M NaCl (Fig. 32a) the overall curve bears a clear resemblance to that of the exchange between two PEO samples and can be regarded as the superposition of (1) an undisturbed adsorption curve for the largest molar mass, and (2) an adsorption (followed by rapid desorption) curve for the smaller fraction in the sample. Since the exchange is rapid, the adsorbed mass is virtually constant during the exchange process; hence, the depression in the curve. When the experiment is repeated at a somewhat lower salt concentration (0.025 M NaCl), see Fig. 32b, the depression in the curve becomes smaller and eventually disappears, because the exchange process cannot keep up with the supply. The same behavior has been found in an exchange experiment with PS samples [99].

Krabi and Cohen Stuart [109] have studied sequential adsorption of carboxylated pullulan and pullulan on solid PS from water by reflectometry. It was shown that at pH 2 higher molar mass pullulans were able to displace carboxylated pullulans up to a maximum of 70%. At pH 9 the carboxylated pullulan was negatively charged and, due to repulsion from the surface, partial desorption of the remaining

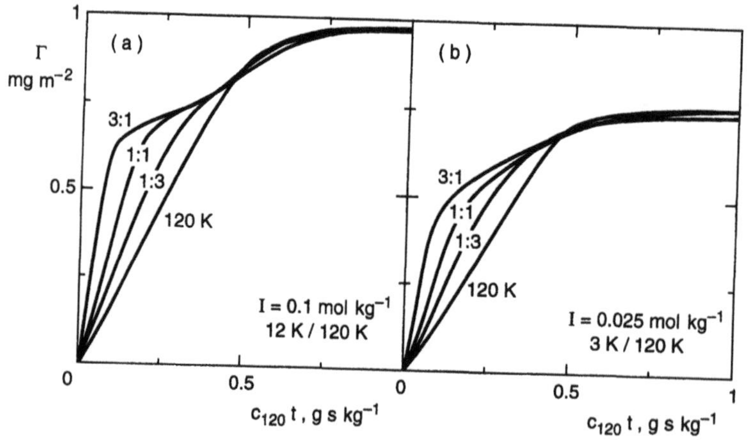

FIG. 32 Adsorption of poly(vinylpyridinium) (PVP^+) (120 K) and of mixtures of PVP^+ (3 K) and PVP^+ (120 K) at an ionic strength of 0.025 M (a), and samples of 12 K and 120 K at an ionic strength of 0.1 M (b) on TiO_2 at pH = 8. Here, t is the time and c_{120} is the concentration of PVP^+ (120 K). (Adapted from Ref. 108.)

adsorbed carboxylated pullulan took place. However, roughly 60% of the remaining carboxylated pullulan did not desorb and it was concluded that this fraction was in a *trapped* state caused by entanglements (pinning) with the adsorbed pullulan. Van Eijk et al. [110] have studied the adsorption kinetics of the semiflexible polysaccharide *xanthan* from aqueous solution on to zirconium oxide. The authors stressed the importance of the presence of helices in solution, leading to a gradual increase in the adsorbed amount after the initial fast, transport-limited adsorption step. The effect was attributed to slow rearrangements of the stiff helices near the surface.

We conclude that the dynamics of the exchange process, at the level of individual segments, can be accelerated by increasing the ionic strength; apparently, the salt ions compete with charged groups on the polymer for forming ion pairs with surface groups and this shortens the lifetime of polymer/surface ion pairs.

F. Pores

In most experiments on the kinetics of polymer adsorption discussed above, a strong effect of the transport from the bulk solution to the subsurface was found. In these cases, surfaces of well-defined geometries, i.e., mainly flat surfaces, have been considered. It is expected that porous systems will behave completely differently, showing adsorption rates that differ by orders of magnitude from, e.g., a flat surface. Although porous systems are highly relevant in many practical situations, systematic experimental or theoretical investigations are scarce and mainly limited to diffusional transport of polymers in the absence of segment–pore surface interactions. In a recent review on polymer solutions in confined geometries [111] some attention was paid to the dynamics of polymer chains in pores.

Diffusional transport of polymers into (cylindrical) pores can be separated into three contributions: (1) transport into the pore volume by diffusion through its free cross-section ("lumen"), (2) diffusion retarded by protruding loops and tails in a layer near the surface, and (3) lateral diffusion of attached molecules. The relative importance of these three contributions will depend on the ratio of the pore size to the polymer size, and on the free energy of the interaction between surface and polymer segments. If the size of the lumen exceeds the size of the polymeric molecule, the polymer can freely diffuse into the cylindrical pore. This contribution to the rate of adsorption will be proportional to the area of the lumen, i.e., $\sim (R_{\text{cyl}} - d_{\text{ads}} - R\text{pol})^2$. It is likely that the adsorbed amount will initially depend on the penetration depth. It is not too difficult to model this situation quantitatively. If polymeric molecules have an affinity for the pore surface, they will adsorb near the entrance of the pore. Subsequently arriving molecules will be strongly retarded by the protruding polymer chains of the adsorbed layer. Further transport must now be accomplished either by diffusion of nonadsorbed polymer molecules partially hampered by tails and loops, or by lateral diffusion of the adsorbed molecules. It may be expected that these processes will be slow, but a systematic theoretical or experimental study has not been performed. We must realize that the adsorbed molecules will relax after adsorption, leading to a decrease in the thickness of the adsorbed layer and an increase in the area of the lumen. If the polymer size exceeds the size of the pores, the bulk conformation has to be changed in order for the molecules to penetrate the pores, which is unfavorable from an entropic point of view. It is expected that molecules can only enter the pores if the entropy loss is sufficiently compensated by interaction with the pore wall. However, adsorption will be extremely slow, i.e., of the order of days and weeks. If single molecules can be attached to the cylindrical wall over the full circumference, longitudinal transport in the pores can only take place by lateral diffusion of (strongly) adsorbed molecules. In practice, pores will be fully blocked by adsorbed molecules and it is hardly imaginable that significant adsorption takes place even in a period of months.

It is not easy to perform measurements of kinetics of polymer adsorption or exchange in porous systems. Pore geometries, even in model systems like controlled-pore glass or Stöber silicas, are usually poorly defined. In situ measurements are difficult to perform. Usually, indirect measurements are performed in which one measures batchwise the time dependence of the concentration of the polymer in intensively stirred bulk solutions. As discussed above, depending on the size of the polymer and the pore radius, adsorption can be extremely slow. Of course, one must realize that even in model porous systems nonideality of the pore geometry, such as the presence of tortuous pore channels with junctions and branches, a pore size distribution, and a non-uniform pore diameter, may thwart the interpretation of experimental results.

The kinetics of the adsorption of PS at a porous silica surface was studied under Θ conditions (cyclohexane at 35°C) by Kawaguchi and coworkers [112, 113]. For small PS molecules ($M = 96{,}400$), with a size much smaller than that of the pores, it took less than 10 h to obtain a (quasi)-plateau value. For large PS ($M = 355{,}000$) the size of the polymer was half the size of the pores and a plateau was obtained only after 35 h. Adsorption of a mixture of the small and large PS molecules showed an interesting behavior. The rate of diffusion of the small molecules into the pores was much faster than that of the large molecules. As a result, the adsorption of the small molecules obtained a maximum in about 8 h (Fig. 33). After that period the large molecules replaced some of the small molecules. Pseudoplateau values were obtained after 25 h. All these experiments were carried out below saturation, and the solution concentration was varied between its initial concentration and

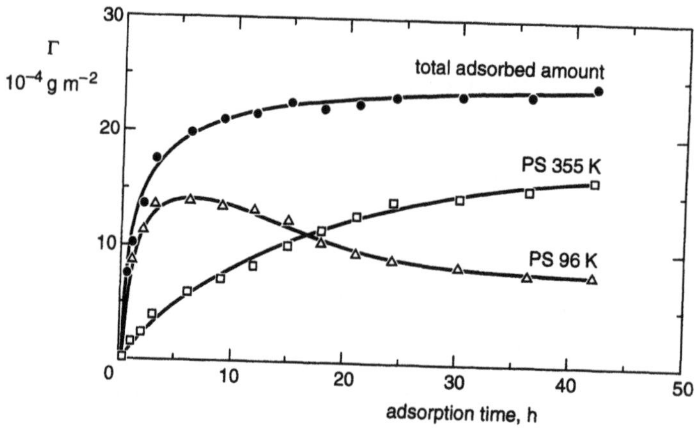

FIG. 33 Adsorption of a mixture of two polystyrenes with different molar masses from cyclohexane (35°C) on silica gel beads. Initial concentrations amount to 3 g L^{-1}. (Adapted from Ref. 113.)

zero. The results were empirically interpreted by distinguishing a rate of adsorption, desorption, and exchange [114] in a "Langmuir" like approach. Subsurface concentrations and concentration gradients inside the pores were not considered.

A polymer chain with repulsive interactions between the segments confined into a cylindrical pore has been studied by *Monte Carlo simulations* by Milchev et al. [115]. Besides the equilibrium conformation of the molecules in the confined geometry, the time-dependent mean-square displacements of the segments was also studied. It is shown that relaxation times scaling with N^2 as well as N^3 play a role in the reptative motion of the chain in these tubes.

Research with respect to the adsorption of polymer in pores is still in its infancy despite the technological importance. In particular, studies of self-exchange and exchange, as a function of molar mass, could contribute much to an insight into adsorption mechanisms. Well-defined porous systems and highly monodisperse polymers are required for a successful approach.

ACKNOWLEDGMENTS

We are indebted to Dr R.A. Gage for critically reading our manuscript and suggesting linguistic improvements.

REFERENCES

1. GJ Fleer, MA Cohen Stuart, JMHM Scheutjens, T Cosgrove, B Vincent. Polymers at Interfaces. London: Chapman & Hall, 1993.
2. JMHM Scheutjens, GJ Fleer. J Phys Chem 83:1619, 1979.
3. JMHM Scheutjens, GJ Fleer. J Phys Chem 84:178, 1980.
4. PG de Gennes. Adv Colloid Interface Sci 27:189, 1987.
5. HP van Leeuwen. In: AJ Bard, ed. "Electroanalytical Chemistry" Vol. 12. New York: Marcel Dekker, 1982, pp 159–238.
6. Z Adamczyk, T Dabros, J Czarnecki, TGM van de Ven. Adv Colloid Interface Sci 19:183, 1983.
7. Z Adamczyk, TGM van de Ven. J Colloid Interface Sci 97:68, 1984.
8. Z Adamczyk, T Dabros, J Czarnecki, TGM van de Ven. J Colloid Interface Sci 97:91, 1984.
9. J Lyklema. Fundamentals of Interface and Colloid Science. Vol. 1: Fundamentals. London: Academic Press, 1991, Ch. 6.
10. JC Dijt, MA Cohen Stuart, GJ Fleer. Macromolecules 25:5416, 1992.
11. HA Kramers. Physica 7:284, 1940.
12. A Johner, JF Joanny. Macromolecules 23:5299, 1990.
13. MA Cohen Stuart, CW Hoogendam, A de Keizer. J Phys: Condens Matter 9:7767, 1997.
14. AN Semenov, JF Joanny. J Phys II (Paris) 5:859, 1995.
15. C Ligoure. Macromolecules 24:2968, 1991.
16. R Zajac, A Chakrabarti. Phys Rev E 52:6536, 1995.
17. E Pefferkorn, A Elaissari. J Colloid Interface Sci 138:187, 1990.
18. MCP van Eijk, MA Cohen Stuart. Langmuir 13:5447, 1997.
19. MPC van Eijk, MA Cohen Stuart, S Rovillard, J de Coninck. Eur Phys J B 1:233, 1998.
20. P Frantz, S Granick. Phys Rev Lett 66:899, 1991.
21. ZL Fu, M Santore. Macromolecules 32:1939, 1999.
22. LK Filippov. J Colloid Interface Sci 181:232, 1996.

23. NL Filippova. J Colloid Interface Sci 195:203, 1997.
24. VN Kislenko. J Colloid Interface Sci 202:74, 1998.
25. R Zajac, A Chakrabarti. J Chem Phys 104:2418, 1996.
26. R Zajac, A Chakrabarti. J Chem Phys 107:8637, 1997.
27. A Milchev, K Binder. Macromolecules 29:343, 1996.
28. LC Jia, PY Lai. J Chem Phys 105:11319, 1996.
29. TD Hahn, J Kovac. Macromolecules 23:5153, 1990.
30. T Haliloglu, DC Stevenson, WL Mattice. J Chem Phys 106:3365, 1997.
31. JS Wang, RB Pandey. Phys Rev Lett 77:1773, 1996.
32. R Hasegawa, MS Doi. Macromolecules 30:3086, 1997.
33. M Kawaguchi. Adv Colloid Interface Sci 32:1, 1990.
34. Y Wang, RG Diermeier, R Rajagopalan. Langmuir 13:2348, 1997.
35. MA Cohen Stuart, T Cosgrove, B Vincent. Adv Colloid Interface Sci 24:143, 1986.
36. RMA Azzam, NM Bashara. Ellipsometry and Polarized Light. Amsterdam: North Holland, 1989.
37. P Schaaf, P Dejardin, A Schmitt. Langmuir 3:1131, 1987.
38. FAM Leermakers, AP Gast. Macromolecules 24:718, 1991.
39. L Heinrich, EK Mann, JC Voegel, GJM Koper, P Schaaf. Langmuir 12:4857, 1996.
40. Z Fu, MM Santore. Colloids Surfaces A 135:63, 1998.
41. JC Dijt, MA Cohen Stuart, JE Hofman, GJ Fleer. Colloids Surfaces 51:141–158, 1990.
42. JC Dijt, MA Cohen Stuart, GJ Fleer. Adv Colloid Interface Sci 50:79, 1994.
43. Z Fu, MM Santore. Langmuir 13:5779, 1997.
44. DJ Kuzmenka, S Granick. Colloids Surfaces 31:105, 1988.
45. HE Johnson, S Granick. Macromolecules 23:3367, 1990.
46. D Ausserré, H Hervet, F Rondelez. Phys Rev Lett 54:1948, 1985.
47. F Rondelez, D Ausseré, H Hervet. Ann Rev Phys Chem 38:317, 1987.
48. I Caucheteux, H Hervet, R Jerome, F Rondelez. J Chem Soc, Faraday Trans 86:1369, 1990.
49. D Parsons, R Harrop, EG Mahers. Colloids Surfaces 64:151, 1992.
50. MS Kelly, MM Santore. Colloids Surfaces A 96:199, 1995.
51. VA Rebar, MM Santore. Macromolecules 29:6273, 1996.
52. VA Rebar, MM Santore. J Colloid Interface Sci 178:29, 1996.
53. VA Rebar, MM Santore. Macromolecules 29:6262, 1996.
54. MM Santore, MJ Kaufman. J Polym Sci, Part B: Polym Phys 34:1555, 1996.
55. M Santore, Z Fu. Macromolecules 30:8516, 1997.
56. MA Bos, JM Kleijn. Biophys J 68:2573, 1995.
57. EM Lee, RK Thomas, AR Rennie. Europhys Lett 13:135, 1990.
58. T Cosgrove, TL Crowley, K Ryan, JRP Webster. Colloids Surfaces 51:255, 1990.
59. FD Blum. Colloids Surfaces 45:361, 1990.
60. T Cosgrove, PC Griffiths. Adv Colloid Interface Sci 42:175, 1992.
61. T Cosgrove, JW Fergie-Woods. Colloids Surfaces 25:91, 1987.
62. J Van Alsten. Macromolecules 24:5320, 1991.
63. ID Robb, R Smith. Polymer 18:500, 1977.
64. MA Cohen Stuart. Polym J 23:669, 1991.
65. MA Cohen Stuart, GJ Fleer. Ann Rev Mater Sci 26:463, 1996.
66. MA Cohen Stuart, MCP van Eijk, JC Dijt, NG Hoogeveen. Macromol Symp 113:163, 1997.
67. HM Schneider, P Frantz, S Granick. Langmuir 12:994, 1996.
68. E Pefferkorn, A Haouam, R Varoqui. Macromolecules 21:2111, 1988.
69. A Haouam, E Pefferkorn. Colloids Surfaces 34:371, 1988.
70. HE Johnson, H-W Hu, S Granick. Macromolecules 24:1859, 1991.

71. A Elaissari, A Haouam, C Huguenard, E Pefferkorn. J Colloid Interface Sci 149:68, 1992.
72. JC Dijt, MA Cohen Stuart, GJ Fleer. Macromolecules 27:3207, 1994.
73. M Ruths, JN Israelachvili, HJ Ploehn. Macromolecules 30:3329, 1997.
74. PJ Barham. Colloid Polym Sci 265:584, 1987.
75. Y Cohen. Macromolecules 21:494, 1988.
76. G Fuller, J-J Lee. ACS Symposium Series 240:67, 1984.
77. J-J Lee, GG Fuller. Macromolecules 17:375, 1984.
78. J-J Lee, GG Fuller. J Colloid Interface Sci 103:569, 1985.
79. GJ Besio, RK Prud'homme, JB Benziger. Macromolecules 21:1070, 1988.
80. SH Chang, IJ Chung. Macromolecules 24:567–571, 1991.
81. MR Munch, AP Gast. J Chem Soc, Faraday Trans 86:1341, 1990.
82. MR Munch, AP Gast. Macromolecules 23:2313, 1990.
83. C Huguenard, R Varoqui, E Pefferkorn. Macromolecules 24:2226, 1991.
84. F Tiberg, M Malmsten, P Linse, B Lindman. Langmuir 7:2723, 1991.
85. CP Tripp, ML Hair. Langmuir 12:3952, 1996.
86. M Xie, FD Blum. Langmuir 12:5669,1996.
87. MA Awan, VL Dimonie, LK Filippov, MS El-Aasser. Langmuir 13:130, 1997.
88. E Pelletier, A Stamouli, GF Belder, G Hadziioannou. Langmuir 13:1884, 1997.
89. K Eskilsson, F Tiberg. Macromolecules 30:6323, 1997.
90. HD Bijsterbosch, MA Cohen Stuart, GJ Fleer. Macromolecules 31:9281, 1998.
91. H Watanabe, M Tirrell. Macromolecules 26:6455, 1993.
92. A Stamouli, E Pelletier, V Koutsos, E van der Vegte, G Hadziioannou. Langmuir 12:3221, 1996.
93. Z Li, W Zhao, Y Liu, MH Rafailovich, J Sokolov, K Khougaz, A Eisenberg, RB Lennox, G Krausch. J Am Chem Soc 118:10892, 1996.
94. JH Maas, MA Cohen Stuart, GJ Fleer. Thin Solid Films 358:234, 2000.
95. P Frantz, DC Leonhardt, S Granick. Macromolecules 24:1868, 1991.
96. HM Schneider, S Granick. Macromolecules 25:5054, 1992.
97. HE Johnson, JF Douglas, S Granick. Phys Rev Lett 70:3267, 1993.
98. JC Dijt, MA Cohen Stuart, GJ Fleer. Macromolecules 27:3219, 1994.
99. JC Dijt, MA Cohen Stuart, GJ Fleer. Macromolecules 27:3229, 1994.
100. HE Johnson, S Granick. Science 255:966, 1992.
101. P Frantz, S Granick. Macromolecules 28:6915, 1995.
102. L Sjöstrom, T Åkesson, B Jönsson. Ber Bunsenges Phys Chem 100:889, 1996.
103. CW Hoogendam, A de Keizer, MA Cohen Stuart, BH Bijsterbosch, JG Batelaan, PM van der Horst. Langmuir 14:3825, 1998.
104. Q Liu, JS Laskowsky. Int J Mineral Process 27:147, 1989.
105. JG Göbel, NAM Besseling, MA Cohen Stuart, C Poncet. J Colloid Interface Sci 209:129, 1998.
106. AWM de Laat, GLT van den Heuvel, MR Böhmer. Colloids Surfaces A 98:61, 1995.
107. C Geffroy, J Persello, A Foissy, P Lixon, F Tournilhac, B Cabane. Colloids Surfaces A: 162:107, 2000.
108. NG Hoogeveen, MA Cohen Stuart, GJ Fleer. J Colloid Interface Sci 182:146, 1996.
109. A Krabi, MA Cohen Stuart, Macromolecules 31:1285, 1998.
110. MCP van Eijk, MA Cohen Stuart, GJ Fleer. Progr Colloid Polym Sci 105:31, 1997.
111. I Teraoka. Prog Polym Sci 21:89, 1996.
112. M Kawaguchi, S Anada, K Nishikawa, N Kurata. Macromolecules 25:1588, 1992.
113. M Kawaguchi, Y Sakata, S Anada, T Kato, A Takahashi. Langmuir 10:538, 1994.
114. VN Kislenko, AA Berlin, M Kawaguchi, T Kato. Langmuir 12:768, 1996.
115. A Milchev, W Paul, K Binder. Macromol Theor Simul 3:305, 1994.

4

Activity Coefficient and Electrostatics at the Solution–Metal Oxide Interface

JAMES A. WINGRAVE University of Delaware, Newark, Delaware

I. EVOLUTION OF THE ACTIVITY COEFFICIENT

A. Chemistry and Thermodynamics

The activity coefficient was introduced by G. N. Lewis [1–4] a century ago. As described by Lewis, the activity coefficient evolves as a rather convenient method for reconciling thermodynamic properties of ideal solutions (which can usually be calculated) with those of real solutions (which are in most cases incalculable). Thus, the activity coefficient provides the bridge by which real solution properties can be expressed in terms of ideal solution properties. Unfortunately, this is often a "bridge to nowhere" due to the difficulty of calculating the activity coefficient.

It will be the goal of this chapter to examine fundamentally the activity coefficient in order to develop a rigorous and yet intuitive understanding from which activity coefficient calculations can be made for solute molecules located in solution and adsorbed at the solution–solid interface or just near a solution–solid interface. Since, the activity coefficient was developed from thermodynamics, the starting point will be an examination of the thermodynamic principles upon which activity coefficients are based. Very shortly, however, the limitation of thermodynamcis to descriptions of bulk phases requires the incorporation of electrostatics into the discussion. Using chemical thermodynamics and electrostatics, it will be possible to examine the role of the activity coefficient for solute molecules throughout solution–solid interfacial systems. By applying thermodynamic, electrostatic, and mathematical rigor to solutes in solution–solid interfacial systems, proper methods for applying and calculating the activity coefficient will be possible. We shall begin with the thermodynamic description of the activity coefficient.

The theories for thermodynamics were initially established almost exclusively to guide the design of machines to convert heat energy more efficiently into mechanical energy [5, 6]. The applications of classical thermodynamics in the eighteenth and early nineteenth centuries were mostly concerned with changes in temperature, heat (entropy), pressure, and volume among bodies of matter whose composition remained essentially unchanged. In other words, before the twentieth century, thermodynamics was used primarily to predict physical, thermal, or mechanical changes, but not chemical changes.

By the end of the nineteenth century this focus had begun to change and applications of classical thermodynamics to chemical changes (referred to as chemical thermodynamics) had become a popular subject of study [5]. In principle, specifying molecules and molecular reactions in a solution phase posed no problem to classical thermodynamics. The chemical potential expressed partial molar energy changes as the molecular composition of a phase was changed. However, evaluation of the chemical potential turned out to be elusive. To appreciate this problem requires a careful examination of the chemical potential.

B. The Chemical Potential

The classical thermodynamic equation for differential Gibbs free energy changes in a phase in which no chemical or compositional changes occur can be written:

$$dG = -S\,dT + V\,dP \tag{1}$$

When chemical or compositional phase changes do occur, extra terms must be added to the simple Gibbs free energy equation above in order to account for the contribution that compositional changes make in the Gibbs free energy of the phase. Each of the added N terms (one term for each of the N components in the phase) is the product of the chemical potential and the (differential) change in the number of moles for that component contained in the phase. The differential Gibbs free energy equation for chemical and physical changes in a phase is written as

$$dG = -S\,dT + V\,dP + \sum_{i=1}^{i=N} \mu_i\,dn_i \tag{2}$$

For purposes of defining the chemical potential for a single solute, Y^{ν_y}, with a valence, ν_y, in a solution, one must determine how the Gibbs free energy of the solution phase depends on the quantity of solute in the solution. From elementary calculus, the change in Gibbs free energy, G, of a solution phase with an incremental change in the amount, dn_Y, of a solute, Y^{ν_y}, can be determined from Eq. (2) as

$$\mu_Y = \left(\frac{\partial G}{\partial n_Y}\right)_{T,P,n_i \neq n_Y} \tag{3}$$

In Eqs (2) and (3) it can be clearly seen that the chemical potential term in these chemical thermodynamic equations rigorously represents the chemical influences on the Gibbs free energy of a solution phase. However, rigorous evaluation and rigorous representation are not the same. While Eq. (3) serves as a thermodynamic definition for the chemical potential [7] for a single solute, Y^{ν_y}, in any real phase, it provides no explicit general equation for G as a function of n_Y, from which an explicit equation for μ_Y can be derived.

Another route for evaluation of thermodynamic variables, such as chemical potential, is to apply Maxwell's formalism to the exact differential of the Gibbs free energy equation (2). Maxwell showed that the following equation:

$$\left(\frac{\partial M}{\partial z}\right)_x = \left(\frac{\partial N}{\partial x}\right)_z \tag{4}$$

can be applied generally to any exact differential equation of the form:

$$du = M\,dx + N\,dz \tag{5}$$

From Eqs (2), (4), and (5) it can be seen that only two Maxwell equations are possible which involve the chemical potential, μ_Y, of the component Y^{ν_y} as

$$\left(\frac{\partial \mu_Y}{\partial T}\right)_{P,n_Y} = -\left(\frac{\partial S}{\partial n_Y}\right)_{P,T,n_i \neq Y} \equiv -\bar{S}_Y \tag{6}$$

$$\left(\frac{\partial \mu_Y}{\partial P}\right)_{T,n_i} = -\left(\frac{\partial V}{\partial n_Y}\right)_{T,P,n_i \neq Y} \equiv \bar{V}_Y \tag{7}$$

Equations (6) and (7) allow determination of the temperature and pressure dependence of μ_Y from experimental measurements of the partial molar entropy (sometimes called latent heat) and partial molar volume, $\bar{S}_Y$ and $\bar{V}_Y$, respectively. Unfortunately, however, Eqs (6) and (7) provide no explicit general equation for μ_Y as a function of n_Y. From the form of Eq. (2) one can see that μ_Y and n_Y will always appear as a product in the same term in every thermodynamic energy equation. Therefore, it will never be possible to derive a Maxwell equation from which the composition dependence of μ_Y can be experimentally determined.

Including Eqs (3), (6), and (7), no thermodynamic equations have ever been derived from thermodynamic equation (2) which provide an explicit, rigorous, and general equation for the phase composition dependence of μ_Y in terms of experimentally measurable variables [8]. In all probability, it was this dilemma that led to the concept and incorporation of the ideal solution into chemical thermodynamics.

C. The Ideal Phase

In order to determine the composition dependence of the chemical potential of a component in a solution, an alternative derivation is needed. Derivation of the explicit equations for the chemical potential on solution composition begin with a discussion of the chemical potential, μ_g, of a component, g, in a gaseous solution. From Eq. (2) the Gibbs–Duhem equation can be derived as

$$-V\,dp + S\,dT + \sum_{i}^{N} n_i\,d\mu_i = 0 \tag{8}$$

Under the constraints of constant temperature and composition of all components except n_g, Eq. (8) becomes an exact differential equation which can be rearranged into integral form as

$$\int_{\mu_g(p_1)}^{\mu_g(p_2)} d\mu_g = \int_{p_1}^{p_2} \bar{V}_g\,dp\,; \qquad \text{iff, } dT = dn_{i \neq g} = 0 \tag{9}$$

In order to integrate the thermodynamic equation (9), an expression for $\bar{V}_g(p)$ is needed. However, no general, rigorous analytical expression for real gases has ever been found for $\bar{V}_g(p)$. One conventional solution to this problem [9, 10] has been to combine the rigorous thermodynamic equation (9) with the state equation for a hypothetical ideal gas to make integration possible as

$$\mu_g^{ID}(p_2) - \mu_g^{ID}(p_1) = \int_{p_1}^{p_2} \frac{RT}{p}\,dp = RT\ln\frac{p_2}{p_1}\ ;\ dT = dn_i = 0 \tag{10}$$

If the chemical potential change, $\mu_g^{ID}(p_2) - \mu_g^{ID}(p_1)$, in Eq. (10) is taken to represent the chemical potential difference between a gas in an ideal, μ_g^{ID}, and a reference, μ_g^*, state, respectively, then Eq. (10) becomes

$$\mu_g^{ID} = \mu_g^* + RT\ln\frac{p^{ID}}{p_g^*} = \mu_g^* + RT\ln\chi_g^{ID} \tag{11}$$

The partial pressure variable, χ_g^{ID}, in Eq. (11) for the gas, g, in an ideal gas solution provides a convenient correspondence to the mole fraction, χ_Y^{ID}, for a solute, Y^{ν_y}, in an ideal liquid solution. The basis for the correspondence between the partial pressure of an ideal gas, χ_g^{ID}, and the mole fraction of a solute, Y^{ν_y}, in an ideal solution, χ_Y^{ID}, comes from Lewis' concept of escaping tendency.

In defining escaping tendency [11], Lewis and Randall use classical mechanics as an analogy, where the mechanical work performed is regarded as the product of a "capacity" and a "potential". This model is the basis for each term in Eq. (2). Therefore, in the last term of Eq. (2), the chemical potential, μ_i, can be seen as the "potential" with the "capacity unit" represented by either the partial pressure, χ_g^{ID} or the mole fraction, χ_Y^{ID}.

Since χ_g^{ID} and χ_Y^{ID} have been determined as interchangeable "capacity units" by analogy between classical mechanics and thermodynamic equation (2), they can be used interchangably in Eq. (11). Therefore, an equation of state for an ideal solution can be written between the ideal chemical potential, μ_Y^{ID}, and solution mole fraction, χ_Y^{ID}, (defined in terms of ideal and standard state molar concentrations, $[Y^{\nu_y}(ID)]$ and $[Y^{\nu_y}(*)]$) of a solute, Y^{ν_y}, as

$$\mu_Y^{ID} = \mu_Y^* + RT\ln\frac{[Y^{\nu_y}(ID)]}{[Y^{\nu_y}(*)]} = \mu_Y^* + RT\ln\chi_Y^{ID} \tag{12}$$

Equations (11) and (12) are based on ideal solutions (gaseous and liquid, respectively) and are, therefore, only approximately accurate for real liquid or gaseous solutions. The inadequacy of the ideal solution model to predict real solution behavior is testimony to the need for the activity coefficient.

D. The Activity Coefficient

In order to deal with the nonideality of real solutions, the mole fraction variable, χ_Y^{ID}, in Eq. (12), was replaced by Lewis [1–4] with a fictitious mole fraction-like variable, α_Y. The variable α_Y was defined by Lewis as the activity of solute, Y^{ν_y}, and can be regarded as the variable for real solutions which replaces χ_Y^{ID} in the ideal solution equation (12). Thus, for real solutions, μ_Y can be written in analogy to Eq. (12) as

$$\mu_Y = \mu_Y^* + RT\ln\alpha_Y \tag{13}$$

Since, in a real solution, χ_Y and α_Y vary *about* proportionally, it was further propsed by Lewis [1–4] that the difference in proportionality between χ_Y and α_Y

could be expressed as a coefficient, f_Y, referred to as the activity coefficient and defined as

$$f_Y \equiv \frac{\alpha_Y}{\chi_Y} \tag{14}$$

Substitution of Eq. (14) into Eq. (13) gives the general Gibbs–Lewis thermodynamic equation for the chemical potential of solute, Y^{ν_y}, in a real solution by incorporating the definition of the activity coefficient, f_Y, as

$$\mu_Y = \mu_Y^* + kT \ln \chi_Y + kT \ln f_Y \tag{15}$$

In analogy with Eqs (10) and (11), the mole fraction ratio, χ_Y, in Eq. (15) can be expanded as

$$\mu_Y = \mu_Y^* + kT \ln \frac{[Y^{\nu_y}]}{[Y^{\nu_y}(^*)]} + kT \ln f_Y \tag{16}$$

If we now restrict Eqs (12) and (16) to apply to ideal and real equimolar ($[Y^{\nu_y}] = [Y^{\nu_y}(ID)]$) solution phases of solute, Y^{ν_y}, respectively, then the equations can be substrated to give

$$\mu_Y = \mu_Y^* + kT \ln \frac{[Y^{\nu_y}]}{[Y^{\nu_y}(^*)]} + kT \ln f_Y \tag{16}$$

$$-\left(\mu_Y^{ID} = \mu_Y^* + RT \ln \frac{[Y^{\nu_y}(ID)]}{[Y^{\nu_y}(^*)]}\right) \tag{12}$$

$$\mu_Y - \mu_Y^{ID} = RT \ln f_Y \qquad \text{iff, } [Y^{\nu_y}(ID)] = [Y^{\nu_y}] \tag{17}$$

In principle, the activity coefficient, f_Y, can be determined from any of the four equations (14–17) above. Practically, however, in none of the above equations does f_Y appear as the only variable which cannot be directly measured by experiment. At least one variable other than f_Y appears in each equation which is unknown and cannot be directly measured by experiment. Derivation of an equation for f_Y will require determination of the chemical potential difference, $\mu_Y - \mu_Y^{ID}$. For this step we will turn to the subject of electrostatics.

II. CHEMICAL POTENTIAL OF A SOLUTE ION

A. Work to Create a Solute Ion

The chemical potential, μ_Y, of a solute ion, Y^{ν_y}, is, by definition, the reversible work involved in creating an ion in a solution at constant temperature and pressure. The work for this ion creation process has two parts [12]; (1) the work necessary to make a hole in the solution for the ion, w_{Y_h}, and (2) the work of adding electrostatic charge to the ion in a solution, w_{Y_e}. In most solutions, $w_{Y_h} \ll w_{Y_e}$, so we can sensibly ignore w_{Y_h} and proceed with the derivation of an explicit equation for w_{Y_e}.

To an uncharged ion in solution, Y^0, electrostatic charge is transported from a reference location where the electrostatic charge has no interaction energy, as shown in Fig. 1. This reference location is considered to be a vacuum at an infinite distance from the solute ion being charged.

As infinitesimal amounts of electrostatic charge, dq, are transferred reversibly to the uncharged solute ion, Y^0, from the reference location, an infinitesimal amount

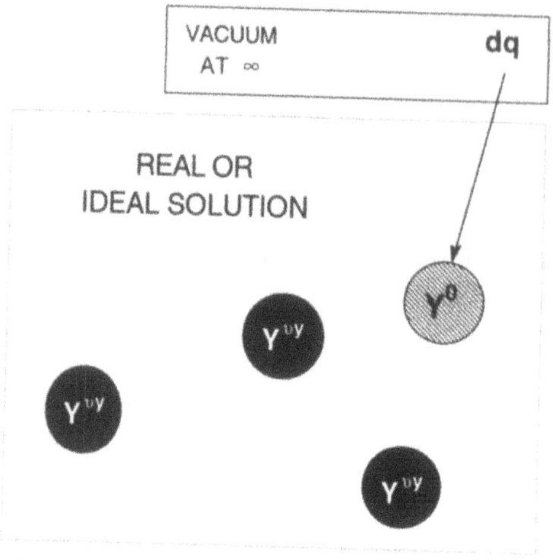

FIG. 1 Ion charging process.

of work, dw_Y, must be done at constant volume to overcome the increasing electrostatic potential, ψ_Y, on the solute ion, Y^0. From the definition of electrostatic potential [13], the work, dw_Y, for this infinitesimal charge transfer process is written as

$$dw_Y = \psi_Y \, dq \tag{18}$$

Invoking the assumption that ionic atmospheres are additive, the electrostatic potential, ψ_Y, on a solute ion is proportional to the amount of added electrostatic charge, q, so

$$\psi_Y = \phi q \tag{19}$$

Substitution of Eq. (19) into Eq. (18) gives

$$dw_Y = \phi^{-1} \psi_Y \, d\psi_Y \tag{20}$$

Equation (20) can be integrated immediately to give an equation, commonly referred to as the Born equation [14]:

$$\int_0^{w_Y} dw_Y = w_Y = \phi^{-1} \int_0^{\psi_Y} \psi_Y \, d\psi_Y = \frac{\phi^{-1} \psi_Y^2}{2} \tag{21}$$

Combining Eqs (21) and (19) gives an expression for the potential energy of a solute molecule as

$$w_Y = \tfrac{1}{2} q \psi_Y \tag{22}$$

From the rigorous equation (22) it can be seen that the reversible work involved in solute ion creation depends on the charge on the ion, q_Y, and the electrostatic potential, ψ_Y, created on the ion surface. It can be seen from Eq. (22) that the solution and solute properties of interest are implicitly contained in the electrostatic potential, ψ_Y, which in turn depends on the electrostatic environment of the solute, Y^{υ_y}.

For the general case of solutions containing differing amounts and types of solute molecules and/or interfaces, only approximate explicit equations for ψ_Y are possible [15]. In later discussions, it will be necessary to find approximations for ψ_Y, so that explicit equations for the activity coefficient can be derived. Of immediate interest is to determine the precise thermodynamic identity of the work depicted in Fig. 1 and expressed in Eqs (17) and (22).

B. Free Energy of a Solute Ion

The energy change for any equimolar process occurring at constant temperature is a work process. If the isothermal, equimolar process is carried out reversibly at constant pressure, the work is Gibbs free energy. If the isothermal process is carried out reversibly at constant volume, the work is Helmholtz free energy.

For the isothermal, equimolar transfer of a solute between two phases, the work involved will depend on the electrostatic interaction of the transferred solute molecule as a function of the spatial distribution of all the solution molecules. This view that molecular interactions in a solution can be regarded solely as electrostatic in origin is a corollary of the Hellman–Feynman theorem which has been succinctly stated by Israelachvili as, " . . . all intermolecular forces are essentially electrostatic in origin . . . once the spatial distribution of the electron clouds has been determined by solving the Schrödinger equation, the intermolecular forces may be calculated on the basis of straightforward classical electrostatics." [16]. Following Hellman and Feynman, it can be concluded that, in the absence of other potential energy sources (i.e., magnetic, atomic, gravitational, etc.), the potential energy of solute–solute and solute–solvent interactions result from electrostatic interactions.

It should be noted that the Hellman–Feynman theorem applies to *all* molecular interactions, not just Coulombic ionic. Therefore, electrostatic equation (38) applies to all types of solutes (ions, dipoles, induced dipoles, hydrogen bonded, etc.) interacting in a solution or at a solid–solution interface.

Due to the distance dependence of electrostatic forces, electrostatic interactions are always presumed to occur at constant volume. Therefore, (electrostatic) interactions among solute and solvent molecules must correspond to Helmholtz and not Gibbs free energies. However, f_Y in Eq. (17) is expressed in terms of the partial molar Gibbs free energy, so it is clear that the electrostatic work equation (21) cannot be substituted directly into Eq. (17). Equation (17) will first have to be rewritten in terms of Helmholtz free energies by beginning with the thermodynamic identity related chemical potential to partial molar Helmholtz free energy for the solute molecule, Y^{ν_y}, as

$$\mu_Y = \bar{A}_Y + p\bar{V}_Y \tag{23}$$

where $\bar{A}_Y$ is the partial molar Helmholtz free energy, p is the externally applied pressure on the solution, and $\bar{V}_Y$ is the solution molar volume of the solute molecule, Y^{ν_y}.

Substitution of Eq. (23) into Eq. (17) gives

$$RT \ln f_Y = \left(\bar{A}_Y - \bar{A}_Y^{\mathrm{ID}}\right) + \left(p\bar{V}_Y - p\bar{V}_Y^{\mathrm{ID}}\right) \tag{24}$$

If the pressure–volume terms in Eq. (24) are approximated with the ideal gas law, they can be seen to be negligible as

$$p\bar{V}_{\mathrm{Y}} - p_0\bar{V}_{\mathrm{Y}^0} \cong RT\left(\frac{\bar{V}_{\mathrm{Y}}}{\bar{V}_{\mathrm{w,vap}}} - \frac{\bar{V}_{\mathrm{Y}^0}}{\bar{V}_{\mathrm{w}^0,\mathrm{vap}}}\right) \ll (\bar{A}_{\mathrm{Y}} - \bar{A}_{\mathrm{Y}}^{\mathrm{ID}}) \tag{25}$$

where $\bar{V}_{\mathrm{w}^0,\mathrm{vap}}$ and $\bar{V}_{\mathrm{w,vap}}$ are the vapor molar volume of the pure solvent and the partial pressure of the solvent over the solution, respectively. Hence, chemical potential and partial molar Helmholtz free energy will henceforth be used interchangably since, as Onsager [17] stated so eloquently, " . . . distinctions between the 'free energies' at constant pressure *viz.* volume etc. . . . of real solutions, are more tedious than important, . . . "!

From Eqs (24) and (25):

$$kT\ln f_{\mathrm{Y}} = \frac{(\mu_{\mathrm{Y}} - \mu_{\mathrm{Y}}^{\mathrm{ID}})}{N_{\mathrm{A}}} \cong \frac{(\bar{A}_{\mathrm{Y}} - \bar{A}_{\mathrm{Y}}^{\mathrm{ID}})}{N_{\mathrm{A}}} = w_{\mathrm{Y}} - w_{\mathrm{Y}}^{\mathrm{ID}} = \frac{1}{2}q(\psi_{\mathrm{Y}} - \psi_{\mathrm{Y}}^{\mathrm{ID}}) \tag{26}$$

where N_{A} is Avogadro's number.

Equation (26) establishes the link between the activity coefficient and the free energy of interaction of a solute ion, Y^{ν_y}, with its surrounding solution. However, it is still not possible to calculate the activity coefficient from Eq. (26). Before activity coefficient evaluation from Eq. (20) is possible, a definition of the electrostatic properties of an ideal solution is needed.

C. Free Energy of a Solute Ion in Real and Ideal Solutions

Derivation of a partial molar Helmholtz free energy equation for an ideal solution will provide a tool by which ideal and real solution behavior can be differentiated. Specifically, we will make use of the fact that the partial molar enthalpy of a real solution will depend on the type and concentration of solutes in a solution while for an ideal solution, the partial molar enthalpy for a solute is independent of the solution composition [18]. As a brief proof of this ideal solution property, consider the defining Eq. (12) for the chemical potential of a solute, Y^{ν_y}, in an ideal solution:

$$\frac{\mu_{\mathrm{y}}^{\mathrm{ID}}}{T} = \frac{\mu_{\mathrm{Y}}^{*}}{T} + k\ln\chi_{\mathrm{y}}^{\mathrm{ID}} \cong \frac{\bar{A}_{\mathrm{y}}^{\mathrm{ID}}}{T} \tag{27}$$

Taking the partial derivative with respect to temperature of Eq. (27):

$$\left[\frac{\partial(\mu_{\mathrm{y}}^{\mathrm{ID}}/T)}{\partial T}\right]_{p,n_i} = \left[\frac{\partial\mu_{\mathrm{Y}}^{*}/T)}{\partial T}\right]_{p} \tag{28}$$

The partial molar enthalpy, $\bar{H}_{\mathrm{Y}}$, for real solution can be expressed in terms of the Gibbs–Helmholtz equation as

$$\left[\frac{\partial(\mu_{\mathrm{Y}}/T)}{\partial T}\right]_{p.n_i} = -\frac{\bar{H}_{\mathrm{Y}}}{T^2} \tag{29}$$

Thus, for ideal solutions, Eqs (28) and (29) can be combined to give

$$\left[\frac{\partial(\mu_{\mathrm{y}}^{\mathrm{ID}}/T)}{\partial T}\right]_{p,n_i} = \left[\frac{\partial(\mu_{\mathrm{Y}}^{*}/T)}{\partial T}\right]_{p} = -\frac{\bar{H}_{\mathrm{Y}}^{\mathrm{ID}}}{T^2} \tag{30}$$

From Eq. (30) it can be seen that the partial molar enthalpy, $\bar{H}_Y^{ID}$, for a solute, Y^{ν_y}, in an ideal solution is equal to a partial derivative term for the standard state chemical potential, μ_Y^*. Since μ_Y^* and all its derivatives must be composition independent, it can be concluded that the partial molar Helmholtz free energy, $\bar{H}_Y^{ID}$, for a solute, Y^{ν_y}, in an ideal solution must be independent of the composition of the ideal solution phase as well. The solution composition independence of $\bar{H}_Y^{ID}$ can now be used to find the functional form of $\bar{A}_Y^{ID}$.

A thermodynamic identity for the partial molar enthalpy, $\bar{H}_Y$, in terms of the Helmholtz free energy, $\bar{A}_Y$, of a solute, Y^{ν_y}, in any solution, real or ideal, can be written:

$$\bar{H}_Y = \bar{A}_Y + p\bar{V}_Y + T\bar{S}_Y \cong \bar{A}_Y + T\bar{S}_Y \tag{31}$$

The approximation in Eq. (31) acknowledges that pressure–volume work makes a negligible contribution to $\bar{H}_Y$ in a solution, as was deduced earlier in Eq. (25). Using another thermodynamic identity for $\bar{S}_Y$ and dropping the approximation sign, Eq. (31) can be written:

$$\bar{H}_Y = \bar{A}_Y - T\left(\frac{\partial \bar{A}_Y}{\partial T}\right)_{T,\bar{V}_Y,} = -T^2\left[\frac{\partial(\bar{A}_Y/T)}{\partial T}\right]_{T,\bar{V}_Y,} \tag{32}$$

Evaluation of the differentials in Eq. (32) requires an explicit expression for $\bar{A}_Y$. However, no general solution to Eq. (32) exists due in large part to the difficulty of obtaining explicit expressions for the density distribution of electrostatic charge around the solute molecule, Y^{ν_y}, which are both general and rigorous.

The relationship between $\bar{A}_Y$ and electrostatic potential, ψ_Y, can be intuitively inferred from continuum mechanics [19] to depend on: (1) the solution temperature, T, (2) the dielectric strength of the solvent, and (3) the spatial distribution or concentration, $\bar{c}_i$, of solute ions around the central solute species, Y^{ν_y}. Further evidence confirming this intuition can be gained from approximate treatments of solutions such as the Debye–Hückel theory [20], which will be discussed later. It would therefore be reasonable to presume that a solution for $\bar{A}_Y$ would contain at least one variable with temperature dependence, $u(T)$, and at least one variable with concentration dependence $v(\bar{c}_i)$ or $\bar{A}_Y[u(T), v(\bar{c}_i)]$ (assuming no solvent change). It will be necessary to return later to determine if the proposed functional form of $\bar{A}_Y[u(T), v(\bar{c}_i)]$ is a satisfactory representation.

Substituting the functional equation, $\bar{A}_Y = u(T) \cdot v(\bar{c}_i)$, into Eq. (32) gives

$$\bar{H}_Y = -T^2\left[\frac{\partial(\bar{A}_Y/T)}{\partial T}\right]_{T,\bar{V}_Y.} = -T^2\left\{\frac{\partial[u(T)\cdot v(\bar{c}_i)/T]}{\partial T}\right\}_{T,\bar{V}_Y} \tag{33}$$

Expanding Eq. (33) with the chain rule gives

$$\begin{aligned}\bar{H}_Y &= -T^2\left\{\frac{\partial[u(T)(\bar{c}_i)/T]}{\partial T}\right\}_{T,\bar{V}_Y}\\ &= uv\left\{1-\left(\frac{T}{v}\right)\left[\frac{\partial v}{\partial T}\right]_{T,\bar{V}_Y}-\left(\frac{T}{u}\right)\left[\frac{\partial u}{\partial T}\right]_{T,\bar{V}_Y}\right\}\end{aligned} \tag{34}$$

Equation (34) would describe the partial molar enthalpy, $\bar{H}_Y$, of a solute in a real solution. From the discussion above, it was shown that for an ideal solution, $\bar{H}_Y^{ID}$ is a special case of $\bar{H}_Y$, which is independent of solution concentration. In Eq. (34), $\bar{H}_Y$ will become composition independent only if the composition independent term, $v(\bar{c}_i)$, vanishes. Setting $v(\bar{c}_i) = $ constant, in Eq. (34), gives the desired equation for $\bar{H}_Y^{ID}$ as

$$\bar{H}_Y^{ID} = u\left\{1 - \left(\frac{T}{u}\right)\left(\frac{\partial u}{\partial T}\right)_{T,\bar{V}_Y}\right\} \tag{35}$$

Equations (32), (34), and (35) can be combined to give

$$(-T^2)\left[\frac{\partial(\bar{A}_Y^{ID}/T)}{\partial T}\right]_{T,\bar{V}_Y} = u\left\{1 - \left(\frac{T}{u}\right)\left(\frac{\partial u}{\partial T}\right)_{T,\bar{V}_Y}\right\} \tag{36}$$

A solution for Eq. (36) can be seen to be

$$\bar{A}_Y^{ID} = u(T) \tag{37}$$

Comparing Eq. (37) with the equation for the partial molar Helmholtz free energy, $\bar{A}_Y = u(T) \cdot v(\bar{c}_i)$, one clearly sees that $\bar{A}_Y^{ID} = u(T)$ or that $\bar{A}_Y^{ID}$ is simply the composition *independent* partial molar Helmholtz free energy for real solutions. The meaning of "composition independent" should not be taken to mean pure solvent or zero solute ion concentration. This subtle but important difference will be examined later.

From the previous discussion it is now possible to specify the functional dependence of the activity coefficient.

$$kT \ln f_Y \cong \frac{\bar{A}_Y(\bar{c}_i, T) - \bar{A}_Y^{ID}(T)}{N_A} = \frac{1}{2}q[\psi_Y(\bar{c}_i, T) - \psi_Y^{ID}(T)] \tag{38}$$

Equation (38) is an implicit equation for the activity coefficient for the solute ion, Y^{ν_y}. Explicit equations for the electrostatic potentials would allow the activity coefficient to be calculated. However, as stated earlier, explicit, rigorous expressions for electrostatic potentials do not exist for solute ions in solutions. Instead, explicit equations will be derived for the activity coefficient employing solution models.

III. ACTIVITY COEFFICIENT EQUATIONS

A. Solute Ions—Debye–Hückel Model

The Debye–Hückel treatment is based on continuum electrostatics. Continuum solutions are composed of ions distributed throughout a continuum solvent, w, which has no molecular structure but has a solvent dielectric strength, ε_w. The solvent dielectric strength is completely independent of the type and/or amount of solute molecules in the continuum solution.

The electrostatic interaction energy for a single solute ion, Y^{ν_y}, in a continuum solution can be envisioned as the work required to transfer an amount of electrostatic charge, $\nu_y e$, to the ion from an infinitely distant vacuum. The electrostatic interaction energy for such a transfer can be determined as a function of the electrostatic charge, $\nu_y e$, and the electrostatic potential, $\psi_Y(r)$ measured radially from the center of the solute ion, Y^{ν_y}.

In a continuum solution, the electrostatic potential, $\psi_Y(r)$, surrounding each solute ion, must be a solution to the Poisson–Boltzmann equation. The Debye–Hückel solution to the linearized Poisson–Boltzmann equation [15, 20, 23] gives the electrostatic potential at a distance, r, from a solute ion, Y^{ν_y}, of diameter, a_Y, and valence, ν_y, in a solvent with dielectric strength, ε_w, as

$$\psi_Y(r) = \left(\frac{\nu_y e}{4\pi\varepsilon_0\varepsilon_w}\right)\left[\frac{\exp(\kappa a_Y)}{1+\kappa a_Y}\right]\left[\frac{\exp(-\kappa r)}{r}\right] \tag{39}$$

At $r = a_Y$, which is the minimum distance of approach for an ion in solution to the solute ion, Y^{ν_y}, the potential would be

$$\psi_Y(a_Y) = \left(\frac{\nu_y e}{4\pi\varepsilon_0\varepsilon_w a_Y}\right)\left(\frac{1}{1+\kappa a_Y}\right) \tag{40}$$

where,

$$\kappa \equiv \left(\frac{2000e^2 N_A}{\varepsilon_w\varepsilon_0 kT} I\right)^{1/2} = \left\{\begin{array}{l}\text{Inverse thickness}\\ \text{of the diffuse}\\ \text{ion atmosphere}\end{array}\right\}[=]\text{m}^{-1} \tag{41}$$

where e is the electronic charge in coulombs, ε_0 is the dielectric permittivity of a vacuum ($8.854 \times 10^{-12}\,\text{C}^2\,\text{J}^{-1}\,\text{m}^{-1}$), I is the ionic strength [24], N_A is Avogadro's number, and the first parenthetical expression in Eq. (40) is the Coulomb electrostatic potential.

Equation (40) represents an explicit equation for the electrostatic potential at the surface of an ion in a real continuum solution. The accuracy of Eq. (40) is limited by the approximations of the Debye–Hückel theory.

In Eq. (22) it was shown that half the product of the electrostatic charge on an ion and the electrostatic potential on the ion surface gives the Helmholtz free energy, $\bar{A}_Y$, for that ion. Therefore, Eqs (22) and (40) can be combined to give an electrostatic equation for the partial molar Helmholtz free energy, $\bar{A}_Y$, of an ion, Y^{ν_y}, as

$$\bar{A}\nu_y = \frac{\nu_y e\psi_Y(a_Y)}{2} = \left[\frac{(\nu_y e)^2}{8\pi\varepsilon_0\varepsilon_w a_Y}\right]\left(\frac{1}{1+\kappa a_Y}\right) \tag{42}$$

A similar expression for the ideal partial molar Helmholtz free energy, $\bar{A}_Y^{\text{ID}}$, cannot be determined directly from the preceding equations because no method for determining the electrostatic potential of an ion in an ideal solution is known. Only if the ionic structure of an ideal solution was known, could the electrostatic potential of an ion in an ideal solution be determined from electrostatic theory. We will return to the thermodynamic method used earlier to determine the functional dependence of $\bar{A}_Y^{\text{ID}}$, taking advantage of the fact that the partial molar enthalpy for real, $\bar{H}_Y$, and ideal, $\bar{H}_Y^{\text{ID}}$, solutions are dependent and independent of solution composition, respectively.

A thermodynamic identity for the partial molar enthalpy, $\bar{H}_Y$, can be written:

$$\bar{H}_Y = \bar{A}_Y + p\bar{V}_Y + T\bar{S}_Y \cong \bar{A}_Y + T\bar{S}_Y \tag{43}$$

The approximation in Eq. (43) acknowledges that pressure–volume work makes a negligible contribution to $\bar{H}_Y$ in a solution, as was deduced earlier in Eq. (29). Using another thermodynamic identity:

$$\bar{H}_Y = \bar{A}_Y - T\left(\frac{\partial \bar{A}_Y}{\partial T}\right)_{T,\bar{V}_Y} = -T^2\left[\frac{\partial(\bar{A}_Y/T)}{\partial T}\right]_{T,\bar{V}_Y} \tag{44}$$

Substituting Eq. (40) into Eq. (32) and differentiating:

$$\bar{H}_Y = \left[\frac{(\nu_y e)^2}{8\pi\varepsilon_0\varepsilon_w a_Y}\right]\left(\frac{1}{1+\kappa a_Y}\right)\left[1+\left(\frac{T}{\varepsilon_w}\right)\left(\frac{\partial\varepsilon_w}{\partial T}\right)_{\bar{V}_Y}+\left(\frac{T}{1+\kappa a_Y}\right)\left(\frac{\partial\kappa}{\partial T}\right)_{\bar{V}_Y}\right] \tag{45}$$

For ideal solutions the ideal partial molar enthalpy, $\bar{H}_Y^{ID}$, must remain invariant with changes in the concentration of ions in the solution. Since κ is the only concentration dependent variable in Eq. (45), one can deduce:

$$\bar{H}_Y^{ID} = \bar{H}_Y(\kappa = 0) \tag{46}$$

From Eqs (45) and (46) the equation for $\bar{H}_Y^{ID}$ can be written:

$$\bar{H}_Y^{ID} = \left[\frac{(\nu_y e)^2}{8\pi\varepsilon_0\varepsilon_w}\right]\left(\frac{1}{a_Y}\right)\left[1+\left(\frac{T}{\varepsilon_w}\right)\left(\frac{\partial\varepsilon_w}{\partial T}\right)_{\bar{V}_Y}\right] \tag{47}$$

Since Eq. (32) is a thermodynamic identity, it must be obeyed for ideal solutions, so

$$\bar{H}_Y^{ID} = -T^2\left[\frac{\partial(\bar{A}_Y^{ID}/T)}{\partial T}\right]_{T,\bar{V}_Y} \tag{48}$$

Equations (47) and (48) can be combined to give

$$T^2\left[\frac{\partial(\bar{A}_Y^{ID}/T)}{\partial T}\right]_{T,\bar{V}_Y} = -\left[\frac{(\nu_y e)^2}{8\pi\varepsilon_0\varepsilon_w}\right]\left(\frac{1}{a_Y}\right)\left[1+\left(\frac{T}{\varepsilon_w}\right)\left(\frac{\partial\varepsilon_w}{\partial T}\right)_{\bar{V}_Y}\right] \tag{49}$$

The solution for Eq. (49) takes the form:

$$\bar{A}_Y^{ID} = \left[\frac{(\nu_y e)^2}{8\pi\varepsilon_0\varepsilon_w a_Y}\right] \tag{50}$$

That Eq. (50) is a solution for Eq. (49) can be confirmed by substitution of the former into the latter and differentiating the result.

By comparison of Eqs (50) and (22), the ideal free energy can be seen as one-half the product of the ion charge, $\nu_y e$, and the ideal solution electrostatic potential, $\psi_Y^{ID}(a_Y)$, at the surface of a single solute ion, Y^{ν_y}, in a pure dielectric solvent:

$$\bar{A}_Y^{ID} = \left[\frac{(\nu_y e)^2}{8\pi\varepsilon_0\varepsilon_w a_Y}\right] = \frac{(\nu_y e)\psi_Y^{ID}(a_Y)}{2} \tag{51}$$

Equations (40) and (51) allow comparison of the electrostatic potentials at the surface of a solute ion, Y^{ν_y}, in a real, $\psi_Y(a_Y)$, and an ideal, $\psi_Y^{ID}(a_Y)$, solution:

$$\psi_Y(a_Y) = \left(\frac{\nu_y e}{4\pi\varepsilon_0\varepsilon_w a_Y}\right)\left(\frac{1}{1+\kappa a_Y}\right) \quad \left\{\begin{array}{l}\text{Electrostatic potential}\\ \textit{with}\text{ screening by}\\ \text{the diffuse ion atmosphere}\end{array}\right\} \tag{52}$$

$$\psi_Y(a_Y) = \left(\frac{\nu_y e}{4\pi\varepsilon_0\varepsilon_w a_Y}\right) \quad \left\{\begin{array}{l}\text{Electrostatic potential}\\ \textit{without}\text{ screening by}\\ \text{the diffuse ion atmosphere}\end{array}\right\} \tag{53}$$

The difference in the electrostatic potentials in Eqs (52) and (53) is the expression, $1/1+\kappa a_Y$, with κ dependent on solute ion concentration as can be seen from Eq. (41). As κ increases with increasing ion concentration, the electrostatic potential of an ion in a real solution will decrease in agreement with Eq. (52). This inverse relationship between concentration and electrostatic potential of solute ions in real solutions is referred to as "screening" of the electrostatic potential by a diffusion ion atmosphere.

The diffusion ion atmosphere can be visualized as a slight concentration of ions of opposite charge (counter ions) around and near an ion in a real solution. Let us return to the ion charging process to examine how a diffuse ion atmosphere forms and how it affects the electrostatic potential of an ion in solution.

Consider the ion charging process as a two-step process: (1) an ion is fully charged without allowing any surrounding ions to move, and (2) the surrounding ions are allowed to move, forming the diffuse ion atmosphere. In Step 1 the ion is charged while solution composition remains unchanged. If solution composition remains invariant, ion charging is independent of solution composition or ideal. Equation (53), which is simply the Coulomb electrostatic potential for an ion in a pure dielectric medium, would describe the ideal electrostatic potential of an ion after step 1. In step 2, counter ions concentrate near the ion surface so that the electrostatic charge on the ion surface (initially ideal solution) is partially offset by the proximity of ions of opposite charge. The summed (or "screened") electrostatic charge of the ideal surface potential and the ion atmosphere is the electrostatic potential for ions in real solutions and is given by Eq. (52).

The reversible work or free energy at constant volume for adding electrostatic charge, $q = \nu_y e$, to ions in real and ideal solutions would be

$$\bar{A}_Y = \frac{1}{2}q\psi_Y = \left[\frac{(\nu_y e)^2}{8\pi\varepsilon_0\varepsilon_w a_Y}\right]\left(\frac{1}{1+\kappa a_Y}\right) \tag{54}$$

$$\bar{A}_Y^{ID} = \frac{1}{2}q\psi_Y^{ID} = \left[\frac{(\nu_y e)^2}{8\pi\varepsilon_0\varepsilon_w a_Y}\right] \tag{55}$$

The absence of screening in Eq. (55) indicates that counter ions play no role in determining the free energy of solute ions in ideal solutions. This is consistent with the derivation requirement that the free energy of ions in ideal solutions be independent of solute concentration.

For real solutions, the role of solute ion concentration on free energy is manifest in the expression $1/1+\kappa a_Y$. This effect can best be understood by again returning to the ion charging process. An increment of work must be expended each time an increment of electrostatic charge is added to the surface of a solute ion. The source of this expenditure of work results from the fact that an ion with some amount of electrostatic charge will resist the addition of even more electrostatic charge of like sign. However, if counter ions move nearer the ion surface each time an increment of charge is added to an ion, the counter ion approach favors the addition of electrostatic charge thereby lowering the amount of energy required to make each incremental electrostatic charge addition. Hence, the increasing counter ion concentration with charge build up at the ion surface slightly decreases the work of charging the ion and simultaneously screens the electrostatic potential interaction among like-charged ions.

Equations (54) and (55) can be substituted into Eq. (26) to give

$$kT\ln f_{\mathrm{Y}} = \frac{(\bar{A}_{\mathrm{Y}} - \bar{A}_{\mathrm{Y}}^{\mathrm{ID}})}{N_{\mathrm{A}}} = \left[\frac{(\nu_{\mathrm{y}}e)^2}{8\pi\varepsilon_0\varepsilon_{\mathrm{w}}a_{\mathrm{Y}}}\right]\left[\left(\frac{1}{1+\kappa a_{\mathrm{Y}}}\right) - 1\right]$$
$$= -\left(\frac{\nu_{\mathrm{y}}e}{2}\right)\left[\frac{\nu_{\mathrm{y}}e}{4\pi\varepsilon_0\varepsilon_{\mathrm{w}}a_{\mathrm{Y}}}\right]\left(\frac{\kappa a_{\mathrm{Y}}}{1+\kappa a_{\mathrm{Y}}}\right) = -\left(\frac{\nu_{\mathrm{y}}e}{2}\right)\psi_{\mathrm{Y}}^{\mathrm{ID}}\left(\frac{\kappa a_{\mathrm{Y}}}{1+\kappa a_{\mathrm{Y}}}\right) \quad (56)$$

Solving for the activity coefficient:

$$f_{\mathrm{y}} = \exp\left[-\frac{(\nu_{\mathrm{y}}e)^2}{8\pi\varepsilon_0\varepsilon_{\mathrm{w}}kT}\left(\frac{\kappa}{1+\kappa a_{\mathrm{Y}}}\right)\right] \quad (57)$$

In Eq. (57) a selection of variables can be grouped together, which have the units of length:

$$q_{\mathrm{B}} \equiv \frac{e^2}{8\pi\varepsilon_0\varepsilon_{\mathrm{w}}kT} [=] \text{length} \quad (58)$$

This group of variables has units of length and was shown by Bjerrum [25, 15] to be the distance at which the potential energy of attraction between two oppositely charged ions equals the kinetic potential energy due to molecular motion, $2kT$. Bjerrum used the sumbol q (q_{B} will be used instead of q as the latter is more commonly used to designate electrostatic force) for this length and regarded oppositely charged ions separated by less than q_{B} to be associated.

Equation (58) can be seen as independent of any ion properties; it is dependent only on temperature and solvent properties. For aqueous solutions and primitive interfacial systems with aqueous solutions at 25°C, Eq. (58) can be evaluated to give a distance of 3.44 Å. For activity coefficient calculations the significance of the distance q_{B} can be appreciated by the fact that it will appear in equations for both solute and surface site ions. The activity coefficient equation for an ion in solution in Eq. (57) is

$$f_{\mathrm{Y}} = \exp\left\{-\left[\frac{(\nu_{\mathrm{y}}e)^2}{8\pi\varepsilon_0\varepsilon_{\mathrm{w}}kT}\right]\left(\frac{\kappa}{1+\kappa a_{\mathrm{Y}}}\right)\right\} = \exp\left(-\frac{\nu_{\mathrm{y}}^2 q_{\mathrm{B}}\kappa}{1+\kappa a_{\mathrm{Y}}}\right) \quad (59)$$

Equation (59) can be further rearranged to separate variables, dependent and independent of ion type, as

$$f_{\mathrm{Y}} = \exp\left(-\frac{\nu_{\mathrm{y}}^2\kappa q_{\mathrm{B}}}{1+\kappa a_{\mathrm{Y}}}\right) = \left[\mathrm{e}^{(-\kappa q_{\mathrm{B}})}\right]^{\left(\frac{\nu_{\mathrm{y}}^2}{1+\kappa a_{\mathrm{Y}}}\right)} \equiv Z^{\left(\frac{\nu_{\mathrm{y}}^2}{1+\kappa a_{\mathrm{Y}}}\right)} \quad (60)$$

where

$$Z \equiv e^{(-\kappa q_B)} \tag{61}$$

Several important insights can be gleaned from Eqs (56), (57), (60), and (61).

1. From the form of Eq. (60), the variable Z can be seen as the solute ion screening variable, approaching unity (no ionic screening) for dilute solutions of low ionic strength where $\kappa \rightarrow 0$, and approaching zero (complete ionic screening) for concentrated solutions of high ionic strength where $\kappa \rightarrow \infty$. The exponent ν_y^2 modifies the screening variable Z for the intensity of the screening through the valence of the screened solute ion.

2. The ideal partial molar enthalpy equation (47) was derived from Eq. (45) under the constraint that in an ideal solution, $\kappa = 0$. The interpretation of this constraint is that the thickness of the Debye–Hückel diffuse ion atmosphere, κ^{-1}, has become infinite, indicative of the absence of electrostatic screening among solute ions. While κ is required to be zero for ideal solutions and real solutions as they approach infinite dilution, it does not restrict $\bar{H}_Y^{ID}$ to solutions at infinite dilution. Since no restraint with regard to solute concentration was employed in the derivation of Eq. (47), one is led to the interesting conclusion that κ is made to go to zero in an ideal solution by requiring the absence of screening of all the ions. This result depicts ideal solutions as differing from real solutions in that ions in the former interact with full electrostatic potential in the absence of a diffuse layer of charged solute ions. From this perspective each solute ion in an ideal solution is deduced to be surrounded by neighboring solute ions whose positions have *not* been altered by electrostatic interactions, i.e., no screening. In this way, ions in ideal solutions are independent of the entropy reduction induced by electrostatic attractive forces among each central ion and its surrounding counterions, the latter making up the diffuse layer of electrostatic charge responsible for ion screening in real solutions.

3. The ionic diameters used in the Debye–Hückel theory have been described as, "... the 'effective diameter' of the ion in solution. Since no independent method is available for evaluating a_i this quantity is an empirical parameter, but the a_i's obtained are of a magnitude for ion sizes." [26]. Values for these effective ionic diameters have been experimentally evaluated and can be found in the literature [26, 27].

4. The derivation of the Debye–Hückel equation for the activity coefficient is based on the linearized Boltzmann equation for electrostatic charge distribution around an ion. This limits the applicability of Eq. (57) to solutes with low surface potentials, which occurs for solution concentrations of monovalent ions of < 0.01 M. However, it is important to note that the method used for deriving activity coefficient equation (25) is based on rigorous thermodynamics and is not limited by the Debye–Hückel theory. If, for example, the Gouy–Chapman equation [22] was used in Eq. (30), more rigorous equations for $\bar{A}_Y$, $\bar{A}_Y^{ID}$, and f_Y would be expected.

5. A central ion, Y^{ν_y}, with valence, ν_y surrounds itself with an ion cloud of net electrostatic charge, $-\nu_y e$. As a result of the negative exponent, the value of the activity coefficient calculated from Eq. (60) for a solute ion can never be greater than unity. However, experimentally determined activity coefficients can exceed unity. The source of this discrepancy results largely from the use of the *linearized* Poisson–Boltzmann equation to obtain the electrostatic potentials for solute ions.

The activity coefficient equation (38) can be seen to be dependent on the "screened" electrostatic potential, $\psi_Y - \psi_Y^{ID}$, for an ion, Y^{ν_y}:

$$kT \ln f_Y = \frac{1}{2}\nu_y e(\psi_Y - \psi_Y^{ID}) \tag{62}$$

Equation (62) is rigorous (assuming Gibbs and Helmholtz free energies are the same for ions in a solid–solution interfacial system) but requires explicit electrostatic potential equations in order to obtain an explicit activity coefficient equation. Since rigorous, explicit electrostatic potential equations for solute and surface site ions have yet to be derived, the approximate electrostatic potential equations, which are solutions to the linearized Poisson–Boltzmann equation, were used here and by Debye–Hückel to give

$$kT \ln f_y = -\frac{(\nu_y e)^2}{8\pi\varepsilon_0\varepsilon_w}\left(\frac{\kappa}{1+\kappa a_Y}\right) \tag{63}$$

From the linearized Poisson–Boltzmann equation the work, $-(\nu_y e)^2/8\pi\varepsilon_0\varepsilon_w(\kappa/1+\kappa a_Y)$, due to ion screening is mathematically always less than zero, which restricts the activity coefficient calculated by Eq. (63) to a range from zero to unity.

Solutions to the Poisson–Boltzmann equation in which the exponential charge distribution around a solute ion is not linearized [15] have shown additional terms, some of which are positive in value, not present in the linear Poisson–Boltzmann equation [28, 29]. From the form of Eq. (62) one can see that whenever the work, $q(\psi_Y - \psi_Y^{ID})$, of creating the electrostatic screening potential around an ion becomes positive, values in excess of unity are possible for the activity coefficient. Other methods that have been developed to extend the applicable concentration range of the Debye–Hückel theory include mathematical modifications of the Debye–Hückel equation [15, 26, 28, 29] and treating solution complexities such as: (1) "ionic association" as proposed by Bjerrum [15, 25], and (2) quadrupole and second-order dipole effects estimated by Onsager and Samaras [30], etc.

6. Derivation of the activity coefficient equation (57) began with the Debye–Hückel electrostatic potential, Eq. (40), which employs only the Coulombic potential for ions. However, other types of electrostatic interactions such as: ion–dipole, dipole–dipole (Keesom), ion–atom, dipole–atom (Debye), atom–atom (London), hydrogen bonding, etc., are operative in real solutions. Of all the electrostatic forces, coulombic interaction is by far the largest in magnitude. In most solutions, particularly electrolyte solutions, all but the Coulombic interactions can be ignored, based on estimates of interaction energies [21, 22]: Coulombic:Keesom:London $\approx$ 800:3:1.

Except at absolute zero, every molecule must have one or more sources of electrostatic potential.Even if a molecule bears no net electrostatic charge, atomic dipoles which result from the motion of electrons in their orbits around a nucleus will give rise to dispersive van der Waals or London interactions. These atomic dipoles insure that a solute molecule has a small but finite electrostatic interaction energy with the surrounding solution molecules. For example, in aqueous ethanol solutions, the dipole–dipole interaction between ethanol and water molecules becomes the primary interaction energy. Ethanol molecules are not ionic, so use of the Debye–Hückel equation (57), based on Coulombic interaction, cannot be used to determine the activity coefficient of an aqueous ethanol solution. At sensible ethanol concentrations, the activity coefficient would be expected to be close to, but not exactly equal to, unity. Thus, improved accuracy and extended applicability of the activity coefficient equations would be expected if free energy expressions substituted into Eq. (26) included both Coulombic and non-Coulombic ion interaction energies.

7. The activity coefficient in Eq. (56) depends on the expression $\bar{A}_Y - \bar{A}_Y^{ID}$. From earlier discussions, it was found that this free energy difference represents a difference in screened and unscreened partial molar Helmholtz free energies. Thus, the activity coefficient can be seen to result from the partial molar free energy, which is due to the screening of the counter ion atmosphere surrounding each ion in any real solution. This definition is consistent with the view that the activity coefficient predicts electrostatic screening in real solutions or the difference between real and ideal solution properties.

8. The free energy of a solute ion is dependent on the work of electrostatic charging, w_{Y_e}, and the work of making a hole in the solution into which the uncharged ion is placed, w_{Y_h}. The latter energy has been ignored throughout the derivation of the activity coefficient equations (56) and (57). The free energies, upon which the activity coefficient is functionally dependent in Eq. (50), *each* contain implicitly the work of hole making, w_{Y_h}. Thus, since the activity coefficient is dependent on the *difference* in free energies of ideal and real solution ions, any error in neglecting w_{Y_h} would be minimized.

B. Interfacial Ions–Primitive Interfacial Model

An explicit rigorous solution for an ion at a solution–solid interface does not exist at present. Explicit solutions for the electrostatic potential experienced by an ion at a solution–solid interface can be obtained only for model interfaces. For the following treatment, the primitive interfacial model [31] will be used. The primitive interface is composed of three parts: (1) a semi-infinite continuum dielectric solid, (2) a semi-infinite solution containing solute molecules in a continuum dielectric solvent, and (3) a Stern layer of ions lying between the two semi-infinite phases, as shown in Fig. 2.

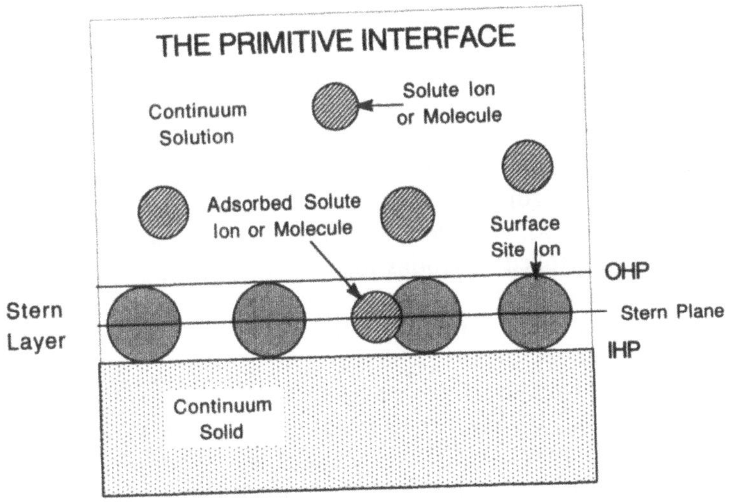

FIG. 2 The primitive interfacial system.

The semi-infinite continuum solid has a dielectric strength, ε_s. The semi-infinite solution is identical to that used in the Debye–Hückel model, consisting of a continuum solvent with dielectric strength, ε_w, and solute ions, Y^{ν_y}, with ionic diameter, a_Y. Hence, the primitive interfacial model can be considered to be at about the same level of electrostatic sophistication as the Debye–Hückel model for electrolyte solutions.

Between the two semi-infinite phases lies a Stern layer which contains continuum solvent with dielectric strength ε_w, surface site ions, and adsorbed solute ions. The thickness of the Stern layer, a_s, corresponds to the diameter, a_s, of a single surface site ion, S^s. The surface site ions may or may not be bound to solute ions. The presence of adsorbed solute ions will not affect a_S, but will change the valence of the surface site. In general, the area of a surface site ion, A_s, will be $A_s \geq a_s^2$. This relationship can be seen qualitiatively in Fig. 2 and understood quantitatively when one consideres that ionic diameters are nominally 2–3 Å, while surface sites are nominally spaced 7–10 Å (the electrostatic surface charge density on a fully ionized surface is of the order of 0.3 coulombs m^{-2} [32], which corresponds to an electrostatic surface site density of 50–100 $Å^2$ per surface site). The volume of a surface site not occupied by a surface site ion is filled with continuum solvent. The surface site area, A_s, for a metal oxide powder with a specific surface area, $\bar{\Sigma}$, cation exchange capacity, X_c, and N_A as Avogadro's number is determined to be

$$A_s = \frac{\bar{\Sigma}}{X_c N_A} [=] \frac{\frac{m^2}{g}}{\left(\frac{\text{mole surface sites}}{g}\right)\left(\frac{\text{surface sites}}{\text{mole surface sites}}\right)} [=] \frac{m^2}{\text{surface site}} \tag{64}$$

The Stern plane (SP) is the planar surface containing the centers of all the surface sites. The plane between the Stern layer and the solid phase is the inner Helmholtz plane (IHP). The plane between the Stern layer and the solution phase is the outer Helmholtz plane (OHP). The IHP is the origin of the x-axis. The x-axis measures distance perpendicular to the IHP with the SP located at $x = \frac{1}{2}a_s$.

The electrostatic potential in the solution phase has a surface electrostatic potential, ψ_s, at the SP. Although the Stern layer contains "discrete" surface sites, the electrostatic surface potential is not discrete but "smeared out" evenly over the SP so that the electrostatic potential varies only perpendicularly with increasing x from the SP ($x \geq \frac{1}{2}a_s$). How the electrostatic potential varies at $x < \frac{1}{2}a_s$ is of no interest and will not be examined.

We can now proceed to determine the activity coefficient for a surface site, f_s, at a primitive interface. From Eq. (26) it can be seen that determination of f_s requires explicit expressions for the real, ψ_s, and ideal, ψ_S^{ID}, electrostatic potentials of the surface site ion. Derivation of explicit equations for the electrostatic potentials, ψ_s and ψ_S^{ID}, at the primitive interface for surface site ions will proceed as for the solute ion except for the boundary constraint imposed on the surface site ions by the two semi-infinite contiguous phases.

The electrostatic potential for a surface site ion at the primitive interace is taken from a model first derived by Wagner [33] and Onsager and Samaras [30] using the method of electrical images. An electrical point charge, e, is positioned in a solution a distance, x, above an interface, and an image charge is placed in the solid a distance, $-x$, below the interface, as shown in Fig. 3.

The electrostatic potential, ψ_p, at any point, p, in the system is the sum of the two potentials resulting from a point charge at x and its image charge at $-x$ or

$$\psi_p = \frac{e}{4\pi\varepsilon_0 e_w r_1} + \frac{(\varepsilon_w - \varepsilon_s)}{(\varepsilon_w + \varepsilon_s)} \frac{e}{4\pi\varepsilon_0\varepsilon_w r_2} \tag{65}$$

The second term on the right-hand side of Eq. (65) is the electrostatic potential at point p due to the image charge. The work, w_p, required to place a unit of electrostatic charge at point p can be deduced from Eq. (22) to be

$$w_p = \frac{1}{2} e\psi_p = \frac{(\varepsilon_w - \varepsilon_s)}{(\varepsilon_w + \varepsilon_s)} \frac{e^2}{8\pi\varepsilon_0\varepsilon_w r_2} \tag{66}$$

Since point p can be anywhere in the system, the work of adding a unit of electrostatic charge at x due to the image charge at $-x$ is given by Eq. (66) when point p is moved to the electrostatic charge, e, by requiring that $r_2 = 2x$, and

$$w_e = \frac{(\varepsilon_w - \varepsilon_s)}{(\varepsilon_w + \varepsilon_s)} \frac{(+1e)^2}{8\pi\varepsilon_0\varepsilon_w r_2}\bigg|_{r_2=2x} = \frac{(\varepsilon_w - \varepsilon_s)}{(\varepsilon_w + \varepsilon_s)} \frac{(+1e)^2}{16\pi\varepsilon_0\varepsilon_w x} \tag{67}$$

Returning to Eqs (22) and (67) the electrostatic potential, ψ_{e°, for an electrostatic charge, e, in a pure dielectric solvent with dielectric strength, ε_w, due to its image charge in a contiguous solid with dielectric strength, ε_s, can be seen as

$$\psi_e = \frac{(\varepsilon_w - \varepsilon_s)}{(\varepsilon_w + \varepsilon_s)} \frac{e}{8\pi\varepsilon_0\varepsilon_w x} \tag{68}$$

Equation (68) gives the electrostatic potential at the surface of a solute ion in a pure dielectric solvent. To find the electrostatic potential, ψ_e, for electrostatic charge, e, in a dielectric solution requires that, ψ_e, be a solution of the Poisson–Boltzmann equation [21, 25]. Owing to the spherical and planar geometry of a charge in an interfacial system, Onsager and Samaras chose cylindrical co-ordinates (ξ, λ) for the Poisson–Boltzmann equation,

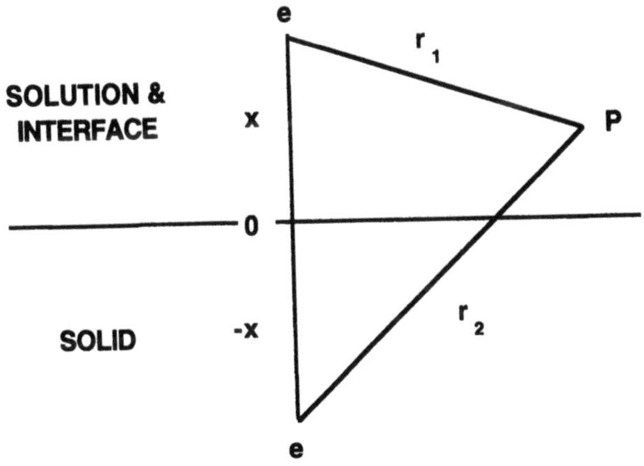

FIG. 3 Wagner [33] and Onsager and Samaras [30] models for computing the electrostatic potential due to a point charge, e, and its image, $e[(\varepsilon_w - \varepsilon_s)/(\varepsilon_w + \varepsilon_s)]$.

$$\nabla^2\psi_e = \frac{\partial^2\psi_e}{\partial^2\xi} + \frac{1}{\lambda}\frac{\partial}{\partial\lambda}\left(\lambda\frac{\partial\psi_e}{\partial\lambda}\right) = \kappa^2(\xi)\psi_e \tag{69}$$

The co-ordinates of electrostatic charge, e, are ($\xi = x, \lambda = 0$). For $\xi < 0, \kappa^2(\xi) = 0$, and for $\xi > 0, \kappa^2(\xi)$ is given by Eq. (41) with the ion concentrations of all the solute ions a function of x, as given by the linearized Boltzmann equation for the ith ion:

$$\begin{aligned}[c_i(\xi)] &= [c_i(\infty)]\exp\left[-\frac{w_i(\xi)}{kT}\right] \\ &= [c_i(\infty)]\exp\left[-\frac{e\psi_i(\xi)}{2\,kT}\right] \cong [c_i(\infty)]\left[1 - \frac{e\psi_i(\xi)}{2\,kT}\right]\end{aligned} \tag{70}$$

Boundary conditions are: ψ_e is continuous and $\partial\psi_e/\partial\xi$ changes by $\varepsilon_0\varepsilon_w$ when $\xi = 0$.

In order to solve Eq. (69) explicitly, Onsager and Samaras [30] found two approximations necessary: (1) the dielectric strength of the solid was set equal to zero, $\varepsilon_w \to 0$; and (2) the dependence of κ on distance from the interface; κ was omitted making κ a constant independent of x. Error estimates of both approximations suggest negligible errors for ions in most real interfacial systems (pp 530–531, Ref. 30). Equations (69) and (70) can be solved for the electrostatic potential, $\psi_e(x)$, at x, or for the more interesting case of the electrostatic potential, $\psi_{S_{IC}}(x)$, induced by an image charge (IC) on the surface of an ion, S^{ν_s}, centered at x, with diameter a_s, the latter result being

$$\psi_{S_{IC}}(x) = \left(\frac{+\nu_s e}{8\pi\varepsilon_0\varepsilon_w x}\right)\left(\frac{e^{+\kappa a_s}}{1+\kappa a_s}\right)e^{-2\kappa x} = \left[\frac{(+\nu_s e)e^{+\kappa a_s}}{8\pi\varepsilon_0\varepsilon_w(1+\kappa a_s)}\right]\frac{e^{-2\kappa x}}{x} \tag{71}$$

The electrostatic potential given by Eq. (71) describes the potential, $\psi_{S_{IC}}(x)$, experienced by an ion, S^{ν_s}, anywhere in a semi-infinite continuum solution or the Stern layer of a primitive interfacial system. If the ion, S^{ν_s}, is moved to the interface the electrostatic potential for the ion can be obtained by setting $x \to \frac{1}{2}a_s$ in Eq. (71) to give

$$\psi_{S_{IC}}\left(x = \frac{1}{2}a_s\right) \equiv \psi_{S_{IC}} = \left(\frac{+\nu_s e}{4\pi\varepsilon_0\varepsilon_w a_S}\right)\left(\frac{1}{1+\kappa a_S}\right) \tag{72}$$

The electrostatic potentials in Eqs (71) and (72) were derived with the "image charge method," which is used to determine electrostatic potentials at any point in any system of two phases with different dielectric strengths. The image charge creates an electrostatic potential which is equivalent to a surface potential described by Onsager and Samaras as "the fictitious surface charge" (pp 531, Ref. 30). This fictitious surface charge produces an electrostatic potential in the Stern layer at $x = \frac{1}{2}a_s$, where the distance a_s is the mean distance of approach for the surface site ion. If all other surface site ions have *no* electrostatic charge, then Eq. (72) completely describes the electrostatic surface potential of a surface site in a primitive interface.

At real solution–solid interfaces, particularly at solution–metal oxide (S–MO) interfaces, surface site ionization and solute ion adsorption will occur, creating a permanent electrostatic surface charge, in general. Rigorous determination of an explicit equation for the adsorption-induced electrostatic surface potential, ψ_{S_σ}, on a surface ion in a real interface would be a very complicated problem in electrostatics and quantum mechanics. However, for the primitive interface an explicit equation for ψ_{S_σ} is possible.

For the primitive interface the permanent surface electrostatic charge will be observed as a net electrostatic charge density, σ_s, located at the **P** where $x = \frac{1}{2}a_s$. The presence of σ_s will induce an electrostatic potential, ψ_{S_σ}, on the surface site ion, S^{ν_s}, in addition to $\psi_{S_{IC}}$. The electrostatic surface charge density, σ_s, at the Stern plane can be found by division of the site-average electrostatic charge, $\bar{\nu}_s e$, by the area per surface site, A_s, defined in Eq. (64) as

$$\sigma_s = \frac{e}{A_s}\bar{\nu}_s [=] \frac{\text{Coulombs}}{\text{m}^2} \tag{73}$$

The relation of σ_s to $\bar{\nu}_s$ and the location of this adsorption-induced electrostatic charge at $x = \frac{1}{2}a_s$ provides an opportunity to write an explicit equation for the adsorption-induced electrostatic surface potential, ψ_{S_σ}, very similar in form to Eq. (72). An uncharged surface site ion in a primitive interface surrounded by surface site ions all with the identical electrostatic charge, $\bar{\nu}_s e$, will create an electrostatic potential, ψ_{S_σ}, at $x = \frac{1}{2}a_s$, so from Eqs (71) and (72) an explicit equation for ψ_{S_σ} can be written:

$$\psi_{S_\sigma} = \left(\frac{\bar{\nu}_s e}{4\pi\varepsilon_0\varepsilon_w a_S}\right)\left(\frac{1}{1+\kappa a_S}\right) \tag{74}$$

The electrical potential, ψs_σ, with electrostatic charge, $+\ \bar{\nu}_s e$, will induce an equal but opposite electrostatic charge of, $-\bar{\nu}_s e$, on the surface site. If the surface site, S^{ν_s}, has a formal electrostatic charge or valence where, $\nu_s = \pm 1, \pm 2, \pm 3$, etc., then the added units of electrostatic charge would give the surface site an effective valence of, $\nu_s - \bar{\nu}_s$). The effective electrostatic charge on the surface site ion would be, $(\nu_s - \bar{\nu}_s)e$, giving the surface site ion at, $x = \frac{1}{2}a_s$, an effective electrostatic potential, ψ_s, of

$$\psi_s \equiv \psi_{S_{IC}} + \psi_{S_\sigma} = \left[\frac{(\nu_s - \bar{\nu}_s)e}{4\pi\varepsilon_0\varepsilon_w a_S}\right]\left(\frac{1}{1+\kappa a_S}\right) \tag{75}$$

Equation (75) can also be derived from the method of images directly, by starting with an uncharged surface site at an uncharged Stern plane where $x = \frac{1}{2}a_s$, then bringing $\nu_s - \bar{\nu}_s$ units of electrostatic charge to the initial uncharged surface site. This procedure will result in Eq. (75) but the intuition for the mean electrostatic mean surface site charge, $\bar{\nu}_s e$, is lost.

Equation (75) expresses the electrostatic potential, ψ_s, experienced by an ion, S^{ν_s}, in the Stern layer of a primitive interfacial system contiguous to a continuum solution. The electrostatic potential, ψ_s, is needed to evaluate the activity coefficient, given in Eq. (26). Also needed is the electrostatic potential, ψ_s^{ID}, experienced by an ion, S^{ν_s}, in the Stern layer of a primitive interfacial system contiguous to an ideal continuum solution. Evaluation of ψ_S^{ID} will follow the partial molar enthalpy method used to find the electrostatic potential of a solute ion in the interior of an ideal solution.

For an ion, S^{ν_s}, in the Stern layer the partial molar Helmholtz free energy, $\bar{A}_S$, for that ion results from the addition of an amount of electrostatic charge, $(\nu_s - \bar{\nu}_s)e$, to the uncharged ion, S^0, to give a resulting electrostatic potential to the ion of ψ_s. The equation for $\bar{A}_S$ for this process can be written from Eqs (26) and (75):

$$\bar{A}_{\mathrm{S}} = \frac{(\nu_{\mathrm{s}} - \bar{\nu}_{\mathrm{s}})e\psi_{\mathrm{s}}}{2} = \left[\frac{(\nu_{\mathrm{s}} - \bar{\nu}_{\mathrm{s}})^2 e^2}{8\pi\varepsilon_0\varepsilon_{\mathrm{w}} a_{\mathrm{s}}}\right]\left(\frac{1}{1 + \kappa a_{\mathrm{s}}}\right) \tag{76}$$

Equations (42) and (76) can be seen to be quite similar. Therefore, the procedures for deriving ideal electrostatic potentials will yield the same equations for the ideal electrostatic potentials differing only in electrostatic charge and numerical constants. In spite of these similarities, a brief derivation will be given. Substitution of Eq. (76) into the thermodynamic identity:

$$\bar{H}_{\mathrm{s}} = \bar{A}_{\mathrm{s}} - T\left(\frac{\partial \bar{A}_{\mathrm{s}}}{\partial T}\right)_{T,\bar{V}_{\mathrm{s}}} = -T^2\left[\frac{\partial(\bar{A}_{\mathrm{s}}/T)}{\partial T}\right]_{T,\bar{V}_{\mathrm{s}}} \tag{77}$$

gives

$$\bar{H}_{\mathrm{s}} = -T^2\left[\frac{(\nu_{\mathrm{s}} - \bar{\nu}_{\mathrm{s}})^2 e^2}{8\pi\varepsilon_0 a_{\mathrm{s}}}\right]\left\{\frac{\partial[\varepsilon_{\mathrm{s}}^{-1} T^{1}(1 + \kappa a_{\mathrm{s}})^{-1}]}{\partial T}\right\}_{T,\bar{V}_{\mathrm{s}}} \tag{78}$$

Equation (78) gives upon differentiation:

$$\bar{H}_{\mathrm{s}} = \left[\frac{(\nu_{\mathrm{s}} - \bar{\nu}_{\mathrm{s}})^2 e^2}{8\pi\varepsilon_0 a_{\mathrm{s}}}\right]\left[\frac{1}{\varepsilon_{\mathrm{w}}(1 + \kappa a_{\mathrm{s}})}\right]\left[1 + \left(\frac{T}{\varepsilon_{\mathrm{w}}}\right)\left(\frac{\partial \varepsilon_{\mathrm{w}}}{\partial T}\right)_{\bar{V}_{\mathrm{s}}} + \left(\frac{a_{\mathrm{s}} T}{1 + \kappa a_{\mathrm{s}}}\right)\left(\frac{\partial \kappa}{\partial T}\right)_{\bar{V}_{\mathrm{s}}}\right] \tag{79}$$

For ideal solutions the ideal partial molar enthalpy, $\bar{H}_{\mathrm{s}}^{\mathrm{ID}}$, must remain invariant with changes in the concentration of ions in the solution. Since κ is the only concentration-dependent variable in Eq. (79), one can deduce

$$\bar{H}_{\mathrm{s}}^{\mathrm{ID}} = \bar{H}_{\mathrm{s}}(\kappa = 0) \tag{80}$$

From Eqs (79) and (80) the equation for $\bar{H}_{\mathrm{s}}^{\mathrm{ID}}$ can be written:

$$\bar{H}_{\mathrm{s}}^{\mathrm{ID}} = \left[\frac{(\nu_{\mathrm{s}} - \bar{\nu}_{\mathrm{s}})^2 e^2}{8\pi\varepsilon_0\varepsilon_{\mathrm{w}} a_{\mathrm{s}}}\right]\left[1 + \left(\frac{T}{\varepsilon_{\mathrm{w}}}\right)\left(\frac{\partial \varepsilon_{\mathrm{w}}}{\partial T}\right)_{\bar{V}_{\mathrm{s}}}\right] \tag{81}$$

Substitution of Eq. (77) into the thermodynamic identity for ideal partial molar enthalpy:

$$\bar{H}_{\mathrm{s}}^{\mathrm{ID}} = -T^2\left[\frac{\partial(\bar{A}_{\mathrm{s}}^{\mathrm{ID}}/T)}{\partial T}\right]_{T,\bar{V}_{\mathrm{s}}} \tag{82}$$

gives

$$T^2\left[\frac{\partial(\bar{A}_{\mathrm{s}}^{\mathrm{ID}}/T)}{\partial T}\right]_{T,\bar{V}_{\mathrm{s}}} = -\left[\frac{(\nu_{\mathrm{s}} - \bar{\nu}_{\mathrm{s}})^2 e^2}{8\pi\varepsilon_0\varepsilon_{\mathrm{w}} a_{\mathrm{s}}}\right]\left[1 + \left(\frac{T}{\varepsilon_{\mathrm{w}}}\right)\left(\frac{\partial \varepsilon_{\mathrm{w}}}{\partial T}\right)_{\bar{V}_{\mathrm{s}}}\right] \tag{83}$$

The solution for Eq. (83) can be seen to be

$$\bar{A}_{\mathrm{s}}^{\mathrm{ID}} = \left[\frac{(\nu_{\mathrm{s}} - \bar{\nu}_{\mathrm{s}})^2 e^2}{8\pi\varepsilon_0\varepsilon_{\mathrm{w}} a_{\mathrm{s}}}\right] \equiv \frac{(\nu_{\mathrm{s}} - \bar{\nu}_{\mathrm{s}})\psi_{\mathrm{s}}^{\mathrm{ID}}}{2} \tag{84}$$

That Eq. (84) is a solution for Eq. (83) can be confirmed by substitution of the former into the latter and differentiating the result.

Equations (81) and (84) give the electrostatic potentials for a surface site ion in the Stern layer (more precisely, the electrostatic potential on the surface of a surface site ion, S^{ν_s}, adsorbed in the Stern layer) contacting a real, ψ_s, and an ideal, ψ_s^{ID}, solution:

$$\psi_s = \left[\frac{(\nu_s - \bar{\nu}_s)^2 e}{4\pi\varepsilon_0\varepsilon_w a_s}\right]\left(\frac{1}{1+\kappa a_s}\right) \quad \left\{\begin{array}{l}\text{Electrostatic potential} \\ \textit{with} \text{ screening by the} \\ \text{diffuse ion atmosphere}\end{array}\right\} \tag{85}$$

$$\psi_s^{ID} = \left[\frac{(\nu_s - \bar{\nu}_s)^2 e}{4\pi\varepsilon_0\varepsilon_w a_s}\right] \quad \left\{\begin{array}{l}\text{Electrostatic potential} \\ \textit{without} \text{ screening by the} \\ \text{diffuse ion atmosphere}\end{array}\right\} \tag{86}$$

The difference in the electrostatic potentials in Eqs (85) and (86) is the same expression for solute ions and surface site ions, $1/1+\kappa a_s$, with κ dependent on solute ion concentration as can be seen from Eq. (41). As κ increases with increasing ion concentration, the electrostatic potential of an ion in a real solution will be decreased by "screening" in agreement with Eq. (85).

Equations (76) and (84) can be substituted into Eq. (26) to give the equation for the activity coefficient:

$$kT\ln f_s = \frac{(\bar{A}_s - \bar{A}_s^{ID})}{N_{\mathrm{A}}} = \left[\frac{(\nu_s - \bar{\nu}_s)^2 e^2}{8\pi\varepsilon_0\varepsilon_w a_s}\right]\left[\frac{1}{1+\kappa a_s} - 1\right]$$
$$= -\left[\frac{(\nu_s - \bar{\nu}_s)^2 e^2}{8\pi\varepsilon_0\varepsilon_w a_s}\right]\left(\frac{\kappa a_s}{1+\kappa a_s}\right) = -\left[\frac{(\nu_s - \bar{\nu}_s)e}{2}\right]\psi_s^{ID}\left(\frac{\kappa a_s}{1+\kappa a_s}\right) \tag{87}$$

Solving for the activity coefficient:

$$f_s = \exp\left[-\frac{(\nu_s - \bar{\nu}_s)^2 e^2}{8\pi\varepsilon_0\varepsilon_w kT}\left(\frac{\kappa}{1+\kappa a_s}\right)\right] \equiv \exp\left[-\frac{(\nu_s - \bar{\nu}_s)^2 \kappa q_B}{1+\kappa a_s}\right]$$
$$= \left[e^{(-\kappa q_B)}\right]^{\left[\frac{(\nu_s - \bar{\nu}_s)^2}{1+\kappa a_s}\right]} \equiv Z^{\left[\frac{(\nu_s - \bar{\nu}_s)^2}{1+\kappa a_s}\right]} \equiv Z^{\epsilon_s} \tag{88}$$

Several observations about the activity coefficient equation for interfacial surface site ions derived for the primitive interfacial model can be made:

1. The activity coefficient equations for a solute ion in the interior of a solution, Eq. (60), and for a surface site ion in the Stern layer of a primitive interface, Eq. (88), are quite similar in mathematical form. In particular, the Bjerrum association length, q_B, is found in both activity coefficient equations. In addition, the activity coefficient can be conveniently expressed in terms of the base activity coefficient, Z, and its exponent, ϵ_s. However, the significance of these two variables, comprising the activity coefficient, is somewhat different for surface site ions vis-à-vis solute ions.

The explicit variables comprising the entire exponent of the activity coefficient, $(\nu_s - \bar{\nu}_s)^2\kappa q_B/1+\kappa a_s$, are related to the two different electrostatic potentials experienced by a surface site ion. The variables $\nu_s - \bar{\nu}_s$ are the electrostatic charge variables for the SP and specify the effect adsorption-induced electrostatic potential has the activity coefficient. Thus, the relative effects of each different surface electrostatic potential can be evaluated for any solid–solution interfacial system of interest.

2. Experimentally determined activity coefficients can exceed unity. The maximum value predicted for primitive interfacial activity coefficients in Eq. (85) is unity. Just as with activity coefficient for solute ions, the source of this discrepancy results largely from the use of the *linearized* Poisson–Boltzmann equation to obtain the electrostatic potentials for solute and surface site ions. Reasons and remedies for this discrepancy for the activity coefficients of surface site ions is the same as discussed for solute ions in Section III.A.

3. Activity coefficient equation (88) for surface site ions has the same functional dependence as Eq. (59) for solute ions in solution. The obvious differences are the ion valences and effective ion diameters. The most important difference is in the mean surface site valence, $\bar{\nu}_s$, which has an implicit dependence on x. The mean surface site valence, $\bar{\nu}_s$, can be viewed as the effective electrostatic charge which each surface site must have to give rise to the electrostatic surface charge, ψ_{s_σ}, at the SP. Any ion, Y^{ν_y}, located at the SP will experience an electrostatic surface potential, ψ_s, resulting from electrostatic charge, $(\nu_y - \bar{\nu}_y)e$. If the ion is now withdrawn from the SP along the positive x-axis, the ion carries its electrostatic charge, $\nu_y e$, but the electrostatic charge due to the mean surface site electrostatic charge, $\bar{\nu}_s e$, must diminish and approach zero at $x = \infty$. The variation of $\bar{\nu}_s$ along the x-axis will be described in Section IV.F.

The primitive interfacial model allowed the derivation of an explicit equation in closed form for the activity coefficient, Eq. (88). The derivation was made possible as a result of many simplifying features. Some of these primitive interfacial model approximations are described below:

1. Smearing out the electrostatic potential over the SP oversimplifies the discreteness of the surface site ions. Solute adsorption at interfaces with discrete surface sites is more readily directed away from like charged surface sites and toward unlike charge. Since the electrostatic potential of the nondiscrete primitive interface is the site-average mean of all discrete surface sites, one would expect that the electrostatic potential for the primitive model to be an overestimate of the electrostatic potential experienced by a discrete surface site.

2. It should be noted that many adsorption theories, discussed in Chapter 1, require the difference in electrostatic potential with distance from the interface (potential gradient) as the basis for solute adsorption. In this chapter one can see that electrostatic potential does play a role in adsorption through the activity coefficient; however, adsorption is more dependent on other forces of greater magnitude such as ionic and covalent bonding between surface sites and solute molecules as expressed through mass action formalism. Furthermore, the complexity of electrostatic potential gradients at real S–MO interfaces makes the basis for purely electrostatic adsorption models tenuous.

3. The dielectric strength of the solvent is considered to be independent of solute. For all but the most concentrated solutions this should be a negligible error except possibly at the interface. The Stern layer contains solvent and surface site ions. The derivation of the electrostatic potential equations implicitely assumed that ε_w is constant up to the SP at $x = \frac{1}{2}a_s$. Owing to the high electrostatic potential near the SP, ε_w could possibly be expected to decrease. This alleged electrostatic potential-induced decrease in ε_w is commonly referred to as dielectric saturation. However, the effect is still poorly understood and may in fact not be significant [34].

4. Solvation forces are the result of solvent molecules adsorbing on surface sites and solute ions. In the primitive interfacial model, the solvent is a continuum so solvation effects do not affect properties of a primitive interface. In real systems, electrostatic properties are seen to be state functions, depending on initial (uncharged ion) and final (charged ion) electrostatic states only. Hence, it is unlikely that solvation effects play any role at all in the electrostatic forces of S–MO interfaces.

5. For the primitive interfacial model all surface site and adsorbed ions are restricted to a single ion layer of thickness, a_s, and surface site area, $A_s \geq a_s^2$. This location of interfacial ions with centers in the SP facilitates the calculation of position-dependent electrostatic properties of the interface. In the primitive interface, surface sites with or without adsorbed metal ions are assigned to the Stern layer with a spherical geometry of diameter a_s.

In reality, surface site ions with and without different adsorbed solute ions almost certainly do not all reside in a single plane parallel to the interface, are not, in general, spherical, and do not all have the same diameter. The error incurred in calculating the activity coefficient caused by these approximations in the primitive interfacial model can be semiquantitatively estimated by examining how the electrostatic potential changes away from the SP by dividing Eqs (71) and (72):

$$\frac{\psi(x)}{\psi\left(x = \frac{a_s}{2}\right)} = \left(\frac{a_s}{2x}\right) e^{\kappa(a_s - 2x)} \tag{89}$$

Equation (89) gives the decrease in electrostatic potential with distance x, measured perpendicularly away from the IHP. This equation is commonly shown incorrectly in the literature with the parenthetical expression, $a_s/2x$, absent. The reciprocal x expression in Eq. (89) is functionally far more important to the distance dependence of the electrostatic potential than is the x-dependent exponential expression.

The change in electrostatic potential with distance according to Eq. (89) is shown in Fig. 4.

To examine the significance of the assumption that solute ions must reside in the Stern layer, consider a solute ion adsorbing by moving along the x-axis toward a surface site ion until the two ion surfaces contact so that the solute ion would be located at $x = \frac{3}{2}a_s$. If solute ion location at $x = \frac{3}{2}a_s$ is taken as an alternative model of adsorption, one can see from Fig. 4 that the electrostatic potential at $x = \frac{3}{2}a_s$ has decreased to about one-third of the SP value. The location of the adsorbed solute in this two-plane surface-site/adsorbed-solute ion complex would experience a different electrostatic potential, $\psi_s(x = (3a_s/2))$, than the surface site ion, $\psi_s(x = (a_s/2))$, on which it is adsorbed by Eq. (71):

$$\psi_s\left(x - \frac{3a_s}{2}\right) = \left[\frac{(\nu_s - \bar{\nu}_s')e\, e^{-2\kappa a_s}}{12\pi\varepsilon_0\varepsilon_w(1 + \kappa a_s)a_s}\right] \tag{90}$$

For a surface site complex of a surface site ion, S^{ν_s}, and a solute ion, Y^{ν_y}, the free energy is calculated as the work of electrostatically charging the surface complexes for the two different methods as follows.

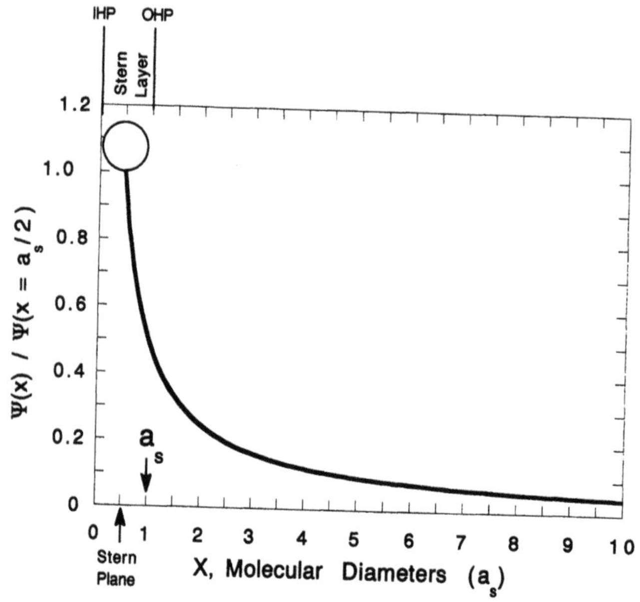

FIG. 4 Electrostatic potential in a primitive interfacial system.

Primitive model

$$w = \frac{[(\nu_s + \nu_y - \bar{\nu}_s)e]^2}{8\pi\varepsilon_0\varepsilon_w(1 + \kappa a_s)a_s} \tag{91}$$

The work of the solute-complexed surface site is a result of the net valence experienced by the solute-complexed surface site of $\nu_s + \nu_y - \bar{\nu}_s$, located at, $x = \frac{1}{2}a_s$.

Two-plane model

$$w = \frac{[(\nu_s - \bar{\nu}_s)e]^2}{8\pi\varepsilon_0\varepsilon_w(1 + \kappa a_s)a_s} + \frac{[(\nu_y - \bar{\nu}_s')e]^2 e^{-2\kappa a_s}}{24\pi\varepsilon_0\varepsilon_w(1 + \kappa a_s)a_s} \tag{92}$$

The work of the solute-complexed surface site is a result of the net valence experienced by the surface site of $\nu_s - \bar{\nu}_s$, located at $x = \frac{1}{2}a_s$, plus the next valence experienced by the solute ion of $\nu_y - \bar{\nu}_s'$, located at $x = \frac{3}{2}a_s$.

The differences between the two models is significant with the latter method being mathematically more complicated. Moreover, the difference between $\bar{\nu}_s$ and $\bar{\nu}_s'$ adds further complexity (see Section IV.F). Since both methods are approximate models, they ignore important properties of interfaces, the determination of which would require unavailable quantum mechanical solutions or experimental evidence. Therefore, the primitive interfacial model, shown in Eq. (91), will be the model of choice due to its simplicity and for lack of defining experimental information about real interfaces.

IV. THE ACTIVITY COEFFICIENT IN COMPLEXATION

A. Solution Complexation

In the following discussion the role of activity coefficients in adsorption will be examined. A more in-depth discussion of adsorption can be found in Chapters 1–3. Since adsorption is nothing more than complex formation at a solution–solid interface, the discussion will begin with the more familiar topic of complexation in solution. Adsorption will then be seen as simply complexation at an interface between solute and surface site ions.

Electrolyte solution complex formation between two solute ions: a metal cation, M^{ν_m}, and a ligand, L^{ν_l}, is shown in Fig. 5.

The chemical equation for the solution complexation reaction shown in Fig. 5 is

$$\mathrm{m}M^{\nu_m} + lL^{\nu_l} = [M_mL_l]^{\nu_c} \quad \text{where} \quad \nu_c = \mathrm{m}\nu_m + l\nu_l \tag{93}$$

The chemical potential equations can be written for all three species in Eq. (93) as shown in Eq. (15):

$$\mu_M = \mu_M^* + kT\ln\chi_M + kT\ln f_M \tag{94}$$

$$\mu_L = \mu_L^* + kT\ln\chi_L + kT\ln f_L \tag{95}$$

$$\mu_{ML} = \mu_{ML}^* + kT\ln\chi_{ML} + kT\ln f_{ML} \tag{96}$$

The chemical potential for the complexation reaction, $\Delta\mu_{ML}$, shown in Fig. 5:

$$\Delta\mu_{ML} \equiv \mu_{ML} - \mathrm{m}\mu_L - l\mu_L \tag{97}$$

Substitution of Eqs (94)–(96) into Eq. (97) gives

$$\Delta\mu_{ML} = (\mu_{ML}^* - \mathrm{m}\mu_M^* - l\mu_L^*) + kT\ln\left(\frac{\chi_{ML}}{\chi_M^m\chi_L^l}\right) + kT\ln\left(\frac{f_{ML}}{f_M^m f_L^l}\right) \tag{98}$$

At equilibrium, $\Delta\mu_{ML} = 0$, so Eq. (83) can be rearranged to give

$$-(\mu_{ML}^* - \mathrm{m}\mu_M^* - l\mu_L^*) \equiv kT\ln(K_{ML}) + kT\ln\left(\frac{\chi_{ML}}{\chi_M^m\chi_L^l}\right) + kT\ln\left(\frac{f_{ML}}{f_M^m f_L^l}\right) \tag{99}$$

Removing logarithms from Eq. (99) yields:

$$K_{ML} = \left(\frac{\chi_{ML}}{\chi_M^m\chi_L^l}\right)\left(\frac{f_{ML}}{f_M^m f_L^l}\right) \tag{100}$$

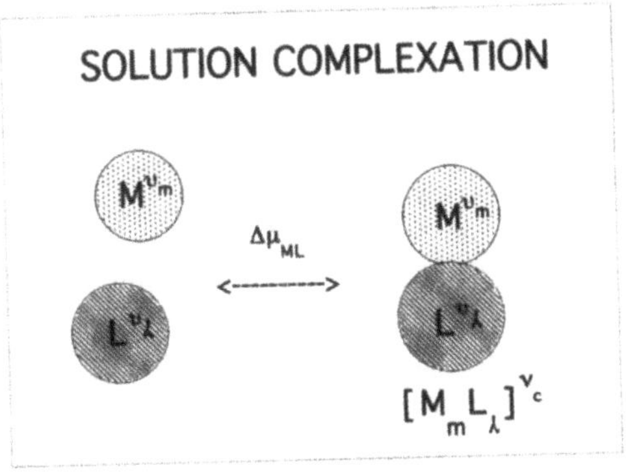

FIG. 5 Solution complexation.

Equation (100) is the mass action equation for the solution complexation reaction shown in Fig. 5. For the activity coefficients in Eq. (96), the rigorous equation (22) can be substituted so that the activity coefficient can be expressed in terms of valence and electrostatic potentials:

$$f_M = e^{\left[-\frac{\nu_m e(\psi_M - \psi_M^{ID})}{2kT}\right]} \tag{101}$$

$$f_L = e^{\left[-\frac{\nu_l e(\psi_L - \psi_L^{ID})}{2kT}\right]} \tag{102}$$

$$f_{ML} = e^{\left[-\frac{\nu_c e(\psi_{ML} - \psi_{ML}^{ID})}{2kT}\right]} \tag{103}$$

Combining Eqs (101)–(103) into Eq. (100):

$$\begin{aligned} K_{ML} &= \left(\frac{\chi_{ML}}{\chi_M^m \chi_L^l}\right)\left(\frac{f_{ML}}{f_M^m f_L^l}\right) \\ &= \left(\frac{\chi_{ML}}{\chi_M^m \chi_L^l}\right)\exp\left\{-\frac{e\left[\nu_c(\psi_{ML} - \psi_{ML}^{ID}) - m\nu_m(\psi_M - \psi_M^{ID}) - l\nu_l(\psi_L - \psi_L^{ID})\right]}{kT}\right\} \end{aligned} \tag{104}$$

Equation (104) is a rigorous mass action equation in terms of explicit functions except for the electrostatic potentials. Electrostatic potentials are, at best, difficult to measure and no rigorous explicit equation expressing the electrostatic potential of a solute ion has yet been developed. Instead, the approximate explicit equation (60) (based on Debye–Hückel theory) for the electrostatic potential of a solute ion will be used to write equations for each of the activity coefficients in Eq. (104):

$$f_M = \left[e^{(-\kappa q_B)}\right]^{\left(\frac{\nu_M^2}{1+\kappa a_M}\right)} \equiv Z^{\left(\frac{\nu_M^2}{1+\kappa a_M}\right)} \equiv Z^{\epsilon_M} \tag{105}$$

$$f_L = \left[e^{(-\kappa q_B)}\right]^{\left(\frac{\nu_L^2}{1+\kappa a_L}\right)} \equiv Z^{\left(\frac{\nu_L^2}{1+\kappa a_L}\right)} \equiv Z^{\epsilon_L} \tag{106}$$

$$f_{ML} = \left[e^{(-\kappa q_B)}\right]^{\left(\frac{\nu_c^2}{1+\kappa a_{ML}}\right)} \equiv Z^{\left(\frac{\nu_c^2}{1+\kappa a_{ML}}\right)} \equiv Z^{\epsilon_{ML}} \tag{107}$$

The activity coefficient exponent, $\nu_Y^2/1 + \kappa a_Y$, for the solute ion, Y^{ν_y}, influences the solute activity coefficient through the solution properties, κa_y, and the ion valence, ν_y. These effects are the familiar Debye–Hückel behavior resulting from screening of the electrostatic potential of a solute ion by the diffuse counter ion layer. Since $Z \leq 1.0$, ϵ_{SO} will vary inversely with the activity coefficient due to solute ion screening by neighboring solute ions. To understand this behavior better it is instructive to examine electrostatic potential screening of solute ions.

A solute ion, Y^{ν_y}, will automatically attract a slight excess of counter ions (solute ions of opposite charge) at equilibrium. Since the electrostatic potential of a solute ion is destructively additive with the electrostatic potential of the counter ions, the electrostatic potential of the solute ion, Y^{ν_y}, will appear to be diminished or screened due to interaction with the electrostatic potentials of the counterions. If the screened fraction of the electrostatic potential is $\S_Y$, then the activity coefficient, f_Y, can be understood to represent that fraction of the electrostatic potential which is *not* screened. This description can be used to heuristically define the screening fraction, $\S_Y$, as the complement of the activity coefficient:

$$\S_Y \equiv 1 - f_Y \tag{108}$$

The activity coefficient quotient for the adsorbed cation mass action adsorption equation (100) can be written in explicit form by substituting Eqs (105)–(107) into Eq. (100) to give

$$K_{ML} = \left(\frac{\chi_{ML}}{\chi_M^m \chi_L^l}\right)\left(\frac{f_{ML}}{f_M^m f_L^l}\right) = \left(\frac{\chi_{ML}}{\chi_M^m \chi_L^l}\right) Z^{\left(\frac{v_c^2}{1+\kappa a_{ML}} - \frac{m v_m^2}{1+\kappa a_M} - \frac{l v_l^2}{1+\kappa a_L}\right)} \tag{109}$$

Equation (105) gives the activity coefficient quotient in explicit terms, allowing direct calculation.

Before leaving the subject of activity coefficients in solution, it will be instructive to discuss the mean activity coefficient, $f_\pm$, for an electrolyte. For an electrolyte, A_aB_b, which ionizes as

$$A_aB_b = A^{+b} + B^{-a} \tag{110}$$

the mean activity coefficient, $f_\pm$, is written [35]:

$$f_\pm = \left(f_A^a f_B^b\right)^{1/(a+b)} \tag{111}$$

First, the Debye–Hückel activity coefficient equation (57) can be used to write explicit expressions for the individual ion activity coefficients:

$$f_A = Z^{\left[\frac{(+b)^2}{1+\kappa a_A}\right]} = Z^{\left(\frac{b^2}{1+\kappa a_A}\right)} \tag{112}$$

$$f_B = Z^{\left[\frac{(-a)^2}{1+\kappa a_B}\right]} = Z^{\left(\frac{a^2}{1+\kappa a_B}\right)} \tag{113}$$

Substituting Eqs (112) and (113) into Eq. (112) gives an explicit equation for $f_\pm$:

$$f_\pm = (f_A^a f_B^b)^{1/(a+b)} = \left\{ Z^{\left[a\left(\frac{b^2}{1+\kappa a_A}\right)\right]} Z^{\left[b\left(\frac{a^2}{a+\kappa a_B}\right)\right]} \right\}^{1/(a+b)}$$
$$= Z^{\left[\frac{ab}{a+b}\left(\frac{b}{1+\kappa a_A} + \frac{a}{1+\kappa a_B}\right)\right]} \tag{114}$$

Note that if the "effective" ionic diameters are about the same, $a_a \cong a_b \cong a$, Eq. (114) reduces to

$$f_\pm = (f_A^a f_B^b)^{1/(a+b)} = Z^{\left[\frac{ab}{a+b}\left(\frac{b}{1+\kappa a_A} + \frac{a}{1+\kappa a_B}\right)\right]}\Bigg|_{a_A \cong a_B \cong a} \cong Z^{\left(\frac{ab}{1+\kappa a}\right)} \tag{115}$$

Since the variables a and b are stoichiometric they will always be positive, so $0 < f_\pm \leq 1$. We can now make use of the principles of solution complexation as a background for adsorption or surface complexation at the S–MO interface.

B. Proton Adsorption–Interfacial Complexation

In Chapters 1–3, adsorption has been discussed in great detail. A poorly understood part of the adsorption process is the role of activity coefficients or electrostatics. In the preceding section, solution complexation was examined in order to establish a basis for surface complexation or adsorption. To further the understanding of the role activity coefficients play in adsorption, explicit activity coefficient equations will be derived for protonated surface sites.

The exchange of protons between pure water (no cations other than protons) and an amphoteric metal oxide interface will produce three different types of surface site: (1) unprotonated anionic sites each with a unit negative valence, (2) monoprotonated nonionic sites each with a zero valence, and (3) diprotonated cationic sites each with a unit positive valence. Given sufficient time to reach equilibrium, the proton exchange will produce a surface in which the number of each type of site and the number of protons in solution becomes constant. To specify the stoichiometry of this equilibrium proton distribution, the total number of surface sites will be desianted as n_s^T, with the number of each specific type of surface site designed as: (1) n_{SO}, for unprotonated anionic sites, (2) n_{SOH}, for monoprotonated nonionic sites, and (3) n_{SOH_2}, for diprotonated cationic sites. The surface site mole fraction can therefore be defined relative to n_s^T as:

$$\theta_{SO} \equiv \frac{n_{SO}}{n_s^T} \tag{116}$$

$$\theta_{SOH} \equiv \frac{n_{SOH}}{n_s^T} \tag{117}$$

$$\theta_{SOH_2} \equiv \frac{n_{SOH_2}}{n_s^T} \tag{118}$$

For the sake of clarity, surface mole fractions will be designated with the lower case Greek symbol, θ, while molar solution concentrations will be designated with the lower case Greek symbol, χ. The exchange of protons between the metal oxide interface and the solution can be visualized as shown in Fig. 6.

From Fig. 6, proton adsorption can be seen to occur by two chemical pathways,

$$SOH_2{}^+ \leftrightarrow SOH + H^+ \tag{119}$$

$$SOH \leftrightarrow SO^- + H^+ \tag{120}$$

Mass action equations are derived for equilibrium chemical reactions by equating the chemical potential of the reactants and products, since by definition the chemcial potential of reactants and products of a chemical reaction must be equal at equilibrium. For the protonation reactions shown in Fig. 6, chemical potentials, μ_H, μ_{SO}, μ_{SOH}, and μ_{SOH_2}, for each of the four chemicals, H^+, SO^-, SOH, and $SOH_2{}^+$, respectively, can be written, following Lewis [1], in terms of their respective standard state chemical potentials (μ_H^*, μ_{SO}^*, μ_{SOH}^*, and $\mu_{SOH_2}^*$), solution molarities, S–MO surface mole fractions (χ_H, $_{SO}$, θ_{SOH}, and θ_{SOH_2}), and activity coefficients (f_H, f_{SO}, f_{SOH}, and f_{SOH_2}) to give

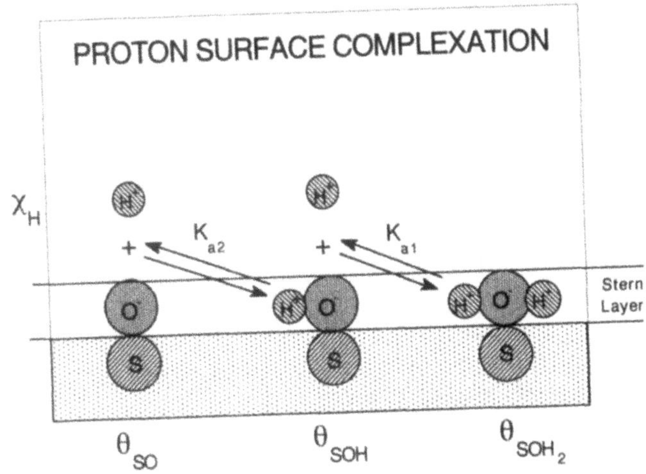

FIG. 6 Interfacial complexation reactions of protons at the solvent–metal oxide interface. Molarity of protons in solution, χ_H. Mole fraction of: (1) unprotonated anionic surface sites, θ_{SO}, (2) monoprotonated nonionic surface sites, θ_{SOH}, and (3) diprotonated cationic surface sites, θ_{SOH_2}.

$$\mu_H = \mu_H^* + kT\ln\chi_H + kT\ln f_H \tag{121}$$

$$\mu_{SO} = \mu_{SO}^* + kT\ln\theta_{SO} + kT\ln f_{SO} \tag{122}$$

$$\mu_{SOH} = \mu_{SOH}^* + kT\ln\theta_{SOH} + kT\ln f_{SOH} \tag{123}$$

$$\mu_{SOH_2} = \mu_{SOH_2}^* + kT\ln\theta_{SOH_2} + kT\ln f_{SOH_2} \tag{124}$$

For the protonation reaction shown in Fig. 6, which is designated by the symbol K_{a1}, the chemical potentials of all the reactants must equal the chemical potentials of all the products in Eq. (119) at equilibrium, so

$$\mu_{SOH_2} = \mu_{SOH} + \mu_H \tag{125}$$

Substitution of Eqs (121), (123), and (124) into Eq. (125) gives upon rearrangement:

$$-(\mu_{SOH}^* + \mu_H^* - \mu_{SOH_2}^*) = +kT\ln\left(\frac{\theta_{SOH}\,\chi_H}{\theta_{SOH_2}}\right) + kT\ln\left(\frac{f_{SOH}f_H}{f_{SOH_2}}\right) \tag{126}$$

The standard state chemical potential terms on the left-hand side of Eq. (126) vary only with temperature and solvent type, so it is conventional to define the mole fraction mass action constant or adsorption constant, K_{a1}, as

$$kT\ln(K_{a1}) \equiv -(\mu_{SOH}^* + \mu_H^* - \mu_{SOH_2}^*) \tag{127}$$

Equations (126) and (127) can be combined to give a mass action equation for the reaction represented in Fig. 1 by the adsorption constant, K_{a1}, as

$$K_{a1} = \left(\frac{\theta_{SOH}\chi_H}{\theta_{SOH_2}}\right)\left(\frac{f_{SOH}f_H}{f_{SOH_2}}\right) \quad \begin{pmatrix}\text{Mass action equation for}\\ \text{diprotonated surface site}\\ \text{ionization}\end{pmatrix} \tag{128}$$

The equation for the other proton adsorption reaction in Fig. 6 can be written:

$$SOH \leftrightarrow SO^- + H^+ \tag{129}$$

By an analogous procedure, using Eqs (121)–(124), the mass action equation for the reaction in Eq. (129), which is represented by the adsorption constant K_{a2} in Fig. 6, can be derived:

$$K_{a2} = \left(\frac{\theta_{SO}\chi_H}{\theta_{SOH}}\right)\left(\frac{f_{SO}f_H}{f_{SOH}}\right) \quad \left(\begin{array}{l}\text{Mass action equation for}\\ \text{monoprotonated surface site}\\ \text{ionization}\end{array}\right) \tag{130}$$

The explicit expressions for the activity coefficient quotients can be written using Eqs (60) and (88). First, activity coefficients in explicit form:

$$f_H = Z^{\left(\frac{\nu_H^2}{1+\kappa a_H}\right)} = Z^{\left(\frac{1}{1+\kappa a_H}\right)} \equiv Z^{\epsilon_H} \tag{131}$$

$$f_{SO} = Z^{\left[\frac{(\nu_s-\bar{\nu}_s)^2}{1+\kappa a_s}\right]} = Z^{\left[\frac{(-1-\bar{\nu}_s)^2}{1+\kappa a_s}\right]} = Z^{\left[\frac{(1+\bar{\nu}_s)^2}{1+\kappa a_s}\right]} \equiv Z^{\epsilon_{SO}} \tag{132}$$

where $\nu_s = -1$ for an anionic surface site, SO^-;

$$f_{SOH} = Z^{\left[\frac{(\nu_s-\bar{\nu}_s)^2}{1+\kappa a_s}\right]} = Z^{\left[\frac{(0-\bar{\nu}_s)^2}{1+\kappa a_s}\right]} = Z^{\left[\frac{\bar{\nu}_s^2}{1+\kappa a_s}\right]} \equiv Z^{\epsilon_{SOH}} \tag{133}$$

where $\nu_s = 0$ for a nonionic surface site, SOH;

$$f_{SOH_2} = Z^{\left[\frac{(\nu_s-\bar{\nu}_s)^2}{1+\kappa a_s}\right]} = Z^{\left[\frac{(1-\bar{\nu}_s)^2}{1+\kappa a_s}\right]} \equiv Z^{\epsilon_{SOH_2}} \tag{134}$$

where $\nu_s = +1$ for a cationic surface site, $SOH_2{}^+$. Several aspects of the activity coefficient equations (131)–(134) deserve further discussion:

1. Activity coefficients for surface sites and protons have been written in terms of two variables: (1) a base coefficient, Z, which is dependent upon the solution ionic strength, solvent type, and temperature; and (2) an exponent, ϵ, dependent mainly upon ion properties such as valence and effective diameter. The base activity coefficient, Z, appearing in all activity coefficient equations, (131)–(134), decreases from unity as the solution ionic strength increases. The activity coefficients exponents, ϵ, are all positive. As discussed previously, the linearized Poisson–Boltzmann equation restricts the range of the primitive interfacial model activity coefficients from zero to unity while experimentally determined activity coefficients can be greater than unity in concentrated solutions.

2. The activity coefficient exponent, $(1 + \bar{\nu}_s)^2/1 + \kappa a_s$, for the anionic surface site, SO^-, influences the surface site activity coefficient through both interfacial (the mean electrostatic surface site charge, $e\bar{\nu}_s$) and solution (the diffuse ion concentration, κa_s) properties. The latter effects are the familiar Debye–Hückel behavior resulting from screening of the electrostatic potential of a solute ion by the diffuse counter ion layer, which has been examined for solution complexation. The effects of interfacial electrostatic charge on the activity coefficient become predominate in the limit of dilute solute ion concentration so that $\kappa a_s \ll 1$. In the dilute solution limit, ϵ_{SO} will vary inversely with the activity coefficient due to surface site ion screening by neighboring surface site ions. To understand this behavior better it is instructive to examine electrostatic potential screening of surface site ions.

When an anionic surface site ion, SO^-, is surrounding by a large number of positive surface site ions, the positive electrostatic potentials are destructively additive with the negative electrostatic potential of the anionic surface site ion. In this way, the electrostatic potential at some distance from the anionic surface site ion will appear diminished or screened due to interaction with the electrostatic potentials emanating from the neighboring cationic surface sites as can be seen in Fig. 7. If the screened fraction of the electrostatic potential is $§_{SO}$, then the activity coefficient, f_{SO}, can be understood to represent that fraction of the electrostatic potential which was *not* screened. This description can be used to define heurestically the screening fraction, $§_{SO}$, as the complement of the activity coefficient:

$$§_{SO} \equiv 1 - f_{SO} \tag{135}$$

As the mean electrostatic surface site charge, $e\bar{\nu}_s$, decreases (becomes more negative) more neighboring surface site ions have a charge similar to that of the anionic surface site, causing two things to happen: (1) $§_{SO}$ decreases because neighboring cationic surface ions that cause screening are diminished, and (2) f_{SO} increases with fewer neighboring cationic surface site ions present to screen the anionic surface site ion. In summary: $\bar{\nu}_s \downarrow$, $\epsilon_{SO} \downarrow$, $§_{SO} \downarrow$, $f_{SO} \uparrow$.

Equation (132) can be seen to predict the screening behavior described above and shown in Fig. 7. As $\bar{\nu}_s$ increases (becomes more positive), the activity coefficient exponent, ϵ_{SO}, increases, which decreases the activity coefficient. This result is reasonable because a more positive environment around an anionic surface site ion, SO^- will increase screening, $§_{SO}$. This behavior can be clearly seen in a few numerical examples in Table 1, where a value of $Z = 0.8$ was chosen arbitrarily so that numerical values could be calculated. In addition, polyvalent electrolytes enhance this effect as seen in Fig. 8.

3. The activity coefficient exponent, $(1 - \bar{\nu}_s)^2/1 + \kappa a_s$, for the cationic surface site, $SOH_2{}^+$, can be seen to decrease as the mean surface site electrostatic charge, $e\bar{\nu}_s$, becomes more positive. As a result, f_{SOH_2} will increase as the interfacial electrostatic potential charge becomes more positive. As the interfacial electrostatic potential increase, bringing positive electrostatic charge to a more positive interface requires more work, which means less screening or increasing f_{SOH_2}. In summary: $\bar{\nu}_s \uparrow$, $\epsilon_{SOH_2} \downarrow$, $§_{SOH_2} \downarrow$, $f_{SOH_2} \uparrow$.

Equation (134) can be seen to predict the screening behavior described above and shown in Fig. 7. As $\bar{\nu}_s$ increases (becomes more positive), the activity coefficient exponent, ϵ_{SOH_2}, decreases, which increases the activity coefficient, f_{SOH_2}. Note that

TABLE 1 Dependence of Activity Coefficient on the Mean Electrostatic Surface Site Charge for Anionic Surface Sites, SO^-

$\bar{\nu}_s$	ϵ_{SO}	$f_{SO}\ (Z = 0.8)$	$§_{SO} \equiv 1 - f_{SO}$
+1.0	2.0	0.64	0.36
+0.5	1.125	0.78	0.22
0.0	0.5	0.89	0.11
−0.5	0.125	0.97	0.03
−1.0	0.0	1.0	0.0

this trend between $\bar{\nu}_s$ and f_{SOH_2} is opposite to the trend between $\bar{\nu}_s$ and, f_{SO}, discussed above. These results for f_{SOH_2} are reasonable because a more positive environment around a cationic surface site, SOH_2^+, will decrease screening, $\S_{SOH_2}$. This behavior can be clearly seen in a few numerical examples in Table 2, where a value of $Z = 0.8$ was chosen arbitrarily so that numerical values could be calculated. Moreover, the enhanced effect of polyvalent electrolytes can be seen in Fig. 8.

4. The prediction of an activity coefficient other than unity for the nonionic, monoprotic surface site, SOH, is somewhat unexpected. The existence of a nonunity activity coefficient can be seen to result from the electrostatic potential induced on SOH by the electrostatic potential of the neighboring surface sites. When the interfacial electrostatic charge $\bar{\nu}_s \rightarrow 0$, then $f_s \rightarrow 1$. If $\bar{\nu}_s$ is finite, then f_{SOH} is predicted to be less than unity by Eq. (129).

The activity coefficient exponent, $\bar{\nu}_s^2/1 + \kappa a_s$, in Eq. (133) for the nonionic surface site, SOH, can be seen to increase as the mean surface site electrostatic charge, $e\bar{\nu}_s$, becomes more positive. As a result, f_{SOH} will decrease as the interfacial electrostatic potential charge increases owing to the inverse relationship between ϵ_{SOH} and f_{SOH}. This behavior can be clearly seen in a few numerical examples in Table 3, where a value of $Z = 0.8$ was chosen arbitrarily so that numerical values could be calculated. In summary: $\bar{\nu}_s \uparrow$, $\epsilon_{SOH_2} \uparrow$, $\S_{SOH_2} \uparrow$, $f_{SOH_2} \downarrow$.

It is interesting to note that the trend between $\bar{\nu}_s$ and f_{SOH} is parabolic and therefore different than the trend for $\bar{\nu}_s$ with either f_{SO} or f_{SOH_2} as shown in Figs. 7 and 8. In addition, the trend in f_{SOH} with $\bar{\nu}_s$ is smaller in magnitude than for either f_{SO} or f_{SOH_2} although polyvalent electrolytes enhance the effects for all three types of surface sites. This trend is reasonable, based on the fact that extremes in electrostatic surface potentials will induce charges in the nonionic surface sites which in turn will manifest as nonunity activity coefficients for the SOH sites.

Substitution of Eqs (131)–(134) into the mass action equations (128) and (130) gives explicit equations for the activity coefficient quotients corresponding the mass action equation for ionization of the cationic diprotic surface site:

$$K_{a1} = \left(\frac{\theta_{SOH}\chi_H}{\theta_{SOH_2}}\right)\left(\frac{f_{SOH}f_H}{f_{SOH_2}}\right) = \left(\frac{\theta_{SOH}\chi_H}{\theta_{SOH_2}}\right)Z^{\left[\frac{\bar{\nu}_s^2}{1+\kappa a_s}+\frac{1}{1+\kappa a_H}-\frac{(1-\bar{\nu}_s)^2}{1+\kappa a_s}\right]} \tag{136}$$

or

$$K_{a1} = \left(\frac{\theta_{SOH}\chi_H}{\theta_{SOH_2}}\right)Z^{\epsilon_{K_{a1}}} \quad \text{where} \quad \epsilon_{K_{a1}} \equiv \frac{1}{1+\kappa a_H} + \frac{2\bar{\nu}_s - 1}{1+\kappa a_s} \tag{137}$$

TABLE 2 Dependence of Activity Coefficient on the Mean Electrostatic Surface Site Charge for Cationic Surface Sites, SOH_2^+

$\bar{\nu}_s$	ϵ_{SOH_2}	f_{SOH_2} $(Z = 0.8)$	$\S_{SOH_2} \equiv 1 - f_{SOH_2}$
+1.0	0.0	1.0	0.0
+0.5	0.125	0.97	0.03
0.0	0.5	0.89	0.11
−0.5	1.125	0.78	0.22
−1.0	2.0	0.64	0.36

TABLE 3 Dependence of Activity Coefficient on the Mean Electrostatic Surface Site Charge for Cationic Surface Sites, SOH

$\bar{\nu}_s$	ϵ_{SOH}	f_{SOH} $(Z = 0.8)$	$\S_{SOH} \equiv 1 - f_{SOH}$
+1.0	0.5	0.89	0.11
+0.5	0.125	0.97	0.03
0.0	0.0	1.0	0.0
−0.5	0.125	0.97	0.03
−1.0	0.5	0.89	0.11

For the ionization of the nonionic monoprotic surface site, the explicit activity coefficient equation can be written as

$$K_{a2} = \left(\frac{\theta_{SO}\chi_H}{\theta_{SOH}}\right)\left(\frac{f_{SO}f_H}{f_{SOH}}\right) = \left(\frac{\theta_{SO}\chi_H}{\theta_{SOH_2}}\right)Z^{\left[\frac{(-1-\bar{\nu}_s)^2}{1+\kappa a_s}+\frac{1}{1+\kappa a_H}-\frac{\bar{\nu}_s^2}{1+\kappa a_s}\right]} \tag{138}$$

or

$$K_{a2} = \left(\frac{\theta_{SO}\chi_H}{\theta_{SOH_2}}\right)Z^{\epsilon_{K_{a2}}} \quad \text{where} \quad \epsilon_{K_{a2}} \equiv \frac{1}{1+\kappa a_H} + \frac{2\bar{\nu}_s + 1}{1+\kappa a_s} \tag{139}$$

In adsorption equations the product of the two proton mass action adsorption constants is used so it will be useful to determine the activity coefficient exponent for the activity coefficient quotient which arises for $K_{a1}K_{a2}$. Multiplying Eqs (135) and (136):

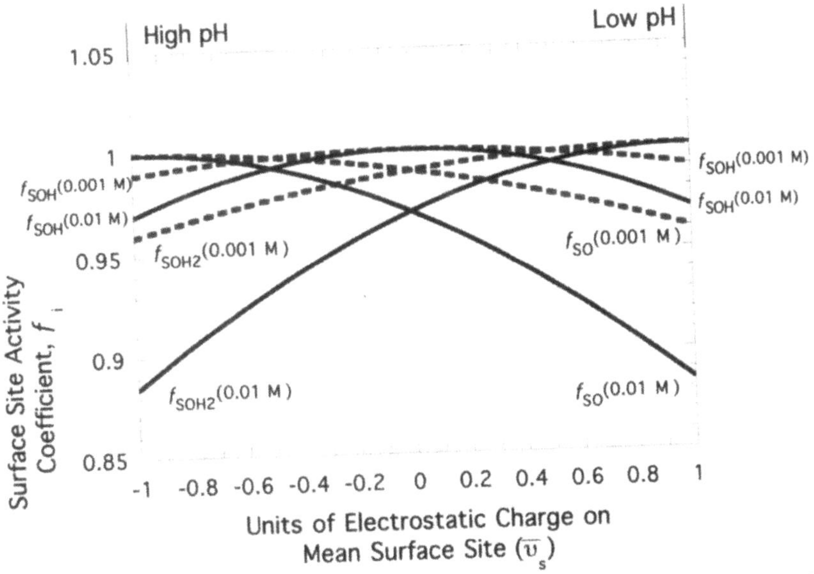

FIG. 7 Dependence of surface site activity coefficients on interfacial electrostatic charge, $\bar{\nu}_s$ and κ, for 1:1 electrolytes at 0.01M and 0.001M concentrations.

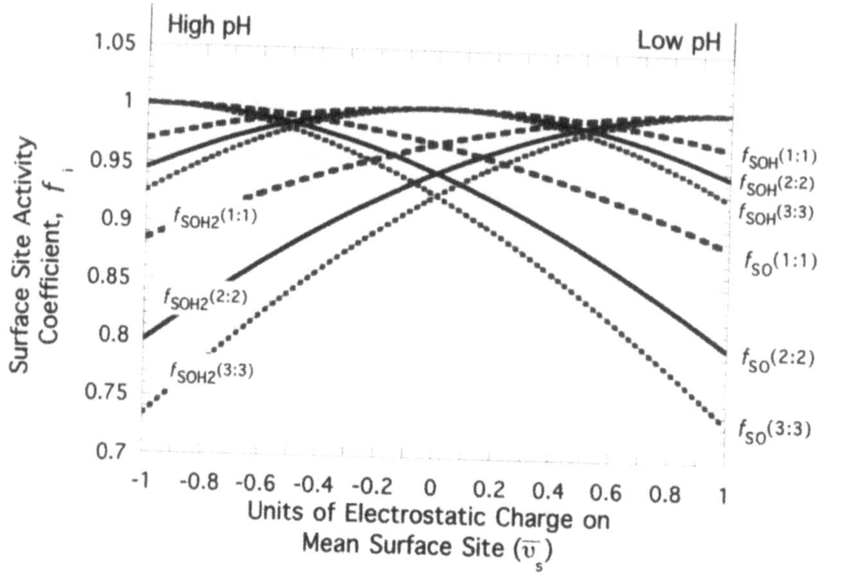

FIG. 8 Dependence of surface site activity coefficients on interfacial electrostatic charge, $\bar{\upsilon}_s$ and κ, for 1:1, 2:2, and 3:3 electrolytes at 0.01M concentration.

$$K_{a1}K_{a2} = \left(\frac{\theta_{SOH}\chi_H}{\theta_{SOH_2}}\right)\left(\frac{f_{SOH}f_H}{f_{SOH_2}}\right)\left(\frac{\theta_{SO}\chi_H}{\theta_{SOH}}\right)\left(\frac{f_{SO}f_H}{f_{SOH}}\right) = \left(\frac{\theta_{SO}\chi_H^2}{\theta_{SOH_2}}\right)\left(\frac{f_{SO}f_H^2}{f_{SOH_2}}\right) \tag{140}$$

Expanding the activity coefficients into explicit equations:

$$K_{a1}K_{a2} = \left(\frac{\theta_{SO}\chi_H^2}{\theta_{SOH_2}}\right) Z^{\left[\frac{(-1-\bar{\upsilon}_s)^2}{1+\kappa a_s} + \frac{2}{1+\kappa a_H} - \frac{(1-\bar{\upsilon}_s)^2}{1+\kappa a_s}\right]} \tag{141}$$

or

$$K_{a1}K_{a2} = \left(\frac{\theta_{SO}\chi_H^2}{\theta_{SOH_2}}\right) Z^{\epsilon_{K_{a12}}} \quad \text{where} \quad \epsilon_{K_{a12}} \equiv \frac{2}{1+\kappa a_H} + \frac{4\bar{\upsilon}_s}{1+\kappa a_s} \tag{142}$$

The exponents are also shorthand notation that will simplify the adsorption equations in Chapter 1.

C. Cation Adsorption–Interfacial Complexation

Since protons are cations any interface which adsorbs protons must also adsorb cations to a greater or lesser degree. A simple surface complexation example will be discussed here to demonstrate how activity coefficients enter into the mass action equation for the cation adsorption process and the role activity coefficients play in the adsorption process. An example adsorption process for a cation, M^{ν_m}, adsorbing at a S–MO interface is shown in Fig. 9.

From Fig. 9, the adsorption process is seen as a chemical reaction where a cation, M^{ν_m}, moves from the interior of a resolution where it experiences a potential,

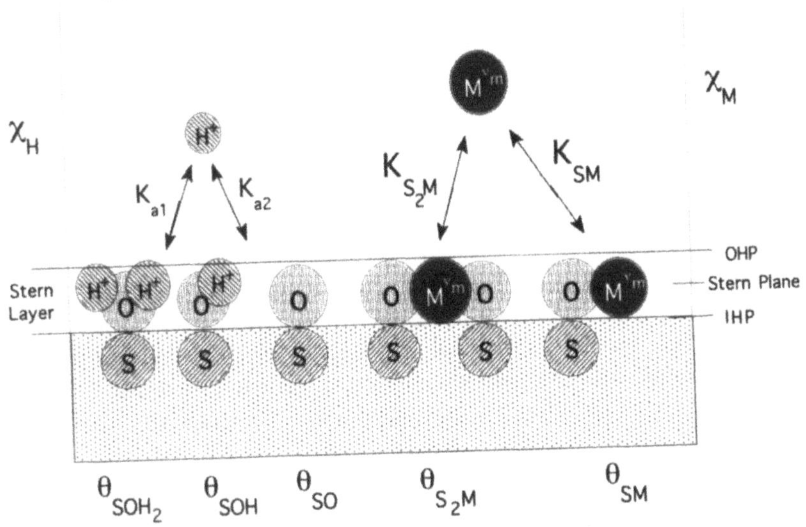

FIG. 9 Adsorption as an interfacial complexation reaction.

ψ_M, to the interface where it experiences the interfacial potential, ψ_s, and forms some type of bond with an anionic surface site ion, SO^-, as

$$SO^- + M^{\nu_m} = SOM^{\nu_m^{-1}} \tag{143}$$

Note that the adsorption bond can be a short- or long-range potential, but whatever the type of bond formed, it is of no consequence to the thermodynamic argument and is indeterminate from thermodynamics alone. Bonding mechanisms must be determined by other methods such as spectroscopy, quantum mechanics, etc.

The mass action equation for the adsorption reaction shown in Fig. 9 is derived by the same method used to derive the mass action equation (100) for solution complexation. The chemical potential equations for the two reactants (cation in solution at solution mole fraction, χ_M, and uncomplexed surface site at surface mole fraction, χ_{SO}) and one product (surface site complexed with a cation at surface mole fraction, θ_{SM}) shown in Fig. 9 can be written as

$$\mu_M = \mu_M^* + kT\ln\chi_M + kT\ln f_M \tag{144}$$

$$\mu_{SO} = \mu_{SO}^* + kT\ln\theta_{SO} + kT\ln f_{SO} \tag{145}$$

$$\mu_{SM} = \mu_{SM}^* + kT\ln\theta_{SM} + kT\ln f_{SM} \tag{146}$$

$$\mu_{S_2M} = \mu_{S_2M}^* + kT\ln\theta_{S_2M} + kT\ln f_{S_2M} \tag{147}$$

The chemical potential for the complexation reaction, $\Delta\mu_{SM}$, shown in Fig. 9 is

$$\Delta\mu_{SM} \equiv \mu_{SM} - \mu_{SO} - \mu_M \tag{148}$$

Substitution of Eqs (144)–(146) into Eq. (148) gives

$$\Delta\mu_{SM} = (\mu_{SM}^* - \mu_{SO}^* - \mu_M^*) + kT\ln\left(\frac{\theta_{SM}}{\theta_{SO}\chi_M}\right) + kT\ln\left(\frac{f_{SM}}{f_{SO}f_M}\right) \tag{149}$$

At equilibrium, $\Delta\mu_{SM} = 0$, so Eq. (149) can be rearranged to give

$$-(\mu^*_{SM} - \mu^*_{SO} - \mu^*_{M}) \equiv kT\ln(K_{SM}) = kT\ln\left(\frac{\theta_{SM}}{\theta_{SO}\chi_M}\right) + kT\ln\left(\frac{f_{SM}}{f_{SO}f_M}\right) \quad (150)$$

Removing logarithms from Eq. (150):

$$K_{SM} = \left(\frac{\theta_{SM}}{\theta_{SO}\chi_M}\right)\left(\frac{f_{SM}}{f_{SO}f_M}\right) \quad (151)$$

Equation (151) is the mass action equation for the adsorption reaction shown in Fig. 9. As before the solution complexation, rigorous expressions for the activity coefficients in Eq. (151) can be substituted from Eq. (38), but the electrostatic potentials in the resulting equations make evaluation difficult. Instead, Eqs (59) and (88) will be used to write activity equations for the three activity coefficient equations in Eq. (151).

For a solute cation, M^{ν_m}, the electrostatic charge was determined in Section III.A to be $\nu_m e$. The equation for the activity coefficient can be derived by substitution of the electrostatic charge, $\nu_m e$, into Eq. (59):

$$f_M = e^{\left(-\frac{\nu_m^2 \kappa q_B}{1+\kappa a_m}\right)} = (e^{-\kappa q_B})^{\left(\frac{\nu_m^2}{1+\kappa a_m}\right)} \equiv Z^{\left(\frac{\nu_m^2}{1+\kappa a_m}\right)} \equiv Z^{\epsilon_m} \quad (152)$$

For the anionic surface site, SO^-, the electrostatic charge was determined in Section III. B to be $(-1 - \bar{\nu}_S)e$, and the activity coefficient was determined in Section IV.B to be

$$f_{SO} = e^{\left[-\frac{(\nu_s - \bar{\nu}_s)^2 \kappa q_B}{1+\kappa a_s}\right]} \equiv Z^{\left[\frac{(-1-\bar{\nu}_s)^2}{1+\kappa a_s}\right]} \equiv Z^{\left[\frac{(1+\bar{\nu}_s)^2}{1+\kappa a_s}\right]} \equiv Z^{\epsilon_{SO}} \quad (153)$$

For the cation–surface site complex, $SOM^{(\nu_m - 1)}$, the net electrostatic charge at a solid–solution interface, where the electrostatic charge on the SP plane was $\bar{\nu}_s e$, was determined in Section III.B to be the sum of electrostatic charges on the anionic surface site, $(-1)e$, and the adsorbed cation, $\nu_m e$, less the electrostatic charge on the SP, $\bar{\nu}_s e$, to give $(\nu_m - 1 - \bar{\nu}_s)e$. The equation for the activity coefficient can be derived by substitution of the electrostatic charge, $(\nu_m - 1 - \bar{\nu}_s)e$, into Eq. (88):

$$f_{SM} = Z^{\left[\frac{(\nu_m - 1 - \bar{\nu}_s)^2}{1+\kappa a_s}\right]} = Z^{\left[\frac{\nu_m^2 - 2(\nu_m + \nu_m\bar{\nu}_s - \bar{\nu}_s) + \bar{\nu}_s^2 + 1}{1+\kappa a_s}\right]} \equiv Z^{\epsilon_{SM}} \quad (154)$$

For the cation–surface site complex, $(SO)_2M^{(\nu_m - 2)}$, the net electrostatic charge at a solid–solution interface, where the electrostatic charge on the Stern plane was $\bar{\nu}_s e$, was determined in Section III.B to be the sum of electrostatic charges on the two anionic surface site, $(-2)e$, and the adsorbed cation, $\nu_m e$, less the electrostatic charge on the SP $\bar{\nu}_s e$, to give $(\nu_m - 2 - \bar{\nu}_s)e$. The equation for the activity coefficient can be derived by substitution of the electrostatic charge, $(\nu_m - 2 - \bar{\nu}_s)e$, into Eq. (88):

$$f_{S_2M} = Z^{\left[\frac{(\nu_m - 2 - \bar{\nu}_s)^2}{1+\kappa a_s}\right]} = Z^{\left[\frac{\nu_m^2 - 2(\nu_m + \nu_m\bar{\nu}_s - 2\bar{\nu}_s) + \bar{\nu}_s^2 + 4}{1+\kappa a_s}\right]} \equiv Z^{\epsilon_{S_2M}} \quad (155)$$

The explicit equation for the activity coefficient quotients for the cation–surface site complex, $SOM^{(-1+\nu_m)}$, can be derived by substitution of Eqs (152)–(154) into the activity coefficient quotient in Eq. (151):

$$\left(\frac{f_{SM}}{f_{SO}f_M}\right) = Z^{\left[\frac{\nu_m^2 - 2(\nu_m + \nu_m\bar{\nu}_s - \bar{\nu}_s) + \bar{\nu}_s^2 + 1 - (-1-\bar{\nu}_s)^2}{1+\kappa a_s} - \frac{\nu_m^2}{1+\kappa a_m}\right]} \equiv Z^{\epsilon_{K_{SM}}} \tag{156}$$

Expanding and collecting terms in the exponent of Eq. (156):

$$\left(\frac{f_{SM}}{f_{SO}f_M}\right) = Z^{\left[\frac{\nu_m(\nu_m - 2 - 2\bar{\nu}_s)}{1+\kappa a_s} - \frac{\nu_m^2}{1+\kappa a_m}\right]} \equiv Z^{\epsilon_{K_{SM}}} \tag{157}$$

where

$$\epsilon_{K_{SM}} \equiv \frac{\nu_m(\nu_m - 2 - 2\bar{\nu}_s)}{1+\kappa a_s} - \frac{\nu_m^2}{1+\kappa a_m} \tag{158}$$

In a similar fashion the explicit equation for the activity coefficient quotient for the cation–double surface site complex, $(SO)_2M^{(-2+\nu_m)}$, can be derived. The chemcial potential for the complexation reaction, $\Delta\mu_{S_2M}$, shown in Fig. 9 is

$$\Delta\mu_{S_2M} \equiv \mu_{S_2M} - 2\mu_{SO} - \mu_M \tag{159}$$

Substitution of Eqs (144), (145), and (147) into Eq. (159), and invoking equilibrium conditions, gives

$$-(\mu_{S_2M}^* - 2\mu_{SO}^* - \mu_M^*) \equiv kT\ln(K_{S_2M}) = kT\ln\left(\frac{\theta_{S_2M}}{\theta_{SO}^2\chi_M}\right) + kT\ln\left(\frac{f_{S_2M}}{f_{SO}^2 f_M}\right) \tag{160}$$

Removing logarithms from Eq. (155):

$$K_{S_2M} = \left(\frac{\theta_{S_2M}}{\theta_{SO}^2\chi_M}\right)\left(\frac{f_{S_2M}}{f_{SO}^2 f_M}\right) \tag{161}$$

Substitution of Eqs (152), (153), and (155) into the activity coefficient quotients in Eq. (161):

$$\left(\frac{f_{S_2M}}{f_{SO}^2 f_M}\right) = Z^{\left[\frac{\nu_m^2 - 2(\nu_m + \nu_m\bar{\nu}_s - 2\bar{\nu}_s) + \bar{\nu}_s^2 + 4}{1+\kappa a_s} - \frac{2(-1-\bar{\nu}_s)^2}{1+\kappa a_s} - \frac{\nu_m^2}{1+\kappa a_m}\right]} \equiv Z^{\epsilon_{K_{S_2M}}} \tag{162}$$

Expanding and collecting terms in the exponent of Eq. (162)

$$\left(\frac{f_{S_2M}}{f_{SO}^2 f_M}\right) = Z^{\left[\frac{\nu_m(\nu_m - 2 - 2\bar{\nu}_s) + (2 - \bar{\nu}_s^2)}{1+\kappa a_s} - \frac{\nu_m^2}{1+\kappa a_m}\right]} \equiv Z^{\epsilon_{K_{S_2M}}} \tag{163}$$

where

$$\epsilon_{K_{S_2M}} \equiv \frac{\nu_m(\nu_m - 1 - \bar{\nu}_s) + (2 - \bar{\nu}_s^2)}{1+\kappa a_s} - \frac{\nu_m^2}{1+\kappa a_m} \tag{164}$$

D. Anion Adsorption–Interfacial Complexation

Anion adsorption at the solid–solution interface occurs in the same way cation adsorption occurs except that anions adsorb almost exclusively on cationic surface sites, SOH_2^+. A simple surface complexation example will be discussed here to demonstrate how activity coefficients enter into the mass action equations for the anion adsorption process and the role activity coefficients play in the adsorption process. An example adsorption process for an anion, L^{ν_l}, adsorbing at an S–MO interface is shown in Fig. 10.

The adsorption process shown in Fig. 10 is seen as a chemical reaction where an anion, L^{ν_l}, moves from the interior of a solution where it experiences a potential, ψ_L, to the interface where it experiences the interfacial potential, ψ_s, and forms some type of bond with a cationic surface site ion, SOH_2^+, as

$$SOH_2^+ + L^{\nu_l} = SOH_2L^{(1+\nu_l)} \tag{165}$$

The chemical potential equations for the two reactants (anion in solution at solution mole fraction, χ_L, and uncomplexed cationic surface site at surface mole fraction, χ_{SOH_2}) and one product (surface site complexed with an anion at surface mole fraction, θ_{SL}) shown in Fig. 10 can be written as

$$\mu_L = \mu_L^* + kT\ln\chi_L + kT\ln f_L \tag{166}$$

$$\mu_{SOH_2} = \mu_{SOH_2}^* + kT\ln\theta_{SOH_2} + kT\ln f_{SOH_2} \tag{167}$$

$$\mu_{SL} = \mu_{SL}^* + kT\ln\theta_{SL} + kT\ln f_{SL} \tag{168}$$

The chemical potential for the complexation reaction, $\Delta\mu_{SL}$, shown in Fig. 10 is

$$\Delta\mu_{SL} \equiv \mu_{SL} - \mu_{SOH_2} - \mu_L \tag{169}$$

Substitution of Eqs (166)–(168) into Eq. (169) gives

$$\Delta\mu_{SL} = (\mu_{SL}^* - \mu_{SOH_2}^* - \mu_L^*) + kT\ln\left(\frac{\theta_{SL}}{\theta_{SOH_2}\chi_L}\right) + kT\ln\left(\frac{f_{SL}}{f_{SOH_2}f_L}\right) \tag{170}$$

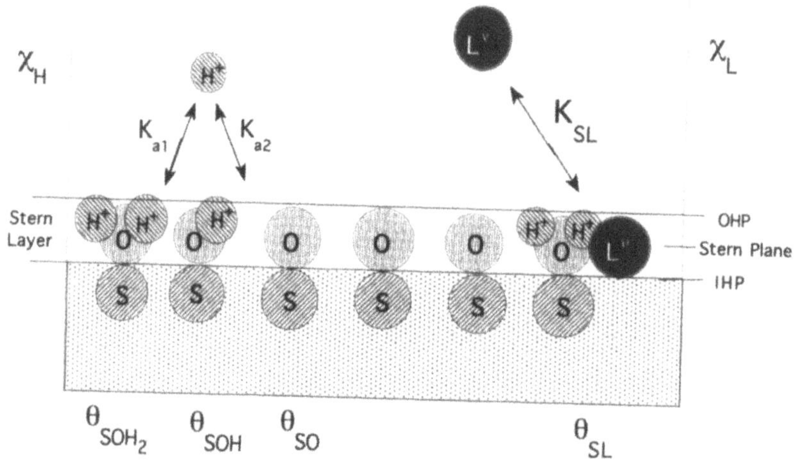

FIG. 10 Anion adsorption as an interfacial complexation reaction.

At equilibrium, $\Delta\mu_{SL} = 0$, so Eq. (170) can be rearranged to give

$$-(\mu^*_{SL} - \mu^*_{SOH_2} - \mu^*_{L}) \equiv kT\ln(K_{SL}) = kT\ln\left(\frac{\theta_{SL}}{\theta_{SOH_2}\chi_L}\right) + kT\ln\left(\frac{f_{SL}}{f_{SOH_2}f_L}\right) \tag{171}$$

Removing logarithms from Eq. (171):

$$K_{SL} = \left(\frac{\theta_{SL}}{\theta_{SOH_2}\chi_L}\right)\left(\frac{f_{SL}}{f_{SOH_2}f_L}\right) \tag{172}$$

Equation (172) is the mass action equation for the adsorption reaction shown in Fig. 10. As before with cation complexation, rigorous expressions for the activity coefficients in Eq. (177) can be substituted from Eq. (38), but the electrostatic potentials in the resulting equations make evaluation difficult. Instead, Eqs (59) and (88) will be used to write activity coefficient equations for the three activity coefficients in Eq. (168).

For a solute cation, L^{ν_l}, the electrostatic charge was determined in Section III.A to be $\nu_l e$. The equation for the activity coefficient can be derived by substitution of the electrostatic charge, $\nu_l e$, into Eq. (59):

$$f_L = e^{\left(-\frac{\nu_l^2\kappa q_B}{1+\kappa a_l}\right)} = (e^{-\kappa q_B})^{\left(\frac{\nu_l^2}{1+\kappa a_l}\right)} \equiv Z^{\left(\frac{\nu_l^2}{1+\kappa a_l}\right)} \equiv Z^{\epsilon_l} \tag{173}$$

For the cationic surface site, $SOH_2{}^+$, the electrostatic charge was determined in Section III.B to be $(+1 - \bar{\nu}_s)e$, and the activity coefficient was determined in Section IV.B to be

$$f_{SOH_2} = e^{\left[-\frac{(\nu_s-\bar{\nu}_s)^2\kappa q_B}{1+\kappa a_s}\right]} \equiv Z^{\left[\frac{(1-\bar{\nu}_s)^2}{1+\kappa a_s}\right]} \equiv Z^{\epsilon_{SOH_2}} \tag{174}$$

For the cation–surface site complex, $SOH_2L^{(1+\nu_l)}$, the net electrostatic charge at a solid–solution interface, where the electrostatic charge on the SP plane was $\bar{\nu}_s e$, was determined in Section III.B to be the sum of electrostatic charges on the cationic surface site, $(1 - \bar{\nu}_s)e$, and the adsorbed anion, $\nu_l e$, to give $(1 + \nu_l - \bar{\nu}_s)e$. The equation for the activity coefficient can be derived by substitution of the electrostatic charge, $(1 + \nu_l - \bar{\nu}_s)e$, into Eq. (88):

$$f_{SL} = Z^{\left[\frac{(1+\nu_l-\bar{\nu}_s)^2}{1+\kappa a_s}\right]} = Z^{\left[\frac{\nu_l^2-2(-\nu_l+\nu_l\bar{\nu}_s+\bar{\nu}_s)+\bar{\nu}_s^2+1}{1+\kappa a_s}\right]} \equiv Z^{\epsilon_{SL}} \tag{175}$$

Substitution of Eqs (173)–(175) into the activity coefficient quotient in Eq. (172):

$$\left(\frac{f_{SL}}{f_{SOH_2}f_L}\right) = Z^{\left[\frac{\nu_l^2+2(\nu_l-\nu_l\bar{\nu}_s-\bar{\nu}_s)+\bar{\nu}_s^2+1-(+1-\bar{\nu}_s)^2}{1+\kappa a_s}-\frac{\nu_l^2}{1+\kappa a_l}\right]}$$

$$\equiv Z^{\epsilon_{K_{SL}}} \tag{176}$$

Expanding and collecting terms in the exponent of Eq. (175):

$$\left(\frac{f_{SL}}{f_{SOH_2}f_L}\right) = Z^{\left[\frac{\nu_l(\nu_l+2-2\bar{\nu}_s)}{1+\kappa a_s}-\frac{\nu_l^2}{1+\kappa a_l}\right]} \equiv Z^{\epsilon_{K_{SL}}} \tag{177}$$

where

$$\epsilon_{K_{SL}} \equiv \frac{\nu_l(\nu_l + 2 - 2\bar{\nu}_s)}{1 + \kappa a_s} - \frac{\nu_l^2}{1 + \kappa a_l} \tag{178}$$

E. Nonionic Adsorption–Interfacial Complexation

While cations adsorb on anionic surface sites and vice versa for anion adsorption, molecules not possessing a full electrostatic charge, N^0, in all certainty adsorb to some degree on all three types of surface sites: anionic, SO^-, nonionic, SOH, and cationic, $SOH_2{}^+$. The extent to which a nonionic molecule complexes with each of the three surface site types: anionic, SON^-, nonionic, SOHN, and cationic, SOH_2N^+, is governed by each of the experimentally determined three mass action constants, anionic, K_{SN}, nonionic, K_{SHN}, and cationic, K_{SH_2N}, respectively. The procedure for deriving the activity coefficients and quotients is the same as for protons, cations, and anions as discussed in Sections IVB, C, and D, so the derivations given below will be abbreviated with less analysis and description. The adsorption processes for a nonionic molecule, N^0, adsorbing at all three types of surface site on a S–MO interface is shown in Fig. 11.

Chemical reactions for the three surface complexation reactions shown in Fig. 11 for adsorption of a nonionic molecule, N^0, with corresponding mass action equations can be written as

$$SO^- + N^0 = SON^- \qquad K_{SN} = \left(\frac{\theta_{SN}}{\theta_{SO}\chi_N}\right)\left(\frac{f_{SN}}{f_{SO}f_N}\right) \tag{179}$$

$$SOH + N^0 = SOHN \qquad K_{SHN} = \left(\frac{\theta_{SHN}}{\theta_{SOH}\chi_N}\right)\left(\frac{f_{SHN}}{f_{SOH}f_N}\right) \tag{180}$$

$$SOH_2{}^+ + N^0 = SOH_2N^+ \qquad K_{SH_2N} = \left(\frac{\theta_{SH_2N}}{\theta_{SOH_2}\chi_N}\right)\left(\frac{f_{SH_2N}}{f_{SOH_2}f_N}\right) \tag{181}$$

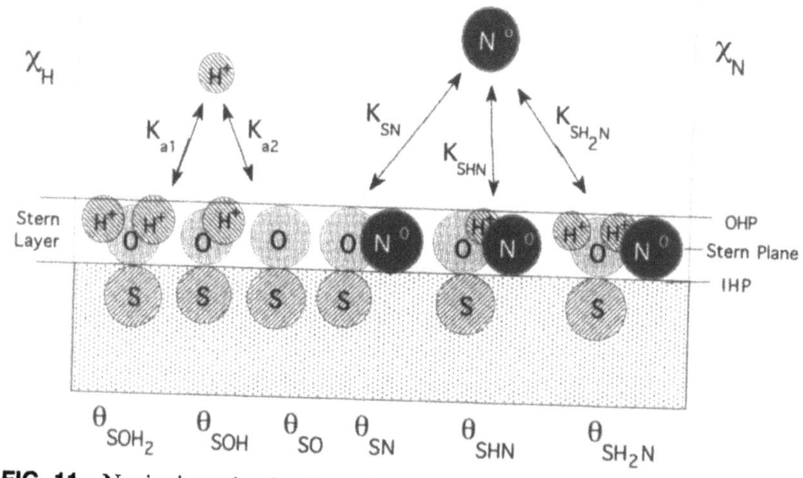

FIG. 11 Nonionic molecule adsorption as an interfacial complexation reaction.

Equations (179)–(181) are the mass action equations for the adsorption reactions shown in Fig. 11. As before with cation and anion complexation, rigorous expressions for the activity coefficients can be substituted from Eq. (38), but the electrostatic potentials in the resulting equations make evaluation difficult. Instead, Eqs. (59) and (88) will be used to write activity equations for the three activity coefficients in the equations above.

For a solute, N^0, the electrostatic charge is obviously zero so substitution of the electrostatic charge, 0, into Eq. (59) gives

$$f_N = e^{\left(-\frac{0^2 \kappa q_B}{1+\kappa a_n}\right)} Z^{\left(\frac{0^2}{1+\kappa a_n}\right)} = 1 \tag{182}$$

Of course, it should be remembered that the only type of electrostatic charge being considered is Coulombic. Many nonionic solutes will interact with solution molecules through other types of electrostatic interactions such as polar, induced dipole, hydrogen bonding, etc. These other electrostatic interactions would lead to a nonunity value for f_N, in general.

The activity coefficient equations (132)–(134) for the three surface sites, SO^-, SOH, and $SOH_2{}^+$, respectively, were derived in Section IV.B as

$$f_{SO} = Z^{\left[\frac{(\nu_s-\bar{\nu}_s)^2}{1+\kappa a_s}\right]} = Z^{\left[\frac{(-1-\bar{\nu}_s)^2}{1+\kappa a_s}\right]} = Z^{\left[\frac{(1+\bar{\nu}_s)^2}{1+\kappa a_s}\right]} \equiv Z^{\epsilon_{SO}} \tag{132}$$

$$f_{SOH} = Z^{\left[\frac{(\nu_s-\bar{\nu}_s)^2}{1+\kappa a_s}\right]} = Z^{\left[\frac{(0-\bar{\nu}_s)^2}{1+\kappa a_s}\right]} = Z^{\left[\frac{\bar{\nu}_s^2}{1+\kappa a_s}\right]} \equiv Z^{\epsilon_{SOH}} \tag{133}$$

$$f_{SOH_2} = Z^{\left[\frac{(\nu_s-\bar{\nu}_s)^2}{1+\kappa a_s}\right]} = Z^{\left[\frac{(1-\bar{\nu}_s)^2}{1+\kappa a_s}\right]} \equiv Z^{\epsilon_{SOH_2}} \tag{134}$$

For the nonionic–anionic surface site complex, SON^-, the net electrostatic charge at a solid–solution interface, where the electrostatic charge on the SP was $\bar{\nu}_s e$, was determined in Section III.B to be the sum of electrostatic charges on the anionic surface site, $(-1-\bar{\nu}_s)e$, and the adsorbed nonionic species, 0, to give $(-1-\bar{\nu}_s)e$. The equation for the activity coefficient can be derived by substitution of the electrostatic charge, $(-1-\bar{\nu}_s)e$, into Eq. (88):

$$f_{SN} = Z^{\left[\frac{(-1-\bar{\nu}_s)^2}{1+\kappa a_s}\right]} = Z^{\left[\frac{1+2\bar{\nu}_s+\bar{\nu}_s^2}{1+\kappa a_s}\right]} \equiv Z^{\epsilon_{SN}} \tag{183}$$

For the nonionic–nonionic surface site complex, SOHN, the net electrostatic charge at a solid–solution interface, where the electrostatic charge on the SP was $\bar{\nu}_s e$, was determined in Section III.B to be the sum of electrostatic charges on the nonionic surface site, $\bar{\nu}_s e$, and the adsorbed nonionic species, 0, to give $\bar{\nu}_s e$. The equation for the activity coefficient can be derived by substitution of the electrostatic charge, $\bar{\nu}_s e$, into Eq. (88):

$$f_{SHN} = Z^{\left[\frac{(\bar{\nu}_s)^2}{1+\kappa a_s}\right]} = Z^{\left[\frac{\bar{\nu}_s^2}{1+\kappa a_s}\right]} \equiv Z^{\epsilon_{SHN}} \tag{184}$$

For the nonionic–cationic surface site complex, SOH_2N^+, the net electrostatic charge at a solid–solution interface, where the electrostatic charge on the SP was $\bar{\nu}_s e$, was determined in Section III.B to be the sum of electrostatic charges on the cationic surface site, $(1-\bar{\nu}_s)e$, and the adsorbed nonionic species, 0, to give $(1-\bar{\nu}_s)e$. The equation for the activity coefficient can be derived by substitution of the electrostatic charge, $(1 - \bar{\nu}_s)e$, into Eq. (88):

$$f_{SH_2N} = Z^{\left[\frac{(1-\bar{\nu}_s)^2}{1+\kappa a_s}\right]} = Z^{\left[\frac{1-2\bar{\nu}_s+\bar{\nu}_s^2}{1+\kappa a_s}\right]} \equiv Z^{\epsilon_{SH_2N}} \tag{185}$$

Explicit equations for the activity coefficient quotients are given in Eqs (179)–(181).

Substitution of Eqs (183), (132), and (182) into the activity coefficient quotient in Eq. (179) gives

$$\left(\frac{f_{SN}}{f_{SO}f_N}\right) = Z^{\left[\frac{(-1-\bar{\nu}_s)^2-(1+\bar{\nu}_s)^2-0}{1+\kappa a_s}\right]} = Z^{\left[\frac{0}{1+\kappa a_s}\right]} = 1 \tag{186}$$

Substitution of Eqs. (184), (133), and (182) into the activity coefficient quotient in Eq. (181) gives

$$\left(\frac{f_{SHN}}{f_{SOH}f_N}\right) = Z^{\left[\frac{\bar{\nu}_s^2-\bar{\nu}_s^2-0}{1+\kappa a_s}\right]} = Z^{\left[\frac{0}{1+\kappa a_s}\right]} = 1 \tag{187}$$

Substitution of Eqs (185), (134), and (182) into the activity coefficient quotient in Eq. (182) gives

$$\left(\frac{f_{SH_2N}}{f_{SOH_2}f_N}\right) = Z^{\left[\frac{(-1-\bar{\nu}_s)^2-(1+\bar{\nu}_s)^2-0}{1+\kappa a_s}\right]} = Z^{\left[\frac{0}{1+\kappa a_s}\right]} = 1 \tag{188}$$

For Eqs (182)–(188) we see some interesting results:

- The activity coefficient equation (182) for the nonionic solute molecule in solution is unity as it should be for the Coulombic model from which the equation was derived. As discussed above, other electrostatic interactions, other than Coulombic, between nonionic solute and solution molecules in real solutions will make the activity coefficient for nonionic molecules non-unity, especially at high solute concentrations.
- For the three nonionic–surface complexes, the activity coefficient equations (183)–(185) are not in general unity because of the interfacial electrostatic charge, $\bar{\nu}_s e$, which induces an electrostatic potential into the nonionic–surface complexes. This result is reasonable and was observed and discussed for the activity coefficient for the proton complexed surface sites in Section IV.B.
- The three activity quotient equations (186)–(188) all reduce to unity. This is noteworthy because the activity coefficients in these activity coefficient quotients are not unity, but they cancel so that unity is the result. From this result it can be concluded that the role of activity coefficients in surface complexation reactions for nonionic solute molecules is negligible.

However, in real systems, activity coefficients may play a small role in nonionic surface complexation reactions. In order for the cancellation to occur in Eqs (186)–(188) requires that the activity coefficient of the surface site and the nonionic complexed surface are the same. For the primitive interfacial model they are the same so the activity coefficient quotient reduces to unity. For nonionic–surface site complexes in real solid–solution interfacial systems the activity coefficients quotients will be close to unity but not exactly, because the activity coefficients of nonionic complexed and uncomplexed surface sites will, in general, be slightly different.

F. Polymer Adsorption–Interfacial Complexation

Owing to the spatial dependencies of electrostatic potentials, derivation of equations describing electrostatic influences on the adsorption process is difficult. The primitive interfacial is a model that solves this problem by regarding all adsorbed species as surface sites in a monolayer (the Stern layer). Further, the surface sites (bonded or not bonded to adsorbed solute molecules or ions) are required to be spherical in geometry with the same diameter. For polymeric adsorbates, these restrictions would be a very poor electrostatic model. For polymer adsorption the same electrostatic and free energy equations can be applied, but the restriction of the entire polymer molecule residing in the Stern layer must be eliminated.

The role of electrostatics and free energy in polymer adsorption is discussed in great detail in Chapters 2 and 3. To determine adsorption for a polymer molecule at a solution–solid interface requires a free energy minimization procedure. Such a procedure requires that the free energy for a polymer molecule can be expressed by an equation which is a function of the polymer location relative to the solid–solution interface. The distance-dependent free energy functions can then be minimized with respect to the distance from the interface thereby predicting the configuration and adsorbed amount of polymer. The application and merits of free energy minimization procedures are discussed in great detail in Chapters 2 and 3, and require no further examination. Instead the focus of the following discussion will be applying thermodynamic and electrostatic principles, already discussed for monomeric adsorbents, to polymeric adsorbents.

In most treatments of polymer adsorption, polymers are modeled as monomer units bonded together. Adsorbed polymer molecules will have some segments adsorbed (in the Stern layer) with other segments in the solution (not in the Stern layer) as tails or loops. Figure 12 shows an adsorbed homopolymer.

For polymer monomer units the free energy will depend on additional factors not affecting simple monomeric solute molecules such as bond angles, bond rotations, movement limited by bonds, etc. However, the free energy for segments in a polymer and a solute molecule will have two common terms in the free energy equation: (1) the molecular concentration, and (2) the activity coefficient. If additional free energy terms representing unique polymer properties such as bond angles, bond rotations, movement limited by bonds, etc. are represetned with the symbol δ then the free energy equations for an adsorbed segment numbered 2 (in the Stern layer), and a segment in solution numbered 3, as shown in Fig. 12, can be written:

$$\mu_2 = \mu_s^* + kT\ln\theta_2 + kT\ln f_2 + \delta_2 \qquad f_2 = Z^{\left[\frac{(\nu_2-\bar{\nu}_s)^2}{1+\kappa a_2}\right]} \equiv Z^{\epsilon_2} \tag{189}$$

$$\mu_3 = \mu_w^* + kT\ln\chi_3 + kT\ln f_3 + \delta_3 \qquad f_3 = Z^{\left[\frac{(\nu_3-\bar{\nu}_s')^2}{1+\kappa a_3}\right]} \equiv Z^{\epsilon_3} \tag{190}$$

The activity coefficient equations (189) and (190) can be seen to vary only in the mean surface site ion valence, $\bar{\nu}_s$ and $\bar{\nu}_s'$, respectively. The similarity between these two equations simplifies mathematical manipulation. However, the important x-axis dependence which differentiates these equations is an implicit function of $\bar{\nu}_s'$. The x-

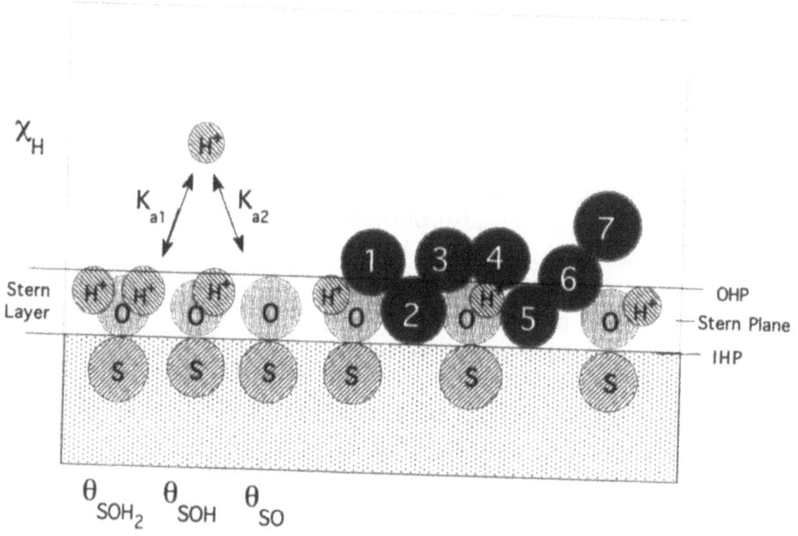

FIG. 12 Adsorbed homopolymer.

axis dependence of $\bar{\nu}_s'$ can be understood from the SP electrostatic potential, ψ_2, experienced by segment 2 at $x = a_s/2$, and the electrosatic potential, ψ_3, experienced by segment 3 at $x = 3a_s/2$.

The SP electrostatic potential, ψ_2, is a result of neighboring ions in the Stern layer having an electrostatic charge density, σ_s, which is electrostatically equivalent to locating $\bar{\nu}_s e$ electrostatic charge on segment 2. The electrostatic potential at ψ_3 must be less than ψ_2, which could be accomplished by placing a lesser amount of electrostatic charge, $\bar{\nu}_s' e$, on segment 3 at $x = 3a_s/2$. On the premise that the effective electrostatic charge and potential for segments 2 and 3 are proportional, Eq. (89) can be used to write:

$$\frac{\psi_3}{\psi_2} = \frac{\bar{\nu}_s'}{\bar{\nu}_s} = \left(\frac{a_s}{2x}\right) e^{\kappa(a_s - 2x)} \bigg|_{x=\frac{3}{2}a_s} = \frac{e^{2\kappa a_s}}{3} \tag{191}$$

Equation (191) gives the x-axis dependence of the mean surface site electrostatic charge at any distance, x, from the interface, $\bar{\nu}_s' e$, in terms of the mean surface site electrostatic charge, $\bar{\nu}_s e$, at the SP. The relationship between these effective electrostatic charges is a simple exponential dependence on the Stern layer thickness and the diffuse ion layer thickness.

V. SUMMARY DISCUSSION OF ACTIVITY COEFFICIENT EQUATIONS

A. Activity Coefficient Behavior

From rigorous thermodynamic arguments, the activity coefficient for any solute has been shown to be dependent on the difference in free energy for that solute in real and ideal solutions which are equimolar and at the same temperature and pressure. Further thermodynamic arguments, utilizing the concentration independence of the partial molar enthalpy, were used for the derivation of the free energy equation of an ideal solute from the free energy equation of a real solute.

To this point, equations for the activity coefficient were rigorous. Due to the lack of partial molar free energy equations for real solutes, Coulombic electrostatic equations were substituted. While the electrostatic equations were rigorous, solutes interact through more than just Coulombic forces. In addition, electrostatic equations have distance dependencies which make application to real solutions and interfaces with unknown and complex structures difficult.

In order to derive explicit activity coefficient equations, solutions were depicted with the Debye–Hückel model, and interfaces with the primitive interfacial model. From these models, explicit activity coefficient equations were derived and used to examine the activity coefficient behavior of solutes in solution and adsorbed at interfaces. In addition, the effect of activity coefficients on adsorption were examined. These equations were used to assess the activity coefficient effects on adsorption in Chapter 1. A few examples of activity coefficient behavior are also given below.

Since the advent of the Debye–Hückel equation, calculation of activity coefficients for solute ions has been possible. The activity coefficient for solute ions is dependent on the solute concentration through the diffuse ion layer thickness, κ^{-1}. Using the equations listed in Table X, the activity coefficients for solute and surface sites ions can now be calculated as a function of solute ion solution concentration and compared, as shown in Fig. 13.

In Fig. 13, calculated activity coefficients for solute cations in Debye–Hückel solutions (lines a and b) for cation–surface site complexes at the primitive interface (lines c, d, e, g, h, and k) are plotted. One can see that activity coefficients for ions in solution and at interfaces are of a similar magnitude. For the activity coefficient of the surface site ions, mostly negative values for the mean surface site electrostatic charge, $\bar{\nu}_s$, were chosen because cation adsorption is strongest at high pH where $\bar{\nu}_s$ would tend to be more negative. A slight positive value of, $\bar{\nu}_s = +0.2$, was included for line p, just for comparison.

Another interesting comparison is the effect of the mean electrostatic surface charge, $\bar{\nu}_s$, on the activity coefficient of surface sites, which is plotted in Fig. 14.

From the data in Fig. 14, the activity coefficients of the monovalent anion– and cation–surface site complexes are symmetrical with $\bar{\nu}_s$ and are nearly unity. For divalent anion– and cation–surface site complexes the activity coefficients deviate significantly from unity and are not the same. As might be expected when surface site charge is counter to the complexed ion, activity coefficients deviate most from unity. This is analogous to lower activity coefficients for solute ions in solution as counter ion concentration increases. The activity coefficient of the monovalent cation–double surface site complex follows the single surface site divalent anion complex behavior more closely than the single surface site divalent cation complex behavior.

Finally, the predicted activity coefficient behavior for nonionic–surface site complexes will be examined. In Fig. 15 the activity coefficient variation with the mean electrostatic surface charge, $\bar{\nu}_s$, is plotted.

From Table 5 one can deduce that the activity coefficient behavior for nonionic–surface site complexes should be the same as for the protonated surface site complexes. A comparison of Figs 7 and 15 confirms this obvious fact. Since nonionic

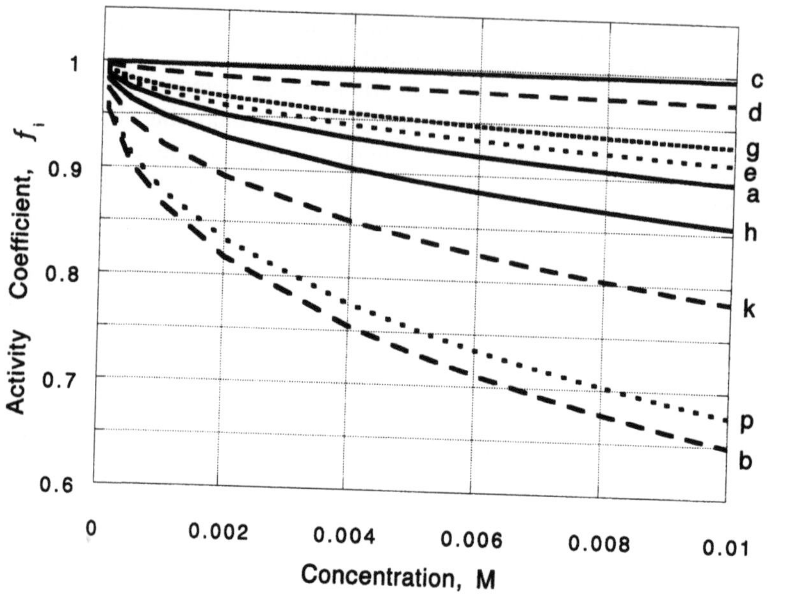

FIG. 13 Calculated activity coefficients for solute and surface site ions at the S–MO interface as a function of a 1:1 solute concentration in solution. Lines in the figure correspond to the following activity coefficients:

Line	Activity coefficient description	ν_m	$\bar{\nu}_s$
a	f_M, cation in solution	+1	–
b	f_M, cation in solution	+2	–
c	f_{SM}, cation–surface site complex	+1	−0.2
d	f_{SM}, cation–surface site complex	+1	−0.5
e	f_{SM}, cation–surface site complex	+1	−0.9
g	f_{SM}, cation–surface site complex	+2	+0.2
h	f_{SM}, cation–surface site complex	+2	−0.2
k	f_{SM}, cation–surface site complex	+2	−0.5
p	f_{SM}, cation–surface site complex	+2	−0.9

molecules add no electrostatic charge to the surface site, this result can be easily understood.

One final point of interest for Fig. 15 is the role which ions in the solution play in the activity coefficients of nonionic–surface site complexes. This observation shows that even though nonionic molecules possess no electrostatic charge, the electrostatic charge induced in such complexes is influenced by the electrostatic charge on solute and surface site ions. The interaction between solute ion electrostatic charge and the induced electrostatic charge of the nonionic–surface site complexes is manifest as deviations of the activity coefficient from unity.

B. Tables of Explicit Activity Coefficient Equations

Activity coefficients have been derived and discussed for ions and electrolytes and for activity coefficient quotients in mass action equations. The following is a summary reference for all the activity coefficient derivations divided into three tables.

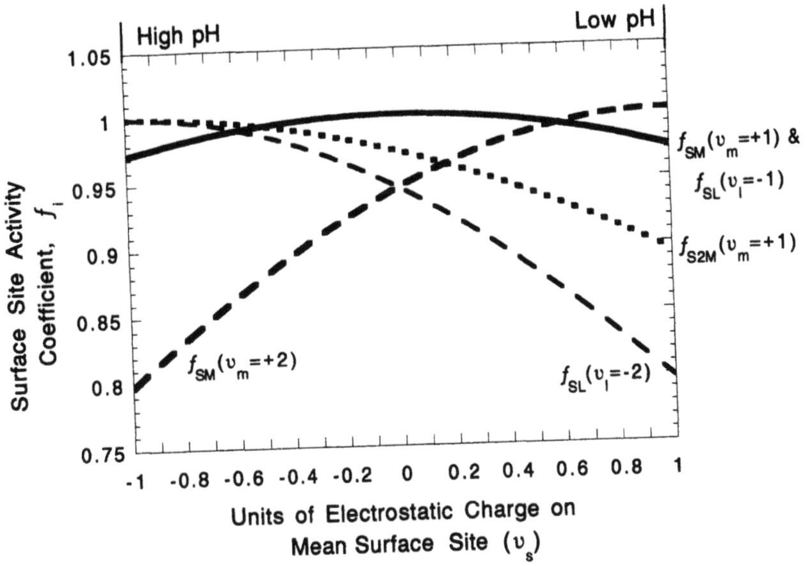

FIG. 14 Role of the mean electrostatic surface charge, $\bar{\upsilon}_s$, on the activity coefficient of anion and cation complexed surface sites.

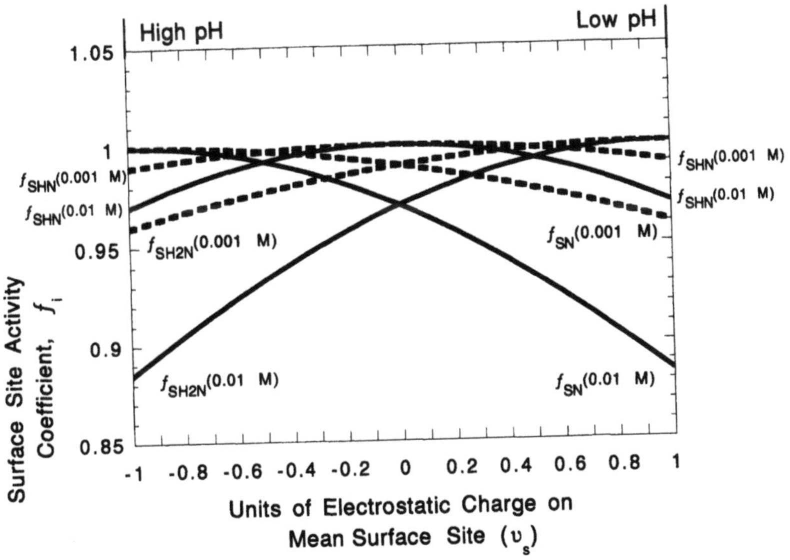

FIG. 15 Role of the mean electrostatic surface charge, $\bar{\upsilon}_s$, on the activity coefficient of surface sites complexed with nonionic molecules.

In Table 4, activity coefficients for individual ions, molecules, and electrolyte salts are given.

In Table 5, activity coefficients for surface sites complexed with different solute ions are given.

In Table 6, activity coefficients for different mass action equations, which occur frequently in adsorption equations, are listed.

For all individual ions, molecules, surface sites, and electrolyte salts, the equations for the activity coefficient, f_i, based on the Debye–Hückel and primitive interfacial models, have a common form:

$$f_{\mathrm{i}} = \mathrm{e}^{\left[-\frac{\nu_i^2 \kappa q_{\mathrm{B}}}{1+\kappa a_i}\right]} = Z^{\left[\frac{\nu_i^2}{1+\kappa a_i}\right]} \equiv Z^{\epsilon_i} \tag{192}$$

where

i = ion, molecule, surface site or electrolyte salt;
ν_i = valence of i;
q_{B} = Bjerrum ion association distance [13, 29];
κ = Debye–Hückel inverse diffuse ion thickness, Eq. (41);
a_{i} = "effective" ionic diameter of i [13, 33];
Z = base activity coefficient;
ϵ_i = exponent base activity coefficient for i.

For individual ions, molecules, and electrolyte salts the general activity coefficient equation is

$$f_{\mathrm{i}} = \mathrm{e}^{\left(-\frac{\nu_i^2 \kappa q_{\mathrm{B}}}{1+\kappa a_i}\right)} \equiv Z^{\left(\frac{\nu_i^2}{1+\kappa a_i}\right)} \equiv Z^{\epsilon_i} \tag{193}$$

From Eq. (193) explictic equations for activity coefficients for individual ions, molecules, and electrolyte salts, based on the Debye–Hückel model, can be seen to vary

TABLE 4 Explicit Activity Coefficients for Individual Ions, Molecules, and Electrolyte Salts

Section	Solute	f_i	ϵ_i
IV.B	Protons, H^+	f_{H}	$\frac{1}{1+\kappa a_{\mathrm{H}}}$
IV.A, IV.C	Cations, $M^{\nu_{\mathrm{m}}}$	f_{M}	$\frac{\nu_{\mathrm{m}}^2}{1+\kappa a_{\mathrm{m}}}$
IV.A, IV.D	Anions, L_l^{ν}	f_{L}	$\frac{\nu_l^2}{1+\kappa a_l}$
IV.E	Nonionic, $N^{\nu_{\mathrm{n}}} = N^0$	f_{N}	$\frac{\nu_{\mathrm{n}}^2}{1+\kappa a_{\mathrm{n}}} = \frac{0}{1+\kappa a_{\mathrm{n}}} = 0$
IV.A	Electrolyte, A_aB_b	$f_{\pm}$	$\frac{ab}{a+b}\left(\frac{b}{1+\kappa a_{\mathrm{A}}} + \frac{a}{1+\kappa a_{\mathrm{B}}}\right)$

TABLE 5 Explicit Activity Coefficients for Surface Sites at the Primitive Interface

Section	Surface site	f_i	ϵ_i
IV.B	Unprotonated anionic, SO^-	f_{SO}	$\frac{(-1-\bar{\nu}_s)^2}{1+\kappa a_s}$
IV.B	Protonated nonionic, SOH	f_{SOH}	$\frac{(0-\bar{\nu}_s)^2}{1+\kappa a_s}$
IV.B	Diprotonated cationic, $SOH_2{}^+$	f_{SOH_2}	$\frac{(1-\bar{\nu}_s)^2}{1+\kappa a_s}$
IV.C	Cation complex, $SOM^{(-1+\nu_m)}$	f_{SM}	$\frac{(\nu_m-1-\bar{\nu}_s)^2}{1+\kappa a_s}$
IV.C	Cation dicomplex, $(SO)_2M^{(-2+\nu_m)}$	f_{S_2M}	$\frac{(\nu_m-2-\bar{\nu}_s)^2}{1+\kappa a_s}$
	Cation tricomplex, $(SO)_3M^{(-3+\nu_m)}$	f_{S_3M}	$\frac{(\nu_m-3-\bar{\nu}_s)^2}{1+\kappa a_s}$
IV.E	Nonionic complex, $SON^{(-1)}$	f_{SN}	$\frac{(-1-\bar{\nu}_s)^2}{1+\kappa a_s}$
IV.E	Nonionic complex, $SOHN^{(0)}$	f_{SHN}	$\frac{(0-\bar{\nu}_s)^2}{1+\kappa a_s}$
IV.E	Nonionic complex $SOH_2N^{(+1)}$	f_{SH_2N}	$\frac{(1-\bar{\nu}_s)^2}{1+\kappa a_s}$
IV.D	Anion complex, $SOH_2L^{(1+\nu_l)}$	f_{SL}	$\frac{(1+\nu_l-\bar{\nu}_s)^2}{1+\kappa a_s}$
	Anion dicomplex, $(SOH_2)_2L^{(2+\nu_l)}$	f_{S_2L}	$\frac{(2+\nu_l-\bar{\nu}_s)^2}{1+\kappa a_s}$

simply by the exponent ϵ_i. Table 4 lists ϵ_i for individual ions, molecules, and electrolyte salts.

For surface sites at the primitive interface the general activity coefficient equation is

$$f_i = e^{\left[-\frac{\nu_i^2 \kappa q_B}{1+\kappa a_i}\right]} \equiv Z^{\left[\frac{\nu_i^2}{1+\kappa a_i}\right]} \equiv Z^{\epsilon_i} \tag{194}$$

From Eq. (194) explicit equations for surface sites at the primitive interface can be seen to vary simply by the exponent ϵ_i. Table 5 lists ϵ_i for surface sites at the primitive interface.

TABLE 6 Explicit Activity Coefficients for Activity Coefficient Quotients

Section	Reaction	Activity coefficient quotient	ϵ_{K_i}
IV.B	$SOH_2{}^+ = SOH + H^+$	$\frac{f_{SOH} f_H}{f_{SOH_2}} \equiv Z^{\epsilon_{K_{a1}}}$	$\epsilon_{K_{a1}} \equiv \frac{1}{1+\kappa a_H} + \frac{2\bar{\nu}_s - 1}{1+\kappa a_s}$
IV.B	$SOH = SO^- + H^+$	$\frac{f_{SO} f_H}{f_{SOH}} \equiv Z^{\epsilon_{K_{a2}}}$	$\epsilon_{K_{a2}} \equiv \frac{1}{1+\kappa a_H} + \frac{2\bar{\nu}_s + 1}{1+\kappa a_s}$
IV.B	$SOH_2{}^+ = SO^- + 2H^+$	$\frac{f_{SO} f_H^2}{f_{SOH_2}} \equiv Z^{\epsilon_{K_{a12}}}$	$\epsilon_{K_{a12}} \equiv \frac{2}{1+\kappa a_H} + \frac{4\bar{\nu}_s}{1+\kappa a_s}$
IV.C	$SO^- + M^{\nu_m} = SOM^{(-1+\nu_m)}$	$\frac{f_{SM}}{f_M f_{SO}} \equiv Z^{\epsilon_{K_{SM}}}$	$\epsilon_{K_{SM}} \equiv \frac{\nu_m(\nu_m - 2 - 2\bar{\nu}_s)}{1+\kappa a_s} - \frac{\nu_m^2}{1+\kappa a_m}$
IV.C	$2SO^- + M^{\nu_m} = (SO)_2M^{(-2+\nu_m)}$	$\frac{f_{S_2M}}{f_M f_{SO}^2} \equiv Z^{\epsilon_{K_{S_2M}}}$	$\epsilon_{K_{S_2M}} \equiv \frac{\nu_m^2 - 4\nu_m - 2\nu_m\bar{\nu}_s - \bar{\nu}_s^2 + 2}{1+\kappa a_s} - \frac{\nu_m^2}{1+\kappa a_m}$
IV.D	$SOH_2{}^+ + L^{\nu_l} = SOH_2L^{(1+\nu_l)}$	$\frac{f_{SL}}{f_L f_{SOH_2}} \equiv Z^{\epsilon_{K_{SL}}}$	$\epsilon_{K_{SL}} \equiv \frac{\nu_l(\nu_l + 2 - 2\bar{\nu}_s)}{1+\kappa a_s} - \frac{\nu_l^2}{1+\kappa a_l}$
IV.E	$SO^- + N^0 = SON^{(-1+0)}$	$\frac{f_{SON}}{f_{SO} f_N} \equiv Z^{\epsilon_{K_{SN}}}$	$\epsilon_{K_{SN}} = 0$
IV.E	$SOH + N^0 = SOHN^{(0+0)}$	$\frac{f_{SHN}}{f_{SOH} f_N} = Z^{\epsilon_{K_{SHN}}}$	$\epsilon_{K_{SHN}} = 0$
IV.E	$SOH_2{}^+ + N^0 = SOH_2N^{(1+0)}$	$\frac{f_{SH_2N}}{f_{SOH_2} f_N} \equiv Z^{\epsilon_{K_{SH_2N}}}$	$\epsilon_{K_{SH_2N}} = 0$

For activity coefficient quotients the general activity coefficient equation is

$$\frac{f_i \cdots}{f_j \cdots} = e^{\left[-\sum_i \frac{\nu_i^2 \kappa q_B}{(1+\kappa a_i)} + \sum_j \frac{\nu_j^2 \kappa q_B}{(1+\kappa a_j)}\right]} \equiv Z^{\left[\sum_i \frac{\nu_i^2}{(1+\kappa a_i)} - \sum_j \frac{\nu_j^2}{(1+\kappa a_j)}\right]} \equiv Z^{\epsilon_{K_{ij}}} \tag{195}$$

From Eq. (195) explicit equations for activity coefficient quotients, based on both the Debye–Hückel and primitive interfacial models, can be seen to vary simply by the exponent $\epsilon_{K_{ij}}$. Table 6 lists $\epsilon_{K_{ij}}$ for activity coefficient quotients.

REFERENCES

1. GN Lewis. Proc Am Acad 43:259, 1907.
2. GN Lewis. Z Physik Chem 61:129, 1907.
3. GN Lewis. Proc Am Acad 37:49, 1901.
4. GN Lewis, M Randall. Thermodynamics. 2nd ed. New York: McGraw-Hill, 1961, Ch. 20.
5. GN Lewis, M Randall. Thermodynamics. 2nd ed. New York: McGraw-Hill, 1961, pp 1–3.
6. CJ Adkins. Equilibrium Thermodynamcis. 3rd ed. London: Cambridge University Press, 1883, Ch. 1.
7. JW Gibbs. The Collected Works of J. Willard Gibbs. New Haven, CT: Yale University Press, 1948, p 93.
8. K Denbigh. The Principles of Chemical Equilibrium. 4th ed. London: Cambridge University Press, 1981, pp 215–216.
9. K Denbigh. The Principles of Chemical Equilibrium. 4th ed. London: Cambridge University Press, 1981, pp 111–116.
10. GN Lewis, M Randall. Thermodynamics. 2nd ed. New York: McGraw-Hill, 1961, pp 44–46.
11. GN Lewis, M Randall. Thermodynamics. 2nd ed. New York: McGraw-Hill, 1961, pp 145–149.
12. JN Israelachvili. Intermolecular and Surface forces. 2nd ed. London: Academic Press, 1991, pp 37–38.
13. EM Pugh, EW Pugh. Principles of Electricity and Magnetism. Reading, MA: Addison-Wesley, 1960, Ch. 2.
14. JN Israelachvili. Intermoleculear and Surface Forces. 2nd ed. London: Academic Press, 1991, Ch. 3.
15. HS Harned, BO Owen. The Physical Chemistry of Electrolyte Solutions. 2nd ed. New York: Reinhold, 1950, Ch. 3.
16. JN Israelachvili. Intermolecular and Surface Forces. 2nd ed. London: Academic Press, 1991, p 11.
17. L Onsager. Chem Rev 13:73–89, 1933.
18. K Denbigh. The Principles of Chemical Equilibrium. 4th ed. London: Cambridge University Press, 1981, pp 252–255.
19. JN Israelachvili. Intermolecular and Surface Forces. 2nd ed. London: Academic Press, 1991, Ch. 4.
20. MC Gupta. Statistical Thermodynamics. New York: John Wiley, 1990, pp 345–362.
21. JN Israelachvili. Intermolecular and Surface Forces. 2nd ed. London: Academic Press, 1991, p 50.

22. JN Israelachvili. Intermolecular and Surface Forces. 2nd ed. London: Academic Press, 1991, p 76.
23. PC Hiemenz. Principles of Colloid and Surface Chemistry. 2nd ed. New York: Marcel Dekker, 1986, pp 681–703.
24. PC Hiemenz. Principles of Colloid and Surface Chemistry. 2nd ed. New York: Marcel Dekker, 1986, pp 686–697.
25. N Bjerrum. Kgl Danske Vidensk Selskab 7:9, 1926.
26. IM Klotz, RM Rosenberg. Chemical Thermodynamics. 4th ed. Malibar, FL: Krieger, 1991, Ch. 20.
27. J. Kielland. J Am Chem Soc 59:1675, 1927.
28. TH Gronwall, VK LaMer, K Sandved. Phys Z 29:358, 1929.
29. VK LaMer, TH Gronwall, LJ Greiff. J Phys Chem 35:2245, 1931.
30. L Onsager, NNT Samaras. J Chem Phys 2:528–536, 1934.
31. P Attard. In: I Prigogne, SA Rice, eds. Advances in Chemical Physics. Vol. XCII. New York: John Wiley, 1996, pp 9–10.
32. JN Israelachvili. Intermolecular and Surface Forces. 2nd ed. London: Academic Press, 1991, p 217.
33. K Wagner. Phys Zeits 25:474, 1924.
34. JN Israelachvili. Intermolecular and Surface Forces. 2nd ed. London: Academic Press, 1991, pp 43, 57.

5
The Dissolution of Insulating Oxide Minerals

WILLIAM H. CASEY, JAN NORDIN, and BRIAN L. PHILLIPS University of California, Davis, Davis, California

I. INTRODUCTION

It is difficult to overstate the importance of oxide mineral growth and dissolution to our lives. Oxide corrosion products build up on high-speed turbine blades, in the piping in nuclear reactors, on the wings of aircraft, and in such mundane settings as an automobile fender. These corrosion products must be periodically removed by chemical dissolution using strong chelating agents or by physical abrasion.

Our understanding of dissolution reactions at oxide mineral surfaces is undergoing a dramatic evolution, begun over a decade ago, as ideas from solution chemistry are quantitatively applied to surfaces. The motivation for this work has been largely geochemical: most of the minerals that make up the Earth are electrically insulating and lend themselves poorly to comparison with electrodes. Furthermore, the understanding of mineral dissolution reactions is important to a wide range of environmental concerns, as geochemists are at the sharp end of predictions about the safety of waste repositories and the fate of toxicants in the environment [1–3]. In most cases, waste repositories are assumed ultimately to leak metals and receive ligands into the aquifers that contain them. The rates of subsequent solute migration are moderated through reactions at oxide mineral surfaces. Likewise, these reactions are central to the cycling of organic matter in the atmosphere [e.g., 4, 5] and to the bacterial transformations of pesticides in soil [e.g., 6, 7]. Finally, oxygen, after all, makes up over 80% of the Earth's crust by volume and all of this is in oxide minerals.

II. SOLUTES AND SURFACES

A. Structural Considerations

This huge volume of oxygen is expressed in a rich array of oxide minerals, most of which contain a relatively unreactive silicate polymer with varying concentrations of covalent crosslinks and charge-compensating cations. Some high-temperature phases, the orthosilicate minerals, have isolated SiO_4 tetrahedra, but more common, especially in the near surface environment, are aluminosilicate minerals with extensive sheets or frameworks of linked silicate or aluminate tetrahedra, or aluminate octahedra. In soils, many of the most reactive minerals are nanocrystalline oxyhydroxides of ferric iron (e.g., Fe_2O_3, hematite), manganese (e.g., MnO_2, birnessite), or aluminum [e.g., $Al(OH)_3$ gibbsite] without a silicate anion at all.

Although the structures of these minerals are well understood, we currently lack techniques capable of determining the structure at the fluid–mineral interface, where the reactions that are important to mineral dissolution occur. For guidance, it can be very helpful to study dissolved complexes as analogs for surface sites. The structures of these dissolved complexes, and the effects of chemistry and structure on the rates of certain reactions, can be studied directly by spectroscopic techniques.

There is an important dichotomy when comparing the surface chemistry of minerals to the chemistry of dissolved multimers. The co-ordination number of metals to oxygens at the surface of these minerals is usually similar to those in a dissolved metal–ligand complex. For example, the Ni(II) metal center is hexaco-ordinated to oxygens in the $Ni(H_2O)_6^{2+}(aq)$ complex (Fig. 1a), at the NiO(s) (bunsenite) surface (Fig. 1b), and even at the surface of the orthosilicate mineral $Ni_2SiO_4(s)$ (liebenbergite; Table 1). The ⟨Ni–O⟩ bond lengths vary slightly from 2.04 to 2.1 Å. A few metals, such as Zn(II), B(III), and Al(III), change co-ordination numbers to oxygen as they pass from a solid to a solute, but most metals retain the general features of the oxygen–co-ordination geometry upon release from the solid.

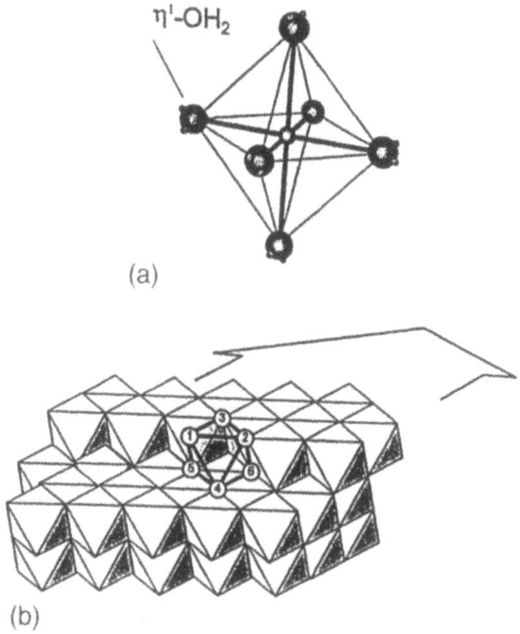

FIG. 1 (a) Schematic representation of a monomeric metal co-ordinated to six waters of hydration in the inner co-ordination sphere. Examples would be the $Ni(H_2O)_6^{2+}(aq)$ complex where all of the co-ordination sites are filled with water molecules (η^1-OH_2 sites) that can undergo Brønsted reactions with the solution to form hydrolyzed species [e.g., $NiOH^+(aq)$]. (b) A retreating monomolecular step on the [100] surface of NiO(s). Numbers indicate the co-ordination numbers of oxygens to Ni(II); e.g., the number 4 identifies a μ_4-oxo site. Brønsted reactions can affect the protonation states of these exposed oxygens.

TABLE 1 Measured Metal–Oxygen Distances (<M–O>, Å) for Hydrated Cations, Oxide Minerals, and Orthosilicate Minerals

Ion	<M–O>	Oxide	<M–O>	Silicate	<M–O>
$Ca(H_2O)_6^{2+}$	2.39–2.46	CaO	2.405	Ca_2SiO_4	2.346–2.392
$Mg(H_2O)_6^{2+}$	2.10	MgO	2.11	Mg_2SiO_4	2.101–2.127
$Be(H_2O)_4^{2+}$	1.67	BeO	1.649	Be_2SiO_4	1.645
$Zn(H_2O)_6^{2+}$	2.08–2.17	ZnO	1.95	Zn_2SiO_4	1.92
$Mn(H_2O)_6^{2+}$	2.18–2.20	MnO	2.22	Mn_2SiO_4	2.185–2.227
$Co(H_2O)_6^{2+}$	2.05–2.08	CoO	2.13	Co_2SiO_4	2.123–2.134
$Ni(H_2O)_6^{2+}$	2.04–2.10	NiO	2.095	Ni_2SiO_4	2.076–2.102
$Al(H_2O)_6^{3+}$	1.87–1.94	α-Al_2O_3	1.86–1.97		

See Ref. 8 for sources.

In contrast, oxygens at mineral surfaces are commonly co-ordinated to a high number of metals and several co-ordinatively distinct sets of oxygens are exposed at a given crystal face. In Fig. 1b, a monomolecular step on the [100] surface of NiO(s) is shown with the co-ordination numbers of oxygens to metals identified. The co-ordination numbers range from μ_6-O to η^1-OH_2. These co-ordination numbers decrease as the mineral reacts with the aqueous solution, often releasing small monomeric complexes to the solution with the oxygens bonded to a single metal center.

Here, we employ the formalism that η^i sites are nonbridging sites with i ligand atoms bonding to the metal, and μ_i-sites are ligand atoms that bridge i metals [9]. These designations allow us to convey structural information with the stoichiometry. Consider the example of the $Al_2(H_2O)_8(OH)_2^{2+}$(aq) complex. Using the formalism: $Al_2(\mu_2\text{-}OH)_2(\eta^1\text{-}H_2O)_8^{4+}$(aq) it is clear that the hydroxyls bridge two Al(III) metals, and the waters are monodentate and nonbridging. Designating terminal water molecules in this way is usually unnecessary, but becomes useful for surfaces.

Low-symmetry solids expose a range of oxygen co-ordination numbers narrower than that of high-symmetry solids. The low-symmetry Bayerite [β-$Al(OH)_3$, a platy mineral] surface, for example, has only terminal η^1-OH_2 and μ_2-OH sites and not the rich array of co-ordination chemistries illustrated in Fig. 1, where μ_2- through μ_6-oxo sites surround each metal. The co-ordination chemistry, of course, varies with the exposed surface, and the protonation states of these oxygens may vary with solution composition as Brønsted reactions modify charge on the oxide surface. Obviously, many properties of a solid surface have no complete analog in solution. For example, the co-ordination numbers of oxygens to metals in solutes, even multimers in highly concentrated solutions, rarely exceed three.

However, by carefully choosing the dissolved complex, one can usually find striking similarities with sites on the oxide surface. Compare, for example, the structure of the $Al_{13}O_4(OH)_{24}(H_2O)_{12}^{7+}$(aq) oligomer (Fig. 2a) with the [010] surface of Bayerite [β-$Al(OH)_3$(s)] (Fig. 2b) and the [001] surface of corundum [α-Al_2O_3] (Fig. 2c). The $Al_{13}O_4(OH)_{24}(H_2O)_{12}^{7+}$(aq) oligomer has the Keggin structure [9, p 816] with edge-shared AlO_6 octahedra surrounding a central, AlO_4 tetrahedron [10, 11]. At the apices of the AlO_6 octahedra are 12 co-ordinated waters (η^1-OH_2) that undergo Brønsted acid–base reactions with the aqueous phase [12]. There are two sets of

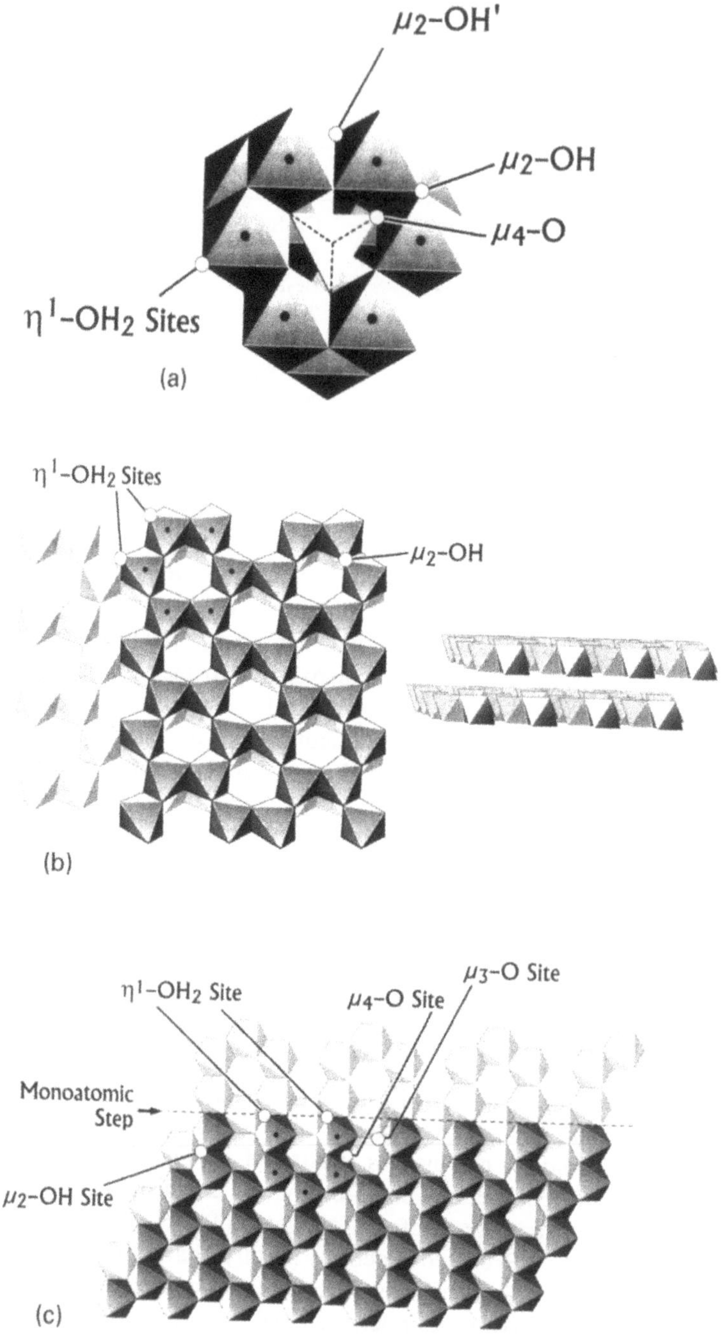

FIG. 2 (a) The $Al_{13}O_4(OH)_{24}(H_2O)_{12}^{7+}$(aq) dissolved complex. The structure has two structurally distinct sets of 12 μ_2-OH sites and four μ_4-O sites and 12 η^1-OH_2 sites. (b) A monomolecular step on Bayerite [β-$Al(OH)_3$]. Solid dots are added to six-membered rings on the uppermost layer of AlO_6 octahedra to make them easier to identify. Also shown is the structure edge-on, to emphasize that the layers are linked to one another via van der Waals' forces. (c) A monomolecular step on the [001] surface of α-Al_2O_3(s). Solid circles are placed on the face-shared AlO_6 octahedra which make up six-membered rings that surround an unoccupied octahedral site.

μ_2-hydroxo groups that differ, depending upon whether they are *trans* or *cis* to the four μ_4-O sites on the apices of the AlO_4 tetrahedron.

Both the dissolved oligomer and the two mineral surfaces (Fig. 2) are dominated by six-membered rings of shared AlO_6 octahedra. The oligomer has edge-shared octahedra, of which each shares an oxygen with an AlO_4 tetrahedron. In Bayerite, truncation of a monomolecular step on the basal surface exposes μ_2-hydroxo groups formed at the shared edges of AlO_6 octahedra. The basal layers are linked to one another via van der Waals' bonds so that the exposed surface is dominated by η^1-OH_2 and μ_2-OH groups that can react with the aqueous phase.

The [001] surface of α-Al_2O_3 also includes six-membered rings, although the structure is much more complicated than that of the lower-symmetry solids or the solutes. The crystal structure shows shared octahedral faces that lie perpendicular to the [001] axis with every other octahedral site vacant; these vacant sites lie at the center of the six-membered rings, shown exposed on the surface. The octahedra are linked to one another via faces and edges, and expose μ_4-O, μ_3-O, μ_2-OH, and η^1-OH_2 sites that can change protonation state through reaction with the aqueous solution. We do not know the exact protonation states of the oxygens, and the μ_4-O sites are structurally distinct from those in the $Al_{13}O_4(OH)_{24}(H_2O)_{12}^{7+}$(aq) oligomer. At the α-Al_2O_3 surface the oxygens are not linked to AlO_4 tetrahedra.

By hypothesis, chemistry determined on individual sites on these oligomers can be extended to the surfaces because it is only local short-range forces that control the kinetics of dissociation and the Brønsted acidities. This hypothesis will certainly be confirmed qualitatively; the extent to which information can be quantitatively extended from solutes to surfaces remains to be discovered. The important point is that dissolved multimers are useful, but incomplete, analogs for the surface sites on oxide minerals.

B. Reactivities of Metal–Oxo and –Hydroxo Complexes

A wide range of oligomers has been prepared from inert metals [13] and these complexes have proven enormously useful in assigning microscopic equilibrium constants for Brønsted reactions and rate coefficients for elementary reactions (Table 2). There are a few useful rules to summarize from this fine work:

1. Metals in these complexes are co-ordinatively saturated, and the mechanisms of ligand exchange and bridge cleavage have a considerable dissociative character.
2. The most important variable controlling reactivity is the metal–ligand [generally metal–oxygen (M–O)] bond strength, which can be determined from electronic structures and from positions in the periodic chart of the elements (see Fig. 3 and Ref. 15). The rates of water exchange from the inner co-ordination sphere of these metals to the bulk solution is a useful measure of M–O bond strength and vary by over a factor of 10^{15}.
3. Electron exchange considerably modifies the M–O bond strengths, introducing *reductive* and *oxidative* pathways for reaction. For example, Co(II) has the electronic structure $t_{2g}^6 e_g^1$ and $Co(H_2O)_6^{2+}$(aq) exchanges waters of hydration with the bulk solution in approximately 10^{-7} s. The oxidized metal, Co(III), has the t_{2g}^6 electronic structure and $Co(H_2O)_6^{3+}$(aq) exchanges waters in approximately 10 s [15, 16].

TABLE 2 Variation in Rates of Hydrolytic Processes with the OH/Cr(III) Ratios in Monomers and Oligomers at 298 K and $I = 1.0$ M.[a] The Abbreviations SBD and DBD Correspond to a Singly Bridged Dimer and Doubly Bridged Dimer, Respectively, e.g., $[Cr(\mu\text{-}OH)Cr]^{5+}$ and $[Cr(\mu\text{-}OH)_2Cr]^{4+}$

Reactant	OH/Cr(III)	$10^5\times$ rate (s^{-1})	Comment
			1. Exchange of water from inner co-ordination sphere to bulk
$Cr(H_2O)_6^{3+}$	0	0.24	Fully hydrated monomer
$Cr(OH)^{2+}$	1.0	18	Hydrolyzed monomer
$[Cr(\mu\text{-}OH)_2Cr]^{4+}$			
trans to μ-OH	1	36	DBD, fully hydrated
cis to μ-OH	1	6.6	DBD, fully hydrated
$[Cr(\mu\text{-}OH)_2CrOH]^{3+}$			
trans to μ-OH	1.5	1260	DBD, hydrolyzed
cis to μ-OH	1.5	490	DBD, hydrolyzed
			2. Intramolecular bridge formation
$[Cr(\mu\text{-}OH)Cr]^{5+} + H_2O \rightarrow DBD + H^+$	0.5	10	SBD → DBD; fully hydrated reactant
$[Cr(\mu\text{-}OH)CrOH]^{4+} + H_2O \rightarrow DBD + H^+$	1.0	40	SBD→DBD; singly hydrolyzed reactant
$[HOCr(\mu\text{-}OH)CrOH]^{3+} + H_2O \rightarrow$ DBD-H	1.5	1140	SBD→DBD; doubly hydrolyzed reactant
$[Cr_4(OH)_6]^{6+} \rightarrow$ closed tetramer	1.5	8700	
$[Cr_4(OH)_7]^{5+} \rightarrow$ closed tetramer	1.75	24,000	
			3. Sulfate anation
$Cr(H_2O)_6^{3+}$	0	1.1	Hydrated monomer
$Cr(OH)^{2+}$	1.0	61	Hydrolyzed monomer
$[Cr(\mu\text{-}OH)_2Cr]^{4+}$	1	35	Hydrated DBD
$[Cr(\mu\text{-}OH)_2CrOH]^{3+}$	1.5	1700	Hydrolyzed DBD

[a] *Source*: Ref. 14.

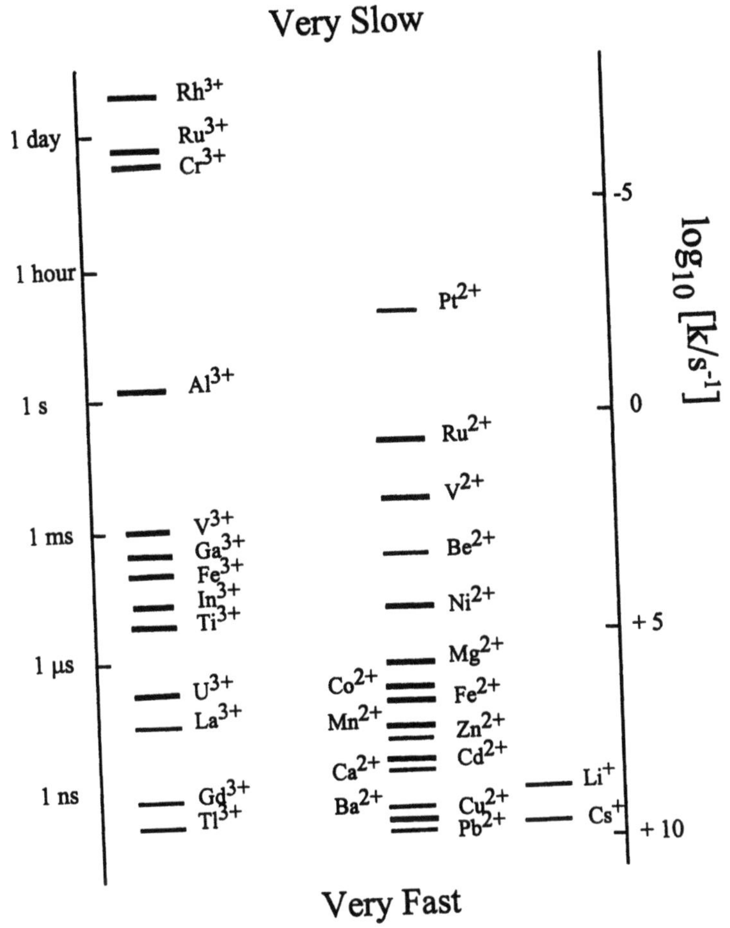

FIG. 3 Rates of exchange (s^{-1}) and characteristic lifetimes of water molecules in the inner co-ordination spheres of the metals. (From Ref. 15.)

4. In general, reactivities decrease and the Brønsted acidities increase with increased co-ordination of oxygens to metals. Within a dissolved complex, η^1-OH_2 sites are more reactive than the μ-hydroxo or μ-oxo sites, and μ_3-oxo bridges are less reactive than μ_2-oxo sites. The differences are large: Richens et al. [17] found no evidence over 2 years for exchange of the μ_2- or μ_3-oxo bridges in the $Mo_3O_4(OH_2)_9^{4+}(aq)$ complex, yet the η^1-OH_2 sites exchange in minutes with the aqueous solution.

5. Oxygen liabilities correlate with: (1) structural positions in the complex, (2) bond lengths, and (3) Brønsted acidities. Waters *trans* to a μ-hydroxo or μ-oxo site are commonly more reactive and acidic than waters *cis* to the bridges. Rates of exchange of the water molecules *cis* or *trans* to the μ-oxo bridges in $Mo_3O_4(OH_2)_9^{4+}(aq)$ differ by a factor of 10^5 [17].

6. Protonation of oxygens is fast and dramatically changes reactivities because the charged protons are so much smaller ($\approx 10^{-5}$ times) than typical M–O bonds. Protonation reactions introduce *proton-* and *base-assisted* pathways for

dissociation, depending upon which site (bridging or nonbridging) is attacked by the proton (see below).

7. The Brønsted acidity of oxygens in the complex depends upon the other ligands also present in the inner co-ordination sphere and can be strongly influenced by intermolecular hydrogen bonding, commonly between the water at one metal center and the hydroxyl bound at another or a hydroxyl bridge. Any adsorbed ligand that can change the Brønsted acidity of a surface oxygen that is also in the inner co-ordination sphere of the metal can accelerate or retard rates of reaction. Likewise, an inner-sphere ligand that attracts protons from adjacent inner co-ordination sphere oxygens changes the reactivities of the M–O bond. These are *conjugate-base pathways* because the ligand (the conjugate base of a weaker acid) associates with a structural proton to modify the rate.

8. Substitution of a stable ligand into the inner co-ordination sphere of a metal can enhance or retard the depolymerization rates. This substitution proceeds to equilibrium, introducing *ligand-assisted* pathways.

C. Catalytic and Inductive Roles of Ligands and Protons

Rate-enhancing protons and ligands are commonly stable constituents in the inner co-ordination sphere of a dissociation complex. Consider how a single-bridged dimer dissociates to release monomers.

In one pathway (marked "1" in Fig. 4) a proton associates with the μ_2-OH bridge, forming a weak, bridging water molecule:

$$\mu_2 - OH + H^+(aq) = \mu_2 - OH_2^+ \tag{1}$$

If this bridge dissociates to release a relatively weak acid, such as fully hydrated metal monomers [e.g., $Fe(H_2O)_6^{3+}(aq)$], the proton *induces* dissociation, but is not *catalytic* because the proton remains in the inner co-ordination sphere of the monomer as an associated hydration water or hydroxyl ion. One molecule of water must enter the inner co-ordination sphere of the underbonded metal as it dissociates from the oligomer to complete the co-ordination sphere. A water molecule may subsequently

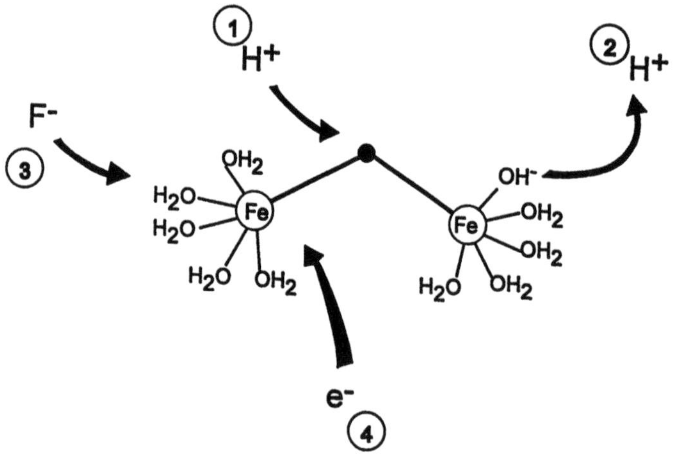

FIG. 4 Pathways to accelerate the dissociation of a bridging oxo- or hydroxo-group.

dissociate to achieve some equilibrium charged state [e.g., $FeOH(H_2O)_5^{2+}(aq)$], such as pathway 2 in Fig. 4. Hydrogen exchange, of course, is usually much faster than ligand exchange, and protons can come on and off the bridge many times before inducing dissociation. Because hydrogen exchange with the solvent is so rapid, it can be difficult to know the protonation state of the product.

If the hydrated metal monomer is a much stronger acid than the water-bridged dimer, the proton is released after dissociation of the bridge and can recycle. For these cases, the proton adsorption is *catalytic* and these pathways may be particularly important for oxide minerals with polymers of oxyanions, such as aluminosilicates and phosphates. There is, for example, no such thing as a η^1-OH_2 site on monomeric Si(IV) as $Si(OH_2)_4^{4+}(aq)$ dissociates immediately to form silicic acid: $Si(OH)_4^0(aq)$, which has co-ordinated hydroxyls.

Inner-sphere ligand substitutions also affect the dissociation rates of multimers. The deprotonation to form stable hydroxyls is, in essence, such a ligand exchange:

$$\eta^1 - OH_2 = \eta^1 - OH^- + H^+(aq) \tag{2}$$

This deprotonation, or *hydroxyl-assisted* pathway is marked "2" in Fig. 4.

Deprotonation of the terminal η^1-OH_2 site weakens the bridging hydroxo- or oxo-groups and accelerates the rates of dissociation (Table 2). The effect is not trivial. This *ligand-directed labilization* by a stable hydroxyl is manifested in accelerated rates of polymerization, anation, water exchanges, and mineral dissolution. Rates of water exchange from the inner co-ordination sphere of $Cr(H_2O)_6^{3+}(aq)$ to the bulk solution, for example, are a factor of 75 times slower than around the deprotonated $Cr(H_2O)_5OH^{2+}(aq)$ complex. Similar increases are observed for other metals [e.g., 16, 18].

Replacement of a hydration water with other stable ligands can also make the distal oxygens more reactive. Substitution of fluoride ion for a hydration water in $Al(H_2O)_6^{3+}(aq)$ to form the $AlF(H_2O)_5^{2+}(aq)$ complex, for example, causes the rate of exchange of each of the other water molecules in the inner co-ordination sphere to increase by a factor of $\approx 10^2$ [19]. The effect is progressive so that rates continue to increase with each additional substitution. This pathway is marked "3" in Fig. 4 and is a type of *ligand-directed labilization*:

$$\eta^1 - OH_2 + F^-(aq) = \eta^1 - F + H_2O(aq) \tag{3}$$

Ligand-directed labilization is well documented for ammonia substitution into $Ni(NH_3)_x(H_2O)_{6-x}^{2+}(aq)$ complexes and for substitution of amine and aminocarboxylate chelates for water molecules to form multidentate Ni(II) complexes (Table 3). Rates of water exchange increases by ≈ 10–10^2 for each increase in the number of nitrogens or carboxyl oxygens co-ordinated to the metal in the complex. Rates of water exchange in a Ni(II) complex co-ordinated to four nitrogens in the aliphatic amino-chelate triethylenetetra-amine (trien) is more rapid than around Ni(II) co-ordinated to two nitrogens in ethylenediamine [$NH_2(CH_2)_2NH_2$, en]. This reactivity trend will be examined later in the context of NiO(s) dissolution.

Finally, because the reactivity of a bridging oxygen depends on the M–O bond strength, transfer of an electron to the metal can dramatically accelerate the dissociation rate. This *reductive* pathway is marked "4" in Fig. 4.

TABLE 3 Rate Coefficients for Exchange of a Single Water Molecule from Around Ni(II)–Ligand Complexes

Complex	k(s^{-1})	Complex	k (s^{-1})
$Ni(H_2O)_6^{2+}$	0.32×10^5	$Ni(NH_3)(H_2O)_5^{2+}$ [a]	2.5×10^5
$Ni(NH_3)_2(H_2O)_4^{2+}$ [a]	6.1×10^5	$Ni(NH_3)_3(H_2O)_3^{2+}$ [a]	25×10^5
$Ni(en)(H_2O)_4^{2+}$ [b]	4.4×10^5	$Ni(dien)(H_2O)_3^{2+}$ [b]	18×10^5
$Ni(trien)(H_2O)_2^{2+}$ [c]	29×10^5	$Ni(en)_2(H_2O)_2^{2+}$ [b]	54×10^5
$Ni(bpy)(H_2O)_4^{2+}$	0.49×10^5	$Ni(bpy)_2(H_2O)_2^{2+}$ [c]	0.66×10^5
$Ni(tpy)(H_2O)_3^{2+}$ [c]	0.52×10^5		
$Ni(ida)(H_2O)_3^{2+}$ [d]	2.4×10^5	$Ni(H_2O)(edta)^{2-}$ [d]	7×10^5
$Ni(H_2O)_5Cl^{+}$ [e]	1.5×10^5	$Ni(H_2O)_3(NCS)_3^{-}$ [e]	11×10^5
$Ni(2,3,2\text{-}tet)(H_2O)_2^{2+}$ [f]	40×10^5	$Ni(12[ane]N_4)(H_2O)_2^{2+}$ [f]	200×10^5

[a]Ammonia.
[b]Tertiary amines.
[c]Pyridines.
[d]Carboxylates.
[e]Inorganic ligands.
[f]Macrocycles.
Source: Refs. 16, 18, and 20.

III. DISSOLVING OXIDE MINERAL SURFACES

Dissolution rates of oxide minerals vary with solution pH, and this variation arises from the adsorption of protons to surface oxygens and the deprotonation of functional groups. In addition, a wide range of ligands has been shown to accelerate oxide mineral dissolution, including fluoride, carboxylates, amines, and phenols [21, 22] and these ligands usually form surface complexes rapidly. These adsorbates exist as stable ligands in the inner co-ordination sphere of surface metals and influence the reaction rate through the reactivity of the distal M–O bonds that link the metal to the structure. These stable ligands are commonly released in a metal–ligand complex that detaches from the surface.

A. Dissociation Water Molecules and Metal Co-ordination Spheres

In all of the hydrolytic pathways for multimer dissociation discussed in the previous section, the rate-controlling step probably involves hydration as one of the metals released as a monomer is not co-ordinatively saturated. The co-ordination number is increased by association with a water molecule, which may subsequently deprotonate to form a stable hydroxo-species. Because water molecules exist in large excess in aqueous solutions, the dependence of the reaction rate on water concentration is ambiguous.

The same process takes place on an oxide surface. The NiO_6 octahedra shown with the wireframe of Fig. 1, for example, ultimately detaches as a Ni(II) monomer [shown in Fig. 5 as $Ni(H_2O)_6^{2+}(aq)$] as the monomolecular step migrates. The co-ordination numbers of oxygens co-ordinated to adjacent Ni(II) atoms are all decreased by one as a monomer detaches. These metals re-establish their inner co-ordination spheres by movement and association with water molecules or adsor-

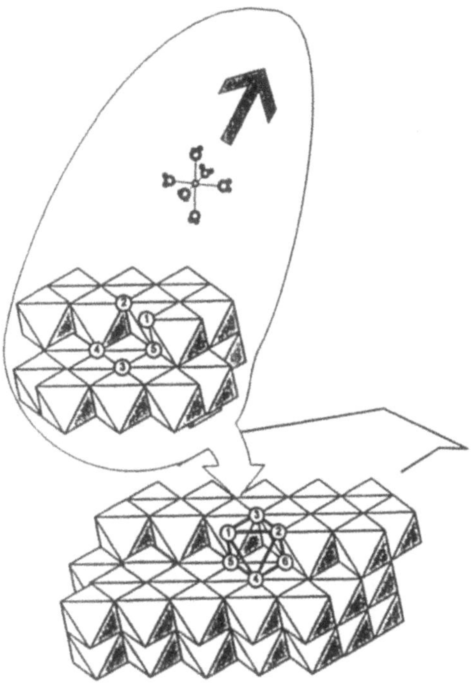

FIG. 5 The retreating monomolecular step NiO(s) from Fig. 2 shown with the changed co-ordination number of oxygens in the inset. The co-ordination sphere of the metals is maintained by movement and dissociation of water molecules to re-establish a steady-state surface.

bates. To maintain a steady-state charge on the surface, these ligands and water molecules may deprotonate [23, 24]. We cannot, of course, identify a priori whether the surface metal or detaching solute gets the structural oxygens, but this identification is unimportant. What is important is that a steady-state surface can only be maintained by movement and dissociation of ligands and water molecules, and this movement proceeds simultaneously with detachment of the surface complex.

It has long been known that the dissolution rates of minerals vary with solution pH (Fig. 6). The careful work of Pulfer et al. [27] and Furrer and Stumm [23] elaborated on these pH variations and expressed simplified forms of the rate laws in terms of real changes in the composition of surfaces [28–31]. Before discussing these rate laws, it is useful to examine first the acid–base chemistry of oxide surfaces.

B. Brønsted Reactions on Oxide Surfaces are Treated Phenomenologically

For surfaces, the protonation/deprotonation reactions are of uncertain stoichiometry, and the reactions are generally written to involve a neutral "average" functional group [32, 33]:

$$> SOH(s) + H^+(aq) => SOH_2^+(s) \tag{4}$$

$$> SOH(s) => SO^-(s) + H^+(aq) \tag{5}$$

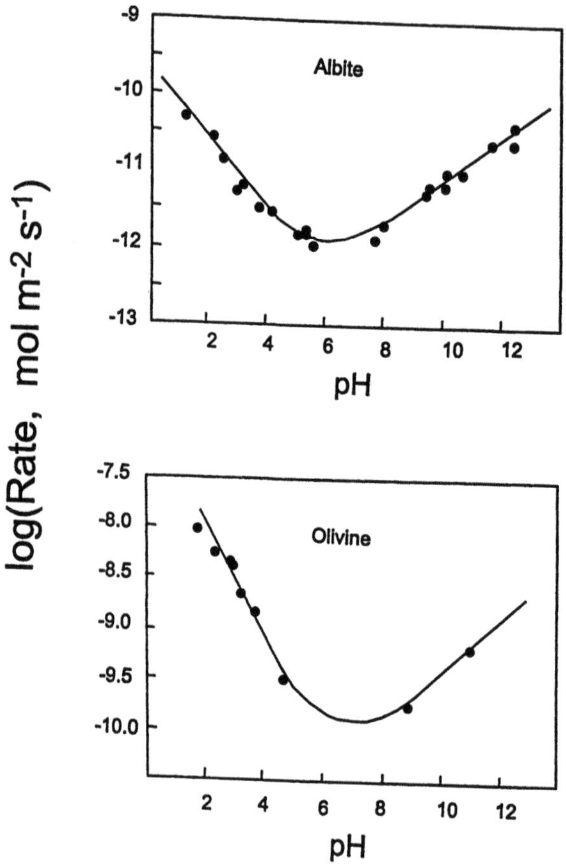

FIG. 6 Variation in dissolution rates of albite ($NaAlSi_3O_8$) and forsterite olivine (Mg_2SiO_4) as a function of solution pH. (From Refs. 25 and 26.)

where $>$ SOH(s) represents a model hydroxyl function group and $>$ SOH_2^+(s) and $>$ SO^-(s) represent the hyperprotonated and deprotonated forms, respectively. The deprotonation equation (5) is in the form of a conventional acid ionization reaction. However, as a result of writing both protonation/deprotonation equations (4) and (5) with the $>$ SOH(s) group as a reactant, Eq. (4) takes the form of a "reversed" acid ionization reaction. Simultaneous use of both "reversed" and conventional acid ionization equations (4) and (5), respectively, poses no problem to the mathematical description of proton reaction with surface groups. However, differences in convention appear in the literature and can be confusing. Before about the mid-1980s, conventional acid ionization reactions are found almost exclusively. After this time, both "reversed" and conventional acid ionization reactions are common. A more complete discussion of these conventions and their chronology are given in Chapter 1 of this book.

The conditional equilibrium constants for such charging reactions have no unique expression, which is to be expected since these equilibria do not correspond to specific microscopic reactions. The phenomenological two-state expression of equilibrium is most common, giving an equilibrium constant for Eq. (4):

$$\dot{K} = \frac{\lambda_{SOH_2^+}\chi_{SOH_2^+}}{(H^+)\lambda_{SOH}\chi_{SOH}} \tag{6}$$

where λ_i is the rational activity coefficient for the ith surface species (discussed in more detail in Chapter 4 of this book) and the parentheses indicate the activity of protons in solution. Calculation of the ratio of rational activity coefficients is virtually always treated with a complexation model [32]. In the constant-capacitance model [33] the conditional equilibrium constant is expressed as

$$K' = \left[K_{a1}^{s}(\text{int}) \cdot e^{\frac{F\psi_s}{RT}}\right]^{-1} \tag{7}$$

where K' is the conditional equilibrium constant (a "reverse" acid ionization constant), K_{a1}^{s} is the "intrinsic" equilibrium constant for the dissociation of $> SOH_2^+(s)$ sites (a conventional acid ionization constant) and represents the proton affinity in an uncharged state, ψ_s is the electric potential (V) at the mean adsorption plane of the surface species and accounts for electrostatic repulsion and attraction to the charged surface, F is the Faraday constant (C/mol), R is the gas constant, and T is temperature in Kelvins.

There are unique properties of surfaces that deserve specific mention. First, the conditional equilibrium constants for these Brønsted reactions vary by factors of 100–1000 with both the solution pH and the total charge concentration on the surface (which also varies with pH; Fig. 7) because more work is required to put a proton on a positively charged surface than on a neutral surface. Likewise, more work is required to remove a proton from a negatively charged surface than from a positively charged or neutral surface. No information about microscopic acid–base reactions is provided by Eqs (6) and (7) and the choice of a diprotic model for the acid–base reactions is arbitrary. By microscopic reactions we mean the actual protonation of co-ordinatively distinct oxygens at the mineral surface. This approach to Brønsted acidity is wholly a phenomenological treatment, although progress is being made in assigning microscopic equilibrium constants to surface oxygens [e.g., 36, 37].

Second, any property of the electrolyte solution that affects the surface potential at the Stern layer of the surface (ψ_s), in turn affects the concentrations of $> SOH_2^+$ and $> SO^-$. These variables include outer-sphere adsorbates that can shield or deshield charges from one another on the surface, and inner-sphere adsorbates that create or neutralize charges. One expects, for example, that background electrolytes with cations of differing hydrated radii (e.g., $Li^+ > Na^+ > K^+ > Cs^+ > Rb^+$ or $Mg^{2+} > Ca^{2+} > Ba^{2+}$) affect the Brønsted acidities of the surface by shielding charges to different extents. This shielding is expressed through the value of ψ_s in Eq. (7).

On a broad scale, separation of the conditional equilibrium constant into contributions from an intrinsic affinity [$K_{a1}^{s}(\text{int})$] and electrostatic repulsion or attraction (ψ_s) accounts for the overall charge properties of most oxides (Fig. 8). Wieland et al. [38] compiled charge data for oxide minerals with well-defined surfaces and showed that, on an areal basis and a logarithmic scale, the oxide minerals and latex beads fall in a similar trend where the concentration of positive charge increases as pH is decreased from the point of neutrality (the point of net zero proton charge, or PZNPC). This PZNPC is interpretable as a measure of the fundamental affinity of the surface for protons apart from electrostatic repulsion and attraction. That these

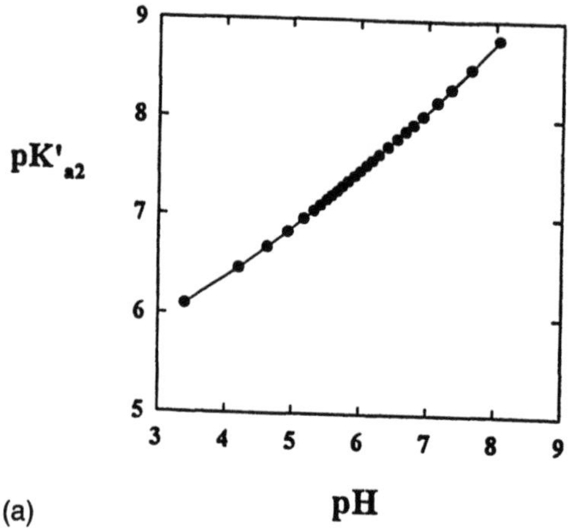

(a)

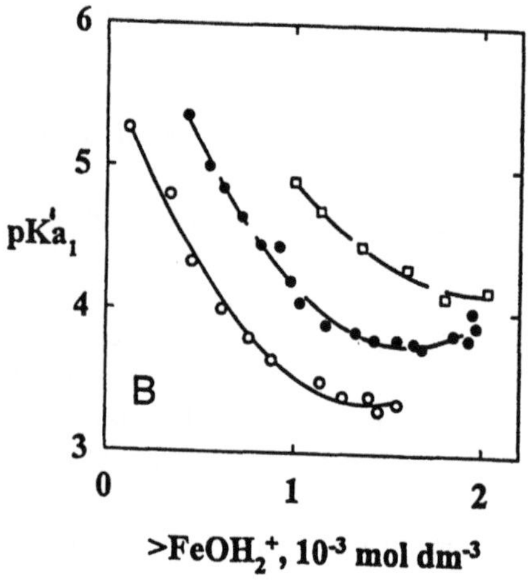

(b)

FIG. 7 (a) Variation with pH of the conditional equilibrium constant for forming negative surface charge on SiO_2(s) by deprotonation of surface groups. (From Ref. 34.) (b) Negative logarithm of the conditional equilibrium constant (pKa_1) for protonation of the surfaces of hematite (dotted line), magnetite (solid line), and goethite (dashed line) at $I = 0.001$ M and 25°C as a function of total charge concentration. (From Ref. 35.) The abscissa corresponds to the bulk concentration of positive charge (mol dm^{-3}) and can be converted into surface concentrations (mol m^{-2}) through knowledge of the specific area of the solid (m^2 g^{-1}) and the total mass (g) of solid in the experiment.

differing materials (even latex beads!) exhibit similar variations in charge concentration at pH $\neq$ PZNPC is equivalent to stating that the dependence on pH is largely attributable to electrostatic effects only.

The orthosilicate minerals fall conspicuously off this trend (Fig. 8), indicating that the charge concentrations are inaccurately known, that they have different electrical properties, or that protonation can affect deeper parts of the near-surface region of the solid. These orthosilicate minerals have no polymerized silicate anion to retain structural integrity at the surface once the divalent metals are leached away. These minerals also exhibit a similar pH dependence in dissolution rates [40]. As one

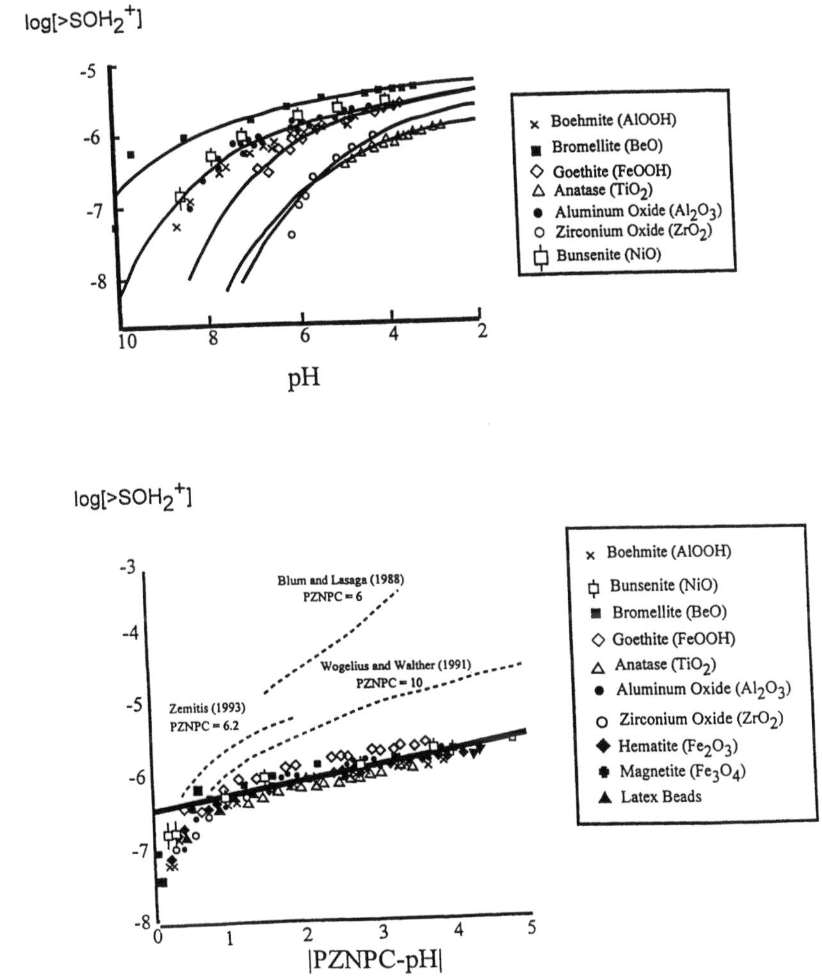

FIG. 8 Logarithm of positive charge concentrations in units of mol m^{-2} on a wide range of oxide and orthosilicate minerals, along with data for latex beads prepared with sulfonate functional groups. (Adapted from Ref. 38; data for bunsenite from Ref. 39.) *Top*: charge concentrations as a function of pH; *Bottom*: charge concentrations as a function of the deviation in pH from the point of zero charge. The lines correspond to estimated charge concentrations for orthosilicate minerals. (Data for silicate minerals from Ref. 25, 60, and 66.)

can see, the protonation of orthosilicate mineral surfaces is either dramatically different from that of simple oxides or much more difficult to study because protons are consumed by side reactions, such as the leaching of metals from the near-surface reaction of the solid.

C. Rates are Proportional to Concentrations of Surface Complexes

Rates of mineral dissolution are properly expressed in terms of the concentration of each metal–ligand surface complex with distinct reactivity and stoichiometry. If hydration steps control the rate of detachment of a surface complex from a mineral, by hypothesis, the overall dissociation of an oligomer or hydrolytic complex on a mineral surface can be resolved into a sum of elementary reactions at far-from-equilibrium conditions:

$$\text{Rate} = \sum_{p,l,o,i} k_{p,l,o,i} \quad x_{S_{p,l,o,i}}^{p+l+o} + k_{H_2O} \tag{8}$$

where $k_{p,l,o,i}$ and k_{H_2O} are rate coefficients for reaction via the proton- (p), hydroxide- (o), and ligand-promoted (l) pathways, and nucleophilic attack by water. In this rate law, the stable ligands modify the labilities of the near-surface M–O bonds, but are stable in the surface complex. Movement of a water molecule is assumed to do the real work in breaking bonds at the surface. Water exists in large and constant concentrations and is left out of Eq. (8).

In using rate laws such as that of Eq. (8) we assume that reacations along various pathways are independent, and therefore additive, and that the rate-controlling step reaction is the hydration and disruption of bonds at or near the metal–ligand complex, and not movement of the reactive solutes through the bulk aqueous phase to (or from) the site(s) of reaction. The subscript p accounts for the number of protons involved in the rate-controlling step, the subscript l accounts for the number of adsorbed ligands, and the subscript i accounts for ligands of similar stoichiometry but different structures, including protonation states. The $x_{S_{p,l,o,i}}$ variables are the mole fractions of surface complex i (although any concentration variable could substitute) and this variable includes a description of the concentrations of adsorbed protons and hydroxide ions. In this formalism, a distinction is made between protons that are adsorbed on to the ligand and those that adsorb on to the mineral surface in addition to the ligand.

The great difference between surface reactions and oligomer dissociation is that it is possible to distinguish analytically the various stoichiometries of oligomers. For dissolved oligomers, the overall rate of dissociation described by a rate law similar to that of Eq. (8) will have contributions from all stoichiometrically distinct species. For example, the dissociation of the μ_2-hydroxo sites in the $(en(H_2O)Cr(\mu\text{-}OH)_2Cr(H_2O)(en)^{4+}(aq)$ complex can be studied separately from the $(H_2O)_4Cr(\mu\text{-}OH)_2Cr(OH_2)_4]^{4+}(aq)$ species.

In contrast, the stoichiometries and concentrations of different reactive complexes at the mineral surface are unknown so we cannot fully evaluate rate laws such as that of Eq. (8). There exist numerous simplifying rate laws, however, that relate rates of dissolution of an oxide mineral to total adsorbed concentrations, and do not distinguish among the stoichiometrically distinct adsorbates, as we discuss below.

D. Protons in Surface Complexes are Particularly Important

For dissolution under conditions where positive surface charge predominates ($pH < PZNPC$) and far from equilibrium, Furrer and Stumm [23] report an immensely useful empirical rate law:

$$\text{Rate} = k_{H^+}\, x^{n}_{SOH_2^+} \tag{9}$$

where: $x_{SOH_2^+}$ is the surface concentration (here on the mole fraction scale) of the positive charge that arises from proton adsorption to the surface, n is a rate order, and k_{H^+} is a rate coefficient in units of mol $cm^{-2}s^{-1}$. A similar rate law is expressed for dissolution under conditions where negative charge predominates on the surface from desorption of protons:

$$\text{Rate} = k_{OH}\, x^{m}_{SO^-} \tag{10}$$

where x_{SO^-} is the concentration (mole fraction) of the negative charge that arises from adsorption of hydroxyl (or desorption of protons) to the surface, m is a rate order, and k_{OH^-} is a rate coefficient in units of mol cm^{-2} s^{-1}.

It is very difficult to determine the equilibrium proton charge on solid surfaces as they dissolve, and that on mixed oxide minerals is virtually impossible to determine [41–43]. Nevertheless, some careful studies (Fig. 9) relate dissolution rates of simple oxide minerals to the concentration of surface-charge data and suggest that the n values scale like the formal charge on the metal ion. This surprising and controversial result is consistent with a simple model where monomolecular steps on the surface of an oxide mineral retreat at a steady state. Because the charge state of the surface must be maintained, removal of a fully hydrated monomer from this step [e.g., $Ni(H_2O)_6^{2+}(aq)$] requires addition of three protons to recover the charge. In this model, protons only induce dissociation and are not catalytic. Furthermore, it requires that monomers detach in a slow step, that protonation affects these detaching species only at the surface and not in solution after detachment, and that newly uncovered, underbonded metals satisfy their coordination sphere by the movement and association of water molecules [24].

The rate order is given by a charge balance of protons at the surface; this result is, however, not observed for complicated silicate minerals. There are many potential reasons for this, including the selective leaching of cations from the surface [e.g., 41, 45, 46], self-poisoning of the reaction [e.g., 47], or migration of protons deep into the mineral. We yet have virtually no fundamental understanding of the dissolution kinetics of complicated mixed oxide and aluminosilicate polymerized structures.

E. Dissolution Rates Correlate with Bond Strengths for Isostructural Oxides

If the effects of M–O bond dissociation can be isolated from other reactions, such as depolymerization of a silicate chain, one expects oxide mineral dissolution rates to correlate with bond strengths. In other words, there should be correlation between dissolution rates at constant pH and the rates of water exchange about the corresponding dissolved cation, since similar dissociation of M–O bonds are involved in both solvent exchange and mineral dissolution. Well-defined conditions mean that:

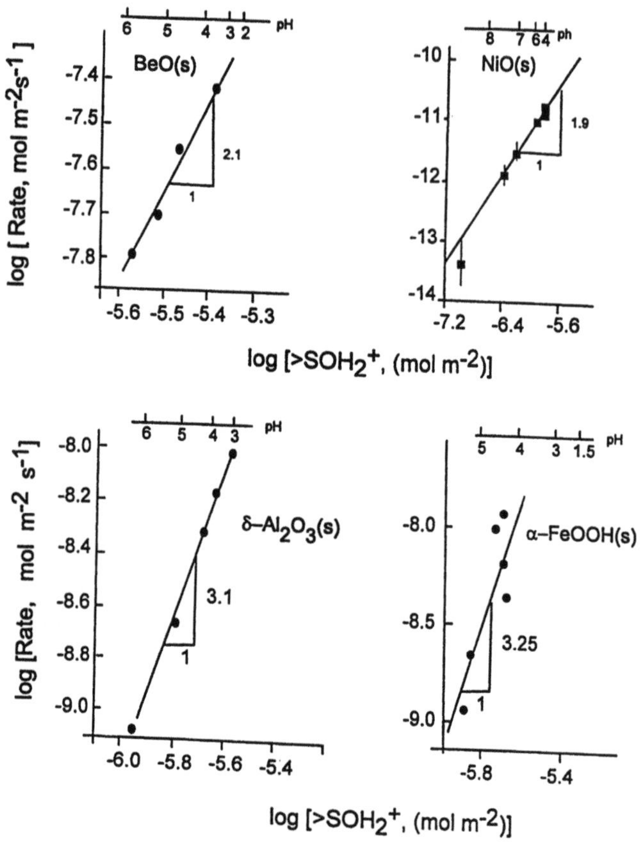

FIG. 9 Variation in dissolution rate of a series of oxide minerals as a function of pH and the density of hyperprotonated sites. (Adapted from Ref. 21; data for bunsenite from Ref. 39 and for boehmite from Ref. 44.)

(1) minerals are compared that do not have a relatively unreactive silicate polymer holding the structure together, (2) experiments are conducted such that the surface compositions are comparable, and (3) electron exchange does not modify the valence state of metals at the mineral surface.

A collection of dissolution rates of orthosilicate minerals is shown in Fig. 10 where the divalent metal is systematically varied but the conditions are otherwise maintained constant. As one can see (Fig. 10) the dissolution rates of orthosilicate minerals correlate strongly with the rate of solvent exchange around the corresponding fully hydrated divalent metal [49]. Calcium can be removed from a surface site more quickly than magnesium; beryllium is very resistant. For minerals containing first-row transition metals, rates vary with the number of valence d-electrons and, hence, the crystal-field energies for the cation. The dissolution rate of liebenbergite [Ni_2SiO_4(s)] is much slower than that of tephroite [Mn_2SiO_4(s)] although the M–O bond lengths differ only slightly. A similar reactivity trend is reported for metal adsorption and desorption from γ-Al_2O_3 [51].

The variation in rate spans a range of 10^5, in a way that is perfectly consistent with the rates of ligand exchange. Both dissolution of the solid surface and ligand

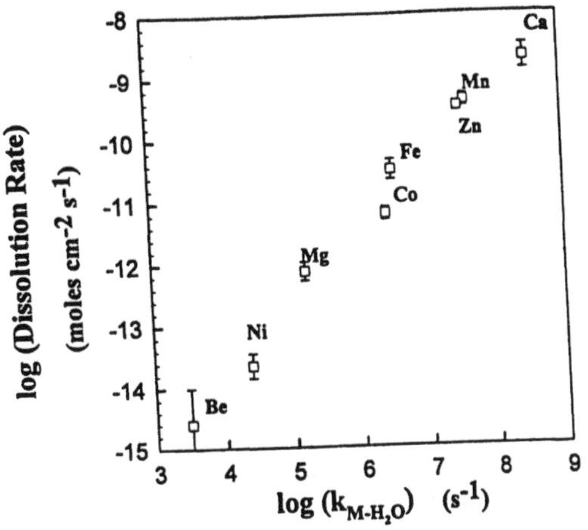

FIG. 10 Dissolution rates of orthosilicate minerals at pH 2 and 25°C plotted against the rate coefficient, which describes exchange of water molecules from the inner co-ordination sphere of the hydrated metal with bulk solution. (From Ref. 48.) For example, Ni^{2+} identifies both the dissolution rates of Ni_2SiO_4(s) (liebenbergite) and the rate of solvent exchange from the inner co-ordination sphere of $Ni(H_2O)_6^{2+}$ (aq) to the bulk solution.

exchange around the monomeric solute are controlled by the strengths of similar bonds to oxygen. The co-ordination number and distances of metals to oxygens at the surface and in the dissolved complex are similar (see Ref. 49, Table 1 and Ref. 50).

F. Rates Vary with Temperature and Solution Compositions

Most researchers studying oxide mineral dissolution express the temperature dependence in terms of an unconventional Arrhenius-like function by plotting the logarithm of rate (i.e., not k_H or k_{OH^-}) versus $1/T$ because it is so much easier to measure dissolution rates than equilibrium adsorbate concentrations. The important geochemistry is illustrated by differentiating the empirical rate laws and by applying the Van't Hoff relation [52, 53], yielding from Eqs (6), (7), and (9):

$$E_{app} = -R\left[\frac{\partial \ln k_{H^+}}{\partial\left(\frac{1}{T}\right)}\right]_{pH} - nR(1 - x_{SOH_2^+})\left[\frac{\partial(\ln K')}{\partial\left(\frac{1}{T}\right)}\right]_{pH} \tag{11}$$

Equation (11) includes not only a variation in the rate coefficient, k_{H^+} with temperature, but also includes a temperature variation in the conditional equilibrium constant, K'. As shown earlier, proton charge densities vary with pH because more work is required to move a proton to a positively charged surface than to a neutral or negatively charged surface. Just as the protonation state of the surface varies with both pH and temperature, so too does the derivative parameter, E_{app}. The temperature dependence of dissolution rate varies with pH due to changes in the adsorption of protons to the surface with temperature and pH (Fig. 11).

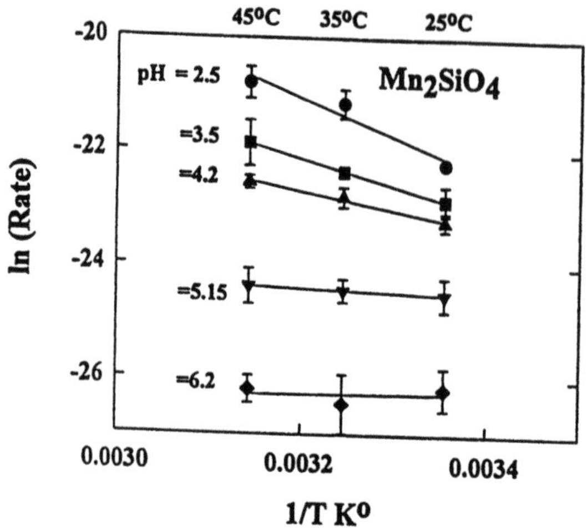

FIG. 11 Because the enthalpies of proton adsorption are large and vary with pH, the derivatives of dissolution rates (here tephroite: Mn_2SiO_4) vary with both temperature and pH. (See Refs. 8, 52 and 53.)

G. Adsorbed Ligands Promote Dissolution By Modifying Bond Strengths

All of the important classes of *ligand-directed labilization* are represented at mineral surfaces, including the deprotonation of a terminal η^1-OH_2 site. One therefore expects a similar increase in dissolution of oxide minerals as the surface sites become deprotonated. For a mineral surface, these deprotonations form negative surface charge [cf. Eqs (5) and (10)] and the rates correspondingly increase with pH or charge. The actual location of the deprotonations is unknown, but a likely scenario is that η^1-OH_2 sites on monomolecular steps deprotonate with increases in pH.

Ligands or ligand functional groups at the surface of minerals can enter the inner co-ordination sphere of a surface metal at a much more rapid rate than the rates of dissolution. Therefore, one finds that ligands such as fluorides or carboxylates enhance the rates of dissolution of oxide minerals. For reasons of experimental convenience, much work on ligand-directed labilization in dissolved metal complexes has been conducted on Ni(II) species (e.g., Table 3). The effect of ammonia, amino-ligands, and aminocarboxylates on Ni(II)–oxygen bonds is particularly well understood and a strong correlation has been shown between the rate coefficients for water exchange and the equilibrium constant for complexation.

Ludwig and coworkers [54, 55] set out to reproduce these reactivity trends for dissolution of NiO(s). They first took care to identify ligands that probably form structurally similar complexes with Ni(II) both in solution and at the surface of NiO(s). They also chose a pH region for experiments where all of the functional groups on the ligand could bind strongly to a surface Ni(II) atom; at low pH values, for example, it is possible to protonate one carboxyl of the EDTA complex so that the denticity of the surface complex is poorly known. Finally, they chose ligands that only form five-membered chelating rings with the surface Ni(II) metal, since it is known that effectiveness in enhancing dissolution increases as the chelate ring decreases in size down to five [e.g., 23]; five-membered chelate rings are more effective than similar six-membered rings.

In these NiO(s) dissolution experiments, the authors established close consistency between the ligand-promoted dissolution rates of NiO(s) and labilizing of water molecules in dissolved Ni(II)–aminocarboxylate complexes. They specifically found rates to increase with the number of $-NH_2$ and –COOH ligand functional groups that could attach to the surface Ni(II) atom, exactly in the same trend as is observed for dissolved Ni(II)–ligand complexes (Table 3; Fig. 12), and the effectiveness of a ligand in enhancing dissolution also scales like the equilibrium constant.

Beyond these largely qualitative experiments, there is very little known about the actual structure and stoichiometries of the adsorbates that enhance dissolution rates. Acidic organic ligands such as carboxylates are negatively charged in natural waters and adsorb to positively charged oxide surfaces as either outer-sphere or inner-sphere complexes. One can speculate a reasonable structure for bidentate complexes, such as ethylenediamine on NiO(s) (Fig. 12), but it becomes difficult to imagine how a larger chelate, such as *trien*, can adsorb to a single site and thereby enhance the labilities of the distal M–O bonds.

Finally, it is clear that adsorbed ligands can influence the dissolution rates of minerals through their effect on the Brønsted acidities. The clearest such influence is when an adsorbed ligand is the conjugate base of a weak acid and removes a proton from a surface oxygen. We saw earlier, for example, that removal of a proton from an η^1-OH_2 on a dissolved, fully hydrated metal causes the rates of exchange of each of the distal oxygens to increase by a factor of about 100. Therefore, any ligand that encourages such deprotonation can enhance the rate of reaction even if it does not bond directly to the metal [16, 18]. Such *conjugate-base* mechanisms undoubtedly operate at mineral surfaces, but we do not have sufficient resolution of the speciation to identify these cases.

H. Electron Exchange Dramatically Reduces Bond Strengths

In discussing the dissociation of dissolved multimers we emphasized that the most important control is the strength of the M–O bond, which is predictable from the electronic structure. For oxides of many trivalent and tetravelent metals, *reductive* processes are virtually the only pathways for dissolving the mineral because the bond strengths to oxygen of a reduced metal are commonly weaker than the oxidized forms. There are, however, corresponding *oxidative* pathways for those minerals [e.g., $UO_2(s)$] that contain a very strong bond between a reduced metal and oxygen that can be weakened by oxidation [e.g., 5, 56, 57]. These repeated processes of reduction and oxidation contribute to the chemical stratification of soils and the elimination of organic matter [e.g., 58, 59].

As one might expect, the rates of reductive dissolution of a mineral depend upon the concentrations of reduced metals at the surface. These concentrations, in turn, depend on the concentrations of electron-donating ligands at the surface and the rates of electron transfer. This subject is immensely complicated because the rates of electron transfer can depend on many variables, including the stoichiometry of the adsorbate, the electronic structure of the metal oxide mineral, and whether light is involved. Many adsorbates that are important in nature are chromophores, and the electron transfer is facilitated by adsorption of light at the appropriate wavelength.

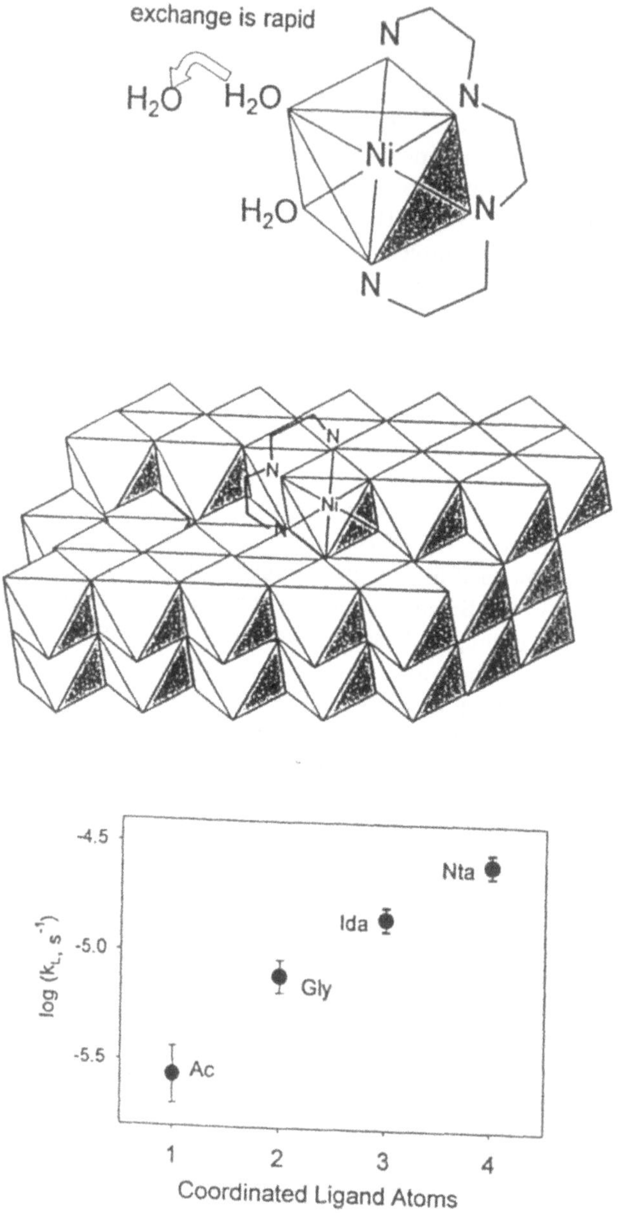

(a)

(b)

(c)

FIG. 12 (a) Labilities of hydration water molecules in the inner co-ordination sphere of Ni(II) increase as the metal is co-ordinated to amine functional groups in the *trien* chelate. (b) A retreating monomolecular step on NiO(s) shown with an adsorbed ethylenediamine molecule as a hypothetical surface complex. The three amino groups co-ordinate to the Ni(II) via exchange for oxygen-containing functional groups and can enhance the labilities of the distal Ni(II)–O bonds. (c) Correlation between the ligand-promoted rate coefficient for dissolution of NiO(s) as a function of the number of ligand atoms available for co-ordination. (From Ref. 55.)

These *photoreductive* processes are central to the elimination of dissolved organic matter in natural waters [e.g., 5] and are the only pathways for degradation of some chelating agents and toxicants.

Photoreductive pathways are also enormously complicated. Consider the example provided by Karametaxas et al. [61] that is shown in Fig. 13. Ethylenediaminetetra-acetate (EDTA) is a hexadentate chelating agent that is used in many industrial processes as a water softener, a metal binder, and as a chemical reagent to eliminate oxide corrosion products. Unfortunately, wide use of EDTA has led to dispersal of this chelating agent in natural waters. The ligand does not hydrolyze at appreciable rates in groundwater without light [62, 63] and can complex and transport toxic metals. In fact, one of the most troubling concerns about the geological storage of toxic waste is the burial with metals of these strong chelating agents [e.g., 64, 65].

One particularly effective pathway for eliminating EDTA is to react it with ferric hydroxide minerals in the presence of light. The ligand undergoes a complicated series of decompositions that are coupled with the reductive dissolution of the ferric hydroxide mineral. In Fig. 13, for example, the concentrations of the dissolved Fe(III)–EDTA complex increase initially with time as the ferrihydrite mineral is dissolved nonreductively. The mineral dissolves via proton- or ligand-assisted pathways to release Fe(III)–EDTA complexes to the aqueous phase. After about 100 h, however, the concentration of this ferric–EDTA complex decreases as it readsorbs on to the surface and undergoes an electron exchange that releases Fe(II) from the dissolving mineral and formaldehyde as a scission product of the aminocarboxylate ligand. Light is key to the reaction because it activates the electron transfer [61]. In this example, mineral dissolution proceeds simultaneously via several pathways (proton-promoted, ligand-protomoted, photoreduction-promoted) that depend sensitively on the solution concentration and incident light flux.

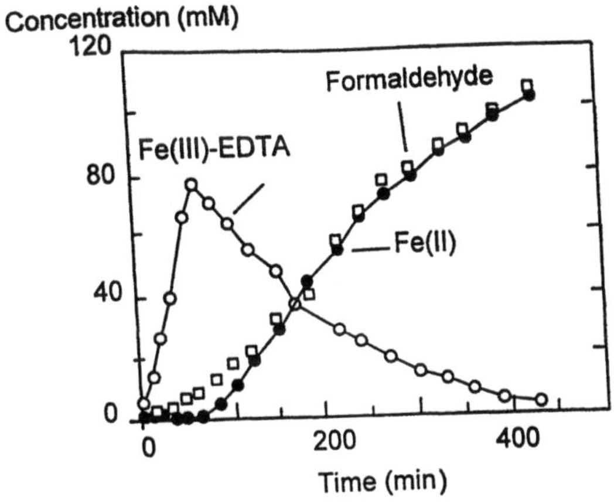

FIG. 13 Concentrations of formaldehyde (CH_2O=●), dissolved Fe(III)–EDTA complexes (○), and dissolved ferrous iron (□) as a function of time in an aerated suspension of γ-FeOOH(s) in the presence of EDTA and irradiated with 0.5 kW m^{-2} of white light. (From Ref. 61.)

IV. THE FUTURE: LIGANDS AND SURFACE SITES AT A MOLECULAR SCALE

The important message from the works discussed in this chapter is that the important information about understanding dissolution lies at the molecular scale. Examination of complex changes in concentration, such as shown in Fig. 13, in the context of Eq. (11), emphasizes the enormous distance between the results of our experiments and the real molecular configurations that control the reaction at a surface. In many cases, we cannot even identify the ligand atoms that adsorb to surface metals or the co-ordination numbers of those metals to bulk oxygens. Knowledge of the denticity and protonation states of the ligand in the surface complex is pure speculation in most cases. The future challenge is to establish methods of assigning actual stoichiometries to surface complexes and measuring their concentrations via methods such as NMR and FTIR. Once these stoichiometries are known, we can bring to bear the powerful methods of computation chemistry in order to simulate these complicated processes.

ACKNOWLEDGMENTS

Support for this research was from the US NSF grants EAR 96-25663, 94-14103, and US DOE DE-FG03-96ER 14629.

REFERENCES

1. KB Krauskopf. Radioactive Waste Disposal and Geology. New York: Chapman and Hall, 1988.
2. KB Krauskopf. Science 249:1231, 1990.
3. DG Brookins. Geochemical Aspects of Radioactive Waste Disposal. New York: Springer-Verlag, 1984.
4. P Behra, L Sigg. Nature 344:419, 1990.
5. B Sulzberger. In: W. Stumm, ed. Aquatic Chemical Kinetics. New York: John Wiley, 1990, pp 401–430.
6. A Torrents, AT Stone. Soil Sci Soc Am J 58:738, 1994.
7. MB McBride. Environmental Chemistry of Soils. Oxford Press, New York, 1994, 403 pp.
8. WH Casey, C Ludwig. (1995). In: AF White, SL Brantley, eds. Chemical Weathering Rates of Silicate Minerals. Reviews in Mineralogy. Vol. 31. Mineralogical Society of America, Washington D.C. pp 87–114.
9. FA Cotton, G Wilkinson. Advanced Inorganic Chemistry. Wiley-Interscience, New York, 1988.
10. G Johansson. Acta Chem Scand 14:771, 1960.
11. B Wehrli, E Wieland, G Furrer. Aqu Sci 52:1, 1990.
12. G Furrer, C Ludwig, PW Schindler. J Colloid Interface Sci 149:56, 1992.
13. J Springborg. Adv Inorg Chem 32:55, 1988.
14. SJ Crimp, L Spiccia, HR Krouse, TW Swaddle. Inorg Chem 33:465, 1994.
15. J Burgess. Ions in Solution: Basic Principles of Chemical Interactions. Chichester: Ellis Horwood, 1988.
16. DW Margerum, GR Cayley, DC Weatherburn, GK Pagenkopf. (1978) In: A Martell, ed. Coordination Chemistry. Vol. 2. ACS Monograph 174. Washington DC: American Chemical Society, pp 1–220.

17. DT Richens, L Helm, P-A Pittet, AE Merbach, F Nicolo, G Chapuis. Inorg Chem 28:1394, 1989.
18. RG Wilkins. Kinetics and Mechanism of Reactions of Transition Metal Complexes. VCH. New York, 1991, 424 pp.
19. BL Phillips, WH Casey, S Neugebauer-Crawford. Geochim Cosmochim Acta 61:3041, 1997.
20. F Basolo, RG Pearson. Mechanisms of Inorganic Reactions. John Wiley, 1967.
21. W Stumm. The Chemistry of the Solid–Water Interface. John Wiley, New York, 1991, 428 pp.
22. MA Blesa, PJ Morando, AE Regazzoni. Chemical Dissolution of Metal Oxides. New York: CRC Press, 1993.
23. G Furrer, W Stumm. Geochim Cosmochim Acta 50:1847, 1986.
24. WH Casey, C Ludwig. Nature 381:506, 1996.
25. AE Blum, AC Lasaga. Nature 331:431, 1988.
26. L Chou, R Wollast. Geochim Cosmochim Acta 48:2205, 1984.
27. K Pulfer, PW Schindler, JC Westall, R Grauer. J Colloid Interface Sci 101:554, 1984.
28. GW Wirth, JM Gieskes. J Colloid Interface Sci 68:492, 1979.
29. SM Budd. Phys Chem Glass 2:111, 1961.
30. G Koch. Berichte der Bunsengesellshaft 69:141, 1965.
31. R Grauer, W Stumm. Colloid Polym Sci 260:959, 1982.
32. G Sposito. The Surface Chemistry of Soils. Oxford Press, New York, 1984.
33. PW Schindler, W Stumm. (1987) In: W Stumm, ed. Aquatic Surface Chemistry. New York: John Wiley, pp 83–107.
34. B Furst. Das coordinationschemishe Adsorptionsmodell: Oberflächenkomplexbildung von Cu(II), Cd(II) und Pb(II) an SiO_2 (Aerosil) und TiO_2. PhD dissertation, Universität Bern, Switzerland, 1976, 80 pp.
35. RD Astumian, M Sasaki, T Yasunaga, ZA Shelly. J Phys Chem 85:3832, 1981.
36. T Hiemstra, P Venema, WH Van Riemsdijk. J Colloid Interface Sci 184:680, 1996.
37. JR Rustad, AR Felmy, BP Hay. Geochim Cosmochim Acta 60:1563, 1996.
38. E Wieland, B Wehrli, W Stumm. Geochim Cosmochim Acta 52:1969, 1988.
39. C Ludwig, WH Casey. J Colloid Interface Sci 178:176, 1995.
40. HR Westrich, RT Cygan, WH Casey, C Zemitis, GW Arnold. Am J Sci 293:869–893, 1993.
41. R Wollast, L Chou. Geochim Cosmochim Acta 56:3113, 1992.
42. AE Blum, AC Lasaga. Geochim Cosmochim Acta 55:2193, 1991.
43. LL Stillings, SL Brantley, ML Machesky. Geochim Cosmochim Acta 59:1473, 1995.
44. J Nordin, P Persson, E Laiti, S Sjöberg. Langmuir 13:4085, 1997.
45. WH Casey, HR Westrich, JF Banfield, G Ferruzzi, GW Arnold. Nature 366:253, 1993.
46. R Hellmann, CM Eggleston, MF Hochella, DA Crerar. Geochim. Cosmochim Acta 54:1267, 1990.
47. EH Oelkers, J Schott, J-L Devidal. Geochim. Cosmochim. Acta 58:2011, 1994.
48. WH Casey, HR Westrich. Nature 355:147, 1992.
49. WH Casey. J Colloid Interface Sci 146:586, 1991.
50. DA Sverjensky. Nature 358:310, 1992.
51. K Hachiya, M Sasaki, T Ikeda, N Mikami, T Yasunaga. J Phys Chem 88:27, 1984.
52. WH Casey, G Sposito. Geochim Cosmochim Acta 56:3825, 1992.
53. PV Brady, JV Walther. Am J Sci 53:2822, 1992.
54. C Ludwig, WH Casey, J-L Devidal. Geochim Cosmochim Acta 60;213, 1996.
55. C Ludwig, WH Casey, PA Rock. Nature 375:44, 1995.
56. JG Hering, W Stumm. In: MF Hochella, AF White, eds. Mineral–Water Interface Geochemistry. Mineralogical Society of America, Reviews in Mineralogy. Washington D.C., Vol. 23, pp 427–465.

57. AT Stone, JJ Morgan. (1987) In: W Stumm, ed. Aquatic Surface Chemistry. John Wiley, New York, pp 221–255.
58. Y Zuo, J Hoigné. Environ Sci Technol 26:1014, 1992.
59. MA Jauregui, HM Reisenauer. Soil Sci Soc Am J 46:314, 1982.
60. CR Zemitis. Dissolution kinetics of three beryllate minerals. MS thesis, University of California, Davis, 1993, (unpublished).
61. G Karametaxas, SJ Hug, B Sulzberger. Environ Sci Technol 29:2992, 1995.
62. B Nowack. Behavior of EDTA in groundwater – a study of the surface reactions of metal–EDTA complexes. PhD dissertation, Swiss Federal Institute of Technology, Zürich, 1996 (unpublished).
63. B Nowack, L Sigg. Geochim Cosmochim Acta 61:951, 1997.
64. JL Means, DA Crerar, JO Duguid. Science 200:1477, 1978.
65. AJ Francis, CJ Dodge, JB Gillow. Nature 356:140, 1992.
66. RA Wogelius, JV Walther. Geochim Cosmochim Acta 55:943, 1991.

6

The Surface Chemistry of Clay Minerals

PATRICK V. BRADY and JAMES L. KRUMHANSL Sandia National Laboratories, Albuquerque, New Mexico

I. INTRODUCTION

The surface chemistry of clay minerals is typically a great deal more complex than that of oxides and hydroxides, in part because there are a seemingly infinite number of clay stoichiometries. Moreover, the reactivity of a metal cation in a clay mineral of a given composition can vary, depending on whether it is exposed and hydroxylated at a truncated edge, incorporated as a component of a basal plane, or existing as a charge-balancer on an exchange site. It is, consequently, impossible to estimate accurately the integrated surface reactivity of a clay mineral solely from its chemical composition. Said differently, the chemical behavior of clay mineral (hydr)oxide components cannot simply be summed to arrive at any realistic estimate of the surface reactivity of the whole.

Because the crystal structure of clay minerals plays a very large role in determining the way a clay will interact with aqueous solutions, a review of clay mineral surface chemistry must, by necessity, begin with a brief outline of the chemical origins of clay mineral structures. From this basis we will outline the role of edge-site chemical interactions and, subsequently, the controls on ion exchange. This overview of clay chemistry will conclude with a survey outlining how the various site interactions determine the macroscopic behavior of a number of clays in natural systems.

II. CRYSTAL CHEMISTRY

The multitude of clay mineral stoichiometries are built from just two building blocks, "T" sheets composed of tetrahedrally co-ordinated cations (principally Si with lesser amounts of Al) and "O" sheets of octahedrally co-ordinated cations. The latter are primarily Mg, Al, and Fe, though in unusual environments cations such as Cr, Ni, and even Li, may occupy a significant fraction of the sites. When Al dominates the

---

This work was supported by the United States Department of Energy under contract DE-AC04-94AL85000.

octahedral sites, the clays are referred to as "dioctahedral". These are found in nature principally where silica and alkali-rich rocks, such as granites, have been weathered by natural solutions. Weathering of Fe- and Mg-rich rocks (e.g., basalts) releases a proportionally greater fraction of divalent cations that subsequently form "trioctahedral" clays wherein the divalent cations make up the octahedrally co-ordinated fraction. While many of the newly formed clays have a neutral charge, others possess a net residual negative charge. In these cases, a cation (commonly Na^+, Ca^{2+}, K^+, or Mg^{2+}) resides in interlayer positions to balance heterovalent substitution (e.g., substitution of trivalent Al for quadrivalent Si) elsewhere in the lattice. Where heterovalent substitution prevails, exchange of the interlayer cations will dominate the ion exchange capacity (typically given in meq per 100 g) of the clay. Where heterovalent substitution is minimal, the surface behavior of the clay will be closer to that of the simple oxides described in earlier chapters. The large number of potential elemental substitutions and packing arrangements of the individual layers obviously allows for a virtually infinite number of possible clay minerals. Fortunately, in nature only a restricted number of these possible combinations occur commonly (Table 1).

It is the unique geometry of the individual T and O sheets which makes possible the multitude of different clay minerals. In the ideal T sheet, shown in Fig. 1, tetrahedral Si forms a two-dimensional plane with oxygen atoms. For the ideal T sheet each oxygen atom in the basal plane is co-ordinated to two Si atoms so that stoichiometrically 1/2 of each basal oxygen atom can be considered to be co-ordinated to each Si atom. Since each Si atom is co-ordinated to three basal oxygen atoms plus a single apical oxygen atom, the net number of oxygen atoms stoichiometrically co-ordinated to each Si atom must be (1/2) (3, basal oxygens) + (1) (1, apical oxygen) = 5/2 oxygen/Si. As a result the stoichiometric formula of the repeat unit of the T layer can be seen to be Si_2O_5.

With a +4 charge on Si and a −2 charge on the oxygen atoms, the net charge on the T layer stoichiometric formula, Si_2O_5, can be seen to be −2 or $Si_2O_5^{2-}$. Alternatively, each SiO_4 tetrahedra can be seen to possess a net charge of −1 because

TABLE 1 Simplified Clay Mineral Nomenclature

Stacking sequence	Al≫Mg	FeMg,Fe > Al
	Dioctahedral	*Trioctahedral*
TO, no interlayer cation	Kaolinite, halloysite	Mg > Fe serpentine group
TO_+TO^-, no interlayer cation	Cookeite (rare)	Chlorite Mg > Fe; clinochlore Fe > Mg; chamosite
TOT, no interlayer cation	Pyrophyllite	Talc
TOT, interlayer cations abundant, but not easily exchanged	Illite	Mg > Fe; vermiculite Fe > Mg; glauconite
	Smectites	
TOT, interlayer cations abundant and readily exchanged	Montmorillonite (Bentonite)	Mg > Fe; saponite Fe > Mg; nontronite

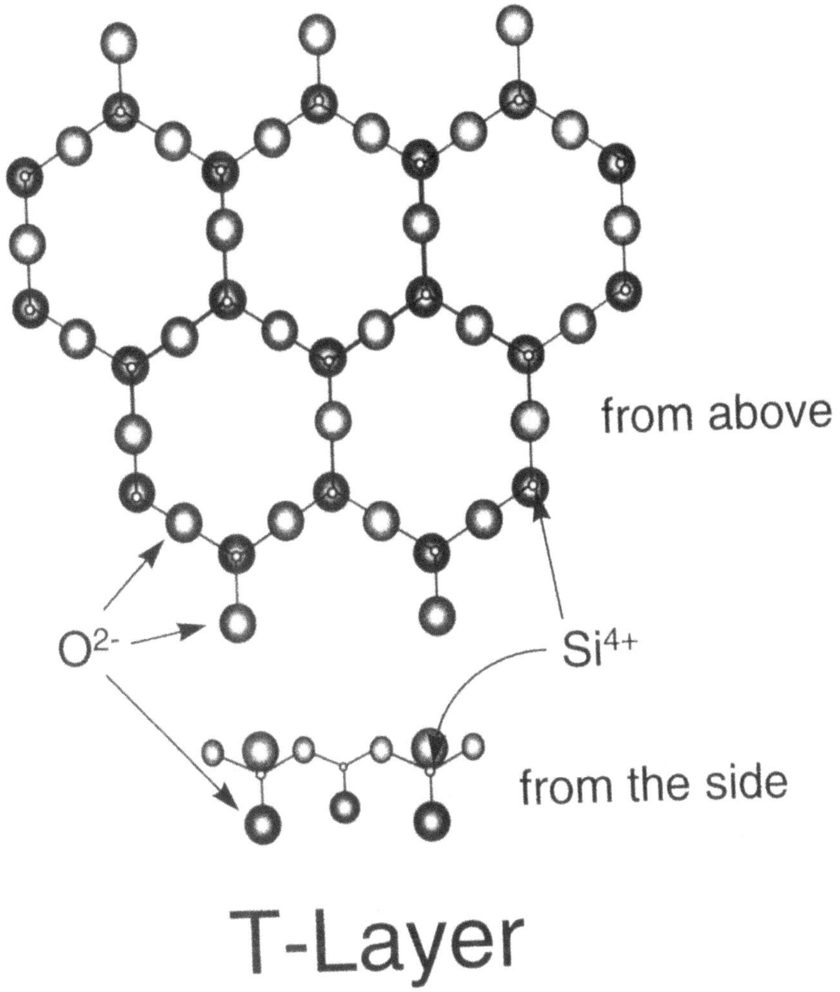

FIG. 1 The SI tetrahedral sheet.

each of the three basal plane oxygens donate a −1 charge, whereas the unshared apical oxygen donates a −2 charge. The apex of the Si tetrahedra in Fig. 1 points downward. It is of great importance that the spacing of apex oxygens in the T sheet happens to correspond almost exactly to some of the oxygen positions in an O sheet. This allows T and O sheets to join together to form the various clay minerals outlined in Table 1.

In the O sheets, shown in Fig. 2, the cations are co-ordinated octahedrally by oxygen atoms. The O layer is dioctahedral if the cations are primarily trivalent cations or trioctahedral if divalent cations predominate. The nomenclature arises from the fact that it takes three divalent cations to equal the charge of two trivalent cations.

For the dioctahedral sheet, each oxygen atom is seen to be co-ordinated to two trivalent cations, Me^{3+}, so that stoichiometrically 1/2 of each oxygen atom must be co-ordinated to a trivalent cation. Alternatively, each trivalent cation can be seen to

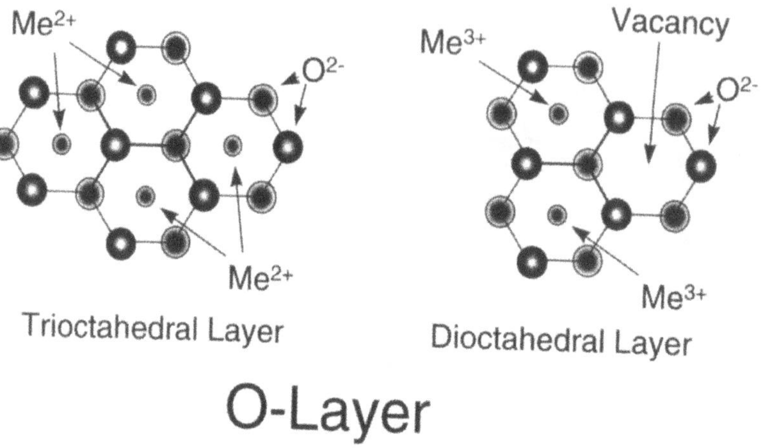

FIG. 2 Trioctahedral and dioctahedral sheets.

be co-ordinated to six oxygen atoms, so the net number of oxygen atoms co-ordinated to each trivalent cation must be (1/2) (6 oxygens) = 3 oxygens/trivalent cation. As a result, the stoichiometry for each trivalent cation in the dioctahedral O layer can be seen to be MeO_3. However, the formula MeO_3 represents just the stoichiometry of each trivalent cation not the overall dioctahedral layer stoichiometry.

As one can see from Fig. 2, every third cation site in the dioctahedral sheet is vacant. Therefore, the repeat unit of the dioctahedral sheet will be the sum of the two cation sites plus a vacant site. However, the formula for each cation site, MeO_3 was derived by considering all three sites: two cation-filled plus one vacant site. Hence, the stoichiometry for the dioctahedral sheet repeat unit can be determined as the sum of the formulas for two trivalent cation sites since the repeat unit for the dioctahedral sheet contains two trivalent cations to give $Me(III)_2O_6$. The net charge on the repeat unit of the diotrahedral O layer with two trivalent cations and six oxygen atoms is simply $2(+3) + 6(-2) = -6$ or $Me\ (III)_2O_6^{6-}$.

For the trioctahedral layer structure shown in Fig. 2, each oxygen atom is co-ordinated to three divalent cations, Me^{2+}, so that stoichiometrically 1/3 of each oxygen atom must be ascribed to the divalent cations to which it is co-ordinated. Since each divalent cation can be seen to be co-ordinated to six oxygen atoms, the net number of oxygen atoms stoichiometrically co-ordinated to each divalent cation must be (1/3) (6 oxygens) = 2 oxygens/divalent cation. As a result the stoichiometry for each divalent cation in the trioctahedral O layer can be seen to be MeO_2. The repeat unit for the trioctahedral layer, to be consistent with the dioctahedral layer, is determined as the sum of three divalent cation sites to give $Me(II)_3O_6$. The net charge on the repeat unit of the trioctrahedral O layer stoichiometric formula, $Me(II)_3O_6$, with three divalent cations and six oxygen atoms is simply $3(+2) + 6(-2) = -6$, or, $Me(II)_3O_6^{6-}$.

The repeat units determined above [tetrahedral, $Si_2O_5{}^{2-}$, dioctahedral, $Me(III)_2O_6^{6-}$ and, trioctahedral, $Me(II)_3O_6^{6-}$] also represent the stoichiometric combining units for clay minerals. By combining these repeat units, clay minerals can be constructed in two steps: (1) substituting the apical oxygens on the Si tetrahedral sheet for oxygens on the octahedral sheet followed by (2) substituting hydroxyls for oxygens in the octahedral sheet. In the first step, two apical oxygens are removed from $Si_2O_5^{2-}$, resulting in $Si_2O_3^{2+}$ units in the tetrahedral layer. When the $Si_2O_3^{2+}$ unit is combined with a trioctahedral or dioctahedral unit the result is $Me_3Si_2O_5(O_4)^{4-}$ or $Me_2Si_2O_5(O_4)^{4-}$, respectively. In the second step the four oxygens in parentheses are replaced with four hydroxyl groups (or four hydrogen ions are combined with the four oxygen atoms in parentheses) to produce either $Me_3Si_2O_5(OH)_4$ or $Me_2Si_2O_5(OH)_4$, respectively.

If the cation in a dioctahedral sheet is trivalent aluminum the mineral is kaolinite with the formula $Al_2Si_2O_5(OH)_4$ Fig. 3. If the cation in a trioctahedral sheet is magnesium, the resulting mineral is serpentine, $Mg_3Si_2O_5(OH)_4$. Because of the structural mismatch between the Mg octahedral layer and the Si tetrahedral layer these minerals are often tubular or fibrous, and in many cases carcinogenic if inhaled in substantial quantities.

If instead, two Si tetrahedral sheets, each with two apical oxygens removed, $Si_2O_3{}^{2+}$, were combined to sandwich a dioctahedral Al layer, $Al_2O_6^{6-}$, the resulting structure would be $Al_2Si_4O_{10}(OH)_2^{2}$. When the two oxygen atoms in parentheses are converted nto hydroxyls, the clay mineral, pyrophyullite, with the formula $Al_2Si_4O_{10}(OH)_2$ results. Similarly, if two Si tetrahedral sheets, each with two apical oxygens removed, $Si_2O_3{}^{2+}$, are combined to sandwich a trioctahedral Mg layer, $Mg_3O_6^{6-}$, the mineral talc, with the structure, $Mg_3Si_4O_{10}(OH)_2$, results once two oxygens are converted into hydroxyls. Kaolinite and the serpentines, because they are a combination of one Si tetrahedral layer and one octahedral layer, are often also referred to as 1:1 clay minerals. By the same token, pyrophyllite and talc are 2:1 clay minerals. Note that in each case the clay mineral is completely charge-balanced.

TO or TOT packages propagate the lattice in the direction perpendicular to the sheet through relatively weak van der Waals' bonds. Unsatisfied charges only occur where the lattice is truncated – at the edges. The chemical reactivity of this region appears to be similar to that of the component oxides. In essence, pH-dependent charge arises from the interaction of protons and hydroxyls (and charge-balancing

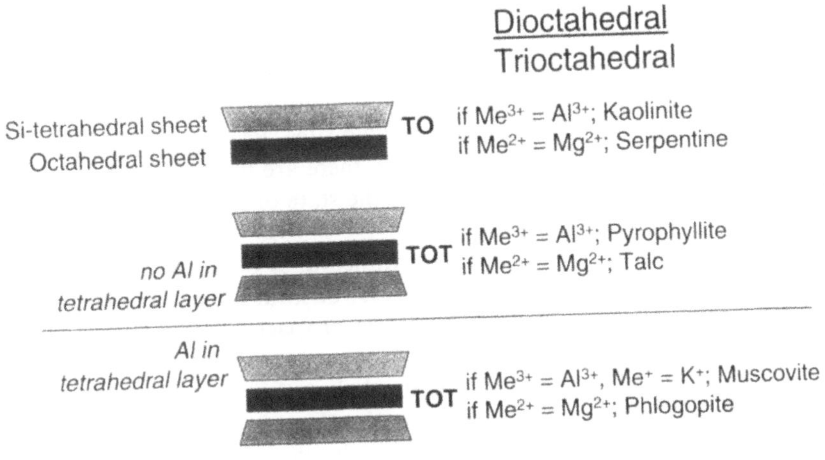

FIG. 3 Common clay mineral structures.

counter ions) with these truncated edges, similar to that which is seen for truncated (hydr)oxide mineral surfaces (see below).

Further complexity arises though due to the substitution of Al for Si in the tetrahedral layer, or where the sum of the cations in the octahedral layers fails to satisfy the charge requirement for an ideal O-type layer. These substitutions result in a "permanent" negative charge because the charge depends on the chemistry of the crystal, as opposed to pH and ionic strength. Although "permanent," such a site still requires charge balance, and this balancing gives rise to much of the most interesting characteristics of clays that will be covered in considerable detail below. The negative charge is typically balanced by the incorporation of a univalent cation, such as K^+ or Na^+ into the clay structure. For example, paired substitution of aluminum and potassium into a pyrophyllite makes muscovite, $KAl_2(AlSi_3)O_{10}(OH)_2$. Paired substitution of aluminum and potassium into talc makes phlogopite, $KMg_3(AlSi_3)O_{10}(OH)_2$. The parentheses in the two chemical formulas specifically indicate that Al is on the tetrahedral sites.

A number of additional clay types require definition before proceeding. Smectites are TOT clays wherein substantial negative charge arising from heterovalent substitution resides in the octahedral layer. Typically, the negative charge is balanced by hydrated monovalent cations in between the layers. In general, the fact that the negative charge on a smectite is somewhat smaller than, say, on a mica, the layers can be easily separated to allow the passage of hydrated cations. Chlorites are made up of TOT stacks that together possess a negative charge that is ultimately balanced by the presence of an intervening metal hydroxide layer [commonly brucite – $Mg(OH)_2$]. The presence of the metal hydroxide layer accounts for the low cation exchange capacity of chlorites and their low tendency to hydrate. Illites are equivalent to hydrated and poorly crystalline micas. Their sizable negative charge is satisfied primarily by K^+, which tends to hold the layers together and prevent swelling.

In contrast to the charge inequalities found at the edge of the crystal, substitutional charge deficiencies extend throughout the lattice. To accommodate this distribution the compensating cations must somehow be distributed throughout the interior of the crystal. Most of the ions involved are too large to fit in either the tetrahedral or octahedral sites of the T and O layers. Instead, they reside between the faces of the sheets in such a way that the overall electrostatic energy of the arrangement is minimized. Such cations are referred to as "interlayer" cations and the positions they occupy are called basal sites. It is reasonable to assume that monovalent cations lie roughly opposite the charge imbalance on the interior of the lattice. Divalent cations must, however, straddle at least two sites. Because, in actuality, a two-dimensional array of charge imbalances exists, there are many possible configurations for the interlayer cations that depend on the state of hydration of the clay and the proportions of the various ions present.

Whether the charge balance occurs in the tetrahedral or octahedral site plays a late role in how clays behave in hydrous systems. A tetrahedral site charge imbalance in a 2:1 (i.e., TOT) clay is relatively close to the interlayer space. Thus, it has a strong electrostatic attraction for the interlayer cation. At room temperature, K^+ (or Rb^+ and Cs^+), but not Na^+ or Ca^{2+}, may over time slowly lose their waters of hydration and allow the layers to move relatively close to each other. At this point the interlayer cations cannot be readily exchanged. This is why illite has an ion exchange capacity significantly smaller than that of the various smectites even though a bulk chemical analysis of the mineral will typically yield more equivalents of K than the total exchange capacity of the typical smectite. When the charge deficiency originates in the octahedral layer (e.g., smectites), the interlayer spaces remain open so that one charge compensating cation can be exchanged for another.

In short, the surface reactivity of clays arises from two distinct causes: (1) exchangable cations on interlayer sites that are incorporated to maintain an overall neutral charge on the crystal lattice, and (2) the broken bonds (e.g., discontinuities) that exist on the edges when the periodic arrangement of atoms is terminated.

III. ION EXCHANGE CAPACITY

Ion exchange, measured in the laboratory, monitors two processes, adsorption on to edge sites and true ion exchange at basal plane sites. We differentiate the two because the crystallochemical origins are quite different. The first arises due to the breaking of bonds. The latter occurs as a result of heterovalent substitution. For edge sites, the identity of the metal cation in the mineral structure is an important determinant of the ensuing sorption reaction; for the case of true ion exchange the identity of the sorbing metal is more important (see below). Ion exchange capacities for common clays are shown in Table 2. The relatively low values for kaolinite, halloysite, illite, and chlorite all probably represent mostly edge reactivity of a well-crystallized fraction, while the higher values probably represent measurements made on minerals that departed significantly from the ideal structure by incorporating an appreciable portion of smectite-like layers possessing appreciable basal plane charge. For vermiculite and smectite it is likely that the broken bonds on the edges only account for roughly 20% of the exchange capacity. In contrast to interactions with the broken bonds on the edges, the actual charge deficiency may lie buried in the interior of the lattice and not at a single exposed oxygen. One result is that the potential for hydrogen bonding is much reduced and the bonds to the charge compensating cations are essentially electrostatic. For the hydrogen ion to compete with the normal groundwater cations in such a setting it must be present in similar concentrations. Typically though this requires inordinately low pH values that would cause the octahedral and tetrahedral sheets themselves to be destabilized and dissolved. Thus, while the displacement of hydrogen ions plays a central role in understanding ion

TABLE 2 Cation Exchange Capacities of Common Clays

Clay mineral	Exchange capacity (meq/100 g)
Kaolinite	3–15
Halloysite.$2H_2O$	5–10
Smectite	80–150
Illite	10–40
Vermiculite	100–150
Chlorite	10–40

Source: Ref. 1.

exchange reactions on edge sites, the hydrogen ion plays almost no role in the exchange reactions involving interlayer cations. From here forward, ion exchange will specifically refer to interactions occurring on basal planes resulting from heterovalent substitution in the lattice.

A. Edge Reactivity

The broken edges of clay minerals are typically assumed to possess roughly the same chemical properties of their (hydr)oxide equivalents [e.g., 2–5]. For example, it is reasonable to first treat the acid–base characteristics and cation sorption properties of the broken edge of an Si tetrahedral sheet on a clay as being equivalent to those of a pure Si (hydr)oxide, such as quartz or silica. By the same approach, Al octahedra exposed at the edge of a clay mineral might be modeled as behaving equivalently to an Al site exposed at the corundum (or gibbsite)–solution interface, acquiring and donating protons as a function of pH and ionic strength similar to that of the latter interface. Because, in fact, there does appear to be some correspondence between (hydr)oxide surface properties and clay edges, it is useful to restate briefly the chemical controls on (hydr)oxide surface charge.

The acid–base properties of metal (hydr)oxide surfaces is most simply explained with a three-site model [6,7]. This model postulates that three surface complexes of differing charge exist in dilute solutions free of adsorbents other than H^+ and OH^-: $>M\text{–}O\text{–}H_2^+$, $>M\text{–}O\text{–}H_2^+$, $>M\text{–}O\text{–}H$, and $>M\text{–}O^-$, where $>M-$ denotes a surface metal site which is bound to the solid through one, two, or three surface oxygens. Proton transfer reactions between the three can be written as

$$>M\text{–}O\text{–}H_2{}^+ \leftrightarrow >M\text{–}O\text{–}H + H^+$$
$$K_{(1)}[>M\text{–}O\text{–}H]\exp(F\psi/RT)a_{H^+}/[>M\text{–}O\text{–}H_{2^+}] \tag{1}$$

$$>M\text{–}O\text{–}H \leftrightarrow >M\text{–}O^- + H^+$$
$$K_{(2)} = [>M\text{–}O^-]\exp(\text{-}F\psi/RT)a_{H^+}/[>M\text{–}O\text{–}H] \tag{2}$$

In Eqs (1) and (2) bracketed terms are site concentrations in mol cm^{-2} $K_{(1)}$ and $K_{(2)}$ are thermodynamic equilibrium constants which are specific to individual oxides, a_{H^+} is the activity of protons in solution, F is Faraday's constant (96,485 C mol^{-1}), and ψ is the electrochemical potential of the surface (in volts). The exponential term models the work entailed in the formation of the respective charged surface species; R is the gas constant (8.314 J mol^{-1} K^{-1}) and T is temperature (K). The total number of surface exchange sites, $[>M\text{–}O\text{–}H] + [>M\text{–};O\text{–}H_2{}^+] + [>M\text{–}O^-]$, is generally close to the value calculated from the mineral surface area and crystallographic constraints. These range from 1 to 6 sites nm^{-2} ($1.7\text{–}10 \times 10^{-6}$ mol m^{-2}). At the pH of zero charge, pH_{zpc}, where $pH = (pK_{(1)} + pK_{(2)})/2$, the surface has zero net charge [$pK_{(1)}$ and $pK_{(2)}$ represent the negative logarithm of the equilibrium constants of Eqs (1) and (2), respectively]. Representative pH_{zpc} values for a number of solids are shown in Table 3.

Surface charge in dilute solutions free of other multivalent adsorbing species is positive at $pH < pH_{zpc}$ and negative at $pH > pH_{zpc}$. Specific adsorption of multivalent cations on to anionic surface sites causes an upward shift in the observed pH_{zpc}. Adsorption of multivalent anions to cationic surface sites causes an analogous

TABLE 3 Hydr(oxide) pH_{zpc} Values

Hydr(oxide)	pH_{zpc}
SiO_2	~2
TiO_2	4.7
BeO	10.2
β-MnO_2	7.0
α-Fe_2O_3	8.5
α-FeOOH	7.3
α-Al_2O_3	9.5
MgO	12.4

Source: Ref. 7.

downward shift in the pH_{zpc}. Combining Eqs (1) and (2) with edge-species association reactions and with expressions for charge and mass balance provides enough information to calculate the distribution of adsorbed species at the mineral–solution interface, and a number of programs such asMINEQL and HYDRAQL [8] exist to perform such calculations if the necessary surface equilibrium constants and total site densities are available.

The three-site (hydr)oxide surface charge model has been applied repeatedly to describe clay-edge reactivity [2–5, 9, 10]. Note though, that with the exception of kaolinite and pyrophyllite, two clays which have only minor basal plane charge, the edge contribution to surface charge is often outweighed by that on the basal planes in most solutions. Exceptions are in relatively high and/or low pH solutions where all of the edge sites become protonated (or deprotonated) and their contribution becomes visible above the high background caused by basal plane reactions [4, 5, 10].

One complication to the metal (hydr)oxide-clay edge site analogy is seen most clearly for the case of kaolinite. Kaolinite is useful as a model mineral from which to examine clay mineral surface chemistry because: (1) it contains Al in octahedral co-ordination and Si in tetrahedral co-ordination with oxygen like most clays, (2) it has no interlayer cations that can be easily leached and complicate the charge balancing procedure, (3) each (hydr)oxide component exists in two distinct structural environments at the surface (i.e., basal planes and edges), and (4) there is minimal substitution of variable-valence cations; hence, minor permanent structural charge. Figure 4 shows pH-dependent kaolinite surface charge as a function of temperature [9]. The surface charge is measured through a potentiometric titration wherein protons (or hydroxyls) are added to a mineral–solution slurry and the pH is monitored. A mass balance of the protons and hydroxyls in the solution allows the respective net excesses at the surface to be calculated. Apportioning protons over the kaolinite surface as a function of pH then provides some constraints over the integrated surface reactivity of kaolinite. Typically, this is done by testing possible surface reaction stoichiometries for goodness of fit using a program such as FITEQL [11]. To fit best the kaolinite surface charge data, two modifications from a simple model of oxide additivity are required:

1. The exposed edge area is required to be at least roughly one-third of the total surface area.

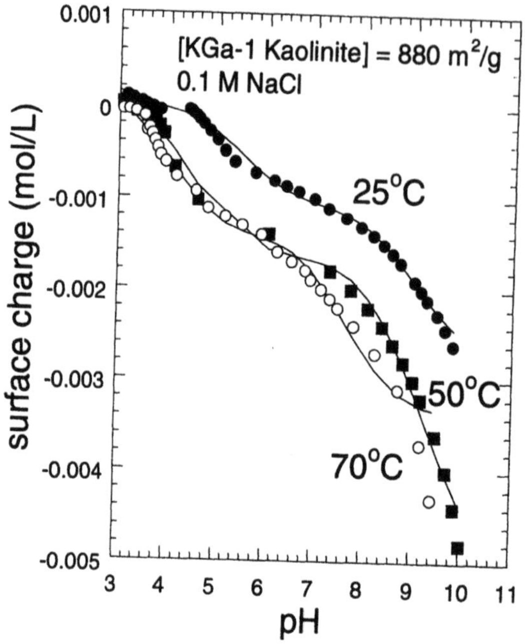

FIG. 4 pH- and temperature-dependent surface charge on kaolinite. (From Ref. 9, with permission from Academic Press.)

2. The Al edge sites were estimated to be a great deal more acidic than those in simple Al (hydr)oxides such as corundum or gibbsite (Si site acidity appears to behave identically to Si sites on pure Si oxides, such as quartz).

Independent evidence pointing to heightened Al site acidity comes from molecular electrostatic potential calculations [9].

1. Metal–Edge Interactions

A complete model of clay mineral edge charge is required to model effectively the interaction of metals with edges as the latter is often appreciably electrostatic, and hence depends on the charge of surface. At low pH, when edges are likely to be neutral, or positively charged, there is less electrostatic interaction between the edges and positively charged metal ions in solution. Instead, metal sorption by edge sites typically occurs at neutral to basic pH, with sorption increasing with increasing pH. Figure 5 illustrates this for the case of cadmium sorption on to kaolinite at 25°C. Sorption on to edge sites is generally modeled as a unidentate reaction:

$$>\text{M–O–H} + \text{Me}^{2+} \leftrightarrow >\text{M–O–Me}^{+} + \text{H}^{+}$$

$$K_{(\text{Me})} = [>\text{M–O–Me}^{+}]\exp(F\psi/RT)a_{\text{H}^{+}}/[>\text{M–O–H}][\text{Me}^{2+}] \quad (3)$$

Recall again that the contribution of any pH-independent exchange reactions must be accounted for to understand the clay–mineral interaction completely. Figure 6 shows Sr adhesion to kaolinite. Note that the sizable and relatively pH-independent fraction of Sr sorbed at $4 < \text{pH} < 8$ probably indicates exchange on to basal planes

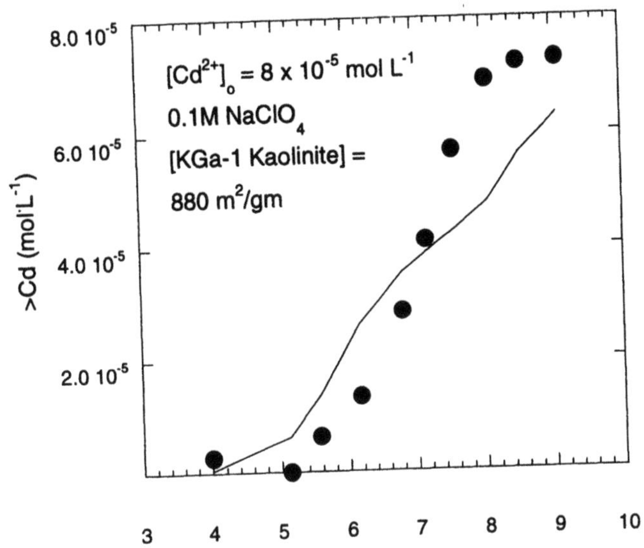

FIG. 5 pH-dependent cadmium sorption. The line is the behavior predicted using as input a unidentate binding of Cd^{2+} with Al sites on the surface. (From Ref. 12.)

(kaolinite has a minor, but measurable, basal plane charge that may be caused by smectite impurities). The pH-independent sorption seen at pH > 8 is most reasonably ascribed to edge interactions resulting from the pH-dependent reactivity of the latter.

A thorough examination of the respective contributions of edges and basal planes to Cd sorption by smectites is that of Zachara et al. [5] who examined sorption on to SWy-1, a smectite routinely used to model soil smectite behavior, as well as a number of soil-derived smectites. The cation exchange capacity, as measured by ^{22}Na isotopic exchange of the various specimens, is shown in Fig. 7. Note, in parti-

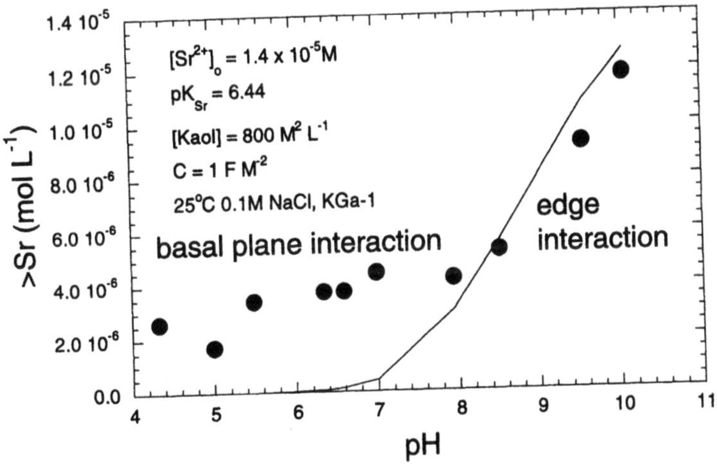

FIG. 6 Sorption of Sr^{2+} to edge and basal sites.

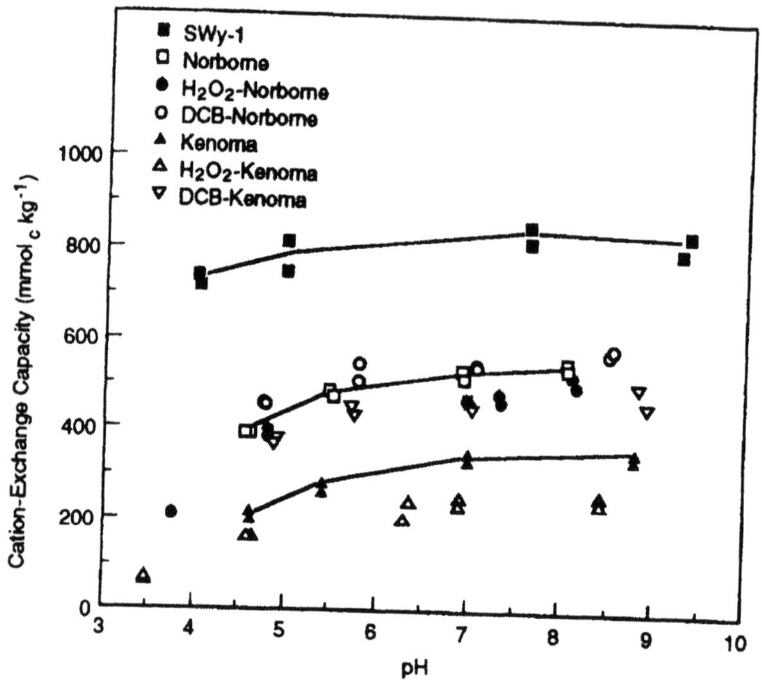

FIG. 7 Cation exchange capacity of specimen, soil, and treated soil smectites. (From Ref. 5, with permission.)

cular, the absence of substantial pH dependence, pointing to the very sizable basal plane contribution. Figure 8 shows sorption on to SWy-1 as a function of pH and Na^+ and Ca^{2+} content of the solution. High Na levels decrease Cd^{2+} interaction from pH 4.5 to 6.5, presumably by competing for basal plane sites. Calcium apparently has an even greater ability to displace cadmium. The line in Fig. 8 is the calculated fraction of sorption attributable to edge sites.

B. Ion Exchange

The ability of ions to displace previously sorbed/exchanged ions is central to the role of clays as nutrient reservoirs and contaminant sinks. For this reason, ion exchange on to clays has been studied since the inception of modern chemistry. Way (1850, 1852) concluded that the relative ability of one ion to displace another from a soil followed the general order: Na < K < Ca < Mg < NH_4. Gedroiz [13] found that using 0.01 N solutions of various ions to displace Ba from a Chernozem soil resulted in the following sequence: Li < Na < NH_4 < K < Mg < Rb < Ca < Co < Al. Two regularities appear in this sequence:

1. For both alkali and alkali earths, the larger the cation, the better is its ability to displace a smaller ion of the same charge.
2. For elements in the same row of the periodic table, the greater the charge, the greater is the ability to displace cations of a lower charge.

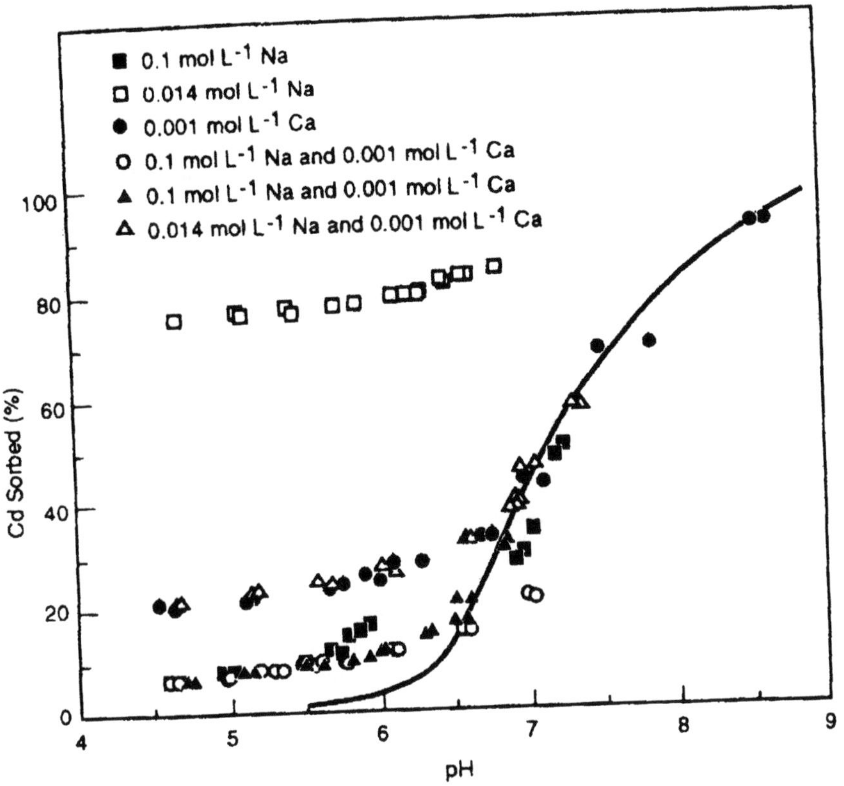

FIG. 8 Cd sorption onto SWy-1 smectite as a function of solution composition. (From Ref. 5, with permission.)

One early study on individual clay minerals [14] revealed that, at equal equivalence, a similar sequence was observed for the alkali ions (e.g., Li < Na < K < Rb < Cs). However, all the alkali earths behaved about the same at levels below 100 meq/L and were slightly more effective than K in displacing the ammonium that originally occupied the exchange sites on the clay. For kaolinite the same pattern was observed, though at lower levels of 30 meq/L. For cations of a given charge the larger cations tend to exchange most readily on to basal planes. The larger radius is inversely correlated with the degree of hydration, and the hydrated radius of a large cation is typically smaller than that of a small cation. Because ion exchange is appreciably electrostatic, and binding depends on the charge density of the solute, larger cations are taken up first.

Although these generalizations are useful for making predictions when "all other things are equal," further investigations revealed that, when concentrations differed or different clays were compared, there was not a general order that could be applied across the board, pointing up the need for a broader mathematical treatment of the problem. The simplest treatment of these follows the by now familiar "mass action" formalism already discussed above. The basic framework considers a series of reaactions of the form:

$$aN^{b+} + bMX \leftrightarrow aNX + bM^{a+} \tag{4}$$

For each reaction there is a corresponding equilibrium expression:

$$[NX]^a[M^{a+}]^b/[MX]^b[N^{b+}]^a = K_{ab} \tag{5}$$

Species in brackets are thermodynamic activities. For the aqueous species the standard state is a 1 molal solution referenced to infinite dilution. For exchange sites, X, activities are typically set equal to the fraction of exchange sites occupied. If all exchange sites were occupied by N^{b+}, [NX] would equal one. If no sites were occupied, [NX] would equal zero. In other words, the sum of all of the exchange sites must equal one. Since each *pair* of interacting species produces only a *single* equilibrium expression, an additional equation is needed if the system of equations is actually to be solved to predict the equilibrium state of a particular system. This equation is the mass balance expression that states that the total site occupancy equals the cation exchange capacity:

$$\text{CEC} = a\text{MX} + b\text{NX} + \cdots \tag{6}$$

If many species are involved the algebra becomes complicated and a computer is typically used to calculate the distribution of species at the clay surface and in solution.

Even in a simplified form, however, this approach to modeling such systems can yield informative results. Consider, for example, the relative behavior of a Na/K exchange pair as compared to a Na/Ca exchange pair. The two equilibrium reactions can be written as

$$[K^+][NaX]/[KX][Na^+] = K_{Na/K} \tag{7}$$

$$[Ca^{2+}][NAX]^2/[CaX][Na^+]^2 = K_{Na/Ca} \tag{8}$$

In the case of the Na/K exchange, Eq. (7) implies that the relative amounts of Na and K exchanged on to the clay should depend *only* on the ratio of the activities of Na and K in solution. Since activity coefficients for Na^+ and K^+ change in a similar manner with increasing ionic strength, it follows that, over a broad concentration range, the ratio of sorbed Na to sorbed K should not change so long as the concentration ratio of Na to K in solution remains fixed. This is a fairly safe approximation, and Eq. (7) works because the charge on the ions exchanging is the same. Nevertheless, deviations (termed nonideal behavior) appear when, for example, two monovalent cations exchange for a divalent cation. For this reason, exchange equations are often modified to include a selectivity coefficient explicitly to account for deviations from ideality. Typically, selectivity coefficients approach unity when the ion exchangers are of the same charge and of similar size. Selectivity coefficients (f_i) are larger if charge and/or radius mismatches are involved, and are used to "correct" the site occupancy term in the exchange equilibria. For example, for the case of the K–Na exchange, the exchange constant in Eq. (7) becomes

$$[K^+]f_{Na}[NaX]/f_{KL}[KX][Na^+] = K_{Na/K} \tag{9}$$

There is a specific preference for divalent cations over monovalent cations in ion exchange that is less apparent when the ionic strength of the solution is great. Figure 9 shows that, for an ion exchanger clay in a solution having a fixed ratio of divalent to monovalent cations, the number of cation-occupied sites decreases with increasing ionic strength. This follows directly from the exchange equations outlined above.

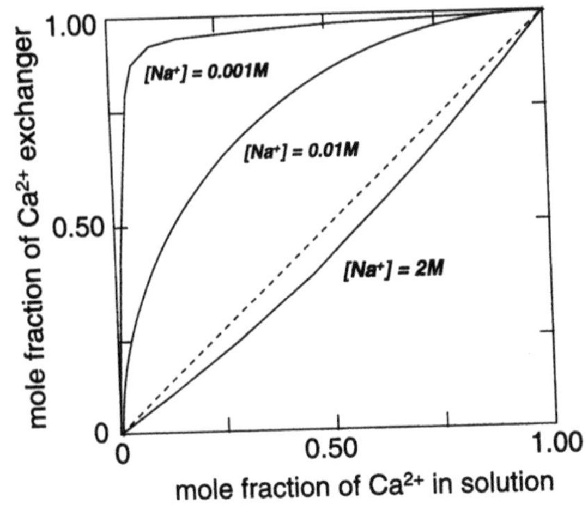

FIG. 9 Ion exchange isotherms describing Na/Ca exchange as a function of Na content. (From Ref. 7, with permission from John Wiley & Sons.)

Using as an example the Ca/Na exchange reaction in Eq. (8), and rewriting the equation in terms of mole fractions of Na and Ca in solution, X_{Na^+} and $X_{Ca^{2+}}$:

$$K_{Na/Ca} = [Ca^{2+}][NaX]^2/[CaX][Na^+]^2 = [X_{Ca^{2+}}]M_t[NaX]^2/[CaX][X_{Na^+}]^2M_t^2 \\ = [X_{Ca^{2+}}][NaX]^2/[CaX][X_{Na^+}]^2M_t \quad (10)$$

where M_t is the total sum of exchangeable ions in solution. One implication of Eq. (9), termed the "constant-charge rule," is that the distribution of heterovalent cations on clay interlayers depends on more than simply the ratio of one to the other in solution. Specifically, the total sum of exchangable ions in solution must be considered as well. One consequence of the constant-charge rule is that the ion exchange sites on clays in contact with low ionic strength groundwaters typically contain a greater proportion of divalent cations. In contact with high ionic strength brines or ocean water, monovalent cations predominate on the clay exchange sites. Consequently, when river-borne clays reach the ocean they typically donate divalent cations (primarily Ca^{2+}) to the ocean before they flocculate. Flocculation, a process that will not be covered much further here (see Ref. 7 for details), occurs when either: (1) an increase in the salinity of the adjacent solution causes a compression of the electric double layer at the clay–solution interface, allowing Van der Waals' interactions to hasten the joining of dispersed solids into larger, more easily settled ones; or (2) multivalent cations bridge and join dispersed particles that subsequently settle.

IV. ADSORPTION OF ORGANIC MATTER

Organic cation exchange on to clay interlayers is, to an extent, largely determined by [15]:

1. Molecular weight – higher molecular weight organic cations sorb more selectively.

2. Hydration energy. Organics that, owing to their molecular configuration, are poorly hydrated, sorb more readily than those that are more hydrated.
3. The identity of the cation that is being replaced.
4. The charge density of the clay mineral itself.

Some clays, because of their high layer charge are sufficiently constrained for space so as to prevent the incorporation of larger organic cations. At the same time, the strongly hydrated nature of the high charge layers may prevent displacement by organic cations.

For many organics that are protonated and possess a positive charge (e.g., amines) the pH of the solution controls the extent of exchange on to the basal plane. At pHs above the pK_a (the negative logarithm of the acidity constant) such molecules will be uncharged and less likely to interact with the permanent negative charge on the basal plane. Instead, maximal adsorption occurs at lower pHs where the protonated form is most common. Note though that, as the selective fraction of ionized groups from an aqueous population of ionized, uncharged groups causes a net buildup of the former at the surface relative to the solution, there is an apparent enhancement of organic protonation at the surface relative to the solution.

The other means by which organic molecules adhere to clays is through: (1) hydrogen bonding to the hydration layers of exchange cations, and (2) by forming ion–dipole linkages between exchange cations and polar functional groups of the organic molecule.

V. K_d VALUES AND ENVIRONMENTAL TRANSPORT

The equations above can be modified to examine specifically the transport of trace contaminants through the environment. Typically, trace levels of contamination do not alter the distribution of the major components sorbed on to the clay. Consequently, the equation describing exchange of a contaminant N can be simplified to

$$[NX]^a/[N^{b+}]^a = K_{ab}[MX]^b/[M^{a+}]^b \tag{11}$$

Since the terms on the right-hand side all are effectively constant, and both terms on the left-hand side are raised to the same exponent, Eq. (10) can be rewritten as

$$[NX]/[N^{b+}] = K_d \tag{12}$$

This K_d or "distribution coefficient" approach fits readily into many reaction/transport codes used for environmental assessments, but is only applicable if the sorption process is actually reversible, and the distribution of the major exchangable components on the clays does not change. The latter point is an important one, neglect of which can lead to serious errors in estimated transport. As a case in point consider the transport of radioactive ^{137}Cs in the Na-rich fluids generated during nuclear waste processing. Tamura [16] measured the K_d describing Cs sorption to five clays. In order of decreasing tendency to sorb Cs, they were; Illite, biotite, hydrobiotite, montmorillonite, and kaolinite. For all minerals, Tamura noted a linear relationship between the log of the Na concentration and log K_d. In general, an increase in Na concentration from 0.01 to 1 M Na caused the K_d values to decrease by about a factor of 70. The decrease in K_d is thought to arise because of increased competition by Na for exchange sites. Beall et al. [17] found that for a suite of trivalent cations (Am, Cf, Cm, Es, La, Sm, and Yb) the K_d decreased by about a factor of 17 for both kaolinite and montmorillonite when the Na concentration was increased from 0.25 to 4 M. This means that the "competing ion" effect is not limited to ion exchange between monovalent cations. In such a case, the use of a dilute system K_d to model the early part of the transport would seriously underestimate the extent of migration. The important point here is that a K_d approach tends to lump together a number of system-specific parameters, particularly the concentration of any ions that might compete with a trace contaminant on ion exchange sites. For this reason it is critical than when K_d values are measured in the laboratory, fluids are used of similar composition to those likely to be found in the field.

VI. DISSOLUTION/GROWTH DURING DIAGENESIS

The changes in clay bulk composition that are caused by ion exchange and edge adsorption at low temperatures are dwarfed by the significantly greater changes that occur when clays are exposed to greater temperatures and salinity during geological burial. The whole process is termed diagnesis (early diagenesis corresponds to 90°–140°C, middle diagenesis goes up to roughly 200°C, and late diagenesis roughly covers the range from 250° to 280°C [18]. The phases formed initially by rock weathering are demonstrably not the most thermodynamically stable for near-surface environments – though kinetics may indeed favor forming poorly crystallized materials with abundant solid solution substitutions. Given the passage of time and increased temperatures these materials start to recrystallize, forming new sheet silicates. At the lower temperatures, ion exchange reactions clearly play important roles, while at higher temperatures the wholesale reconstruction of the lattice dominate the picture.

The literature on clay diagenesis is vast because of the effect of the latter on petroleum production. The generalizations that emerge [19] are: (1) unless destabilizing pore fluids migrate from elsewhere in the subsurface, kaolinites retain their crystallographic make up to the upper limit of middle diagenesis, though significant recrystallization may occur; (2) chlorites behave similarly, with the added twist that over the middle diagenesis range the Fe/Mg ratio in the lattice typically decreases; and (3) smectites undergo a wholesale crystallographic restructuring toward the upper limit of lower diagenesis that clearly exceeds the simple exchange of ions in response to changing pore fluid chemistries. This reforming is complex and reflects both an increase in the amount of charge imbalance arising from Al for Si substitutions in tetrahedral sites and the displacement of other exchangable cations by K. The net effect is that the layers collapse and the CEC decreases. In Al-rich clays the end product is illite. If the original clay was rich in Mg and Fe the result is a regular mixture of saponite and vermiculite layers known as corrensite.

In the geological column, aluminous clays greatly predominate over those rich in Mg and Fe. Thus, the coupled effects of Al/Si substitutions and K enrichment have been most heavily studied for this system and a combination of laboratory and field studies give semiquantitative estimates of transformation rates. Perry and Hower [20] first noted, that, in the thick wedge of sediments along the gulf coast of North America, there was a progressive decrease in smectite with depth that they attributed to an increase in temperature from 50°C (where the disappearance begins) to 100°C (where smectite is 80% converted). A number of experimental studies followed, though to accelerate the rate to a point where experiments could be completed in less than a year hydrothermal conditions, often of 250°C or higher, were required. The hydrothermal results were then extrapolated back down to the temperature of the conditions of interest by using an Arrhenius relation. Eberl and Hower [21] demonstrated that synthetic K-montmorillonite at 80°C would take slightly less than a million years to achieve the 80% collapse noted in the gulf coast sediments; Na apparently slows the transformation rate. Roberson and Lahann [22], showed that montmorillonites from gulf coast sediments in contact with a more realistic solution Na-to-K ratio of 23 could require in excess of 10 million years to achieve the 80% conversion. Calcium and Mg were found specifically to inhibit conversion by, respectively, factors of 10 and 30, relative to Na.

VII. ION HYDRATION AND SWELLING

Recall that interlayer hydration of clays, such as montmorillonite, occurs primarily because of the tendency for exchange cations to bring water molecules with them, and that high charge and/or small radius cations are the most hydrated. Nevertheless, because fewer divalent cations than monovalent cations are needed to satisfy the same amount of permanent negative charge on the basal plane of a clay, the actual amount of hydration-induced swelling observed is less for exchange of divalent cations than for monovalent cations. The drop in permeability of some clay-rich soils when Na^+ exchanges for Ca^{2+} and Mg^{2+} is generally ascribed to this process. Once interlayers exceed several water molecules thick, osmotic forces begin to predominate and changes in the salinity of the adjacent fluid cause water to diffuse out of clay interlayers (if the salinity of the adjacent fluid is high) or into clay interlayers (if the salinity of the adjacent fluid is low), causing, respectively, a decrease or increase in clay bulk volume. In effect, the tendency to swell depends on both the mineral and solution. Smectites that possess high interlayer basal charge are less likely to swell then smectites having less basal charge. The tendency of Na-smectites to swell relative to Ca-smectites is routinely taken advantage of when $CaSO_4$ is added to moderate the swelling of sodic soils.

VIII. ^{137}Cs AND THE KINETICS OF EXCHANGE

Ion exchange reactions are typically very fast, going to completion within seconds if the exchange sites are exposed to solution. The kinetics of ion exchange become important in practical applications when physical obstacles, such as collapsed interlayers prevent the rapid movement of exchangable cations. A classic example of the latter situation, which also illustrates how clay chemistry can impact human and environmental health, is the transport of ^{137}Cs. Cesium-137 is produced during the generation of nuclear power and the production of nuclear weapons. Wide swaths of eastern Europe were exposed to moderately high levels of ^{137}Cs contamination from the radioactive plume generated by the Chernobyl accident (a nice movie of the modeled path of the ^{137}Cs plume can be viewed on the worldwide web at www.jaeri.go.jp/~speed/simulation.html). High levels of ^{137}Cs contamination were also added to a number of soils and groundwaters as a secondary result of the weapons production activities of the US DOE and its predecessors. Cesium-137 decays to ^{137}Ba and emits a β particle. The latter causes tissue damage in humans. Consequently, considerable effort has been expended to minimize the environmental transport of ^{137}Cs or, if possible, to remove it altogether from the environment.

Cesium is arguably one of the simplest of potentially exchangable ions; it is weakly hydrated, forms no insoluble mineral phases, and forms few strong complexes with other ligands. Consequently, its exchange behavior, and in theory, its environmental transport should be reasonably straightforward from a chemical standpoint. In actuality, the subject remains a fertile area of research, primarily because of uncertainties surrounding the interaction between ^{137}Cs and clay minerals.

To begin with, uptake of ^{137}Cs by smectite-containing clays is the primary mechanism by which it is removed from soil solutions and groundwater. Typically though, once ^{137}Cs interacts with a given smectite, extreme measures are required to separate the two. In others words, a hysteresis is observed and sorption occurs much more readily than desorption. Typically rapid uptake on to clays (seconds to hours) is followed by slow uptake of Cs over days to weaks. Typically, K_d values are much higher when the background exchangable cation is Ca^{2+}, as opposed to K^+. Whereas changes in the composition of the adjacent fluid affect the first reaction, the second population of sites remains somewhat independent [23]. This is thought to reflect the bimodal distribution of Cs between exposed edge sites and interior "high energy" sites. The sequence of steps that cause this is schematized in Fig. 10. Absorption of Cs^+ to Al sites exposed at clay edges occurs, presumably through some combination of electrostatic forces [24, 25]. Migration of Cs^+ to sites of permanent, basal charge

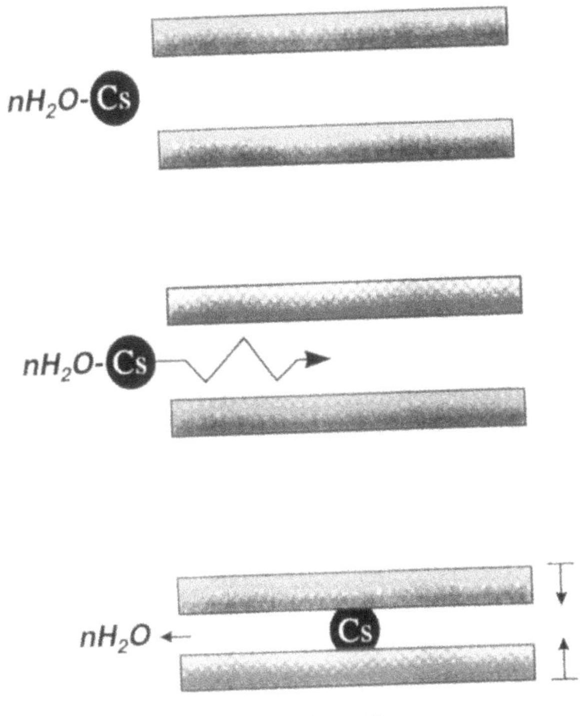

FIG. 10 Cs exchange and clay collapse.

occurs followed by dehydration of the Cs ion and collapse of the clay structure around the ion. This collapse has two important implications for the environmental reactivity of ^{137}Cs: (1) It locks in a long residence time for ^{137}Cs in the interlayers, and, consequently, (2) leaves particle movement as a primary means of ^{137}Cs transport in the subsurface.

Note that ^{137}Cs can be removed from collapsed interlayer sites – it just requires that an ion such as NH_4^+ be the displacing agent. Indeed, in situ techniques for leaching Cs from soils typically rely on the introduction of NH_4^+ into the subsurface as the means for freeing up Cs from clay interlayers. A case can certainly be made for simply leaving the ^{137}Cs-contaminated soils to remediate themselves; ^{137}Cs has a 30-year half-life. Consequently, a wait of several decades will result in severely diminished radiological hazards, assuming no displacement of ^{137}Cs. Obviously, such a scenario would be better if the potential use of ammonia-rich fertilizers for agriculture were avoided.

ACKNOWLEDGMENTS

PVB appreciates the efforts of Professor H. C. Helgeson, the fellow who introduced him to clay chemistry many years ago, as well as the support of DOE/BES-Geosciences and the US Nuclear Regulatory Commission. This work was supported by the United States Department of Energy under contract DE-AC04-94AL85000.

REFERENCES

1. RE Grim. Clay Mineralogy. 2nd ed. McGraw-Hill, 1968.
2. SA Carroll. The dissolution behavior of corundum, kaolinite, and andalusite: A surface complex reaction model for the dissolution behavior of aluminosilicate minerals in diagenetic and weathering environs. PhD thesis, Northwestern University, 1989.
3. P Fletcher, G Sposito. Clay Mineral 24:375–391, 1989.
4. JM Zachara, SC Smith. Soil Soc Am J 58:762–769, 1994.
5. JM Zachara, SC Smith, JP McKinley. CT Resch. Soil. Soc Am. J 57:1491–1501, 1993.
6. PW Schindler, W Stumm. In: W Stumm, ed. Aquatic Surface Chemistry: Chemical Processes at the Particle–Water Interface. John Wiley, 1987, pp 83–110.
7. W Stumm, JJ Morgan. Aquatic Chemistry. Wiley-Interscience, 1996.
8. C Papelis, KF Hayes, JO Leckie. HYDRAQL: A program for the computation of chemical equilibrium composition of aqueous batch systems including surface-complexation modeling of ion adsorption at the oxide/solution interface. Technical Report 306. Stanford University, 1988.
9. PV Brady, RT Cygan, KL Nagy. Colloid Interface Sci 183:356–364, 1996.
10. E Wieland, W. Stumm. Geochim Cosmochim Acta 56:3339–3355, 1992.
11. AL Herbelin, JC Westall. FITEQL – A computer program for determination of chemical equilibrium constants from experimental data. Oregon State University, 1994.
12. PV Brady, RT Cygan, KL Nagy. In: EA Jenne, ed. Adsorption of Metals by Geomedia. Academic Press.
13. K Gedroiz. On the adsorptive power of soils. US Dept of Agriculture, 1922. (Transl. by S. Waksman.)
14. P. Schachtschabel. Kolloid-Beihefte 51:199–276, 1940.
15. MB McBride. Environmental Chemistry of Soils, 1993.
16. T Tamura. Underground Waste Management and Environmental Implications. Vol. Memoir 18: American Association of Petroleum Geologists, 1972, pp 318–330.

17. GW Beal, BH Kettelle, RG Haire, GD O'Kelley. In: S Fried, ed. Radioactive Waste in Geologic Storage. American Chemical Society, pp 201–213.
18. CE Weaver, EV Eslinger, HW Yew. In: CE Weaver, ed. Shale–Slate Metamorphism in the Southern Appalachians. Elsevier, 1984, pp 142–152.
19. CE Weaver. Geothermal alteration of clay minerals and shales: diagenesis. Office of Nuclear Waste Isolation, 1979.
20. EAJ Perry, J Hower. Clays Clay Minerals 18:165–177, 1970.
21. DL Eberl, J Hower. Geol Soc Bull Am 87:1326–1330, 1976.
22. HE Roberson, RW Lahann. Clays Clay Minerals 29:129–135, 1981.
23. RNJ Comans, M Haller, P DePreter. Geochim Cosmochim Acta 55:433–440, 1991.
24. RT Cygan, KL Nagy, PV Brady. In: EA Jenne, ed. Adsorption of Metals by Geomedia. Academic Press.
25. Y Kim, RT Cygan, RJ Kirkpatrick. Geochim Cosmochim Acta 60:1041–1052, 1996.

7
Nonaqueous Liquid–Mineral Oxide Interface

GEIR MARTIN FØRLAND Bergen College, Bergen, Norway

HARALD HØILAND University of Bergen, Bergen, Norway

I. INTRODUCTION

To our knowledge, the first quantitative adsorption study from nonaqueous solvents was carried out by Davis in 1907 [1]. He studied the adsorption of iodine from five different nonaqueous solvents on charcoal. It was concluded that the adsorption was determined by a rapid surface condensation process and a slow diffusion process, and a solid solution diffusion to the interior of the charcoal. The experimental procedure was adsorption by the batch method, analysing the solution before and after adsorption had taken place.

Then, as now, by far the most of the adsorption studies were carried out in aqueous solution. The early studies were largely performed by agricultural chemists, being basically interested in the properties of soil. Theories were developing based on these measurements, and the first adsorption isotherm, sulfuric acid adsorbed by metastannic acid, was published by van Bemmelen [2]. The history of solute–solvent adsorption isotherms has been reviewed by Forrester and Giles [3], where it is also apparent that the vast majority of adsorption studies have been carried out in aqueous solution. However, in modern chemistry some of the most important industrial applications relate to adsorption or adhesion from a nonaqueous environment [4]. The technological applications include electrophoretic image processing, monodisperse colloid production, magnetic tape recording, high-performance ceramic processes, and enhanced oil-recovery processes. In most of these applications it is not adsorption of molecular species that is of interest, but rather nonaqueous colloidal dispersions used to coat various materials. Details can be found in a paper by Novotny (4).

Coating technologies are outside the scope of this work. Here we shall be concerned with normal molecular adsorption studies from nonaqueous solution. Important questions concern the driving force of the various adsorption processes, the topography of the solid surface and how it relates to the adsorption process, the role of the solvent and other additives in the adsorption process, and the possibility of chemical reactions between the solid and the adsorbed molecules. Figure 1 is an illustration of the problems encountered; the topic is divided into successively smal-

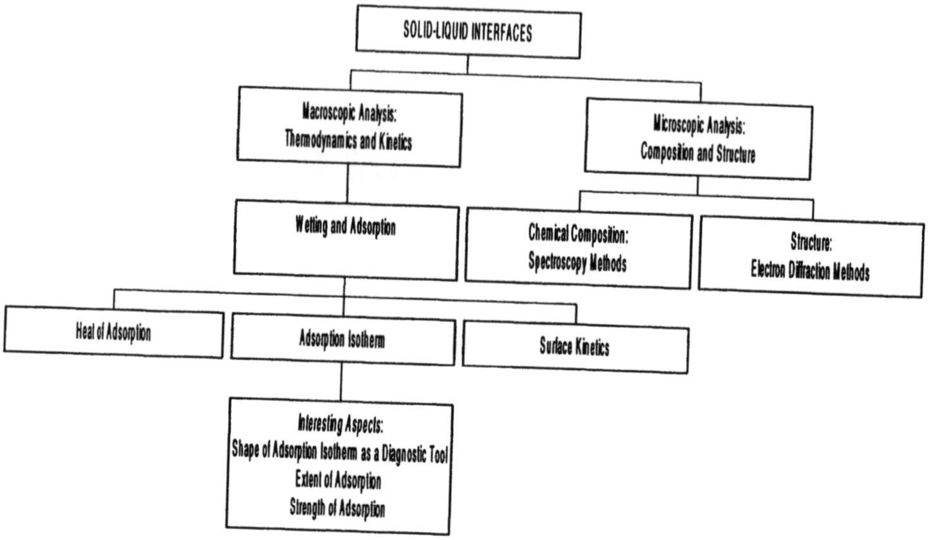

FIG. 1 Diagram showing a dividing into more specific categories of the working area of solid–liquid interfaces.

ler sections of disciplines. Within this framework two main divisions appear. The systems can be treated either microscopically or macroscopically.

Microscopic investigation provides information about the composition and structure of the surface [5]. The structure of the outermost atomic layer can be investigated by several electron diffraction methods like low energy electron diffraction, scanning tunneling microscopy, and field ion microscopy to mention some. Of great interest to surface chemists is also the chemical composition of an interface. Some of the experimental methods used to gain information of the composition includes spectroscopic techniques like Auger electron spectroscopy, low-angle reflection infrared spectroscopy, and secondary ion mass spectroscopy. One technique which provides information of the electronic structure of the surface is electron spectroscopy for surface analysis. This is a specifically designed spectroscopic technique for the chemical analysis of surfaces.

The macroscopic investigation of surfaces is based on chemical thermodynamics, and includes studies of quantities like pressure, volume, surface area, temperature, and chemical composition. The physical aspect of a simple system containing a solid in contact with a pure liquid is manifested in the phenomena of wetting and spreading. However, the majority of practical applications are concerned with systems where two or more components are present in the liquid phase. Therefore, it is necessary to consider the distribution of the components at the interface and relate this information to the observed physical behavior of the system. Study of the adsorption phenomenon is one way to obtain information on the distribution of the various components at the interfaace. Adsorption obviously occurs because it results in a lower free energy of the system, and the main question is what kinds of interactions between the adsorbate molecules and the surface adsorbent molecules are responsible for this lowering. In a broad sense it can be divided into two distinct groups. First, there is the general van der Waals type of interactions, leading to what is termed physical adsorption. The second group involves formation of definite chemical bonds between sorbent and sorbate molecules and is termed chemisorption.

Within this chapter we are mostly concerned with physical adsorption. Our first task is to explore some simple and frequently used equations for adsorption isotherms. We will focus on the acidity/basicity of the adsorbate and adsorbent as a criterion for the adsorbate–adsorbent interaction energy, and make a special case of the adsorption of alcohol molecules from apolar solvents. The alcohol molecules self-associate and form molecular clusters in apolar and inert solvents. This aggregation process may influence the adsorption process and will be encountered when discussing and analyzing the alcohol adsorption isotherms.

II. ADSORPTION FROM DILUTE SOLUTIONS

In this section, we intend to present some simple and straightforward aspects of adsorption of nonelectrolytes from binary liquid solutions on to mineral surfaces. Owing to the vast literature on this subject, we limit the presentation to a description of some frequently observed adsorption isotherms applicable to experimental data represented by the adsorption of polar molecules on to mineral oxide surfaces from dilute nonaqueous solutions.

A common experimental approach is to determine the degree of adsorption from batch methods. These methods are based on analyzing the adsorbate solution before and after adsorption has taken place, thus determining the decrease in adsorbate bulk concentration. The number of moles of solute species adsorbed per gram of adsorbent is often denoted q_2, and it is given by $(\Delta C_2 V/w)$. Here, ΔC_2 is the change in concentration of the solute due to adsorption, V is the volume of the solution, and w is the weight of the adsorbent in grams; q_2 is called the surface excess of component 2. At constant temperature, $q_2 = f_t(C_2)$ is called the adsorption isotherm function. Knowledge of this function is often valuable in practical adsorption technology, and several simple analytical expressions have been presented to describe the adsorption isotherms.

The most common type of adsorption isotherm is that known as the L-isotherm. The frequently used Langmuir adsorption equation describes this functional form [6]. This simple isotherm function builds on three important assumptions. The adsorbed molecules are randomly localized on to equal energy sites on the surface of the adsorbent. Furthermore, the adsorption occurs in a monolayer, and one assumes that no lateral interactions take place among the adsorbate molecules. Although the Langmuir adsorption equation was derived for gas adsorption, it has frequently been used for adsorption of nonelectrolytes from dilute solutions [7–9]. The physical significance of this model is likely to be of limited value due to violation of the assumptions on which the Langmuir isotherm is based. Badiali et al. [10] present a related lattice model developed for adsorption of nonelectrolytes from solutions. In this model, the adsorbed monolayer is regarded as an ideal two-dimensional solution consisting of a mixture of equally sized solute and solvent molecules [11]. Lateral interactions among the molecules in the monolayer film cancel if one assumes that this interaction is independent of composition in the adsorbed monolayer. For solute interaction A and solvent interaction B, such an adsorption process can be written as: A (solute in solution, n_2) + B (adsorbed solvent, n_1^s) = A′ (adsorbed solute, n_2^s) + B′ (solvent in solution, n_1).

The equilibrium constant is

$$K = x_2^s a_1 / x_1^s a_2 \tag{1}$$

where a denotes the activity in the solution, and x^s denotes the mole fraction at the surface. Since the treatment is restricted to dilute solutions, a_1 is regarded as a constant, and one obtains $b = K/a_1$. This gives

$$x_2^s = ba_2/(1 + ba_2) \tag{2}$$

Since x^s is the mole fraction of surface sites occupied by solute molecules, Eq. (2) can also be written as

$$\theta = ba_2/(1 + ba_2) \tag{3}$$

Here, θ is the fraction of the surface sites on the adsorbent occupied by solute molecules. This equation has an analytical expression identical to the Langmuir equation.

The number of moles adsorbed per gram adsorbent, q, is proportional to θ with a proportionality constant equal to n^s. Therefore, q can be written as

$$q = n^s ba_2/(1 + ba_2) \tag{4}$$

This equation can be directly fitted to the experimentally determined isotherm data in the dilute region by replacing a_2 with C_2, the solute concentration. The constants can be determined by linearizing Eq. (4) in the form:

$$q/C_2 = n^s b - bq \tag{5}$$

A plot of q/C_2 against q should give a straight line with the slope $-b$, and the intercept $n^s b$. At sufficiently high solute concentrations, q approaches the limiting value n^s, which describe the adsorption capacity. The term b determines the initial slope of Eq. (4) and is a measure of the intensity of the adsorption or the adsorbent–adsorbate interaction energy.

Quite often, Eq. (5) will be slightly curved, normally convex to the axes. This indicates that the Langmuir model assumptions, particularly the coverage independence of the heat of adsorption, are not fulfilled. For such heterogeneous surfaces, b varies systematically with the adsorption coverage. The adsorption isotherm equation can then be written:

$$F(C_2, T = \int_0^{\infty} f(b)\theta(C_2, b, T)\,\mathrm{d}b \tag{6}$$

Here, $f(b)$ is an unknown distribution function for b, and $\theta(C_2, b, T)$ is the adsorption isotherm function.

In a special case, when the variation in b is attributed entirely to the variation in the heat of adsorption, the solution to the Langmuir adsorption equation can be reduced to the Freundlich adsorption equation [12]:

$$F = aC_2^{1/n} \tag{7}$$

The Freundlich equation, unlike the Langmuir equation, does not become linear at low concentrations, but remains convex to the concentration axis. It does not show a saturation or limiting value. For adsorption of nonelectrolytes from nonaqueous solutions on to solid surfaces this empirical equation is limited in its usefulness. It will normally not fit data in the dilute concentration range. In a practical sense this means that if the data fit the Freundlich equation, it is likely, but not proven that the surface is heterogeneous.

III. OXIDE MINERALS AND ORGANIC DISPERSIONS

The minerals can be classified in various ways, for instance in terms of chemical composition, in terms of crystal lattice, crystal size, and porosity, or in terms of surface polarity. Chemical composition has been the most common basis for the classification. However, the development of X-ray diffraction measurements in the first half of this century made it possible to determine crystal structures, and a finer classification of the minerals was developed based on chemical composition as well as the atomic structure of the mineral.

For the oxide compounds, the oxygen is bound to one or more metal atoms. They are relatively hard and dense, and the bond type in the structure is ionic. Surface hydroxyl groups are formed by the reaction of the oxygen atoms on the mineral surface with the atmospheric moisture. The hydroxyl groups are amphoteric in nature. This means that a strong base can complex with the acidic proton, or the oxygen can be basic in reacting with an acid. The density of these surface hydroxyls determines the relative basicities of minerals. The acidic/basic properties can also be inferred from the isoelectric point (IEP) of the minerals in aqueous solutions. The lower the IEP, the more acidic is the mineral. Silica has a low density of hydroxyls (3–4 OH/nm^2) and is an acidic oxide with an IEP value as low as 2.5 [13]. On the other hand, alumina has a high density of hydroxyls (10–15 OH/nm^2) and is a basic oxide with an IEP value as high as 8.5–9.0 [13]. Even though the surface protonation may be absent when the mineral oxides become dispersed in organic solvents, the acidity/basicity of the mineral is a highly important property for the adsorption process, and it often contributes to the formation of stable surface complexes at the hydroxyl groups.

The adsorption properties for hydrous oxides are highly dependent on the moisture content, and thus to the degree of heating. Up to 10 molecular layers of water may exist on kaolinite at high humidities [14]. Part of the bound water will evaporate upon heating, and careful drying of the mineral should normally be carried out before any adsorption study. The infrared (IR) spectra of kaolinite at the OH stretching wavenumber region and at various temperatures can be seen in Fig. 2. The lowering of absorbance with increasing temperatures represents the loss of H-bonded water molecules.

Some naturally occurring mineral oxides exhibit a high degree of isomorphous substitution of metal ions. These complex oxides make the interpretation of the adsorption study very difficult, and a detailed pretreatment of the mineral is often required before any adsorption study. With inert nonaqueous solvents, the need for a detailed pretreatment of the mineral is substantially reduced. This can be explained from the fact that the protonation and deprotonation of the mineral surface in aqueous solutions must be controlled by adjusting the pH, while this adjustment is not necessary in most organic solutions due to the lack of surface protonation. On the other hand, the molecular behavior of many amphiphilic solute molecules in

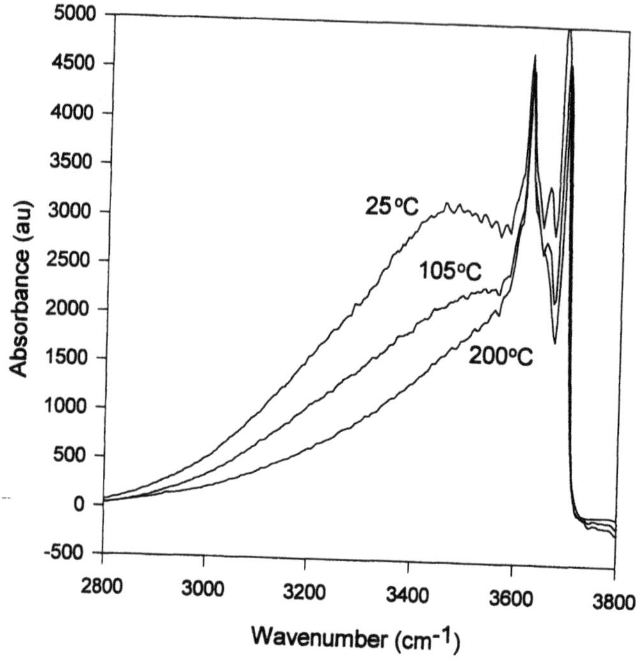

FIG. 2 Infrared absorption spectra of kaolinite showing the loss of bonded water at the mineral surface as the temperature is increased.

apolar solvents is often highly complex, and a more detailed knowledge of the solution behavior is therefore required in order to obtain a more reliable interpretation of the adsorption isotherm.

IV. EVALUATION OF ADSORPTION PROCESSES

The nature of the adsorbent and the solute molecules and their mutual interaction play an important role concerning the adsorption strength, which is often related to the orientation and structure of the adsorbed layer. Moreover, the solvent may also strongly influence the adsorption process. It can weaken the adsorption of solute molecules and it can compete with the solute molecules for available sites on the adsorbent. In order to evaluate the strength of adsorption between the hydroxyl functional groups on a mineral oxide surface and an organic probe molecule, one can use the Lewis concept of electron acceptor (Lewis acid) and electron donor (Lewis base). A variety of physical properties have previously been proposed as measures of acidity and basicity of the components involved in the system [15–17]. The most frequently used properties describing the relative rankings of solvent acidity–basicity are the donor–acceptor approach of Gutmann [17]. The electron donor number (DN) is an indication of the tendency of a molecule to give away electrons in a reaction between donor and acceptor molecules. Likewise, the electron acceptor number (AN) is the tendency of a molecule to accept electrons in a corresponding reacation. Thus, the donor–acceptor approach as used in adsorption is based on the separation of electric charges that takes place at the interface when a liquid adsorbate molecule comes in contact with a solid surface. An electron pair and an empty orbital can give rise to a bond that holds together a donor–acceptor complex. The ability of any surface to interact specifically with functional groups on a probe molecule through acidic and basic sites can then be estimated from their respective AN and DN values. The use of the donor–acceptor approach in solid–liquid interactions gives an interesting aspect for improved understanding of the adsorption processes. It gives us the ability to choose conditions for desired bond polarities within a functional group by applying an appropriate molecular environment, and it offers a method to predict the adsorption strength of a given system.

Both the DN and AN values are determined empirically. The electron DN is taken as the molar enthalpy value for the reaction of the donor species with antimony pentachloride ($SbCl_5$) as a reference acceptor in a 10^{-3} molar solution of 1,2-dichloroethane. The DN values range mainly from 0 to 60 kcal/mol. The AN values are dimensionless numbers related to the capacity for sharing electron pairs from standard donating molecules. The numbers can be determined from ^{31}P NMR chemical shift measurements in triethylphosphine (Et_3PO) upon addition of the respective acceptor. The normalized AN scale ranges from 0 for hexane to 100 for 1:1 components formed between $SbCl_5$ and (Et_3PO).

The electron donor and acceptor properties of minerals are often determined using zetametry or inverse gas chromatography. In zeta-potential measurements, the particles are submitted to a constant electric field. One measures the particle migration, which depends on the value of the electric charge acquired by the particle in the liquid medium, and it is considered to be directly proportional to the donor–acceptor properties of the solid surface. Thus, this concept considers the acid–base interaction to be predominant. The relative ranking of particles based on acid–base properties has been done by measuring the point of zero charge (PZC). Labib and Williams [18] measured zeta potentials in a series of organic liquids in order to determine the donor–acceptor properties of several solids. Siffert et al. [19] have published a method which obtains a direct measure of the DN and AN values of solids from zeta-potential measurements in a series of organic liquids with different donicities.

Inverse gas chromatography provides accurate thermodynamic data for solid surfaces [20]. It can be used to determine the DN and AN values of oxides and to establish adsorption isotherms [20, 21]. In this method, the solid is placed in a gas chromatographic column as the stationary phase, and probe gases are introduced into an inert carrier gas. The measured quantity is the retention time, which is related to the adsorption interaction energy and the equilibrium between adsorbed and nonadsorbed gas molecules.

A qualitative study of the adsorption, in order to determine the functional group which takes part in the adsorption process, can be done by IR spectroscopy measurements. The OH stretching vibration of an hydroxylated surface serves in prinicple as a spectroscopic probe of the molecular environment, and IR spectroscopy can be used to determine the chemical functionality and the donor acceptor properties of an oxide surface upon adsorption of various probe molecules. Noller and Kladnig [22] found that the changes in the OH valence frequency values ($\Delta\nu_{OH}$-values), upon adsorption of a variety of organic donor molecules on to acidic silica, were linearly related to the donor number of the adsorbate.

V. ADSORPTION OF AMPHIPHILIC MOLECULES FROM DILUTE NONAQUEOUS SOLUTIONS

The functional form of an adsorption isotherm will serve as a diagnostic tool for elucidating some characteristics of the adsorption process, i.e., the initial slope of the isotherm serves as a measure of the adsorbent affinity for the adsorbate molecules, and any plateau section gives the adsorption capacity of the mineral toward the adsorbate molecules.

Thus, some important aspects to consider include the extent to which solvent is adsorbed, the significance of the adsorption limit or plateau sections found in most isotherms, the orientation of the adsorbed molecules, and the existence of physical and/or chemical adsorption. One should be aware that the data obtained from studies of molecular adsorption represent information on a macroscopic level. Information on the molecular adsorption mechanism will, therefore, require additional investigation of the system. In order to elucidate some areas concerning adsorption from dilute nonaqueous solvents on to mineral oxides, we will discuss more closely the adsorption of amphiphilic molecules from organic solvents on to various such oxides.

Polar solvents may interact strongly with a mineral oxide surface. In principle, the adsorption of the solvent must be considered. Claesson [13] studied the adsorptiotn of fatty acids by silica from solvents of various polarities. The results show that polar solvents compete with the solute for available sites on the surface, while nonpolar solvents show little competition. The polarity of the solvent is often determined from the measured dielectric properties. Krishnakumar and Somasundaran [13] studied surfactant adsorption on to silica and alumina from solvents with various dielectric properties. The aim of the study was to look at the effect of adsorbent and surfactant acidities and solvent polarity on the adsorption properties of the surfactant molecules. They used anionic and cationic surfactants as adsorption probes. The results show that polar interactions control the adsorption from solvents of low dielectric properties while hydrocarbon chain interactions with the surface play an important role in determining adsorption from solvents of higher dielectric properties. It was also found that an acidic surfactant interacts strongly with a basic adsorbent, and vice versa. One should be aware that the polarity of a molecule as measured from the dielectric properties is not always correlated with the ability of the molecules to form ion pairs. For example, dimethylformamide and nitromethane have almost equal dielectric constants. However, the extent of ion pairing in nitromethane is much greater than that in dimethylformamide. Thus, the solvent acidity and basicity are the physical properties which can best characterize the ability of the solvent to compete with the solute for available sites on the mineral surface.

Apolar solvents display low donor and acceptor values, and do not take part in the formation of stable donor–acceptor complexes at the oxide surface. The solvent can be asumed to interact with the mineral or with the adsorbate molecules only through weak dispersion forces. This means that an apolar and inert solvent will not affect the adsorption strength differently for various adsorbate molecules on to hydroxylated mineral surfaces. Thus, a characterization of the chemical functionalities of a mineral oxide surface can be done from adsorption studies of polar or amphiphilic molecules from inert and apolar solvents. Also, a need for detailed treatment of the mineral surface before adsorption will be substantially reduced in relation to a corresponding adsorption from aqueous solution where electrostatic charges on the surface may continuously change throughout the adsorption process.

In one of the early adsorption studies, Crisp [24] examined the adsorption properties of various polar and nonpolar alcohol molecules on to alumina from benzene solutions. He found that the Langmuir adsorption isotherm is practically applicable over the whole range of adsorption sites available. In dilute solutions, only the polar group takes part in the adsorption while the hydrocarbon chain remains in the solvent. This indicates a reasonable uniformity in energy of the adsorption sites, and that the energy of adsorption (i.e., the energy between the adsorbed alcohol molecules and the hydroxyl groups on the solid surface) is high compared to the energy associated with lateral interactions among the hydrocarbon chains of the adsorbed molecules. He also found that the energy of adsorption is very similar for the various alcohol molecules despite the great difference in the alcohol chain length. Also, the adsorption of poly(vinyl alcohol) molecules on to kaolin from a dimethyl sulfoxide solution [25] shows an isotherm shape which can be mathematically described by the Langmuir adsorption isotherm. The poly(vinyl alcohol) molecules form dense coils in water, while they are rather expanded in many organic solvents, and from the strong adsorbent–adsorbate interaction one might expect that these molecules also become adsorbed with their hydroxyl groups oriented toward the mineral, gaining their stability from hydrogen bonding and dispersion forces.

In general, the acidity and basicity of the adsorbate probe molecules and the density of hydroxyl groups on the surface of the oxide can be used to evaluate the adsorption interaction strength and sometimes also the degree of adsorption. An acidic probe molecule will interact strongly with a basic mineral and vice versa. This behavior has been demonstrated by Pugh [26]. The adsorption of several acidic and basic probe molecules from cyclohexane solutions was studied in order to characterize the surface of various types of ceramic powders. The classification of the acidic and basic probe molecules was done in terms of their AN and DN values, and the classification of the oxide powders was done from their IEPs in water. Generally, the adsorption data followed a Langmuir type. The acidic probes were most strongly adsorbed on to the powders, and especially on to the basic powders (the powders with high IEP values). Moreover, the highest adsorption capacities were observed for the basic powders towards the acidic probe molecules.

In a recently published work [27], Ardizzone et al. studied the adsorption of pyridine and aniline on to α-alumina and iron powder from *n*-hexane solutions. The aim of the work was to obtain the adsorption isotherms and to analyze the adsorption behavior in view of a potential use as corrosion inhibitors in nonaqueous solvents. The results showed a strong adsorbent–adsorbate interaction at low amine concentration, especially for the adsorption on to α-alumina. The adsorption isotherms are presented in Fig. 3. In the case of the aluminum oxide, the adsorption isotherms have a Langmuirian shape with a long plateau section, but with a final upward bend. The shape of the isotherms belonging to the iron powder are sigmoidal and present two successive plateau regions and a final upward bend. The ΔG values obtained indicate physisorption processes with short-range van der Waals' interactions between the adsorbate and the adsorbent. Both solutes show higher affinity toward the oxide than toward the iron metal. This fact was interpreted as a result of better donor/acceptor properties between the amines and the hydroxylated surface sites of alumina compared to the surface sites on the iron.

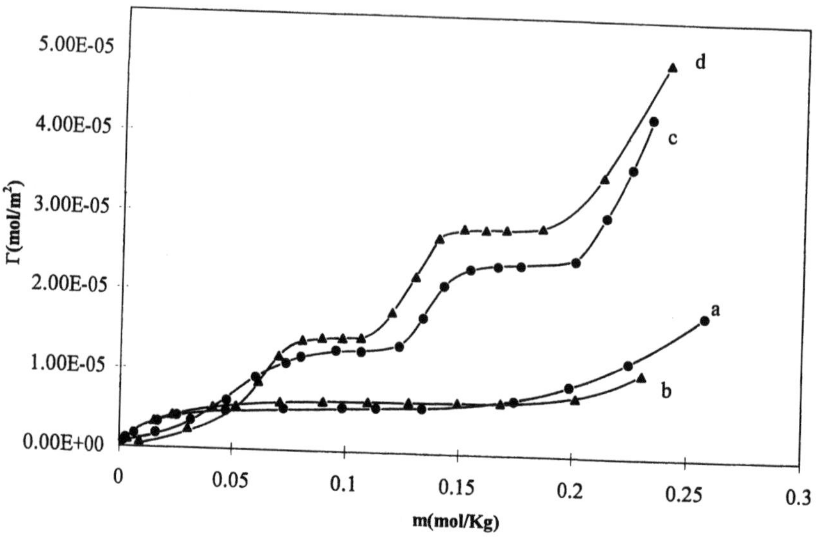

FIG. 3 Adsorption isotherms of aniline and pyridine onto alumina and iron. (a) Aniline on α Al_2O_3; (b) pyridine on α Al_2O_3; (c) aniline on Fe; (d) pyridine on Fe. (From Ref. 27.)

Systems where the mineral oxide is well defined, and the inert solvents do not influence the adsorption of amphiphilic probe molecules, can be considered as relatively simple ones. However, even a well-defined and simple system requires a comprehensive and careful treatment of the adsorption data in order to be described properly. One phenomenon to be considered when adsorption of amphiphilic molecules occurs from an apolar and inert solvent is the ability of the solute molecules to self-associate in an apolar solution. This is the case for alcohol molecules dissolved in inert apolar liquids, and it complicates their adsorption properties in that different aggregates may have different affinities for the surface. This can be illustrated by discussing in some detail, the adsorption properties of some alcohol molecules on to kaolin and alumina.

The adsorption of short-chain aliphatic alcohol molecules on to the clay mineral kaolin has been elucidated by considering the heterogeneous surface sites of the clay and the self-association of alcohol molecules in apolar decane solutions [28]. The adsorption isotherms are shown in Fig. 4. The isotherms of ethanol, propanol, and butanol are of L-type, showing large initial slopes that readily decrease with increasing alcohol concentration, while the isotherm of methanol appears as a stepwise function.

The adsorption data suggest high affinity between the mineral and the alcohol molecules at low alcohol concentrations. It also indicates that monolayer adsorption takes place for ethanol, propanol, and butanol molecules, while a multilayer adsorption takes place for methanol. At the plateau section of the monolayer isotherms, the adsorption density decreases with increasing chain length of the alcohol. The low area available for each molecule indicates that the alcohols are more or less perpendicularly oriented on to the mineral surface. As previously mentioned, the adsorption strength of alcohol molecules is normally correlated with the acidity of the surface and through the donor–acceptor properties of the alcohol molecules.

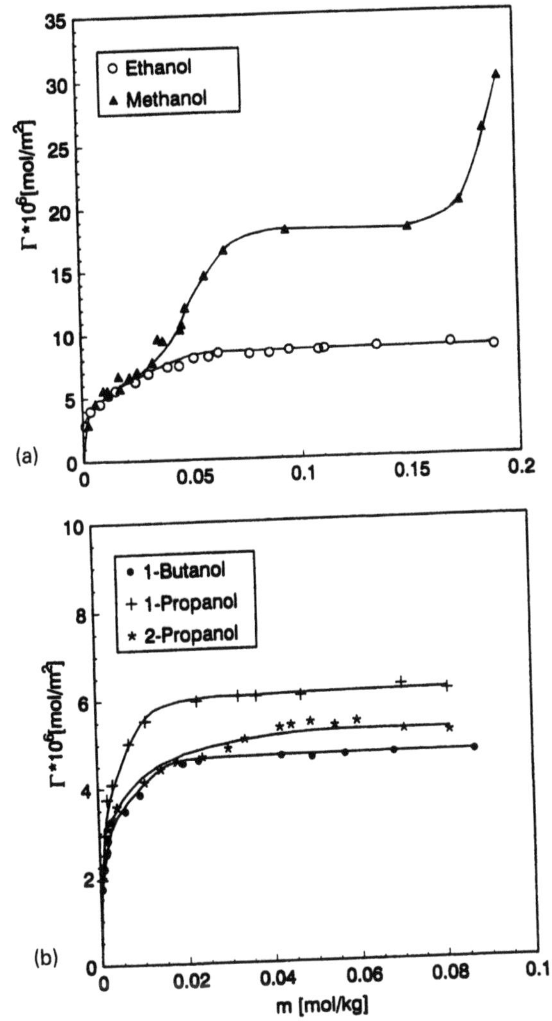

FIG. 4 Adsorption isotherms of alcohol molecules on to kaolin from decane solutions: (a) methanol and ethanol; (b) propanol, isopropanol, and butanol. (From Ref. 28.)

Figure 5 shows some scanning electron micrograph (SEM) images of the kaolin used in this study. The low value for the PZC of kaolin [29] shows that the clay is a highly acidic mineral. For the alcohol molecules, the high AN and DN values represent strong amphoteric properties. These probe molecules can act as Lewis bases and Lewis acids, depending on the acidity/basicity of the solid surface. Thus, an alcohol molecule will act as a Lewis base toward a hydroxylated group on the mineral surface of kaolin.

The L-type isotherms can usually be fitted by the Langmuir equation, Eq. (4). However, it is not uncommon that one observes a curve convex to both axes when a linearization of the Langmuir equation, Eq. (5), is used. This is also the case here, and it indicates that the basic assumption for using Langmuir is not fulfilled. The poor fit of the Langmuir equation can be because a variety of energy sites are present

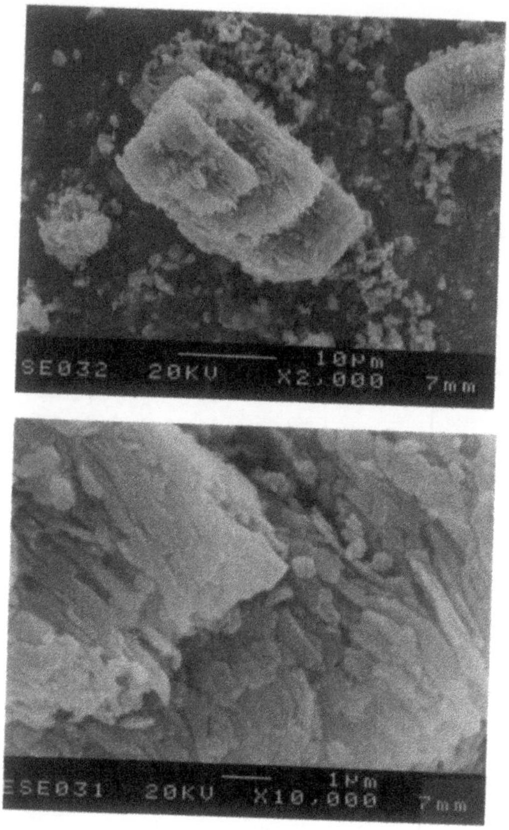

FIG. 5 SEM images of kaolin.

on the mineral surface. One should be aware that the edges of the kaolin platelets contain more than one type of OH group. Also, the basal kaolin surface is heterogeneous owing to isomorphous substitution of Si^{4+} by Al^{3+}. The high surface reactivity exhibited by clay minerals is a consequence of both the high specific surface area and the irregular stacking of crystalline layers. More details on kaolin and other clay structural features can be found in Chapter 6 of this volume.

In order to make further progress, we assume that the functional adsorption sites of kaolin can, with some ambiguity, be divided into two main groups. The most important functional sites for the alcohol adsorption at low alcohol concentration are assumed to be the hydroxylated groups exposed on the edges of the kaolin platelets. The second group is assumed to be the functional sites on the basal siloxane surface. So, even though the adsorption sites on the kaolin will represent a variety of energy sites, we have chosen to fit mathematically a two-term Langmuir adsorption isotherm to the data obtained for ethanol, propanol, and butanol. The two-term Langmjir equation represents adsorption on to a surface containing two distinctly different energy sites.

One other possible explanation which can contribute to the deviation from linearity is the effect of self-association of the alcohol molecules in the hydrocarbon solution. A schematic presentation of some alcohol aggregates is seen in Fig. 6. Studies of primary alcohols in aliphatic hydrocarbon solvents have shown that the

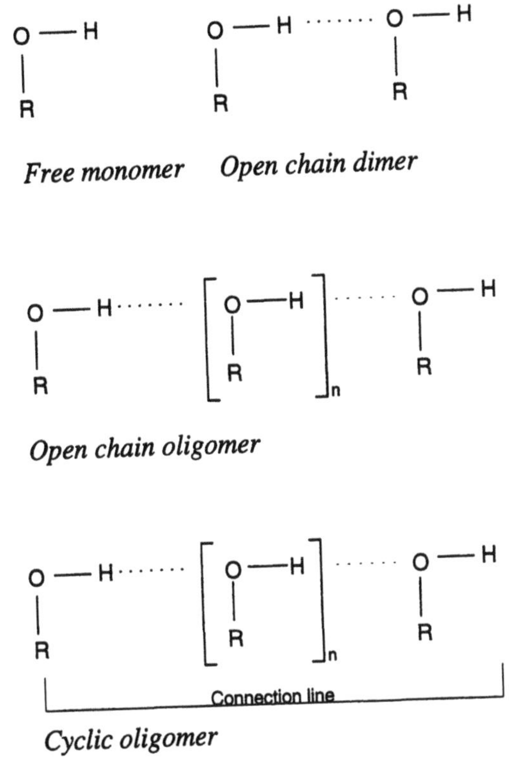

FIG. 6 Schematic presentation of various alcohol species.

alcohol molecules self-associate by hydrogen bonding to form mainly tetramers [30, 31]. Even though no generally agreed conclusion for the association process has emerged, a number of studies emphasize cyclic aggregates to be the most important structures [31–35]. Such aggregates will have no unsaturated valencies for forming hydrogen bonds. Although this does not rule out the possibility that hydrogen-bonded alcohol aggregates adsorb on to a mineral oxide surface, it suggests that monomer adsorption is the most important process owing to the higher polarity of the molecule.

An estimate for the association constants of aliphatic alcohols can be derived from the literature, and Table 1 contains the monomer–tetramer equilibrium constants for various systems [31]. The high values of the equilibrium constants show that, even at the low alcohol concentration region used here, a large proportion of the alcohol molecules take part in the association process. An equilibrium constant of 220 calculated on the molal scale implies that, at alcohol concentrations as low as 0.1 molal, about 26% of the alcohol molecules will be H-bonded. This shows quite clearly that alcohol association should be taken into account. Therefore, all the monolayer isotherms were modified to take into account the self-association of the alcohol molecules in the hydrocarbon solution by suggesting that only the free alcohol monomers adsorb on to the clay surface.

Not surprisingly, the results show a noticeably improved fit for all systems when going from a one-term to a two-term Langmuir equation. One should be

TABLE 1 Monomer–Tetramer Equilibrium Constants at 25°C for Various *n*-Alcohols in Saturated Hydrocarbon Solvents

Solute	Solvent					
	Hexane	Heptane	Iso-octane	Decane	Dodecane	Cyclopentane
Methanol	100					110
Ethanol	155		215	220		
Propanol			175			
Butanol			180			
Pentanol		215	205	235		
Hexanol			215		180	
Heptanol			200			
Octanol			215	300		

Source: Ref. 27.

aware that the mathematical modeling by a two-term Langmuir equation can always be well fitted to a convex distribution function, and does not necessarily provide any information about the adsorption mechanism. However, from the point of view of surface chemistry, the kaolin may very well display adsorption sites with contrasting energies, and a two-term Langmuir equation will in this case be useful. The introduction of self-association into the adsorption models does not lead to improvements in the fits, and only minor changes are obtained in the Langmuir parameters. An explanation why the derived Langmuir parameters are only mildly sensitive to the inclusion of association can be given from the fact that the association of alcohol molecules cannot have any impact on the initial slope and the plateau value of the isotherms.

Another example of alcohol adsorption will be given from the data obtained for adsorption of benzyl alcohol on to alumina and well-crystallized kaolin (KGa-1) from carbon tetrachloride solutions. A SEM image of the alumina can be seen in Fig. 7. The mineral is a high-area form of aluminum oxide. Elemental analysis by X-ray fluorescence spectroscopy shows that the alumina surface contains more than 99.9% aluminum oxide. X-ray diffraction measurement indicates an amorphous surface structure. The adsorption isotherms for benzyl alcohol on to the alumina and the kaolin can be seen in Fig. 8. The isotherms are plotted on a log–log form in order to elucidate the region at low alcohol concentration. One can see that the shape of the isotherms represents typical multilayer adsorption with well-defined monolayer plateau sections. The plateaus represent the adsorption capacity of the minerals for monolayer adsorption. The data for the adsorption density at the plateaus correspond to an available area for each alcohol molecule of 0.54 nm^2. At higher alcohol concentrations, the increase in the adsorption densities represents a multilayered stacking of alcohol.

The sufficiently large area available for each alcohol molecule at the plateau section, and the readily increasing adsorption density in the multilayer region with increasing alcohol concentration, suggest that the alcohol becomes adsorbed in a vertical orientation with the benzene ring lying in the plane of the interface. If this is the case, the formation of multilayers may occur by a stacking of the benzene rings. The stability of such a configuration will be attained by sharing of π-electrons.

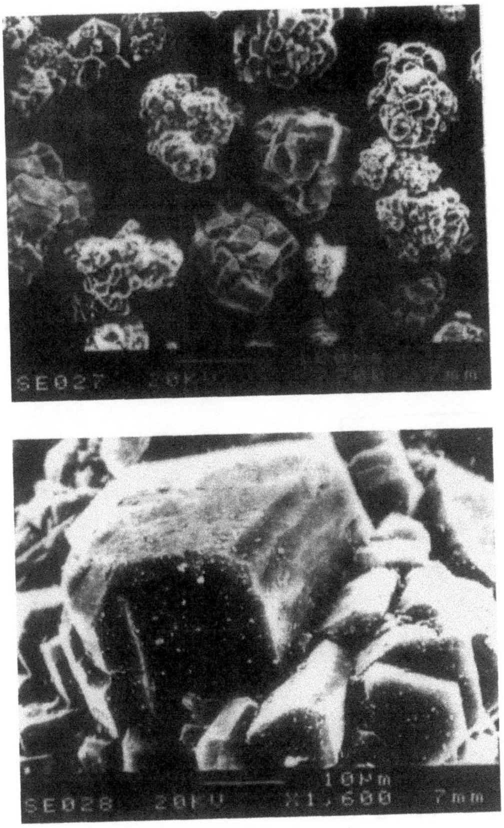

FIG. 7 SEM images of alumina.

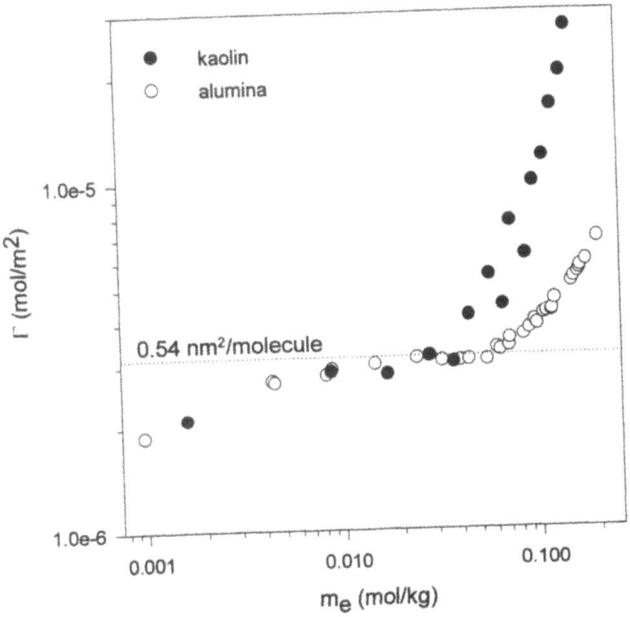

FIG. 8 Adsorption isotherms of benzyl alcohol onto kaolin and alumina.

For an adsorption process where the first layer is formed by H-bonding between the hydroxylated groups on the mineral surface and the OH group of the alcohol molecules, followed by subsequent layers formed by π-electron interactions, the energy of the first adsorbed layer will most likely be much higher than the energy of the subsequent layers. This fits the adsorption data. The well-defined plateau sections which are obtained at a low alcohol concentration and last over a rather large concentration range indicate that the affinity of the minerals for the first adsorbed layer is much higher than for the subsequent layers.

Benzyl alcohol is shown to form open-chain oligomers as well as cyclic oligomers in carbon tetrachloride solutions [36]. Figure 9 shows the concentration profiles

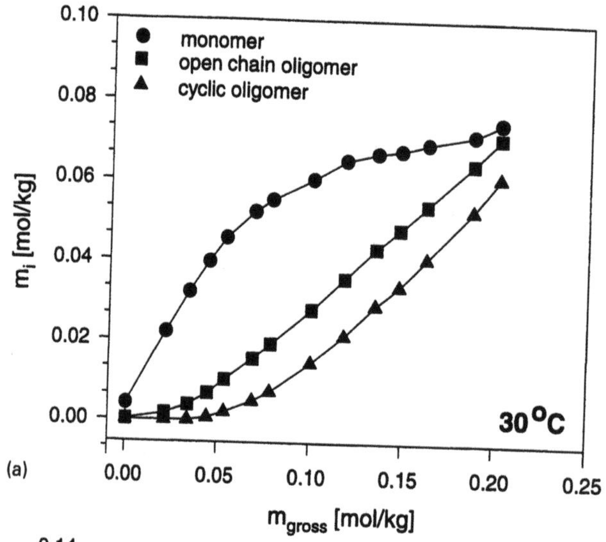

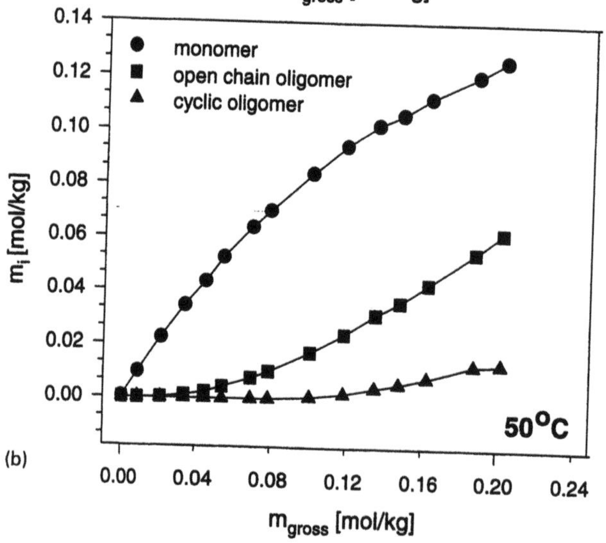

FIG. 9 Concentration profiles of the benzyl alcohol species. Molality of the various alcohol species plotted against the molality of the total alcohol content at (a) 30°C and (b) 50°C.

of the various alcohol species in CCl_4 solutions. The result indicates that about 40% of the alcohol molecules are associated at 30°C and at an alcohol concentration as low as 0.1 molal. The loss of unsaturated valencies for forming hydrogen bonds in a cyclic alcohol aggregate limits their ability to adsorb on to the mineral oxide surface. However, this does not rule out the possibility that open-chain hydrogen-bonded alcohol aggregates adsorb on to such a surface. An open-chain aggregate has the ability of forming two hydrogen bonds and should, therefore, be able to compete with an alcohol monomer for available sites on the oxide surface. It is possible that the adsorbed alcohol monomers can be exchanged by open-chain oligomers. If this is the case, increased adsorption will be measured. However, this process will not take place until the total alcohol content reaches the limit where open-chain aggregates form. From the concentration profiles in Fig. 9, one can see that nearly all alcohol molecules exist as monomeric species in the concentration range up to about 0.04 mol/kg. Above this concentration, the number of open-chain alcohol aggregates increases rapidly. The observed increased adsorption starts at equilibrium concentrations of alcohol between 0.04 and 0.05 mol/kg. Thus, there appears to be a correlation between increased adsorption, shown in Fig. 8, and formation of open-chain aggregates, shown in Fig. 9.

VI. CONCLUDING REMARKS

The theories of adsorption processes and the resulting adsorption isotherms are normally derived without specifically considering the solvent. Thus, in principle, the theories do not distinguish between aqueous and nonaqueous solutions. However, concerning the mechanisms of adsorption there is one significant difference. The low permittivity of most nonaqueous solvents excludes ionization. The adsorbent will be uncharged, but even more significantly, the surface of the solid will not be modified by dissociation or ionization of the surface groups. Electrostatic forces will not be a driving force of the adsorption processes. This leaves hydrogen bonding as a favored mechanism of adsorption in nonaqueous media. In nonaqueous solvents, solute adsorption may result in a competition between intermolecular hydrogen bonding and the adsorption process itself. It is apparent that hydrogen-bonded aggregates are less likely to adsorb than the monomolecular species, and this will influence the adsorption isotherm.

REFERENCES

1. OCM Davis. J Chem Soc 91:1666, 1907.
2. JM van Bemmelen. J Pr Chem 23:324, 1881.
3. SD Forrester, CH Giles. Chem Ind 318–328, 1972.
4. V Novotny. Colloids Surfaces 24:361–375, 1987.
5. K Christmann. In: H Baumgärtel, EU Franck, W Grünbein, eds. Introduction to Surface Physical Chemistry. Darmstadt: 1991, pp 34–83.
6. I Langmuir. J Am Chem Soc 40:1361, 1918.
7. CC Travis, EL Etnier. J Environ Qual 10:8–17, 1981.
8. M Tschapek, C Wasowski. J Soil Sci 27:175–182, 1976.
9. ICR Holford, RWM Wedderburn, GEG Mattingly. J Soil Sci 25:242–255, 1974.
10. JP Badiali, L Blum, ML Rosenberg. Chem Phys Lett 129:149, 1986.

11. AW Adamson. In: AW Adamson, ed. Physical Chemistry of Surfaces. 5th ed. New York: John Wiley, 1990, p. 422.
12. H Freundlich. Colloid and Capillary Chemistry. London: Methuen, 1926.
13. S Krishnakumar, P Somasundaran. Langmuir 10:2786, 1994.
14. JJ Jurinak. Soil Sci Soc Am J 27:270, 1963.
15. TB Lloyd. Colloids Surfaces A: Physicochem Eng Aspects 93:25, 1994.
16. RJ Drago, BB Wayland. J Am Chem Soc 87:3571, 1965.
17. V Gutmann. The Donor–Acceptor Approach to Molecular Interactions. New York: Plenum Press, 1978.
18. ME Labib, R Williams. J Colloid Interface Sci 97:356, 1984.
19. B Siffert, J Eleli-Letsango, A Jada, E Papirer. Colloids Surfaces A: Physicochem Eng Aspects 92:107–111, 1994.
20. DR Lloyd, TC Ward, HP Schreiber (eds). Inverse Gas Chromatography. ACS Symposium Series 391. Washington DC: American Chemical Society, 1989.
21. CR Hegedus, IL Kamel. J Coatings Technol 65:23, 1993.
22. H Noller, W Kladnig. Catal Rev Sci Eng 13:149, 1976.
23. S Claesson. Rec Trav Chim 65:571, 1946.
24. DJ Crisp. J Colloid Sci 11:356, 1956.
25. T Kashmoula, J. Schurz. Paper Darmst 29:45, 1975.
26. RJ Pugh. Ceram Trans 12:375, 1990.
27. S Ardizzone, H Høiland, C Lagioni, E Sivieri. J Electroanal Chem 447:17, 1998.
28. M Førland, KJ Børve, H Høiland, A Skauge. J Colloid Interface Sci 171:261, 1995.
29. Z Zhou, WD Gunter. Clays Clay Minerals 40:365, 1992.
30. M Costas, D Patterson. J Chem Soc, Faraday Trans 1 81:635, 1985.
31. GM Førland, FO Libnau, OM Kvalheim, H Høiland. Appl Spectrosc 50:1264, 1996.
32. AN Fletcher, CA Heller. J Phys Chem 71:3742, 1967.
33. T Hoffman. Fluid Phase Equil 55:271, 1990.
34. G Brink, I Glasser. J Phys Chem 82:1000, 1978.
35. M Iwahashi, Y Hayashi, N Hachiya, H Matsuzawa, H Kobayashi. J Chem Soc, Faraday Trans 89:707, 1993.
36. GM Førland, Y Liang, OM Kvalheim, H Høiland, A Chazy. J Phys Chem 101:6960, 1997.

8

Surface Modification of Inorganic Oxide Surfaces by Graft Polymerization

YORAM COHEN, WAYNE YOSHIDA, VAN NGUYEN, and NIANJIONG BEI
University of California, Los Angeles, Los Angeles, California

JENG-DUNG JOU Printronix, Inc., Irvine, California

I. INTRODUCTION

Modification of inorganic oxide surfaces is used in a variety of practical applications and in the study of interfacial phenomena. There are a multitude of surface modification methods, including modification by reactive polymer interlayers [1], diblock copolymerization [2], and polymer adsorption [3]. Recent years have also seen advancement in layer-by-layer deposition methods for preparation of multilayer thin films on surfaces [4]. In the textile industry, excimer laser irradiation is used in preparing highly absorbing and oriented synthetic fibers [5]. The semiconductor industry uses the metal organic chemical vapor deposition [6], physical vapor deposition [7], and reactive ion etching plasma etching processes [8] for surface modification of silica wafers.

In particular, recent years have been marked by a growing interest in surface modification of inorganic substrates by a covalently bonded polymer phase (Table 1). The grafted polymer alters the surface chemistry while retaining the basic mechanical strength and geometry of the solid substrate. The unique feature of a covalently bonded polymer layer is that it allows tailoring and manipulation of the interfacial properties. In particular, by the appropriate choice of a soluble polymer, the modified substrate can be made highly compatible with the liquid medium. Polymer dissolution is prevented owing to the covalent attachment of the polymer to the surface. Applications of polymer-modified solid substrates include control of filler/polymer in polymer composites [9], support packing for gas and liquid chromatography [10–12], biocompatible surfaces [13–15], colloid stability [16], and modified inorganic separation membranes [17, 18].

Two popular methods of graft polymerization onto inorganic oxide supports (silica, alumina, and zirconia) are *free-radical polymerization* [16–34] and *anionically initiated polymerization* [12, 35–37]. Studies have demonstrated that free-radical graft polymerization onto surface-active vinyl alkoxysilanes is an efficient and controllable method for obtaining a high polymer graft yield in the dense "brush" regime [10, 17,

TABLE 1 Surface Modification of Inorganic Oxide Substrates

Substrate	Surface activation method	Polymerization method	Grafted monomer	Graft yield (mg/m^2)	Reference
Aerosil 200	Silylation (GPTMS)[a] + ACPA[b]	Free-radical polymerization	Acrylonitrile	1.79	[37]
			NVC[e]	1.69	
			Styrene	1.31	
			MMA[f]	2.26	
Nonporous silica	Silylation (γ-APS)[c] + ACPA	Free-radical polymerization	MMA	1.0–4.8	[26]
Aerosil 200	Chlorosilylation + hydroperoxide	Free-radical polymerization	MMA	1.73	[31]
Aerosil 200	Silylation (TSPA)[d] + hydroperoxide	Free-radical polymerization	MMA	1.91	[31]
			Styrene	1.25	
			NVC	1.43	
Novacite nonporous	Silylation	Free-radical polymerization	VP[g]	3.5	[47]
			VAc[h]		
Nonporous silica	Silylation	Free-radical polymerization	VP	2.0–4.0	[20]
Cellulose	–	Free-radical polymerization	MA[i]	–	[48]
Porous silica	–	Condensation	Alkyl chains	5.6–7.4	[8]
Nonporous silica	–	Condensation	Dimethylsiloxane	0.54–4.65	[49]
Nonporous silica	Silylation	Free-radical polymerization	PVP	1.27–6.99	[21]

Aerosil 130	Siloxane	Anionic	Styrene	0.63	[12]
Aerosil 200				0.63	
Aerosil 300				0.58	
Aerosil 380				0.61	
Nonporous silica	Siloxane	Anionic	Styrene	4.6	[50]
Nonporous silica	Chloride	Anionic	Styrene	4.33	[51]
Aerosil 200	Chloride	Anionic	Styrene	0.8	[36]
Porous silica	Silylation	Free-radical polymerization	MPS	1.1–3.3	[30]
Nonporous silica	Silylation	Free-radical polymerization	Styrene	0.4–2.1	[16]
Aerosil 200	Silylation	Free-radical polymerization	Styrene MMA	2.0–6.4 (St) 2.2–4.3 (MMA)	[32]

[a]GPTMS: 3-glycidoxypropyltrimathoxysilane.
[b]ACPA: 4,4′-azobis(4-cyanopentanoic acid).
[c]γ-APS: γ-aminopropyltriethoxysilane.
[d]TSPA: 4-trimethoxysilyltetrahydrophthalic anhydride.
[e]NVC: *N*-vinylcarbazole.
[f]MMA: methyl methacrylate.
[g]VP: vinylpyrrolidone.
[h]VAc: vinyl acetate.
[i]MA: methylacrylate.

38]. This grafting method, via free-radical polymerization, has been shown to be most suitable for many applications in membrane separation [17,18] and size exclusion chromatography [10, 11, 39–43].

Free-radical graft polymerization in solution is relatively simple to carry out experimentally. However, it involves simultaneous formation of both homo- and grafted polymers. The grafted polymer is covalently bonded to the surface through two concurrent mechanisms: polymer grafting and graft polymerization (Fig. 1). Polymer grafting involves the chemical bonding of homopolymer radicals to active surface sites to form grafted chains. This approach allows for the attachment of monodisperse surface chains. However, since the homopolymer radicals must diffuse to the solid surface, diffusional limitations can severely reduce the contribution of polymer grafting to the overall graft yield, especially when porous substrates are employed. Graft polymerization differs from polymer grafting in that it involves the growth of surface chains by propagation (the addition of one monomer molecule at a time). In graft polymerization, diffusion and steric effects are diminished since the smaller monomer units can readily diffuse to the solid surface. Graft polymerization thus allows the formation of a denser and more uniform surface coverage [17, 19–21]. Therefore, to control and optimize the graft density and grafted chain length, it is necessary to understand the relative contributions of polymer grafting and graft polymerization to the overall graft yield [21, 33].

Although various polymers can be used to modify an inorganic oxide substrate, this chapter focuses on poly(vinylpyrrolidone) (PVP) and poly(vinyl acetate) (PVAc). PVP grafting can be used to form a nontoxic surface that possesses superior hydrophilic and biocompatible properties. A number of studies have shown that surface-bonded PVP can also enhance the performance of silica resins in size exclusion chromatography [11, 44], inorganic membranes of silica and zirconia [17, 18], and wetting of synthetic fibers and biocompatible hydrogens [14, 15, 45]. A grafted layer of PVAc is especially versatile since it can render an inorganic oxide substrate hydrophobic or hydrophilic, depending on the post-grafting treatment [18, 21, 29, 46]. In this chapter, the main features of the free-radical grafted polymerization of the above two polymers onto silica are reviewed with emphasis on various factors affecting the engineering of a graft-polymerized surface.

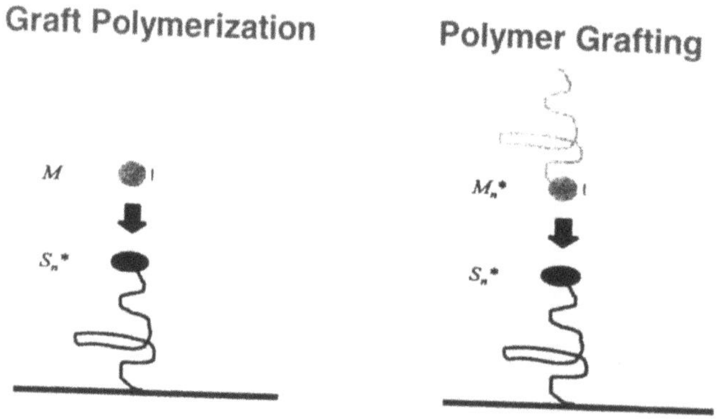

FIG. 1 Chemical modification processes: graft polymerization and polymer grafting. M: Monomer, M_n^*: Active Polymer Chain of Size n, S_n^* : Active surface Chain of Size n.

II. SURFACE MODIFICATION

Chemical modification of inorganic oxides has become a standard approach to controlling surface chemistry in a variety of applications, including gas and liquid chromatography, polymer composites, colloidal stabilization, biocompatibility, and modified inorganic membranes [9, 10, 13, 16–18, 20, 33, 38, 52]. Among various surface modification methods (Table 1), free-radical graft polymerization is of particular interest since it can be carried out under mild reaction conditions, and the reaction can be easily controlled. Clearly, knowledge of the kinetics involved is essential for proper control and optimization of the grafting process. The free-radical graft polymerization procedure consists of two main steps: a silylation reaction producing a network of surface-bonded vinylsilane molecules and a subsequent surface polymerization of a vinyl monomer producing a chemically and terminally anchored polymer phase. In the subsequent sections, an overview of a typical experimental procedure is presented, focusing on the techniques of pretreatment, silylation, hydrolysis, and graft polymerization.

A. Pretreatment

Surface pretreatment is a crucial step which determines the surface activity for the subsequent graft polymerization. Most inorganic oxide surface modification procedures rely on surface silylation whereby silane compounds with the desired functionality are reacted with surface hydroxyls. For example, the surface of amorphous silicas consists of silanol ($\equiv$SiOH) and/or siloxane ($\equiv$Si$-$O$-$Si$\equiv$) groups, with reported hydroxyl concentration, on a hydroxylated amorphous silica, of 4.6 hydroxyls/nm^2 [53].

Before silylation, the ceramic substrate must be cleaned and hydroxylated. Silica substrates can be treated with a dilute acid solution to remove traces of iron ions and organic solvents. The substrate is then rinsed with deionized water for a prescribed period to hydrolyze, if necessary, any surface siloxane groups into silanol groups. The final pretreatment step involves drying in order to remove excess surface water. Although the average surface density on a hydroxylated amorphous silica is 4.6 hydroxyls/nm^2 [53], the actual surface distribution of silanol and siloxane groups depends on the temperature history of the substrate and the type of solvent employed in the hydrolysis [54].

B. Surface Silylation

Before graft polymerization, the surface is activated with a vinylsilane as shown for the vinyltrialkoxysilane example in Fig. 2. Since the vinyl groups of the attached silane molecules provide surface anchoring points for the later grafted chains, the ability to regulate the silylation coverage provides a method of controlling the subsequent polymer surface density (number of chains/area). It is noted that the use of trialkoxysilanes allows multiple surface attachments and/or silane crosslinking, which produces a multilayer coverage [55] as illustrated in Fig. 3.

Chemical bonding of alkoxysilanes to a hydroxylated substrate can be accomplished in an aqueous or anhydrous environment [19]. In aqueous silylation, the presence of water molecules leads to the formation of polysilanes, which may limit the development of a uniform surface coverage owing to steric effects. In addition,

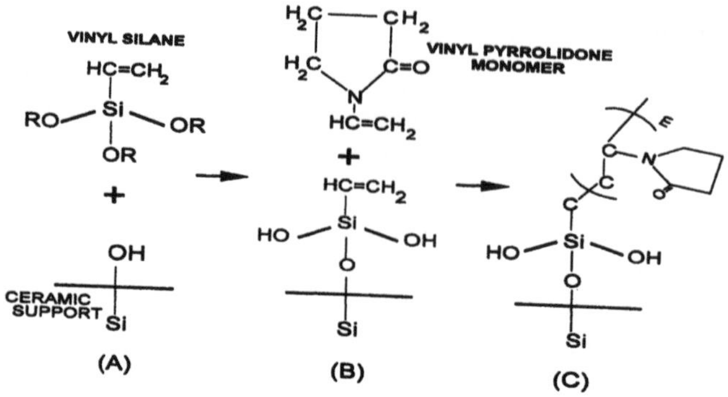

FIG. 2a Illustration of vinylpyrrolidone graft polymerization onto silica: (A) surface silylation; (B) graft polymerization; (C) terminally anchored polymer chain.

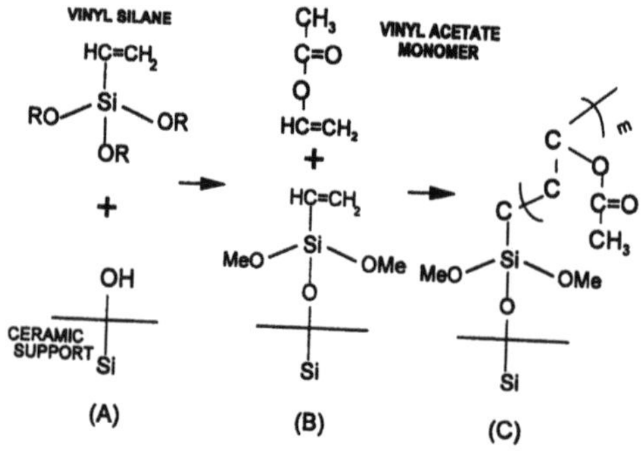

FIG. 2b Illustration of vinyl acetate graft polymerization onto silica: (A) surface silylation; (B) graft polymerization; (C) terminally anchored polymer chain.

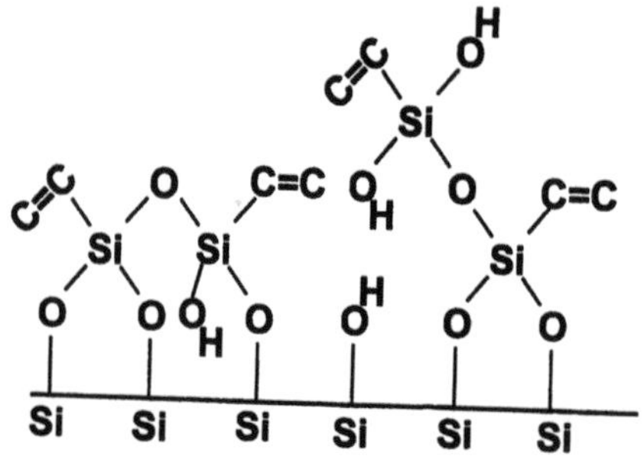

FIG. 3 Schematic diagram of surface anchored polysilane network.

the silylation yield is strongly affected by the pH of the reaction medium. Anhydrous silylation can be achieved by using pure silane or anhydrous organic solvents with properly dried substrates. A high-yield multilayer coverage may be achieved by allowing adsorption of a small amount of surface water onto the dried substrates [34, 56].

In general, the net surface silylation reaction can be written as

$$\equiv Si-OH + R_nSiX_{4-n} \rightarrow Si-O-SiX_{(3-n)}R_n + XH \quad (1)$$

where R represents a nonhydrolyzable group with the desired functionality to be imparted to the silica surface; n can take on values from 1 to 3; and X represents a hydrolyzable group (an alkoxy, acyloxy, amine, or chlorine), which can react with a surface silanol. When the silane contains two or more hydrolyzable groups and sufficient water is present, a network of surface-bonded polysilanes can form. Since the reaction as written above consumes a hydroxyl group, surface silane coverage is limited by the initial concentration of hydroxyl groups on the surface. For example, the upper limit of surface silylation coverage on silica, for silanes with one hydrolyzable group, is 4.6 molecules/nm^2. (i.e., the maximum surface hydroxyl concentration on silica). Notwithstanding, surface densities of vinylsilanes in excess of 20 molecules/nm^2 can be achieved, as is the case when polysilanes form by the reaction of silanes with hydrolyzable groups of already-bonded trialkoxysilanes [34]. It is important to emphasize that the vinyl groups of the surface silanes provide active sites on to which polymer chains are grown by the sequential addition of the selected vinyl monomer. Therefore, the surface density of the later grafted polymer is dictated by the initial surface silane concentration, which is in turn fixed by the previous pretreatment step.

C. Hydrolysis

The use of vinyl alkoxysilanes with two or three alkoxy groups is particularly useful when the graft polymerization process is conducted in an aqueous solution. Hydrolysis of the remaining alkoxy groups, after silylation, increases the hydrophilicity of the silylated layer, thus facilitating a more efficient wetting of the support by the aqueous monomer solution. Surface hydrolysis can be accomplished by using a basic solution (pH $\approx$9.5) as shown in Fig. 4. For graft polymerization in organic solvents, hydrolysis is not required since it is necessary to maintain the hydrophobicity of the surface.

D. Graft Polymerization

The final part of the surface modification procedure is monomer grafting via free-radical polymerization. Schematic illustrations of the graft polymerization of vinylpyrrolidone (VP) and vinyl acetate (VAc) on to silica are shown in Figs 2a and 2b, respectively. In both cases, an initiator is used to initiate the grafting reaction by creating free radicals that initiate polymer chain growth.

In general, free-radical graft polymerization is carried out under reaction conditions similar to those used for free-radical homopolymerization. The reaction is performed in a temperature-controlled environment in which radical scavenging contaminants and chain transfer agents are carefully excluded (in a jacketed glass reactor under a nitrogen blanket). Although a bulk monomer graft polymerization

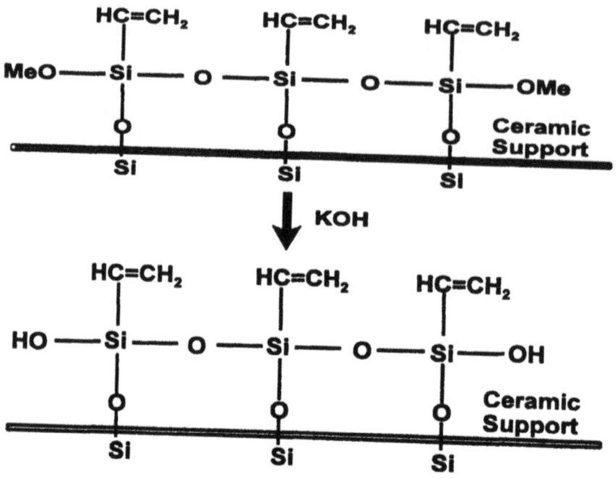

FIG. 4 Hydrolysis of an alkoxysilane-modified silica surface with potassium hydroxide.

can be performed, solution polymerization under mild conditions is preferred since good mixing can be attained even at high conversions. The experimental grafting process is typically started by the direct immersion of the silylated particles, wafer, or porous tube into the reaction solution. After heating the reaction mixture to a desired temperature, graft polymerization is initiated by the addition of the initiator. Since the rates of polymer grafting and graft polymerization depend on the diffusion of monomer molecules and homopolymer radicals toward surface chain radicals, it is particularly important to keep the reaction mixture well mixed at all times.

III. KINETICS OF FREE-RADICAL GRAFT POLYMERIZATION

The kinetics of free-radical graft polymerization involves the following species: solvent, initiator, monomer, surface vinylsilane, free and grafted polymer radicals, and terminated homo- and surface polymers. The consumption of monomer via covalent attachment to the silane-modified surface occurs by two processes: homopolymerization and graft polymerization, with the sum of these two rates representing the overall rate of polymerization. Similar to classical free-radical homopolymerization, the reaction sequence in a free-radical graft polymerization can be divided into four main parts: (1) initiation, (2) propagation, (3) chain transfer, and (4) termination.

A detailed kinetic description of free-radical graft polymerization was provided by Chaimberg and Cohen [33], who accounted for the effect of chain length on various reaction rate coefficients. The above study, illustrated for graft polymerization of VP onto silica, with hydrogen peroxide initiator, utilized a restricted initiation mechanism which could not account for a reaction order greater than unity with respect to monomer concentration. A higher reaction order can be obtained by employing a more elaborate initiation scheme, such as the cage-complex model [57] or the high initiation rate mechanism [58–62]. To illustrate the kinetics, the subsequent sections focus on examples of the graft polymerization of VP and VAc onto silica. The discussion presented here is restricted to the long-chain hypothesis (invariance of kinetic parameters with respect to chain length), an approach which is sufficient to capture the main features of most free-radical polymerization reactions.

A. Graft Polymerization Kinetics

As stated earlier, the kinetics of free-radical graft polymerization can be conveniently divided into the following four parts:

(1) Initiation:

Initiator decomposition

$$I_2 \xrightarrow{k_d} 2I\cdot \qquad (2)$$

Chain initiation

$$I\cdot + M \xrightarrow{k_{im}} M_1\cdot \qquad (3)$$

$$I\cdot + S \xrightarrow{k_{is}} S_1\cdot \qquad (4)$$

(2) Propagation:

Homopolymerization

$$M_1\cdot + M \xrightarrow{k_{pmm}} M_2\cdot \qquad (5)$$

$$\vdots$$

$$M_n\cdot + M \xrightarrow{k_{pmm}} M_{n+1}\cdot \qquad (6)$$

Graft polymerization

$$S_1\cdot + M \xrightarrow{k_{psm}} S_2\cdot \qquad (7)$$

$$\vdots$$

$$S_n\cdot + M \xrightarrow{k_{psm}} S_{n+1}\cdot \qquad (8)$$

(3) Chain transfer:

Chain transfer with monomer

$$M_n\cdot + M \xrightarrow{k_{trmm}} H_n + M_1\cdot \qquad (9)$$

$$S_n\cdot + M \xrightarrow{k_{trsm}} G_n + M_1\cdot \qquad (10)$$

Chain transfer with surface-active site

$$\mathrm{M}_n\cdot + \mathrm{S} \overset{k_{\mathrm{trms}}}{\rightarrow} \mathrm{H}_n + \mathrm{S}_1\cdot \tag{11}$$

$$\mathrm{S}_n\cdot + S \overset{k_{\mathrm{trss}}}{\rightarrow} Gn + S_1 \tag{12}$$

(4) Termination

$$\mathrm{M}_m\cdot + \mathrm{M}_n\cdot \overset{k_{\mathrm{tcmm}}}{\rightarrow} \mathrm{H}_{m+n} \tag{13}$$

$$\mathrm{M}_m\cdot + \mathrm{M}_n\cdot \overset{k_{\mathrm{tdmm}}}{\rightarrow} \mathrm{H}_m + \mathrm{H}_n \tag{14}$$

$$\mathrm{M}_m\cdot + \mathrm{S}_n\cdot \overset{k_{\mathrm{tcms}}}{\rightarrow} \mathrm{G}_{m+n} \tag{15}$$

$$\mathrm{M}_m\cdot + \mathrm{S}_n\cdot \overset{k_{\mathrm{tdms}}}{\rightarrow} \mathrm{H}_m + \mathrm{G}_n \tag{16}$$

$$\mathrm{S}_m\cdot + \mathrm{S}_n\cdot \overset{k_{\mathrm{tcss}}}{\rightarrow} \mathrm{G}_{m+n} \tag{17}$$

$$\mathrm{S}_m\cdot + \mathrm{S}_n\cdot \overset{k_{\mathrm{tdss}}}{\rightarrow} \mathrm{G}_m + \mathrm{G}_n \tag{18}$$

where I_2 and $\mathrm{I}\cdot$ are the initiator and initiator radical, $\mathrm{M}\cdot$ and $\mathrm{S}\cdot$ are the free and surface-bonded polymer radicals, M and S are the monomer and surface-active site, and H_n and G_n are the terminated homo- and grafted polymers, respectively. The initiator decomposition rate coefficient is given by k_{d}, Eq. (2). The monomer and surface site initiation rate constants are denoted by k_{im} and k_{is}, Eqs (3) and (4). The homopolymer propagation rate coefficient is given by k_{pmm}, Eqs (5) and (6). The grafted polymer propagation rate coefficient k_{psm}, Eqs (7) and (8), is for the reaction of a growing grafted polymer with a monomer. The reaction rate coefficients for chain transfer reactions of a growing homopolymer ($\mathrm{M}_n\cdot$) and a growing grafted polymer ($\mathrm{S}_n^{\cdot}$) with a monomer are given by k_{trmm} and k_{trsm}, Eqs (9) and (10). Similarly, the reaction rate coefficients for chain transfer reactions of a growing homopolymer and a growing grafted polymer with a surface site are given by k_{trms} and k_{trss}, Eqs (11) and (12). The combination and disproportionation rate coefficients are given by k_{tcmm} and k_{tdmm} for termination reactions between two growing homopolymer chains [Eqs (13) and (14)], k_{tcms} and k_{tdms} for termination reactions between a growing homopolymer chain and a growing grafted chain [Eqs (15) and (16)], and k_{tcss} and k_{tdss} for termination reactions between two growing grafted chains [Eqs (17) and (18)].

Experimental studies of the free-radical graft polymerization of VP onto vinylsilane-modified silica indicated that polymer chains are grafted onto a small fraction of the initial surface sites [20]. The low graft efficiency (contributed by both polymer grafting and graft polymerization) suggests that not all surface vinyl groups are available for grafting; some surface sites may be shielded by the neighboring grafted chains. Since only a small fraction of the surface-active sites is converted, the surface density of vinyl groups is essentially constant during the grafting process ($S \approx S_0$).

The rate at which monomer is incorporated into the grafted polymer (gmol/l.min) is given by

$$R_G = PG + GP \tag{19}$$

where PG and GP denote the rates of grafted polymer formation due to polymer grafting and graft polymerization, respectively. These two rates are given by

$$PG = DP_h k_{tcms}[\mathrm{M}\cdot][\mathrm{S}\cdot] \tag{20}$$

$$GP = k_{psm}[\mathrm{S}\cdot][\mathrm{M}] \tag{21}$$

where both PG and GP are in the unit of moles of monomer/volume solution/time; DP_h is the number-average degree of homopolymerization. This factor must be included since each homopolymer radical coupling at the surface contains DP_h monomer units.

The graft efficiency (f), which is defined as the fraction of the initial surface sites converted into grafted chains, is related to the polymer graft yield G (g surface monomer/m^2 surface) and the number-average molecular weight of the grafted polymer M_n by

$$f = \frac{G}{M_n S_a} \tag{22}$$

where S_a is the initial surface density of vinylsilane (mol vinylsilane/m^2 surface).

Although a simple initiation scheme has been demonstrated in this section, it is known that more elaborate mechanisms, which account for the complex initiator–monomer interactions, are available in the literature as discussed in the next section.

B. Initiation Mechanisms

Over the past two decades, numerous investigators have confirmed that the rate of free-radical homopolymerization of vinyl monomers can be greater than first-order dependence on monomer concentration [57]. A number of studies have rationalized this phenomenon by regarding the initiation efficiency as being proportional to the monomer concentration [63–66]. Although this assumption predicts an overall kinetic order in accord with experimental observations, it provides no insightful description of the molecular mechanism of the initiation process. This section discusses the complex-cage model and the high initiation rate mechanism, which are particularly suitable for the graft polymerization of VP and VAc onto silica.

1. Complex-Cage Mechanism

The early studies of Gee and coworkers [67,68], Matheson [69], Burnett and Loan [70] and Allen et al. [71] on free-radical homopolymerization with peroxide initiators proposed the concept of complex-cage structures that can account for a kinetic order of up to 1.5 power dependence on the monomer concentration. Noyes [72] broadened the scope of the cage-effect theory by developing a hierarchal cage structure. This latter approach assumed that the initiator can form a compact cage. In the compact-cage regime, the distance between fragments in an initiator is less than one molecular diameter, and the time scale of fragment recombination rate is greater than the time scale of monomer diffusive displacement rate. When the initiator fragments have diffused to a distance greater than one molecular diameter, they are in the diffused-cage regime where the time scales of fragment recombination and monomer diffusion are comparable in their magnitude. Tertiary recombination is also possible for fragments from different initiator molecules. More recent studies have proposed that the diffuse-cage initiator can associate not only with a monomer [57], but also with surface-active sites (the vinyl groups in the case of vinylalkoxy-silane molecules) to form a monomer–initiator associate or a surface site–initiator associate [21] where the bonding between the diffuse-cage initiator and the monomer or the surface site is weak. The weakly bonded monomer– and surface site–initiator associates can then decompose to yield an initiator radical and a corresponding free or surface polymer radical, respectively. The various steps in the complex-cage model are summarized in the next subsection.

(a) Proposed Mechanism for Hydrogen Peroxide Initiator,

Formation of a caged hierarchy:

$$I_2 \underset{k_1'}{\overset{k_1}{\rightleftharpoons}} (I_2) \overset{k_2}{\rightarrow} (I_2^*) \overset{k_d^*}{\rightarrow} 2I\cdot \tag{23}$$

$$(I_2^*) \overset{k_r}{\rightarrow} I_2 \tag{24}$$

Formation of monomer–initiator and surface–initiator associates:

$$(I_2^*) + M \overset{k_{am}^*}{\rightarrow} (IMI) \tag{25}$$

$$(I_2^*) + S \overset{k_{as}^*}{\rightarrow} (ISI) \tag{26}$$

Dissociation of Monomer–initiator and surface–initiator associates:

$$(IMI) \overset{k_{bm}}{\rightarrow} I\cdot + M_1\cdot + H_2O \tag{27}$$

$$(ISI) \overset{k_{bs}}{\rightarrow} I\cdot + S_1\cdot + H_2O \tag{28}$$

$$(IMI) \overset{k_{cm}}{\rightarrow} Q_m \tag{29}$$

$$(ISI) \overset{k_{cs}}{\rightarrow} Q_s \tag{30}$$

Primary chain initiation:

$$I\cdot + M \overset{k_{im}}{\rightarrow} M_1\cdot \tag{31}$$

$$I\cdot + S \overset{k_{is}}{\rightarrow} S_1\cdot \tag{32}$$

where (I_2) and (I_2^*) designate the compact- and diffuse-cage initiators. The diffuse-cage initiator decomposes to yield the primary initiator radical $I\cdot$, which reacts with a monomer M and a surface vinyl group S to initiate new polymer radicals $M\cdot$ and $S\cdot$. The forward and reverse reaction-rate coefficients for the formation and recombination are given by k_1 and k_1' for the compact-cage initiator [Eq. (23)], and k_2 and k_r for the diffuse-cage initiator [Eqs (23) and (24)]. The reaction rate coefficient for the thermal decomposition of the diffuse-cage initiator is given by k_d^*, Eq. (23). The reaction rate coefficients for the monomer–initiator and surface–initiator associate formation are denoted by k_{am}^* and k_{as}^*, Eqs (25) and (26). The dissociation rate coefficients of the monomer–initiator and surface–initiator associates are given by k_{bm} and k_{bs}, Eqs (27) and (28). The reaction rate coefficients for the deactivation of IMI and ISI into nonreactive products Q_m (in the solution) and Q_s (on the surface) are designated by k_{cm} and k_{cs}, Eqs (29) and (30). Reactions (29) and (30) are included for generality as suggested by Hunkeler [57]. Finally, the primary chain initiation-rate coefficients for the monomer and surface-active site are given by k_{im} and k_{is}, Eqs (31) and (32).

(b) Analysis of Poly(vinylpyrrolidone) Polymerization Data.

With the complex-cage initiation scheme, Eqs (23)–(32), the total initiator concentration $[I_t]$ is given by the sum of concentrations of the undissociated initiator $[I_2]$, compact-caged initiator $[(I_2)]$, and diffuse-cage initiator $[(I_2^*)]$:

$$[I_t] = [I_2] + [(I_2)] + [(I_2^*)] \tag{33}$$

where

$$[I_2] = \varphi_0[I_t] \tag{34}$$

$$[(I_2)] = \varphi_1[I_t] \tag{35}$$

$$[I_2^*)] = \varphi_2[I_t] \tag{36}$$

in which φ_0, φ_1, and φ_2 are mole fractions of the undissociated, compact-, and diffuse-cage initiators, respectively, such that

$$\varphi_0 + \varphi_1 + \varphi_2 = 1 \tag{37}$$

To obtain the overall rate of polymerization, it is convenient to define the overall rate coefficient for the thermal bond rupture k_d as $k_d = \varphi_2 k_d^*$ [57]. Similarly, it can be defined that $k_{am} = \varphi_2 k_{am}^*$, where k_{am} and k_{am}^* are the overall and actual association rate coefficients.

Based on the complex-cage scheme, Eqs (23)–(32), combined with the propagation, chain transfer, and termination reactions given by Eqs (5)–(18), the overall rate of polymerization R_p can be written in terms of lumped kinetic parameters as

$$R_p = (K_T + K_M[M])^{\frac{1}{2}}[M][I_t]^{\frac{1}{2}} \tag{38}$$

where

$$K_T = \left[k_{pmm}^2\left(f_d\frac{k_d}{k_{tmm}}\right)\right] \tag{39}$$

and

$$K_M = \left[k_{pmm}^2\left(f_m\frac{k_{am}}{k_{tmm}}\right)\right] \tag{40}$$

where K_T and K_M are the lumped rate coefficients for the thermal bond rupture [Eq. (23)] and monomer-enhanced decomposition [Eq. (27)] of the diffuse-cage initiator; the corresponding efficiencies are given by f_d and f_m. The overall rate of polymerization is thus proportional to the total initiator concentration to the one half power and has an effective rate order ranging from 1 to 1.5 with respect to the monomer concentration. At high temperatures and/or high conversions, where the thermal decomposition dominates [the first term on the right-hand side of Eq. (38) is large relative to the second term], the overall rate of polymerization is linearly proportional to the monomer concentration. Monomer-enhanced decomposition, which dominates at low temperatures and/or high monomer concentrations, however, yields a 1.5 rate order.

When empirical representation of rate data is sought, Eq. (38) can be rewritten as

$$R_p = K_P[I_t]^{\frac{1}{2}}[M]^{\alpha} \tag{41}$$

where K_P denotes the lumped polymerization rate coefficient, and α is the overall kinetic order with respect to the monomer. When the initiator decomposes primarily through thermal bond rupture, α approaches unity. An apparent reaction order greater than unity signifies the presence of monomer-enhanced decomposition.

(c) Case Study: Vinyl(pyrrolidone) Polymerization with Hydrogen Peroxide Initiator.

The experimental and calculated [Eq. 38] monomer conversion data for VP in the range 70°C–90°C are shown in Fig. 5. The initial VP monomer, surface vinylsilane, and hydrogen peroxide initiator concentrations for the above system are $[M_0] = 2.81$ M, $S_a = 31$ μmol/m^2, and $[I_0] = 0.043$ M, respectively. It is evident from Fig. 5 that monomer conversion increases with temperature, and the complex-cage model is capable of accurately describing the overall rate of polymerization. According to Eq. (38), the present model predicts that the kinetic order (with respect to the monomer concentration) is an exclusive function of the relative chemical interactions between the monomer–initiator and initiator–initiator pairs (hydrogen bonding interactions in the case of the VP/hydrogen peroxide system). At high monomer concentrations and/or low reaction temperatures, where both free and grafted poly-

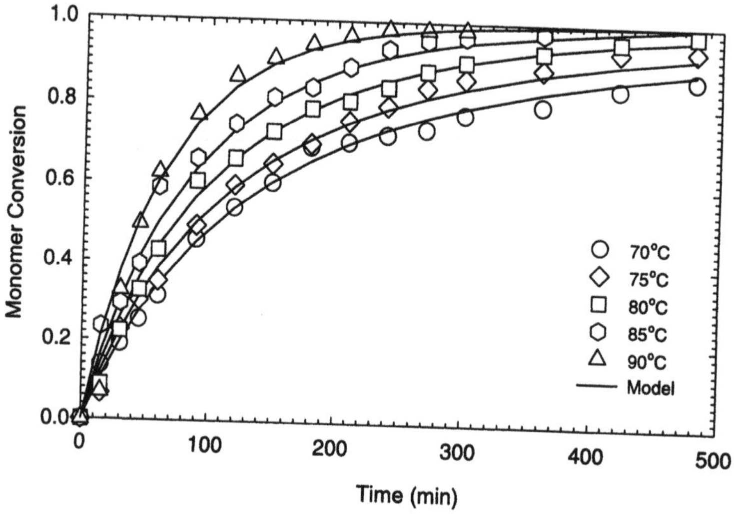

FIG. 5 Monomer conversion data for aqueous vinylpyrrolidone graft polymerization. The initial vinylpyrrolidone monomer, surface vinylsilane, and hydrogen peroxide initiator concentrations are $[M_0] = 2.81$ M, $S_a = 31$ μmol/m^2, and $[I_0] = 0.043$ M, respectively.

mer radicals are initiated primarily due to the monomer-enhanced decomposition, the present model predicts a 1.5 kinetic order. Conversely, at low monomer concentrations (toward the end of the polymerization reaction) and/or high reaction temperature, the kinetic order approaches unity, indicating a dominating chain initiation resulting from thermal bond rupture.

The fitted lumped rate coefficients for the thermal bond rupture K_T and monomer-enhanced decomposition K_M [Eq. (38)] are shown in Fig. 6. Although both rate constants increase with temperature according to the Arrhenius relationship, K_T is more temperature sensitive. Clearly, the increase in chain initiation rate with temperature is attributed mainly to thermal bond rupture. At high temperatures, the thermal energy possessed by fragments of the diffuse-cage initiator increases, producing a higher displacement frequency away from their equilibrium bond distance. This increasing displacement frequency, coupled with an increase in concentration of the diffuse-cage initiator at the expense of the compact-cage initiator, enhances the probability of the thermal bond rupture process. On the other hand, the increase in the apparent monomer association rate constant is due solely to the increasing concentration of the diffuse-cage initiator, rather than the probability of the actual monomer association process, which becomes more energetically unfavorable as the reaction temperature rises. The negative temperature dependence of the actual association phenomenon is anticipated since the probability of coupling between a monomer and a diffuse-cage initiator drastically decreases with the increasing frequency of displacement of the initiator fragments.

Although Fig. 5 illustrates excellent fits of the complex-cage model, Eq. (38), it is noted that equally good fits of the experimental conversion data are obtained from the emprical relation given by Eq. (41) in which the overall rate constant K_P and reaction order α are treated as adjustable parameters. Although such an empirical approach is useful for data analysis and reactor scale up, it does not provide information regarding the detailed reaction mechanism.

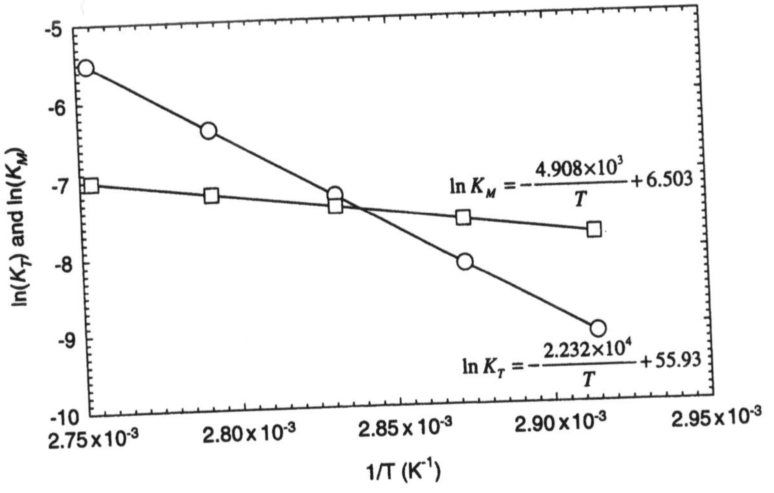

FIG. 6 Lumped rate constants for thermal bond rupture (K_T; l/gmol.min)^2 and monomer-enhanced decomposition (K_M; l/gmol.min^2) for vinylpyrrolidone graft polymerization.

2. High Initiative Rate Mechanism

The kinetic order of the free-radical polymerization of some vinyl monomers can exceed a 1.5 power dependence on the monomer concentration. Such behavior is observed, for example, in the polymerization of VAc with azo-initiators [46, 58–62]. To account for this high reaction order, Ito [58–62] proposed that, under high reaction temperatures and/or low monomer concentrations, the initiator radicals can also be involved in termination reactions with the polymer radicals. The chain initiation and propagation reactions of free and grafted polymer radicals, the chain transfer reactions with monomer and surface, and the termination reactions between polymer radicals remain identical to those given by Eqs (3)–(18).

(a) Initiator Decomposition and Additional Termination Reactions for 2,2′-Azobis-2,4-dimethylvaleronitrile (ABVN) Initiator.

$$A \xrightarrow{k_d} 2I\cdot + B \tag{42}$$

$$2I\cdot \xrightarrow{k_{ii}} Q + T \tag{43}$$

$$I\cdot + M_n\cdot \xrightarrow{k_{tim}} P_n \tag{44}$$

$$I\cdot + S_n\cdot \xrightarrow{k_{tis}} Y_n \tag{45}$$

where A denotes the undissociated initiator, B is a decomposition byproduct, and Q and T are (neutral) recombination products between two initiator radicals. The initiator radical $I\cdot$, besides initiating new free and grafted chains, can also terminate with another initiator radical, Eq. (43), or participate in termination reactions with a growing homopolymer chain $M\cdot$, Eq. (44), or a growing grafted chain $S\cdot$, Eq (45). The corresponding rate coefficients for the above three termination reactions, Eqs (43)–(45) are given by k_{ii}, k_{tim}, and k_{tis}, respectively. Finally, P_n and Y_n denote the terminated homo- and grafted polymers.

Under normal reaction conditions, the primary initiator radicals are highly reactive and most monomers are efficient radical scavengers. Consequently, most initiator radicals are readily removed from the solution by reacting with the monomer, Eq. (3). Termination reactions beween the initiator radical and the polymer radicals, Eqs (44) and (45), are usually neglected in the derivation of kinetic relations. However, at high temperatures and/or low monomer concentrations, the initiator radical concentration increases significantly and the monomer is incapable of acting as an efficient radical scavenger. This phenomenon leads to a high-rate termination scheme as given by Eqs (42)–(45). In this case, the competition between the monomer and polymer radicals for the initiator radical results in a rate of initiation which depends on the monomer concentration. The termination process no longer occurs solely via normal bimolecular reactions between macroradical species.

(b) Analysis of PVAc Polymerization Data.

Analysis of the polymerization scheme, given by Eqs (3)–(18) and Eqs (42)–(45), subject to the standard pseudo steady-state assumption for radical species, leads to the following expression for the rate of polymerization:

$$R_p = k_{pmm}[M][M\cdot] + k_{im}[M][I\cdot] \tag{46}$$

where

$$[M\cdot] = \frac{k_{im}}{k_{tim}}[M] \tag{47}$$

$$[I\cdot] = \left\{\left[\frac{2fk_d[A]}{k_{ii}} + \left(\frac{k_{im}[M]}{k_{ii}}\right)^2\right]^{\frac{1}{2}} - \frac{k_{im}[M]}{k_{ii}}\right\} \tag{48}$$

The propagation is the main rate of monomer consumption [58, 60], and thus the first term on the right-hand side of Eq. (46) dominates, leading to the following approximation:

$$R_p = \frac{k_{im}k_{pmm}}{k_{tim}}[M]^2 \tag{49}$$

The above quadratic kinetic order [Eq. (49)] was reported by Ito [58] for the free-radical polymerization of VAc with the ABVN initiator at 50°C. Recent data for VAc polymerization, under reaction conditions such that Eq. (49) applies, are shown in Fig. 7. The fitted lumped rate coefficient $k_p = k_{im}k_{pmm}/k_{tim}$ increases with temperature according to the Arrhenius relationship (Fig. 8). The above kinetic observations are consistent with the high initiation rate model, which is applicable at low monomer concentrations and/or high initiator concentrations.

C. Graft Polymerization of Vinylpyrrolidone onto Silica

1. Polymer Graft Yield

Control and optimization of the grafting process can be achieved once the kinetic rate parameters are known. This can be illustrated with data for VP graft polymer-

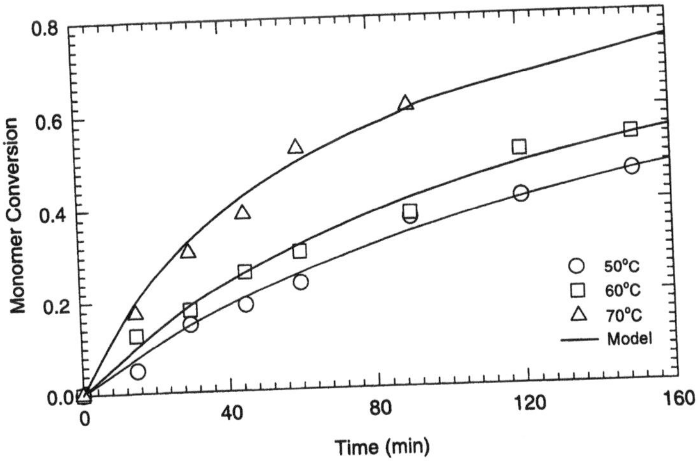

FIG. 7 Monomer conversion data for aqueous vinyl acetate graft polymerization with ABVN initiator. The initial vinyl acetate monomer, surface vinylsilane, and ABVN initiator concentrations are $[M_0] = 4.33$ M, $S_a = 10^{-3}$ M, and $[I_0] = 0.03$ M, respectively.

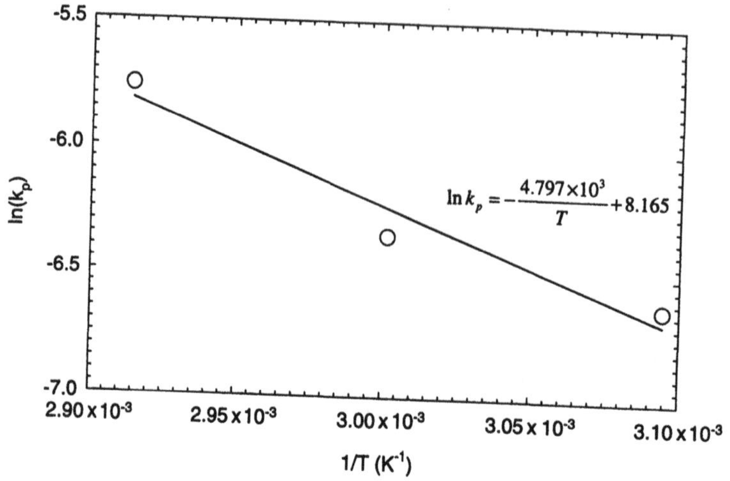

FIG. 8 Lumped rate coefficient $k_p(l/\text{gmol.min})$ for vinyl acetate polymerization with ABVN initiator.

ization provided by Chaimberg and Cohen [33] and more recently by Jou [21], using the cage-complex initiation mechanism, Eqs (23)–(32). A typical fit of the cage-complex model to experimental graft yield data is shown in Fig. 9. Due to the sensitivity limits of thermogravimetric analysis, coupled with the relatively minute amount of the grafted polymer, experimental graft yield data are more scattered (with an average error of less than 15%) relative to monomer conversion data. Nevertheless, analogous to the monomer conversion data, an increasing trend with temperature is still obvious. Although not shown here, the present model also predicts an increase in the polymer graft yield at higher initial monomer and/or surface vinylsilane concentrations.

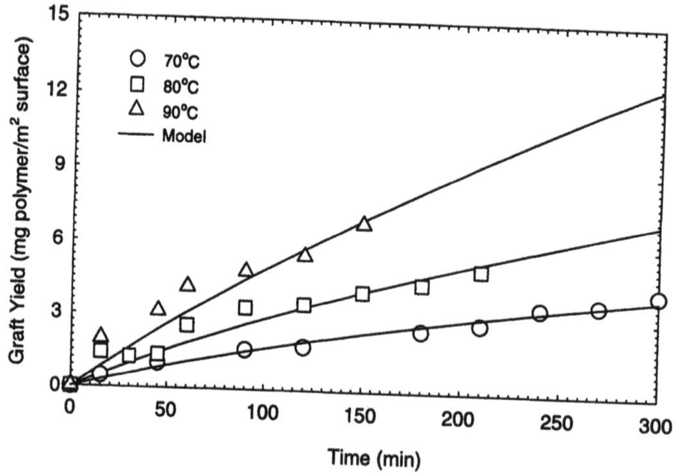

FIG. 9 Experimental and calculated poly(vinylpyrrolidone) graft yield data. The initial vinylpyrrolidone monomer, surface vinylsilane, and hydrogen peroxide initiator concentrations are $[M_0] = 4.68$ M, $S_a = 46\,\mu\text{mol/m}^2$, and $[I_0] = 0.043$ M, respectively.

2. Polymer Grafting and Graft Polymerization

The complete kinetic model is nonlinear, and the kinetic rate parameters of specific interest are k_{tcms} [Eq. (15)] and k_{psm} [Eqs (7) and (8)], which are associated with the polymer grafting and graft polymerization, respectively (Fig. 10). Analysis of VP graft polymerization data reveals that both k_{tcms} and k_{psm} increase with temperature as predicted by the Arrhenius equation [21]. In addition, the rate coefficient for polymer grafting is less temperature sensitive and is roughly four orders of magnitude larger than for graft polymerization. As an illustration, the percentage contribution of polymer grafting to the total polymer graft yield for the VP/silica system is provided in Fig. 11. The polymer grafting rate coefficient k_{tcms} for PVP increases with temperature, consistent with the increasing mobility of the polymer radicals, even though their mobility is partially offset by the increasing viscosity of the reaction medium at high conversions. In addition, as the reaction temperature rises, owing to a higher rate of polymerization, both rates of monomer consumption and polymer radical formation increase. In other words, an increase in the reaction temperature yields a higher ratio of concentrations of polymer radical to monomer. The same argument holds with respect to monomer conversion. Therefore, although both rates of polymer grafting and graft polymerization increase with temperature, the rate of polymer grafting is more temperature sensitive owing to the combined effects of the augmented chain mobility and increased concentration of growing homopolymer. Ultimately, graft polymerization is still the dominant process for the growth of surface chains (Fig. 11), but its contribution relative to polymer grafting declines with rising reaction temperature. On the other hand, since the rate of graft polymerization increases more rapidly with increasing monomer concentration, the percentage contribution of polymer grafting to the total graft yield decreases with increasing initial monomer concentration.

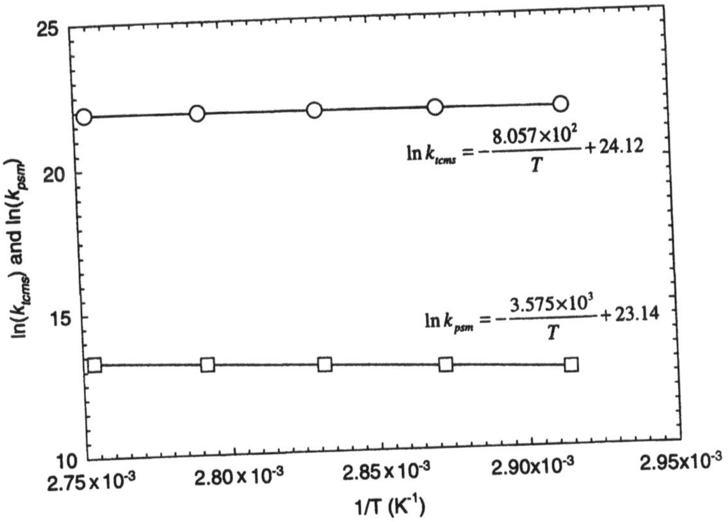

FIG. 10 Rate coefficients for polymer grafting (k_{tcms}, l/gmol.min) and graft polymerization (k_{psm}, l/gmol.min) of poly(vinylpyrrolidone).

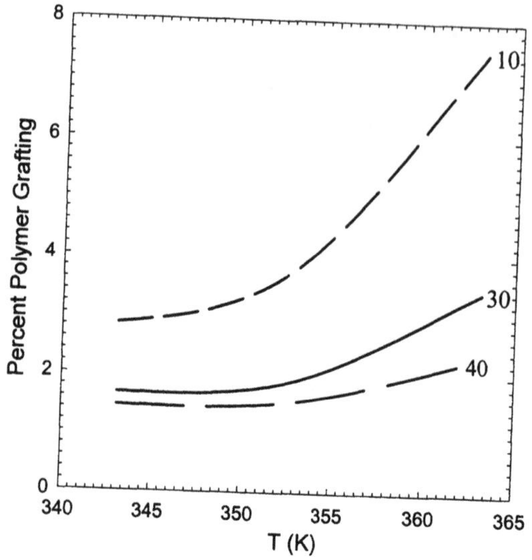

FIG. 11 Percentage contributions of polymer grafting to the overall polymer graft yield of poly(vinylpyrrolidone). The three curves correspond to initial vinylpyrrolidone monomer concentrations of 10, 30, and 40% by volume. The initial hydrogen peroxide initiator concentration is $[I_0] = 0.043$ M.

3. Molecular Weight of Grafted Polymer

The calculated instantaneous number-average degree of graft polymerization decreases slightly with increasing monomer conversion (Fig. 12). From the expressions for polymer grafting [Eq. (20)] and graft polymerization [Eq. (21)], it is obvious that the average grafted chain length is governed by the relative magnitudes of the rate of polymer grafting, which should depend on the length of free macroradicals coupling at the surface, and the rate of graft polymerization. The average grafted polymer chain length is also affected by the initial number of surface vinyl groups [21]. Although the total graft yield always increases with temperature and/or initial monomer concentration (Fig. 9), monomer molecules added to the surface are distributed among all surface chains. Therefore, the average grafted molecular weight could conceivably remain unaffected with modest increases in the reaction temperature and/or initial monomer concentration, given a simultaneous increase in the concentration of utilized surface vinylsilanes. However, as a dense grafted phase is formed, the initiation of new chains is likely to be limited by the diffusion of reactive species to the solid surface through an already bonded polymer layer. As a consequence, a further increase in the reaction temperature and even in initial monomer concentration is unlikely to lead to any additional increase in graft efficiency (fraction of the initial vinylsilanes on to which polymer chains are grafted), but it will result in an increase in the molecular weight of the grafted polymer.

D. Graft Polymerization of Vinyl Acetate onto Silica

The polymer graft yield for VAc graft polymerization on to inorganic oxides follows a similar trend as previously illustrated for VP. As an illustration, the polymer graft

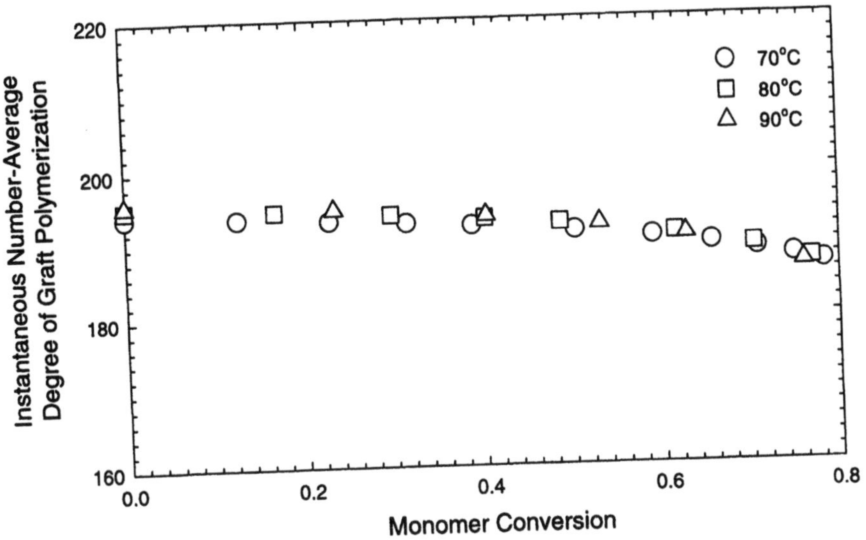

FIG. 12 Instantaneous number-average degrees of poly(vinylpyrrolidone) graft polymerization calculated at discrete monomer conversions. The initial vinylpyrrolidone monomer, surface vinylsilane, and hydrogen peroxide initiator concentrations are $[M_0] = 4.68$ M, $S_a = 46\,\mu mol/m^2$, and $[I_0] = 0.043$ M, respectively.

yield data for the VAc/silica system are shown in Fig. 13 for three different temperatures and an initial ABVN initiator concentration of $[I_0] = 0.03$ M [46]. Clearly, reaction temperature has a marked effect on polymer graft yield. For example, an increase in reaction temperature from 60° to 70°C resulted in a 32% increase in the polymer graft yield at a reaction time of 2.5 h. The trend of increasing polymer graft yield with temperature has also been observed in several other studies [29, 31, 47]. As discussed earlier for the graft polymerization of VP on to silica, polymer chain mobility increases with rising temperature, thereby increasing the rate of polymer grafting. It is noted, however, that the contribution of polymer grafting to the total graft yield remains significantly lower relative to that of graft polymerization.

It has also been shown that, for the VAc/silica system, polymer graft yield increases significantly with initial monomer concentration. Since the rate of graft polymerization is directly proportional to the monomer concentration [Eqs (7) and (8)], whereas the rate of polymer grafting only increases slightly with monomer concentration [Eq. (15)], graft polymerization becomes more favorable at higher initial monomer concentrations. In addition, with increasing monomer concentration, termination reactions with polymer radicals [Eqs (44) and (45)] are expected to become less significant given the reduction in the concentration of available initiator radicals due to monomer scavenging [Eq. (3)]. In other words, an increase in initial monomer concentration, while keeping the initial initiator and surface vinylsilane concentrations constant, would increase grafted chain length.

The number of available sites for polymer grafting is dictated by the initial vinylsilane surface density. However, the initial surface vinylsilane concentration should have the same degree of rate enhancement for both polymer grafting and graft polymerization since both rates are proportional to the grafted polymer radical concentration ($S\cdot$), whose magnitude is in turn proportional to the initial concen-

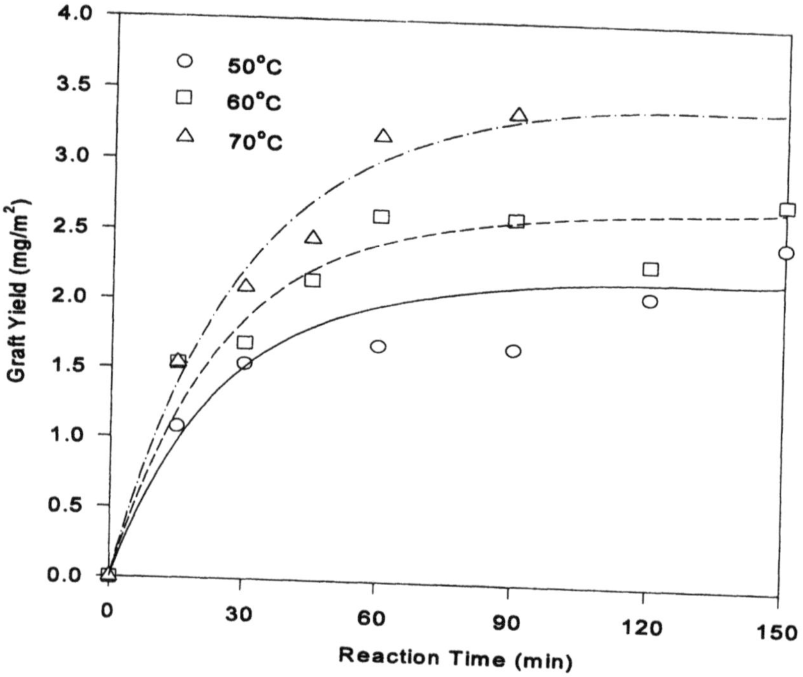

FIG. 13 Experimental and calculated poly(vinyl acetate) graft yield data. The initial monomer, surface vinylsilane, and ABVN initiator concentrations are $[M_0] = 4.33$ M, $S_a = 10^{-3}$ M, and $[I_0] = 0.03$ M, respectively.

tration of surface vinylsilane molecules. The effect of surface vinylsilane concentration on the graft density and grafted chain length is expected to be similar to that observed for the PVP/silica system. The graft polymerization of VAc, however, is known to be susceptible to chain transfer reactions which are likely to affect the grafted polymer molecular weight and its distribution. Clearly, additional studies are needed for such systems in order to map their detailed kinetics, and thus to optimize the desired surface properties.

IV. PHYSICAL PROPERTIES AND HYDRODYNAMICS OF GRAFTED POLYMER

The ultimate purpose of surface modification by graft polymerization, as discussed previously, is to alter surface properties of the substrate so as to manipulate its physicochemical interactions with the surrounding fluid medium. Knowledge of surface properties such as molecular weight, surface density, and chemical structure is essential to predicting solute–substrate and solvent–substrate interactions, which in turn govern the macroscopic behavior (adsorption, wetting) of the modified surface.

Surface density of grafted polymers can be characterized by the distance D between two nearest anchoring points on the surface. This distance is related to grafting efficiency (f) and initial surface concentration of available anchoring sites (S_a) by

$$D = (fS_aN_A)^{-\frac{1}{2}} \tag{50}$$

where N_A is the Avogadro number. Equation (50) is based on the approximation of evenly spaced square grids of anchoring points on the surface. Since each surface vinyl group presents a potential anchoring point, the number of sites available for grafting should increase with the extent of silylation. Therefore, a high graft density (or a small D) can be obtained by allowing the formation of a polysilane network during the silylation reaction [an increase in S_a in Eq. (50). Of course, the actual fraction of initial surface sites on to which polymer chains are grafted also depends on the grafting efficiency, Eq. (50). The grafting efficiency, in principle, can be determined by experimental analysis of the initial number of surface vinyl silanes, the polymer graft yield, and the number-average molecular weight of the grafted polymer, Eq. (22).

In the dense grafting regime, as the graft density increases to a point where the distance between two neighboring chains becomes less than their radius of gyration [75], the polymer coils begin to overlap. In order to maximize their entropy, the grafted chains extend away from the surface, forming a layer that resembles the bristles on a closely packed brush. This regime of dense surface chains is termed a brush layer. For grafted polymer layers with a uniform segment distribution, according to the scaling analysis of de Gennes [72] for a good solvent condition, the terminally anchored chains can be considered to be in a stretched (or brush) configuration when the following criterion is met:

$$\sigma = \left(\frac{a}{D}\right)^2 > DP_g^{-6/5} \tag{51}$$

where σ is the fractional surface coverage of the grafted polymer, a is the size of a monomer unit, and DP_g is the number-average degree of graft polymerization. The fractional surface coverage for the free-radical graft polymerization is typically low ($\sigma < 2\%$).

Various theoretical approaches [72–79] have been developed to predict the brush height, density distribution, and solute partitioning. In this chapter, however, a simple scaling analysis [72] for a good solvent condition will suffice to illustrate calculation of the grafted polymer thickness from the graft density and grafted chain length. For grafted polymer layers obeying the polymer brush criterion given by Eq. (51), the volume fraction ϕ_N of the grafted polymer scales with the fractional surface coverage as

$$\phi_N \propto \sigma^{2/3} \tag{52}$$

and the brush height can be approximated by

$$h = DP_g a\sigma^{1/3} \tag{53}$$

Equation (53) suggests that the maximum height of the brush, if the chains were fully extended, is given by $h = DP_g a$. It is noted that an equivalent brush height can be obtained from either a denser surface coverage with shorter chains (a larger σ and a smaller DP_g) or a less dense grafted layer with longer chains (a smaller σ and a larger DP_g). A denser grafted layer of shorter chains may yield a brush layer if chain spacing is sufficiently small such that Eq. (51) is satisfied. A less dense polymer layer of longer chains may also result in a brush layer if the radius of gyration of each chain is sufficiently large for chain overlap to occur. In the limit of low surface density, where the distance between neighboring chains is sufficiently large so that there is negligible overlap, the height of the polymer layer can be estimated by the Flory radius for a single chain [73]. Clearly, in this regime the definition of an average polymer thickness is somewhat vague since chain spacing is too high to define a uniform polymer layer.

Studies have shown that it is also important to consider a continuous segment concentration profile from the solid surface to the edge of the brush layer, rather than a simple step concentration profile of height h as implied in the above scaling analysis. For example, the theoretical self-consistent field (SCF) analysis of Milner et al. [75] predicts that, given a SCF-calculated equilibrium brush height h^*, the polymer volume fraction $\phi(z)$ in the brush layer follows a parabolic distribution with respect to the distance z from the surface:

$$\phi(z) = C[(h*)^2 - z^2] \tag{54}$$

where C is a proportionality constant related to the effective potential coefficient and the excluded-volume parameter as described by Milner et al. [75]. Equation (54) predicts a polymer segment profile varying smoothly from the surface ($z = 0$) to the edge of the brush layer ($z = h*$).

Although the grafted polymer can be easily deformed as a consequence of changing solvent power or applied shear rate, the chains remain covalently attached to the surface. The polymer conformational change is particularly important when polymer chains are grafted inside porous substrates, such as in silica size exclusion chromatography resins and ceramic membranes. The deformation of the grafted polymer layer can be quantified in terms of the effective hydrodynamic thickness L_H, determined from permeability measurements in a column packed with nonporous inorganic oxide particles which have been previously grafted with the selected polymer. The hydrodynamic permeability measurement is expected to be especially sensitive to long polymer chains, a fact which is well known from L_H measurements of adsorbed polymers [10, 77, 80, 81]. For a packed bed of polymer-modified porous particles, assuming a simple cylindrical pore geometry, the hydrodynamic thickness of the grafted layer, in a given solvent, can be determined from

$$L_H = R_E\left[1 - \left(\frac{k_g}{k}\right)^{1/4}\right] \tag{55}$$

where k and k_g are the solvent permeabilities in a column packed with the nongrafted and polymer-grafted particles, respectively. The effective pore radius R_E is given by

$$R_E = \frac{D_p\varepsilon}{12(1-\varepsilon)} \tag{56}$$

where D_p is the particle diameter, and ε is the porosity of the packed column [82]. In Eq. (55), the solvent permeabilities k and k_g can be obtained from flow measurements in a column packed with native inorganic oxide particles and a column packed with the polymer-grafted particles, respectively. These permeabilities, which are specific to a particular solvent/polymer/substrate system, are calculated using Darcy's law:

$$k = Q\frac{\mu L}{A\Delta P} \tag{57}$$

where Q denotes the solution volumetric flow rate, L and A are the column length and cross-sectional area, ΔP is the pressure difference over the column, and μ is the solution viscosity. The interpretation of the above representation is that a decrease in column permeability due to the addition of a polymer-grafted phase is equivalent to a decrease in the effective pore diameter, or an increase in the effective hydrodynamic thickness of the polymer-grafted layer. If the surface density of the grafted polymer is sufficiently high, or if the chain length is small relative to the pore radius, the column permeability is expected to be invariant with respect to shear stress in the interstitial pores of the column [10].

An illustration of the above approach is demonstrated by the hydrodynamic behavior of three porous PVP-grafted silica tubes as reported by Castro et al. [83] (Fig. 14). A decrease in the hydrodynamic thickness of the grafted polymer layer with pore–wall shear stress is evident for the substrate with the longest grafted chains (Tube II). For short chain surfaces (Tubes I-A and I-B), however, the percentage change in hydrodynamic polymer layer thickness is smaller.

A simple analysis of the flow inside a polymer-grafted pore can be performed to illustrate the effect of polymer graft density and grafted chain length on the pore velocity profile. Considering a simple cylindrical pore geometry, the flow field can be divided into a core region (outside the brush layer) and a brush layer region. In the core region, the axial velocity profile is described by the classical equation of motion:

$$\frac{1}{r}\frac{\mathrm{d}}{\mathrm{d}r}\left(r\frac{\mathrm{d}v^{(0)}}{\mathrm{d}r}\right) + \frac{1}{\mu}\frac{\mathrm{d}P}{\mathrm{d}x} = 0 \tag{58}$$

where $v^{(0)}$ is the fluid velocity in the core region, $\mathrm{d}P/\mathrm{d}x$ is the pressure drop along the pore axis, r and x are the radial and axial positions, and μ denotes the fluid viscosity.

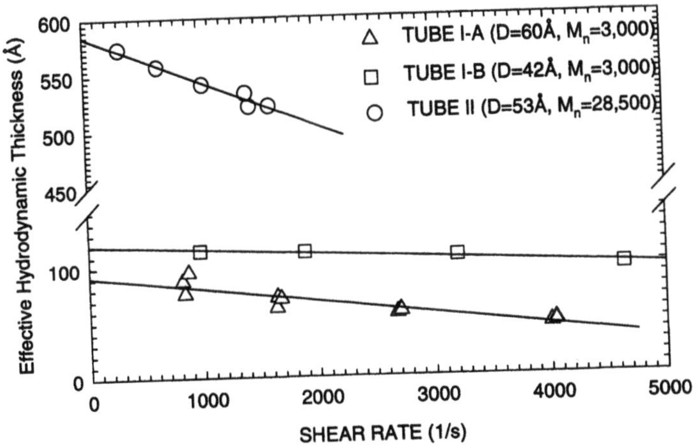

FIG. 14 Polymer hydrodynamic thicknesses of poly(vinylpyrrolidone)-grafted silica membranes (from Ref. 34). D and M_n designate the distance between nearest anchoring points and the grafted polymer molecular weight, respectively.

Axial flow in the brush layer can be determined from the Debye–Brinkman equations [45, 80]:

$$\frac{1}{r}\frac{\mathrm{d}}{\mathrm{d}r}\left(r\frac{\mathrm{d}v^{(1)}}{\mathrm{d}r}\right)-\frac{v^{(1)}}{\kappa}+\frac{1}{\mu}\frac{\mathrm{d}P}{\mathrm{d}x}=0 \tag{59}$$

in which $v^{(1)}$ is the fluid velocity in the brush layer. The solvent permeability coefficient κ, which determines the magnitude of fluid drag due to the presence of grafted chains, can be estimated from sedimentation data for a given polymer/solvent system [45]. Qualitatively, it can be shown that the drag term [the second term in Eq. (59)] increases as the polymer chains extend into the core region of higher fluid velocities. In order to solve the coupled equations (58) and (59), the following boundary conditions can be set:

$$\frac{\mathrm{d}v^{(0)}}{\mathrm{d}r}\Big|_{r=0}=0 \tag{60}$$

$$v^{(0)}\big|_{r=r^*}=v^{(1)}\big|_{r=r^*} \tag{61}$$

$$\mu\frac{\mathrm{d}v^{(0)}}{\mathrm{d}r}\Big|_{r=r^*}=\mu\frac{\mathrm{d}v^{(1)}}{\mathrm{d}r}\Big|_{r=r^*} \tag{62}$$

$$v^{(1)}\big|_{r=R}=0 \tag{63}$$

where $r=0$ is at the pore center, $r^*=R-h$ is the radial position at the edge of the grafted layer (h is the brush height), and $r=R$ is at the pore wall. Equation (60) expresses symmetry of the velocity profile at the pore center. Equations (61) and (62) specify the continuity of velocity and shear stress at the edge of the brush layer. Finally, Eq. (63) denotes the no-slip condition at the solid pore wall.

Analytical solutions of the coupled differential equations (58) and (59), subject to the boundary conditions given by Eqs (60)–(63), are possible when the permeability coefficient κ is assumed to be a constant [80]. In general, however, a numerical solution is required since κ may vary with the radial position from the surface owing to the non-uniform segment distribution within the brush layer, Eq. (54). The above approach was recently used by Castro et al. [83] to analyze the performance of ceramic-supported polymer membranes. Figure 15 shows the computed velocity profiles of water flowing inside the pores of the PVP-grafted silica tubes of Fig. 14. Clearly, a brush layer made of long chains significantly displaces the velocity profile in the core region, and the presence of an axial flow is obvious in the brush layer. Although the axial velocity in the brush layer is small, the shear rate at the edge of the brush layer is sufficiently high, leading to a certain degree of polymer deformation, as evident in the behavior of the hydrodynamic thickness with changing pore–wall shear rate. As noted earlier, the grafted polymer conformation is affected not only by the applied shear rate but also by solvent quality. In a good solvent, in order to minimize the free energy, the grafted chains stretch away from the solid surface, as shown schematically in Fig. 16. This chain extension (or swelling) further decreases the effective pore side, and thus reduces solvent flow through the pore. In a poor solvent, on the other hand, the grafted chains will collapse and coil up against the surface (Fig. 17), resulting in a larger effective pore diameter (or a smaller hydrodynamic thickness) and thus a higher rate of solvent permeation. The above behavior is consistent with permeability measurements of

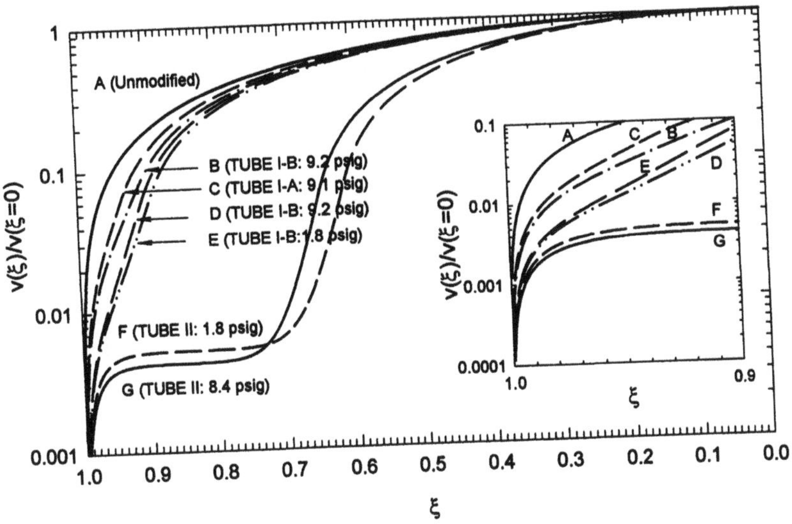

FIG. 15 Velocity profiles inside cylindrical pores of tubular unmodified (curve A) and poly(-vinylpyrrolidine)-modified silica membranes shown in Fig. 13 (from Ref. [34]). ξ is defined by $\xi = r/\mathrm{Re}$ and $\xi = 0$ is at the pore center.

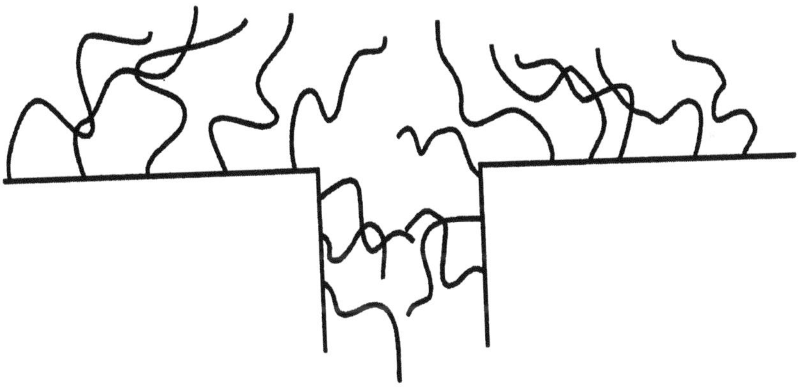

FIG. 16 Polymer brush configuration in a good solvent.

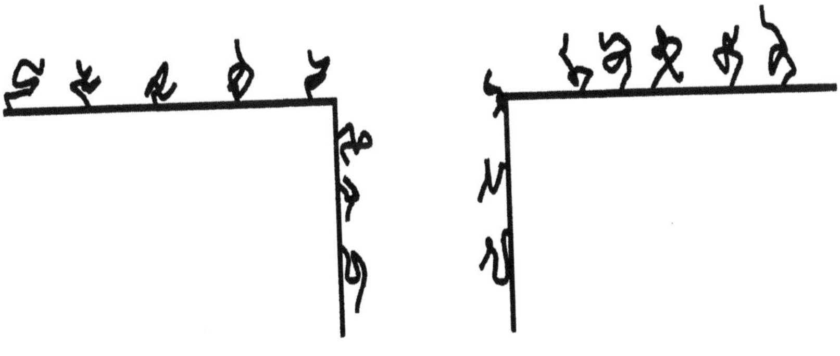

FIG. 17 Polymer brush configuration in a poor solvent.

different solvents through a porous PVP-grafted silica disk (Table 2) [83]. Clearly, increasing solvent power results in a smaller apparent pore size and lower solvent permeability.

V. GRAFT POLYMERIZATION APPLICATIONS

A. Size Exclusion Chromatography

Free-radical graft polymerization can be used to modify silica resins for size exclusion chromatography (SEC) [84]. In SEC, solute molecules such as proteins, polymers, and polypeptides, dissolved in an appropriate solvent, are separated as they elute through a packed column of porous polymer-modified particles, such as PVP-grafted silica pellets. The use of graft-polymerized silica resins is especially attractive owing to their excellent thermal, mechanical, and chemical stability [84]. The masking of silica resins by grafted PVP was shown effectively to prevent the retention of solute molecules, which would otherwise be readily adsorbed onto native silica substrates [84]. In addition, it has also been demonstrated that the porosity of the native silica resins, and thus their exclusion limits and permeability, can be effectively controlled by varying graft density, grafted chain length, solvent quality, and column temperature [84]. An example of the SEC calibration curves of PVP-grafted silica SEC resins using poly(ethylene oxide) standards is shown in Fig. 18 [85]. The initial pore diameters for the N4000-10 and N1000-10 silica resins are 4000 and 1000 Å, respectively. It is noted that by employing a higher initial monomer concentration (and thus a higher rate of polymerization), it is possible to use a larger initial pore size. In this manner, for example, the PVP-grafted N4000-10 (VP 30%) can be tailored to give an effective pore size similar to that of the grafted N1000-10 (VP 5%) SEC resin. Similar observations hold for calibration curves for the N4000-10 (VP 30%) and N4000-10 (VP 10%) SEC resins.

TABLE 2 Solvent Permeabilities for a Porous PVP-Grafted Silica Membrane

Solvent	Permeability[a] $\times 10^{12}$ (cm^2)	Permeability reduction[b] (%)	Unmodified permeability ratio ($k_{solvent}/k_{water}$)	Modified permeability ratio ($k_{solvent}/k_{water}$)
Water	15.4	47.8	1.00	1.00
Ethanol	14.5	40.4	0.827	0.941
1-Propanol	18.2	45.0	1.12	1.18
1-Butanol	16.4	52.3	1.17	1.06
Toluene	26.5	24.0	1.18	1.72
Cyclohexane	31.4	14.0	1.24	2.04

[a]Permeability k as defined in Eq. (57).

[b]Permeability decrease $= \frac{(k_{modified} - k_{modified})}{k_{unmodified}} \times 100\%$.

Source: Castro [34].

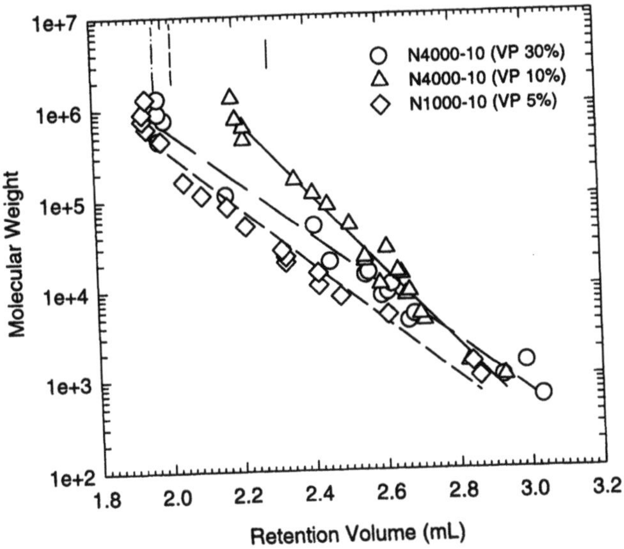

FIG. 18 SEC calibration curves for poly(vinylpyrrolidone)-grafted silica resins (from Ref. 85). Resins N4000-10 (VP 30%), N4000-10 (VP10%), and N1000-10 (VP 5%) have native pore diameters of 4000, 4000, and 1000 Å, respectively, and were prepared with initial monomer concentrations of 30, 10, and 5% by volume, respectively.

B. Microfiltration Using PVP-Modified Silica Membranes

In recent years, there has been a growing interest in ceramic membranes as a result of their excellent temperature resistance and structural strength [17, 18, 85]. Unfortunately, as with most membrane processes, permeate flux decline due to membrane fouling presents a serious barrier to achieving desired separation goals. Membrane fouling can be reduced by graft polymerization of the ceramic surface. The grafted polymer helps to reduce surface adsorption and extend the membrane life.

A fouling-resistant membrane was demonstrated by Castro [34] for the separation of an oil/water emulsion, using a 2000 Å pore diameter tubular silica membrane grafted with PVP. The membrane tested had relatively isolated grafted chains (with a graft yield of 0.277 mg/m^2, a chain spacing of 44.8 Å, and a Flory radius of 31 Å). The transmembrane pressure was approximately 8 psig and flow rates well in the region of full turbulence were tested (Re = 5300 and 9300). Figure 19 shows that the PVP-modified membrane had a hydraulic resistance similar to the resistance of the unmodified membrane, but exhibited a markedly slower increase in the hydraulic resistance curve. After 3 h at a Reynolds number of 5300, the resistance of the modified membrane stabilized while the unmodified membrane continued to foul. Further investigation revealed that at very high Reynolds numbers the modified membrane would begin to foul due to the pore blockage, but at a much slower rate compared to the corresponding unmodified membrane. It is emphasized that the above experiment was performed with a sparse brush layer to maximize permeability. It is postulated that the antifouling character of the membrane would significantly increase with a denser brush coverage and/or longer grafted chains. However, longer chains would decrease the permeability by reducing the effective pore diameter, while dense surface

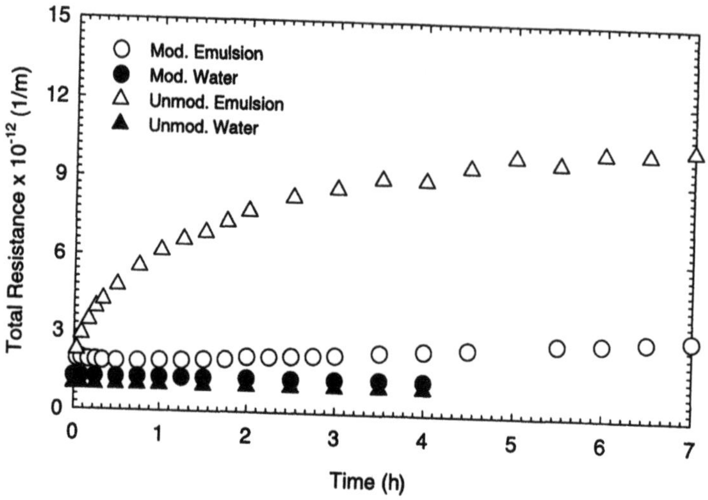

FIG. 19 Fouling resistance of unmodified and poly(vinylpyrrolidone)-grafted tubular silica membranes (from Ref. 34). The native membrane has a pore diameter of 2000 Å. The polymer graft yield, grafted molecular weight, and silane coverage for the modified membrane are 0.277 mg/m^2, 3340 g/mol, and 12.9 $\mu mol/m^2$, respectively.

coverages of short chains would not necessarily cause a significant decrease in permeability unless the pore diameter was very small.

C. Removal of Trichloroethylene from Water by Pervaporation

Pervaporation is a novel separation technique which is useful for the dehydration of organic compounds and the recovery of volatile organic solvents from aqueous mixtures. Pervaporation is usually performed using a polymeric membrane; however, the physical characteristics that many polymers possess make polymeric membranes unreliable and subject to physical deformation and swelling. In contrast, inorganic oxide membranes possess good structural stability, but poor chemical selectivity and are thus relegated to simple size separation. Graft-polymerized ceramic membranes can combine the structural stability of the inorganic oxide backbone with the high chemical selectivity of a grafted polymer layer. As an example, recent studies [18] have shown that pervaporation with PVAc-grafted silica membranes is an effective method for removing dilute trichloroethylene (TCE) from water. The grafted PVAc reduces the average pore size of the native silica membrane to a suitable pervaporation pore size. The TCE enrichment factor (the ratio of TCE concentration in the permeate to that in the feed) ranged from 69 to 106, giving TCE as a pure phase, along with a residual saturated aqueous phase which can be recycled to the feed (Fig. 20). For the dilute aqueous solutions (TCE concentration of less than 700 ppm), the enrichment factor was observed to be independent of the TCE feed concentration. Since the transport of TCE across the PVAc-modified silica membrane is controlled by the feed-side concentration boundary layer, pervaporation selectivity is significantly affected by the tube-side Reynolds number.

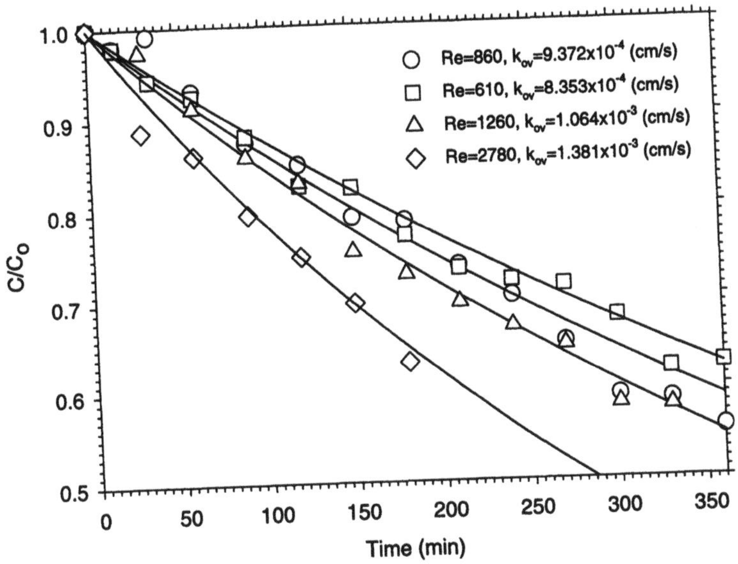

FIG. 20 Relative concentration data for TCE in aqueous feed solution for TCE pervaporation (from Ref. 18). The poly(vinyl acetate)-grafted silica membrane has a graft yield of 0.17 mg/m^2. The initial TCE concentration in the feed (C_0) ranges from 500 to 700 ppm. Re and k_{ov} denote the tube-side Reynolds number and the overall mass-transfer coefficient, respectively. (Note: retentate stream was recirculated back to the feed reservoir.)

VI. CONCLUSIONS

As demonstrated in this chapter, the modification of inorganic oxide surfaces by graft polymerization is an efficient and controllable process that combines the chemical versatility of the organic polymer with the structural integrity of the inorganic oxide substrate. Grafted chains, which are covalently attached to the surface, impart attributes that are entirely different from the chemical properties of the native inorganic oxide substrate. Given knowledge of the kinetics of graft polymerization and the characteristics of the resulting surface, suitable polymers and reaction conditions can be chosen to tailor the substrate to specific needs. This is a key step toward the creation of modified inorganic oxide surfaces with customizable chemical resistance, biocompatibility, and chemical selectivity. It is emphasized that the examples of SEC, microfiltration, and pervaporation, discussed in this chapter, are not the limits for the use of polymer-grafted oxide surfaces, but rather their thresholds in practical applications.

REFERENCES

1. D Beyer, TM Bohanon, W Knoll, H Ringsdorf. Langmuir 12:2514, 1996.
2. W Wang, HP Schreiber, Y Yu, A Eisenberg. J Polym Sci, Part B: Polym Phys 35:1793, 1997.
3. T Hwa, D Cule. Phys Rev Let 79;4930, 1997.
4. W Chen, TJ McCarthy. Macromolecules 30:76, 1997.
5. T Eahners. Opt Quantum Electron 27:1337, 1995.

6. H Moffat, KF Jensen. J Cryst Growth 77:108, 1986.
7. GI Font, ID Boyd. J Vac Sci Technol 15:320, 1997.
8. A Vidal, E Papirer, W Jiao, J Donnet. Chromatographia 23:121, 1987.
9. NV Yablokova, YA Aleksandrov, OM Titova. Vysokmol Soyed A28:1908, 1986.
10. Y Cohen, P Eisenberg, M Chaimberg. J Colloid Interface Sci 148:579, 1992.
11. I Krasilnikov, V Borisova. J Chromatogr 446:211, 1988.
12. E Papirer, VT Nguyen. Polym Lett 10:167, 1972.
13. D Cohn, AS Hoffman, BD Ratner. J Appl Polym Sci 29:2645, 1984.
14. BD Ratner, AS Hoffman. J Appl Polym Sci 81:3183, 1974.
15. H Inoue, S Kohama. J Appl Polym Sci 29:877, 1984.
16. R Laible, K Hamann. Adv Colloid Interface Sci 13:63, 1980.
17. RP Castro, Y Cohen, HG Monbouquette. J Membr Sci 115:179, 1996.
18. JD Jou, W Yoshida, Y Cohen. J Membr Sci 162:269, 1999.
19. M Chaimberg. Free-radical Graft Polymerization of Vinyl Pyrrolidone. PhD dissertation, University of California, Los Angeles, 1989.
20. M Chaimberg, Y Cohen. Ind Eng Chem Res 30:2534, 1991.
21. JD Jou. Graft Polymerization and Application to Membrane Pervaporation PhD dissertation, University of California, Los Angeles, 1998.
22. BB Wheals. J Chromatogr 107;402, 1975.
23. B Monrabal. Advances in Chromatography. Proceedings of Chromatography Symposium, Houston, TX, II, 117 (1979).
24. B Monrabal. In: J Crazes, ed. Liquid Chromatography of Polymer and Related Materials III. Chromatographic Science Series 19. New York: Marcel Dekker, 1981.
25. RS Parnas, M Chaimberg, V Taepaisitphongse, Y Cohen. J Colloid Interface Sci 129:441, 1989.
26. G Boven, ML Oosterling, G Challa, A Schouten. Polymer 31:2377, 1990.
27. G Boven, R Folkersma, G Challa, A Schouten. Polym Commun 32:51, 1991.
28. Y Cohen. US Patent 5035803, 1991.
29. T Browne, M Chaimberg, Y Cohen. J Appl Polym Sci 44:671, 1992.
30. E Carlier, A Guyot, A Revillon, M Darricades, R Petiaud. Reactive Polym 16:115, 1992.
31. N Tsubokawa, H Ishida. J Polym Sci, Part A: Polym Chem 30:2241, 1992.
32. N Tsubokawa, T Kimoto, K Koyama. J Colloid Polym Sci 271:940, 1993.
33. M Chaimberg, Y Cohen. AIChEJ 40:294, 1994.
34. RP Castro. Development of Ceramic-supported Polymeric Membranes. PhD dissertation, University of California, Los Angeles, 1997.
35. E Papirer, VT Nguyen. Angew Makromol Chem 128:31, 1973.
36. J Horn, R Hoene, K Hamann. Makromol Chem Suppl 1:329, 1975.
37. N Tsubokawa, A Kogure, Y Sone. J Polym Sci, Part A: Polym Chem 28;1923, 1990.
38. RP Castro, Y Cohen, HG Monbouquette. J Membr Sci 84:151, 1993.
39. J Kohler. Chromatographia 21:573, 1986.
40. AE Ivanov, L Zigis, MF Turchinskii, VP Kopev, PD Reshetov, VP Zubov, LN Kastnkina, NZ Lonskaya. Mol Genet Mikrobiol Virusol 11:39, 1987.
41. K Komiya, Y Kato. Can Patent 1293083, 1987.
42. K Komiya, Y Kato. Chem Abstr 117:163125, 1991.
43. Y Cohen, P Eisenberg. In: RS Harland, RK Prud'homme, eds. Polyelectrolyte Gels: Properties, Preparation, and Applications. Washington, DC: American Chemical Society, 1992.
44. JG Heffernam, DC Sherrington. J Appl Polym Sci 29:3013, 1984.
45. MA Cohen Stuart, FHWH Waajen, T Cosgrove, B Vincent, TL Crowley. Macromolecules 17:1825, 1984.
46. Y He. Graft Polymerization of Vinyl Acetate onto Silica and the Potential Application in Membrane Modification. MS thesis, University of California, Los Angeles, 1995.

47. M Chaimberg, R Parnas, Y Cohen. J Appl Polym Sci 37:214, 1989.
48. I Casinos. Angew Makromol Chem 221:33, 1994.
49. J Auroy. J Colloid Interface Sci 50:187, 1992.
50. J Edwards, S Lenon, AF Toussaint, B Vincent. ACS Symp Ser 240;281, 1984.
51. K Bridger, D Fairhurst, B Vincent. J Colloid Interface Sci 68:190, 1979.
52. M Chaimberg, R Parnas, Y Cohen. J Appl Polym Sci 37:2921, 1990.
53. RK Iler. The Chemistry of Silica. New York: John Wiley, 1979.
54. LT Zhuravlev. Langmuir 3:316, 1987.
55. KMR Kallury, PM Macdonald, M Thompson. Langmuir 10:492, 1994.
56. RE Majors, MJ Hopper. J Chromatogr Sci 12:767, 1974.
57. D Hunkeler. Macromolecules 24:2160, 1991.
58. K Ito. J Polym Sci, Part A 10:1481, 1972.
59. K Ito. J Polym Sci, Polym Chem Ed 13:521, 1975.
60. K Ito. J Polym Sci, Polym Chem Ed 15:2037, 1977.
61. K Ito. J Polym Sci, Polym Chem Ed 18:701, 1980.
62. K Ito. Macromolecules 13:193, 1980.
63. RT Woodhams. The Kinetics of Polymerization of Vinyl Pyrrolidone. PhD dissertation, Polytechnic Institute of Brooklyn, NY, 1954.
64. FW Billmeyer. Textbook of Polymer Science. 3rd ed. New York: Wiley-Interscience, 1984.
65. JA Biesenberger, DH Sebastian. Principles of Polymerization Engineering. New York: Wiley-Interscience, 1983.
66. GG Odian. Principles of Polymerization. 3rd ed. New York: John Wiley, 1991.
67. G Gee, EK Rideal. Trans Faraday Soc 32:666, 1936.
68. G Gee, AC Cuthbertson, EK Rideal. Proc R Soc London A170:300, 1939.
69. MS Matheson. J Chem Phys 13:584, 1945.
70. GM Burnett, LD Loan. Trans Faraday Soc 51:214, 1955.
71. PW Allen, FM Merrett, J Scanlan. Trans Faraday Soc 51:95, 1955.
72. RM Noyes. J Am Chem Soc 77:2042, 1955.
73. PG de Gennes. Macromolecules 13:1069, 1980.
74. MA Carignano, I Szleifer. Macromolecules 28:3197, 1995.
75. ST Milner, TA Witten, ME Cates. Macromolecules 21:2610, 1988.
76. RS Parnas, Y Cohen. Macromolecules 24:4646, 1991.
77. RS Parnas, Y Cohen. Rheol Acta 33:485, 1994.
78. S Patel, M Tirrell, G Hadziioannou. Colloids Surfaces 31:157, 1988.
79. JH van Zanten. Macromolecules 27:5052, 1994.
80. R Varoqui, P Dejardin. J Chem Phys 66:4395, 1977.
81. P Gramain, P Myard. Macromolecules 14:180, 1981.
82. RB Bird, W Stewart, E Lightfoot. Transport Phenomena. New York: John Wiley, 1960.
83. RP Castro, HG Monbouquette, Y Cohen. J Membr Sci 179: 207, 2000.
84. Y Cohen, RS Faibish, M Rovira-Bru. In: E Pefferkorn, ed. Surface Interaction and Size Exclusion Chromatography. New York: Marcel Dekker, 1999.
85. C Chen. Behavior of Polyvinylpyrrolidone-grafted Silica Resin in Size Exclusion Chromatography. MS thesis, University of California, Los Angeles, 1992.

9
Metal Oxide Membranes

ANTONIO HERNÁNDEZ, PEDRO PRÁDANOS, JOSÉ IGNACIO CALVO, and LAURA PALACIO Universidad de Valladolid, Valladolid, Spain

I. INTRODUCTION

In most books dealing with or related to industrial separations, membrane processes are usually considered a new and promising field of recent development. In fact, if we consider the plethora of industrial processes where separation by means of membranes have become a standard procedure, we can conclude that now this field is fully developed.

In these applications, membranes are used to achieve an effective and low-cost separation of several compounds. A membrane can be defined as a semipermeable barrier, which separates two phases and is selective to one or more of the compounds in one of the phases. Selectivity is the key factor of the membrane and it governs the kind of applications and processes where membranes can be used [1]. The most outstanding of the currently used membrane processes are shown in Fig. 1, along with some more traditional separation processes, their ranges, and selective factors.

It can be seen in this figure that membranes have found a wide range of applications corresponding to quite different membrane processes. It is clear that the membranes developed for such a range of processes must be necessarily very different in their structure, properties, and manufacture. A useful classification of membranes can be made based on different criteria [2]. Among them, a classification based on structural aspects is of key importance as far as it determines the functional properties of the membrane as a barrier.

A first distinction can be to separate porous membranes from nonporous ones. The latter being those presenting molecular interstices under the limit of detection of the most powerful microscopes. A porous membrane presenting equalized pores should be called homoporous, while membranes with pores in a more or less wide range should be called heteroporous.

On the other hand, membranes can be symmetric or asymmetric. Asymmetric membranes have a porous support and a thin skin layer which gives selectivity. If the two layers are made of different materials, the membrane is called composite. A more or less complete classification of membranes, along with the main features of each kind of membrane, is shown in Fig. 2.

Referring to the membrane material (or materials for composite ones), many interesting materials have been tested and developed in the some 40 years of investigation and applications of membrane processes [3]. Mostly polymeric materials

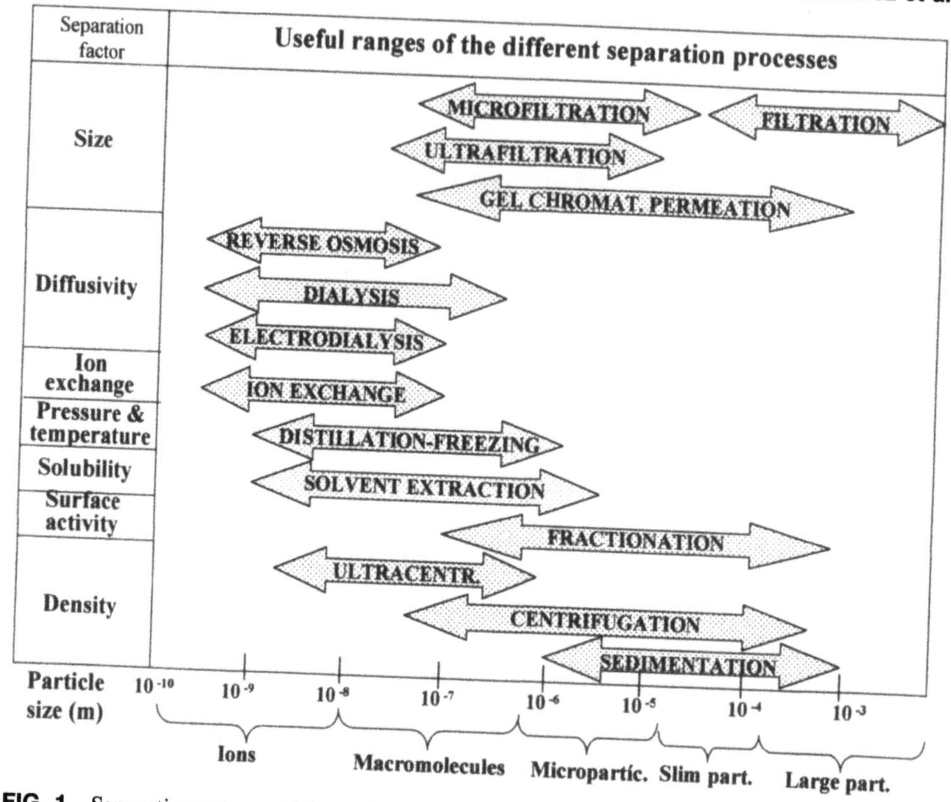

FIG. 1 Separation ranges of the main membrane processes as compared with other typical methods.

have been used from the very beginning (the cellulose acetate asymmetric membrane of Loeb and Sourirajan [4] is frequently considered an outstanding milestone in membrane technology). Nevertheless, some other materials have proven reliability for membrane manufacturing. Among those inorganic materials some are relatively new, but their development is actually emerging to cover a broad range of potential applications [5].

Most of the interest paid to the inorganic membranes is due to the advantages they offer over polymeric materials [5]. Properties such as a high resistance to several chemicals including corrosive products, a wide tolerance to a broad range of pH and temperatures, along with a high mechanical resistance to pressure, give an idea of the enormous versatility of inorganic membranes. Of particular importance is the ability to resist repeated sterilization cycles, which make them very suitable when dealing with biotechnological species. These evident advantages are offset by an initial higher cost and fragility. In any case, their higher stability makes possible a more efficient cleaning, giving a comparatively longer operation life to inorganic membranes. A sketch of the main advantages and some drawbacks of inorganic membranes are shown in Table 1.

To give a complete review of the historical background [6] of inorganic membranes is beyond the scope of this chapter. Nevertheless, we think that some historical notes should be of interest to make understandable to the reader the initial steps

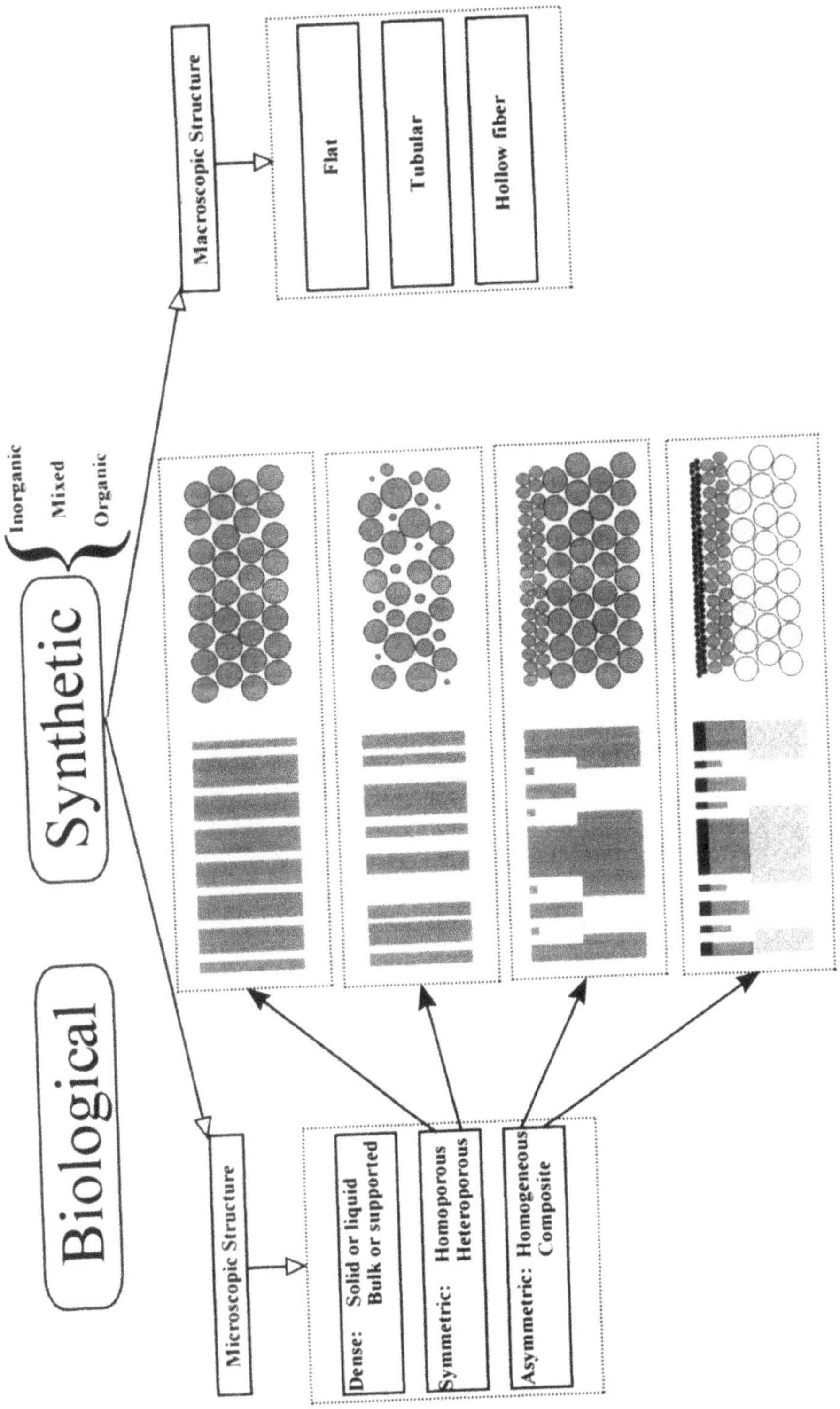

FIG. 2 Classification of membranes according to different aspects.

TABLE 1 Advantages and Disadvantages of Inorganic Membranes

	Advantages				Disadvantages	
Thermal	High-Temperature stability		Strong cleaning agents and temperatures can be used	Low fouling levels	Sealing is not easy at high temperatures	
Chemical	High compatibility in aggressive media	Easy electrochemical activity	Strong cleaning agents and temperatures can be used	Low fouling levels		
Mechanical	High stability under high pressures	No creeps	Good control on pore dimensions	Low fouling levels	Brittle easily	Special needs of configuration or support media
Economical	Long lifetime				High initial or reposition inversions needed	

and efforts in the development of inorganic membranes for industrial uses. Inorganic membranes were first applied in the separation of uranium isotopes for the nuclear industry [7]. Even now, uranium enrichment is the single most important field of application for inorganic membranes, based on the number of plants in use. Gaseous diffusion technology using UF_6 (the most practical volatile compound of uranium) across a porous membrane with a Knudsen mechanism as the gas transport regime, allows separation of the main isotopes (^{235}U and ^{238}U) of natural uranium. Combining a large number of such separation steps the mined uranium can be concentrated to 3% of the ^{235}U required in the power generating plants (over 1000 steps are usually needed) or even to higher contents for nuclear weapons.

It is clear that this work, performed from the 1940s and 1950s, led to the development of several inorganic membranes as the most convenient materials to resist the chemically aggressive vapors of uranium. Corning Glass (manufacturers of the Vycor glass membranes) and Union Carbide in the USA, and Le Carbone Lorraine and SFEC (a subsidiary of the French Commisariat a l'Energie Atomique, CEA) in France, were leading companies in this pioneering period along with USSR researchers. Not many details are known on the membranes used for easily understandable reasons. Typically tubular membranes were developed, comprising a macroporous support and a ceramic separating layer [8].

The delay or even abandonment of the nuclear plants in the developed world along with the potential of the accumulated know-how led the inorganic membrane manufacturers to explore the use of these membranes for other membrane processes, such as microfiltration (MF) or ultrafiltration (UF), from 1980 to 1990.

A major advance was the Ucarsep® UF membrane patented by Union Carbide in 1973 and made from a nonsintered ceramic oxide layer (usually ZrO_2) deposited on to a carbon porous tubular support (6 mm inner diameter). The worldwide license of these membranes was sold to SFEC, who added the sintering of the ZrO_2 layer to obtain a permanent attachment to the support. With the Carbosep® trademark, these membranes were marketed by SFEC from 1980, and later, the company was sold to Rhône-Poulenc, which created a subsidiary, Tech Sep, still having the ZrO_2 UF membrane as its main inorganic product.

Ceraver, in the early 1980s, developed a quite different membrane, made from α-Al_2O_3 supported on α-Al_2O_3, having tubular multichannel geometry and the possibility of backflushing, with application to the MF range. Their membranes were sold under the trademark of Membralox® and included, from 1988, UF membranes with ZrO_2 separating layers. A similar approach was given by Norton, developing a MF membrane with an α-Al_2O_3 layer on α-Al_2O_3. They included multichannel geometry from 1988–1989, and are sold by Millipore as Ceraflo® filters.

Le Carbone Lorraine, another company involved in the French nuclear program, developed MF and UF membranes with a tubular carbon support, made of carbon fibers coated with CVD carbon, and a carbon separating layer.

Finally, asymmetric alumina membranes, made by the anodization of aluminum sheets, were developed for MF and UF, and marketed as Anopore®, by Anotec, now a part of Whatman.

As ever more membrane companies enter the field of inorganic filters, new manufacturing techniques have been developed and a much broader exploration of the potential of inorganic membranes has been seen. A variety of basic processes, including the coupling of catalytic reactors and membrane separations, is emerging, with an increasing number of companies, universities, and industrial laboratories being involved in this process. A register of inorganic membrane providers can be found in Scott [9] or Mulder et al. [10].

From this brief historical perspective, it can be seen that metal oxides were the basis of inorganic membrane development. In this chapter we will try to examine, in a tutorial rather than complex form, the main features of the metal oxide based membranes, as an important part of the field of inorganic membranes. In this way, some paragraphs will be initially devoted to the fundamentals of membrane synthesis, covering the most common membrane making techniques. Structural, functional, and electrical membrane characterization methods will then be introduced.

Finally, we will analyze membrane fouling, a major problem that complicates the widespread use of membrane technology.

II. SYNTHESIS

It is clear that metal oxide membranes can be and have been developed by using many different methods, which have a direct influence on the material properties and the resulting porous structure. Some general methods for manufacturing inorganic membranes will be given here [6, 11, 12]. Most of these methods can be used to make metal oxide membranes as well as other inorganic membranes, while others are more specific. In any case, many metal oxide membranes are deposited on to other inorganic porous supports; thus, the fundamentals of synthesis of general inorganic filters should be known if the characteristics of metal oxide membranes are to be understood. Many examples will be given, referring mostly to metal oxide membranes.

A. Nonceramic Membranes

Zirconia, cerates, or bismuth oxides in very thin solid layers are permeable to ionic forms of hydrogen or oxygen; thus, they have been used especially in membrane reactors. In order to obtain significant permeabilities these membranes should be very thin, making it necessary to deposit them on to an appropriate porous support. Other membranes that could be considered dense are obtained by filling a porous membrane with a liquid that remains immobilized and controls permselectivity [13, 14].

Membranes with pores having pore diameters in the nanometer range can be obtained by pyrolysis. Molecular sieves can be prepared by controlled pyrolysis of thermoset polymers [poly(vinylidene chloride), poly(furfuryl alcohol), cellulose, cellulose triacetate, polyacrylonitrile (PAN), and phenol formaldehyde] to obtain carbon membranes, or of silicone rubbers to obtain silica filters. For example, carbon molecular sieves can be obtained by pyrolysis of PAN hollow fibers in an inert atmosphere, which leads to dense membranes whose pores are opened by oxidation, initially at 400°–500°C and finished at 700°C [15]. These membranes are used to separate O_2/N_2 mixtures. Le Carbone-Lorraine deposits a resin into a tubular macroporous substrate and then by pyrolysis creates a thin (< 1 μm) carbon active layer. Silicon rubber tubes can be pyrolyzed in an inert atmosphere at temperatures around 700°C followed by oxidation in air at temperatures from 500° to 900°C [16]. The membranes are composed almost completely of SiO_2 with pores having a maximum porosity of 50% and diameters from 5 to 10 nm. The permeabilities for He, H_2, O_2, and Ar range from 0.5 to 5×10^{-9} m s^{-1} Pa.

Membranes very regular in shape and size can be produced by the track-etching method [17, 18]. Here, a thin (~ 10 μm) solid layer is bombarded by highly energetic particles (fission fragments from a nuclear reactor or accelerated ions, for instance). These particles leave damage that after a chemical etching is enlarged. This method has been extensively used for polymeric membranes. For inorganic membranes it was originally used to obtain mica membranes, but recently it has been used with other more useful materials. For example, recrystallized aluminum foils have been track-etched to obtain pore sizes from 0.5 to 8 μm that have been made narrower by coating with aluminum oxide, yielding pores down to 2 nm [9].

A particular type of metal oxide membrane produced by anodic oxidation of aluminum seems to be especially interesting and easy to model. This anodic oxidation in a certain acid electrolyte results in a structure comprising a uniform column array of hexagonally close packed alumina cells each containing a cylindrical pore. According to O'Sullivan and Wood [19] the cell and pore sizes are controlled by the anodizing voltage, whereas the film thickness is determined by the current density and the anodizing time. The porous films are detached from the aluminum substrate by a controlled reduction of the anodizing voltage, which results in a skin layer formed by smaller pores, as decribed elsewhere [20–22]. The resulting membranes have 0.02 μm size pores in the skin layer. This membrane can be subsequently converted to 0.1 or 0.2 μm size pores by dissolving the fine pore layers and thinning the walls of the major capillary structure [23]. The resulting membranes are highly porous and exhibit a capillary pore structure with very uniform pore sizes. Field emission scanning electronic microscopy (FESEM) images of the A01 membrane are shown in Fig. 3. Atomic-force microscopy (AFM) images of the A002 membrane are shown in Fig 4. Both these membranes are made by Anotec.

Thin zirconia (~ 20 μm) membranes with pore diameters below 5 nm can be deposited on to porous carbon. Tech Sep uses macroporous (> 0.3 μm) carbon tubes embedded in acetone which are filled with a ZrO_2 solution in water–methanol. The presence of acetone precipitates zirconia on the carbon substrate.

B. Ceramic Membranes

The most developed and useful inorganic membranes are ceramic ones [24, 25]. Ceramics can be defined as polycrystalline consolidated materials based on compounds of elements in groups III and VI of the periodic table.

The raw materials used are:

1. Alumina (Al_2O_3) is a part of most of the ceramic pastes. It can appear in several more or less hydrated crystalline forms: γ, δ, θ, and α; γ-alumina has a hexagonal structure and is obtained by controlled drying at low temperatures (400°–500°C), while α is trigonal and is obtained by high temperature (> 1000°C) drying. The other forms are dried at intermediate temperatures [26, 27].

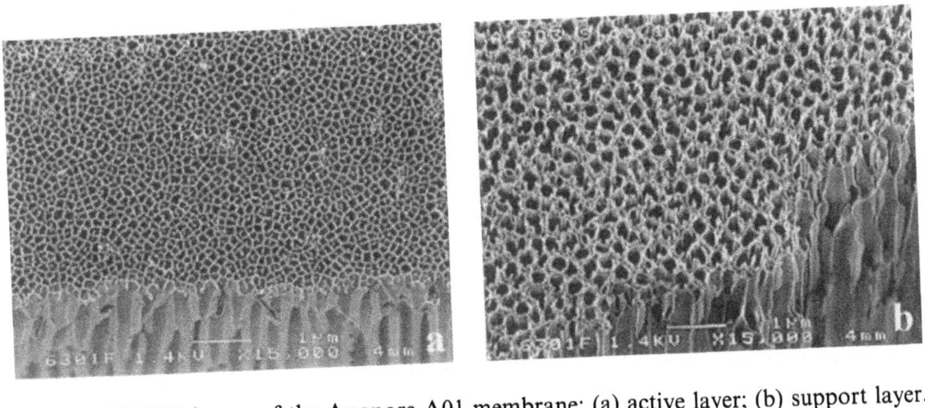

FIG. 3 FESEM image of the Anopore A01 membrane: (a) active layer; (b) support layer.

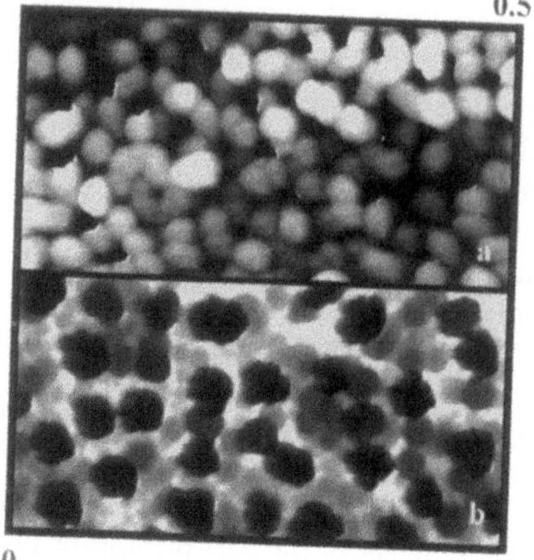

FIG. 4 AFM image of the Anopore A002 membrane: (a) active layer; (b) support layer.

2. Silica (SiO_2) is one of the most common constituents of ceramic pastes; it can appear as different alotropic forms.
3. Calcium carbonate ($CaCO_3$) acts as a melting facilitator in ceramic pastes and increases porosities when high temperature sintering is used because of CO_2 liberation.
4. Clay is a common membrane substance which contain principally SiO_2 (~ 45 wt%) and Al_2O_3 (~ 38 wt%) along with other metallic (Ti, Fe, Mg, Ca, K, Na, etc.) oxides each under 1 wt% usually. The proportions of its constituents and impurities along with the crystalline structure determines different kinds of clays and their resultant membrane properties.
5. Talc and steatite are very similar to clays with very similar plasticity. They are SiO_2 and MgO systems differing in their metal oxide ratios and hydration.
6. Cordierite ($2MgO.Al_2O_3.5SiO_2$) is not naturally abundant so it is almost always synthesized.
7. Finally, it is possible to obtain ceramic membranes from other inorganic compounds from groups III and VI as, for example, zirconia (ZrO_2), titania (TiO_2), silica–palladium ($SiO_2 + PdCl_2$), and lanthanum oxichloride (LaOCl).

Structure, porosity, and pore size are directly determined by the method used to transform them. Namely, wide pores can be obtained by traditional ceramic fabrication processes while narrower pores require more modern techniques as, for example, the sol–gel method. An outline of these techniques is shown in Fig. 5 [24].

1. Traditional Ceramic Techniques

From the point of view of membrane manufacture, these techniques are used mainly to make macroporous support media that can be used as microfiltration membranes. Nevertheless, they provide poor performance resulting from wide pore size distribu-

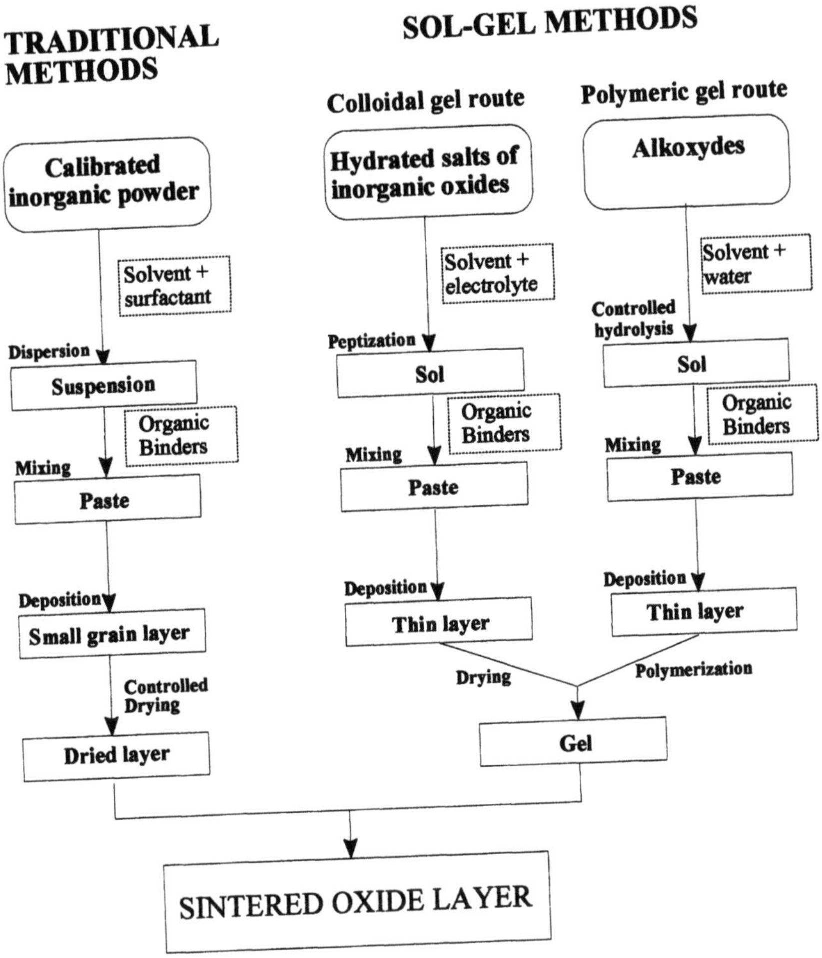

FIG. 5 Sketch of the methods used to make ceramic membranes.

tions and many faults leading to cracks or pinholes. To improve the quality of these membranes a selective layer is usually deposited by using a sol–gel method.

The classic techniques for making ceramics consist in the deposition or molding of a powder suspension. Thus, the main steps include: powder preparation, molding or deposition, drying, and sintering (see Fig. 5).

The most important factors determining the final product are the grain size and the additives, which control microstructure. When starting from a pure oxide, several additives (frequently organic) are needed and should be expelled by a thermal treatment. The main types of additives used are:

1. Defloculants which keep the suspension stable.
2. Binders that increase viscosity.
3. Plasticizers to make the product flexible.
4. Lubricants to allow extrusion, if necessary.

5. Solvents which homogenize the whole system of additives plus oxide to form a suspension of fine particles of oxide, usually called barbotine or slip. Partial evaporation will lead to a ceramic paste.

The optimal properties of the barbotine and paste are determined by both the desired microstructure of the final product and the molding procedure to be used [28, 29]. There are four main molding techniques:

1. Extrusion to make tubular membranes (with a mono- or multi-channel geometry).
2. Pressing to make tubular or flat membranes.
3. Tapping to make both tubes and flat sheets.
4. Slide coating to make flat membranes.

When extrusion is used, the wet paste is dried at low temperatures (< 100°) after extrusion and then sintered under temperatures and times that will determine texture and mechanical properties of the final product [30–32].

When pressing, the paste can be molded by applying isostatic pressure to obtain a flat product or uniaxial if tubular membranes are required. Sometimes a thermal treatment is simultaneously applied followed by a final sintering. A scanning electron microscopy (SEM) image of a cordierite membrane support is shown in Fig. 6. This support was manufactured from fine cordierite particles (< 120 μm in size), mixed with poly(vinyl alcohol) (PVA) (25 wt%) as binder [33]. Water was used as solvent and eliminated by drying and so the resulting paste was uniaxially pressed

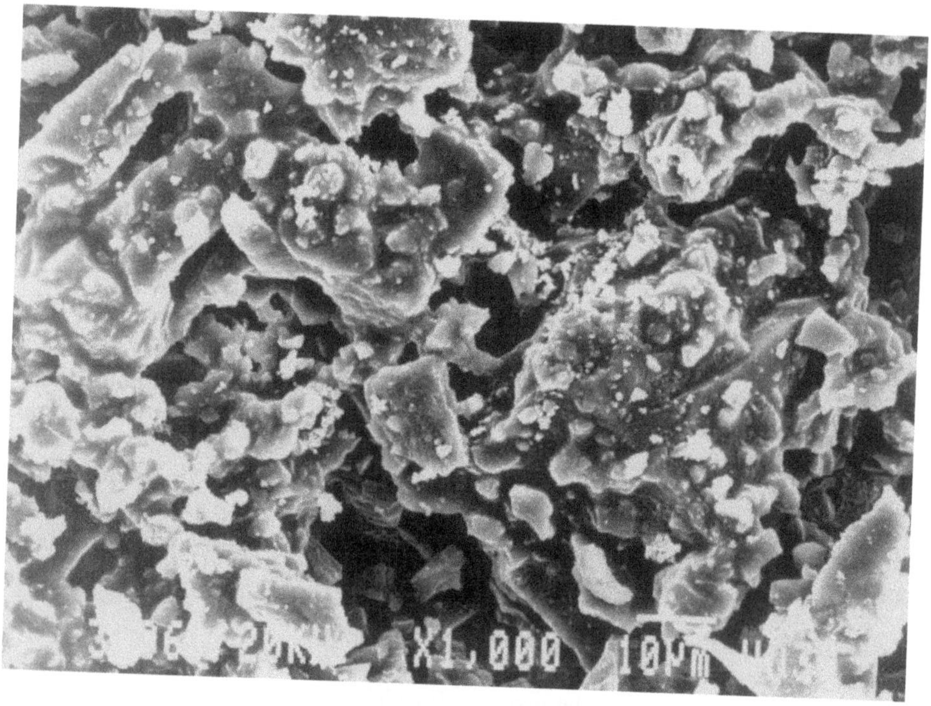

FIG. 6 SEM image of a membrane support made of cordierite.

heated to 1280°C at 4°C/min and then held for 2h [34].

Tapping molding begins with a thick paste which is poured into a porous mold. Here, an addition of carbonate helps to have a fluid enough paste without too much water that could lead to cracks during sintering.

Slide molding consists in the deposition of the slip layer on to a flat surface. Ultrasound is frequently used to disgregate the slip. As mentioned, one or several layers of metallic oxides ($Al^2 O^3$, TiO^2, ZrO^2, etc.) can be deposited on a support obtained by these traditional methods to yield really useful membranes. These composite membranes are usually obtained by slide deposition on to a traditional ceramic [35, 36], improving adherence by allowing the oxides to penetrate slightly into the macropores of the support, by centrifugation (spin-coating) for instance, followed by drying and sintering. In Fig. 7, SEM and AFM images of a TiO_2 active layer deposited on to a cordierite support (see Fig. 6) are shown. The active layer was obtained from a defloculated slip produced by mixing 5 wt% of titania, 20 wt% of PVA (12 wt% aqueous solution) as binder, and a 75 wt% of water, deposited by centrifugation and sintered initially at 1050°C for 2 h and finally at 300°C for 1 h [34].

2. Sol–Gel Techniques

This technique produces ceramic materials from a solution. The starting particles are smaller (a few nanometers) than for traditional techniques since the particles are obtained by hydrolysis of organometallic compounds or metallic salts to oxides. The ceramic materials so obtained are more dense with a higher resistance to compression. As a result, imperfections are smaller and less frequent than with other methods. A few cracks of less than 1 µm are formed, while traditional materials can have pinholes of up to 1 mm.

After controlled hydrolysis, all particles have very similar sizes (2–3 nm usually) [37], that are stabilized (by addition of carefully chosen chemicals) due to their charge or to the presence of an organic polymer in the so-called sol phase. At appropriate concentrations these particles are linked to form a homogeneous three-dimen-

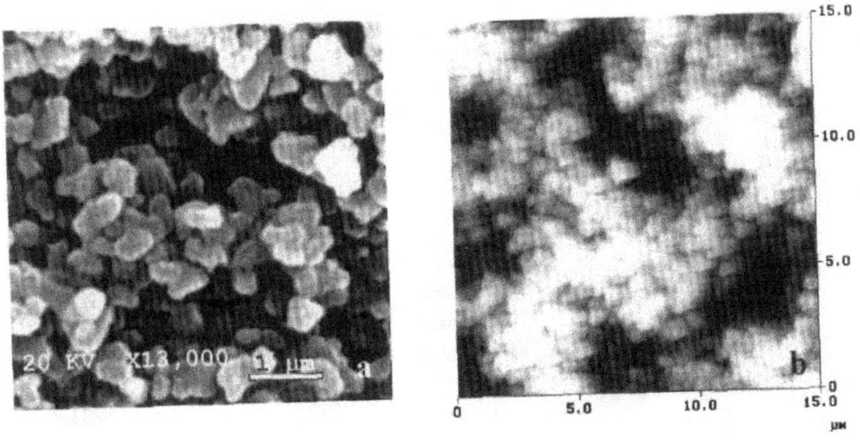

FIG. 7 Two images of a titania active layer: (a) SEM; (b) AFM.

appropriate concentrations these particles are linked to form a homogeneous three-dimensional solid that separates from the liquid in a gel phase. After drying and sintering the final ceramic membrane is obtained.

The sol–gel technique can follow two routes, depending on the nature of starting materials (see Fig. 5):

1. The colloidal route starts from a metallic salt that after hydrolysis gives a colloidal suspension of metallic oxides and hydroxides. After solution, the metallic salt gives a metallic cation (M^{n+}), surrounded by water molecules. The cation can be considered to be linked to OH_2, OH^-, or O^{2-}. The effective groups that surround the cations depend on the relative charge-to-radius ratio and on the balance between pH (conditioned by the OH^- and M^{n+} concentration ratio $[OH^-]/[M^{n+}]$) and the cationic zeta potential (ζ, i.e., the potential appearing at the shear surface of the colloidal particle). So, it can be assumed that low charge cations, $n < 4$, are complexed with OH_2 or OH^- groups, while OH^- or O^{2-} groups are present if $n > 5$; for $n = 4$ all these groups can be present. When $[OH^-]/[M^{n+}] \leq \zeta$ the colloidal particles (hydrated cations) precipitate and after peptization transform to a sol phase. Peptization is accomplished by the addition of diluted acid that donates protons (hydroxyl groups in the case of alkaline addition) which break the precipitate owing to the electrostatic repulsion from dispersed particles. The sol–gel transition consists in the formation of increasing clusters that, due to collisions, condense to form a gel. These steps are schematized in Fig. 8 [24].

   The characteristics of the ceramic gel so obtained depend on the zeta potentials of the sol particles, which is controlled by changing the pH and salt concentration or ionic strength. In effect, dense gels are obtained if the sol phase is stable, which happens if the ionic strength and the absolute value of ζ-potential is high enough (i.e., for pH away from the isoelectric point or zero charge pH). These gels are appropriate to have low-porosity membranes with small pores. On the other hand, if the sol phase has low zeta potentials, high porosities and wider pores are obtained.
2. The polymeric gel route starts from organometallic compounds, normally alkoxides linked by oxygen to the metal. Here, the alkoxide is hydrolyzed and condensed to form a viscous solution of organometallic polymers. By adequate control of the hydrolysis the polymer clusters increase in the sol phase until gelification.

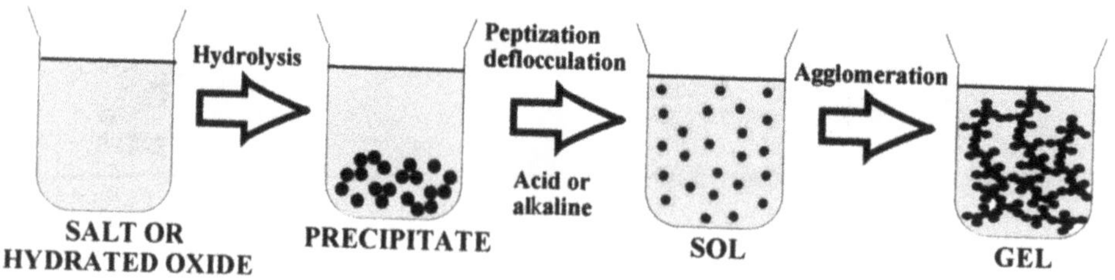

FIG. 8 Representation of the sol–gel technique.

In both techniques, several route specific additives are needed, including some to improve sol rheology to allow easy molding. Normally, the slide deposition on an appropriate support layer is used, followed by an in situ gelification, drying, and sintering. In both cases, organic compounds are burnt during sintering. Some examples are:

1. Zirconia membranes can be obtained by the colloidal route from Zr oxychloride. Adequate control of sintering temperatures can give particle sizes from 6 to 12 nm, depending on the crystallization process. This produces nanofiltration membranes with a molecular weight cut-off as low as 750 Da [24]. They can also be made from Zr alkoxides that completely hydrolyze giving hydrated oxides. By this method, membranes with pore sizes from 6 to 70 nm have been obtained [38–40]. If the Zr alkoxides are partially hydrolyzed with acetylacetone an organometallic compound is obtained and, using the polymeric gel route, will give membranes with pore sizes down 4 nm [24].
2. Nonsupported titania membranes have been obtained by Anderson et al. [41], using the colloidal suspension route. The membranes obtained by this method have pore sizes from 3.8 to 20 nm, depending on the crystalline phase reached (depending on the sintering temperature) by titania. Also through the polymeric gel route, titania membranes with pore sizes of 2 nm can be obtained by following the same steps described for zirconia membranes [24].
3. Alumina membranes with pore sizes under 5 nm can be made from bohemite, γ-AlOOH, suspensions peptized in aqueous solutions of HNO_3 or HCl [42, 43], by following the colloidal gel route. Pore size increases with sintering temperature, giving pores as wide as 78 nm when alumina totally transforms to α-Al_2O_3 [44].
4. Membranes of LaOCl can be manufactured from aqueous solutions of lanthanum chloride. After an adequate sintering step, LaOCl crystallizes at 400°C and is very stable up to 800°C, giving pore sizes as small as 1.5 nm [24]. These membranes are used as catalytic reactors.
5. Silica membranes have only recently been synthesized. Silica forms sols and gels very easily through both routes. Larbot et al. [45] reported a SiO_2 membrane (5 μm thick) prepared through a colloidal route. This membrane was supported on an alumina layer (0.2 μm pore size) and had pore sizes from 6 to 10 nm. The polymeric route can start from a tetramethoxysilane leading to crack-free layers up to 20 nm with grain sizes of 6×2 nm [24].
6. Binary mixtures of oxides have also been used to make membranes. Electrically conductive RuO_2–TiO_2 active layers on alumina supports have been manufactured [46], with pores of 10–20 nm when made through a colloidal route and of 5 nm for polymeric gelification. Binary mixtures of Al_2O_3–TiO_2, Al_2O_3–CeO_2, or Al_2O_3–ZrO_2 have also been used [47] by the colloidal gel route. In all these cases, alumina seems to diminish cracks and defects. TiO_2–SiO_2 membranes can be made by the polymeric gel route [5].

III. STRUCTURAL CHARACTERIZATION

As mentioned, membranes can be classified according to their material components and structure as being porous or dense. What can be understood as dense membranes depends on the scale to which they are studied. The performance of dense membranes (permeability and selectivity) is determined by the intrinsic properties of the material by solution-diffusion through the molecular interstices in the membrane material. When solution-diffusion is not the main transport mechanism, the membranes are considered porous.

Some liquid or gas membranes are composed of solid matrices whose porous structure may play a role in the actual transport. On the other hand, solid dense membranes are usually fixed on a porous support whose structure can penetrate into the dense layer, leading to defects in the dense material and corrections to the expected flow. Even if these defects are not present the transport through the support material can play a significant role in process features.

For porous membranes, the selectivity is mainly determined by the dimension of the pores, and the material has only an effect through phenomena such as adsorption and chemical stability under the conditions of actual application and membrane cleaning [1]. The prediction of the process performances of these membranes for industrially relevant separations ultimately rests on the development and application of effective procedures for membrane characterization.

Most membrane manufacturers characterize their products by a single pore size or a molecular weight cut-off value. These data are usually obtained by measuring the rejection of various macromolecules or particles of increasing hydrodynamic diameter or molecular weight. Nevertheless, it is clear that this single value does not specify the separation properties of the membrane nor its structure. While the molecular weight cut-off can be accepted as a datum useful for preliminary selection, all membranes must be assumed to contain size-distributed pores.

There are several independent methods for determining pore statistics [1, 48–52]. The major ones used to investigate the actual structure of membranes are summarily presented below [53]:

1. Microscopical methods. These methods include several electronic microscopy techniques such as SEM (scanning electron microscopy), TEM (transmission electron microscopy), FESEM (field effect scanning electron microscopy), etc. Also, scanning probe techniques such as AFM (atomic-force microscopy) and STM (scanning tunneling microscopy) could be included in this group.
2. Methods where a fluid is intruded into the pores without changing its phase and displacing another fluid that was filling the pores. The penetrating fluid can flow through the membrane or just fill it; in the first case the resulting flow is measured versus pressure while in the second case the volume of the penetrating fluid needs to be measured. Flow measuring techniques are, for isntance, gas–liquid or liquid–liquid displacement techniques, while mercury porosimetry is a volumetric method.
3. Techniques where there is a phase change in the fluid which penetrates the pores. In adsorption–desorption methods the amount of an inert gas which condenses into the pores is measured versus pressure. In permporometry the flux of a noncondensable gas is measured while other condensable gas

remains in a portion of pores, depending on pressure. Finally, in thermoporometry the variations of the solidification temperature of a liquid filling the pores is measured.

It is clear that these methods can be included in two main groups. Some of them (those developed for the characterization of general porous materials) obtain morphological properties directly, while others give parameters related to the membrane permeation (those designed specifically to characterize membrane materials) [54]. These techniques will be described in some detail here. Of course, they are clearly not the only methods that can be used, but certainly are the most useful and widespread. Other characterization methods can be considered complementary and will not be treated here. Some of them can also be used to characterize pore size distributions while others are useful for characterizing other properties of membrane materials.

Methods that can also be used for determining pores and pore sizes in filters include: nuclear magnetic resonance (NMR) measurements, wide-angle X-ray diffraction, small-angle X-ray scattering, electrical conductance, etc. Glaves and Smith [55] have demonstrated the determination of pore size in water-saturated membranes using NMR spin–lattice relaxation measurements. The NMR measurements must be calibrated using a known pore population material, but nevertheless no specific pore geometry has to be assumed. Wide-angle X-ray diffraction allows determination of the degree of crystallinity of the membrane material, which can be related to the pore sizes [56]. Small-angle X-ray scattering gives important structural information on pores producing the distinct heterogeneities of the electronic density [57]. Finally, electrical conductance measurements have been used by Bean et al. [58] to obtain pore sizes of mica sheets.

Other techniques can give important physical and chemical parameters, other than pore size distributions, of porous materials: ESCA (electron spectroscopy for chemical analysis), FTIR (Fourier transform infrared spectroscopy), or contact angle determinations [59, 60].

It is clearly interesting to know which are the principal features of each technique, in order to be able to choose adequately the more interesting technique for a given membrane and purpose. Of principal concern would be the range of pore sizes that each technique covers. Another important factor is the number of samples required to perform a membrane characterization. While liquid displacement and microscopy methods need only a small quantity of membrane, the techniques of gas adsorption and mercury porosimetry require a considerable amount of sample area, depending on the porsity of the membrane. Moreover, SEM and mercury intrusion can destroy (or at least lead to a change in performance of) most of the membrane materials (especially mercury intrusion, which adds the highly contaminant character of mercury).

A. Microscopical Methods

In this section, we will consider the characterization methods based on microscopy techniques that together with the appropriate image analysis allows evaluation of the membrane parameters without any previous assumption on the pore geometry, or the grain structure, when dealing with ceramic metallic oxides.

Microscopy techniques were used very early in the characterization of membrane filters [61, 62]. Visual inspection of pore structure is an invaluable tool for a deep knowledge of the filters themselves. Nevertheless, as filter development entered the range of submicrometer pores, i.e., pores with sizes under 1 μm, optical microscopy was not useful to achieve a real picture of the membrane structure, given that resolution is limited by the light diffraction pattern. Only the development of nonoptical microscopy techniques circumvented this problem. Primarily, electron microscopy [63] and recently probe microscopy [64] have given a gentle push to the microscopic characterization of membranes, so that we can now have information of membrane surfaces covering the full range of membrane filters.

Transmission electron microscopy operates by flooding the sample with an electron beam, most commonly at 100–200 keV, and detecting the image generated by both elastically and inelastically scattered electrons passing through the sample. TEM operates in the magnification range from 600× to 10^6×. Similarly, a fine beam of medium-energy electrons (5–50 keV) causes several interactions with the material, producing secondary electrons used in the SEM technique. SEM equipment is able to achieve magnifications ranging from 20× to 10^5×, giving images marked by a great depth of field, thus leading to considerable information about the surface texture of the particles.

Usually, SEM samples are covered by a thin metallic layer (normally a gold film of some hundreds of angstroms), increasing the production of secondary electrons and improving, therefore, image contrast [65]. For TEM, a more complicated procedure is required. If only the surface is to be analyzed, a replica technique can be used by coating the membrane with a carbon film and then removing the membrane material, by dissolving the membrane for example, and analyzing the replica. This procedure is, in fact, not often used for inorganic membranes owing to the difficult procedure of selective dissolution of only the membrane material.

The maximum resolution of TEM is ~ 0.3–0.5 nm, while SEM has 10 times greater resolution and in both cases a high electron beam energy is applied. In the early 1980s the FESEM technique, which nowadays achieves very high resolution (up to 0.7 nm) even at low beam energy (accelerating voltage 1.5–4 kV), was developed and used to observe surface pores of ultrafiltration membranes [66].

Other techniques do not require surface pretreatment. Scanning probe microscopy was developed by Binnig et al. [64]. AFM is a newly developed characterization technique which shows promise for development and application in the field of microscopic observation and characterization of various surfaces. A very small tip scans the surface and moves vertically according to its interaction with the sample, similarly to what is done in STM. Both techniques differ in the method used to detect interactions. In STM the tip is close to the sample (both being electrically conducting) allowing a current to flow by tunnel effect and the sample or tip moves to keep this current constant. In AFM the tip is placed on a cantilever whose deflection can be detected by the reflection of a laser beam appropriately focused. This allows analysis of nonconducting materials, which makes the method more convenient to study membrane materials [67–69]. Several procedures can be used for AFM:

1. Contact mode AFM: measures the sample topography by sliding the probe tip across the sample surface. The tip–sample distance is maintained in the repulsive range of the van der Waals forces.

2. Noncontact mode AFM: similarly, the topography of the sample is measured by sensing the van der Waals attractive forces between the surface and the probe tip held above the surface. While worse resolution than contact mode is achieved, the risk of sample damage is avoided or minimized.
3. Tapping or intermittent contact mode AFM is a variation of the contact mode and operationally similar to the noncontact mode, so that it features the best characteristics of both methods. The cantilever is oscillated at its resonant frequency with high amplitude (over 100 nm) allowing it to touch the sample during the oscillation. This method maintains the high resolutions achieved in the contact mode, but minimizes the surface damage, as far as it eliminates the lateral friction forces.

The previously mentioned techniques give account of the sample topography. Moreover, other properties of the surfaces can be obtained by analyzing different forces between the sample and tip. For example, the phase contrast provides information about differences in surface adhesion and viscoelasticity [70]. Magnetic and electric force microscopy [71] both measure magnetic (or electric) force gradient distribution above the sample surface. Surface potential microscopy measures differences in local surface potential cross the sample surface. Force modulation measures [72] relative elasticity/stiffness of surface features, and lateral force microscopy [73, 74] analyzes the frictional force between the probe tip and the sample surface. Finally, electrochemical microscopy [75] measures the surface structure and properties of conducting materials immersed in electrolyte solutions with or without potential control. In many of these techniques, appropriate treatment of the measured forces is necessary to eliminate the contribution of the topographical images.

As an example, Fig. 9 shows images obtained by SEM, FESEM, and AFM of an A01 membrane. Many other examples can be found in the literature. For instance, Marchese [24] showed a three-dimensional AFM image of a ZrO_2 membrane, Hsieh et al. [76] used surface SEM images to obtain particle sizes and trans-

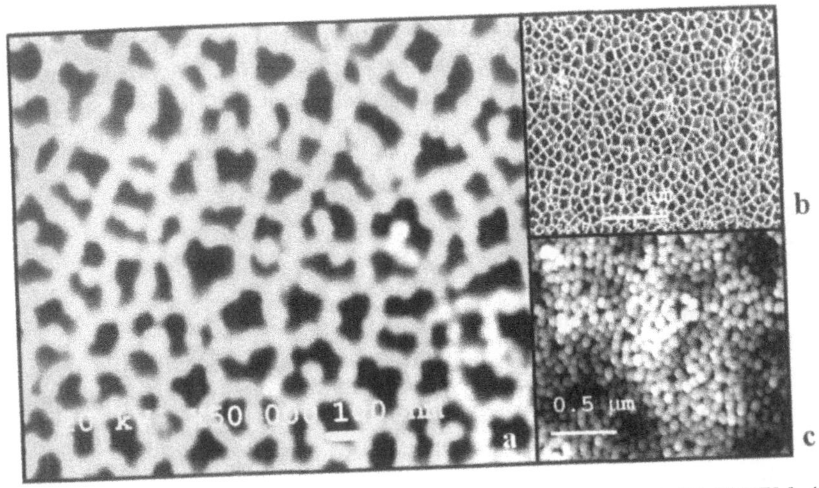

FIG. 9 Images of the active layer of A01 membrane: (a) SEM; (b) FESEM; (c) AFM.

versal SEM images to measure thicknesses of the layers of an alumina membrane, Larbot et al. [77] analyzed FESEM transversal pictures of a commercial UF ZrO_2 (on a porous support of zirconia too) membrane (Kerasep® of Tech-Sep), while Raman et al. [78] used TEM images of a Membralox® alumina support on which they deposited silica whose thickness was determined.

By using the appropriate software, image analysis can be carried out to obtain different structural parameters from the membrane images [79–82], for example, the surface pore density or number of pores per surface unit, N_T, the porosity or porous surface fraction, Θ, and the statistical distribution of, for instance: pore areas, A_p, and pore perimeters, P_p, the equivalent or Feret pore diameter [83]:

$$d_p = 2\sqrt{\frac{A_p}{\pi}} \tag{1}$$

and the pore shape factor:

$$s_p = 4\pi\frac{A_p}{P_p^2} \tag{2}$$

According to these definitions, the equivalent pore diameter, d_p, is the diameter that a pore of area A_p should have if it had a circular section in the surface, while the pore shape factor, s_p, is the ratio between the actual pore area and the area of a circle with the same perimeter. Actually s_p should be unity if the pore sections on the membrane surface were perfectly circular.

In Fig. 10, the pore size distribution for the active layer of the A01 membrane is shown as obtained from FESEM images (Fig. 9b).

In the case of AFM images, we obtain information on height which permits the study of pore entrances in greater detail. In this case, it is possible to use horizontal and vertical sections of the surface [84]. The simultaneous use of three-dimensional images and sections greatly facilitates identification of the entrance of individual pores. However, the diameter deep in the membrane may not be determined directly by surface AFM due to convolution between the tip shape and the pore. Some other parameters are specific for AFM and are usually implemented by the on- and/or off-line analysis software included with the apparatus. In particular, roughness can be analyzed. This analysis is based on determining the heights of the tip over a baseline or reference level, Z. A statistical treatment of such heights allows definition of typical roughness parameters, as the average roughness, R_a:

$$R_a = \frac{1}{n}\sum_{i=0}^{n} |Z_i - Z_m| \tag{3}$$

where n is the total number of points in the image matrix.

In Fig. 11, an AFM micrograph is shown of an ultrafiltration tubular membrane made by TAMI from zirconium oxide deposited on to a multichannel alumina support. Both active layer and support are shown. The corresponding values for roughness are $R_a = 99 \pm 8$ nm for the ZrO_2 layer and $R_a = 59.6 \pm 0.3$ nm for the alumina support.

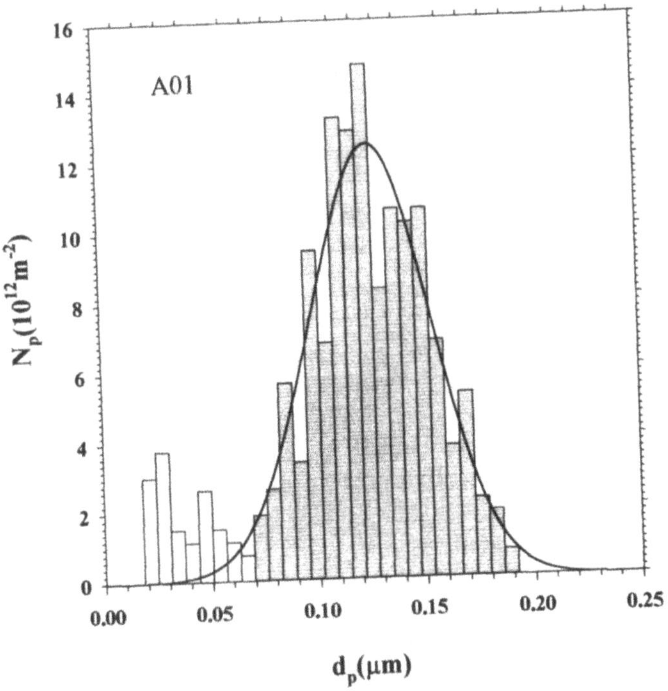

FIG. 10 Pore size distribution of A01 membrane by image analysis of FESEM images.

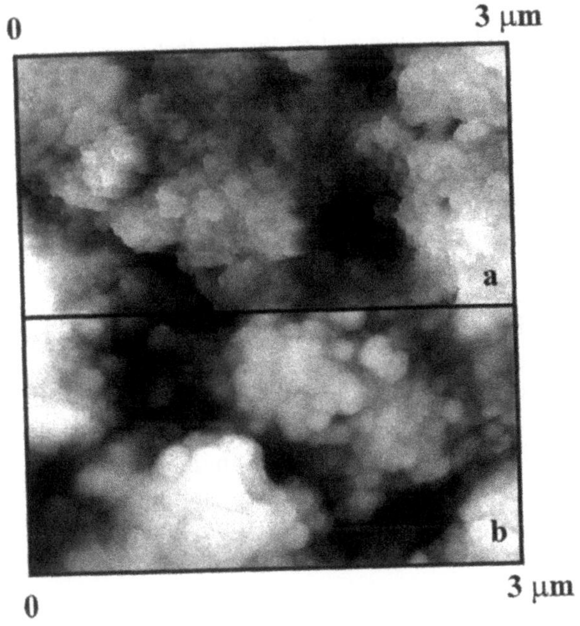

FIG. 11 AFM image of a TAMI UF membrane: (a) active layer; (b) support.

B. Penetration Without Phase Change

When a gas–liquid or liquid–liquid interface is made to move inside a capillary tube, the surface properties affect the liquid–solid interface. However, taking into account that the solid cannot change its shape and size, it will be only the liquid that changes in such a way that, the pressure Δp necessary to intrude or expel the fluid from the capillary is given by the Young–Laplace equation as follows:

$$\Delta p = \frac{2\gamma \cos\theta}{r_p} \tag{4}$$

where θ is the contact angle, γ is the surface tension, and r_p is the capillary radius.

Bechhold [85] was the first to use Eq. (4) to evaluate pore sizes by measuring the pressure necessary to blow air through a water-filled membrane whose contact angle was assumed to be zero. The bubble-point method allows determination of the maximum pore size present in the pore distribution, corresponding to the minimum pressure necessary to blow the first observed air bubble. This method is frequently used as quality control to check the integrity of filters.

Stepwise increments of the applied pressure allows one to calculate [86] the number of pores corresponding to each diameter present in the pore distribution. The method has been recently improved for both liquid–gas interfaces [87, 88] and liquid–liquid ones [89, 90], allowing the evaluation of pore sizes corresponding to a wide range of porous materials.

When a gas–liquid interface is used, a wetted sample is subjected to increasing pressure applied to the gas source. As the pressure of gas increases, it will reach a point where it can overcome the surface tension of the liquid in the largest pores and will push the liquid out. Increasing the pressure still further allows the gas to flow through smaller pores, according to Eq. (4). The range of applicability of this method depends on the characteristics of the liquid soaking the membrane, due to both the surface tension of the gas–liquid interface and to the contact angle of this interface in contact with the solid material. So, to obtain an extensive range of applicability, the liquid needs to have the lowest possible surface tension and a contact angle as near as possible to zero. There exists a range of organic liquids having low surface tensions ranging from 15 to 20 mN/m at temperatures between 293 and 313 K. They have also nearly zero contact angles with the most common membrane materials, allowing analysis of pore sizes below 0.1 μm, with applied pressures of about 10 bar.

Given that the contact angle depends on the liquid–membrane interaction, different liquids will be preferred for different membrane materials. In effect, liquids with low dielectric constant (hydrophobic liquids) should be selected when dealing with hydrophobic membranes, while high dielectric constant liquids should be preferred for hydrophilic materials. This can be somewhat inconvenient when having different materials whose hydrophilicities differ or are unknown. In this case, a single liquid should be selected as a standard. Normally, those exhibiting both hydrophilic and hydrophobic functional groups (as, e.g., hydrocarbonated compounds having a polar functional group) should be selected. This is the case, for instance, of alcohols or halogenated compounds.

Of course, the liquid used must be chemically compatible with the membrane material, which is usually the case when dealing with inorganic membranes. In addition, the liquid should have a low vapor pressure at the working conditions of temperature and pressure, since liquid evaporation during the measurements can lead to erroneous results. Finally, it is worth mentioning that, since these measurements are dynamically performed, it is convenient to choose liquids having viscosities as low as possible to avoid the influence of the measurement speed on the results.

A good candidate as wetting liquid, very frequently used in this kind of porometric analysis, is isopropyl alcohol, which has a surface tension (air–alcohol) of 20.86 mN/m and a vapor pressure of 5.8 kPa at 298 K. There exist, of course, commercial liquids specially designed for these kind of determinations, having very optimized properties. For example, the Porofil® (patented by Coulter) has a surface tension of 16 mN/m with a vapor pressure of 400 Pa at 298 K.

When different pores of diverse sizes are opened, the volume flow of gas, J_V, increases accordingly until all the pores are emptied. By monitoring the applied pressure and the flow of gas through the sample when liquid is being expelled, a wet run is obtained. If the sample is then tested dry (without liquid in its pores), a dry run follows. Computer-controlled instruments with software-driven experiment execution, data acquisition, processing, storage, and plotting are now commercially available.

The volume flow for the wet run, J_V^w, and for the dry one, J_V^d, versus the applied pressure, allows evaluation of several statistical parameters [91–95]. From these data, the cumulative and differential distributions of relative number of pores and flow can be obtained. If the absolute number of pores and porosity are to be evaluated, a model for the gas flow through the pores must be assumed.

The model for gas flow should be determined by the relation between the mean free path of the gas molecules and the pore size. In a first view, three simple models can be proposed to describe gas flow: the Hagen–Poiseuille viscous flow, the Knudsen molecular one, and some transitional regimes between both. For a given gas, the question of the limiting radii to pass from one to the other of these regimes has been frequently discussed [96–98]. This problem can be bridged if we look for the limiting pore size where the representation of $(J_V/\Delta p)$ for both the regimes intersect. For air ($M_W = 29 \times 10^{-3}$ kg/mol; $\eta = 1.904 \times 10^{-5}$ Pa s) at $T = 313$ K with a downstream pressure $p = 1$ bar, this happens at a pore diameter of 0.96 µm [95, 99]. For pores around this limiting diameter, the flow should be better described by a smooth curve. Given that there is not an easy way to establish this transitional regime, a reasonably good approximation can be determined by using Knudsen flow for pores below the limiting diameter, Hagen–Poiseuille flow for those over it, and a smooth curve for those pores of intermediate diameter.

The pore size distribution can also be estimated by using two immiscible liquids, one of them soaking the membrane structure and the other a permeating liquid, which expels the wetting one [89, 100]. This technique allows determination of pore sizes from 5 to 140 nm by using immiscible liquids with appropriately low surface tensions. This method permits characterization of ultrafiltration membranes. The main advantage of this technique is the absence of high applied pressures (always below 10 bar). Examples of typically used pairs of liquids are: distilled water–water saturated isobutanol ($\gamma = 1.7$ mN/m), distilled water–mixture of isobutanol, methanol, and water (5:1:4 v/v) ($\gamma = 0.8$ mN/m), or distilled water–mixture of isobutanol, methanol, and water (15:7:25 v/v) ($\gamma = 0.35$ mN/m) [89].

Since the permeating fluids are liquids, the question of the flow regime does not apply, but the pore numbers require a viscous flux model. If usual assumptions on the pore geometry are done, Hagen–Poiseuille (for cylindrically shaped pores) or Carman–Kozeny (for ceramic particulate media) models should be applied.

In the case of hydrophilic membranes (as most of the inorganic membranes are), when using liquids like those already mentioned, it is reasonable to suppose zero contact angle (or nearly). The same calculation procedure described for gas–liquid interface can be used to calculate the differential flux and pore number distributions.

Another characterization method that can be included with those based on the Laplace equation is mercury intrusion porosimetry. The method (also proposed by Washburn [101]) was developed by Ritter and Drake [52] and was applied for the first time to the characterization of membrane filters by Honold and Skau [102]. It has been shown to be a reliable method for the characterization of pore size distributions, pore structure, and specific surface areas. Here, a Hg–air interface appears inside each pore. Thus, Eq. (4) is also followed. However, in this case, Hg does not wet practically any kind of sample (the corresponding contact angles ranging from 112° to 150°) [52, 101].

Plots of intruded and/or extruded volumes versus pressure, usually called porograms, can be used to obtain the pore size distributions. The porograms can show a great variety of shapes, depending on the characteristics and distribution of pores and voids, if present in the sample. However, two common features are always present in porograms [52]:

1. Hysteresis is always obtained, i.e., the extrusion path does not follow the intrusion curve.
2. Moreover, after completion of an intrusion–extrusion cycle some portion of mercury is always retained by the sample (pore entrapment) avoiding loop closing. This phenomenon usually ceases after the second pressurization–depressurization run.

Both phenomena have been frequently attributed to "ink-bottle" pores [103]. This explanation indicates a very wide distribution of "ink-bottle" pores. A possible more realistic point of view to explain hysteresis and entrapment is the assumption of a network of different sized intersecting pores. Andreoutsopoulos and Mann [104] have calculated the consequences of assuming a two-dimensional square network of cylindrical intersecting pores. Their predictions agree fairly well with the actual hysteresis and entrapment behavior of catalytic materials where pelleted structure can be assumed. Lowell and Shields [105] have shown that coincidence between intrusion and extrusion curves is possible, at least in the second and subsequent cycles, if the contact angle θ is adjusted to distinguish between advancing (θ_i) and receding (θ_e) contact angles. Nevertheless, the first cycle in the porogram cannot be toally closed due to the mercury entrapment which varies greatly (from nearly zero to almost 100%) [104].

Mercury intrusion porosimetry has been used for a long time as an experimental standard technique for the charcterization of pore and void structure. The usual application range is from 0.002 to 1000 μm pore size. Nevertheless, to achieve the lower range, really high pressures are necessary, up to 4.5 kbar, which increases dangerously the risk of distorting the porous structure. However, this is not a critical factor for many inorganic membranes whose compressibility is low.

Most of the commercially available porosimeters include certain common features. First, the sample is evacuated and then the penetrometer is backfilled with Hg in the low-pressure port. The second step of the low-pressure analysis is the collection of the data at pressures up to the last low pressure point specified. When the low-pressure analysis is complete, the high-pressure measurement is carried out up to the maximum pressure. Pore volume data are calculated by determining the volume of Hg remaining in the penetrometer stem. When the maximum pressure is achieved, the extrusion curve starts by reducing slowly the applied pressure. Commercial instruments can work in one or both modes: incremental and continuous. In the former the pressure, or amount of Hg introduced, is increased step by step and the system is allowed to stabilize before the next step. In the continuous mode the pressure is increased continually at a predetermined rate [106].

The question on what contact angle should be used to correlate applied pressures and pore sizes remains open, and it should be taken into account that errors of 1° out of 140° should lead to errors in all pore radii of 1.4% [52]. A contact angle of $\theta = 130°$ seems to be valid for a wide range of materials and it is normally set as default value in most commercial equipment [52], along with a Hg–air surface tension, $\gamma = 0.474\,N/m$ [107]. Nevertheless, some authors calibrate porosimetry results in order to reproduce previous independent calculations on the pore size distributions. This approach has been followed by Liabastre and Orr [101] for Nuclepore membranes, comparing with computerized image analysis of scanning electron microscopy images, leading to a contact angle of 126.3°. This method can induce very inexact estimations of pore size distributions if the reference method is not adequately chosen. SEM photographs are usually not useful for callibration since they refer to surface characteristics of the membrane that can be different from those of the bulk structure [79].

Liquid intrusion/extrusion method should apply ideally to membranes containing cylindrical pores. When this is not the case, the pore size distributions obtained refer actually to equivalent or effective pore sizes. In Fig. 12 the air–liquid displacement distribution is presented along with the Hg-porosimetry one for the A01 membrane. A $TiO_2/\alpha\text{-}Al_2O_3$ layer on a stainless steel woven wire cloth made by Synthesechemie, whose pores are far from being cylindrical, as shown in Fig. 13, gives a different pore size distribution by air–liquid displacement and Hg-porosimetry distributions, as shown in Fig. 14.

Other examples of application of these methods can be found: Palacio et al. [34] show air–liquid penetration and Hg-porosimetry results for the same TiO_2 membrane whose images are shown in Figs 6 and 7; Hsieh et al. [76] show how Hg porosimetry can be used to detect the four layers of an alumina membrane, and Luyten et al. [108] use Hg porosimetry to measure pore size distribution of both the reaction bonded Al_2O_3 selective layer and the $\gamma\text{-}Al_2O_3$ mesoporous support.

C. Penetration with Phase Change

Adsorption phenomena have long been recognized as a surface phenomena, taking place at the gas–solid or liquid–solid interface, where the adsorbate molecules fix themselves to the solid surface as a result of several attractive forces. The nature of

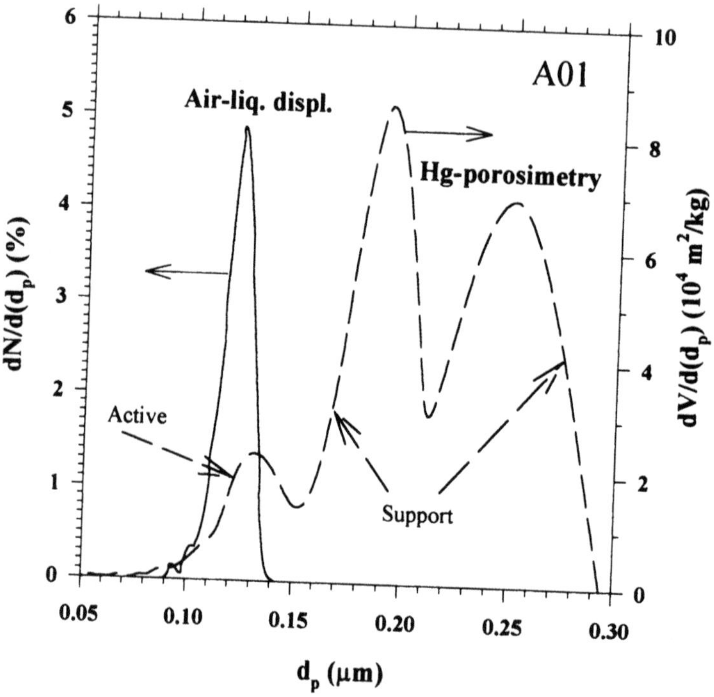

FIG. 12 Air–liquid displacement and Hg porosimetry pore size distributions of the A01 membrane.

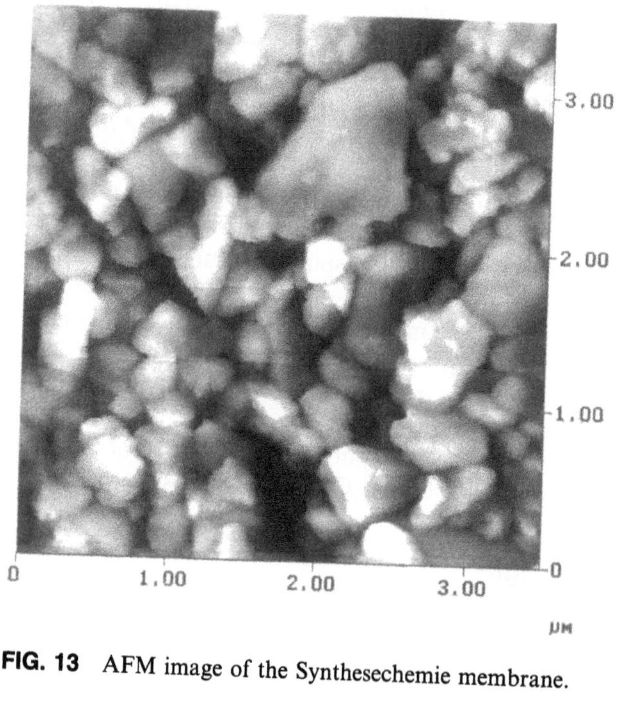

FIG. 13 AFM image of the Synthesechemie membrane.

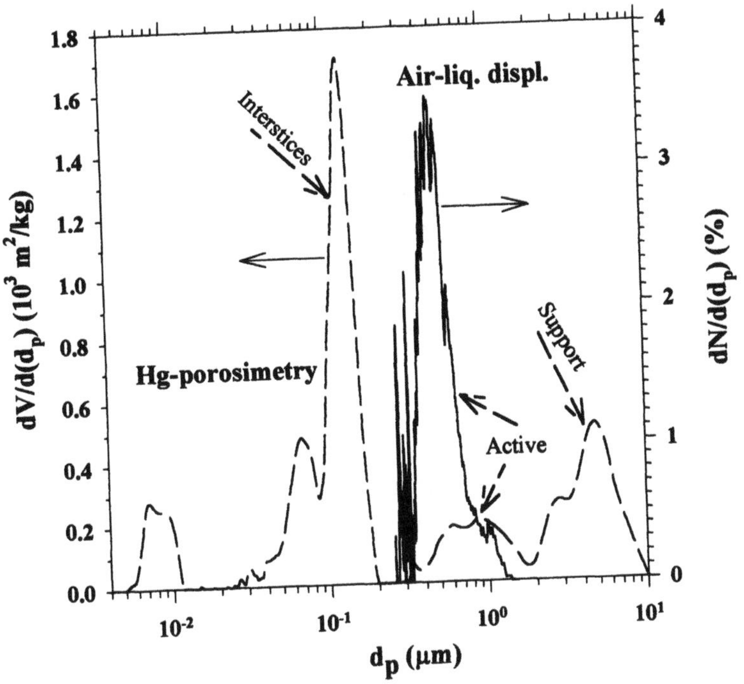

FIG. 14 Air–liquid displacement and Hg porosimetry pore size distributions of the Synthesechemie membrane.

these forces can be very diverse, but it is conventional to speak of physisorption or chemisorption, depending on whether the forces are of a physical (Van der Waals forces) or a chemcial (bonding forces) nature, respectively.

The study of these systems is carried out by determining the so-called adsorption isotherm. An adsorption isotherm consists in the evaluation of the amount of adsorbed gas as a function of the equilibrium pressure at a constant temperature. The physical adsorption is the predominant process for inert gases at temperatures below their critical point. Normally, in the adsorption isotherm, pressure is expressed as the relative pressure $p_r = p/p_0$ (p_0 is the saturation pressure at a given temperature) and the amount absorbed expressed as a mass of gas or its volume at standard temperature and pressure.

When dealing with porous solids, it is common to classify the pores according to their size into three main categories. This classification lacks precision and universality but a tentative one, usually accepted, includes: micropores with sizes below 2 nm, mesopores ranging from 2 to 50 nm, and finally macropores whose sizes are larger than 50 nm. The reason for this classification is the applicability of the Kelvin equation with N_2 at 77 K as adsorbate.

The shape of the adsorption isotherm is related to the inner structure of the adsorbate in such a way that, according to the IUPAC recommendations, five main groups can be distinguished [109]. From the point of view of porous materials whose distribution of pores are to be determined with this technique, only the isotherms of types I (micropores) and IV and V (mesopores) are relevant.

When these methods are used to determine pore size distributions (physisorption), mainly N_2 and sometimes Ar are chosen as adsorbates. Other gases can be used when special characteristics of the materials are to be studied. For example, it can be interesting to study the hydrophilic/hydrophobic character or the dependence of the material accessible for adsorption on the molecular size of the gas. In these cases, vapors from water, hdyrocarbon compounds, alcohols, etc., can be used. For most hydrophilic membranes such as those made from metal oxides, it is clear that when using water vapor, the hydrated structure should be characterized and thus compared with the dry structure that can be obtained by using an inert gas.

The following paragraphs will be devoted to the methods for determination of the pore size distributions of membranes. As previously mentioned, these materials can be divided into two groups (presence of micropores or mesopores). However, as a first step we will consider the determination of the total adsorption surface of these materials by gas adsorption, not only because this parameter is an important structural characteristic, but also due to the relevance of the models used in its determination as the basis of some of the usual methods to calculate pore size distributions.

The total surface area of a membrane where adsorption can occur may be determined by using the Brunauer–Emmett–Teller (BET) method, which is generally regarded as the standard procedure. Thus, a relatively simple expression is obtained:

$$\frac{p}{V(p_0 - p)} = \frac{1}{V_m C} + \frac{C-1}{V_m C}\frac{p}{p_0} \tag{5}$$

where V_m is the total volume adsorbed in a monolayer per mass unit of sample, V is the volume adsorbed per mass unit of the sample, and C is a parameter related to the molar adsorption enthalpy [68].

Accordingly, a plot of $p/[V(p_0 - p)]$ against p_r should give a straight line, allowing evaluation of both C and V_m from the slope and ordinate intercept of the plot. In practice, such a simple model accurately serves only for relative pressures in the range 0.05–0.3. In fact, what ought to be done is to check the linear range to be used; in most cases, p_r from 0.5 to 0.1 should lead to correlation coefficients over 0.9999. Finally, by taking into account the molecular size of the adsorbate A_m (16.2 $Å^2$ for a nitrogen molecule), the specific adsorption area of the membrane should be

$$S = \left(\frac{V_m}{v_g}\right) N_A A_m \tag{6}$$

where v_g is the molar volume of the gas at standard temperature and pressure, and N_A is Avogadro's number.

The BET plots for both a cylindrical pore membrane, A002, and a membrane with a more complex porous structure, TAMI UF, are shown in Fig. 15.

Referring to the pore size distribution, both micropores and mesopores can be analyzed. Although the adsorption isotherm near the condensation pressure ($p_r = 1$) gives important information on macropores, such an analysis is not practical for accurate measurements, because serious condensation on the apparatus walls begins near saturated vapor pressure.

Capillary condensation of the adsorbate molecules into the adsorbant pores is the key mechanism of adsorption in mesopores. In this case, the molecules adsorbed on a solid surface behave as being in the liquid state. This is true for all adsorption

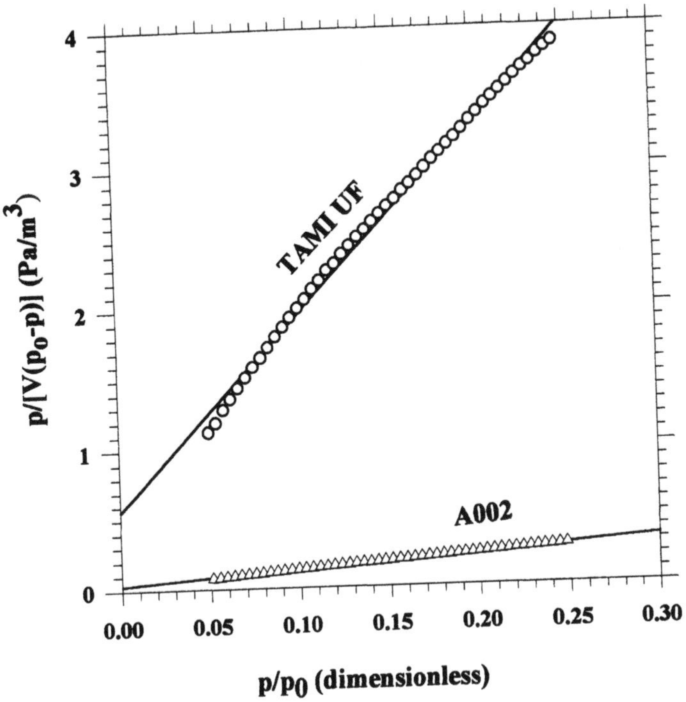

FIG. 15 BET plots for the A002 and TAMI UF membrane. Images of these membranes are presented in Figs 4 and 11, respectively.

layers except the first ones (strictly the first one within the frame of the BET model). Thus, the interface between the gas and the already adsorbed molecules is a vapor–liquid equilibrium. This equilibrium is determined by the surface curvature which, when referring to a porous material, depends mainly on the geometry and size of the pores. This dependency is accounted for by the Kelvin equation which, for a liquid perfectly wetting the solid ($\cos\theta = 1$), can be simplified to

$$\ln\left(\frac{p}{p_0}\right) = \left(-\frac{a\gamma v_1}{RTr_k}\right) \tag{7}$$

where γ is the surface tension, v_1 is the molar volume of the liquid, $a = 2$ for the desorption processes or $a = 1$ for the adsorption process, and r_k is the Kelvin radius (i.e., the pore radius minus the thickness of the adsorbed layer, t). The thickness t is usually evaluated by applying the Halsey phenomenological correlation [110] which gives t in angstroms as

$$t = 0.354\left[\frac{5}{\ln\left(\frac{p_0}{p}\right)}\right]^{(1/3)} \tag{8}$$

Once p_r data are converted into the corresponding values of r_p, the differential distribution of specific volume of the pores with r_p radius, dV_p/dr_p, along with the surface and pore number distributions, can be obtained by the method of Dollimore and Heal, assuming cylindrical pores [109, 111]. This can be done both for adsorption and desorption results. Nevertheless, it should be noted that the adsorption isotherm can only be used if the total specific area of the sample is known. In this case we can use it as obtained from the BET method.

An example of the volume differential distribution obtained from gas desorption is shown in Fig. 16 for membrane A002.

In the range of micropores, the Kelvin equation loses its validity. Several approximations for adsorption isotherm analysis in the micropore range have been used. The most useful alternative is to use the so-called micropores method (or MP method) by Mikhail et al. [112]. It is based on the de Boer *t*-plot where adsorption isotherms are plotted in terms of the statistical adsorbate film thickness instead of the reduced pressure [113].

Mikhail et al. considered that the *t*-plots start to deviate from the initial slope (which should give the total BET surface area) due to the presence of micropores. The subsequent slopes can then be converted into microporous surface areas or the equivalent micropore volumes corresponding to each increment in the abscissa (Δt).

As with the mesopores' analysis, a major point in having reasonable results is the use of an adequate *t*-plot for the given adsorbate. In this case, values of t can be evaluated from an empirical relation given by Harkins and Jura [114]:

$$t = \left[\frac{13.99}{0.034 + \ln\left(\frac{p_0}{p}\right)} \right]^{(1/2)} \tag{9}$$

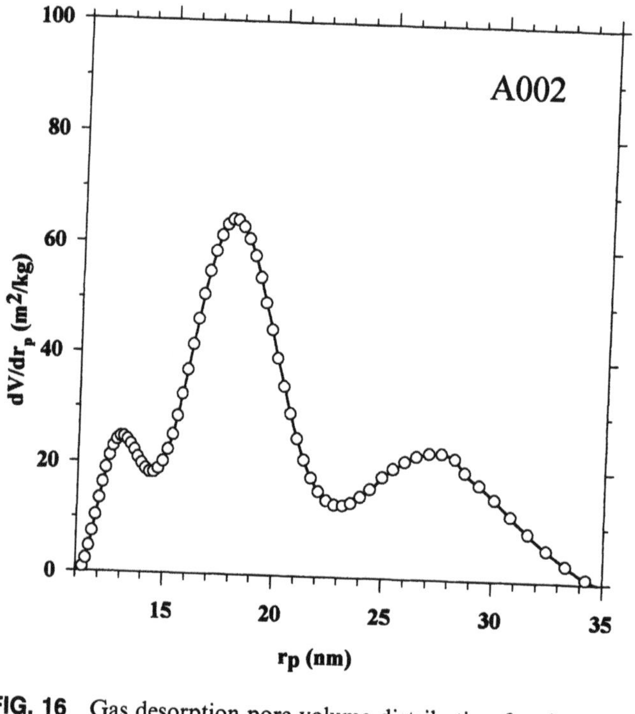

FIG. 16 Gas desorption pore volume distribution for the A002 membrane.

which gives better results in the analysis of micropores than the equation of Halsey [115].

There exists a range of commercially available apparatus that allow determination of the adsorption isotherms of solids with gases or vapors. Taking into account the measurement method, they can be divided into three groups: those which measure the volume and pressure of a gas at equilibrium (static or continuous); those measuring in a continuous manner based on chromatographic methods; and those determining the variation of the solid mass with a microbalance. Some equipment names are: Belsorp 28 from Bel Co.; Sorpty 1750 from Carlo Erba; Fisons Sorptomatic 1900 Series; Horiba SA6200 Series; ASAP 2000, Accusorb 2100E, Digisorb 2600, and Gemini from Micromeritics; Nova 1000 and Nova 1200 from Quantachrome; Omnisorp series from Coulter Ltd, etc.

The micropores' method is frequently used to determine pore size distributions below 20 Å in inorganic membranes. For example, Larbot et al. [77] analyzed the pore size distribution of a TiO_2 membrane with a mean pore size of 8 Å. They also studied how pore size distribution width changes due to small variations of the sol–gel synthesis method. Other methods to analyze micropores are sometimes used too; for instance, Kumar et al. [116] used the Horváth–Kawazoe method to select the appropriate sinterizing (calcination temperature) technique to prepare zeolite membranes.

Another technique where there are phase changes in the pores is permporometry which is based on the principles of the capillary condensation of a vapor inside the membrane pores and the permeability of another noncondensable gas through these pores. As we saw above, capillary condensation is a process which can be modeled with the Kelvin equation, in such a way that control of the relative vapor pressure of a condensable gas allows determination of the size of the pores open to flow of the noncondensable gas.

If flux measurements are started when the relative vapor pressure of the condensable gas equals unity, all the pores in the membrane should be closed (i.e., filled with condensed liquid), avoiding any diffusive flux of the noncondensable gas through the membrane. When the relative vapor pressure is slightly below unity the liquid contained in the largest pores starts to vaporize thereby opening these pores. The Kelvin equation allows correlation of the relative pressure with the size of the pores opened to flux (Kelvin radius, r_k). The measured flow of the noncondensable gas can be easily translated in terms of pore number once the appropriate gas transport model is taken into account. By decreasing steadily the relative pressure until all the pores are opened, both the differential and integral pore size distributions can be obtained.

In this kind of determination, as in the case of the mesopores' analysis through adsorption–desorption measurements, it must be taken into account that the size available to flux is not the real size of the pore. So it is necessary to add the thickness of the adsorbate layer, t, which is a function of relative pressure, as shown above. A good approximation for determining t is again the Halsey correlation.

Once a geometric model for the pore and a model for the gas transport through the pores is chosen, this method (as the liquid displacement technique) allows determination of the absolute distribution (differential or integral) of the number of pores active to flux for the membrane. The model used should depend on the working conditions (type of gas, temperature, pore sizes, etc.). In most of the cases where this technique was used, a diffusive Knudsen model [1, 2] seem to be adequate.

Since this technique is based on the Kelvin equation, it is applicable only to mesopores, limited by the gas and working pressure. However, these limits can be considered as ranging from 2 to 50 nm. The method has been used and compared with other ultrafiltration membrane characterization methods, with very good results [54, 117–119].

Some examples where permporometry is applied to inorganic membranes are: Fain [120] who determined the pore size distribution of alumina membranes made by both sol–gel and anodic deposition methods, by using CCl_4 vapor as the condensable gas; and Cao et al. [121] who analyzed sol–gel γ-alumina membranes by using O_2 as the condensable phase.

Finally, another method where phase changes can be analyzed to obtain pore size distributions is thermoporometry. This method is based on the dependence of the melting or solidification temperature of a substance on the interface curvature. Thus, when a fluid is introduced inside a porous material, since the pores change the surface curvature of the fluid, the determination of the resulting distribution of melting or solidification temperatures can give us information about the size distribution present in the analyzed sample. As a first approximation, pore sizes can be related to freezing and melting temperatures empirically. Such a phenomenological correlation can be obtained by using well known porous substances as test substances. However, it is also possible to obtain equations based on equilibrium thermodynamics, relating those parameters.

It can be shown that

$$\frac{dV}{dr_p} = \Lambda \frac{(\Delta T)^2}{W_a} y \tag{10}$$

where Λ is a constant depending on the sensitivity of the differential scanning calorimeter, the speed of the temperature variation, the mass of the sample, the density of the liquid, and how temperature depends on pore radii. The latent heat of phase change, W_a, depends on the decrease in phase equilibrium temperature in such a way that, in $J \times g^{-1}$ [122] it is given by

$$\begin{aligned} W_a &= -5.56 \times 10^{-2} \Delta T^2 - 7.43 \Delta T - 332 \\ &\quad \text{(solidification)} \\ W_a &= -0.155 \times 10^{-2} \Delta T^2 - 11.39 \Delta T - 332 \\ &\quad \text{(melting)} \end{aligned} \tag{11}$$

for water, and by

$$\begin{aligned} W_a &= -8.87 \times 10^{-3} \Delta T^2 - 1.76 \Delta T - 127 \\ &\quad \text{(solidification)} \\ W_a &= -2.73 \times 10^{-2} \Delta T^2 - 2.94 \Delta T \\ &\quad \text{(melting)} \end{aligned} \tag{12}$$

for benzene.

It is worth noting that the cooling or melting speed must be low enough ($\sim$ 1–6 K/h) for the three phases to remain in constant equilibrium and the temperature to be the same throughout the sample.

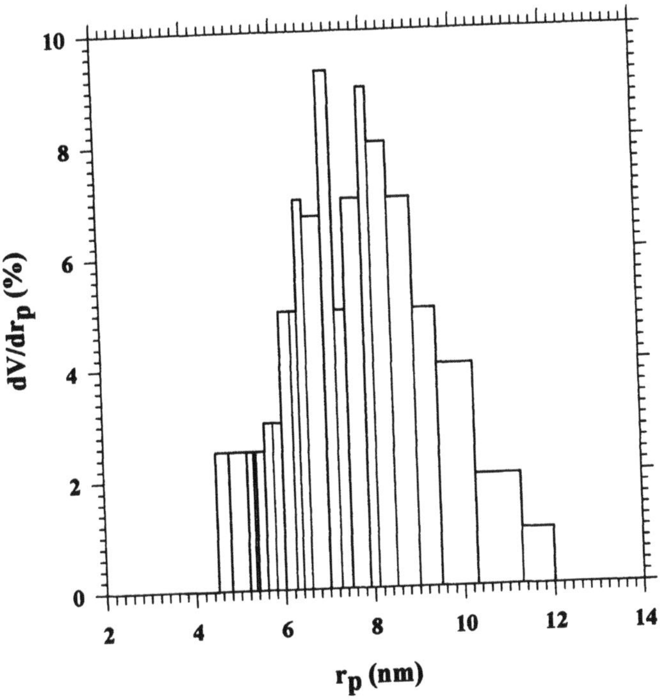

FIG. 17 Thermoporometric pore size distribution of a γ-alumina plug.

This method has been used to characterize porous materials and specifically ultrafiltration membranes, giving good results for pore size distributions in the range 2–30 nm [118, 123–125]. In Fig. 17 the pore size distribution of a γ-alumina plug calculated from results obtained by Brun et al. [126], by using benzene thermoporometry, is shown. By N_2 desorption a quite similar distribution was obtained with an average pore size of 7.2 nm.

IV. FUNCTIONAL CHARACTERIZATION

Appropriate elucidation of structure is not only relevant to describing sieving effects, but also to studying solute–material interactions, as far as the corresponding interfaces are placed inside the pores. Thus, electrically determined membrane properties act on the solutes inside the pores and transport is affected by these properties (zeta potential, surface charges, etc.) in such a way that make necessary a detailed knowledge of pore geometry to correlate interactions adequately with their effects on flux.

The methods based on solute retention are extraordinarily interesting from the application point of view. The factors to be considered to translate actual geometrical or structural characteristics into solute retention performances are especially complex (friction, elasticity, electrical or chemical interaction, hydrophilicity or hydrophobicity, diffusion into the solid matrix, etc.). In the case of metallic oxides, structure can change as well, when the membrane is in an aqueous environment due to hydration. A first approximation should consider sieve factors as the only or the most relevant phenomena by avoiding consideration of friction or elasticity (mechanically or chemically determined) appearing in the hydrodynamic interaction of solute molecules and membrane material. This is equivalent to considering an equivalent pore size to include such factors. In this way, if a polydisperse solution is filtered through a membrane and its molecular weight and/or size distributions are analyzed up and downstream, an equivalent pore size distribution should be determined by sieving effects. On the other hand, the correlation of solute molecular weight and size depends on both the solute–solute and solute–solvent interactions. These require colloidal properties to be held constant by fixing solution properties such as concentration, solvent type, pH, ionic strength, etc.

The process of mass transfer through the membrane is determined by several factors:

1. Adsorption and fouling of the membrane surfaces [127].
2. Formation of a concentration polarization layer due to solute accumulation on the high-pressure side of the membrane with retrodiffusion and possible gelification [128].
3. Other limiting factors such as osmotic pressure [129].

Among these factors the most difficult to quantify and thus to control are adsorption and fouling. In order to determine as much as possible about the actual structural character of the membrane, low interacting solutions should be used under conditions giving low fouling levels. However, since it is impossible in practice to avoid fouling completely, its influence on the porous structure should always be evaluated. This can be accomplished by measuring the decline of flow rate with time and the pure water permeability measured before and after operation [130]. Appropriate models can then be used to take into account the changes due to fouling of the porous matrix. In order to achieve reproducible results one should maintain a stationary flux after fouling, as described below.

The concentration polarization layer can be analyzed by using several different models but none gives generally good results, for all membranes (reverse osmosis, nanofiltration, ultrafiltration, or microfiltration ones) or under all hydrodynamic conditions on the retentate surface. The so-called film model gives good results in spite of its simplicity for membranes in the reverse osmosis to ultrafiltration ranges [131]. Some precautions should be followed to ensure its applicability, namely: (1) mass transfer through the membrane system is limited by the concentration polarization layer more than by the whole membrane, (2) Cross flow (tangential flow) filtration devices should be used, and (3) the retentate should be recirculated on to the membrane at a high speed.

As a consequence of all these factors, the solute should be: (1) used at low concentrations, and (2) be invariant in physicochemical properties over a broad molecular weight range.

In order to study nano- and ultra-filtration membranes some linear polymers could be chosen such as polyethylene glycols or dextranes. However, their linearity makes size designation difficult which is not the case for globular macromolecules such as proteins. In the case of microfiltration membranes: particles of latex, silica, alumina, etc., can be used.

A. Retention and Mass Transfer Coefficients

An important characteristic of membranes is the extent to which they restrict the flow of solutes. Two measures of solute retention behavior are the retention coefficient, R, and the mass transfer coefficient, K_m. An examination of methods to measure R and K_m provides insight into the phenomenon of membrane solute retention.

As solution passes through a membrane, accumulation of solute at the membrane surface occurs as a result of one or more possible mechanisms, i.e., partial or total size exclusion of solute molecules from pores, electrostatic repulsion of solute molecules by a membrane, chemical reaction, adsorption of solute molecules, etc. The configuration of the concentrated solute region contiguous to the solution–membrane interfaces makes it amenable to analysis by the film-theory model [132].

The film is modeled as a thin boundary layer of solution lying between the bulk liquid solution at $y = 0$, and the retentate membrane surface at $y = \delta$. At steady state a concentration gradient for all solutes will exist in the film with $c_{b/f}$ representing the concentration for the solute, b, anywhere within the film, f, such that, $0 \leq f \leq \delta$. Once the flow through a membrane reaches steady state, the concentration, $c_{b/\delta}$, of solute, b (frequently referred to as polarization concentration), in contact with the membrane at $y = \delta$, will be greater than the bulk or feed solution concentration, $c_{b/0}$, at $y = 0$, such that $c_{b/0} < c_{b/\delta}$. Four conditions are assumed for the two solution phases: (1) eddy motion in the film is assumed negligible so that (2) all solute transport in the film can be assumed to occur by diffusive and/or convective laminar flow, (3) steady-state mass transport in the film is perpendicular to the film, and (4) in the bulk liquid solution eddy motion is assumed sufficient to eliminate any concentration gradients.

A material balance of the solute, b, over a differential volume element in the film [133] gives the net or actual mass flux relative to stationary coordinates, $F_{b/f}$, of solute b through the boundary film as

$$F_{b/f} = v_f c_{b/f} - D_b \frac{\partial c_{b/f}}{\partial y} \tag{13}$$

Equation (13) describes the *net mass flux*, $F_{b/f}$, of solute b relative to stationary co-ordinates [132] at any position y $(0 \leq y \leq \delta)$ within the film as the resultant of two fluxes: (1) the *convective mass flux*, $v_f c_{b/f}$, relative to stationary co-ordinates of solute b toward the membrane with concentration $c_{b/f}$ and solution velocity v_f, and (2) the *diffusive mass flux*, $D_b (\partial c_{b/f}/\partial y)$, relative to stationary co-ordinates of solute b away from the membrane along the solute concentration gradient $(\partial c_{b/f}/\partial y)$, with diffusion coefficient, D_b. All fluxes in this chapter can be assumed relative to stationary co-ordinates [132], so for the sake of brevity, reference to stationary co-ordinates will be omitted henceforth. The reader will also note that fluxes are vector quantities [132] and the convective and diffusive fluxes at steady state will move solute in opposite directions, hence the opposing signs for the convective and diffusive flux terms relative to stationary co-ordinates in Eq. (13).

To solve Eq. (13) for flow through a membrane requires additional information which can be obtained from boundary conditions for the steady-state mass fluxes at the retentate membrane surface where $y = \delta$. If molecules of solute b moved into the membrane at the velocity of the average solution molecule in the film, v_f, then the mass flux of solute b would be the convective mass flux, $v_f c_{b/\delta}$. However, at steady-state flow, solute b diffuses away from the membrane, so instead a net mass flux, $F_{b/\delta}$, of solute b actually occurs. The ratio of these two mass fluxes, $F_{b/\delta}/v_f c_{b/\delta}$, is the fractional amount by which the concentration of solute b is reduced as it enters the membrane due to back-diffusion. Conversely, the fraction, R, of solute b rejected by the membrane which is simply unity minus the fraction of solute flowing into the membrane or

$$R = \begin{bmatrix} \text{fraction} \\ \text{of solute} \\ \text{rejected} \end{bmatrix} [=] 1 - \begin{bmatrix} \text{fraction of} \\ \text{solute into} \\ \text{membrane} \end{bmatrix} [=] 1 - \left[\frac{\text{net solute flux}}{\text{convective flux}}\right]$$

$$R \equiv 1 - \frac{F_{b/\delta}}{v_f c_{b/\delta}} \tag{14}$$

where the symbol [=] means "has units of". As solution enters the membrane at $y = \delta$, the concentration of solute b must be reduced by the mass flux ratio, $F_{b/\delta}/v_f c_{b/\delta}$, from $c_{b/\delta}$ to $c_{b/\delta}(F_{b/\delta}/v_f c_{b/\delta})$. If one now assumes that the solution which enters the membrane at $y = \delta$ remains essentially unchanged as it traverses the membrane and emerges as permeate with concentration $c_{b/p}$, then $c_{b/p} = c_{b/\delta}(F_{b/\delta}/v_f c_{b/\delta})$ or $F_{b/\delta} = v_f c_{b/p}$. Substituting this result into Eq. (14) eliminates the net mass flux from the definition of R:

$$R \equiv 1 - \frac{c_{b/p}}{c_{b/\delta}} \tag{15}$$

Equation (15) defines the true membrane retention coefficient, R. However, R is a function of the solute concentration, $c_{b/\delta}$, in contact with the retentate membrane surface which, unlike the bulk feed solute concentration, $c_{b/0}(c_{b/0} < c_{b/\delta})$, is not easily measured. Hence, a more readily determined retention coefficient can be defined as

$$R_0 \equiv 1 - \frac{c_{b/p}}{c_{b/0}} \tag{16}$$

Equation (16) defines the observed or apparent solute retention R_0 as a function of the actual solute concentrations on both faces of the membrane. Unlike the true retention function, R, the observed or apparent retention function, R_0, can be dependent on solution pressure, recirculation speed, and/or molecular weight [134].

Solution of Eq. (13) is now possible if one first notes that mass conservation requires solute flux to be constant everywhere in the film and membrane so that $F_{b/f} = F_{b/\delta}$. Based on this observation, Eqs (13) and (14) can be combined to eliminate $F_{b/f}$ and $F_{b/\delta}$ and give the differential equation for flow in the film and membrane as

$$v_f c_{b/\delta}(1 - R) = v_f c_{b/f} - \frac{D_b \partial c_{b/p}}{\partial y} \tag{17}$$

When Eq. (15) is substituted into Eq. (17), R can be eliminated to give upon rearrangement the differential equation:

$$D_b \frac{\partial c_{b/f}}{\partial y} = v_f c_{b/f} - v_f c_{b/\delta}(1 - R) \tag{18}$$

Equation (19) can now be integrated across the boundary film from the film–bulk solution interface at $y = 0$, with solute b concentration $c_{b/0}$ to the film–membrane interface at $y = \delta$, with solute b concentration $c_{b/\delta}$:

$$\int_0^\delta dy = \left(\frac{D_b}{v_f}\right) \int_{c_{b/0}-c_{b/p}}^{c_{b/\delta}-c_{b/p}} \frac{d(c_{b/f} - c_{b/p})}{c_{b/f} - c_{b/p}} \tag{19}$$

$$\delta = \left(\frac{D_b}{v_f}\right) \ln\left(\frac{c_{b/\delta} - c_{b/p}}{c_{b/0} - c_{b/p}}\right) \tag{20}$$

Equation (20) can be solved for the flow rate of solution through the membrane as

$$v_f = \left(\frac{D_b}{\delta}\right) \ln\left(\frac{c_{b/\delta} - c_{b/p}}{c_{b/0} - c_{b/p}}\right) = K_{b/m} \ln\left(\frac{c_{b/\delta} - c_{b/p}}{c_{b/0} - c_{b/p}}\right) \tag{21}$$

where Eq. (21) serves to define the mass transfer coefficient, $K_{m/b}$ [132].

Before proceeding further with Eq. (21) it is instructive to examine the precise mathematical definition of solution velocity and its relationship to both the volume and mass fluxes relative to stationary co-ordinates. The solution velocity is rigorously defined as [132]:

$$v \equiv \frac{\sum_i c_i v_i}{\sum_i c_i} [=] \frac{\left(\frac{g}{cm^3}\right)\left(\frac{cm}{s}\right)}{\left(\frac{g}{cm^3}\right)} [=] \frac{cm}{s} \tag{22}$$

An alternative analysis of the same units used in Eq. (22) gives

$$\frac{\left(\frac{g}{cm^3}\right)\left(\frac{cm}{s}\right)}{\left(\frac{g}{cm^3}\right)} [=] \frac{\left(\frac{g}{s\,cm^2}\right)}{\left(\frac{g}{cm^3}\right)} [=] \frac{cm^3}{s\,cm^2} [=] J \tag{23}$$

From the unit analysis in Eqs (22) and (23) it would appear that the mass averaged velocity v (units of cm s^{-1}) and the volume flux J (units of $cm^3\,s^{-1}\,cm^{-2}$) of a flowing solution can be used interchangeably. However, for porous media such as a membrane, this is not the case. For steady-state flow in a retentate–membrane–permeate system of constant cross-sectional area, the mass average velocity of solution on the retentate, v_r, and permeate, v_p, side of the membrane will be the same (assuming solution compressibility is negligible) but not in the membrane, v_m. In a membrane the cross-sectional area available for flow for most porous geometries is proportionally limited to the porous fraction of the membrane, Θ_m. Hence, the solution flow velocity, v_m, in the membrane must increase since the volume flux remains constant but only a fraction, Θ_m, of the membrane cross-sectional area is available to flow, so

$$v_r = v_p = v_m \Theta_m \tag{24}$$

For steady-state flow in a retentate–membrane–permeate system of constant cross-sectional area, the volume flow through any perpendicular cross-section of the system must be constant (assuming solution compressibility is negligible). Since flux is the volume flow per unit of cross-sectional area the volume flux of solution must be constant in the retentate, J_r, the membrane, J_m, and in the permeate, J_p, so

$$J_r = J_p = J_m \tag{25}$$

Equations (24) and (25) can be combined to give the relationship between mass averaged solution velocity and volume flux in the retentate, permeate, and membrane:

$$v_r = v_p = v_f = v_m \Theta_m = J_r = J_p = J_m = J \tag{26}$$

For the mass flux, Eq. (22) can be rearranged as

$$v \sum_i c_i \equiv \sum_i c_i v_i [=] \left(\frac{g}{cm^3}\right)\left(\frac{cm}{s}\right) [=] \frac{g}{cm^2\,s} [=] F \tag{27}$$

Equation (27) defines the solution mass flux, F, which, due to mass conservation, must be constant throughout the retentate, permeate, and membrane. Since mass fluxes are vector quantities, the summation term, $\sum_i c_i v_i$, in Eq. (27) is seen to be the vector sum of mass fluxes of all the molecules in the solution. The mass flux, F_b, for a single solute molecule, b, is $c_b v_b$. Since conservation of mass fluxes also applies to each type of molecule, the mass flux for the solute b must be constant throughout the retentate, permeate, and membrane, respectively:

$$F_b = F_{b/r} = F_{b/p} = F_{b/m} = c_b v_b \tag{28}$$

While Eq. (28) is valid for F_b, it is difficult to evaluate experimentally due to the velocity, v_b, of solute b. A more useful equation for F_b, in terms of the mass averaged solution velocity, v, can be derived from Eqs (13) and (17):

$$F_b = v c_{b/p} \tag{29}$$

Combining Eqs (26), (28), and (29):

$$F_b = F_{b/r} = F_{b/p} = F_{b/m} = c_{b/p} v = c_{b/p} J \tag{30}$$

Further evidence that the steady-state mass flux of a solute is constant throughout the retentate, permeate, and membrane is provided by Eq. (30). The far right-hand expression in Eq. (30) is the product of two quantities, which are both constant for a retentate–permeate–membrane system.

Returning to Eq. (21), direct evaluation of $K_{b/m}$ is difficult due to $c_{b/\delta}$, the solute concentration at the membrane retentate surface. In principle, $c_{b/\delta}$ could be experimentally determined by using feed solutions of high solute concentration so that $c_{b/\delta} \cong c_{b/0}$. However, solutions with high solute concentrations have physical properties which make data from membrane permeation experiments difficult to collect and evaluate, i.e., high viscosities, solute–solute (activity coefficient) interactions, aggregate formation, etc. Instead, a graphical method is preferable after modification of Eq. (21).

If Eqs (15), (16), and (26) are substituted into Eq. (21) to eliminate concentrations, the resulting equation upon rearrangement [135] is

$$\ln \frac{1 - R_0}{R_0} = \ln \frac{1 - R}{R} + \frac{v_r}{K_{b/m}} = \ln \frac{1 - R}{R} + \frac{J}{K_{b/m}} \tag{31}$$

When the concentration, $c_{b/0}$, of solute b approaches that for which gellation in the film occurs, the concentration gradient in the film decreases, making solution flow through the membrane mostly convective. Under these near-gelation conditions, $c_{b/p}/c_{b/\delta}$, and hence, R becomes nearly independent of J, so that a plot of $\ln[(1 - R_0)/R_0]$ versus J approximates a straight-line plot with slope of $1/K_{b/m}$, and an ordinate intercept from which R_{max} can be determined since, $R \to R_{max}$ as $J \to 0$. However, experimental problems associated with near-gelation conditions may occur which are associated with high solute concentrations as discussed above. An alternative method to determining $K_{b/m}$ and R_{max} is possible using dimensionless numbers.

The mass transfer coefficient can be written in analogy with the heat transfer coefficient as a combination of dimensionless numbers [136]:

$$\mathrm{Sh} = A(\mathrm{Re})^{\alpha}(\mathrm{Sc})^{\beta} \tag{32}$$

The dimensionless numbers in Eq. (32) are defined as: $\mathrm{Sh} = K_{b/m}d_h/D_b$, $\mathrm{Re} = v\rho d_h/\eta$, and $\mathrm{Sc} = \eta/(\rho D_b)$. Substitution of these explicit expressions for the dimensionless numbers into Eq. (32) and solving for $K_{b/m}$:

$$K_{b/m} = \left[A(d_n)^{\alpha-1}\left(\frac{\rho}{\eta}\right)^{\alpha-\beta}(D_b)^{1-\beta}\right]v^{\alpha} \equiv \Phi v^{\alpha} = \Phi J^{\alpha} \tag{33}$$

From Eq. (33) one can see that, for flow of solutions of different concentrations (as long as the solution viscosity is essentially viscosity independent) of the same single solute through the same membrane at differing flow rates, $K_{b/m}$ becomes the product of a constant, Φ, represented by the variables in square brackets, and the solution velocity in the film, v. Equation (33) can now be substituted into Eq. (31):

$$\ln\frac{1 - R_0}{R_0} = \ln\frac{1 - R}{R} + \frac{v^{1-\alpha}}{\Phi} = \ln\frac{1 - R}{R} + \frac{J^{1-\alpha}}{\Phi} = \ln\frac{1 - R}{R} + \frac{J}{\Phi v^{\alpha}} \tag{34}$$

From Eq. (34) a plot of $\ln[(1 - R_0)/R_0]$ versus $J^{1-\alpha}$ (using α as a fitting parameter) would give a straight-line plot with slope of $1/\Phi$ and an ordinate intercept from which R_{max} can be determined. Since Φ can be seen to be constant by Eq. (33) for any solute–membrane combination of constant cross-sectional area, plots of Eq. (34) should provide better linearity than plots of Eq. (31). Once Φ is determined, values of R can be calculated. A plot for lipase cross-flow filtered at a constant applied pressure through an A002 membrane is shown in Fig. 18. Note that the linearity is almost perfect until the laminar-to-turbulent flow transition occurs in the film.

Several values have been proposed for the variables A, α, and β. The exponents in Eqs (33) and (34) have been shown to depend on the flow regime [137, 138]. For the exponent α the best straight-line plots are obtained for $1/3 \leq \alpha \leq 1/2$ in the laminar flow regime and $\alpha = 0.8$ for turbulent flow.

B. Operative Pore Size Distribution

In order to evaluate the pore size distributions of a membrane which partially retains a solute, two model assumptions are made: (1) retention is due exclusively to molecular sieving or exclusion, and (2) for each molecular weight there is a fraction of pores which totally retain solute, with the remaining pores allowing free solute passage [139]. Consistent with Eq. (30) a mass (flux) balance for each molecular

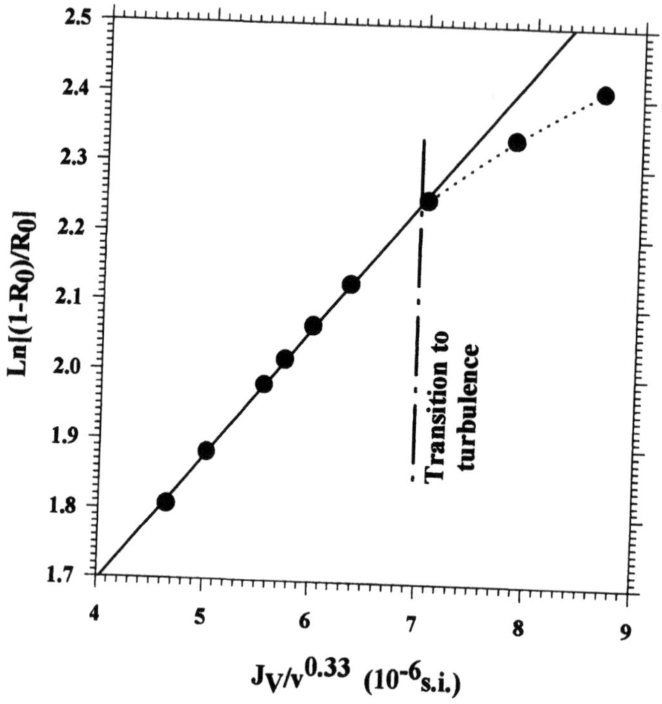

FIG. 18 $\ln[(1-R_0)/R_0]$ versus J_V/V^{α} for an aqueous solution of 0.1wt% of lipase through an A002 membrane at a constant applied pressure.

weight can be written in terms of the net mass flux through the membrane pores which transmit solute, $F_{b/p}$, as

$$F_{b/p} = Jc_{b/p} = F_{b/t} = J_t c_{b/\delta} \tag{35}$$

In Eq. (35), J_t is the volume flow through the pores transmitting solute b with concentration $c_{b/\delta}$. Substituting Eq. (15) into (35) gives, upon rearrangement:

$$J_t = J\frac{c_{b/p}}{c_{b/\delta}} = J(1-R) : \frac{J_t}{J} = (1-R) \tag{36}$$

For solutions of low to moderate concentration of solutes, the volume flux of solution in the transmitting pores, J_t, would be nearly the same as for solvent flux through those same transmitting pores, $J_{w/t}$, so that [140]:

$$J_{w/t} \cong J_t = J\frac{c_{b/p}}{c_{b/\delta}} = J(1-R) \tag{37}$$

The total volume flux of solvent, J_w, through all the membrane pores can be determined from a mass (flux) balance. The total mass flux through all pores in the membrane, F [as defined in Eq. (28)], through which a solution of solvent and solute b are flowing can be written as the (vector) sum of the solvent mass flux, F_w, and the solute mass flux, F_b:

$$F = F_w + F_b \tag{38}$$

Deriving an expression for J_w begins with the derivation of equations for each mass flux in Eq. (38) in terms of respective volume fluxes. Using cgs units to guide the derivation, the mass fluxes can be written:

$$F[=]\frac{\text{g solution}}{\text{s cm}^2}[=]\frac{\text{mL solution}}{\text{s cm}^2}\times\frac{\text{g solution}}{\text{mL solution}}[=]J\rho_{\text{solution}} \tag{39}$$

$$F_w[=]\frac{\text{g solvent}}{\text{s cm}^2}[=]\frac{\text{mL solvent}}{\text{s cm}^2}\times\frac{\text{g solvent}}{\text{mL solvent}}[=]J_w\rho_{\text{solvent}} \tag{40}$$

$$F_b[=]\frac{\text{g solute}}{\text{s cm}^2}[=]\frac{\text{mL solution}}{\text{s cm}^2}\times\frac{\text{g solute}}{\text{mL solution}}[=]Jc_{b/p} \tag{41}$$

Substituting Eqs (39)–(41) into Eq. (38):

$$J_w = \frac{J}{\rho_w}(\rho_p - c_{b/p}) \cong J(1 - c_{b/p}) \tag{42}$$

A ratio of volume fluxes for solvent in the transmitting and total pores can be determined as

$$\frac{J_{w/t}}{J_w}[=]\frac{\text{water flux through large pores}}{\text{water flux through all pores}}[=]\begin{array}{l}\text{fractional of pores}\\ \text{large enough to}\\ \text{transmit solute}\\ \text{molecules}\end{array} \tag{43}$$

Substituting Eqs (37) and (39) into (40):

$$\frac{J_{w/t}}{J_w} = \frac{J\rho_w(1-R)}{J(\rho_p - c_{b/p})} = \left(\frac{c_{b/p}}{c_{b/\delta}}\right)\left(\frac{\rho_w}{\rho_p - c_{b/p}}\right) \cong \left(\frac{c_{b/p}}{c_{b/\delta}}\right) \tag{44}$$

Equation (44) can be used to determine the fraction of transmitting pores. A plot of $J_{w/t}/J_w$ versus solute molecular weight gives the accumulated fraction of flow passing through the large transmitting pores. The derivative of such a graph provides the flux carrying molecules of any given molecular weight.

Alternatively, if the abscissa is changed from molecular weight to pore diameter, d_p, by considering the radius of gyration of a solute molecule, [141–144], the differential pore size distribution, $d(J_{w/t}/J_w)/d(d_p)$, can be obtained. Additionally, if the pores are modified by chemisorbing molecules on to the pore walls, then pore distributions of adsorbate-modified membranes can be determined. When studying modified or unmodified membranes, care must be taken to properly account for phenomena such as adsorbed layers on pore walls and rigidity of the permeating solute molecule.

Finally, the differential flow fraction can be correlated with the differential pore fraction through the Hagen–Poiseuille equation. If the pores are modeled as cylinders extending through the membrane, then the solvent flux through the larger transmitting pores, $J_{w/t}$, can be expressed as a product of the volume flow rate, Q_t (mL s^{-1}), in a single transmitting pore and the cross-sectional density of transmitting pores, N_t (pores cm^{-2}), for pore diameters ranging from d_p to $d_p + d(d_p)$. In a like manner the flux of solvent in all pores, J_w, can be written as the product of the volume flow rate of the average pore, Q (mL s^{-1}), and the cross-sectional density, N (pores cm^{-2}), of all pores. The ratio of the two volume fluxes can be expressed as

$$\frac{J_{\mathrm{w/t}}}{J_{\mathrm{w}}} = \frac{Q_{\mathrm{t}} N_{\mathrm{t}}}{QN} \tag{45}$$

Taking the differential with respect to d_p gives [144]:

$$\frac{\mathrm{d}(J_{\mathrm{w/t}}/J_{\mathrm{w}})}{\mathrm{d}(d_{\mathrm{p}})} = \left(\frac{Q_{\mathrm{t}}}{Q}\right)\frac{\mathrm{d}(N_{\mathrm{t}}/N)}{\mathrm{d}(d_{\mathrm{p}})} + \left(\frac{N_{\mathrm{t}}}{N}\right)\frac{\mathrm{d}(Q_{\mathrm{t}}/Q)}{\mathrm{d}(d_{\mathrm{p}})} \tag{46}$$

Since Q is constant with respect to d_p, it can be represented by a normalization constant, K. From the Hagen–Poiseuille equation, $Q_t \propto d_p^4$, so Eq. (46) can be written:

$$\frac{\mathrm{d}(J_{\mathrm{w/t}}/J_{\mathrm{w}})}{\mathrm{d}(d_{\mathrm{p}})} = \left(\frac{d_{\mathrm{p}}^4}{K}\right)\frac{\mathrm{d}(N_{\mathrm{t}}/N)}{\mathrm{d}(d_{\mathrm{p}})} + \left(\frac{4N_{\mathrm{t}} d_{\mathrm{p}}^3}{NK}\right) = \left(\frac{d_{\mathrm{p}}^4}{K}\right)\left[\frac{\mathrm{d}(N_{\mathrm{t}}/N)}{\mathrm{d}(d_{\mathrm{p}})} + \left(\frac{4N_{\mathrm{t}}}{N d_{\mathrm{p}}}\right)\right] \tag{47}$$

or

$$\frac{\mathrm{d}(J_{\mathrm{w/t}}/J_{\mathrm{w}})}{\mathrm{d}(d_{\mathrm{p}})} \cong \left(\frac{d_{\mathrm{p}}^4}{K}\right)\frac{\mathrm{d}(N_{\mathrm{t}}/N)}{\mathrm{d}(d_{\mathrm{p}})} \tag{48}$$

since

$$\frac{\mathrm{d}(N_{\mathrm{t}}/N)}{\mathrm{d}(d_{\mathrm{p}})} \gg \left(\frac{4N_{\mathrm{t}}}{N d_{\mathrm{p}}}\right) \quad \text{and} \quad \frac{J_{\mathrm{w/t}}}{J_{\mathrm{w}}} \equiv f_{\mathrm{d}} \tag{49}$$

In Fig. 19 the results obtained by the preceding analysis method are shown for the A002 membrane where both the volume flux and relative pore number distributions are plotted. Other anodic alumina membranes were analyzed by Baker and Strathmann [146], by retention of water solutions of polyethylene glycols and proteins as well as with benzene solutions of polystyrene. For these membranes an average pore size of 10 nm was determined.

C. Electrical Properties

As mentioned above, electrical properties of membranes modify retention over and above pure sieving effects. Thus, an analysis of properties like zeta potential, surface charges, etc., is necessary in order to predict the separation properties of the membranes. The most straightforward method for studying the electrical characteristics of membranes is direct measurement of the electrokinetic phenomena involved.

When a pressure gradient is imposed across a membrane, the solution close to the solid surface at the pore wall remains immobile while the rest of the solution filling the pores moves past. This movement creates an electrical potential drop across the membrane, ΔV. The flow-induced potential drop leads to a dynamical contribution to the total electric potential profile, $\phi(r, z)$. Under stationary conditions the electric potential within a pore, $\psi(r)$, varies only radially. Both contributions to the total potential inside a pore can be written as

$$\phi(r, z) = \frac{RT}{z_+ \mathcal{F}} \psi(r) + V(z) \tag{50}$$

where r and z are the radial and axial co-ordinates in a cylindrical pore, and $\mathcal{F}$ is the Faraday constant. The ratio of the z-dependent potential drop, ΔV, to the applied

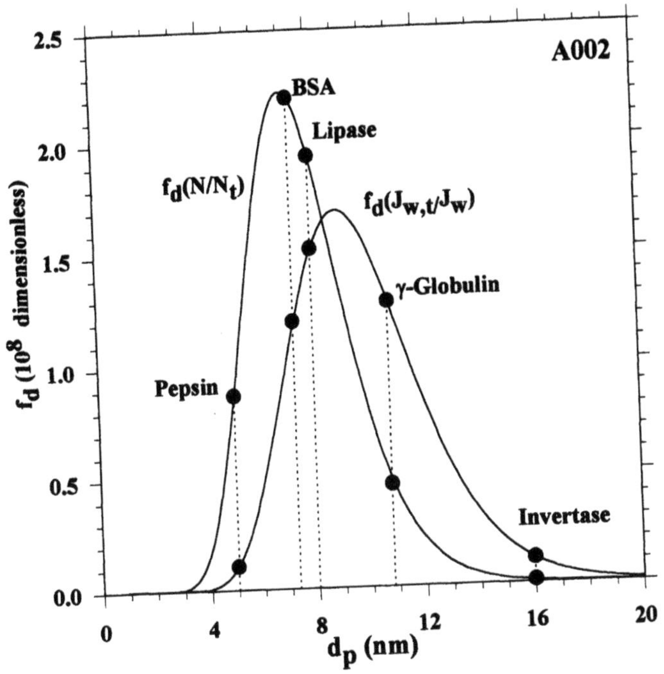

FIG. 19 Pore number and flux relative distribution functions for the A002 membrane obtained by the retention method with several proteins.

pressure difference, Δp, under zero current conditions, is called the streaming potential, ν_p, which can be evaluated from the phenomenological equations [147–149]:

$$\left.\begin{aligned} J_v &= L_{11}\Delta p + L_{12}\Delta V \\ I &= L_{21}\Delta p + L_{22}\Delta V \end{aligned}\right\} \tag{51}$$

where J_v is the volume flow and I is the electric current, both per unit of total membrane area. The L_{ij} are the phenomenological coefficients that fulfill the Onsager reciprocity law. The streaming potential can be written as

$$\nu_p = \left(\frac{\Delta V}{\Delta p}\right)_{I=0} = -\frac{L_{21}}{L_{22}} \tag{52}$$

and evaluated by simultaneous solution of the Nernst–Planck, Navier–Stokes, and Poisson–Boltzmann equations corresponding to the solution actually moving through the pore, which is treated as consisting of point charges according to the Gouy–Chapman surface model [147–149].

Using Eq. (52), ν_p can be calculated from [148–152]:

$$L_{21} = -\frac{\Theta \varepsilon RT}{\Delta z \eta_s z_+ \mathcal{F}}\left(\zeta - \frac{2}{r_p^2}\int_0^{r_p} \psi(r) r \,\mathrm{d}r\right) \tag{53}$$

and

$$L_{22} = \left[\frac{2\mathscr{F}^2 c\Theta}{r_p^2 RT\Delta z} \int_0^{r_p} \left\{ \nu_+ z_+^2 D_+ e^{-\psi(r)} + \nu_- z_-^2 D_- e^{-z_- \psi(r)/z_+} \right\} r \, dr \right.$$
$$\left. - \frac{2cRT\Theta}{r_p^2 z_+ \Delta z} \int_0^{r_p} \left\{ \nu_+ z_+ e^{-\psi(r)} + \nu_- z_- e^{-z_- \psi(r)/z_+} \right\} \{\zeta - \psi(r)\} r \, dr \right] \quad (54)$$

where ν_+ and ν_- are the stoichiometric coefficients, θ is the membrane porosity for pores of radius r_p (the fraction of the surface occupied by these pores), c is the electrolyte concentration, Δz is the pore length, and $\zeta = \psi(r_p)$ is the zeta potential. One should note that neither Θ nor Δz are needed to calculate ν_p with Eqs (52)–(54) since the factor $\Theta/\Delta z$ will cancel.

The streaming potential is not given by an analytical function because the solution of the Poisson–Boltzmann equation is needed to determine the $\psi(r)$ profile and it can only be obtained in general by numerical methods [147–153].

When an aqueous electrolyte solution passes through a microporous membrane with walls having an electrostatic charge per square meter, σ_0, some of the ions bind strongly to the pore surfaces. In this way, the pore surface charge is modified from its initial value due to binding of ions from solution. We will consider [154, 155] that the counter ions are tightly held on the pore surface as a result of: (1) the difference in the static electric potential between the pore walls and the bulk solution, along with (2) specific bonding forces that bind the ions to the solid surface sites. The centers of these bound ions define what is called the Stern surface. According to this model [154–156] the solid–liquid interface is considered as consisting of:

1. The pore surface free of adsorbed ions with a total charge in coulombs and an electrostatic potential difference from the bulk solution given by Σ_0 and ψ_0, respectively.
2. The Stern surface with an electrostatic charge σ_S and an electrostatic potential ψ_S.
3. A diffuse layer including the adjacent solution whose distribution of ions is determined by the requirement of an overall electrical neutrality with a total charge Σ_d and an electrostatic potential ψ_d at the diffuse layer boundary.

In the Stern model, the Stern surface is located at a distance, δ, from the actual solid surface of the pore. The distance δ is conventionally taken as a counter ion radius, $r_\pm$ or $\delta = r_\pm$.

Separating the static and mobile portions of the solution moving through a pore is the shear plane, which is also considered to be the boundary for the diffuse layer. Conventionally, all three planes have been considered approximately coincident at the distance, $\delta = r_\pm$, from the actual solid surface of the pore: (1) the Stern surface, (2) the boundary of the diffuse layer, and (3) the shear plane. With this approximation the electrostatic potential of this plane (the Stern surface, the boundary of the diffuse layer, and the shear plane) located at r_p is

$$\psi(r_p) = \psi_S = \psi_d = \zeta \tag{55}$$

Therefore, the Stern surface is a cylinder of radius r_p while the pore wall itself is a concentric microcapillary of radius $r_p + \delta$.

Electroneutrality requires that a charge balance in a pore must sum to zero. Alternatively, a charge balance over all charged pore surfaces must be zero, so

$$\Sigma_0 + \Sigma_S + \Sigma_d = 0 \tag{56}$$

Considering a membrane as a collection of parallel cylinders perpendicular to the macroscopic membrane surface, the Stern surface, and the membrane pores of the same length Δz, so that the electrostatic charge per unit of pore surface area is

$$\left(\frac{r_p + \delta}{r_p}\right)\sigma_0 + \sigma_S = \sigma_d \tag{57}$$

where σ_0 and σ_S are the surface charge densities, in units of $\mathrm{C\,cm^{-2}}$ on the solid pore wall surface and the Stern layer, respectively, while σ_d is the electrostatic charge (of opposite sign) in the diffuse layer necessary to balance σ_0 and σ_S.

As shown above, the streaming potential or any other combination of the phenomenological coefficients can be calculated [154, 155] as a function of the solution viscosity, η_s, the dielectric constant, ε, both ionic diffusivities, D_+ and D_-, the ratio between the mean hydrodynamic radius and the Debye length, r_p/λ, and the surface charge density, σ_d, which determines the electrical charge profile, $\psi(r)$, through the Poisson–Boltzmann equation.

If the bulk solution values are taken for η_s, ε, D_+, and D_-, we can obtain σ_d for each bulk electrolyte concentration by fitting the experimental results with the chosen electrokinetic parameter. At the same time the corresponding profile of the radial part of the electric potential, $\psi(r)$, and the potential, ζ, can be calculated.

If an homogeneous mechanism of adsorption of anions is assumed, the resulting immobile or adsorbed charge per unit surface should follow [156, 157], the so-called Langmuir isotherm, i.e.

$$\sigma_S = \frac{z_- e N_s \chi_- e^{-\frac{\Delta G^*_{ads}}{RT}}}{1 + \chi_- e^{-\frac{\Delta G^*_{ads}}{RT}}} \tag{58}$$

where e is the elemental charge, N_s is the total number of adsorption sites accessible to the anions, χ_- is the molar fraction of anions in the bulk solution, and ΔG^*_{ads} is the standard state molar Gibbs free energy of adsorption.

On the other hand, a Freundlich dependency, i.e., a power dependence like:

$$\sigma_S(\chi_-) = a\chi_-^b \tag{59}$$

should correspond to an exponentially decreasing molar free energy of adsorption distribution [158]. In this case, the maximum number of occupiable sites should be

$$N_s = \frac{a}{z_- e}\left(\frac{\sin(\pi b)}{\pi b}\right) \tag{60}$$

and the average adsorption free energy:

$$\langle \Delta G_{ads} \rangle = -\frac{RT}{b} \tag{61}$$

It is clear that the electrical properties of membranes depend strongly on their surfce material. For instance, the electrical behavior of an A02 membrane can change substantially if it is modified with a nickel layer. SEM images of an unmodified A02 membrane, along with two recovered A02 membranes with thin and thick coating layers, are shown in Fig. 20. It can be seen that low coating results in a widening of the pores probably due to the change in alumina structure during sintering, while thicker coatings substantially decrease pore sizes by overcoming the widening resulting from high-temperature treatment. In effect, this can be seen in Fig. 21 where the resulting air–liquid displacement pore size distributions are presented. Even for overlapping pore size distributions (curves a and b), the resulting surface charge density changes due to the different wall materials, as shown in Fig. 22. The actual surface charge density along with the corresponding adsorption parameters are shown in Table 2.

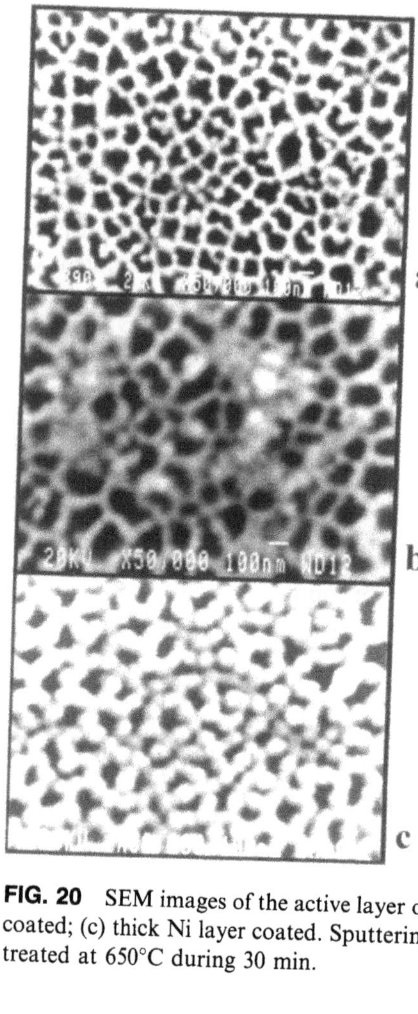

a
b
c

FIG. 20 SEM images of the active layer of A02 membranes: (a) unmodified; (b) thin Ni layer coated; (c) thick Ni layer coated. Sputtering times were 280 and 840 s, respectively. Both were treated at 650°C during 30 min.

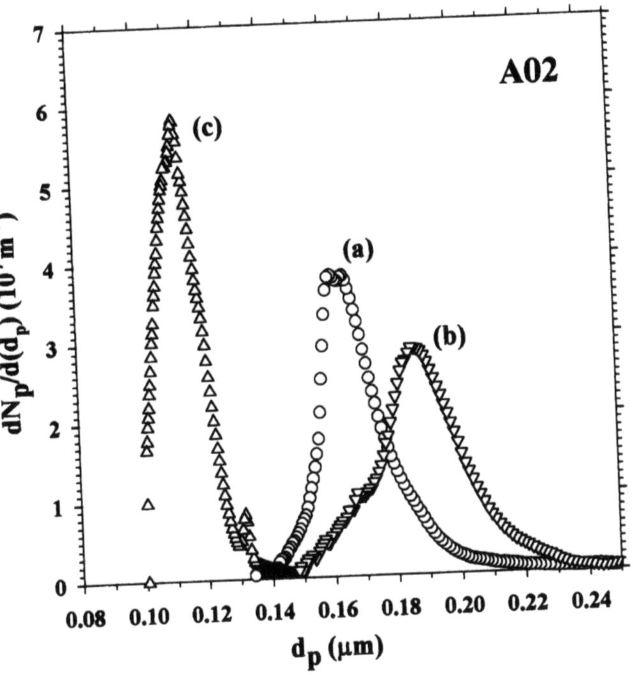

FIG. 21 Air–liquid displacement pore size distributions for unmodified and modified A02 membranes corresponding to the images shown in Fig. 20.

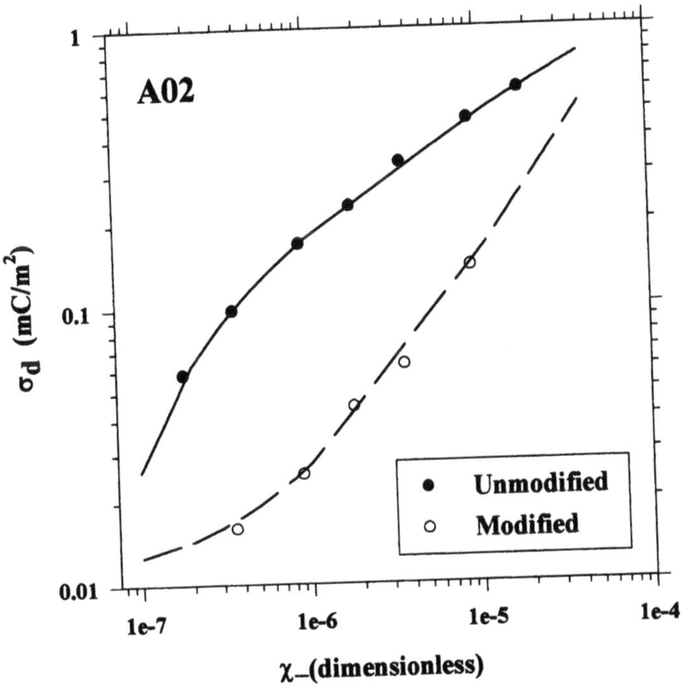

FIG. 22 Surface charge density versus NaCl concentration of membranes a and b whose images and pore size distributions are shown in Figs 20 and 21.

TABLE 2 Surface Charge Density and Adsorption Parameters for A02 and Ni–A02 Membranes as Obtained from Streaming Potential Measurements

	σ_0 (μC m^{-2})	$\langle \Delta G^*_{ads} \rangle$ (kJ mol^{-1})	N (sites m^{-2})
A02	$-(130 \pm 30)$	$-(8.5 \pm 0.5)$	$-(2.0 \pm 1.3) \cdot 10^{15}$
Ni–A02	(11 ± 6)	$-(2.8 \pm 1.5)$	$-(4.5 \pm 2.5) \cdot 10^{17}$

V. FOULING

As mentioned, the permeate flux can be considered to be controlled by several hydraulic resistance mechanisms. Initially, they were attributed only to the membrane itself [159]. It is now known that other time-dependent terms related to the solute–membrane interactions have to be considered. These can be attributed to several phenomena such as concentration polarization, gelation, deposition, adsorption, pore blocking, etc. Among these processes which increase membrane resistance, the most relevant ones (the irreversible ones) can be represented with only four kinetic models [160, 161]: (1) complete blocking, (2) intermediate blocking, (3) standard blocking, and (4) cake filtration models.

If it is assumed that each particle arriving at the membrane blocks some pore or pores completely with no superposition of particles, we have the so-called "complete blocking" phenomenon. It leads to a time law for the permeate flux which is given by

$$\ln J_V = -k_b t + \ln J_{V,0} \tag{62}$$

where

$$k_b = \sigma J_{V,0} \tag{63}$$

σ being the fraction of membrane surface blocked per unit volume of membrane.

The other three models can be expressed by the following equation:

$$\frac{J_{V,0}}{J_V} = (1 + kt)^n \tag{64}$$

When a particle settles among other particles previously deposited on the membrane surface and partially blocks some pores or directly blocks some membrane area, the resulting model corresponds to the so-called "intermediate blocking" phenomenon. In this case $n = 1$ and

$$k = k_i J_{V,0} = \sigma J_{V,0} \tag{65}$$

Alternatively, it is possible that each particle arriving at the membrane is deposited on to the internal pore walls, leading to a decrease in the pore volume. If the membrane is supposed to consist in a collection of identical cylindrical pores, we have the so-called "standard blocking" phenomenon. In this case, $n = 2$ and

$$k = \tfrac{1}{2} A_0 J_{V,0} k_s \tag{66}$$

where A_0 is the cross-section of membrane area and

$$k_s = \frac{2C}{LA_0} \tag{67}$$

where C is the volume of particles deposited per unit of filtrate volume, and L is the pore length.

Finally, a particle can deposit on other particles already deposited and already blocking some pores so that there is no room for direct contact with the membrane surface. In such a way, we have the so-called "cake filtration" phenomenon. In this case $n = 1/2$ and

$$k = 2A_0^2 J_{V,0}^2 k_c \tag{68}$$

It is clear from Eq. (62) that a plot of $\ln(J_V)$ versus time should give a straight line, if complete blocking occurs, from whose slope the constant k_b can be obtained. To obtain similar plots for the other blocking models, Eq. (64) can be modified to

$$\ln J_V = -k_i V + \ln J_{V,0} \quad \text{(for the intermediate blocking)} \tag{69}$$

$$J_V^{1/2} = J_{V,0}^{1/2} - \frac{k_s}{2} J_{V,0}^{1/2} A_0 V \quad \text{(for the standard blocking)} \tag{70}$$

$$\frac{t}{V} = \frac{k_c A_0^2}{2} V + \frac{1}{J_{V,0}} \quad \text{(for the cake model)} \tag{71}$$

When these plots are used, the results allow one to state that [162]:

1. Flux time decline is always divided into two successive steps separated by a short transition zone.
2. When retention is appreciable a first step of pore blocking appears, accounting for the major part of the permeate flux decline.
3. The nature of the slower second step in fouling depends on the ratio of the protein size to the mean pore size. When this relation is far over unity, the retained proteins form a cake on the top of the active side of the membrane. While otherwise, the inner surfaces of the unblocked pores are recovered by adsorbed proteins, leading to a standard blocking model.
4. The corresponding kinetic constants – k_i, k_c, and k_s – follow defined tendencies versus concentration and recirculation velocities. In particular, k_i and k_c increase for increasing feed concentrations, while k_i and k_s decrease for increasing recirculation velocities.
5. When the solute is not totally retained the kinetic constant for the slower fouling step, or standard model, k_s increases with the molecular weight or size of the foulant molecules.

The referred dependencies of the kinetic constants are shown for membrane A002 and several proteins in Fig. 23. Only the standard fouling constant is shown versus the feed recirculation speed for three proteins and for a constant recirculation speed versus molecular gyration radii.

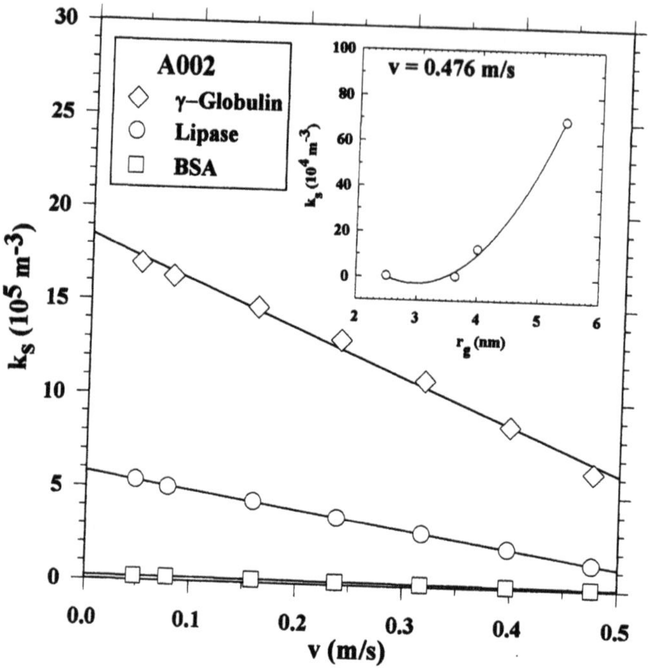

FIG. 23 Kinetic fouling constant for the standard step of the A002 membrane versus recirculation speed and protein gyration radii.

The volume fluxes for the models can also be expressed in terms of resistances:

$$J_v = \frac{\Delta P}{\eta R_t} = \frac{\Delta P}{\eta (R_m + R_c)} \quad \text{(for cake)}$$
$$J_v = \frac{\Delta P}{\eta R_t} = \frac{\Delta P}{\eta (R_m)} \quad \text{(for other models without cake, } R_c = 0) \tag{72}$$

where R_m includes all membrane resistance sources except for flow resistance due to cake formation, and R_t is the total resistance to flow from all sources. Since the volume fluxes are inversely related to the resistances, an equation can be written which expresses this inverse relationship and incorporates Eq. (64) to give

$$\frac{J_{v,0}}{J_v} = \frac{R_m + R_c}{R_{m,0}} = (1 + kt)^n \quad \text{(for cake)}$$
$$\frac{J_{v,0}}{J_v} = \frac{R_m + R_c}{R_{m,0}} = \frac{R_m}{R_{m,0}} = (1 + kt)^n \quad \text{(for other models without cake, } R_c = 0) \tag{73}$$

where $J_{v,0}$ and $R_{m,0}$, are the zero time volume flux and resistance, respectively. Eliminating volume fluxes:

$$R_t \equiv R_m + R_C = R_{m,0}(1 + kt)^n \quad \text{(for cake)}$$
$$R_m = R_{m,0}(1 + kt)^n \quad \text{(for other models without cake, } R_c = 0) \tag{74}$$

Thus, when R_t is plotted verus t, the slope is indicative of the fouling mechanism: (1) the standard model signifying internal pore blocking is indicated by an increasing slope, while (2) the internal and cake models, indicative of external pore blocking, are indicated by a decreasing slope.

When fouling mechanism occurs by complete external pore blocking, a different mathematical relationship is expected, based on Eq. (62):

$$\frac{J_{v,0}}{J_v} = \frac{R_m}{R_{m,0}} = e^{k_b t} \tag{75}$$

$$R_m = R_{m,0} e^{k_b t} \quad \text{(complete blocking model)} \tag{76}$$

thus leading to an increasing slope on an R_t versus t graph.

In the case of the TiO_2 membrane, the two proteins, lysozyme and bovine serum albumin (BSA), which should easily pass through the membrane, have in fact quite different flow behaviors. In Fig. 24, BSA can be seen to give some intermediate but mostly standard fouling while lysozyme gives a resistance which increases strongly with time, thus leading to total retention and zero final flux. For lysozyme, this behavior can be attributed to strong interactions inside the pores as well as formation of lysozyme aggregates. These features can also be seen if we look at the resulting images of their surfaces after fouling, as shown in Fig. 25. These images can be compared with the clean one of this membrane shown in Fig. 7. The material deposition resulting after fouling changes also the electrical properties

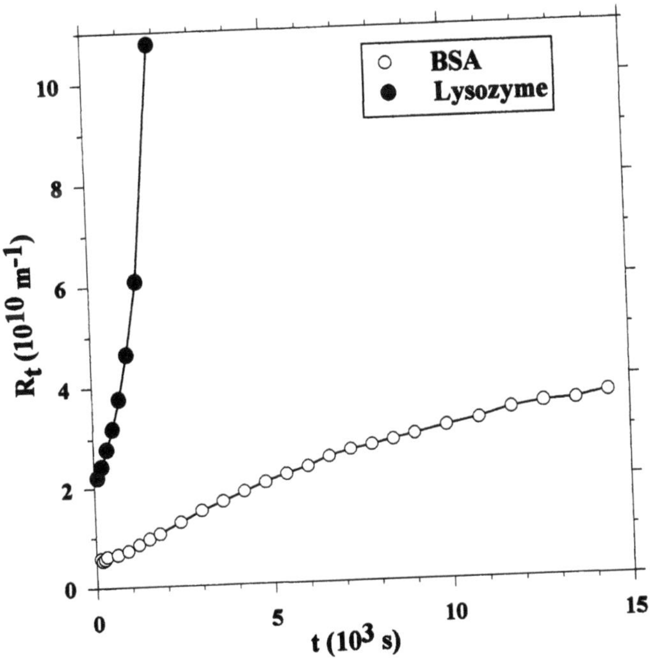

FIG. 24 Time evolution of a TiO_2 membrane resistance when fouled with BSA and lysozyme.

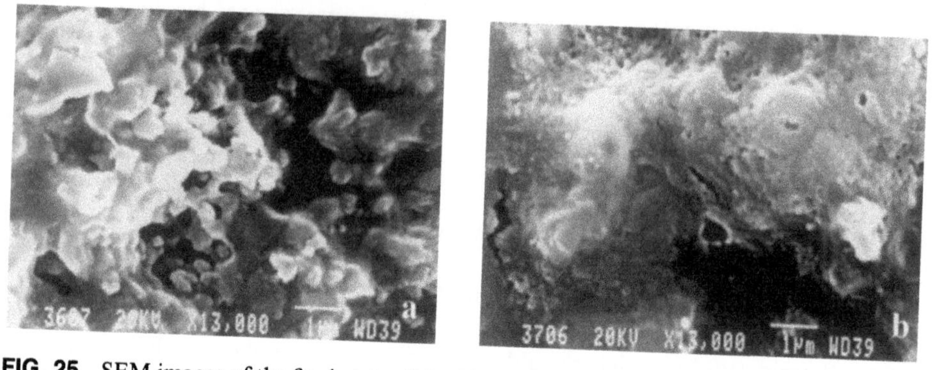

FIG. 25 SEM images of the final state of the TiO_2 membrane after fouling: (a) with BSA; (b) with lysozyme.

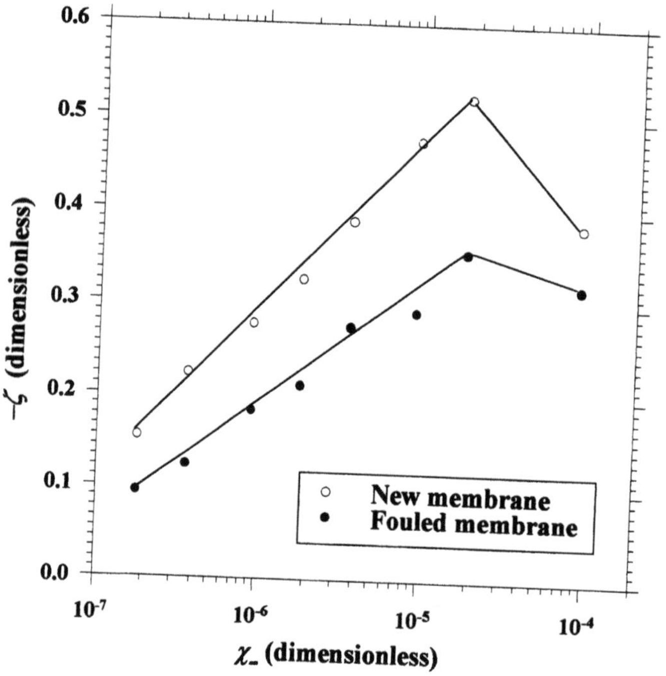

FIG. 26 Zeta potential versus the Cl^- molar fraction for a TiO_2 clean and BSA fouled membrane.

of the membrane, as shown in Fig. 26. Zeta potential is less negative due to the deposition of positively charged (at neutral pH) BSA molecules.

REFERENCES

1. M Mulder. Basic Principles of Membrane Technology. Dordrecht: Kluwer Academic, 1991.
2. ST Hwang, K Kammermeyer. Membranes in Separations. Malabar, FL: Krieger, 1984.
3. HK Lonsdale. J Membr Sci 10:81, 1982.
4. S Loeb, S Sourirajan. US Patent, 1960.
5. RR Bhave. Inorganic Membranes: Synthesis, Characteristics and Applications. New York: Van Nostrand Reinhold, 1991.
6. J Gillot, RR Bhave. In: RR Bhave, ed. Inorganic Membranes: Synthesis, Characteristics and Applications. New York: Van Nostrand Reinhold, 1991.
7. HD Smythe. Atomic Energy for Military Purposes. Princeton, NJ: Princeton University Press, 1945.
8. J Charpin, P Rigny. Inorganic membranes for separative techniques: from uranium isotope separation to non-nuclear fields. Proceedings of the First International Conference on Inorganic Membranes, ICIM89, Montpellier, France, 1990.
9. K Scott. Handbook of Industrial Membranes. Oxford: Elsevier, 1995.
10. M Mulder, J Tholen, W Maaskant. European Membrane Guide. Enschede, The Netherlands: Alinea, 1998.
11. AJ Burggraaf, L Cot. Fundamentals of Inorganic Membrane Science and Technology. Amsterdam: Elsevier, 1996.
12. AJ Burggraaf. Key points in understanding and development of ceramic membranes. Proceedings of the Third International Conference on Inorganic Membranes, ICIM94, Worcester, MA, 1994.
13. GP Pez. US Patent, 1986.
14. BR Dunbobbin, WR Brown. Gas Sepn Purif 1:23, 1987.
15. JE Koresh, F Siffer. J Chem Soc, Faraday Trans I, 82:2057, 1986.
16. KH Lee, SJ Khang. Chem Eng Commun 44:121, 1986.
17. R Fleisher, P Price, P Walker. Science 149:383, 1965.
18. P Price, P Walker. US Patent 3 303 085, 1967.
19. JP O'Sullivan, GC Wood. Proc Roy Soc, Ser A 317:511, 1970.
20. RC Furneaux, WR Rigby, AP Davidson. Nature 337:147, 1989.
21. WR Rigby, ER Cowieson, NC Davies, RC Furneaux. Trans Inst Met Finish 68:95, 1990.
22. P Prádanos, A Hernández, JI Arribas. AIChE J 40:1901, 1990.
23. A Hernández, JI Calvo, P Prádanos, L Palacio, ML Rodríguez, JA de Saja. J. Membr Sci 137: 89, 1997.
24. J Marchese. Membranas: Procesos con Membranas. San Luis, Argentina: Editorial Universitaria de San Luis, 1995.
25. AJ Burggraaf, K Keizer, RR Bhave. In: RR Bhave, ed. Inorganic Membranes: Synthesis, Characteristics and Applications. New York: Van Nostrand Reinhold, 1991.
26. AJ Burggraaf, K Keizer, BA van Hassel. Solid State Ionics 32/33:771, 1989.
27. RJR Uhlhorn, K Keizer, AJ Burggraaf. In: T Matsuura, S Sourirajan, eds. Advances in Reverse Osmosis and Ultrafiltration. Ottawa: National Research Council Canada, 1989.
28. K Pflanz, N Stroh, R Riedel. Flat ceramic ultrafiltration membranes and modules coated by the sol–gel technique. Proceeding of the Third International Conference on Inorganic Membranes, ICIM94, Worcester, MA, 1994.

29. L Cot. New developments in inorganic membranes: fundamental and application aspects. Proceeding of the Third International Conference on Inorganic Membranes, ICIM94, Worcester, MA, 1994.
30. AG Pincus, LE Shipley. Ceram Ind 92:123, 1969.
31. EP MacNamara, JE Comefora. J Am Ceram Soc 28:25, 1945.
32. AL Salamara, JS Reed. Am Ceram Soc Bull 58:618, 1979.
33. GY Onoda. J Am Ceram Soc 58:1975, 1976.
34. L Palacio, P Prádanos, JI Calvo, G Kherif, A Larbot, A Hernández. Colloids Surfaces A 138:291, 1998.
35. V Thoraval, B Castelas, JP Joulin, A Larbot, L Cot. Key Eng Mater 61/62:213, 1991.
36. C Guizard, A Julbe, A Larbot, L Cot. In: BI Lee, EJ Pope, eds. Chemical Processing of Ceramics. New York: Marcel Dekker, 1994.
37. HT Rijnte. In: BG Unsen, ed. Physical and Chemical Aspects of Adsorbents and Catalysis. London: Academic Press, 1970.
38. C Guizard, N Cygankewiecz, A Larbot, L Cot. J Non-Cryst Solids 82:86, 1986.
39. L Cot, C Guizard, A Larbot. Ind Ceram 8:143, 1988.
40. A Larbot, J-P Fabre, C Guizard, L Cot. J Am Ceram Soc 72:257, 1989.
41. MA Anderson, ML Gieselman, Q Xu. J Membr Sci 39:243, 1988.
42. BE Yoldas. J Mater Sci 10:1856, 1975.
43. AFM Leenaars, AJ Burggraaf. J Colloid Interface Sci 105:27, 1985.
44. A Larbot, JA Alary, C Guizard, L Cot, J Gillot. Int J High Technol Ceram 3:145, 1987.
45. A Larbot, A Julbe, C Guizard, L Cot. J Membr Sci 44:289, 1989.
46. C Guizard, F Legault, N Idrissi, L Cot, C Gavach. J Membr Sci 41:127, 1989.
47. RJR Uhlhorn, MHBJ Huis in't Veld, K Keizer, AJ Burggraaf. Sci Ceram 14:551, 1988.
48. WSW Ho, KK Sirkar. Membrane Handbook. New York: Van Nostrand Reinhold, 1992.
49. K Kamide, S Manabe. In: AR Cooper, ed. Ultrafiltration Membranes and Applications. New York: Plenum Press, 1980.
50. B Rasneur. Porosimetry (Characterization of Porous Membranes). Summer School on Membrane Science and Technology, Cadarache, France, 1984.
51. RE Kesting. Synthetic Polymeric Membranes, A Structural Perspective. 2nd ed. New York: John Wiley, 1985.
52. S Lowell, JE Shields. In: B Scarlett, ed. Powder Surface Area and Porosity. Powder Technology Series. New York: John Wiley, 1987.
53. SS Kulkarni, EW Funk, NN Li. In: WSW Ho, KK Sirkar, eds. Membrane Handbook. New York: Van Nostrand Reinhold, 1992.
54. FP Cuperus. Characterization of Ultrafiltration Membranes. PhD thesis, Technical University of Twente, Twente, The Netherlands, 1990.
55. CL Glaves, DM Smith. J Membr Sci 46:167, 1989.
56. K Sakai. J Membr Sci 96:91, 1994.
57. K Kaneko. J Membr Sci 96:59, 1994.
58. CP Bean, MV Doyle, G Entine. J Appl Phys 41:1454, 1970.
59. PC Hiemenz. Principles of Colloid and Surface Chemistry. New York: Marcel Dekker, 1986.
60. L Palacio, JI Calvo, P Prádanos, A Hernández, P Väisänen, M Nyström. J Membr Sci 152:189, 1999.
61. G Hansmann, H Pietsch. Naturwiss 36:250, 1949.
62. O Marti, V Elings, M Haugan, CE Bracker, J Schneir, B Drake, AC Gould, J Gurley, L Hellemans, K Shaw, AL Weisenhorn, J Zasadzinski, K Hansha. J Microsc 152:803, 1988.
63. RL Riley, JO Gardner, U Merten. Science 143:801, 1964.
64. G Binnig, CF Quate, C Gerber. Phys Rev Lett 12:930, 1986.

65. C Riedel, R Spohr. J Membr Sci 7:225, 1980.
66. KJ Kim, AG Fane, CJD Fell, T Suzuki, MR Dickson. J Membr Sci 54:89, 1990.
67. AK Fritzsche, AR Arevalo, MD Moore, VB Elings, K Kjoller, CM Wu. J Membr Sci 68:65, 1992.
68. P Prádanos, ML Rodríguez, JI Calvo, H Hernández, F Tejerina, JA de Saja. J Membr Sci 117:291, 1996.
69. JI Calvo, P Prádanos, A Hernández, WR Bowen, N Hilal, RW Lovitt, PM Williams. J Membr Sci 128: 7, 1997.
70. KL Babcock, CB Prater. Application Note No. 11, Santa Barbara, CA: Digital Instruments, 1995.
71. P Grütter, HJ Mamin, D Rugar. Scanning Probe Microscopy II. New York: Springer-Verlag, 1991.
72. P Maivald, HL Butt, SAC Gould, CB Prater, B Drake, JA Gurley, VB Elings, PK Hansma. Nanotechnology 2:103, 1991.
73. MG Heaton, CB Prater, KJ Kjoller. Application Note No. 5, Santa Barbara, CA: Digital Instruments, 1995.
74. CD Frisbie, A Rozsnyai, A Noy, MS Wrighton, CM Lieber. Science 265:2071, 1994.
75. R Baum. Chem Eng News 72:6, 1994.
76. HP Hsieh, RR Bhave, HL Fleming. J Membr Sci 39:221, 1988.
77. A Larbot, JP Fabre, C Guizard, L Cot. J Membr Sci 39:203, 1988.
78. NK Raman, CJ Brinker, L Delattre, SS Prakash. Sol–gel strategies for amorphous inorganic membranes exhibiting molecular sieving characteristics. Proceeding of the Third International Conference on Inorganic Membranes, ICIM94, Worcester, MA, 1994.
79. JI Calvo, A Hernández, G Caruana, L Martínez. J Colloid Interface Sci 175:138, 1995.
80. R Swenson, JR Attle. Am Lab 11:50, 1979.
81. A Hernández, JI Calvo, P Prádanos, L Palacio, ML Rodríguez, JA de Saja. J Membr Sci 137:89, 1997.
82. F Martínez Villa. Contribución a la Caracterización Morfológico-Estructural y Funcional de Membranas Poliméricas Microporosas. Estudio y Modelización de Flujos y Permeabilidades. PhD thesis, Universidad de Valladolid, Valladolid, 1987.
83. L Zeman. J Membr Sci 71:233, 1992.
84. A Hernández, JI Calvo, P Prádanos, F Tejerina. Colloids Surfaces A 138:391, 1998.
85. H Bechhold. Z Phys Chem 64:328, 1908.
86. P Grabar, S Nikitine. J Chim Phys 33:721, 1936.
87. H Steinhauser, H Scholz, A Hübner, C Hellinghorst. The role of supports for pervaporation composite membranes. Proceedings of ICOM '90, Chicago, 1990.
88. S Pereira-Nunes, KV Peinemann. J Membr Sci 77:25, 1992.
89. G Capannelli, F Vigo, S Munari. J Membr Sci 15:289, 1983.
90. S Munari, A Bottino, P Moretti, G Capannelli, I Becchi. J Membr Sci 41:69, 1989.
91. RA Wenman, BV Miller. In: PJ Lloyd, ed. Particle Size Analysis 1985. New York: John Wiley, 1987.
92. K Venkataraman, WT Choate, ER Torre, RD Husung, HR Batchu. J Membr Sci 39:259, 1988.
93. H Batchu, JG Harfield, RA Wenman. Fluid Particle Sep J 2:5, 1989.
94. Coulter Porometer II. Reference Manual. Coulter Electronics Ltd, Luton, UK, 1991.
95. A Hernández, JI Calvo, P Prádanos, F Tejerina. J Membr Sci 112:1, 1996.
96. RW Schofield, AG Fane, CJD Fell. J Membr Sci 53:159, 1990.
97. KK Sirkar. In: WSW Ho, KK Sirkar, eds. Membrane Handbook. New York: Van Nostrand Reinhold, 1992.

98. H Kreulen, GF Versteeg, CA Smolders, WPM van Swaaij. Determination of mass transfer rates in wetted and non-wetted microporous membranes. Proceedings of ICOM '90, Chicago, 1990.
99. A Hernández, JI Calvo P Prádanos, F Tejerina. Pore size distributions of microfiltration membranes by the bubble point extended method. Proceedings of Euromembrane 95, Bath, UK, 1995.
100. G Capanelli, I Becchi, A Bottino, P Moretti, S Munari. In: KK Unger, J Rouquesol, KSW Sing, K Kral, eds. Characterization of Porous Solids. Amsterdam: Elsevier, 1988.
101. AA Liabastre, C Orr. J Colloid Interface Sci 64:1, 1978.
102. E Honold, EL Skau. Science 120:805, 1954.
103. HM Rootare, CF Prenzlow. J Phys Chem 71:2733, 1967.
104. GP Andreoutsopoulos, R Mann. Chem Eng Sci 34:1203, 1979.
105. S Lowell, JE Shields. J Colloid Interface Sci 83:273, 1981.
106. T Allen. Particle Size Measurement. Vol. 2. London: Chapman & Hall, 1997.
107. HM Rootare. Aminco Lab News 24(3):155, 1968.
108. J Luyten, J Cooymans, P Diels, J Sleurs. Proc. Third Euro-ceramics, Faenza Edittrice, Faenza, Italy, V1:657, 1993.
109. SJ Gregg, KSW Sing. Adsorption, Surface Area and Porosity. London: Academic Press, 1982.
110. J Seifert, G Emig. Chem Eng Technol 59:475, 1987.
111. D Dollimore, GR Heal. J Appl Chem 14:109, 1964.
112. RSh Mikhail, S Brunauer, EE Bodor. J Colloid Interface Sci 26:45, 1968.
113. BC Lippen, HH de Boer. J Catal 4:319, 1965.
114. WD Harkins, G Jura. J Amer Chem Soc 66:1362, 1944.
115. Omnisorp 100/360 Series Manual. Coulter Electronics Ltd, (Beckman Coulter), USA, 1991.
116. K-NP Kumar, K Keizer, AJ Burggraaf, T Okubo, H Nagamoto. J Mater Chem 3:1151, 1993.
117. A Mey-Marom, MG Katz. J Membr Sci 27:119, 1986.
118. C Eyraud, M Bontemps, JF Quinson, F Chatelut, M Brun, B Rasneur. Bull Soc Chim France 9–10:I-237, 1984.
119. M Katz, G Baruch. Desalination 58:199, 1986.
120. DE Fain. A dynamic flow-weighted pore size distribution. Proceedings of First International Conference on Inorganic Membranes, Montpellier, France, 1990.
121. GZ Cao, J Meijerink, HW Brinkman, AJ Burggraaf. J Membr Sci 83:221, 1993.
122. M Brun, A Lallemand, JF Quinson, C Eyraud. Thermochim Acta 21:59, 1977.
123. FP Cuperus, D Bargeman, CA Smolders. J Membr Sci 66:45, 1992.
124. CA Smolders, E Vugteveen. Polym Mater Sci Eng 50:1777, 1984.
125. AP Broek, HA Teunis, D Bergeman, ED Sprengers, CA Smolders. J Membr Sci 73:143, 1992.
126. M Brun, JF Quinson, C Eyraud. Actual Chim (October):21, 1979.
127. E Mathiason. J Membr Sci 16:23, 1983.
128. V Gekas, B Hallström. J Membr Sci 30:153, 1987.
129. P Prádanos, J De Abajo, JG de la Campa, A Hernández. J Membr Sci 108:129, 1995.
130. G Jonsson, P Prádanos, A Hernández. J Membr Sci 112:171, 1996.
131. S Nakao, T Nomura, S Kimura. AIChE J 25:615, 1979.
132. RB Bird, WE Stewart, EN Lightfoot. Transport Phenomena. New York: John Wiley, 1960, pp 495–504, 519–529, 658–668.
133. PLT Brian. In: U Merten, ed. Desalination by Reverse Osmosis. Cambridge, MA: MIT Press, 1966, Ch. 5.
134. S Nakao, S Kimura. J Chem Eng Jpn 14:32, 1981.
135. G Jonsson, CE Boessen. Desalination 21:1, 1977.

136. VL Vilker, CK Colton, KA Smith. AIChE J 27:632, 1981.
137. V Gekas, G Trägårth, B Hallström. Ultrafiltration Membrane Performance Fundamentals. Lund: Swedish Foundation for Membrane Technology, 1993.
138. GB Van der Berg, IG Rácz, CA Smolders. J Membr Sci 47:25, 1989.
139. MS Le, JA Howell. Chem Eng Res Des 62:373, 1984.
140. R Nobrega, H de Balmann, P Aimar, V Sánchez. J Membr Sci 45:17, 1989.
141. M Sarbolouki. Sep Sci Technol 17:381, 1982.
142. M Bodzek, K Konieczny. J Membr Sci 61:131, 1991.
143. DR Lu, SJ Lee, K Park. J Biomater Sci, Polym Ed 3:127, 1991.
144. S Nakatsuka, AS Michaels. J Membr Sci 69:189, 1992.
145. P Prádanos, A Hernández. Biotechnol Bioeng 47:617, 1995.
146. RW Baker, H Strathmann. J Appl Polym Sci 14:1197, 1970.
147. GB Westerman-Clark, L Anderson. J Electrochem Soc 130:839, 1983.
148. E Donath, A Voigt. J Colloid Interface Sci 109:122, 1986.
149. L Martínez, MA Gigosos, A Hernández, F Tejerina. J Membr Sci 35:1, 1987.
150. CL Rice, R Whitehead. J Phys Chem 69:4017, 1965.
151. S Levine, JR Marriot, G Neale, N Epstein. J Colloid Interface Sci 52:136, 1975.
152. W Olivares, T Croxton, D McQuarrie. J Phys Chem 84:867, 1980.
153. L Martínez, A Hernández, A González, F Tejerina. J Colloid Interface Sci 152:325, 1992.
154. A Hernández, L Martínez, MV Gómez. J Colloid Interface Sci 158:429, 1993.
155. J Benavente, A Hernández, G Jonsson. J Membr Sci 80:285, 1993.
156. RJ Hunter. Zeta Potential in Colloid Science. Principles and Applications. London: Academic Press, 1981.
157. AW Adamson. Physical Chemistry of Surfaces. New York: John Wiley, 1982.
158. A Hernández, F Martínez, A Martín, P Prádanos. J Colloid Interface Sci 173:284, 1995.
159. S Nakao, T Nomura, S Kimura. AIChE J 25:615, 1979.
160. J Hermia. Trans Inst Chem Eng 60:183, 1982.
161. WL McCabe, JC Smith, P Jarriott. Unit Operations of Chemical Engineering. New York: McGraw-Hill, 1967.
162. P Prádanos, A Hernández, JI Calvo, F Tejerina. J Membr Sci 114:115, 1996.

10
Vibrational Interactions at the Vapor–Metal Oxide Interface

GEORGI N. VAYSSILOV University of Sofia, Sofia, Bulgaria

I. INTRODUCTION

Vibrational interactions between molecules and a solid surface play a substantial role in many processes occurring on the surface. A large number of investigations show that the vibrational energy exchange can determine the direction and the rate of processes such as adsorption, desorption, laser-induced surface transformations, surface diffusion, chemical transformations of adsorbates, etc. [1–5]. The development of new experimental techniques gives an opportunity for detailed study of different surface processes, and in some cases for direct measurement of the molecular dynamics at surfaces for extremely short times. This is a permanent challenge for the development of the theoretical concepts for vibrational interactions on solid surfaces.

This chapter presents the main theoretical approaches for description of the vibrational energy exchange between a vapor particle and an oxide surface in the nonelastic scattering of the particle from the surface and in the adsorption complex as well as its influence on surface diffusion, dissociation, and desorption. The complete description of the vibrational interactions at the vapor–metal oxide interface includes consideration of energy exchange between different vibrational and translational degrees of freedom of the gas particle, surface vibrations (phonons) of the absorbent, and the bond in the adsorption complex formed. In addition, the electronic degrees of freedom both of the molecule and of the solid participate in most of the surface processes, but in some cases these effects can be accounted for via changes in the concrete parameters of the model.

A. Some General Equations

The understanding of the different theoretical approaches for description of the vibrational interactions considered in this chapter, requires some general knowledge of the stochastic methods and reaction-rate theory used. Here, some of the equation applied later in the chapter are introduced and some necessary simple results are derived.

1. Generalized Langevin Equation (Application to the Harmonic Oscillator)

The original *Langevin equation* considers a Markovian stochastic process [6, 7] with simple constant friction η in the field of an external fluctuation force $F(t)$. For a harmonic oscillator this equation has the form:

$$\mu\frac{\partial^2 x}{\partial t^2} = -\mu\omega_0^2 x - \eta\frac{\mathrm{d}x}{\mathrm{d}t} + F(t) \tag{1}$$

where $x(t)$ is the deviation of the oscillator's length from its mean value x_0 in the moment t, and μ and ω_0 are its reduced mass and vibrational frequency, respectively. The main characteristic of the *fluctuation force* $F(t)$ is its independence of the variable $x(t)$.

Including the *memory friction* $\gamma(t)$ instead of η transforms the equation into its generalized form:

$$\mu\frac{\partial^2 x}{\partial t^2} = -\mu\omega_0^2 x - \int_0^t \gamma(t-\tau)\frac{\mathrm{d}x(\tau)}{\mathrm{d}\tau}\,\mathrm{d}\tau + F(t) \tag{2}$$

The Fourier transform of the above *generalized Langevin equation* is

$$\left[\mu\left(\omega_0^2 - \omega^2\right) + i\omega\gamma(\omega)\right]x_\omega = F_\omega \tag{3}$$

where x_ω and F_ω are the Fourier transforms of the oscillator deviation $x(t)$ and fluctuation force F(t), and i is the imaginary unit. Now, using the independence of x and F, an expression for the spectral density of the oscillator deviation can be found:

$$\langle xx\rangle_\omega = \frac{\langle FF\rangle_\omega}{\mu^2\left(\omega_0^2 - \omega^2\right)^2 + \omega^2\gamma^2(\omega)} \tag{4}$$

The description both of the external fluctuation force and the friction term for real systems is usually difficult or impossible.

Very helpful in this situation is the fluctuation–dissipation theorem, which connects the fluctuation $\langle FF\rangle_\omega$ and $\gamma(\omega)$ dissipation functions due to equilibrium between the incoming and outgoing energy flows to or from the system:

$$\langle FF\rangle_\omega = k_\mathrm{B}T\gamma(\omega) \tag{5}$$

If the spectral function of the dissipation from the system, $\gamma(\omega)$, is known, the deviation spectral density, $\langle xx\rangle_\omega$, can be derived directly from Eqs (4) and (5). For example, if we consider the linear frequency dependence of the friction function $\gamma(\omega) = \gamma_\mathrm{c}.\omega$, this equation tranforms into:

$$\langle xx\rangle_\omega = \frac{k_\mathrm{B}T\gamma_\mathrm{c}\omega}{\mu^2\left(\omega_0^2 - \omega^2\right)^2 + \omega^4\gamma_\mathrm{c}^2} \tag{6}$$

2. Fokker–Planck Equation

The *Fokker–Planck equation* is a differential equation describing various stochastic processes. It concerns the probability $p(x, t)$ of finding our system at the point x at time t in the potential field $U(x)$:

$$\frac{\partial p(x,t)}{\partial t} = \frac{\partial}{\partial x}\left[\frac{p(x,t)}{\eta}\frac{dU(x)}{dx}\right] + D\frac{\partial p(x,t)}{\partial x} \tag{7}$$

Here, η and D denote friction and diffusion coefficients for the motion of the system in potential $U(x)$. In a general case these coefficients could depend on the variables x or t. A stationary solution of this equation gives

$$p_s(x) = N\exp\left(-\frac{U(x)}{\theta}\right)dx \tag{8}$$

where N is the normalization factor and $\theta = \eta D$ accounts for the effects of energy dissipation and diffusion on the distribution. Using this expression, we can define the mean equilibrium potential energy of the system:

$$\langle U\rangle_e = N\int_0^\infty U(x)\exp\left(-\frac{U(x)}{\theta}\right)dx \tag{9}$$

For the one-dimensional parabolic potential $U_0(x)$, this equation results in a more clear evaluation of the parameter θ as a measure of the mean potential energy of the system:

$$\theta_0 = 2\langle U_0\rangle_e \tag{10}$$

Let us now apply the Fokker–Planck equation approach to bond cleavage in a diatomic molecule in a dissipative environment. A typical potential energy profile $U(x)$ along the reaction co-ordinate x is presented in Fig. 1. It has one deep minimum at the distance corresponding to the ordinary chemical bond between the two atoms x_0 and a shallow one at larger distance x_d, usually due to stabilization of dissociated parts by the surroundings (solvent or adsorbent). The maximum between these points determines the potential energy barrier of the reaction:

$$\Delta U = U(x_b) - U(x_0) \tag{11}$$

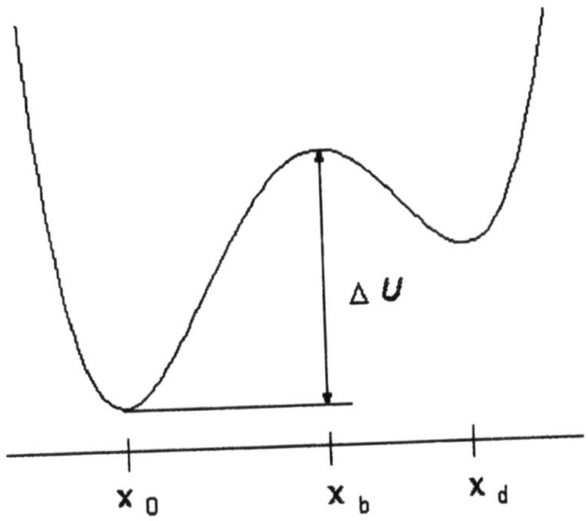

FIG. 1 Potential energy profile for the dissociation of a bond in a molecule adsorbed on a solid surface along the reaction co-ordinate x.

The expression for the rate constant obtained by this formalism is:

$$k = k_0 \cdot \exp\left(-\frac{\Delta U}{\theta}\right) \tag{12}$$

where k_0 is a constant with respect to θ [8]. Near the initial state x_0, the potential energy can be reasonably approximated with the harmonic potential with respect to x (defined as deviation from x_0):

$$U_0(x) = \tfrac{1}{2}\mu\omega_0^2 x^2 \tag{13}$$

Since the total minimum of the potential $U(x)$ around the initial state x_0 determines the highest value of the probability $p_s(x)$ in this area, the influence of the form of the potential in the other parts of the potential profile both on the mean equilibrium potential energy and the parameter θ can be neglected. Hence, we can estimate θ with the value of θ_0 corresponding to the harmonic potential:

$$\theta \approx \theta_0 = 2\langle U_0 \rangle_e \tag{14}$$

and the rate constant becomes

$$k = k_0 \cdot \exp\left(-\frac{\Delta U}{2\langle U_0 \rangle_e}\right) \tag{15}$$

If the dissociating molecule is in a gas thermostat, its mean equilibrium potential energy $2\langle U_0 \rangle_e$ is $1/2k_B T$ and Eq. (15) simply gives the *Arrhenius equation*:

$$k = k_0 \cdot \exp\left(-\frac{E_a}{RT}\right) \tag{16}$$

where $E_a = N_A \Delta U$ is the activation energy, $R = k_B N_A$ is the gas constant, and N_A is the Avogadro number. In another environment of the molecule, with a different mechanism of energy exchange, dissipation or absorption, the mean potential energy can depend on the specific interactions in the system and their description. For example, for a quantum oscillator with

$$\langle U_0 \rangle_e = \frac{1}{2}\hbar\omega_0 \coth\left(\frac{\hbar\omega_0}{2k_B T}\right) \tag{17}$$

the probability is

$$p(x) = N \exp\left[-\frac{\mu\omega_0^2 x^2}{\hbar\omega_0 \coth\left(\frac{\hbar\omega_0}{2k_B T}\right)}\right] \tag{18}$$

which is in accordance with the result from equilibrium quantum-statistical thermodynamics [9]. The expression for the rate constant for dissociation of a diatomic molecule treated as a quantum oscillator, following from Eq. (15):

$$k = k_0 \exp\left[-\frac{\Delta U}{\hbar\omega_0 \coth\left(\frac{\hbar\omega_0}{2k_B T}\right)}\right] \tag{19}$$

is different from the classical Arrhenius equation (16) due to the dependence of the exponential term on the vibrational frequency of the oscillator ω_0. As expected, the quantum effects are important for reactions at lower temperatures and for breaking of bonds with high values of ω_0 (Fig. 2a). Figure 2b shows that temperature hardly affects the dissociation rate of bonds with vibrational frequencies above 2500 cm^{-1}.

3. Kramers Equation

The classical transition state theory (TST) gives the following equation for the rate constant of a chemical reaction:

$$k_{TST} = A_{TST} \cdot e^{-\frac{\Delta U}{k_B T}} \tag{20}$$

where A is the pre-exponential constant ($A_{TST} = \omega_0/2\pi$; ω_0 is the vibrational frequency of the dissociating bond) and ΔU is the energy barrier height. Usually, the most important factor for a chemical transformation is the barrier height, connected roughly with the energy necessary for breaking of the chemical bonds. Often, however, the interaction of the reacting molecule with the surroundings during barrier

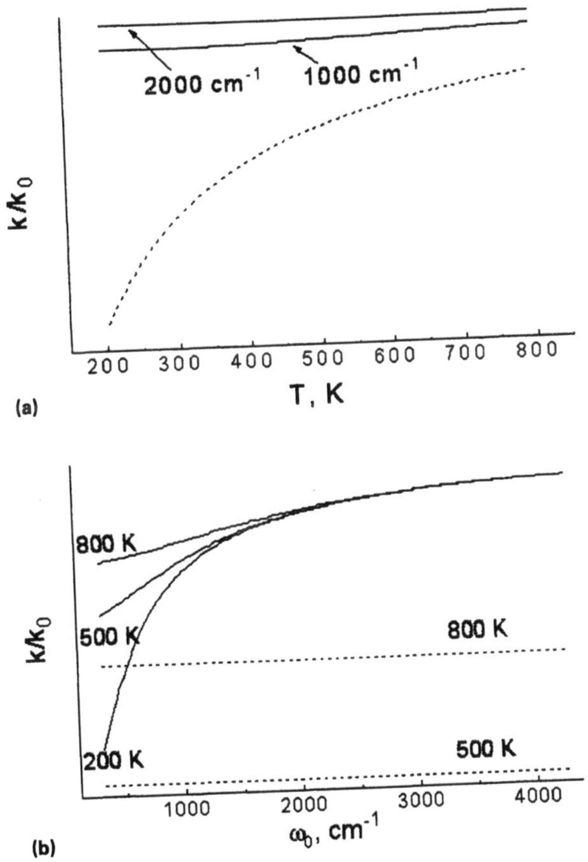

FIG. 2 Rate constant for dissociation of quantum (solid lines) and classical (dashed lines) harmonic oscillators: (a) temperature dependence for different vibrational frequencies ω_0 = 1000 and 2000 cm^{-1}; (b) frequency dependence for different temperatures.

crossing can slow or suppress the reaction due to the energy loss into the environment. This friction effect is substantial for chemical transformations in solutions and on surfaces.

A convenient method for accounting for the energy exchange during the reaction on the rate constant was developed by Kramers [10]. His theory is based on the Langevin equation (1) and allows correction of the pre-exponential term of the rate constant equation (20) with respect to the results obtained by classical TST. Here, a generalized version of Kramers' theory, proposed by Grote and Hynes [11, 12], is briefly presented. Let us consider the generalized Langevin equation (2) for the motion of the reacting system along the reaction co-ordinate x around the top of the energy barrier x_b (Fig. 1):

$$\mu\frac{\partial^2 x}{\partial t^2} = \mu\omega_b^2 x - \int_0^t \gamma(t-\tau)\frac{dx(\tau)}{d\tau}d\tau + F(t) \qquad (21)$$

where $x(t)$ is the deviation of the system from the maximum of the potential barrier x_b at the moment t, and the parameter ω_b is the imaginary "frequency" at the top of the barrier, characterizing the shape of the potential curve around the barrier:

$$\omega_b = -\left(\frac{\partial^2 U(x)}{\partial x^2}\right)_{x=x_b} \qquad (22)$$

The memory friction $\gamma(t)$ and thermal noise $F(t)$ from Eq. (20) are connected by the *classical fluctuation–dissipation theorem*:

$$\langle F(t)F(0)\rangle = k_B T\gamma(t) \qquad (23)$$

The expression for the rate consant obtained in this way is

$$k = \left(\frac{\xi}{\omega_b}\right)\cdot\frac{\omega_0}{2\pi}\cdot e^{-\frac{\Delta U}{k_B T}} = \left(\frac{\xi}{\omega_b}\right)\cdot k_{TST} \qquad (24)$$

The parameter ξ is the so-called transmission frequency which depends on the method for description of the friction term $\gamma(t)$. In the original work of Kramers, a Markovian process is considered, i.e.,

$$\gamma(t) = \eta\delta(t) \qquad (25)$$

and the expression for ξ_{Kr} is

$$\frac{\xi_{Kr}}{\omega_b} = \sqrt{1+\left(\frac{\eta}{2\mu\omega_b}\right)^2} - \frac{\eta}{2\mu\omega_b} \qquad (26)$$

As seen, for the absence of friction ($\eta = 0$), Kramers' equation (24) reduces to the TST expression (20). Using the generalized Langevin equation, the memory-renormalized transmission frequency ξ obeys the following equation:

$$\frac{\xi}{\omega_b} = \sqrt{1+\left(\frac{\hat{\gamma}(\xi)}{2\mu\omega_b}\right)^2} - \frac{\hat{\gamma}(\xi)}{2\mu\omega_b} \qquad (27)$$

Here, $\hat{\gamma}(\xi)$ denotes the Laplace transform of the memory friction. The above equation can be easily transformed into

$$\left(\frac{\xi}{\omega_b}+\frac{\hat{\gamma}(\xi)}{2\mu\omega_b}\right)^2=\left[\sqrt{1+\left(\frac{\hat{\gamma}(\xi)}{2\mu\omega_b}\right)^2}\right]^2$$

which is identical with

$$\left(\frac{\xi}{\omega_b}\right)^2+2\frac{\xi}{\omega_b}\cdot\frac{\hat{\gamma}(\xi)}{2\mu\omega_b}+\left(\frac{\hat{\gamma}(\xi)}{2\mu\omega_b}\right)^2=1+\left(\frac{\hat{\gamma}(\xi)}{2\mu\omega_b}\right)^2$$

From this equation, the following form of Eq. (27) is obtained:

$$\xi=\frac{\omega_b^2}{\xi+\dfrac{\hat{\gamma}(\xi)}{\mu}} \tag{28}$$

which is more convenient for some practical applications.

B. Vibration of Atoms in a Solid

The atoms of the metal oxides, as in all solids, participate in thermal vibrations around their equilibrium positions. These vibrations are determined by the structure of the solid and the interactions between the atom or ions in the lattice and, in general, their detailed description is very complicated. However, there are differetn simplified models for simulation of the collective vibrations of the atoms. Let us consider a linear chain of atoms in a solid at a distance a between neighbors (Scheme 1a). In this case the shorter wavelength will be

$$\lambda_{min}=6a \tag{29}$$

which corresponds to the highest frequency mode ω_{max} with the relation:

$$\omega_{max}=\frac{2\pi\nu}{\lambda_{min}}=\frac{\pi\nu}{a} \tag{30}$$

where ν is the velocity of sound in the material. The dispersion curve (connecting the wave vector $\mathbf{q}$ with vibrational frequency ω) in this case results in two symmetrical parts with a minimum at $|\mathbf{q}|=0$ increasing up to ω_{max} for $|\mathbf{q}|=\pi/2a$ (Fig. 3a).

If the chain consists of alternately bonded atoms of two types, which is similar to the situation in metal oxides, two types of collective vibrations should be observed – common vibrations of both types of atoms (Scheme 1b) and separate vibrations of each type with opposite phase (Scheme 1c). The first part (acoustic) is similar to the case of a homoatomic chain, while the second type of vibrational modes involves higher energy and is connected with optical transitions in the solid. In contrast to the dispersion curve for acoustic vibrations, the optical vibration frequency decreases with $|\mathbf{q}|$ and its highest value is at $|\mathbf{q}|=0$ (Fig. 3b).

The most popular approximation for the frequency distribution of the three-dimensional lattice vibrations was proposed by Debye. The number of vibrations z, with wavelength longer than or equal to λ, in a three-dimensional crystal of volume V is

$$z=\frac{4\pi V}{\lambda^3} \tag{31}$$

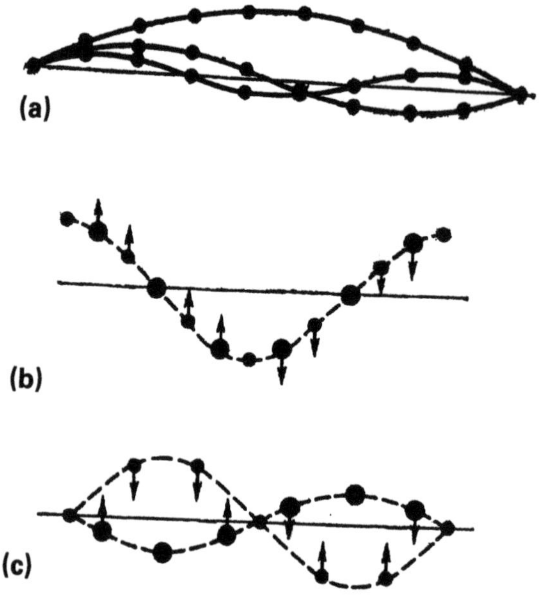

SCHEME 1 Model of normal vibrational modes of a linear chain consisting of one (a) or two (b,c) types of atoms. For the two-atomic chain, modes (b) correspond to acoustic and (c) to optical branches.

Substitution of $\lambda = 2\pi v/\omega$ and differentiation of the equation gives an expression for the frequency distribution $g(\omega)$ defined as the derivative of z with respect to ω:

$$g(\omega) = \frac{dz}{d\omega} = \frac{3V}{2\pi^2 v^3}\omega^2 \tag{32}$$

Debye introduced the characteristic frequency ω_D, as the highest allowed vibrational frequency in the solid. This frequency is connected with the total number of vibrations $3N$:

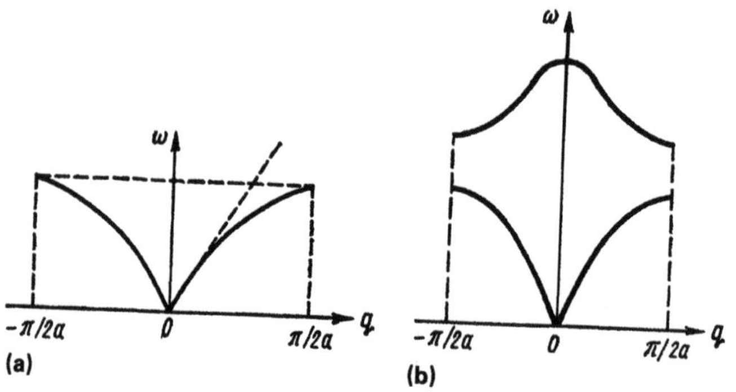

FIG. 3 Disperson curves for vibrations of homoatomic (a) and alternatively bonded heteroatomic linear chains (b) – acoustic (lower line); optical (upper line).

$$3N = \int_0^{\omega_D} g(\omega)\mathrm{d}\omega \tag{33}$$

which gives

$$\omega_\mathrm{D}^3 = \frac{6\pi^2 v^3 N}{V} \tag{34}$$

Now the frequency distribution $g(\omega)$ can be rewritten by combining Eqs (32) and (34):

$$g(\omega) = 9N\frac{\omega^2}{\omega_\mathrm{D}^3}H(\omega_\mathrm{D} - \omega) \tag{35}$$

where $H(\omega_\mathrm{D} - \omega)$ is the Heaviside function (1 for $\omega < \omega_\mathrm{D}$ and 0 for $\omega > \omega_\mathrm{D}$).

Each mode of the collective atomic vibration of the solid can be considered as a quantum oscillator with the same frequency ω and the energy:

$$E_n = (n + 1/2)\hbar\omega \quad (n = 0, 1, \ldots) \tag{36}$$

The minimal quanta of vibrational energy $\hbar\omega$ corresponding to the oscillators in the solid are called phonons. The phonon distribution in the solid is governed by Bose–Einstein statistics:

$$f(E) = \left[\exp\left(\frac{\hbar\omega}{k_\mathrm{B}T}\right) - 1\right]^{-1} \tag{37}$$

where $2\pi\hbar$ is the Planck constant. From this distribution function the mean energy of the normal vibrational mode, with frequency ω, can be obtained:

$$\bar{E} = \hbar\omega \cdot f(E) = \hbar\omega \cdot \left[\exp\left(\frac{\hbar\omega}{k_\mathrm{B}T}\right) - 1\right]^{-1} \tag{38}$$

Using Debye's frequency distribution and the expression for the mean energy of a vibrational mode, we find the total energy of all vibrations with frequencies between ω and $\omega + \mathrm{d}\omega$:

$$\mathrm{d}E = \bar{E} \cdot \frac{g(\omega)}{3N}\mathrm{d}\omega = 3 \cdot \hbar\omega\left[\exp\left(\frac{\hbar\omega}{k_\mathrm{B}T}\right) - 1\right]^{-1}\frac{\omega^2}{\omega_\mathrm{D}^3}H(\omega_\mathrm{D} - \omega) \cdot \mathrm{d}\omega \tag{39}$$

The vibrational energy spectrum of the solid $E(\omega) = \mathrm{d}E/\mathrm{d}\omega$, obtained in the above equation, is important not only for the vibrational energy exchange processes, but also for different thermal properties of the solids. The asymptotic behavior of this spectrum could be considered in the limits of lower and higher temperatures, with respect to the Debye temperature of the solid, defined as

$$\theta_\mathrm{D} = \frac{\hbar\omega_\mathrm{D}}{k_\mathrm{B}} \tag{40}$$

For $T \gg \theta_D$ the exponent can be expressed as a Taylor series:

$$E(\omega) = 3\frac{\hbar\omega}{\left[1 + \frac{\hbar\omega}{k_B T} + \left(\frac{\hbar\omega}{k_B T}\right)^2 + \cdots\right] - 1}\frac{\omega^2}{\omega_D^3}H(\omega_D - \omega)$$

Since the Heaviside function allows variation of ω only between 0 and ω_D, the terms of second and higher orders in the denominator can be neglected and the vibrational energy spectrum transforms into

$$E(\omega) = 3k_B T\frac{\omega^2}{\omega_D^3}H(\omega_D - \omega) \tag{41}$$

The absence of the Planck constant in this expression accounts for the classical behavior of the vibrations at higher temperatures. Conversely, at temperatures close to the Debye temperature of the solid θ_D, the quantum effects are important (Fig. 4).

Debye's model gives only an approximate description of the vibrational properties of real solids, especially for solids containing different atoms or having certain lattice structures. However, it is very convenient in situations where an analytical expression for the distribution functions is necessary, because more rigorous models give analytical solutions for one- or two-dimensional systems only. Since the spectra of surface vibrations are much more complicated, this model is often used. There are also some empirical combinations of Debye and Einstein distributions (in the classical limit):

$$E_{mix}(\omega) = k_B T\left[3\frac{\omega^2}{\omega_D^3}H(\omega_G - \omega) + \frac{N_E}{N}\delta(\omega_E - \omega)\right] \tag{42}$$

where ω_E corresponds to the single vibrational mode of the solid, assumed in the Einstein model. The new highest frequency in the Debye distribution ω_G (instead of ω_D) and the ratio N_E/N allow variation of the contributions of the two model

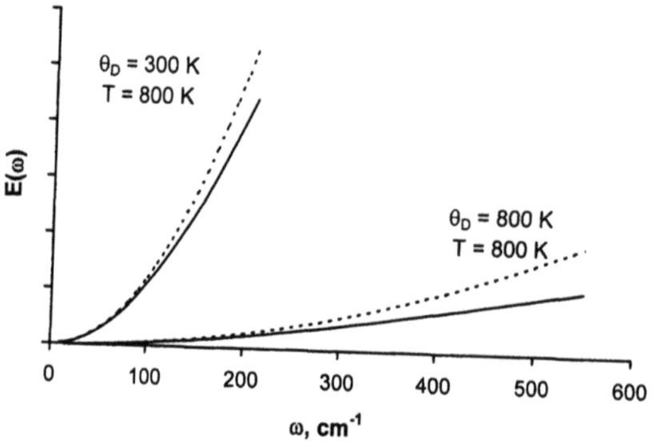

FIG. 4 Classical (dashed curves) and quantum (solid curves) vibrational energy spectrum according to Debye's distribution.

spectra in the mixed vibrational energy spectrum and give better agreement with the experiment.

II. NONELASTIC COLLISION OF THE VAPOR PARTICLE ON THE SURFACE

The nonelastic scattering of gas particles from the metal oxide surface is a typical process in which vibrational energy exchange plays an important role. In this case the kinetic energy of the colliding molecule (formed by two or more atoms) exchanges with its internal vibrational degrees of freedom and/or with the vibrational energy of surface phonons of the oxide.

A. Exchange Between Translational and Intramolecular Vibrational Energy

As a result of the collision of the vapor molecule with the surface, energy can be exchanged in two directions (Scheme 2):

- Translational energy is transferred to the intramolecular vibrations, and excitation of higher vibrational levels occurs (Scheme 2a). If the interaction with surface phonons is neglected, the translational energy loss is exactly equal to the vibrational excitation energy.
- Sufficient intramolecular vibrational energy is transferred to the translational energy of the particle to overcome the energy barrier for adsorption on the surface (Scheme 2b).

This situation is usually applied to dissociative adsorption of molecules, where the adsorption barrier is high. In the case of nondissociative adsorption, vibrational energy transfer to the restricted translational modes results in easier desorption.

1. Vibrational Excitation of Scattered Molecules

A convenient theoretical model for description of these processes has been proposed by Brenig and coworkers [1, 13, 14]. For the first process, they consider the total

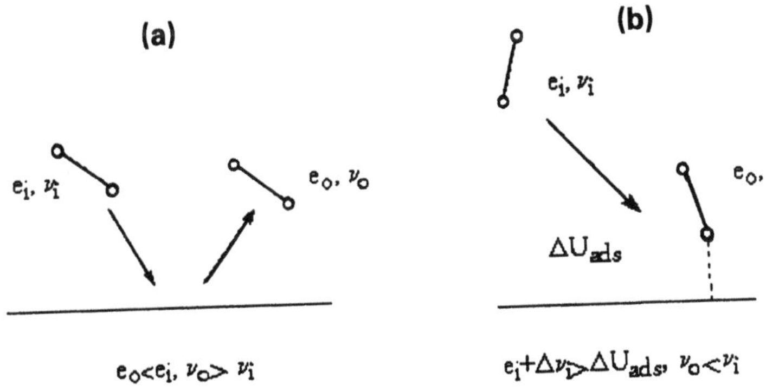

SCHEME 2 Energy exchange during molecule collision at the surface. Subscripts i and o correspond to initial and final values, respectively, and ΔU_{ads} denotes the adsorption barrier. (a) Vibrational excitation of scattered molecules; (b) vibrationally assisted adsorption.

probability $R(\varepsilon', \nu', \varepsilon, \nu)$ for the scattering of a molecule having initial (and final) translational energy $\varepsilon(\varepsilon')$, with corresponding intramolecular vibration states $\nu(\nu')$, as a sum of transition probabilities of direct scattering $P(\varepsilon', \nu', \varepsilon, \nu)$ and trapping–desorption $Q(\varepsilon', \nu', \varepsilon, \nu)$. The probability for direct scattering obeys the following master equation:

$$P(\varepsilon', \nu'; \varepsilon, \nu) = p(\varepsilon', \nu'; \varepsilon, \nu) + \int_{-\infty}^{0} \sum_{\nu''} P(\varepsilon', \nu'; \varepsilon, \nu) p(\varepsilon', \nu'; \varepsilon, \nu) d\varepsilon'' \tag{43}$$

where p is the transition probability of the molecule in a single scattering process, which satisfies the detailed balance equation:

$$p(\varepsilon', \nu'; \varepsilon, \nu) \exp\left(-\frac{\varepsilon + \varepsilon_\nu}{k_B T_s}\right) = (\varepsilon', \nu'; \varepsilon, \nu) \exp\left(-\frac{\varepsilon' + \varepsilon_{\nu'}}{k_B T_s}\right) \tag{44}$$

Here, ε, ν is the internal vibrational energy in state ν, and T_s is the surface temperature. After integration of P over the translational energies and summation over vibrational states of the scattered molecules, the expression for the sticking probability S can be found:

$$S(\varepsilon, \nu) = 1 - \int_{0}^{+\infty} \sum_{\nu'} P(\varepsilon', \nu'; \varepsilon, \nu) d\varepsilon' \tag{45}$$

which can be transformed into

$$S(\varepsilon, \nu) = \int_{-\infty}^{0} \sum_{\nu'} S(\varepsilon', \nu') p(\varepsilon', \nu'; \varepsilon, \nu) d\varepsilon' \tag{46}$$

The other contribution to the transition probability accounting for the trapping–desorption process $Q(\varepsilon', \nu', \varepsilon, \nu)$ is presented as a product of the sticking probability of the incoming molecule $S(\varepsilon, \nu)$ and the desorption probability of the outgoing molecule $D(\varepsilon', \nu')$. The expression for desorption probability D is based on the assumption of thermal equilibrium between sticking and desorption of the molecules on the surface:

$$D(\varepsilon, \nu) = \frac{S(\varepsilon, \nu)}{\langle S \rangle_s} \frac{\exp\left(-\frac{\varepsilon + \varepsilon_\nu}{k_B T_s}\right)}{k_B T_s Z(T_s)} \tag{47}$$

Here, $\langle S \rangle_s$ denotes the integral:

$$\langle S \rangle_s = \int_{0}^{+\infty} \sum_{\nu} S(\varepsilon, \nu) \frac{\exp\left(-\frac{\varepsilon + \varepsilon_\nu}{k_B T_s}\right)}{k_B T_s Z(T_s)} d\varepsilon \tag{48}$$

and $Z(T_s)$ is the partition function for the internal states of the molecule. As seen from Eq. (47), the distribution of the internal vibrational states of desorbed molecules deviates from Boltzmann distribution for T_s because of the dependence of $S(\varepsilon, \nu)$ on the initial internal state. The desorbed molecules are vibrationally heated when the sticking probability is an increasing function of the internal vibrational energy, and vice versa. Equations (46) and (47) give an exact means of determining the trapping–desorption transition probability, Q from the transition probability of a single scattering process p. Now, if we consider only the ground ($\nu = 0$) and the first exited ($\nu = 1$) internal vibrational states of the colliding molecule, the expression for p is

$$p(\varepsilon', \nu'; \varepsilon, \nu) = \frac{\exp\left[-\frac{(\varepsilon + \varepsilon_\nu - \varepsilon' - \varepsilon_{\nu'} - \mu_{\nu\nu'})}{2\sigma^2} K_{\nu\nu'}\right]}{\sqrt{2\pi\sigma}} \tag{49}$$

where $\mu_{\nu\nu'}$ is the average energy transfer from the molecule to the surface degrees of freedom, σ is the width of the distribution, and $K_{\nu\nu'}$ is the transition matrix element between the initial and final vibrational states. From eq. (44), an expression for σ can be found

$$\sigma^2 = k_B T_s(\mu_{\nu\nu'} + \mu_{\nu'\nu}) \tag{50}$$

The transition matrix, K, with zero or small off-diagonal elements, presumes a Gaussian form of the single transition probability, p. A more specific model of the energy exchange between translational and internal degrees of freedom of the molecule and solid surface can be offered. Two main mechanisms of the energy transfer into the oxide surface can be considered, namely (1) generation of electron–hole pairs in conductor oxides, and (2) excitation of surface phonon modes at insulator surfaces. In this model the effect of both channels of molecular scattering is roughly estimated by four parameters $\mu_{\nu\nu'}$, but in numerical calculations they are usually considered equal.

2. Vibrationally Assisted Adsorption

As mentioned earlier, the vibration–translation coupling can help in passing the adsorption energy barrier via transfer of vibrational energy of incoming vibrationally excited molecules to translation energy. This effect was observed for processes with higher activation energy, which implies that the potential energy surface for the interaction between a gas molecule and a solid surface should be considered. However, some simple comparisons based on the proposal for Boltzman distribution for the vibrational levels in the colliding molecules can be examined. For the ratio between occupations P of the ground ($\nu = 0$) and first excited ($\nu = 1$) levels this distribution gives

$$\frac{P(\nu = 1)}{P(\nu = 0)} = \exp\left(-\frac{\hbar\omega_0}{k_B T}\right) \tag{51}$$

As expected, the occupation of the excited states in molecules with higher frequency vibrations such as hydrogen ($\omega_0 = 4160\,\text{cm}^{-1}$) is lower compared to that of nitrogen monoxide ($\omega_0 = 1906\,\text{cm}^{-1}$) or carbon monoxide ($\omega_0 = 2143\,\text{cm}^{-1}$), Fig. 5a. However, the contribution of the vibrational energy of these excited molecules to the total energy is considerable, compared to the translation energy (Fig. 5b). Conversely, many more molecules with lower vibrational frequency (energy) are in the excited state, with relatively low vibrational excitation energy.

This qualitative picture can account for the general factors influencing the process, if the potential energy barrier for adsorption of the molecule on the surface

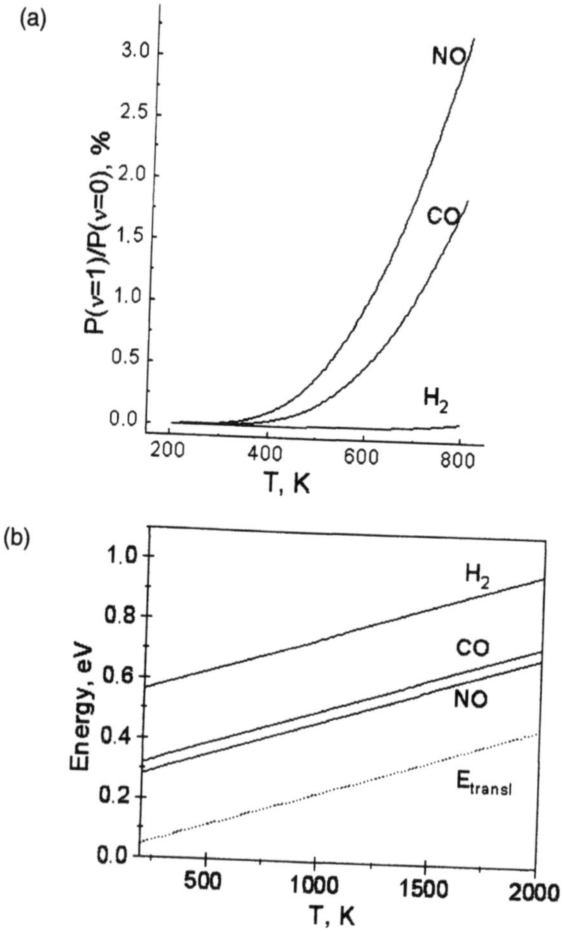

FIG. 5 (a) Ratio between occupations of the first excited and ground state for H_2, CO, and NO molecules calculated from Eq. (51). (b) Mean translational energy in the gas phase E_{transl} (dashed line) and total energy (solid lines) of the molecules with contribution from the vibrational excitation energy.

is known. Since vibrationally assisted adsorption takes place when the barrier is lower than $E_{tr} + \hbar\omega_0$, the most important parameter for the process is the excited state occupation, such as for NO and CO versus H_2. However, for higher adsorption energy barriers, the value of the vibrational excitation energy determines whether the adsorption will proceed at a certain temperature.

B. Exchange of Translational Energy of the Molecule and Vibrational Energy of Surface Phonons

This energy exchange is dependent on the kinetic energy of the gas particles scattered from the solid surface, which can be experimentally measured. In the most simple model when the surface is rigid and the intramolecular vibrations do not affect the collision, only mixing of the normal E_z and parallel E_p components of the molecule's translation energy can be observed after scattering. The magnitude of this energy exchange strongly depends on the surface curvature and is called the corrugation energy $\Delta E_z^{corr} = -\Delta E_p^{corr}$. If the surface geometry follows the equation $Z(x) = Z_0 + \beta\cos(2\pi/c)$, the corrugation energy is

$$\Delta E_z^{corr} = -\frac{8\pi^2\beta^2}{c^2} E_i \cos(2\varphi_i) \tag{52}$$

where E_i and φ_i are the impact beam energy and angle (with respect to the surface normal). This expression shows that ΔE_z^{corr} is negative for $\varphi_i < 45°$ and positive for $\varphi_i > 45°$, with the highest value at 90°.

Participation of the surface phonons in the energy transfer process with translational degrees of freedom can be accounted for by two theoretical approaches based on the theory of the generalized Langevin equation and on quantum or semiclassical solutions of the Schrödinger equation.

1. Methods Based on the Schrödinger Equation

Jackson [15, 16] proposes a quantum-mechanical theory for nonelastic scattering of particles as a result of interaction with surface phonons. Translational degrees of freedom of the gas particle are presented as a time-dependent wave packet. The wave functions describing the scattering of the particles satisfy a Schrödinger-type equation with a potential of interaction between the gas particle and the solid surface, which depends both on time and temperature. In this model the Hamiltonian of the system is

$$\mathbf{H} = \mathbf{H}_m + \mathbf{H}_b + \mathbf{V}_{mb} \tag{53}$$

where $\mathbf{H}_m$ is the operator of interaction of the gas particle with the static surface, $\mathbf{H}_b$ is the Hamiltonian of vibrations of the solid lattice (phonon bath), and the operator of the interaction of the particle with the phonons $\mathbf{V}_{mb}$ is represented as a linear function of the operators of excitation and annihilation of phonons in the solid.

As a result of such separation of the Hamiltonian, the wave function of the system is presented as a product:

$$\Psi = e^s \psi_m \phi_b \tag{54}$$

of the molecular ψ_m and phonon ϕ_b wave functions, and S is a function both of molecular and phonon variables. As a result an equation of the Schrödinger type is obtained for the wave function ψ_m. Depending on the method of presentation of this wave function, the solution can be obtained by a quantum or a semiclassical approach. The result obtained by this method represents the probability for the exchange of a given amount of energy ΔE between the gas particle and phonons, defined as

$$P_n(\Delta E) = \sum \langle \psi_n(t)|m\rangle\langle m|\psi_n(t)\rangle \delta(\Delta E + E_m - E_n) \tag{55}$$

where $|n\rangle$ and $|m\rangle$ are the initial and final state of the lattice, and E_m and E_n are the energies of the these states. Using the presentation of the wave function Ψ, after some transformtions the following equation is obtained:

$$P_n(\Delta E) = \frac{1}{2\pi\hbar} \int_{-\infty}^{\infty} Q(t_1) \exp\left(\frac{1}{\hbar} \Delta E t_1\right) \mathrm{d}t_1 \tag{56}$$

In this equation $Q(t_1)$ is a complex exponential function of the characteristics of phonons, the structure of the surface, and the potential of the interaction between surface oscillations and the gas particle, as well as of co-ordinates and momentum of the particle. As a result the application of this approach requires considerable simplification of the interaction potential. This limits its opportunities for study of more complex systems and for inclusion of additional interactions under consideration (such as intramolecular vibrations). Both quantum and semiclassical variants of the method show reasonable agreement with experimental investigations for some simple systems [15, 16].

Surface phonons also affect the energy exchange among the different components of the gas molecule during the collision. The trapping probability is defined as the total probability that the energy lost to phonons δE is equal to or greater than the initial z component of beam energy E_z^i. In this approach the parallel components of the beam energy are conserved and the energy transfer with phonons involves only the normal component of the beam energy E_z. However, the more detailed picture also includes the contribution of the corrugation energy $\Delta E_z^{\mathrm{corr}}$ [Eq. (52)] to the normal component of the beam energy.

Another quantum-mechanical approach separates the system Hamiltonian in a different way [17]. In the adiabatic approximation, the time-dependent Schrödinger equation for the motion of the atomic nuclei of the solid and the gas particle, excluding the electron degrees of freedom, can be represented as

$$i\hbar \frac{\mathrm{d}\Phi}{\mathrm{d}t} = \mathbf{H}\Phi = [\mathbf{T}(q, q_s) + \mathbf{V}_{g,s}(q, q_s) + \mathbf{V}_{ss}(q, q_s)]\Phi \tag{57}$$

where $\mathbf{T}$ is the kinetic energy operator of the atoms of the solid and the gas, $\mathbf{V}_{ss}(q_s)$ is the operator of interatomic interactions in the solid, and $\mathbf{V}_{g,s}$ is responsible for the interaction of the gas particle with the solid surface. Equation (57) has the following solution:

$$\langle t'; q|\Phi\rangle = \int U(t', q'; t, q)\langle t; q|\Phi\rangle|dq \tag{58}$$

where

$$U(t', q'; t, q) = \left\langle q' \middle| \exp\left[-\frac{i}{\hbar}\mathbf{H}(t'-t)\right] \middle| q \right\rangle \tag{59}$$

For the sake of simplification the period $t' - t$ is divided into N small parts δt and a full system of $N-1$ states $q_i (i = 1, \ldots, N-1)$ is introduced. As a result the integral, Eq. (59), is transformed into

$$U(t', q'; t, q) = \int \left\langle q' \middle| \exp\left[-\frac{i}{\hbar}H\delta t\right] \middle| q_1 \right\rangle \left\langle q_1 \middle| \exp\left[-\frac{i}{\hbar}\mathbf{H}\delta t\right] \middle| q_2 \right\rangle \cdots$$
$$\cdots \left\langle q_{N-1} \middle| \exp\left[-\frac{i}{\hbar}\mathbf{H}\delta t\right] \middle| q \right\rangle \mathrm{d}q_1 \ldots \mathrm{d}q_{N-1} \tag{60}$$

The application of this method requires the use of Monte Carlo methods for calculation of the multidimensional integrals, Eq. (60), which unfortunately results in oscillating functions of the type exp(iF) [17]. In order to avoid this problem different mathematical approaches are proposed, but it is still unclear whether they are appropriate for the dynamics of processes on solid surfaces.

One possibility for simplification of this problem is to consider the motion of the atoms of the solid as classical, which substantially decreases the number of degrees of freedom q in Eq. (58). In this way the time evolution of the quantum system is connected with the classical equations of motion of the atoms of the solid, which presupposes that the matter move much slower than the striking gas molecules:

$$\mu_s \frac{d^2 q_s}{dt^2} = -\nabla_{q_s}\langle\Phi|\mathbf{V}_{g,s}|\Phi\rangle - \nabla_{q_s}\mathbf{V}_{SS}(q_s) \tag{61}$$

The participation of the internal degrees of freedom of the gas molecule (intramolecular oscillations) in the quantum part of the equation still does not provide solutions for the system. Hence, the approach described, as well as all its variants, is generally applied to different peculiarities of the nonelastic scattering of monoatomic or nondissociating gas molecules from solid surfaces.

2. Methods Based on Generalized Langevin Equation

Another group of methods for calculation of the energy exchange between vibrations of the solid and the gas molecules is based on the application of the *generalized Langevin equation* [18–20]. The main assumptions of this approach are [17]:

- The striking molecule interacts only with the atoms from the solid surface which it strikes (primary zone – Scheme 3), while the neighboring atoms in the secondary zone do not interact directly with the molecule, so their equations of motion do not depend on its co-ordinates (x and z).
- Interactions of the atoms from the primary zone with their neighbors (secondary zone) as well as the interactions inside the secondary zone itself are harmonic.
- The solid is considered as an infinite one.

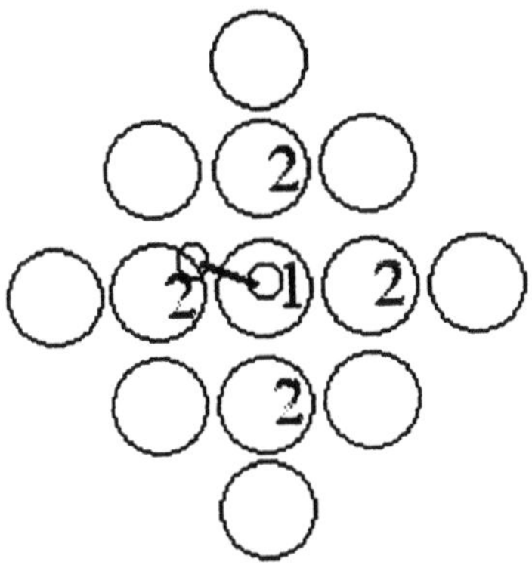

SCHEME 3 Primary (1) and secondary (2) zones of surface atoms.

Introduction of the solution of these equations into the equation of motion for the primary zone atoms gives a generalized Langevin equation (2) with some additional terms. If we consider the primary zone consisting of a single surface atom with equilibrium position y_e, its equation of motion is

$$\mu\frac{\partial^2 y}{\partial t^2} = -\frac{\partial V_{gs}}{\partial y} - \frac{\partial V_{ss}}{\partial y} + \mu\gamma(0)y - \mu\gamma(t)y(0) - \int_0^t \gamma(t-\tau)\frac{dx(\tau)}{d\tau}d\tau + F(t) \quad (62)$$

In this model the potential interaction between the molecule and the surface atoms V_{gs} is a function of the co-ordinates (x and z) of the gas molecule, the primary zone atom y and the equilibrium values for the atoms in the secondary zone Y_e, obeying the classical equation of motion for the molecule:

$$\mu\frac{\partial^2 z}{\partial t^2} = -\frac{\partial V(x, z, y, Y_e)}{\partial z} \quad (63)$$

The interatomic interaction potential V_{ss} on the surface depends only on y. When the primary zone consists of more than one atom, the interaction between them is aharmonic. However, for a one-atom primary zone, the potential is modeled as harmonic with vibrational frequency ω_e. The fluctuation force $F(t)$, acting on the atom, and the dissipation $\gamma(t)$ due to the interactions with the bath are connected by the classical fluctuation–dissipation theorem, Eq. (23).

The application of this method requires knowledge of the explicit form of at least one of the two functions $\langle F(t)F(0)\rangle$ or $\gamma(t)$, in order to find a solution of the equation. Variants of the approach, developed up to the present time are based on different ways of modeling of the dissipation term $\gamma(t)$, connected with the secondary zone of atoms [18–20]. Adelman and Garrison use Debye's model for phonons of the solid and obtain an equation for the dissipation term which, can be solved numerically. Doll and Dion propose $\gamma(t)$ as a linear combination of conveniently chosen functions, where the coefficients are determined by numerical self-consistency. Another possibility is to model the microscopic interactions in the lattice of the solid in order to derive a dissipation term. Tully presents the friction as a white noise or positionally autocorrelated function of a Brownian oscillator, including both oscillation and dissipation terms.

The application of the above approaches to the problem of vibrational energy exchange during molecule collisions with oxide surfaces raises some problems connected not only with correct derivation of the dissipation and stochastic terms, but also with the initial assumptions of the model. The first problem concerns the choice of the primary zone. While adsorption occurs (more or less) at a definite surface position, the collision is a very fast process, far from equilibrium and with different surface atoms. The second question is the different time scales for the exchange of the collision energy, Z_c between the molecule and the primary zone, and the energy dissipation to the secondary zone with characteristic time $\eta^{-1} \approx (\pi\omega_D/6)^{-1}$. For $\tau_c \ll \eta^{-1}$, the effect of collision is limited to motion of the primary zone atoms, while for the opposite extreme (fast dissipation) the system is reliably described with the classical Langevin equation (1) instead of the generalized one.

III. VIBRATIONAL INTERACTIONS OF ADSORBED MOLECULES

In this section the description of the general case – molecules adsorbed on flat oxide surface is considered. The complete description of adsorption on real oxide surfaces is a rather complicated process both due to the complexity of crystallographic and geometrical structure of the solid surface and the chemical behavior and orientation of the approaching molecule. Even on perfect single-crystalline surfaces the adsorption could occur at metal or oxygen atoms. An appreciation for the role of surface structure on electrostatic potential can be seen for the $TiO_2(110)$ surface, calculated in Ref. 21 (Fig. 6). At this surface and other similar nonpolar surfaces such as $Cu_2O(111)$ and NiO, carbon monoxide is adsorbed at the cationic metal sites. Conversely, the adsorption of acidic molecules such as SO_2 or CO_2 proceeds at the surface oxygen on the MgO(111) surface.

Another difficulty is the different adsorption properties of different crystallographic faces of the oxide surface and the significant contribution made by steps, edges, corners, etc. Appearance of specific functional groups on many metal oxide surfaces can also completely change the adsorption kinetics and structure. The interaction of the adsorbed molecules with surface functional groups is a subject with a variety of spectral techniques. In this situation the supporting oxide surfaces affects the properties of the functional groups, but their chemical behavior remains similar in general. For this reason the interaction (both chemical and vibrational) of the adsorbed molecules with such groups is not conveniently described as the interaction between two molecules with the effects of the rest of the surface accounted for by changing some of the parameters.

In addition to surfce heterogeneity, the adsorption characteristics of real surfaces are affected by other types of defects such as dislocations, vacancies, flaws, and

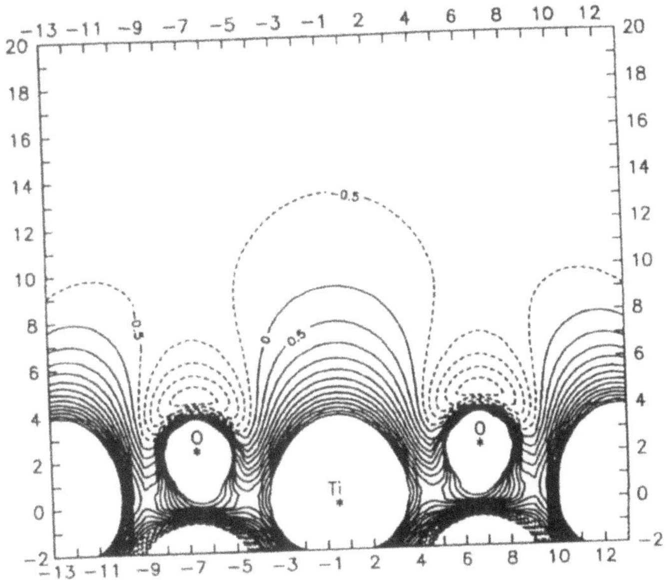

FIG. 6 Electrostatic potential of $TiO_2(110)$ surfaces, calculated by *ab initio* Hartree–Fock calculation. (From Ref. 21; courtesy of Elsevier Science B.V.)

strongly bonded impurities. All these defect sites generate local vibrational modes, which influence the vibrational interaction between the surface and adsorbed molecules, resulting in asymmetric broadening of the spectral bands. However, the theoretical treatment of the vibrational energy exchange is significantly complicated by this site heterogeneity. Experimentally the contribution to vibrational properties of the adsorption system due to site heterogeneity is usually studied by spectral techniques.

A. Vibrations of Adsorbed Molecules

Considerable changes in the intramolecular vibrations of the adsorbed molecule result from adsorption. The first is the transformation of some of the translation and rotational degrees of freedom into vibrational modes, the so-called frustrated translation and rotational modes. The frustrated translation mode perpendicular to the oxide surface corresponds to the vibration of the new adsorption bond. Depending on the geometry and the interactions in the adsorption complex, this bond is formed between one or more atoms from the adsorbate and the surface. The simplest case is one-end adsorption of the molecule, i.e., only one of the atoms is bonded to a surface atom (see Scheme 4) and the frequency of the frustrated translation is equal to the frequency of the adsorption bond. When two or more atoms are connected in the adsorption bond, it is convenient to consider the frustrated translation as the center of mass oscillation of the whole molecule. The frustrated rotational modes usually have a lower frequency, but they are important in the interactions with neighboring surface atoms, in surface diffusion, and in lateral interactions with other adsorbate molecules. In addition to these new vibrational modes, some of the gas-phase vibrations are also changed due to the hindering of the solid surface.

Let us consider adsorption of ammonia on a MgO(100) surface with the N–Mg bond perpendicular to the surface (Scheme 4). The calculated vibrational frequency along the Mg–N direction (corresponding to the adsorption bond) is 310 cm^{-1} [22]. As seen from the potential energy curves in Fig. 7, two of the rotational degrees of

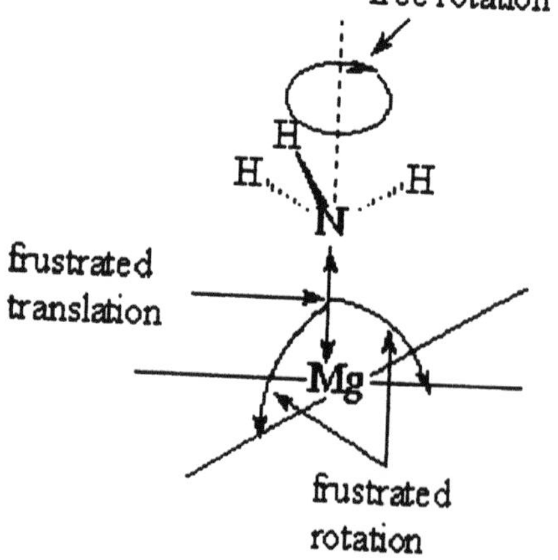

SCHEME 4 Frustrated degrees of freedom after adsorption of ammonia on magnesium oxide surface, perpendicular to the surface.

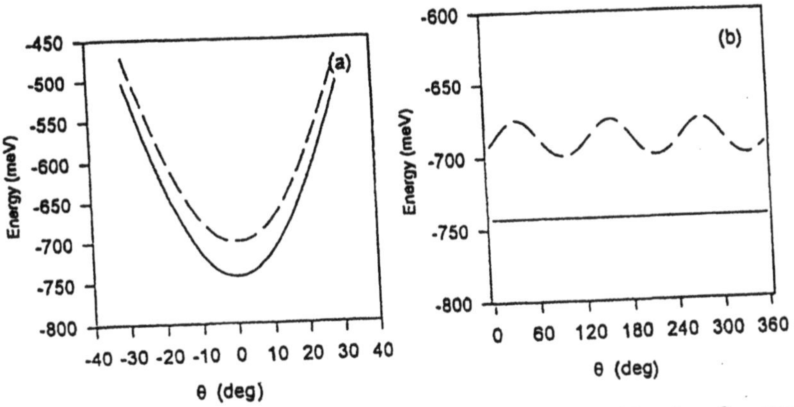

FIG. 7 Potential energy curves for the rotational degrees of freedom for ammonia adsorbed on MgO(110) surface. Solid line corresponds to single admolecule; broken line – to dimmer. (a) Frustrated modes (bending); (b) free rotation around the adsorption bond (perpendicular to the surface). (From Ref. 22; courtesy of Elsevier Science B.V.)

freedom are frustrated at the surface (Fig. 7a) – for rotations around the axis perpendicular the adsorption bond. In the adsorption complex these vibrations are connected with bending motion of the adsorbed molecule towards the surface, and their frequencies (204 and 207 cm^{-1}) are near to the value of the frustrated translational mode. The third rotational mode around the Mg–N axis is almost free (solid line in Fig. 7b) with a frequency 30 cm^{-1}.

Introduction of a second adsorbed ammonia molecule in the neighboring position generates a weak potential barrier in this rotation (broken line in Fig. 7b), which decreases with increase of the distance between the two adsorbed species. The most influenced intramolecular vibration after the adsorption is the "umbrella," connected with molecular inversion of the three N–H bonds in gas-phase ammonia. Stronger attraction of the nitrogen atom to the surface and repulsion of the hydrogens converts the symmetric double-well inversion potential of the gas molecule into a distorted single well (Fig. 8). This change in the potential curve modifies the distances between the vibrational levels for this mode.

B. Vibrational Relaxation

One of the main methods for investigation of the processes occurring on solid surfaces is vibrational spectroscopy in its different variants [23, 24]. As in infrared spectra, the position and the form of the observed bands depend substantially on the interaction of intramolecular vibrations with the molecular environment. In this way conclusions can be made with regard to the ways in which energy exchange occurs in the adsorbate–solid surface system. Several mechanisms for dissipation of vibrational energy of adsorbed molecules are considered [25–30]:

- Radiation mechanism (by photon emission) – classical mechanism for vibrational relaxation of a molecule in vacuum, which is characterized by a relatively long lifetime of vibrationally excited states. The rate of relaxation for adsorbed molecules, however, is much higher, which supposes faster pathways for energy exchange such as the following:

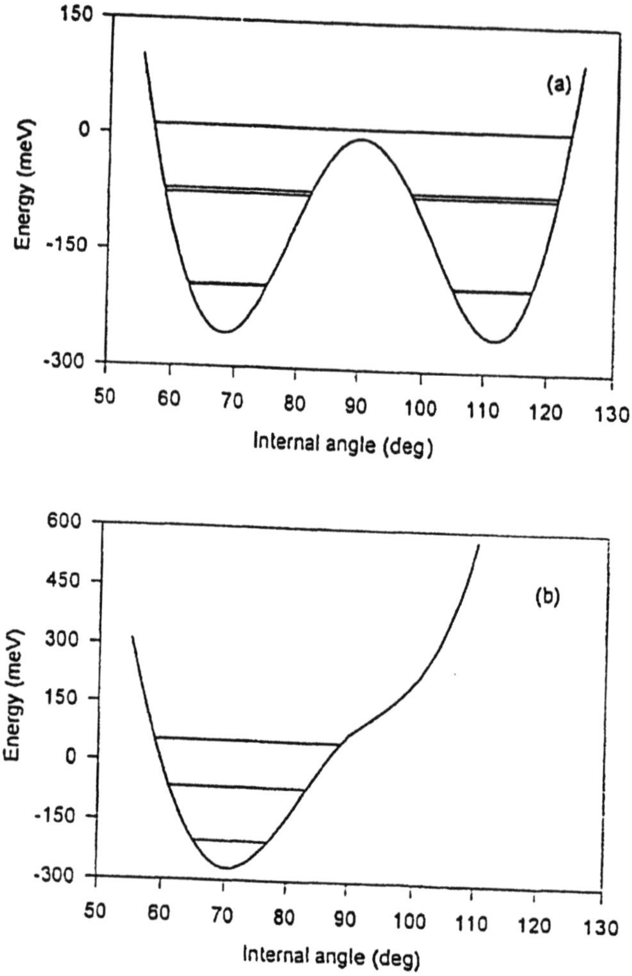

FIG. 8 Potential curve and vibrational states for internal inversion of free (a) and adsorbed (b) ammonia molecule. (From Ref. 22; courtesy of Elsevier Science B.V.)

- energy exchange with the phonons of the adsorbent;
- excitation of electrons from the conduction band of metal adsorbents (electron–hole pair);
- energy transfer into the adsorbent–adsorbate bond.

Theoretical methods for the description of different mechanisms for vibrational energy transfer are directed mainly to their participation in the relaxation of vibrationally excited adsorbed molecules and, respectively, their influence on the form and position of the spectral bands. Some authors [25, 26] propose that the main mechanism for vibrational relaxation of polar molecules adsorbed on conductor surfaces (specifically metals) is excitation of electron–hole pairs, and the interaction with vibrations of the surface atoms of the adsorbent does not significantly influence the form of the spectral bands in this case. On the other hand, for insulator surfaces, it is reasonable to include phonon relaxation. In this section some theoretical approaches for the description of vibrational relaxation by energy transfer into surface vibrations of the adsorbent will be discussed.

The results from the general theory for the vibrational spectrum of a localized harmonic oscillator, linearly coupled with a noninteracting boson continuum (phonons, photons, electron–hole pairs), can be used to estimate the contribution of different relaxation processes at surfaces. The spectral function of the oscillator obtained by normal-mode analysis at zero temperature is [25]

$$\rho(\omega) = \frac{1}{\pi} \frac{2\omega_0^2 \Gamma(\omega)}{\left[\omega^2 - \omega_0^2 - \omega_0 \Delta\omega(\omega)\right]^2 + [\omega_0 \Gamma(\omega)]^2} \tag{64}$$

with frequency shift $\Delta\omega$ and band width Γ:

$$\Delta\omega(\omega) \equiv p \int_0^{\infty} \frac{2\omega' |\lambda(\omega')|^2 \rho_s(\omega')}{\omega^2 - \omega'^2} d\omega \tag{65}$$

$$\Gamma(\gamma) = \pi \rho_s(\omega) |\lambda(\omega)|^2 \tag{66}$$

In the above equations ω_0 is the frequency of a noninteraction oscillator, $\lambda(\omega)$ is the coupling function between the oscillator and the external field, and $\rho_s(\omega)$ is the field density of states. With Eqs (67)–(74) it will now be possible to predict the spectral behavior for localized harmonic oscillators linearly coupled with different model boson continua.

For single phonon–vibrational interactions, the field density of states $\rho_s(\omega)$ can be approximated with the classical version of the Debye vibrational spectral density of a solid (Section I.B) shown in the second row of Scheme 5:

$$\rho_{1p}(\omega) = \begin{cases} 3\omega^2/\omega_D^3 & \text{for } \omega < \omega_D \\ 0 & \text{for } \omega > \omega_D \end{cases} \tag{67}$$

This dependence of $\rho_{1p}(\omega)$ on ω results because there are no surface vibrational modes with frequencies above the Debye frequency, ω_D.

For the frequency spectrum of a molecule adsorbed on a Debye solid the frequency shift is determined from Eq. (65) to be

$$\Delta\omega(\omega) \cong \frac{3\lambda_{ph}^2 \omega_D}{2\omega^2} \tag{68}$$

In Eq. (68), $\Delta\omega(\omega) > 0$, so, owing to the single phonon–vibrational interaction the oscillator band maximum is predicted to undergo a blue shift. Taking into account that the vibrational frequency of the oscillator ω_0 is usually higher than ω_D (as shown in Scheme 5), $\rho_s(\omega_0) = \rho_{1p}(\omega_0 > \omega_D) = 0$. The zero value of $\rho_s(\omega)$ for ω around the oscillator frequency ω_0 results in zero broadening of the vibrational band due to the single phonon–vibrational interaction:

$$\Gamma(\omega > \omega_D) = \pi \rho_s(\omega > \omega_D) |\lambda(\omega > \omega_D)|^2 = 0 \tag{69}$$

The characteristics of the oscillator band due to multiphonon coupling with the solid can be inferred from the single-phonon equations immediately above. The dominant role in multiphonon interactions is played by the n-phonon contribution for which n satisfies the requirement $(n-1)\omega_D < \omega_0 < n\omega_D$. With this requirement,

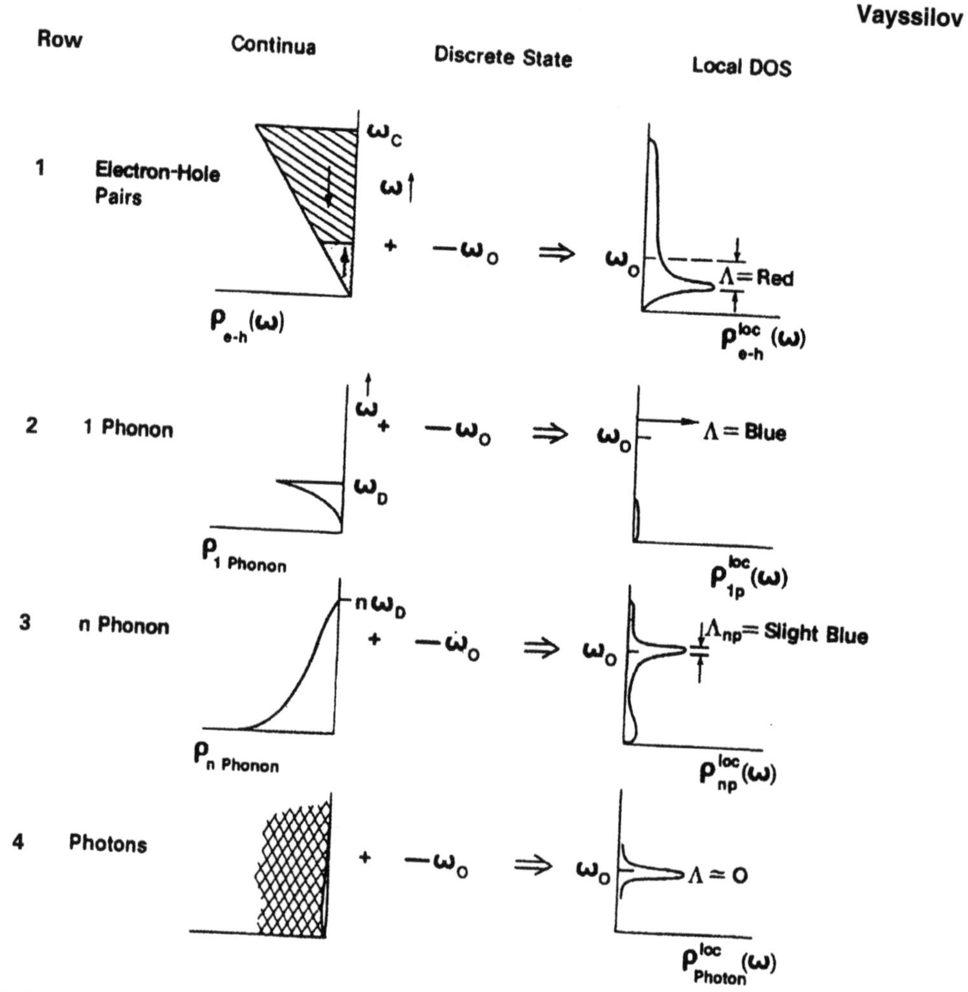

SCHEME 5 Scheme for the characteristic broadening and shift connected with the coupling of the localized oscillator with electron–hole pairs, phonons, and photons. (From Ref. 25; courtesy of Elsevier Science B.V.)

one can see that n phonon-oscillator coupling is possible since ω around ω_0 corresponds to a nonzero value of the n-phonon density of states $\rho_{\mathrm{nph}}(\omega = \omega_0)$ (the upper frequency limit for n-phonon state distribution is $n\omega_{\mathrm{D}}$). Hence, from Eq. (67) and the fact that $\rho_{\mathrm{nph}}(\omega \leq n\omega_{\mathrm{D}}) \neq 0$, one concludes that $\Gamma(\omega \leq n\omega_{\mathrm{D}}) \neq 0$ or that band broadening would be expected for n phonon–oscillator coupling.

The frequency shift of an oscillator coupled with an n-phonon surface $\Delta\omega_{\mathrm{nph}}(\omega)$ can be estimated by Eq. (68); hence, it will be proportional to the coupling constant regardless of how many phonons are linked to the oscillator. It is presumed that: (1) the coupling constants for oscillator coupling with a single and multiple phonon solid are related as $\lambda_{\mathrm{nph}} = \lambda_{\mathrm{ph}}^{n}$, and (2) the coupling constant λ_{ph} is small. From Eq. (68) and these assumptions, one concludes that $\Delta\omega_{\mathrm{nph}}(\omega) \ll \Delta\omega_{\mathrm{ph}}(\omega)$ or that a small frequency shift (smaller than for single-phonon coupling) will occur when an oscillator couples with n-phonon modes of the solid as shown in the third row of the Scheme 5. As a summary of the oscillator–phonon coupling, the main contribution to the line broadening is multiphonon processes while the single-phonon interaction dominates in the band position shift.

The electron–hole pair density of states below the Fermi frequency ($\varepsilon_F/\hbar$) can be approximated by

$$\rho_S(\omega) \cong \hbar^2 \rho_F^2 \omega \tag{70}$$

where ρ_{ε_F} is the electron density of states at the Fermi level. Following the presumption that all the states within the bonded continuum are equally accessible, the coupling constant simplifies to

$$\lambda(\omega) = \begin{cases} \lambda_{e-h} & \text{for } \omega \le \omega_c \\ 0 & \text{for } \omega > \omega_c \end{cases} \tag{71}$$

with $\hbar\omega_c$ of the order of the Fermi energy. Substituting these expressions into Eqs (65)–(67) the band shift and width are

$$\Delta\omega(\omega) \equiv \frac{2}{\pi} C\omega_c \left[-1 + \frac{1}{2}\frac{\omega}{\omega_c} \ln\left|\frac{\omega_c + \omega}{\omega_c - \omega}\right| \right] \tag{72}$$

$$\Gamma(\omega) = \begin{cases} C\omega & \text{for } \omega \le \omega_c \\ 0 & \text{for } \omega > \omega_c \end{cases} \tag{73}$$

where the constant C is associated with the magnitude of the interaction:

$$C \equiv \pi\hbar^2 \rho_{\varepsilon_F}^2 |\lambda_{e-h}|^2$$

and the spectrum due to the interaction is shown in the first row of Scheme 5.

Since the Fermi energy is typically between 1 and 10 eV while the oscillator vibrational energy $\hbar\omega$ is of the order of 0.2 eV, a red shift is always observed, i.e.,

$$\Delta\omega(\omega) \cong -2C\omega_c/\pi$$

The dominant term in the ratio between the shift and band expansion due to the oscillator coupling with electron–hole pairs does not depend on the coupling constant λ_{e-h}:

$$\frac{\Delta\omega(\omega)}{\Gamma(\omega)} \cong -\frac{\pi\omega_c}{2\omega} \tag{74}$$

Following the above estimation of both frequencies, this ratio shows that the magnitude of the frequency shift is 10–100 times larger than the corresponding change in the band width. The coupling of the oscillator with photons (electromagnetic field) results in broadening of the spectral line, but without shift of the band maximum (row 4 in Scheme 5).

This simplified scheme shows the characteristic broadening and shift connected with the coupling of surface phonons and electron–hole pairs, but it does not account for the influence of the temperature and aharmonicity effects, which are very important for oscillator interaction with phonons [25].

1. Interaction with Surface Phonons–Harmonic Model

For the analysis of the vibrational interaction in the adsorption complex, the oscillations of a bond between two atoms of a molecule adsorbed on a solid surface are considered. The adsorbent surface is modeled as a number of oscillators forming the heat surface phonons. The adsorbed molecule is bonded with one of the oscillators and interacts through this bond with the solid. As an initial model, all oscillators in the sytem are presented with a harmonic approximation. Some aharmonic effects will be considered in the next part of this section.

The magnitude of the energy transfer by means of interactions between surface phonons and bond vibrations of the adsorbed molecule is determined by the force balance equation for oscillations of the bond $x(t)$:

$$\mu \frac{\partial^2 x}{\partial t^2} = -\mu \omega_0^2 x - \int_0^t \gamma(t-\tau) \frac{\mathrm{d}x(\tau)}{\mathrm{d}\tau} \mathrm{d}\tau + F_i(t) \tag{75}$$

where ω_0 is the frequency of the oscillator in a free molecule and $f_i(t)$ is the interaction force between the adsorbed molecule and the solid surface. The vibrational interaction force depends on the bond lengths for both oscillators of the adsorbed molecule and the surface. If $y(t)$ denotes an arbitrary deviation of the bond length of the solid surface oscillator at time t, interacting with the adsorbed particle from the equilibrium value y_0, the interaction force $F_i(t)$ can be presented by series expansion of x and y:

$$F_i(t) = F_i(x(t), y(t)) = F_i(x_0, y_0) + \left(\frac{\partial F_i}{\partial x}\right)_{x_0} x + \left(\frac{\partial F_i}{\partial y}\right)_{y_0} y + \cdots \tag{76}$$

The first term in Eq. (76) corresponding to the equilibrium positions is zero. Since the deviations of both oscillators are relatively small compared to the equilibrium values at x_0 and y_0, the second and higher degree terms in this expansion can be neglected. For the sake of convenience, the following symbols are introduced:

$$\left(\frac{\partial F_i}{\partial x}\right)_{x_0} = \mu(\omega_0^2 - \omega_a^2) \tag{77}$$

$$\left(\frac{\partial F_i}{\partial y}\right)_{y_0} = \mu \omega_i^2 \tag{78}$$

Thus, Eq. (75) becomes

$$\mu \frac{\partial^2 x}{\partial t^2} = -\mu \omega_a^2 x - \int_0^t \gamma(t-\tau) \frac{\mathrm{d}x(\tau)}{\mathrm{d}\tau} \mathrm{d}\tau + \mu \omega_i^2 y(t) \tag{79}$$

Equation (79) represents the generalized Langevin equation for an oscillator with specific frequency ω_a, which is the real vibrational frequency of the adsorbed molecule and the fluctuation force $F(t)$:

$$F(t) = \mu \omega_i^2 y(t) \tag{80}$$

which is dependent on the characteristic interaction frequency ω_i. This approach produces the stochastic term $F(t)$ which is necessary for application to real systems, as mentioned in Section I.A.1. The exact value of the vibrational frequency ω_a could be obtained either by experimental measurements of the adsorbed molecule vibrational spectrum or by theoretical calculations. The other vibrational parameter ω_i is connected with molecule–solid surface vibrational interaction. As follows from its definition, Eq. (78), the exact value of the interaction frequency can be calculated by the second derivative of the interaction potential energy with respect to y. When the adsorbed molecule bond is parallel to the solid surface, the derivative of the interaction force $F(t)$ with respect to solid surface oscillator deviation y has the same magnitude as the derivative with respect to the adsorbed molecule oscillator deviation x, i.e.,

$$\omega_i^2 = |\omega_0^2 - \omega_a^2| \tag{81}$$

Following the scheme described in Section I.A.1, Eq. (79) gives the expression for the spectral density of the adsorbed molecule oscillator deviation:

$$\langle xx \rangle_\omega = \frac{\mu^2 \omega_i^4 \langle uu \rangle_\omega}{\mu^2 (\omega_a^2 - \omega^2)^2 + \omega^2 \gamma^2(\omega)} \tag{82}$$

using the spectral density of the autocorrelation function of the fluctuation force:

$$\langle FF \rangle_\omega = \mu^2 \omega_i^4 \langle yy \rangle_\omega \tag{83}$$

From the above equation, the expression for the frequency-dependent friction coefficient $\gamma(\omega)$ of the system can be found by the *quantum fluctuation–dissipation theorem* [6–8], similar to the classical version, Eq. (5):

$$\langle FF \rangle_\omega = \hbar\omega \coth\left(\frac{\hbar\omega}{2k_B T}\right) \gamma(\omega) \tag{84}$$

The quantum version of this theorem is one of the relatively simple ways to include quantum effects in a description of the vibrational interactions. Using Eqs (83) and (84) the memory friction $\gamma(\omega)$ is obtained as

$$\gamma(\omega) = \frac{1}{\hbar\omega} \tanh\left(\frac{\hbar\omega}{2k_B T}\right) \mu^2 \omega_i^4 \langle yy \rangle_\omega \tag{85}$$

and the equation for the spectral density of the autocorrelation function of the adsorbed molecule oscillator deviation transforms into

$$\langle xx \rangle_\omega = \frac{\omega_i^4 \langle yy \rangle_\omega}{\omega_a^2 - \omega^2)^2 + \omega^2 \left[\frac{\mu \omega_i^4}{\hbar\omega} \tanh\left(\frac{\hbar\omega}{2k_B T}\right) \langle yy \rangle_\omega \right]^2} \tag{86}$$

The spectral density of the autocorrelation function of the solid surface oscillator deviation $\langle yy \rangle_\omega$ can be derived by the relationship between the energy spectrum $E_s(\omega)$ of the surface phonons and $\langle yy \rangle_\omega$:

$$E_s(\omega) = \mu_s \omega_s^2 \langle yy \rangle_\omega \tag{87}$$

where μ_s and ω_s are the reduced mass and vibrational frequency of the surface oscillator, respectively. The energy spectral density of the surface phonon vibrations can be determined by experimental measurements or by different theoretical methods. In this simple method Debye's model of the vibrational energy spectrum of the solid is used (Section I.B). To estimte the phonon effect on the vibrations of a molecule adsorbed on an oxide surface, these models are appropriate and often used in semiquantitative studies. Furthermore, the specific surface phonon distribution can be corrected by suitably modeling the values of the vibrational parameters of the solid. From the Debye distribution function [Eq. (39)] and Eq. (87), the spectral density of the surface oscillator $\langle yy\rangle_\omega$ as an explicit function of ω is found:

$$\langle yy\rangle_\omega = \frac{3\omega^2}{\mu_s\omega_s^2\omega_D^3}\frac{\hbar\omega}{\exp\left(\frac{\hbar\omega}{k_BT}\right)-1}H(\omega_D-\omega) \tag{88}$$

The final result for the spectral density $\langle xx\rangle_\omega$ for the adsorbed molecule is

$$\langle xx\rangle_\omega = \frac{C\omega^2\omega_a^2 b_1(\omega)H(\omega_D-\omega)}{(\omega_a^2-\omega^2)^2+\omega^6C^2b_2^2(\omega)H(\omega_D-\omega)} \tag{89}$$

where C is a constant:

$$C = \frac{\mu}{\mu_s}\frac{3\omega_i^4}{\omega_s^2\omega_D^3} \tag{90}$$

and $b_1(\omega)$ and $b_2(\omega)$ are the following functions of ω:

$$b_1(\omega) = \frac{\hbar\omega}{\exp\left(\frac{\hbar\omega}{k_BT}\right)-1},\ b_2(\omega) = \frac{1}{\exp\left(\frac{\hbar\omega}{k_BT}\right)+1} \tag{91}$$

In the classical case ($h'\omega \ll k_BT$) the last two functions reduce to constants $b_1 = k_BT$ and $b_1 = 1/2$. It should be oted that the dissipation term (the second one in the denominator) in Eq. (89) depends on the sixth power of ω, similar to the dissipation processes, associated with the emission of photons from particles. In the present case the adsorbed molecule dissipates energy through phonon emission into the solid. Their frequencies are limited to ω_D because higher frequency phonons cannot exist in the solid according to the model assumptions.

In Fig. 9 the spectral density $\langle xx\rangle_\omega$ of the adsorbed molecule bond vibrations for oxides with different Debye cut-off frequency $\omega_D = 100$–$400\ \mathrm{cm}^{-1}$ are presented. The curves have a maximum which shifts to the higher frequencies and the function value at the maximum becomes smaller with increasing ω_D.

The mean potential vibrational energy of the adsorbed molecule bond can be obtained by integration of the spectral density $\langle xx\rangle_\omega$ in the entire range of the frequencies ω:

$$U_0 = \frac{\mu\omega_a^2}{2\pi}\int_0^\infty \langle xx\rangle_\omega\,\mathrm{d}\omega \tag{92}$$

When the Debye model of the solid is applied, the integral in the above equation can be separated into two parts – for ω from 0 to ω_D, and between ω_D and ∞. In these regions the peculiarities of the Heaviside function can be used and the oscillator vibrational energy U_0 is found to be composed of two terms:

$$U_0 = \frac{1}{4}\hbar\omega_a\coth\left(\frac{\hbar\omega_a}{2k_BT}\right)H(\omega_a-\omega_D)+\frac{1}{2\pi}\int_0^{\omega_D}\frac{C\omega^2\omega_a^2b_1(\omega)}{(\omega_a^2-\omega^2)^2+\omega^6C^2b_2^2(\omega)}\mathrm{d}\omega \tag{93}$$

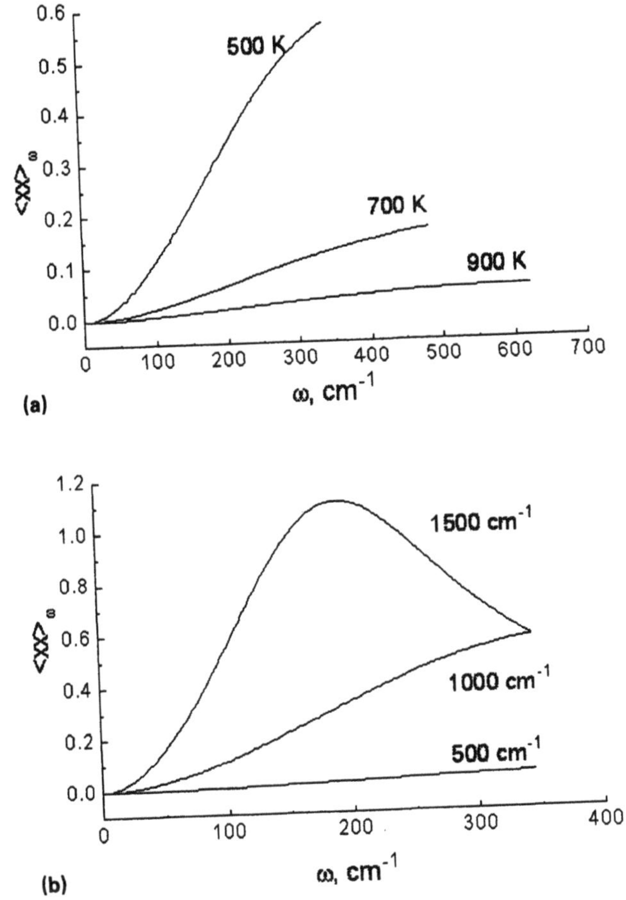

FIG. 9 Spectral density $\langle xx \rangle_\omega$ of the bond vibrations of molecules adsorbed on oxide surfaces with different Debye temperatures θ_D (a) and with different values of the interaction frequency ω_i (b).

The first term predominates in this expression when the vibrational frequency of the adsorbed molecule is higher than the Debye cut-off frequency of the oxide surface, which is the usual situation. This part in the vibrational energy is the contribution from the molecular vibrational mode itself with frequency ω_a. The second term accounts for the contribution from the low-frequency spectral region in the vibrational energy, corresponding to the vibrational energy transfer between the adsorbed molecule and the oxide surface. Thus, for each definite case it can be estimated how different parameters of the adsorption system change the energy of the adsorbed molecule.

The influence of the Debye cut-off frequency on the vibrational energy of the oscillator and interaction parameters ω_i is shown in Fig. 10. When bond vibrations interact strongly with surface phonons the oscillator energy is high.

However, the energy decreases with the rise of θ_D, i.e., catalysts having a higher value of the Debye cut-off frequency have less effect on the vibrational energy of the bond.

It should be mentioned that the magnitude of the vibrational energy of the oscillator generated by interaction with surface phonons is greater than the thermal vibration energy of the oscillator.

2. Interaction with Surface Phonons–Aharmonicity

In the adiabatic approximation, the Hamiltonian of the system **H** can be presented as a superposition of the lattice part $\mathbf{H}_l$, Hamiltonian $\mathbf{H}_p$, describing the adsorbed molecule vibrations in the mean potential of the surface phonons $\langle V(x, x_a)\rangle$ and the interaction Hamiltonian $\mathbf{H}_{int}$, defined by:

$$\mathbf{H}_l = \sum_{\lambda} \omega_\lambda b_\lambda^+ b_\lambda$$
$$\mathbf{H}_p = \left(\frac{p^2}{2m}\right) + \langle V(x, x_a)\rangle \tag{94}$$
$$\mathbf{H}_{int} = V(x, x_a) - \langle V(x, x_a)\rangle$$

where b_λ and b_λ^+ are the creation and annihilation operators of a phonon λ with corresponding frequency ω_λ, and x and x_a denote co-ordinates of the phonons and adsorbed molecules. The potential V is averaged over the phonon co-ordinates:

$$\langle V(x, x_a)\rangle = \mathrm{Sp}_l(\rho_l V(x, x_a)) \tag{95}$$

where

$$\rho_l = \exp(-\beta\mathbf{H}_l)Z_l^{-1}, \quad Z_l = \mathrm{Sp}[\exp(-\beta\mathbf{H}_l)]$$

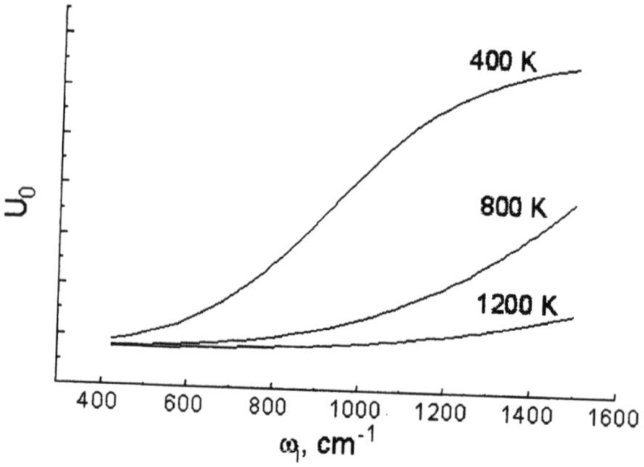

FIG. 10 Vibration energy of the bond in the adsorbed molecule versus interaction frequency ω_i for surfaces with Debye temperatures $\theta_D = 400$, 800, and 1200 K.

If the potential $\langle V(x, x_a)\rangle$ is weakly aharmonic, the Hamiltonian $\mathbf{H}_p$ can be considered in harmonic approximation (b and b^+ are the creation and annihilation operators of the adsorbed molecule vibrational mode):

$$\mathbf{H}_p = \omega_a b^+ b \tag{96}$$

and the interaction part, consisting only of the terms that are linear with respect to x:

$$\mathbf{H}_{int} = x_a V_h(x) \tag{97}$$

In this case the linear response theory gives the following expression for the polarization operator, determining the position and the form of the spectral band:

$$\Pi(\omega) = -i \int_0^{\infty} \exp(i\omega t) \cdot \langle [\langle 0|\mathbf{H}_{int}(t)|1\rangle\langle 1|\mathbf{H}_{int}(0)|0\rangle] \rangle dt \tag{98}$$

where 0 and 1 denote the ground and the first excited state of the molecule vibrations, and the time-dependent operator is introduced with

$$\mathbf{H}_{int}(t) = e^{i\mathbf{H}_1 t}\mathbf{H}_{int}e^{-i\mathbf{H}_1 t} \tag{99}$$

The real and the imaginary parts of the polarization operator $\Pi(_a)$ give the shift $\Delta\omega$ and the half-width Γ of the observed vibrational band. In this case the band width is

$$\Gamma = \operatorname{Im}\Pi(\omega_0) = \left(1 - \exp\left(\frac{-\hbar\omega_0}{k_B T}\right)\right) \int_{-\infty}^{\infty} e^{i\omega_0 t} \cdot \langle [\langle 0|\mathbf{H}_{int}(t)|1\rangle\langle 1|\mathbf{H}_{int}(0)|0\rangle] \rangle dt \tag{100}$$

Let us now consider the frustrated translational mode of an adsorbed molecule which creates oscillations of the whole molecule perpendicular to the surface, using the Morse potential for the interaction between the molecule and the oxide surface atoms. The core of the integral in Eq. (100) then transforms into

$$\begin{aligned}\langle [\langle 0|\mathbf{H}_{int}(t)|1\rangle\langle 1|\mathbf{H}_{int}(0)|0\rangle] \rangle = D_a^2\{C_1^2[\exp(4\alpha^2\langle x(t)x(0)\rangle) - 1] +\\ 4C_2^2[\exp(\alpha^2\langle x(t)x(0)\rangle) - 1] - 4C_1^2C_2^2[\exp(2\alpha^2\langle x(t)x(0)\rangle) - 1]\}\end{aligned} \tag{101}$$

where D_a is the adsorption energy, and the constants

$$\alpha^2 = \frac{m\omega_0^2}{2D_a}, \quad C_2 = \frac{D_a}{\omega_0}\sqrt{\frac{4D_a}{\omega_0} - 3}, \quad C_1 = C_2\left(2 - \frac{\omega_0}{2D_a}\right) \tag{102}$$

The autocorrelation function of the surface phonons $\langle x(t)x(0)\rangle$ in the harmonic approximation is

$$\langle x(t)x(0)\rangle = \frac{1}{2m_s}\int_0^{\omega_D} \frac{\rho_s(\omega)}{\omega}\left[\left(1 - e^{\frac{-\hbar\omega_0}{k_B T}}\right)^{-1} e^{-i\omega t} + \left(e^{\frac{\hbar\omega_0}{k_B T}} - 1\right)e^{i\omega t}\right] d\omega \tag{103}$$

Here, m_s is the reduced mass of the surface atoms and $\rho_s(\omega)$ is the surface phonon spectral density. The above expressions account for multiphonon processes of the second order of the interaction Hamiltonian and are not limited by a specific model of the phonon spectral density [26]. The contributions of the first or higher order phonon excitations to the band width changes can be derived from the integrals in Eq. (100) using a series expansion of $\langle x(t)x(0)\rangle$. When $\omega_a \gg \omega_D$, assuming Debye's distribution of the surface phonons, the band width is found:

$$\Gamma = \frac{2\pi D^2}{\sqrt{\omega_D \omega_a}} \exp\left[-\frac{\omega_a}{\omega_D}(\alpha - 1)\right] \tag{104}$$

where α is the solution of the equation:

$$\frac{\exp \alpha}{\alpha} = \left[1 - \exp\left(-\frac{\hbar\omega_D}{k_B T}\right)\right]\frac{m_s D}{3m\omega_a} \tag{105}$$

3. Contribution of Frustrated Degrees of Freedom

The aharmonicity is also important for the form of the vibrational spectral band of the frustrated translational degrees of freedom which determine the bond between adsorbed molecule and the surface atoms of the adsorbent. Langreth and Persson consider CO adsorbed through the carbon atom perpendicular to the surface atom M, assuming that the distance between atoms in the adsorbed molecule is constant (intramolecular vibrations are neglected) [29]. The Hamiltonian of the system which includes the adsorption bond and the phonon continuum of the solid can be presented as composed of an harmonic $\mathbf{H}_h$ and an aharmonic part – the interaction Hamiltonian $\mathbf{H}_{int}$. According to the authors, aharmonicity of the bond C–M is due to the interaction of carbon atom with the nearest neighbors of atom M from the solid surface. They propose two possibilities for modeling the functions characterising this aharmonicity: (1) quadratic functions of the distance l between the carbon atom and the surface atoms (minimal anharmonicity), and (2) Morse potential. In both approaches experimental data are used for estimation of the constant values in the expressions for the interaction potential.

The harmonic term $\mathbf{H}_h$ is also separated into two components. The first one is responsible for the stretching vibrations of the C–M bond and phonon vibrations perpendicular to the surface which are coupled with it. The second is for bending modes of the bond towards the surface and the modes of the adsorbent parallel to the surface. On the basis of this Hamiltonian method the dipole–dipole correlation function, which is connected with the response of the system to the field created by infrared light (IR) radiation, can be obtained. The spectral function of perpendicular vibrations $\rho^{\perp}(\omega)$ is of the following form:

$$\rho^{\perp}(\omega) = \frac{\omega^4 \rho_s^{\perp}(\omega)}{\{\omega^2 - \omega_s^2[1 + J_1(\omega)]\}^2 + [\omega_s^2 J_2(\omega)]^2} \tag{106}$$

where $\rho_s^{\perp}(\omega)$ is the spectral function for vibrations of the pure surface of the solid, ω_s is the vibrational frequency of the bond C–M for the static solid surface, and functions $J_1(\omega)$ and $J_2(\omega)$ represent the real and imaginary parts of the integral:

$$J(\omega) = \mu_{co}\omega^2 \int_0^{\infty} \frac{2\omega}{\omega^2 - \omega_s^2} \rho_s^{\perp}(\omega) \frac{d\omega_s}{2\pi} \tag{107}$$

From the equations for parallel vibrations the spectral function $\rho^{\parallel}(\omega)$ is derived:

$$\rho^{\parallel}(\omega) = \frac{\gamma_0}{(\omega - \omega_s)^2 + \left(\frac{\gamma_0}{2}\right)^2} \frac{z}{2\mu_{CM}\omega_s} \tag{108}$$

with parameter γ_0, connected with the damping in the system and force constant z.

In the limit of weak flucutations (both thermal and quantum), the band maximum shift and the band half width are found:

$$\Delta\omega = \lambda_2 + \frac{2\lambda\gamma_0^2}{\gamma_0^2 + \lambda^2}\left[\exp\left(\frac{\hbar\omega_0}{k_B T}\right) - 1\right]^{-1}$$

$$\Gamma = \frac{2\lambda^2\gamma_0}{\gamma_0^2 + \lambda^2}\left[\exp\left(\frac{\hbar\omega_0}{k_B T}\right) - 1\right]^{-1} \tag{109}$$

where λ is the coupling constant.

Another problem connected with the frustrated modes at the surface is their influence on the intramolecular vibrations of the adsorbed molecule. Nitzan and Persson [30] suppose that the high-frequency intramolecular vibrations of adsorbed molecules interact with the phonons of the solid via low-frequency oscillations of the adsorbent–adsorbate bond, the latter being regarded as frustrated translational and rotational degrees of freedom of the adsorbed molecule. In the quantum case the Hamiltonian of the system can be expressed as

$$\mathbf{H} = \omega_A a^+ a + \omega_B b^+ b + \lambda_a a^+ a b^+ b + \sum_k \omega_k c_k^+ c_k + \sum_k (\mathbf{V}_k c_k^+ b + \mathbf{V}_k^* c_k b^+) \tag{110}$$

where a^+, a; b^+, b, and c^-, c are operators of creation and annihilation for the high-frequency (**A**) and low-frequency (**B**) modes of the adsorbate. The modes of the thermal bath (surface phonons), and ω_A, ω_B, and ω_k are the frequencies of the corresponding oscillations. Aharmonicity is taken into account in the coupling of modes **A** and **B**, characterized by the parameter λ_a as well as in the coupling of the mode **B** with the adsorbent's phonons, described by operators $\mathbf{V}_k$. The evolution of the system is governed by the equations of motion:

$$\begin{aligned} \dot{a} &= -i\omega_A a - i\lambda_a a b^+ b \\ \dot{b} &= -i\omega_B b - i\lambda_a a^+ a b - i\sum_k \mathbf{V}_k^* c_k \\ \dot{c} &= -i\omega_k c_k - i\mathbf{V}_k b \end{aligned} \tag{111}$$

After simplifications and estimation of some functions involved in the solution of Eqs (111), one finds that the shape of the band of intramolecular oscillations is Lorentzian:

$$\rho(\omega) \approx \frac{\frac{1}{2}\Gamma}{[\omega - (\omega_A + \Delta\omega^2)] + (1/2\Gamma)^2} \tag{112}$$

with the band shift $\Delta\omega$ and half-width Γ equal to

$$\Delta\omega = \lambda_a \left[\exp\left(\frac{\hbar\omega_B}{k_B T}\right) - 1\right]^{-1}$$

$$\Gamma = \frac{2\lambda_a^2}{\lambda_{ph}} \exp\left(\frac{\hbar\omega_B}{k_B T}\right)\left[\exp\left(\frac{\hbar\omega_B}{k_B T}\right) - 1\right]^{-2} \tag{113}$$

where λ_{ph} characterizes the coupling between low-frequency frustrated mode **B** and the adsorbent's phonons. The value of this parameter depends on the type of the coupling operators $\mathbf{V}_k$ and the surface phonon distribution. Due to the simplifications applied, the above result is valid for all temperatures but only in the high friction limit wherein vibrational energy is transferred from the adsorbed molecule into the phonons ($\lambda_{ph} \gg \lambda_a$).

At high temperatures, quantum effects can be neglected and a system of classical equations of motion for description of these vibrational interactions is valid, independent of the high friction limit:

$$\mu_A \ddot{x}A + \omega_A^2 x_A + \lambda_a' x_B^2 x_A = 0$$

$$\mu_B \ddot{x}_B + \omega_B^2 + \lambda_a' \frac{\mu_A}{\mu_B} x_A^2 x_B + \eta \dot{x}_B = F(t) \tag{114}$$

Here, $x_A(t)$ and $x_B(t)$ are the deviations of the length of oscillations **A** and **B** from the equilibrium value at the moment t, μ_A, and μ_B are the reduced masses of these oscillators. The parameter λ_a' in the above equations is proportional to λ_a [from Eq. (110)]:

$$\lambda_a' = \frac{2\mu_B\omega_B\omega_A}{\hbar}\lambda_a$$

The random force $F(t)$ is related to the friction coefficient η by the fluctuation–dissipation theorem, Eqs (23) and (25). On the basis of the assumption $\omega_A \gg \omega_B$, the authors [30] substitute x_A^2 in the second equation (114) in the system with the average value:

$$\langle x_A^2 \rangle = \frac{k_B T}{\mu_A \omega_A^2} \tag{115}$$

and obtain an expression for the spectral density of the autocorrelation function of the deviation of oscillator **B** from the equilibrium value:

$$\langle \tilde{x}_B(\omega)\tilde{x}_B(\omega') \rangle = \frac{2\pi\eta k_B T}{\mu_B\left[\left(\bar{\omega}_B^2 - \omega^2\right)^2 + \omega^2\eta^2\right]}\delta(\omega + \omega') \tag{116}$$

where $\tilde{x}(\omega)$ designates the Fourier transform of the deviation $x(t)$, and $\bar{\omega}_b$ the expression:

$$\bar{\omega}_B = \omega_B + \lambda_a' \frac{\mu_A}{\mu_B}\langle x_A^2 \rangle$$

Considering x_B as a stochastic term, the authors solve the other equation in the system (114) for $x_A(t)$:

$$x_A(t) = x_A(0) \exp\left[-i\omega_A t - i\frac{\lambda_a'}{2\omega_A}\int_0^t \partial\tau [x_B(\tau)]^2 \right] \tag{117}$$

After simplification of the right-hand side of the expression (117) the values of parameters $\Delta\omega$ and Γ of the Lorentzian curve, Eq. (113), corresponding to the oscillations of the mode **A**, are estimated for some boundary cases, i.e., long times for, $\eta \ll \omega_B$ or $\eta \gg \omega_B$. Although the initial representation of the Hamiltonian of the system, consisting of two oscillators and the surface vibrations of the solid (in the quantum case), include the coupling between phonons and low-frequency modes **B** in general form, in the final result this coupling is represented only by the constants λ. In the classical consideration the influence of the surface vibrations is entirely neglected in the initial equations of motion.

C. Interaction Between Adsorbed Molecules

The alteration of the position and the shape of the vibrational spectral bands of adsorbed species with coverage is a well known phenomenon. The most simple effect of the neighboring adsorbed molecule is the steric hindrance which influences mainly the frustrated degrees of freedom, as shown in figure 7 (section III.A for NH_3 adsorption on MgO). For higher coverage this hindrance leads to a change of the adsorption geometry of the molecules. Another structural factor affecting the vibrational properties of the adsorbed molecule is changes in the oxide surface structure due to the presence of other adsorbed species. The role of structural changes becomes more important at coverage near or above one monolayer when the vapor molecules weakly adsorbed in the second layer strongly influence the observed vibrational spectra.

Coupling among the vibrating dipoles of adsorbed molecules influences the dynamic contributions of lateral interactions. Each adsorbed molecule experiences a dynamic electric field composed of the incident optical electric field and the field of neighboring oscillating dipoles. If the surrounding molcules have different vibrational frequencies, this contribution will be zero. For example, when the difference in vibrational frequency is due to adsorption of different molecules (even due to isotope changes of one atom), the peaks for both frequencies appear separately. In the case of adsorption of the same molecules on an heterogeneous surface, a slight difference between their vibrational frequencies will be observed. These small frequency differences lead to inhomogeneous broadening of the band but not a systematic shift. Coupling among the vibrational modes of molecules in identical adsorption states generates dispersion of the vibrations, i.e., dependence of the observed vibrational frequency on the wave vector **q**. For a simple quadratic framework of adsorbed molecules, the dispersion expression for the frequency ω is

$$\omega^2(\mathbf{q}_{\parallel}) = (\omega_0 + \Delta\omega)^2 + 4\beta\omega_0(\cos q_x a + \cos q_y a) \tag{118}$$

This dispersion causes a rise in the single IR-active mode with the frequency shifted to higher wavenumbers and the magnitude of the shift increases with coverage.

The static contribution to lateral interactions among adsorbed molecules includes two possible mechanisms: (1) direct interaction due to the electrostatic interactions of the static dipoles, and (2) indirect interaction via the oxide surface. If two neighboring molecules are adsorbed at positions $\mathbf{r}_0$ and $\mathbf{r}_1$ at the surface with the interaction energy between them $U_{\text{int}}(\mathbf{r}_0, \mathbf{r}_1)$, the frequency shift $\Delta\omega$ of the local vibrations of one of the molecules in the field of the second is proportional to

$$\Delta\omega \approx x_0^2 \left(\frac{\partial^2 U_{\text{int}}}{\partial x_0^2} \right)_{x_0 = x_1 = 0} \tag{119}$$

where x_0 and x_1 are vibrational co-ordinates of the two molecules. In addition to the shift, the frequency splits into two peaks with the difference between the maxima β equal to

$$\beta \approx x_0^2 \left(\frac{\partial^2 U_{\text{int}}}{\partial x_0 \partial x_1} \right)_{x_0 = x_1 = 0} \tag{120}$$

The above equations account for the direct adiabatic contribution to the vibrational interactions. The electrostatic interaction energy depends on the dipole moments of both molecules μ and the distance $r = |r_0 - r_1|$:

$$U_{\text{int}} = 2\mu(x_0)\mu(x_1)r^{-3} \tag{121}$$

In this case the frequency shift $\Delta\omega_\perp$ and split parameter $\beta_\perp$ for vibration perpendicular to the surface are

$$\Delta\omega_\perp \approx 2x_0^2 \frac{\partial^2 \mu'(x)}{\partial x^2} U_{\text{int}}$$
$$\beta_\perp \approx \left(\frac{\mu'}{\mu} \right)^2 U_{\text{int}} \tag{122}$$

where μ' is the dynamic dipole moment. Both parameters for the vibrations parallel to the surface are proportional to $(x_0/r)^2 U_{\text{int}}$. As expected, the magnitude of the electrostatic interaction strongly depends on the distance between two adsorbed molecules, which is connected with the structure of the oxide surface (distance between two neighboring adsorption position) and coverage.

The indirect interaction is determined mainly by the bonding of the molecules on the surface. An estimate of the magnitude of this effect is the asymptotic expression for the interaction energy $U_{\text{int}}(\mathbf{r}_0, \mathbf{r}_1)$:

$$U_{\text{int}} \approx B(x_0)B(x_1) \frac{\cos(2k_F r)}{r^5} \tag{123}$$

where function $B(x)$ is determined by the nature of the surface bonding. If the Fermi level crosses or touches the surface state zone (the oxide behaves as conductor), the distance dependence of the interaction energy transforms into

$$U_{\text{int}} \approx B(x_0)B(x_1) \frac{\cos(2k_F r)}{r^2} \tag{124}$$

That is, the indirect interaction, Eq. (121) decreases more slowly with r than does the static dipole–dipole interaction, Eq. (124).

Let us consider in more detail the contributions of the static and dynamic interactions between adsorbed CO molecules on NiO [31]. As seen from the coverage evolution of the CO valent vibrational mode (Fig. 11), the band at low coverage is at about 2152 cm^{-1} and continuously shifts and narrows to a sharp peak at 2136 cm^{-1} for maximum coverage (CO/surface Ni^{2+} ions = 0.5). The latter peak corresponds to CO weakly adsorbed with the carbon atom bonded to cationic nickel centers. The form of the spectral band implies that at higher coverage CO is adsorbed on identical sites [on NiO(100) surface in the case considered] and the adsorption layer is highly ordered. For smaller amounts of adsorbed CO the adsorption layer is mostly inhomogeneous, which causes asymmetric broadening of the band due both to the differences in: (1) the local Ni(100) surface structure, and (2) the heterogeneous surroundings resulting from other adsorbed moelcules. For this reason, at low coverage, CO neighboring molecules oscillate with different frequencies and dynamic interactions are not observed, i.e., the spectral band at about 2152 cm^{-1} simply results from the vibration of isolated CO molecules on the surface. Also, in the limit of zero coverage, the influence of different surface defects becomes more important in increasing the band half-width.

In order to distinguish between the static and dynamic contributions to adsorbate–adsorbate lateral interactions at maximum coverage, experiments with diluted isotopic mixtures of ^{12}CO in ^{13}CO were performed [31]. At low surface coverage, where $\theta \rightarrow 0$, the ^{12}CO stretching peak occurs at 2152 cm^{-1} in Scheme 6, independent of whether the ^{12}CO:^{13}CO ratio is 99:1 or 12:88. At infinite surface dilution, neighboring CO molecules are too far apart to incur either static or dynamic inter-

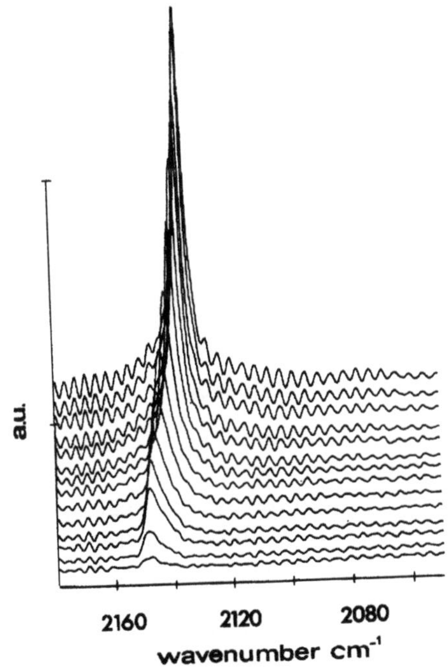

FIG. 11 Infrared spectra of decreasing amounts of carbon monoxide on nickel oxide in the C–O stretching vibrations region. (From Ref. 31; courtesy of Elsevier Science B.V.)

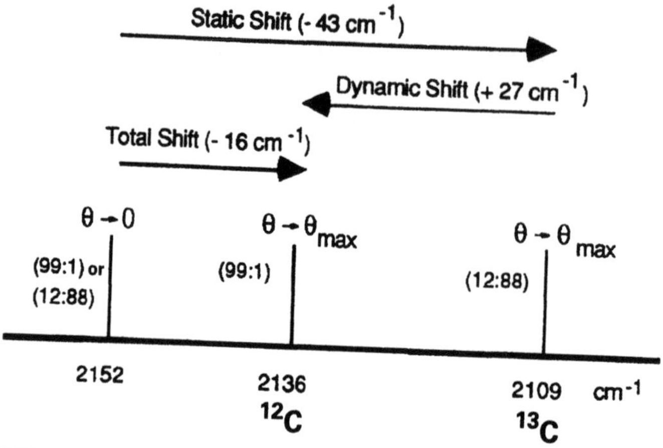

SCHEME 6 Illustration of the static and dynamic shifts of CO (^{12}C:^{13}C) valent vibrational mode due to adsorbate–adsorbate interactions on NiO surface. (From Ref. 31; courtesy of Elsevier Science B.V.)

actions. Hence, for $\theta \to 0$ the ^{12}CO stretching peak found at $2152\,cm^{-1}$ is independent of the ^{13}CO surface concentration and therefore of both static and dynamic interaction.

As the concentration of CO approaches complete coverage, or $\theta \to \theta_{max}$, the stretching peak for ^{12}CO can be seen to be dependent on the ^{13}CO surface concentration in Scheme 6. Since ^{12}CO and ^{13}CO molecules have different vibrational frequencies, the dynamic effect of the ^{13}CO molecules on the shift of the ^{12}CO frequency will be absent. Hence, for the ratio 12:88 (where ^{13}CO molecules predominate over ^{12}CO molecules), the peak shift for ^{12}CO from $2152\,cm^{-1}$ (where ^{12}CO molecules are too dilute to interact with their neighbors) to $2109\,cm^{-1}$ is essentially the result of static interaction of ^{13}CO molecules with the ^{12}CO molecules. On the other hand, when the ^{12}CO molecule is surrounded by molecules with the same mass and vibrational frequency (ratio 99:1) both static and dynamic interaction occurs and the peak shifts from 2152 cm to $2136\,cm^{-1}$. In this way the static and total frequency shifts are determined from experiment to be −43 and $-16\,cm^{-1}$, respectively, so that the dynamic shift can be calculated as $27\,cm^{-1}$ (Scheme 6).

D. Influence of Vibrational Interactions on Surface Diffusion

Diffusion of adsorbed particles on the surface of the adsorbent is one of a number of processes occurring on solid surfaces [4, 32–40] including heterogeneous catalytic reactions. In general, diffusion represents the consecutive processes of breaking old bonds and forming new bonds between the adsorbed particle and the surface atoms of the adsorbent. Diffusion may be influenced by the vibrational interactions of these bonds with the surface oscillations of the solid [4, 35].

According to some qualitative estimates of the influence of the surface vibrations on the diffusion of the adsorbate [37], the surface phonons slightly change the rate of surface diffusion in comparison with the steady surface (when the surface atoms are considered as fixed at their positions). However, a more detailed description of the interactions during migration of adsorbed atoms or molecules on the surface (see following section) showed that the surface diffusion rate is affected by the motion of the surface atoms.

1. Potential Energy Surface Approach

For light atoms adsorbed on the surface, surface atomic motion accelerates surface diffusion due to an increase in the tunneling probability caused by approach of the nearest-neighbor surface atoms to the adsorbed atom [32]. The potential energy surface for an atom adsorbed on a solid surface, at a distance r measured perpendicularily from the surface and a distance q away from its equilibrium position, is represented as a sum:

$$\mathbf{V}(r, q) = \mathbf{V}_0(r) + \mathbf{V}_\mathrm{h}(q) + \mathbf{V}_\mathrm{C}(r, q) \tag{125}$$

of the potential of the interaction between the adsorbed particle and surface atoms in their equilibrium position $\mathbf{V}_0(r)$, the harmonic potential of surface oscillations $\mathbf{V}_\mathrm{h}(q)$, and the correction potential $\mathbf{V}_\mathrm{C}(r, q)$, characterizing the changes in $\mathbf{V}(r, q)$, when the atoms from the adsorbent surface are not at their equilibrium position. The interaction potential for the static surface $\mathbf{V}_0(r)$ is modeled as a set of two-atomic Morse potentials with parameters estimated by the extended Hückel method. In a simple model, the correction term $\mathbf{V}_\mathrm{C}(r, q)$ can be represented as a linear function of the deviation of surface atoms q based on the phonon mode of the solid. Using the model of the potential energy surface described above, Jaquet and Miller found that, with phonon-assisted tunneling the surface diffusion of hydrogen atoms is two to five times that predicted by transition state theory (TST) [32]. Another important finding was that the rate constant increases rapidly with the decrease in frequency of the chosen phonon mode.

Quantum effects can also be included artificially by a tunneling correction in the expression for the rate constant of the process, proportional to the diffusion coefficient D [33]. In this way, the diffusion coefficient is obtained as a modification of the Arrhenius temperature dependence of the rate constant:

$$D = D_0 \exp(-E_\mathrm{a}/k_\mathrm{B}T + \Delta/k_\mathrm{B}T^2) \tag{126}$$

where D_0 and E_a are the classical pre-exponential term and the activation energy in the Arrhenius equation, respectively, and Δ is a parameter for quantum correction. Monte Carlo numerical calculations for diffusion of hydrogen isotopes, carried out on the basis of classical TST with the above additions show that taking into account the adsorbate surface atom oscillations leads to a several fold increase in the migration rate of adsorbed atoms on the surface. Similarly to many other studies [4, 34–37], the calculations by this model led to the conclusion that tunneling of adsorbed hydrogen atoms, assisted by surface phonons, is the main mechanism of surface diffusion at low temperatures.

2. Stochastic Approach

The vibrational influence on the rateof surface diffusion can be considered in terms of Kramers' theory for the rate of a chemical reaction (Section I.A.3). As mentioned earlier, in this theory the form of friction coefficient plays an important role in characterizing the energy dissipation from the dissociating bond to the surroundings. Zhdanov proposed a form of this coefficient for the motion of a particle near the dissociation threshold of the Morse potential; the friction coefficient at the dissociation threshold is practically the same for any potential including the harmonic one [38]. A significant peculiarity of Kramers' theory is the consideration of only the dissipative effects of the surroundings, which always leads to lower values of the rate constant than those obtained by the TST. In contrast to it, quantum effects, connected with the surface vibrations of the adsorbent and included in the rate constant by the above-mentioned methods, accelerate the process more than 10 times in some cases. A reasonable description of phonon influence on the rate of surface diffusion of adsorbates should take into account both effects.

Tsekov and Ruckenstein considered the dynamics of a mechanical subsystem interacting with crystalline and amorphous solids [39, 40]. Newton's equations of motion were transformed into a set of generalized Langevin equations governing the stochastic evolution of the atomic co-ordinates of the subsystem. They found an explicit expression for the memory function accounting for both the static subsystem–solid interaction and the dynamics of the thermal vibrations of the solid atoms. In the particular case of a subsystem consisting of a single particle, an expression for the friction tensor was derived in terms of the static interaction potential and Debye cut-off frequency of the solid.

E. Vibrational Contribution to Bond Breaking in the Adsorbed Molecule

1. Compensation Effect Formalism

The formalism of the compensation effect [41], defined as the dependence of the pre-exponential term k_0 on the activation energy E_a in the Arrhenius equation (16):

$$\ln k_0 = \frac{E_a}{R \cdot \theta} + \ln Z, \text{ i.e., } \ln k = \ln Z + \frac{E_a}{R}\left(\frac{1}{\theta} - \frac{1}{T}\right) \tag{127}$$

can be used to determine the influence of surface oscillations on the rate of heterogeneous catalytic processes. In the above equations θ is the isocatalytic temperature, which is characteristic for a specified set of solid surfaces. As one can see from Eq. (127) θ is the temperature at which the reaction rate is independent of the activation energy. For surface reactions the compensation effect results from the vibrational energy exchange in the adsorption complex [42]. Let us consider the classical equations of motion for a damped oscillator with frequency ω_0, interacting with an external oscillator with frequency ω_e. The energy transfer per unit time (power) in this system is

$$\Delta P = \Delta P_{\text{res}} \frac{\omega_e^2/\tau^2}{(\omega_0^2 - \omega_e^2])^2 + \omega_e^2/\tau^2]} \tag{128}$$

where τ is the relaxation time, defined as $\tau = \mu/\eta$, and ΔP_{res} is energy transfer at resonance. Equations can be written for the energy transfer efficiency:

$$S_0 = \Delta P/\Delta P_{\text{res}}$$

and the quality factor:

$$Q = \tau.\omega_0$$

for the energy transfer from the external oscillator (the surface) to the reaction (adsorbed) molecule. The total efficiency accounting for different orientations and positions of the molecule with respect to the surface can be obtained:

$$S = \int_{1/2}^{\infty} S_0 \mathrm{d}Q = \omega_0 \frac{\omega_e}{\omega_0^2 - \omega_e^2}\left[\pm\frac{\pi}{2} - \arctan\frac{\omega_0\omega_e}{2(\omega_0^2 - \omega_e^2)}\right] \tag{129}$$

The total energy transfer efficiency can be substituted into the rate constant equation, giving

$$\ln k = \ln Z + \sum_i (\omega_0)_i \frac{\omega_e}{\omega_0^2 - \omega_e^2}\left[\pm\frac{\pi}{2} - \arctan\frac{\omega_0\omega_e}{2(\omega_0^2 - \omega_e^2)}\right] - \frac{E_a}{RT} \tag{130}$$

Here, Z is interpreted as a collision number and the summation in the second term is over all vibrational states of the reacting oscillator up to the frequency corresponding to the energy barrier of the reaction ΔU. Performing the summation over these vibrational energies $E_i = hc(\omega_0)i$, Eq. (130) transforms into

$$\ln k = \ln Z + \frac{\omega_e}{\omega_0^2 - \omega_e^2}\left[\pm\frac{\pi}{2} - \arctan\frac{\omega_0\omega_e}{2(\omega_0^2 - \omega_e^2)}\right]\frac{\Delta U}{hc} - \frac{E_a}{RT} \tag{131}$$

since $E_a = N_A \Delta u$, an expression for the isocatalytic temperature is found:

$$\theta = \frac{hcN_A}{R}\frac{\omega_0^2 - \omega_e^2}{\omega_e}\left[\pm\frac{\pi}{2} - \arctan\frac{\omega_0\omega_e}{2(\omega_0^2 - \omega_e^2)}\right]^{-1} \tag{132}$$

This treatment shows that the isocatalytic temperature and the magnitude of the compensation effect strongly depend on the difference between the two vibrational frequencies. The form of the above function is presented in Fig. 12. A simple expression for θ can be derived using a series expansion around the resonsance, i.e., when, in the oxide, vibrational spectra present a frequency near the frequency of the adsorbed gas molcule:

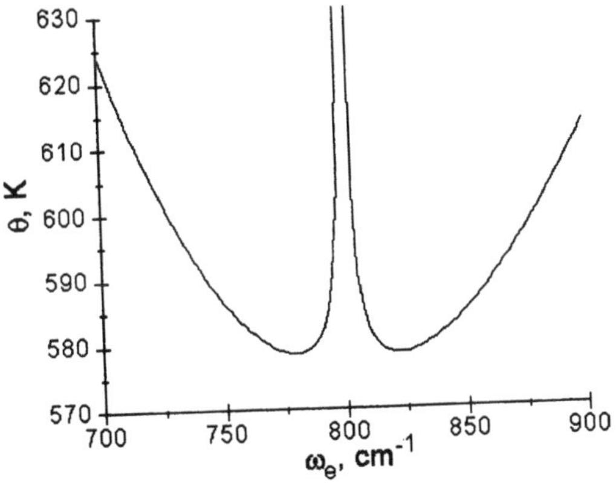

FIG. 12 Isocatalytic temperature θ versus frequency of the surface vibrations ω_e for $\omega_0 = 800\,\mathrm{cm}^{-1}$ according to Eq. (132).

$$\theta_{\rm res} = \frac{hcN_{\rm A}}{R}\frac{\omega_0}{2} \tag{133}$$

Larsson tabulated experimentally determined values of the isocatalytic temperature for specific catalytic systems from measurements on mixed oxide catalysts [42, 43].

2. Generalized Langevin Equation Formalism

A more accurate method for modeling the effect of the vibrational interactions in the adsorption complex on the rate of surface reactions is based on the formalism of the *generalized Langevin equation*. The interaction between intramolecular vibrations on the adsorbed molecule $x(t)$ and the surface phonons of the adsorbent $y(t)$ is considered in harmonic approximation as described in Section III.B.1. As seen from the general considerations in Section I, the stochastic processes, due to interactions of the system with the surroundings, affect both pre-exponential and exponential terms of the reaction rate constant. The change in the pre-exponential term accounts mainly for the energy dissipation from the dissociating bond around the energy barrier, while the influence via the exponential term changes the effective activation energy of the reaction. Both effects are considered below for dissociation of a diatomic molecule, adsorbed on an oxide surface.

(a) Pre-exponential Term. The influence of vibrational energy exchange on the pre-exponential term of the rate constant is investigated by the Grote–Hynes theory [11, 12] (Section I.A.3) for barrier crossing reactions. This effect, generated by friction in the system, leads to a decrease in the rate constant as compared to the result of the TST. In order to find the memory-renormalized transmission frequency ξ for a reaction, we can use the expression for the memory friction $\gamma(\omega)$, obtained with the harmonic approximation, Eqs (85) and (88). For the sake of simplicity, the expression for the classical limit is taken:

$$\gamma(\omega) = \mu C\omega^2 H(\omega_{\rm D} - \omega) \tag{134}$$

The temprature dependence of the friction can be obtained by reverse Fourier transform of Eq. (134), which after the Laplace transform gives

$$\hat{\gamma}(\xi) = \frac{\mu C}{2}\xi\left[\omega_{\rm D} - \xi\cdot\arctan\left(\frac{\omega_{\rm D}}{\xi}\right)\right] \tag{135}$$

where constant C is defined in Eq. (90). Now, using Eq. (28) we found the equation for the memory-renormalized transmission frequency ξ for Debye's distribution of surface phonons:

$$\xi^2 = \frac{\omega_{\rm b}^2}{1 + \frac{C}{2}\left[\omega_{\rm D} - \xi\cdot\arctan\left(\frac{\omega_{\rm D}}{\xi}\right)\right]} \tag{136}$$

This equation can be solved self-consistently for a given set of parameters, and the dissipation-corrected pre-exponential factor A, defined in Eqs (20) and (25) can be found. Calculations for the influence of the interaction frequency ω_i on the pre-exponential term A show that in the region of weak interactions A is equal to the value without friction $A_{\rm TST}$. When ω_i becomes greater, the value of A decreases abruptly by about 10-fold for all curves [44, 45]. This means that the dissociation rate constant is about 10 times lower due to the memory friction between bond vibrations and phonons.

The term A depends stepwise on the curvature of the barrier top ω_b and for very sharp barriers A is almost equal to A_{TST}. The position of the barrier step moves to the lower values of ω_i, Eq. (81), for weaker interactions between bond and solid vibrations. Another conclusion is that catalysts with Debye cut-off freqency ω_D greater than a threshold value (different for different values of ω_b) do not affect the pre-exponential term A while for ω_D below this limit A decreases considerably. This sharp transition takes place at higher ω when the energy barrier of the reaction has a flat top.

(b) Exponential Term. The energy dissipation effects on the term A, discussed above, delay the dissociation of the adsorbed molecule through the influence of friction (dissipation of vibrational energy from the dissociating bond into the solid). However, the fluctuations of the bond oscillations generated by energy transfer from surface phonons into the molecule act in the opposite direction, speeding up the reaction. This effect is taken into account by a change of the exponential term in the Arrhenius equation. As shown in Section I.A.2, the derivation of the expression (15) for the reaction rate constant using Fokker–Planck equation formalism allows consideration of the stochastic behavior of the system to some extent. The effects of the interaction with the surroundings are included in the mean equilibrium potential energy of the reacting molecule $\langle U_0 \rangle_e$, which participates in the denominator of the exponential term in the modified Arrhenius equation [45, 46]. An application of this equation to the description of the vibrational interctions on solid surfaces requries a knowledge of the potential energy of the adsorbed molecule, described in Section III.B.1.

Two different measures for the effective rise of the mean equilibrium potential energy $\langle U_0 \rangle_e$ toward $k_B T/2$ will be examined. The correction parameter α is introduced by the equation:

$$\ln k/k_0 = -\Delta U/\alpha k_B T \quad \text{and} \quad \alpha = 2\langle U_0 \rangle_e / k_B T \tag{137}$$

This parameter is convenient for estimation of the changes in the temperature dependence of the rate constant for dissociation since the equilibrium potential energy of the adsorbed molecule does not depend linearly on the temperature. Another useful parameter is the effective temperature of the vibrating bond T_{eff} defined by

$$T_{eff} = 2\langle U_0 \rangle_e / k_B \tag{138}$$

which should directly replace the temperature in the modified Arrhenius equation. As seen by comparing the two methods, the correction parameter α gives the ratio between the effective and the real temperature.

The temperature dependence of parameters α and T_{eff} is shown in Fig. 13. It can be seen that α will be greater than unity for all Debye temperatures of the model oxide surface, i.e., the phonon–vibration interaction will accelerate the dissociation of the bond according to Eq. (137). Also, from Fig. 13a, it can be seen that higher values of the Debye temperature decrease the rate of the reaction.

At low temperatures the dependence of T_{eff} on T is quite complicated, as can be seen from the following expression obtained from Eqs (138) and (93):

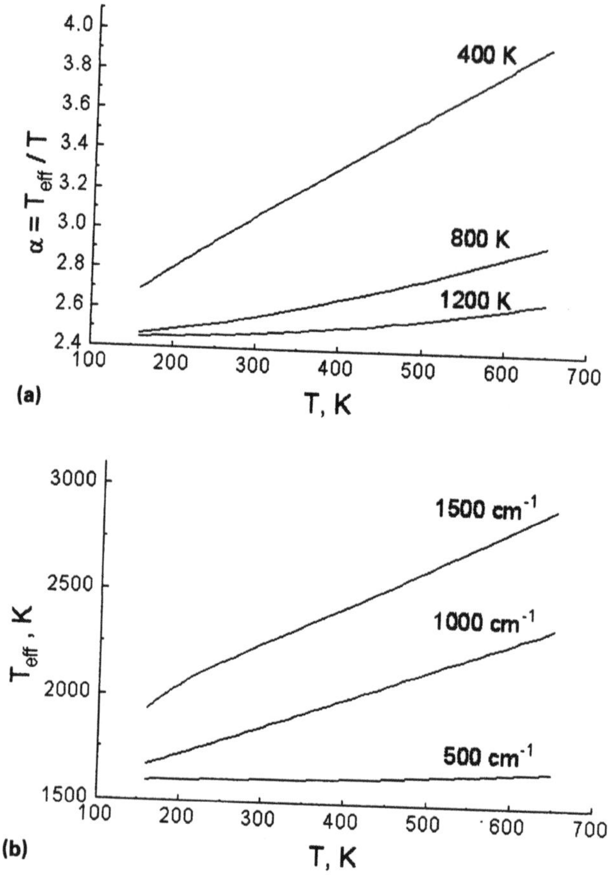

FIG. 13 (a) Temperature dependence of the correction parameter α for oxides with different Debye temperatures θ_D. (b) Temperature dependence of the effective temperature T_{eff} for different values of the interaction frequencies ω_i.

$$T_{eff} = \frac{1}{2}\frac{\hbar\omega_a}{k_B}\coth\left(\frac{\hbar\omega_a}{2k_BT}\right)H(\omega_a - \omega_D) + \frac{1}{\pi k_B}\int_0^{\omega_D}\frac{C\omega^2\omega_a^2 b_1(\omega)}{\left(\omega_a^2 - \omega^2\right)^2 + \omega^6 C^2 b_2^2(\omega)}\,d\omega \tag{139}$$

The constant C and functions b_1 and b_2 were defined in Eqs. (90) and (91). However, at higher temperatures ($k_BT \gg \hbar\omega_D; \hbar\omega_a$), T_{eff} increases linearly with T (Fig. 13b), which corresponds to the simplified classical form of Eq. (139):

$$T_{eff} = T + \frac{4}{\pi}\int_0^{\omega_D}\frac{C\omega^2\omega_a^2 T}{4\left(\omega_a^2 - \omega^2\right)^2 + \omega^6 C^2}\,d\omega \tag{140}$$

The influence of the characteristic interaction frequency ω_i (which is present in the constant C to fourth order) on the effective temperature is much more complicated. However, the calculations, presented in Fig. 13b, show that ω_i determines the slope of the temperature dependence of T_{eff}. For low values of ω_i, the effective temperature is almost independent of T, i.e., the temperature dependence of the dissociation rate constatn is the same as in the Arrhenius equation but with a much lower activation energy, E_a.

IV. VIBRATIONALLY ASSISTED DESORPTION

The desorption of adsorbed molecules from the solid surface is a process of adsorption bond breaking where the vibrational interactions are substantial [3, 5, 47–49]. Quantum stochastic description of the desorption could be based on the stochastic Pauli equation for vibrations of the bond between adsorbed particle and the surface of the adsorbent [3]:

$$\frac{\partial P_n}{\partial t} = -\sum_m W_{n\to m} P_n(t) + \sum_m W_{m\to n} P_m(t) - \int W_{n\to\varepsilon} P_n(t) \partial\varepsilon \tag{141}$$

In this equation, $P_n(t)$ is the probability that the oscillator will be in state n at the moment t, $W_{n\to m}$ is the rate of transfer between two states of the oscillator (from n to m) and ε is the designation for the continuum of states (corresponding to energy greater than the dissociation one), in which the bond is considered broken. Using the method for the average time of the first transition, the authors find an expression for the average time of desorption as a function of the initial state distribution of the oscillator $P_n(0)$:

$$\langle\tau\rangle = \sum_n \sum_m (\mathbf{W}^{-1})_{mn} P_m(0) \tag{142}$$

where $\mathbf{W}$ is a matrix, composed of the rates of transfer between the states of the oscillator $W_{n\to m}$. These rates are determined by the Hamiltonian of the system composed of the Hamiltonian of the lattice $\mathbf{H}_l$ and the Hamiltonian of the adsorbed particle $\mathbf{H}_p$, which takes into account the average interaction between the particle and the surface:

$$\mathbf{H}_p = \mathbf{K}_p + \langle \mathbf{V}(r, q) \rangle \tag{143}$$

and the Hamiltonian connected with fluctuation interactions $\mathbf{H}_i$

$$\mathbf{H}_i = \mathbf{V}(r, q) - \langle \mathbf{V}(r, q) \rangle \tag{144}$$

where the potential $\mathbf{V}(r, q)$ is modeled as a Morse function, and the angular brackets indicate thermal averaging. In order to make the model practically applicable for real systems, some simplifications are introduced – the process of vibrational energy exchange is considered as markovian, the term $\mathbf{H}_i$ is assumed to be small compared to the other terms in the total Hamiltonian of the system, and so forth.

Another possibility for stochastic description of desorption is the use of the generalized Langevin equation as used in the case of vibrational energy exchange for the scattering of particles from a solid surface and for vibrational interactions of adsorbed molecules (Sections II and III). This approach is applied by Guan et al. [48] for the determination of desorption and the relaxation probability of vibrationally excited carbon monoxide molecules adsorbed on model ionic and covalent surfaces. They present the solid as a one-dimensional chain of oscillators, for which fluctuation and friction forces act internally. The adsorbed molecule is bound by the carbon atom at the external end of this row. All interactions are considered harmonic, excluding the bond of the carbon atom with the solid surface, which is modeled by a Morse potential. Calculations with model parameters show that the surface vibrations for both surfaces accelerate predesorption and relaxation of the vibrationally excited adsorbate:

$$CO(\nu = 1)_{ads} \rightarrow CO(\nu = 0)$$

$$CO(\nu = 1)_{ads} \rightarrow CO(\nu - 0)_{ads}$$

in comparison with the case of the static surface in the classical model.

However, the quantum treatment of these processes increases the role of the energy transfer from the model heat bath to the reaction co-ordinate, resulting in thermal desorption of excited carbon monoxide adsorbates. The other frustrated translational or rotational modes may also play an intermediate role in energy transfer from excited carbon monoxide stretching mode into the adsorption bond mode (CO–surface). These interactions are important because direct coupling between the carbon monoxide internal stretching mode and the frustrated translational mode, responsible for desorption, is very weak. Taking into account these interactions, this leads to drastic acceleration of desorption [5, 49].

V. SUMMARY

The main groups of theoretical methods for description of energy exchange with participation of vibrational degrees of freedom are discussed for three types of surface processes: (1) scattering of gas particles from a solid surface, (2) vibrational relaxation of adsorbed molecules, and (3) processes connected with breaking of a bond in an adsorption complex (surface diffusion, desorption, and dissociation of the adsorbates). Most of these methods lead to good agreement between theoretical predictions and experimentally determined results for the phenomenon described. However, each of the approaches involve different ways of modeling the interactions in the system and simplifications for obtaining solutions, which limits their applicability (without substantial alternation) for description of other processes taking place at solid surfaces. In the last part of this work, theoretical approaches are presented that consider both the interaction between the adsorption bond and surface oscillations. These interactions are important to the processes of surface diffusion and desorption. In addition, these interactions are important in the participation of intramolecular vibrations in the exchange of energy, particularly for heterogeneous catalytic processes on solid catalysts. Different types of limitations in application of these methods are also discussed.

REFERENCES

1. H Kasai, A Okiji, W Brenig. J Electron Spec Rel Phenom 54/55:153, 1990.
2. RW Verhoeff, D Kelly, CB Mullins, WH Weinberg. "The direct dissociative chemisorption of methane and ethane on Ir(110); isotope effects and vibrationally assisted chemisorption." Proceedings of 8th International Conference on Solid Surfaces, The Hague, 1992.
3. S Efrima, C Jedrzejek, KF Freed, E Hood, H Metiu. J Chem Phys 79:2436, 1983.
4. JG Lauderdale, DG Truhlar. J Am Chem Soc 107:4590, 1985.

5. T Uzer, JT Muckerman. In: E Yurtsever, ed. Frontiers of Chemical Dynamics. NATO ASI Series. Dordrecht: Kluwer, 1994, pp 267–291.
6. HL Friedman. A Course in Statistical Mechanics. Englewood Cliffs, NJ: Prentice-Hall, 1985.
7. D Chandler. Introduction to Modern Statistical Mechanics. Oxford: Oxford University Press, 1987.
8. CW Gardiner. Handbook of Stochastic Methods. Berlin: Springer, 1985, p 184.
9. LD Landau, EM Lifshitz. Statistical Physics. Part I. Moscow: Nauka, 1976.
10. HA Kramers. Physica 7:197, 1940.
11. RF Grote, JT Hynes. J Chem Phys 77:3736, 1982.
12. JT Hynes. Ann Rev Phys Chem 36:573, 1985.
13. W Brenig, H Kasai. Surface Sci 213:170, 1989.
14. S Kütchenhoff, W. Brenig, Y Chiba. Surface Science 245:389, 1991.
15. B Jackson. J Chem Phys 88:1383, 1988.
16. B Jackson. J Chem Phys 94:5126, 1991.
17. AE DePristo, A Kara. Adv Chem Phys 77:163, 1990.
18. SA Adelman, BJ Garrison. J Chem Phys 65:3751, 1976.
19. JD Doll, DR Dion. J Chem Phys 65:3762, 1976.
20. JC Tully. J Chem Phys 73:1975, 1980.
21. G Pacchioni, AM Ferrari, PS Bagus. Surface Sci 350:159, 1996.
22. A Lakhlifi, S Picaud, C Girardet, A Allouche. Chem Phys 201:73, 1995.
23. N Sheppard. Ann Rev Phys Chem 39:589, 1988.
24. GE Ewing. Int Rev Phys Chem 10:391, 1991.
25. JW Gadzuk, AC Luntz. Surface Sci 144:429, 1984.
26. OM Braun, AI Volokitin, VP Zhdanov. Usp Fiz Nauk 158:421, 1989.
27. G Volpihac, F Achard. J Chem Phys 94:7850, 1991.
28. M Morin, P Jakob, NJ Levinos, YJ Chabal, AL Harris. J Chem Phys 96:6203, 1992.
29. DC Langreth, M Persson. Phys Rev B 43:1353, 1991.
30. A Nitzan, BNJ Persson. J Chem Phys 83:5610, 1985.
31. EE Platero, D Scarano, A Zecchina, G Meneghini, R De Franceschi. Surface Sci 350:113, 1996.
32. R Jaquet, WM Miller. J Chem Phys 89:2139, 1985.
33. KF Freed. J Chem Phys 82:5264, 1985.
34. SM Valone, AF Voter, JD Doll. J Chem Phys 85:7480, 1986.
35. TN Truong, DG Truhlar. J Chem Phys 88:6611, 1988.
36. G Wahnstrom. Chem Phys Lett 163:401, 1989.
37. KD Dobbs, DJ Doren. J Chem Phys 97:3722, 1992.
38. VP Zhdanov. Kinet Katal 28:247, 1987.
39. R Tsekov, E Ruckenstein. J Chem Phys 100:1450, 1994.
40. R Tsekov, E Ruckenstein. J Chem Phys 101:2130, 1994.
41. F Constable. Proc Royal Soc A 108:355, 1923.
42. R Larsson. Chem Scr 27:371, 1987.
43. R Larsson. Catal Today 4:235, 1989.
44. R Tsekov, G Vayssilov. J Phys Chem 96:3452, 1992.
45. GN Vayssilov. Adv Colloid Interface Sci 43:51, 1993.
46. G Vayssilov, R Tsekov. Chem Phys Lett 188:497, 1992.
47. HJ Kreuzer, DN Loney. Chem Phys Lett 78:50, 1981.
48. Y Guan, JT Muckerman, T Uzer. J Chem Phys 93:4383, 1990.
49. F Dsegilenko, E Herbst. J Chem Phys 100:9205, 1994.

11

Thermodynamics of Adsorption at the Vapor–Metal Oxide Interface

JÓZSEF TÓTH University of Miskolc, Miskolc-Egyetemváros, Hungary

I. THERMODYNAMIC INCONSISTENCIES IN GAS/SOLID ADSORPTION

A. Basic Equation of Surface Thermodynamics

The first law of thermodynamics applied to a normal three-dimensional one-component system can be written in the following form:

$$dU = TdS - pdV + \mu dn \tag{1}$$

where U is the internal energy (J), T is the temperature (K), S is the entropy (J K^{-1}), p is the pressure (Pa = J m^{-3}), V is the volume (m^3), μ is the chemical potential (J mol^{-1}), and n is the amount of the component (mol).

For a two-dimensional adsorbed layer, a new variable, the surface area A (m^2), should be introduced:

$$dU^s = TdS^s - pdV^s + \mu^s dn^s - \pi dA \tag{2}$$

where π is the free energy (J m^{-2}) of the surface covered by the adsorbed layers, and the suffix s refers to the fact that Eq. (2) takes the pure adsorbate as a distinct system into account.

Since a one-component system is discussed the free enthalpy (J) can be defined as

$$G^s = n^s \mu^s \tag{3}$$

Integrating Eq. (2) at constant intensive properties, T, p, μ^s, and π, we obtain

$$U^s = TS^s - pV^s - \pi A + \mu^s dn^s \tag{4}$$

Inserting Eq. (3) into Eq. (4) we have

$$G^s = U^s - TS^s + pV^s + \pi A \tag{5}$$

Equations (4) and (5) exactly define the inside energy and the free enthalpy of the adsorbed layer, respectively. For definition of the thermodynamic equilibrium between the gas phase and the adsorbed layer, Eq. (4) should be differentiated:

$$dU^s = TdS^s + S^s dT - pdV^s - V^s dp - \pi dA - Ad\pi + \mu^s dn^s + n^s d\mu^s \tag{6}$$

Substituting Eq. (2) into Eq. (6) we obtain

$$d\mu^{s} = -s^{s}dT + v^{s}dp + \frac{A}{n^{s}}d\pi \tag{7}$$

where s^{s} and v^{s} are the molar entropy ($J\,K^{-1}\,mol^{-1}$) and volume ($m^{3}\,mol^{-1}$), respectively.

If the thermodynamic equilibrium is established, then the chemical potentials must be equal:

$$d\mu^{g} = d\mu^{s} \tag{8}$$

where suffix g refers to the gas phase. Taking Eq. (7) into account the condition of thermodynamic equilibrium can be defined:

$$v^{g}dp - s^{g}dT = v^{s}dp - s^{s}dT + \frac{A}{n^{s}}d\pi \tag{9}$$

From Eq. (9) it follows that

$$\left(\frac{\partial\pi}{\partial p}\right)_{T} = (v^{g} - v^{s})\frac{n^{s}}{A} \tag{10}$$

Equation (10) is the basic relationship of surface equilibrium thermodynamics; therefore, the problems, contradictions, errors, and inconsistencies, etc., found in the literature are directly or indirectly in connection with Eq. (10).

B. Problems and Inconsistencies Relating to Basic Equation of Surface Thermodynamics

The most frequent errors are connected with the exact interpretation of π present in Eq. (10). In order to define π exactly and with *general validity* let us consider the Helmholtz free energy of the adsorbate:

$$A^{s} = U^{s} - TS^{s} \tag{11}$$

or in differential form:

$$dA^{s} = dU^{s} - TdS^{s} - S^{s}dT \tag{12}$$

The notion of π was introduced in Eq. (2); therefore, let us insert it into Eq. (12) so we obtain

$$-\left(\frac{\partial A^{s}}{\partial A}\right)_{V^{s},n^{s},T} = \pi \tag{13}$$

Equation (13) shows that, at constant temperature (in the case of an isotherm), π is equal to the free energy of the surface ($J\,m^{-2}$) covered by a definite amount of the adsorbed layer. Despite this fact there are some experts and scientists who regard π as the tangential pressure or tangential tension of the adsorbed layer. This statement is valid for *totally mobile layers* only, which may be the layers adsorbed on liquid surfaces. However, on solid surfaces, especially on oxide surfaces, where the attractive interactions between the adsorbent and adsorptive molecules are great, the concept of a totally mobile layer is unacceptable and the only exact definition of π is Eq. (13).

The most important contradictions and thermodynamic inconsistencies are in connection with the well known and widely used classical isotherm equations and with the enthalpy functions corresponding to these isotherm relationships. Some of these classical equations are summarized in Table 1.

All these equations suppose monolayer adsorption on a homogeneous surface, that is, the coverage is defined as

$$\theta = n^{s}/n^{s}_{m} \tag{14}$$

where n^{s} is the adsorbed amount (mol) at equilibrium pressure p (kPa). The constant B interprets the interaction energies between molecules adsorbed, and K is defined as

$$K = k_{B}^{-1}\exp(U_0/RT) \tag{15}$$

where U_0 is the adsorptive potential and k_B is defined by deBoer [2] as

$$k_B = 2346(MT)^{1/2}10^{5} \tag{16}$$

where M is the molecular mass of the adsorptive. It is easy to see that both limiting values:

$$\lim_{p\to\infty}\theta = 1; \quad \text{i.e.,} \quad \lim_{\theta\to 1} p = \infty \tag{17}$$

and

$$\lim_{\theta\to 1}\Delta H_{i} = \infty \quad \text{or} \quad \lim_{\theta\to 1}\Delta H_{d} = \infty \tag{18}$$

correspond to all isotherm equations in Table 1. From their mathematical forms it follows that these isotherm equations are able to describe, according to the Brunauer–Emmet–Teller (BET) classification, only isotherms Type I and/or Type V. That is, the classical isotherm equations have limiting adsorption n^{s}_{m} (at the plateau) representing the completion of the monolayer adsorption. This supposition is expressed by limiting values [Eqs (17) and (18)]. Below it will be proven that this supposition contradicts Eq. (10), that is, the classic relationships are thermodynamically inconsistent.

Since Eq. (10) takes the *difference* between the two molar volumes into account, it is evident that n^{s} means the *excess* amount in the adsorbed phase, which is the only *directly* measurable parameter. It is also evident that n^{s}/A means the *excess* surface concentration in relation to the gas (bulk) phase. The situation is quite analogous to adsorption from a two-component liquid mixture. To demonstrate the inconsistencies of limiting values [Eqs (17) and (18)] let us investigate the changes in n^{s}, π, and $(v^{g} - v^{s})$ in the range of very high pressures (densities), that is, when $p \to \infty$. After Menon [3, 4] the excess character of n^{s} can also be defined by the densities of the gas phase (ρ^{g}, mol m^{-3}) and by that of the adsorbed phase (ρ^{s}). The *total* amount of the adsorptive (n^{s}_{t}) present in the sorbed phase is

$$n^{s}_{t} = V^{s}\rho^{s} \tag{19}$$

where V^{s} is volume of the sorbed phase (m^{3}).

If there were no adsorption, the amount of gas still present in the s phase would have been $V^{s}\rho^{s}$. Therefore, the *excess* (measured) amount is

$$n^{s} = V^{s}\rho^{s} - V^{s}\rho^{g} = V^{s}(\rho^{s} - \rho^{g}) \tag{20}$$

By substitution of Eq. (19) into Eq. (20) we have

$$n^{s} = n^{s}_{t} - V^{s}\rho^{g} \tag{21}$$

TABLE 1 Classical Isotherm Equations and the Corresponding Integral and Differential Enthalpies

Isotherm equations	Type of adsorption	Integral enthalpies (ΔH_i)	Differential enthalpies (ΔH_d)
Langmuir (L): $p = \frac{1}{K}\frac{\theta}{1-\theta}$	Localized, no interactions	$RT\left\{\frac{1}{\theta}\ln\frac{1}{1-\theta} - 1\right\}$	$RT\left(\frac{\theta}{1-\theta}\right)$
Fowler–Guggenheim (FG): $p = \frac{1}{K}\frac{\theta}{1-\theta}\exp(-B\theta)$	Localized, with interactions	$RT\left\{\frac{1}{\theta}\ln\frac{1}{1-\theta} - (1+B\theta)\right\}$	$RT\left(\frac{\theta}{1-\theta} - 2B\theta\right)$
Volmer (V): $p = \frac{1}{K}\frac{\theta}{1-\theta}\exp\left(\frac{\theta}{1-\theta}\right)$	Mobile, no interactions	$RT\left(\frac{\theta}{1-\theta}\right)$	$RT\left\{\left(\frac{1}{1-\theta}\right)^2 - 1\right\}$
deBoer–Hill (BH): $p = \frac{1}{K}\frac{\theta}{1-\theta}\exp\left(\frac{\theta}{1-\theta} - B\theta\right)$	Mobile, with interactions	$RT\theta\left\{\frac{1}{1-\theta} - B\right\}$	$RT\left\{\left(\frac{1}{1-\theta}\right)^2 - (1+2B\theta)\right\}$

Source: Ref. 1.

Equation (21) shows that the measured adsorbed amount (n^s) *decreases* at high pressure since ρ^g increases while n_t^s is nearly constant as can be seen in Fig. 1. Due to the decrease in n^s at pressures exceeding surface saturation, the measured isotherm, n^s versus p, must exhibit a maximum followed by a straight-line section with a downward slope, as shown in Fig. 1. Such types of isotherms were experimentally determined and published in the literature [4–7].

The changes in π at high pressures can be estimated if Eq. (20) is substituted into Eq. (10):

$$\left(\frac{\partial \pi}{\partial p}\right)_T = \frac{(v^g - v^s)(\rho^s - \rho^g)V^s}{A} = \frac{(\rho^s - \rho^g)^2 V^s}{\rho^s \rho^g A} \tag{22}$$

By making use of the reciprocal relationship between molar densities and molar volumes in Eq. (22), one can see that changes in π with pressure must always be positive. At low p the density of the adsorbed phase ρ^s far exceeds that of the vapor phase ρ^g, making the density difference, $\rho^s - \rho^g$, relatively large. Therefore, the change in π with p will be expected to change significantly at low p. As p increases, the density difference will decrease so that π will become independent of p, as can be seen in Fig. 1.

Evidently, the types of functions shown in Fig. 1 may only be valid if the temperature is *above* the critical temperature (T_c) of the adsorptive. It should again be emphasized that Fig. 1 is a *schematic* representation of functions because

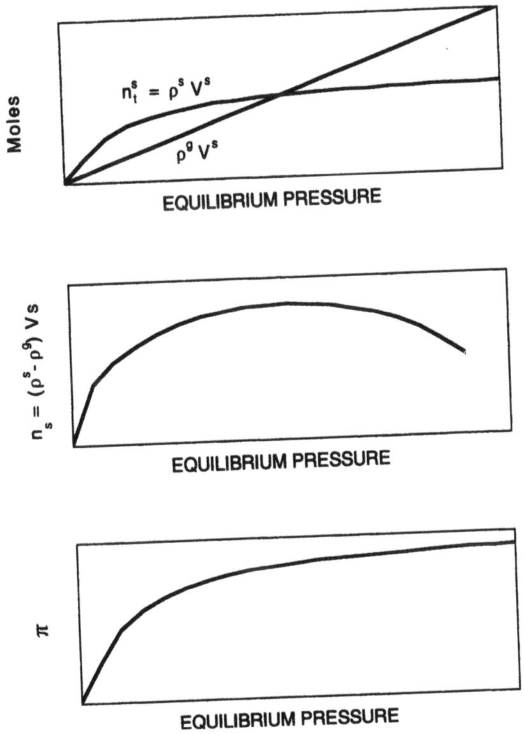

FIG. 1 Schematic representation of changes in $(\rho^s - \rho^g)$, π, and n^s as functions of equilibrium pressure extended to very high values (>3000 bars).

at very high pressures the structure of the solid adsorbents (pore sizes, lattice structures, porosities, specific surface areas) may also change. Therefore, more complicated functions than those shown in Fig. 1 can be measured.

Despite these facts, Fig. 1 and the measurements published in the literature [5–7] prove that an extrapolation to infinite pressure [see Eqs (17) and (18)] is thermodynamically inconsistent, that is, the application of the Langmuir and other classical isotherm equations at high pressures is incorrect and, therefore, these equations need some modification, which will be discussed later in this chapter.

Unfortunately, the classical isotherm equations may lead to contradictions and unreal physical values when they are applied at lower pressure ranges (2–8 MPa) too. This fact is proven by data shown in Fig. 2 where p_m is plotted against p_0 (p_m is the equilibrium pressure when the total monolayer capacity is completed and p_0 is the saturation pressure of the adsorptive).

The calculation of data in Fig. 2 has been made by means of data collected in the handbook of Valenzuela and Myers [8], which includes more than 100 simple gas isotherms measured on adsorbents with known specific surface areas determined by the producers with the controlled BET method. The p_m values were calculated from the BET surface area and from the cross-sectional area occupied by one molecule of the adsorptive investigated. The data in Fig. 2 represent 13 isotherms of ethane, ethylene, carbon dioxide, and propane measured on two activated carbons and silica gel at 10 temperatures below the critical temperatures. It is easy to see that in every case:

$$p_m < p_0 \tag{23}$$

that is, the total monolayer capacity is completed before the condensation of the adsorptive takes place. The uniform method of treament of this problem is proven by the points (p_m, p_0) which in Fig. 2 are, with excellent regression factor 0.998, on the same proportional line.

The open boxes in Fig. 2 refer to CO_2, and the point where the vertical dotted line starts belongs to a pressure which is slightly greater than the critical pressure of

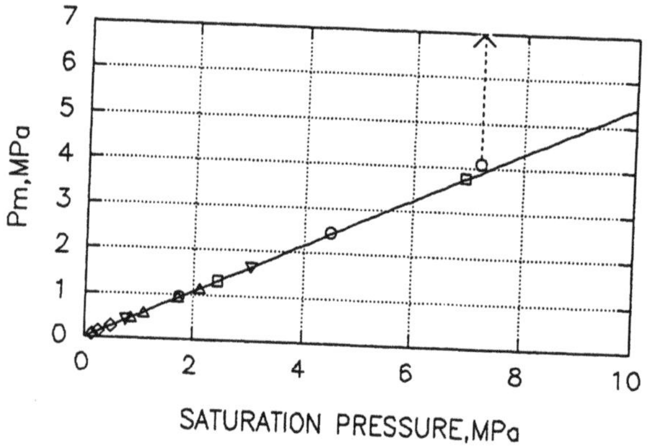

FIG. 2 Pressure (p_m) plotted against the saturation pressure (p_0) of different adsorptives such as ethane, ethylene, carbon dioxide, and propane. Isotherms are measured on BPL- and Columbia-activated carbon and on PH-100 silica gel in the temperature range −60.5°C to 60°C. (From Ref. 8.) For more details about the adsorbents see Ref. 8.

CO_2, i.e., the temperature is also slightly higher than critical ($T_c + dT = T$). At this temperature T the adsorptive is a typical gas, such that the extrapolation to infinite pressure ($p \to \infty$), according to the classical isotherm equations, is physically possible. However, this extrapolation results in a much greater monolayer capacity (specific surface area) than that determined by the BET method, and such a "jump" from the proportional line to infinity shown by the dotted line in Fig. 2 is physical nonsense. Obviously, this statement is valid for all adsorptives represented in Fig. 2 and for isotherms collected in Valenzuela and Myers' handbook. Figure 2 and its physical explanations also prove that the limiting cases defined by Eq. (17) cannot be valid.

As to the unreal limiting values of enthalpies [see Eq. (18)], in the past 10–15 years it has been proven both theoretically and experimentally that the enthalpies and entropies have finite values at total monolayer coverage. For example, in adsorption from solutions a *total excess* coverage is always formed, and, evidently, the changes in enthalpies (entropies) can be measured exactly. The experimental data prove [9] that the enthalpy of a total monolayer coverage can never become infinite. Since the infinite or finite character of thermodynamic functions is independent of the nature of the adsorptive system (gas/solid, vapor/solid, liquid/solid) the supposition of the classical isotherm equation concerning limiting values [Eqs (17) and (18)] should be rejected.

Summarizing the discussions written above, the basic cause of thermodynamic inconsistence of classic isotherm equations can be defined: the exact basic relationship of surface thermodynamics [Eq. (10) or Eq. (22)] demand the calculation with *excess* amount adsorbed. On the contrary, the classical relationships neglect this requirement (i.e., they calculate with the absolute adsorbed amount) and they do that also at very high pressures, especially at the limiting cases [Eqs (17) and (18)]. Having established the applicability and role of Eq. (10) in the adsorption process, it is now possible to write it in a more simplified form. Suppose that in Eq. (10):

$$v^g \gg v^s \tag{24}$$

that is, the molar volume of the sorbed phase can be neglected when compared to that of the gas phase. This omission implicitly includes

$$\rho^g \ll \rho^s \tag{25}$$

In this case, Eqs (10) and (22) transform into Eq. (26):

$$\left(\frac{\partial \pi}{\partial p}\right)_T = v^g \frac{n^s}{A} \tag{26}$$

Supposing that the gas law for ideal gases is valid for the gas phase, so we have

$$\left(\frac{\partial \pi}{\partial p}\right)_T = \frac{RT}{A}\frac{n^s}{P} \tag{27}$$

Equation (27) is the Gibbs equation, widely used in the practice and theory of adsorption. It is also evident that, by application of Eq. (27), the omissions leading to this relationship must be taken into account, otherwise all relationships corresponding to Eq. (27) contradict Eq. (10), that is, they are thermodynamically inconsistent.

C. Basis for Uniform Interpretation of Adsorption

Since Eq. (10) is the basic equation of surface thermodynamics, this relationship must serve as a basis for the uniform interpretation of adsorption. However, before application of Eq. (10) for this purpose, the limiting conditions of this equation should be determined. It has been shown [3–4] that at great equilibrium pressures (when $p \to \infty$) the excess character of adsorption must be taken into account. It is also very important to determine the limiting conditions when $p \to 0$ because, as will be shown, the neglect of this requirement also leads to inconsistent equations.

If $p \to 0$, then Eq. (27) is valid because the condition $v^g \gg v^s$ is fulfilled and the gas phase can be regarded as a perfect (ideal) system. From Eq. (27) the limiting condition as $p \to 0$ is

$$\lim_{p\to 0} \frac{n^s}{p} = \lim_{p\to 0} \left(\frac{\partial \pi}{\partial p} \right)_T \tag{28}$$

where RT/A is constant.

If it is assumed that the adsorption free energy π approaches a constant as $p \to 0$, then

$$\lim_{p\to 0} \left(\frac{\partial \pi}{\partial p} \right)_T = 0 \tag{29}$$

From virial treatments of adsorption, it is known [1, 8] that, in the limit $p \to 0$, the quotient v^s/p approaches the Henry law constant (H):

$$\lim_{p\to 0} \frac{n^s}{p} = H \tag{30}$$

Comparing Eqs (28)–(30) one finds a contradiction. Evidentally, adsorption free energy π is not independent of p as $p \to 0$ so that Eq. (29) must be thermodynamic nonsense. Clearly, π must change as the "first molecules" adsorb as p just becomes finite. As a result, Eq. (28) can be written:

$$\lim_{p\to 0} \frac{n^s}{p} = \lim_{p\to 0} \left(\frac{\partial \pi}{\partial p} \right)_T = \text{finite value} = H > 0 \tag{31}$$

Equation (31) gives the thermodynamic basis for why all isotherm equations (and, evidently, all measured isotherms) must begin (at $p \approx 0$) with a proportional section commonly referred to as Henry's law region.

Since Eq. (10) is a differential equation, the determination of the integration limits is very important as well because badly selected limits can lead to inconsistent relationships. Such inconsistencies were discussed for Eq. (17), the following forms of Eq. (10) may serve as a basis for uniform interpretation of adsorption:

$$\pi(p) = \frac{1}{A} \int_p^{p_\infty} v^g n^s \mathrm{d}p - \frac{1}{A} \int_p^{p_\infty} v^s n^s \mathrm{d}p \tag{32}$$

The upper limit of integration, p_∞, may be as great as is required by the adsorption problem to be solved; however, the following conditions should be met. The surface area A does not change as a function of pressure. For calculation of v^g a state equation must be available which is applicable at high pressures, up to p_∞. For p_∞ the virial equation of state can be used in the following form:

$$v^{g} = \frac{RT}{p}\{1 + b(p/p_c) + c(p/p_c)^2 + d(p/p_c)^3 + e(p/p_c)^4\} \tag{33}$$

where p_c is the critical pressure of the adsorptive, b, c, d, and e are the virial constants calculable from the $Z = Z(p)$ function of the compressibility factor. The measured isotherm, the function $n^s = n^s(p)$ should also have an explicit form, as will be discussed in the next section.

The most complicated problem is to determine the function $v^s = v^s(p)$ because its values are not measurable quantities. In most cases the supposition $v^g \gg v^s$ is applied, so we have from Eq. (32):

$$\pi(p) = \frac{1}{A}\int_{p}^{p_\infty} v^{g} n^{s} \mathrm{d}p \tag{34}$$

The last, but simplest solution to the integral equation (34) derives from the integral form of the Gibbs equation (27). So we have

$$\pi(p) = \frac{RT}{A}\int_{p}^{p_\infty} \frac{n^{s}}{p} \mathrm{d}p \tag{35}$$

Equation (35) is applicable for the case when the gas phase behaves as a perfect gas. This approximation is usually valid if the pressure does not exceed about 500 kPa. Summarized, Eq. (35) contains the following assumptions and conditions: (1) $v^g \gg v^s$, i.e., the excess character of adsorption is not taken into account, (2) p_∞ is not greater than 500 kPa, and (3) $\lim_{p\to 0}(n^s/p)$ is a finite value and it is greater than zero.

In spite of these conditions and approximations, Eq. (35) provides a new possibility for correction of the inconsistent classical equations and a uniform interpretation of isotherm equations.

II. ELIMINATION OF THERMODYNAMIC INCONSISTENCIES OF ADSORPTION EQUATIONS

A. A New Form of the Gibbs Adsorption Equation

Let us write Eq. (35) in the form of the coverage defined by Eq. (14), that is

$$n^{s} = \theta n^{s}_{m} \tag{36}$$

Substituting Eq. (36) into Eq. (35) we obtain:

$$\pi(p) = \frac{RT}{\phi_m}\int_{p}^{p_\infty} \frac{\theta}{p} \mathrm{d}p \tag{37}$$

where

$$\phi_m = A/n^{s}_{m} \tag{38}$$

So ϕ is equal to the surface area of the adsorbent covered by 1 mol of the adsorptive at $\theta = 1$ ($m^2\,mol^{-1}$). Let us introduce a new function, defined as follows:

$$\psi(\theta) = \frac{\theta}{p}\frac{dp}{d\theta} \quad \text{or} \quad \psi(n^s) = \frac{n^s}{p}\frac{dp}{dn^s} \tag{39}$$

Inserting Eq. (39) into Eq. (37) we have

$$\pi(\theta) = \frac{RT}{\phi_m}\int_{\theta}^{\theta_\infty} \psi(\theta)d\theta \tag{40}$$

where, evidently, θ_∞ corresponds to equilibrium pressure p_∞. Let us select the upper limit of integration in Eq. (37) so that $p_\infty = p_m$, i.e., p_m is equal to the equilibrium pressure when the total adsorption capacity (n_m^s) is completed ($\theta = 1$). In this case:

$$\pi(\theta) = \frac{RT}{\phi_m}\int_{\theta}^{1} \psi(\theta)d\theta \tag{41}$$

The expression RT/ϕ_m in Eq. (41) is the change in free energy of the surface covered by adsorbed amount n_m^s if it behaved as a two-dimensional ideal (id) gas:

$$\pi_{id} = RT/\phi_m \tag{42}$$

Taking Eq. (42) into account we obtain from Eq. (41) the following equations:

$$\pi_r(\theta) = \pi_{id}\int_{\theta}^{1} \psi(\theta)d\theta \tag{43}$$

or

$$\psi(\theta) = d\pi_r/d\theta \tag{44}$$

where

$$\pi_r = \pi/\pi_{id} \tag{45}$$

Equation (43) is a special new form of the Gibbs equation and it provides a concrete physical meaning to function $\psi(\theta)$ because its integral value is equal to the change in *relative* free energy of the surface (π_r) when the coverage changes from θ to 1.

B. Conditions of Thermodynamic Consistence Relating to Isotherm Equations

The function $\psi(\theta)$ and its thermodynamic definition [Eq. (43)] make it possible to determine the conditions of thermodynamic consistence relating to isotherm equations. Evidently, this determination is valid under conditions which must also be fulfilled by Eq. (43). These conditions are as follows:

1. The excess character of adsorbed amounts are neglected, that is, $v^g \gg v^s$ and $\rho^s \gg \rho^g$.
2. The function $\pi_r(\theta)$ in Eq. (43) must have a finite value at $\theta = 1$. It should be a thermodynamic nonsense if the change in free energy of the surface tends

to infinity when the total coverage is established. The totally covered surface always has a definite free energy.

3. As has already been shown the limiting value [Eq. (31)] should exist.

The conditions (1) and (2) mentioned above can be defined alone with the function $\psi(\theta)$ in the following way.

If

$$\lim_{\theta=1} \psi(\theta) = \text{finite value} \tag{46}$$

then condition (2) is met.

From Eq. (39) it follows that, if $\psi(\theta) = 1$, then

$$dp/p = d\theta/\theta \tag{47}$$

Equation (47) is the differential equation of a proportional line, i.e., it is the differential equation of the Henry isotherm. It means that condition (3) is fulfilled when

$$\lim_{\theta=0} \psi(\theta) = 1 \tag{48}$$

Equations (46) and (48) are the two conditions of the thermodynamic consistence of isotherm equations and of all adsorption relationships where the thermodynamic functions can be expressed in any form of the function $\psi(\theta)$.

C. Implicit and General Isotherm Differential Equations Fulfilling the Requirements of Thermodynamic Consistence

These general isotherm equations must include the function $\psi(\theta)$ because this is the best possibility for controlling the fulfilment of limiting conditions, Eqs (46) and (48). For this reason, Eq. (39) provides an excellent possibility for deriving two differential equations. The first is

$$\frac{dp}{p} = \frac{\psi(\theta)}{\theta} d\theta \tag{49}$$

and the second:

$$\frac{d\theta}{\theta} = \frac{1}{\psi(p)} \frac{dp}{p} \tag{50}$$

Equation (50) could be defined because

$$\psi(\theta) = \psi(p) \tag{51}$$

where θ and p mean, as elsewhere, the conjugated pairs of isotherm equations in question. Equation (51) expresses also the fact that the relationship in Eq. (39), i.e., $(\theta/p)(dp/d\theta)$, can be expressed both as a function of θ and of p. Nevertheless, the existence of Eq. (51) will be proved below by a concrete example. Before this, it is necessary to emphasize that the equality [Eq. (51)] has great importance in the cases when only one of the two functions ψ can be explicitly expressed for p or θ. Equation (51) provides the possibility for calculating the function ψ numerically which does not have to be explicit. Another (evident) property of function $\psi(\theta)$ or $\psi(p)$ is that their values are directly connected to the shapes of the adsorption isotherms, as shown in Figs 3 and 4.

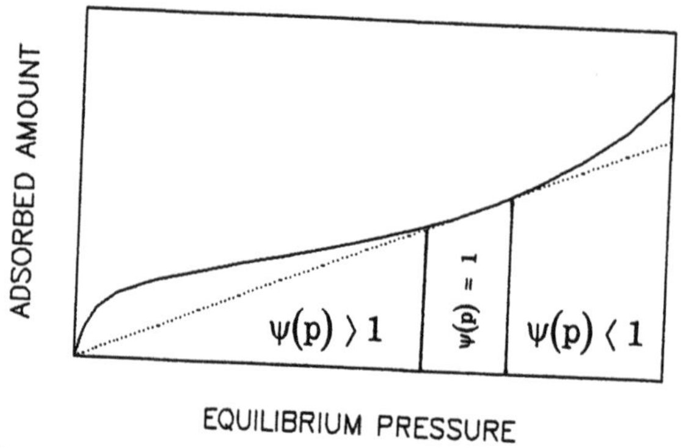

FIG. 3 Values of function $\psi(p)$ corresponding to different domains of Type II isotherms.

It is obvious that $\psi(p) > 1$, where the isotherm concave from below $(\theta/p > \mathrm{d}\theta/\mathrm{d}p)$; $\psi(p) = 1$, where the isotherm has a linear proportional section (its extrapolated part passes through the origin or, according to one of the two conditions of thermodynamic consistence, the isotherms begin with a Henry section; and $\psi(p) < 1$, where the isotherm is convex from below $(\theta/p < \mathrm{d}\theta/\mathrm{d}p)$. So knowing the physical and mathematical properties of functions ψ, let us integrate Eqs (49) and(50). According to Eq. (43) the upper limits of integration are p_{m} and $\theta = 1$, so we have

$$\int_{p}^{p_{\mathrm{m}}} \frac{\mathrm{d}p}{p} = \int_{\theta}^{1} \frac{\psi(\theta)}{\theta} \mathrm{d}\theta \tag{52}$$

and

$$\int_{\theta}^{1} \frac{\mathrm{d}\theta}{\theta} = \int_{p}^{p_{\mathrm{m}}} \frac{\mathrm{d}p}{\psi(p)p} \tag{53}$$

At value p_{m} the gas phase has to behave like an ideal (perfect) gas. Let us express Eqs (52) and (53) for p and θ, respectively, so we obtain:

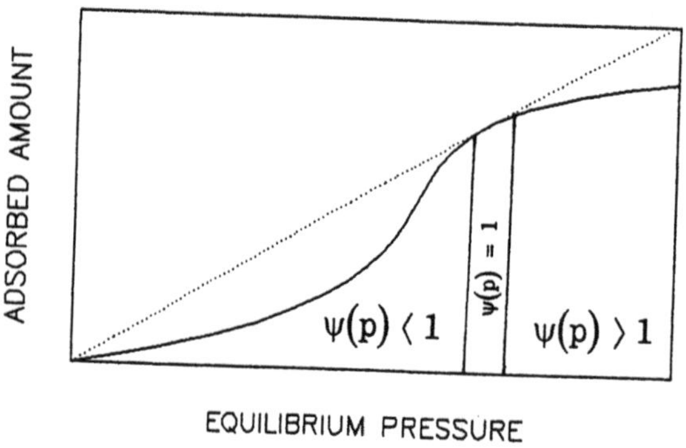

FIG. 4 Values of function $\psi(p)$ corresponding to different domains of Type V isotherms.

$$p = p_{\mathrm{m}} \exp\left\{ -\int_{\theta}^{1} \frac{\psi(\theta)}{\theta} \mathrm{d}\theta \right\} \tag{54}$$

and

$$\theta = \exp\left\{ -\int_{p}^{p_{\mathrm{m}}} \frac{\mathrm{d}p}{\psi(p)p} \right\} \tag{55}$$

Both equations are isotherm relationships and both are thermodynamically consistent if the conditions (46) and (48) are fulfilled. The concrete application and solution of Eqs (54) and (55) are discussed in the next part of this chapter.

D. Solution and Application of Isotherm Differential Equations

Let us discuss the application and solution of the differential equations for the concrete example of a Langmuir (L) equation having two forms, one expressed for p another for θ:

$$p = \frac{1}{K} \frac{\theta}{1-\theta} \quad \text{and} \quad \theta = \frac{p}{\frac{1}{K} + p} \tag{56}$$

Based on Eq. (39) the functions $\psi(\theta)$ and $\psi(p)$ can be explicitly calculated:

$$\psi_{\mathrm{L}}(\theta) = 1/(1-\theta) \tag{57}$$

and

$$\psi_{\mathrm{L}}(p) = Kp + 1 \tag{58}$$

It is easy to see that $\psi(\theta) = \psi(p)$ leads to the original L equation (56). Before inserting Eq. (57) into Eq. (54) or Eq. (58) into Eq. (55), the limiting values of both equations should be determined:

$$\lim_{\theta=0} 1/(1-\theta) = 1 \tag{59}$$

$$\lim_{\theta=1} 1/(1-\theta) = \infty \tag{60}$$

$$\lim_{p=0} (Kp+1) = 1 \tag{61}$$

$$\lim_{p \to p_{\mathrm{m}}} (Kp+1) = \text{finite value} \tag{62}$$

It is evident that function $\psi_{\mathrm{L}}(\theta)$ does not meet both requirements of thermodynamic consistence [only Eq. (48) is fulfilled]. However, function $\psi_{\mathrm{L}}(p)$ fulfils both requirements; it can, therefore, be substituted into Eq. (55) and it is certain that after solving Eq. (55) we obtain a thermodynamically consistent isotherm equation:

$$\theta = \exp\left\{ -\int_{p}^{p_{\mathrm{m}}} \frac{\mathrm{d}p}{(Kp+1)p} \right\} \tag{63}$$

since

$$\int_{p_m}^{p} \frac{dp}{(Kp+1)p} = \ln\left(\frac{p}{1/K+p}\right) + \ln\left(\frac{1/K+p_m}{p_m}\right) \tag{64}$$

Integration of Eq. (63) yields the modified Langmuir (ML) equation:

$$\theta = \frac{\chi_{ML} p}{1/K + p} \tag{65}$$

where

$$\chi_{ML} = 1 + 1/Kp_m \tag{66}$$

The ML equation is thermodynamically totally consistent, because

$$\psi_{ML}(\theta) = \frac{\chi_{ML}}{\chi_{ML} - \theta} \tag{67}$$

that is,

$$\lim_{\theta=0} \psi_{ML}(\theta) = 1 \quad \text{and} \quad \lim_{\theta=1} \psi_{ML}(\theta) = \frac{\chi_{ML}}{\chi_{ML} - 1} = \text{finite value} \tag{68}$$

and the validity of Eqs (61) and (62) is unchanged. This consistence can also be seen from Eq. (65). Namely,

$$\lim_{p=p_m} \theta = 1 \tag{69}$$

It is evident that, by substituting Eq. (67) into Eq. (54), we obtain the same ML equation.

It is also easy to see that integration of Eq. (63) with infinite upper limit, i.e., $p_m = \infty$, leads to the original L equation (56), which is thermodynamically inconsistent as is shown above. This statement can be generalized and stated in other words: the classic isotherm equations (see Table 1) and others suppose the fulfilment of initial conditions $v^g \gg v^s$, i.e., $\rho^s \gg \rho^g$, but they apply in Eq. (54) or (55) to an infinite upper limit of integration ($p_m = \infty$), which contradicts the initial conditions mentioned above. The thermodynamic inconsistence of the original L and other isotherm equations in Table 1 can also be proved with Eq. (43), namely, for the L equation:

$$\pi_{r,L} = \int_{\theta}^{1} \frac{d\theta}{1-\theta} = \left[\ln\left(\frac{1}{1-\theta}\right)\right]_{\theta}^{1} = \infty \tag{70}$$

which is thermodynamic nonsense. The same infinite change in free energy of the surface can be proved for most of the known isotherm equations.

E. Transformations of the Well Known but Thermodynamically Inconsistent Classical Isotherm Equations into Consistent Relationships

These transformations are also based on Eq. (54) or (55) and all transformations are quite similar to that of the L equation discussed in the previous section. So the basic relationships and the results of integration, that is, the modified forms of the classic isotherm equations are discussed here. First, the equations shown in Table 1 are integrated. The following function $\psi(\theta)$ belongs to the *Fowler–Guggenheim* (FG) equation:

$$\psi_{\mathrm{FG}}(\theta) = \frac{1}{1-\theta} - B\theta \tag{71}$$

where

$$B = \frac{c\omega}{RT} \tag{72}$$

In Eq. (72) ω is defined as the interaction energy per pair of molcules of nearest neighbors, and c is a constant. It is evident that in the case of $B = 0$, i.e., when there do not exist any interactions between molecules adsorbed, Eq. (71) transforms into function $\psi_{\mathrm{L}}(\theta)$ [see Eq. (57)]. It has shown that function $\psi_{\mathrm{L}}(\theta)$ is thermodynamically inconsistent; therefore, the only consistent function $\psi(\theta)$ for the modified FG (MFG) equation is the following relationship:

$$\psi_{\mathrm{MFG}} = \frac{\chi_{\mathrm{FG}}}{\chi_{\mathrm{FG}} - \theta} - B\theta \tag{73}$$

Inserting Eq. (73) into Eq. (54) after integration we have

$$p = \frac{1}{K}\frac{\theta}{\chi_{\mathrm{FG}} - \theta}\exp(-B\theta) \tag{74}$$

where

$$\chi_{\mathrm{FG}} = 1 + \frac{1}{Kp_{\mathrm{m}}\exp(B)} \tag{75}$$

Equation (74) fulfils both requirement of thermodynamic consistence. It is expected that the MFG equation (74), which considers explicitly the lateral interactions between molecules adsorbed, represents also the two-dimensional condensation. The formation of a liquid adsorbed layer can be expressed mathematically by the fact that the isotherm equation $p = \varphi(\theta)$ has a local maximum and minimum, i.e., two values of θ exist where

$$\frac{\mathrm{d}\varphi(\theta)}{\mathrm{d}\theta} = 0 \tag{76}$$

The values of θ, where Eq. (76) is fulfilled, is denoted by θ_{c}. Let us differentiate Eq. (74) and calculate the values of θ_{c}, where Eq. (76) is satisfied, so we have

$$B(\theta_{\mathrm{c}}) = \frac{\chi_{\mathrm{FG}}}{\theta_{\mathrm{c}}(\chi_{\mathrm{FG}} - \theta_{\mathrm{c}})} \tag{77}$$

Equation (77) determines all values of B at which two-dimensional condensation can take place. Equation (77) has an absolute minimum value:

$$\theta_{\mathrm{c,min}} = 0.5\chi_{\mathrm{FG}} \tag{78}$$

By inserting Eq. (78) into Eq. (77) this minimum value of function $B(\theta_{\mathrm{c}})$ can be calculated:

$$B_{\mathrm{min}} = 4/\chi_{\mathrm{FG}} \tag{79}$$

It means that at all values of B which are greater than $4/\chi_{FG}$ two-dimensional condensation takes place.

Since $\psi(p) = 1$ [or $\psi(\theta) = 1$] is a distinct part of the isotherm (it separates the concave and convex domains and determines the limit of the Henry section) it is important to calculate the values B where

$$\psi_{MFG}(\theta_s) = 1 \tag{80}$$

Applying Eq. (80) for Eq. (73) we obtain:

$$B(\theta_s) = \frac{1}{\chi_{FG} - \theta_s} \tag{81}$$

These functions and limiting values are shown in Fig. 5.

Equations (81) and (77) are represented with dotted and solid lines in Fig. 5, so it is very easy to determine the limits of values B determining the types of isotherms.

The MFG equation describes isotherm Type I when $0 < B < 1/\chi_{FG}$, isotherms Type V when $1/\chi_{FG} < B < 4/\chi_{FG}$, and two-dimensional condensation takes place when $B > 4/\chi_{FG}$. The different types of isotherms and the corresponding functions $\psi(p_r)$ are shown in Fig. 6.

For the sake of better comparability of functions ψ the relative pressure, i.e., the functions $\psi(p_r)$ are applied in Fig. 6 and in the subsequent figures. This representation is theoretical and numerically entirely equivalent to that of the function $\psi(p)$ because the values of function $\psi(\theta)$ or $\psi(p)$ [see Eq. (39)] are independent of the dimension of the pressure and it is also shown that $\psi(\theta) = \psi(p)$. The only *formal* difference between the two representations is that in Eqs (52)–(54) and in their solutions instead of p_m its relative value $p_{r,m} = p_m/p_0$ is present. Henceforward, in the text the absolute pressures are applied, emphasizing that this uniform interpretation of adsorption is valid both below and above the critical temperature.

As is shown in Table 1 the *Volmer equation* interpets isotherms where the adsorbed layers are mobile, but interactions between adsorbed molecules do not take place. The function $\psi(\theta)$ of the Volmer equation has the following form:

$$\psi_V(\theta) = \left(\frac{1}{1-\theta}\right)^2 \tag{82}$$

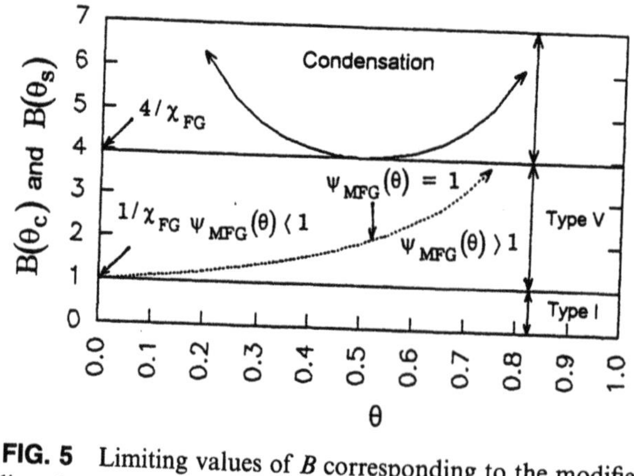

FIG. 5 Limiting values of B corresponding to the modified FG equation; $\chi_{FG} = 1.01$. Solid line: function $B(\theta_c)$; dotted line: function $B(\theta_s)$.

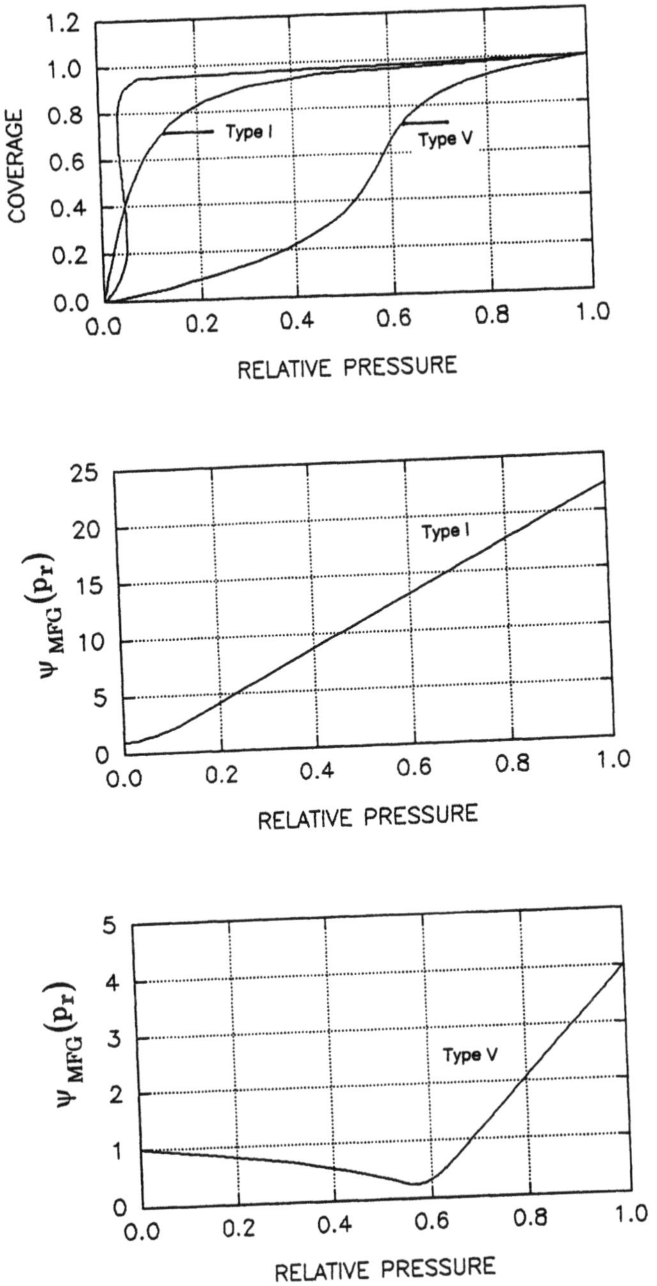

FIG. 6 Types of isotherms defined and described by the modified FG equation (upper diagram) and the corresponding functions $\psi_{MFG}(p_r)$ (lower diagrams). Isotherm Type I: $K = 10$, $B = 0.8$, $\chi_{FG} = 1.0449$, isotherm Type V: $K = 0.30$, $B = 3.0$, $\chi_{FG} = 1.660$; condensation type: $K = 2.0$, $B = 5.0$, $\chi_{FG} = 1.0034$.

The Volmer equation is also thermodynamically inconsistent, because

$$\lim_{\theta=1} \psi_V(\theta) = \infty \tag{83}$$

We obtain the consistent form of the function $\psi_V(\theta)$ similar to the MFG equation:

$$\psi_{MV}(\theta) = \left(\frac{\chi_V}{\chi_V - \theta}\right)^2 \tag{84}$$

By inserting Eq. (84) into Eq. (54) after integration we obtain the modified Volmer (MV) equation:

$$p = \frac{1}{K}\frac{\theta}{\chi_V - \theta}\exp\left(\frac{\theta}{\chi_V - \theta}\right) \tag{85}$$

where

$$\chi_V = 1 + \frac{1}{Kp_m \exp[1/(\chi_V - 1)]} \tag{86}$$

or

$$K = [p_m(\chi_V - 1)]^{-1} \exp[1/(\chi_V - 1)] \tag{87}$$

The types of isotherms described by the MV equation can easily be determined with the help of the function $\psi_V(\theta)$. From Eq. (84) it follows that $\psi_V(\theta) < 1$, i.e., a convex part of the isotherm can only occur when

$$\chi_V < (\chi_V - \theta) \tag{88}$$

Because the relation (88) can never be valid the isotherms described by the modified Volmer equation are always of Type I, that is, $\psi_V(\theta) \geq 1$. This mathematical statement means that the mobility of the adsorbed molecules cannot alone cause a change in the type of isotherms; such a change may happen when interactions exist between the adsorbed molecules.

The last classical isotherm equation shown in Table 1 is the *deBoer–Hill* (BH) *equation* having the following function $\psi(\theta)$:

$$\psi_{BH}(\theta) = \left(\frac{1}{1 - \theta}\right)^2 - B\theta \tag{89}$$

The BH equation is also inconsistent because

$$\lim_{\theta=1} \psi_{BH}(\theta) = \infty \tag{90}$$

Similar to the FG and Volmer equations the consistent form of Eq. (89) is the following relationship:

$$\psi_{MBH}(\theta) = \left(\frac{\chi_{BH}}{\chi_{BH} - \theta}\right)^2 - B\theta \tag{91}$$

By inserting Eq. (91) into Eq. (54) after integration we have the modified deBoer–Hill (MBH) equation:

$$p = \frac{1}{K}\frac{\theta}{\chi_{BH} - \theta}\exp\left\{\frac{\theta}{\chi_{BH} - \theta} - B\theta\right\} \tag{92}$$

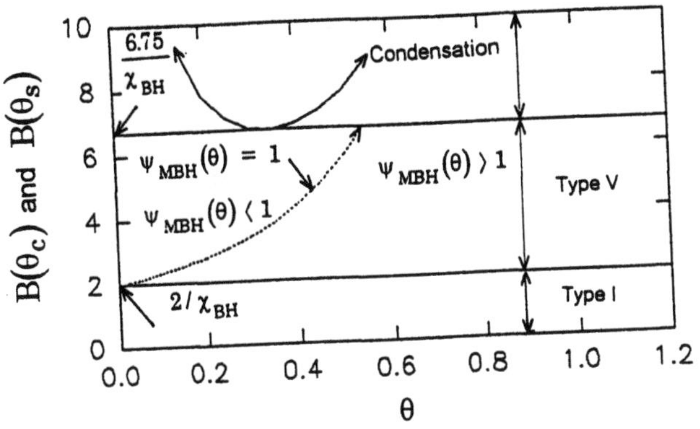

FIG. 7 Limiting values of B corresponding to the modified BH equation; $\chi_{BH} = 1.01$. Solid line: function $B(\theta_c)$; dotted line: function $B(\theta_s)$.

where

$$\chi_{BH} = 1 + \frac{1}{Kp_m \exp\{1/(\chi_{BH} - 1) - B\}} \tag{93}$$

or

$$K = [p_m(\chi_{BH} - 1)]^{-1} \exp\{1/(\chi_{BH} - 1) - B\} \tag{94}$$

In a similar way to the MFG equation the different types of isotherms described by the MBH equation can be determined. The corresponding limits of values for B are shown in Fig. 7, that is, the MBH equation describes isotherm Type I when $0 < B < 2/\chi_{BH}$, type V when $2/\chi_{BH} < B < 6.75/\chi_{BH}$, and two-dimensional condensation takes place when $B > 6.75/\chi_{BH}$. So it is shown that the interactions and mobility together can cause the same changes in types of isotherms as is done by the interactions only. This is the similarity between the MFG and MBH equations.

The modified and original classical isotherm equations shown in Table 1 can only be applied to homogeneous solid surfaces, which are very rare among oxide surfaces; however, these consistent, "homogeneous" isotherms provide the basis for deriving isotherm equations applicable to heterogeneous oxide surfaces.

III. MONOLAYER ISOTHERM EQUATIONS APPLICABLE TO HETEROGENEOUS OXIDE SURFACES

A. "Semiconsistent" Isotherm Equations (the F, MGF, and MS Equations)

In previous parts of this chapter it has been shown that

$$\lim_{\theta=1} \psi(\theta) = \text{finite value} \quad \text{and} \quad \lim_{\theta=0} \psi(\theta) = 1$$

or

$$\lim_{p=p_m} \psi(p) = \text{finite value} \quad \text{and} \quad \lim_{p=0} \psi(p) = 1$$

are the two conditions of thermodynamic consistence [see Eqs (46), (48), and (51)]. There are some well known and widely used isotherm equations which, in spite of the application of Eq. (54) or (55), do not meet one of the conditions mentioned above. These equations are nominated "semiconsistent"; however, they are applicable, in a definite range of equilibrium pressures, to heterogeneous oxide surfaces. This is the reason why these relationships are discussed here.

The *Freundlich* (F) *equation* is the first among these semiconsistent relationships and it has the following form:

$$n^{s} = Cp^{1/n} \tag{95}$$

where C and n are empirical constants. The function $\psi(\theta)$ provides some physical reality to power n, namely,

$$\psi_{F}(\theta) = \psi_{F}(p) = n, \qquad \text{where} \quad n \neq 1 \tag{96}$$

that is, the F equation interprets isotherms having a proportional function π_r versus θ and the slope of this function is equal to n [see Eq. (44)]. It is also evident that Eq. (96) can never meet the condition:

$$\lim_{p=0} \psi(p) = 1$$

therefore, it is a semiconsistent equation. Let us substitute Eq. (96) into Eq. (55) so after integration we obtain

$$\theta = \left(\frac{p}{p_{m}}\right)^{1/n} \tag{97}$$

The value of n (the slope of the function π_r versus θ) determines the type of isotherm described by the F equation, namely, when $n > 1$ the isotherm is of Type I and when $n < 1$ the isotherm is of Type III [$\psi(p) > 1$ or $\psi(p) < 1$].

The following important semiconsistent relationship is the *generalized Freundlich* (GF) *equation*, which has the following form:

$$\theta = \left\{\frac{p}{1/K_{GF} + p}\right\}^{1/n} \tag{98}$$

The power n refers to the heterogeneity of the surface and to the interactions between molecules adsorbed. This fact is reflected in Fig. 8, proving that Eq. (98) describes isotherms Type I, III, and V.

The function $\psi(p)$ is the following:

$$\psi_{GF}(p) = (K_{GF}p + 1)n \tag{99}$$

By inserting Eq. (99) into Eq. (55) after integration we have the modified GF (MGF) equation:

$$\theta - \left\{\frac{\chi_{GF}p}{1/K_{GF} + p}\right\}^{1/n} \tag{100}$$

where

$$\chi_{GF} = 1 + \frac{1}{K_{GF}p_{m}} \tag{101}$$

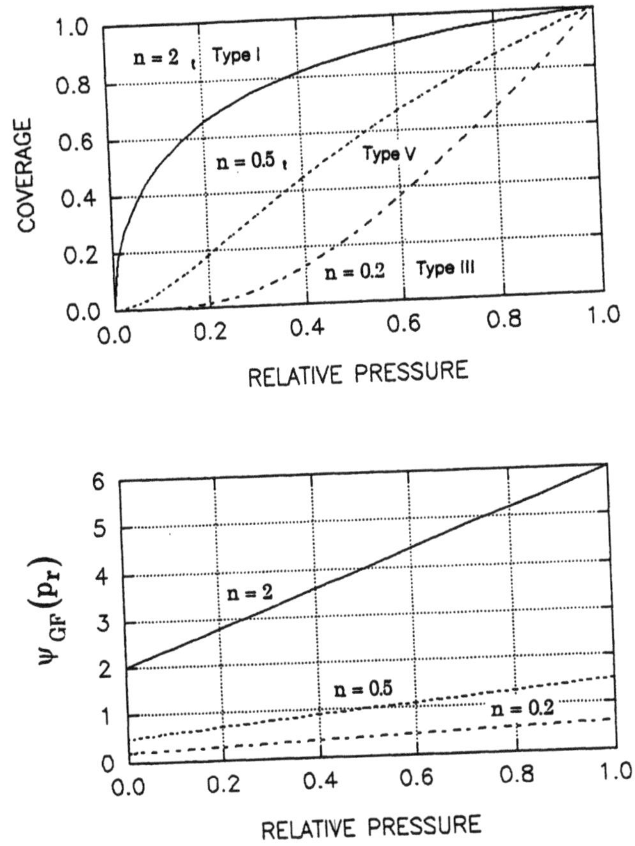

FIG. 8 Modified GF equations (upper diagram) and the corresponding functions $\psi_{GF}(p_r)$ (lower diagram) $K_{GF} = 2.0$; $\chi_{GF} = 1.5$.

From Eq. (99) it follows that the function $\psi_{GF}(p)$ is a linear relationship with a slope of $K_{GF}n$ and with intersection point n. Since this function, depending on the values of K_{GF} and n, may be equal to or greater or less than unity, the isotherms described by the GF equations are of Type I, III, and V. These functions are shown in Fig. 8.

Perhaps the most important semiconsistent isotherm equation is the *Sips* (S) *equation*, which has the following form:

$$\theta = \frac{p^n}{1/K_S + p^n} \tag{102}$$

where again the power n takes the heterogeneity and the interactions between adsorbed molecules into account. Sips has shown [10] that his relationship corresponds to an energy distribution function of Gaussian type. Despite this fact the original S equation is an inconsistent relationship because

$$\psi_S(p) = (K_S p^S p^n + 1)\frac{1}{n} \tag{103}$$

and

$$\psi_S(\theta) = 1/[n(1-\theta)] \tag{104}$$

It is easy to see that neither condition of consistence is fulfilled. The application of Eq. (55) leads to a semiconsistent relationship because after integration we have

$$\theta = \frac{\chi_S p^n}{1/K_S + p^n} \tag{105}$$

where

$$\chi_S = 1 + \frac{1}{K_S p_m^n} \tag{106}$$

The modified Sips (MS) equation fulfils only one of the two conditions of consistence:

$$\lim_{\theta=1} \psi_{MS}(\theta) = \frac{\chi_S}{\chi_S - \theta^n} = \text{finite value} \tag{107}$$

The effect of power n on the shape of Eq. (105) can be seen in Fig. 9.

In this sense the MS equation is similar to the MGF equation because both relationships correspond to isotherms Type I, III, and V.

The three semiconsistent equations (MF, MGF, and MS) discussed in this section have a common property, namely, none of them meet the requirement:

$$\lim_{\theta=0} \psi(\theta) = 1$$

that is, these equations theoretically cannot describe the Henry (initial) section of the isotherms. Since the proportional parts of the measured isotherms are very often small the semiconsistence of the MF, MGF, and MS equations rarely disturbs the practical use of these relationships. On the other hand, these equations can be applied for calculation of the total adsorption capacity (n_m^s) because the values of χ_x assure that at $p = p_m$ the coverage is equal to 1. However, such types of calculation are influenced by other important properties, which are discussed in Section IV.

B. Consistent Isotherm Equations (the MJ, T, FGT, VT, and BHT Equations)

In this section, we discuss such isotherm equations which, after application of Eq. (54) or (55), fulfil both conditions of consistence.

Jovanovic [11] supposed that the uncovered part of the adsorbent surface decreases exponentially as a function of the equilibrium pressure:

$$1 - \theta = \exp(-K_J p) \tag{108}$$

or in a usual form:

$$\theta = 1 - \exp(-K_J p) \tag{109}$$

The original Jovanovic (J) equation is a semiconsistent relationship because of the two conditions only one is met:

$$\psi_J(p) = \frac{\exp(K_J p) - 1}{K_J p} \tag{110}$$

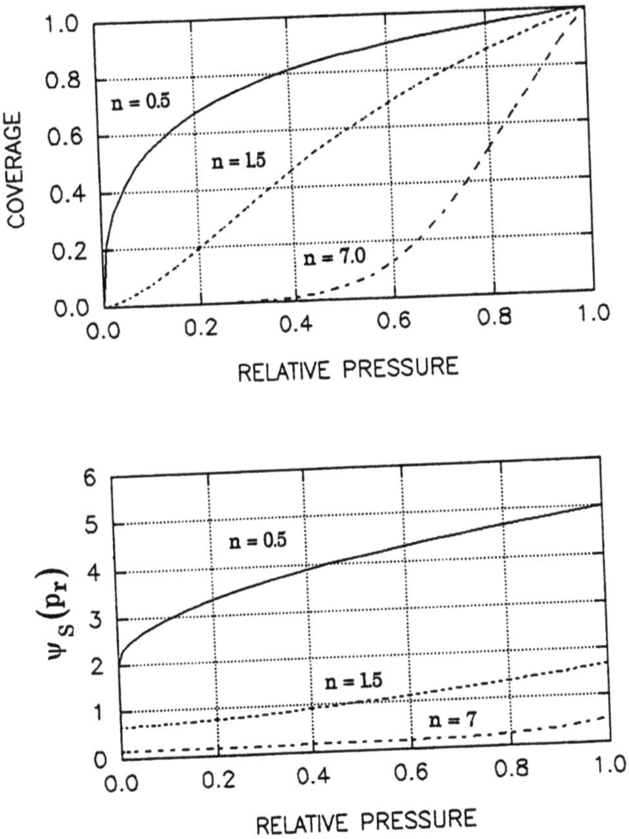

FIG. 9 Modified S isotherm relationships (upper diagram) and the corresponding functions $\psi_S(p_r)$ (lower diagram) $K_S = 1.5$.

and according to the L'Hospital rule:

$$\lim_{p \to 0} \psi_J(p) = 1$$

The other condition is not fulfilled because

$$\lim_{\theta=1} \psi_J(\theta) = \lim_{\theta=1} \frac{\theta/(1-\theta)}{\ln[1/(1-\theta)]} = \infty \tag{111}$$

By inserting Eq. (110) into Eq. (55) after integration we obtain the consistent modified Jovanovic (MJ) equation:

$$\theta = \frac{1 - \exp(-K_J p)}{\chi_J} \tag{112}$$

where

$$\chi_J = 1 - \exp(-K_J p_m) \tag{113}$$

The isotherms (112) corresponding to different values of K_J are shown in the upper panel of Fig. 10. The lower panel refers to the $\psi_{MJ}(p_r)$ functions. It is very interesting

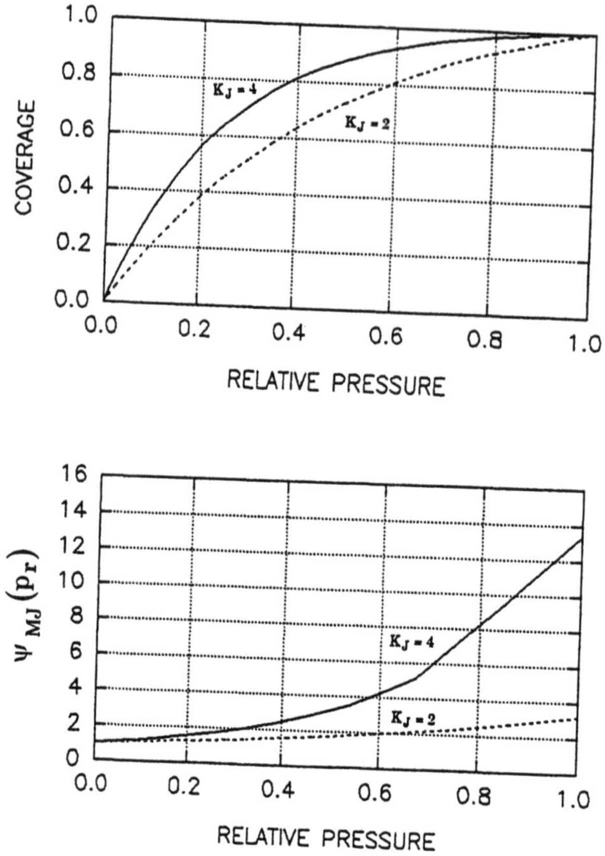

FIG. 10 Modified Jovanovic equation with different values of K_J (upper diagram) and the corresponding functions $\psi_{MJ}(p_r)$ (lower diagram).

that the MJ equation expresses the interactions between molecules adsorbed with a function $\psi(p)$ having concave form from below.

Tóth also derived an equation [12] which, after application of Eq. (54), is totally consistent. The starting point of Tóth's theory is the observation that a heterogeneous surface takes up more adsorptive, at the same relative equilibrium pressure, than a homogeneous adsorbent with specific surface area identical to that of the heterogeneous adsorbent. Consequently, the total monolayer capacities measured on the two adsorbents must be equal. These two requirements regarding the heterogeneity of the adsorbent can be taken by one parameter, m, into account. Namely,

$$\theta^m > \theta \quad \text{if} \quad 0 < m < 1 \tag{114}$$

As has been shown the function $\psi(\theta)$ of the ML equation has the following form:

$$\psi_L(\theta) = \frac{\chi_L}{\chi_L - \theta}$$

Taking the requirement defined by relationship (114) into account we obtain:

$$\psi_T(\theta) = \frac{\chi_T}{\chi_T - \theta^m} \tag{115}$$

Inserting Eq. (115) into Eq. (54) after integration we have the consistent Tóth (T) equation:

$$\theta = \frac{(\chi_T)^{1/m} p}{(1/K_T + p^m)^{1/m}} \tag{116}$$

where

$$\chi_T = 1 + \frac{1}{K_T p_{\rm m}^m} \tag{117}$$

The function $\psi(p)$ has the following form:

$$\psi_T(p) = K_T p^m + 1 \tag{118}$$

The isotherms described by Eq. (116) and the corresponding functions $\psi_T(p_r)$ are shown in Fig. 11.

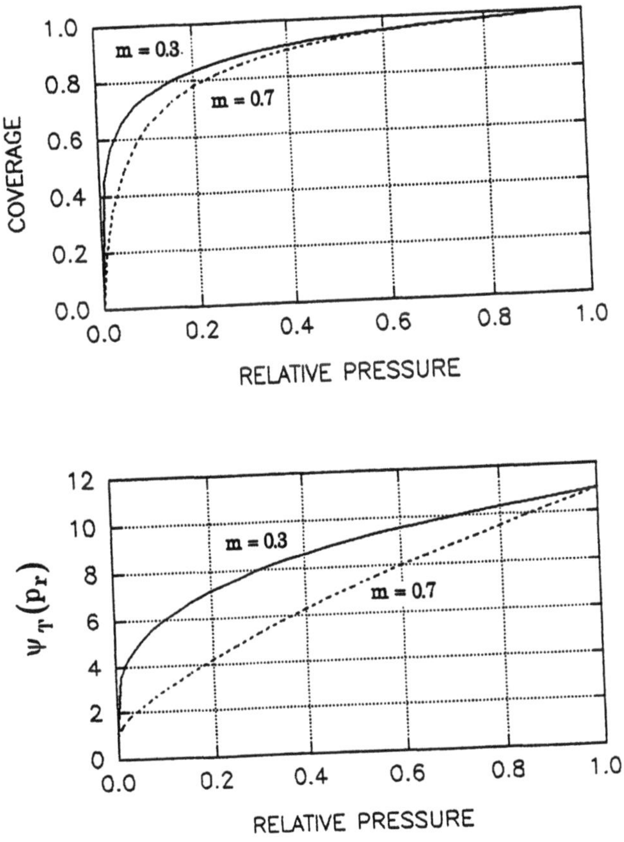

FIG. 11 Tóth's isotherms (upper diagram) and the corresponding functions $\psi_T(p_r)$ at different values of m; $K_T = 10.0$.

The constant K_T present in Eq. (116) and all the similar constants of totally consistent isotherm equations applicable to heterogeneous surfaces can be explained in the following way. As it is known the physical interpretation of the Henry constant (H), according to the deBoer–Hobson theory, is defined by Eq. (15), that is,

$$H = k_{\mathrm{B}}^{-1} \exp(U_0/RT)$$

In this sense U_0 means the adsorptive potential of the heterogeneous surface at very low equilibrium pressure, i.e., when $p \to 0$; under this condition the T equation transforms into the following form:

$$\frac{\theta}{p} = (\chi_T K_T)^{1/m} = H \tag{119}$$

Taking the definition of the Henry constant (H) and Eq. (117) into account we have

$$K_T = k_{\mathrm{B}}^{-m} \exp(mU_0/RT) - p_{\mathrm{m}}^m \tag{120}$$

Equation (120) means that K_T depends not only on the value of m, but is also influenced by the upper limit of integration p_{m}.

A deeper explanation of constant m characterizing the heterogeneity of the adsorbent can be given by a comparison of the following functions $\psi(p)$:

Freundlich: $\psi_{\mathrm{F}}(p) = n = \text{constant}$ (121)

Tóth: $\psi_T(p) = K_T p^m + 1$ (122)

Langmuir: $\psi_{\mathrm{L}}(p) = K_{\mathrm{L}} p + 1$ (123)

It is easy to see that Eqs (121)–(123) differ in the values of m only. Namely, if $m = 1$, then $\psi_T(p) = \psi_{\mathrm{L}}(p)$ and if $m = 0$, then $\psi_T(p) = \psi_{\mathrm{F}}(p) = \text{constant}$. It means that $m = 1$ represents the total homogeneous surface and $m = 0$ means a "total inhomogeneous" surface, which appears in the fact that the original Freundlich equation does not have a limiting value when p tends to infinity. All values of m between zero and unity represent different heterogeneities of the adsorbent existing between the Freundlich and Langmuir theories.

As will be shown below the value of m in the T equation can also express the mobility of the layers and the interactions between the adsorbed molecules. Similar to the ML equation all modified relationsips (MFG, MV, and MBH) applicable to homogeneous surfaces can be transformed into equations applicable to heterogeneous surfaces. This transformation means only that the coverage has to raise to the mth power with values $0 < m < 1$. In this way we obtain the following relationships; derivation of these does not require further explanation because all the mathematical steps and methods have been discussed above.

We have the *Fowler–Guggenheim–Tóth* (FGT) *equation* by the following way:

$$\psi_{\mathrm{MFG}}(\theta) = \frac{\chi_{\mathrm{FG}}}{\chi_{\mathrm{FG}} - \theta} - B\theta \tag{124}$$

$$\psi_{\mathrm{FGT}}(\theta) = \frac{\chi_{\mathrm{FGT}}}{\chi_{\mathrm{FGT}} - \theta^m} - B\theta^m, \quad 0 < m < 1 \tag{125}$$

$$p = \left(\frac{1}{K_{\mathrm{FGT}}}\right)^{1/m} \frac{\theta}{(\chi_{\mathrm{FGT}} - \theta^m)^{1/m}} \exp\left(-\frac{B\theta^m}{m}\right) \tag{126}$$

where

$$K_{\mathrm{FGT}} = \frac{p_m^{-m}}{\chi_{\mathrm{FGT}} - 1} \exp(-B) \quad \text{or} \quad \chi_{\mathrm{FGT}} = 1 + \frac{1}{K_{\mathrm{FGT}} p_{\mathrm{m}}^{m} \exp(B)} \tag{127}$$

These equations are applicable to immobile layers, with interactions on heterogeneous surfaces.

The *Volmer–Tóth* (VT) *equation*:

$$\psi_{\mathrm{MV}}(\theta) = \left(\frac{\chi_{\mathrm{V}}}{\chi_{\mathrm{V}} - \theta}\right)^2 \tag{128}$$

$$\psi_{\mathrm{VT}}(\theta) = \left(\frac{\chi_{\mathrm{VT}}}{\chi_{\mathrm{VT}} - \theta^m}\right)^2 \quad 0 < m < 1 \tag{129}$$

$$p = \left(\frac{1}{K_{\mathrm{VT}}}\right)^{1/m} \frac{\theta}{(\chi_{\mathrm{VT}} - \theta^m)^{1/m}} \exp\left\{\frac{\theta^m}{m(\chi_{\mathrm{VT}} - \theta^m)}\right\} \tag{130}$$

where

$$K_{\mathrm{VT}} = \frac{p_{\mathrm{m}}^{-m}}{\chi_{\mathrm{VT}} - 1} \exp\left(\frac{1}{\chi_{\mathrm{VT}} - 1}\right)$$

These relationships describe adsorption of mobile layers without interactions on heterogeneous surfaces.

Finally, the *deBoer–Hill–Tóth* (BHT) *equations* are applicable to mobile layers with interactions on heterogeneous surfaces:

$$\psi_{\mathrm{MBH}}(\theta) = \left(\frac{\chi_{\mathrm{BH}}}{\chi_{\mathrm{BH}} - \theta}\right)^2 - B\theta \tag{131}$$

$$\psi_{\mathrm{BHT}}(\theta) = \left(\frac{\chi_{\mathrm{BHT}}}{\chi_{\mathrm{BHT}} - \theta^{\mathrm{m}}}\right)^2 - B\theta^m \quad 0 < m < 1 \tag{132}$$

$$p = \left(\frac{1}{K_{\mathrm{BHT}}}\right)^{1/m} \frac{\theta}{(\chi_{\mathrm{BHT}} - \theta^{\mathrm{m}})^{1/m}} \exp\left\{\frac{\theta^m}{m(\chi_{\mathrm{BHT}} - \theta^{\mathrm{m}})} - \frac{\mathrm{B}\theta^m}{m}\right\} \tag{133}$$

$$K_{\mathrm{BHT}} = \frac{p_{\mathrm{m}}^{-m}}{\chi_{\mathrm{BHT}} - 1} \exp\left\{\frac{1}{\chi_{\mathrm{BHT}} - 1} - B\right\} \tag{134}$$

C. Applicability of T Equation Instead of All Consistent Equations Describing Isotherms of Type I

It is obvious that the FGT and BHT equations describe isotherms of Types I and V and that the two-dimensional condensation of adsorbed layers takes place between the same limits of values for B as has been shown for the MFG and MBH equations. However, the transformation of MFG, MV, and MBH equations into FGT, VT, and BHT equations by introducing the power m led to very important changes in the functions $\psi(p_{\mathrm{r}})$ belonging to isotherms of Type I, i.e., when $\psi(p_{\mathrm{r}}) > 1$.

These changes are shown in Figs 12–14. From these figures the following conclusions can be drawn. The initial convex part of functions $\psi(p_{\mathrm{r}})$ transforms into a concave form similar to the whole form of function $\psi_{\mathrm{T}}(p_{\mathrm{r}})$. As a result of

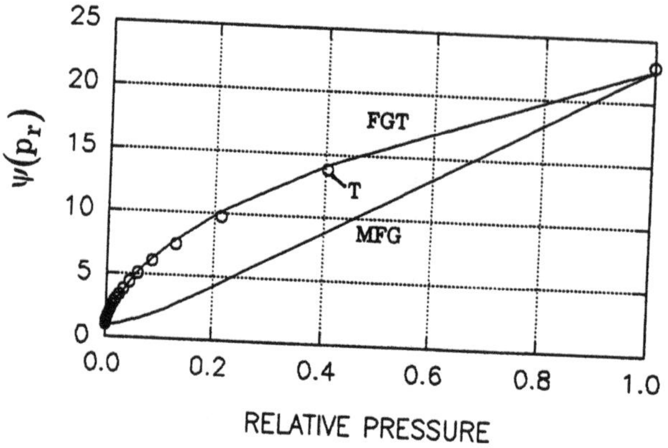

FIG. 12 Function $\psi_{MFG}(p_r)$ can be transformed into function $\psi_{FGT}(p_r)$ (solid lines). Function $\psi_{FGT}(p_r)$ is equivalent to function $\psi_T(p_r)$: O; $\chi_{FG} = 1.045$; $m = 0.5$; $B = 0.8$. The fitted function $\psi_T(p) = K_T p_r^m + 1$; $K_T = 21.811$; $m = 0.5921$.

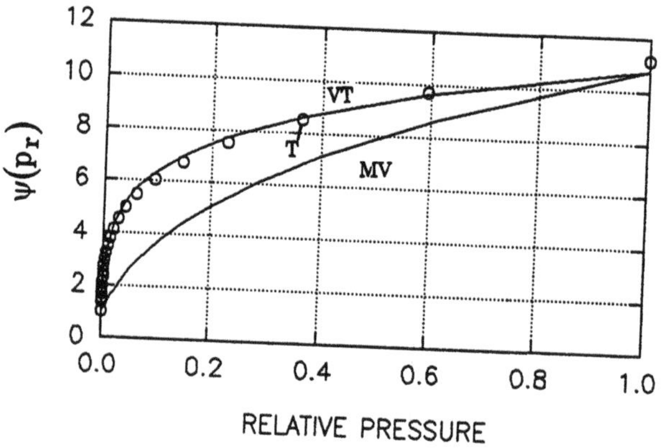

FIG. 13 Function $\psi_{MV}(p_r)$ can be transformed into function $\psi_{VT}(p_r)$ (solid lines). Function $\psi_{VT}(p_r)$ is equivalent to function $\psi_T(p_r)$: O; $\chi_{MV} = \chi_{VT} = 1.439$; $m = 0.5$. The fitted function $\psi_T(p_r) = K_T p_r^m + 1$; $K_T = 10.153$; $m = 0.2956$.

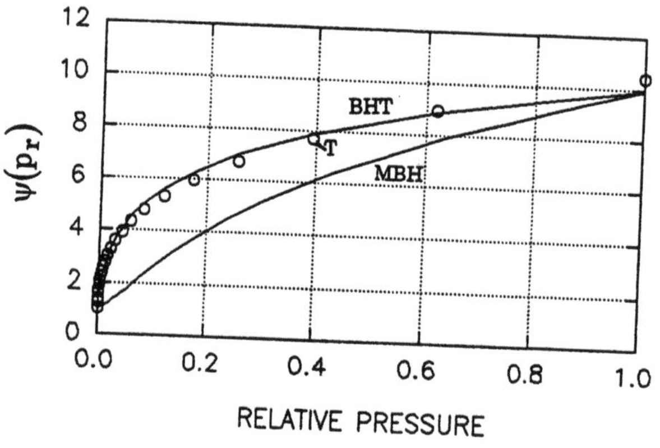

FIG. 14 Function $\psi_{MBH}(p_r)$ can be transformed into function $\psi_{BHT}(p_r)$ (solid lines). Function $\psi_{BHT}(p_r)$ is equivalent to function $\psi_T(p_r)$: O; $\chi_{BH} = \chi_{BHT} = 1.439$; $B = 0.8$. The fitted function $\psi_T(p_r) = K_T p_r^m + 1$; $K_T = 9.4284$; $m = 0.3649$.

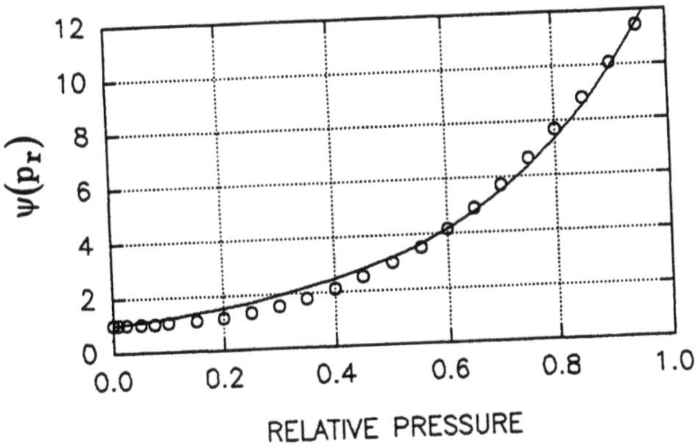

FIG. 15 Comparison of the functions $\psi_{MJ}(p_r)$ (solid line) and to this function fitted $\psi_T(p_r)$: O; $K_{MJ} = 4.000$; $K_T = 11.881$; $m = 2.5765$.

these changes the functions $\psi(p_r)$ of the FGT, VT, and BHT equations can be exactly fitted by function $\psi(p_r)$ of the T equation, i.e., the T equation can be applied instead of the FGT, VT, and BHT equations. It also means that the value of m in the T equation expresses not only the heterogeneity of the surfaces but also takes the interactions and the occasional mobility of the adsorbed layer into account.

This fact is also shown in Fig. 15 where at m values > 1 the T equation can be applied instead of the J equation.

The correctness of these model calculations shown in Figs 12–15 will be proven by the practical applications and calculations in Section V. The great advantage of the wide-ranging applicability of the T equation is its relatively simple form. The disadvantage is its limited applicability to isotherms of Type I.

IV. MULTILAYER ISOTHERM EQUATIONS APPLICABLE TO HETEROGENEOUS OXIDE SURFACES

A. The BET Equation and its Modified Forms

The derivation of the original BET equation can be found in several books and papers; therefore, only the well-known form of this relationship is presented here:

$$n^s = \frac{n^s_m c p_r}{[1 + (c - 1)p_r][1 - p_r]} \tag{135}$$

where p_r is the equilibrium relative pressure, n^s_m is the total monolayer capacity, and c is defined as the difference between the adsorptive potential (U_0) and the heat of condensation (λ) of the adsorptive:

$$RT \ln c = U_0 - \lambda \tag{136}$$

A relatively new representation of the BET equation is its separation into a mono- and multi-layer component isotherm [13]:

$$n^s = n^s_1(p_r) + \frac{n^s_1(p_r)p_r}{[1 - p_r]} \tag{137}$$

where $n_1^s(p_r)$ is the monolayer adsorption isotherm equal to the Langmuir equation:

$$n_1^s(p_r) = \frac{n_m^s c p_r}{1 + (c - 1)p_r} \tag{138}$$

The second expression in Eq. (137) describes only the multilayer adsorption. So it is evident that the total adsorbed amount (the overall multilayer adsorption isotherm) can be described as a sum of the monolayer adsorption and a multilayer one. This representation can be seen in Fig. 16.

It is also known that the BET theory is an oversimplificated picture of the multilayer adsorption for the following reasons: (1) the solid oxide surfaces are not homogeneous, (2) the interactions between molecules adsorbed are not taken into account, and (3) the heats of adsorption in the second and other layers are not equal to the negative heat of condensation. Despite these simplifications the BET equation (135) is widely used for calculation of the specific surface areas of absorbents. The possibility and reality of this calculation can be explained by the fact that the different simplifications of the theory are practically counteracting and so the extrapolation of Eq. (138) to $p_r \to 1$ yields an approximate real value of n_m^s. However, this possibility of calculation does not influence the very limited applicability of the BET equation in the range of equilibrium relative pressure (0.05–0.35). Because of this narrow range there are some modifications of the BET equation; one of the most important was first derived by Anderson [14]. The essence of Anderson's theory is based on the real assumption that the heats of adsorption in the second and higher layers are not equal to the heat of condensation:

$$RT \ln k = \lambda'/RT \tag{139}$$

where

$$Q_s = \lambda + \lambda' \tag{140}$$

In Eq. (140), Q_s is the heat of adsorption of the multilayer adsorption. Taking Eq. (139) into account the value of k appears in the original BET equations [(137) and (138)] [13]:

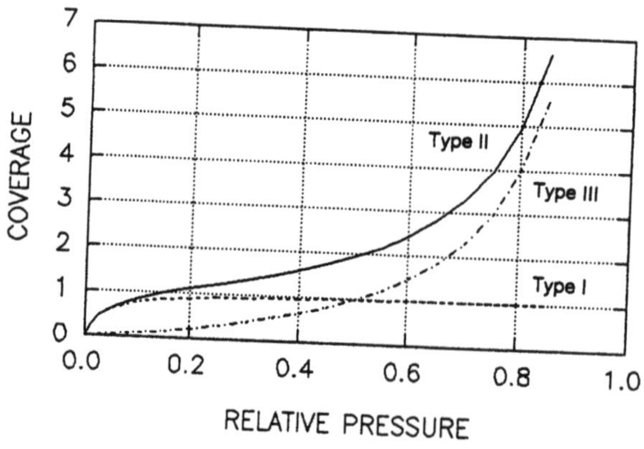

FIG. 16 According to the BET theory a multilayer isotherm of Type II is the sum of a monolayer isotherm of Type I (Langmuir isotherm) and a multilayer isotherm of Type III; $c = 30$.

$$n_1^s(p_r) = \frac{n_m^s c p_r}{1 + (c - k)p_r} \tag{141}$$

and

$$n^s = n_1^s(p_r) + \frac{n_1^s(p_r)kp_r}{(1 - kp_r)} \tag{142}$$

The upper limit of applicability of this modified BET equation is in the relative pressure range $\simeq 0.6$–0.8. In order to distinguish this modified form of the BET equation from the original one the name k-BET is proposed for Eq. (142).

Both the original and the k-BET equation suppose that the mono- and muilti-layer adsorption begins together and immediately when p_r is infinitesimal greater than zero. In contrast with this supposition there is experience to prove that isotherms of Types I and II exist having definite and pure monolayer parts, that is, the multilayer adsorption begins only at a definite relative pressure between 0.05–0.4. This fact essentially influences the value of the total monolayer capacity (n_m^s) calculable from the isotherms (see details in Section V). This is the reason why it is necessary to define such a form of Eq. (142) which includes the equilibrium relative pressure where the multilayer adsorption begins ($p_{r,e}$). It is easy to understand that this form is [13]:

$$n^s = n_1^s(p_r) + \frac{n_1^s(p_r)k\Delta p_r}{(1 - k\Delta p_r)} \tag{143}$$

$$n^s = \frac{n_1^s(p_r)}{1 - k\Delta p_r} \tag{144}$$

where

$$\Delta p_r = p_r - p_{r,e} \tag{145}$$

and it is always valid that $p_r > p_{r,e}$, i.e., Eq. (143) can only be applied to the *multilayer* part of the isotherm. The proposed name for this double-modified BET equation is $p_{r,e} - k -$ BET.

In the frame of the uniform interpretation of gas/solid adsorption it is necessary to calculate the function $\psi(p_r)$ of the BET equation for comparison with those of the monolayer isotherm equations. After application of Eqs (39) and (51) we have

$$\psi_{\mathrm{BET}}(p_r) = \frac{(1 - c)p_r^2 + (c - 2)p_r + 1}{(c - 1)p_r^2 + 1} \tag{146}$$

In the upper part of Fig. 17 a BET equation with its mono- and multi-layer component isotherms are shown and in the lower part its function $\psi(p_r)$ calculated by Eq. (146) is represented. It is obvious, but from a practical point of view it is very important, that the maximum value of Eq. (146) and the decreasing part of that exactly demonstrate the dominantly *multilayer* character of adsorption. This importance of function $\psi(p_r)$ relating to multilayer adsorption is discussed in detail in Section V.

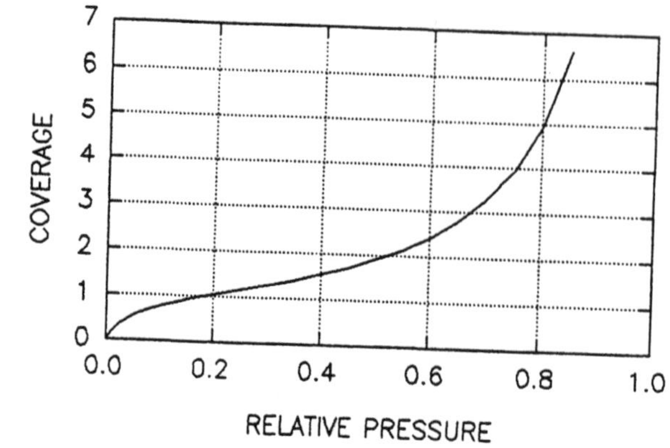

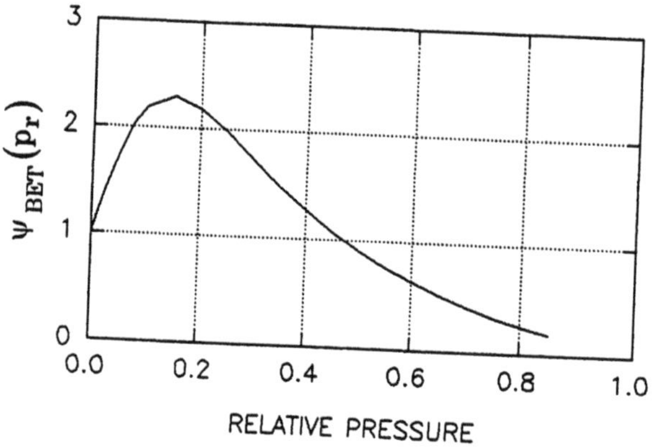

FIG. 17 The BET isotherm (upper diagram) and its function $\psi_{\mathrm{BET}}(p_r)$ (lower diagram); $c = 20$.

B. New Forms of Multilayer Isotherm Equations Applicable to Oxide Surfaces

The BET equation and its modified forms are applicable for calculation of the specific surface area (see Section V), but they cannot describe and explain the monolayer part of the isotherm existing *before* the beginning of the multilayer adsorption (the domain of equilibrium relative pressure $0 < p_r < p_{r,e}$).

The main reason for this fact is that the monolayer composite isotherms of the original and modified BET equations suppose a total homogeneous surface and this simplification can only be eliminated if the heterogeneity of the adsorbent is taken into account.

For that reason let us write the equation k-BET in a form directly following from Eq. (142):

$$n^s = \frac{n_1^s(p_r)}{1 - kp_r} \tag{147}$$

Equation (138) proves that $n_1^s(p_r)$ in Eq. (142) is the Langmuir equation. This fact provides the possibility of a phenomenological change in Eq. (147), namely, in this equation we can write – instead of the Langmuir equation – another monolayer equation applicable to heterogeneous surfaces. In this sense, $n_1^s(p_r)$ may be any of the consistent monolayer isotherm equations discussed in Section III. The only requirement is to write these monolayer equations with equilibrium *relative* pressures. This possibility is discussed in connection with Fig. 6 and with all figures where functions $\psi(p_r)$ are shown. As was shown in Section III the T equation (116) can express all the properties of the monolayer adsorption (heterogeneity, interactions, and occasional mobility or immobility); therefore, it is proposed to insert the T equation into Eq. (147), so we obtain:

$$n^s = \frac{n_m^s(\chi_T)^{1/m} p_r}{(1/K_{T,r} + p_r^m)^{1/m}} \frac{1}{1 - kp_r} \tag{148}$$

where $K_{T,r} = K_T p_0^m$ and $\chi_T = 1 + 1/K_{T,r}$.

The definition of $K_{T,r}$ and χ_T in Eq. (148) means that the upper limit of integration in Eq. (52) is equal to the saturation pressure, i.e., $p_m = p_0$ or $p_{r,m} = 1$ if Eq. (52) is written in relative pressures. This upper limit of integration corresponds to the supposition that – according to the BET and other *multilayer* theories – the total monolayer capacity (n_m^s) is completed at p_0. The proposed name for Eq. (148) is the TM equation. Where m refers to the multilayer adsorption.

Equation (148) can also be separated into mono- and multi-layer component isotherms:

$$n^s = n_1^s(p_r) + \frac{n_1^s(p_r) kp_r}{1 - kp_r} \tag{149}$$

or

$$n^s = \frac{n_1^s(p_r)}{1 - kp_r} \tag{150}$$

where, evidently

$$n_1^s(p_r) = \frac{n_m^s(\chi_T)^{1/m} p_r}{(1/K_{T,r} + p_r^m)^{1/m}}$$

In the frame of the uniform interpretation it is also necessary to calculate the function $\psi(p_r)$ of Eq. (149). So we have

$$\psi_{TM}(p_r) = \left\{ \frac{1}{1 + K_{T,r} p_r^m} + \frac{kp_r}{1 - kp_r} \right\}^{-1} \tag{151}$$

In the upper panel of Fig. 18 are shown the multilayer isotherm (149) and its component isotherms and in the lower panel is represented the function $\psi_{TM}(p_r)$.

V. PRACTICAL APPLICATIONS

Sections I–IV dealt with theoretical, thermodynamical, and mathematical problems of adsorption on heterogeneous surfaces. This last section includes only practical problems and so the readers of this chapter can be convinced of the usefulness of the theoretical considerations discussed above.

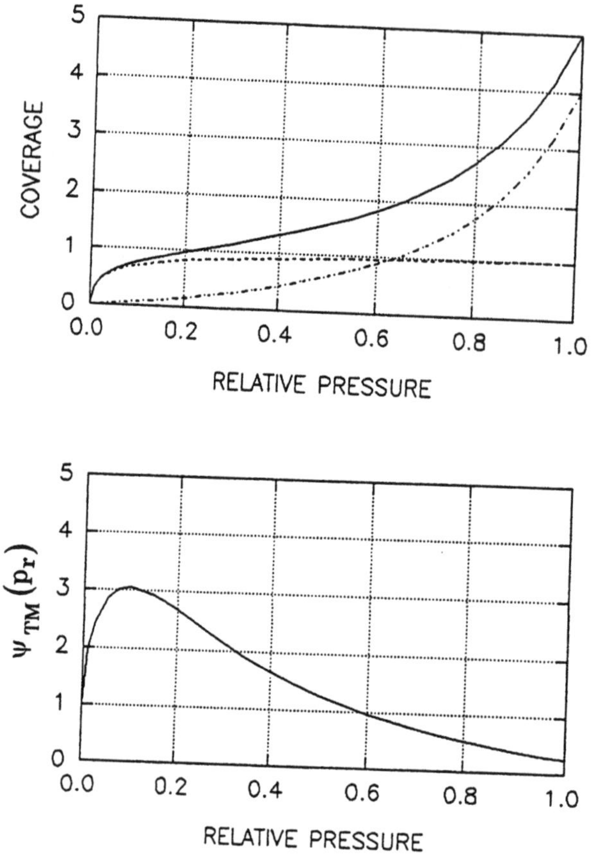

FIG. 18 The TM isotherm and its component isotherms (upper diagram) and the function $\psi_{TM}(p_r)$ (lower diagram); $m = 0.5$; $K_T = 10.0$; $k = 0.8$.

A. Selection of the Most Appropriate Isotherm Equation to Describe and Explain the Measured Isotherms of Types I and II

After measuring an adsorption isotherm on a heterogeneous surface the first question immediately emerges: which equation is the most appropriate to describe and explain the measured data? In most cases this selection is made without deeper consideration and with neglect of the real physical properties of the adsorbates. The uniform interpretation offers a more exact possibility for selecting the appropriate equation: it is proposed to calculate the function $\psi(p)$ of the measured isotherm. This calculation requires only the determination of the differential function (dn^s/dp) of the measured isotherm, which is a simple problem with help of a PC. The only requirement of this PC calculation is the sufficient number of measured data in a definite range of equilibrium pressures. (It is about six or seven data in every relative pressure steps 0.2.) The calculation of function $\psi(p)$ has two important advantages. The first one appears in the case of isotherms of Type I because they are very similar (all are concave from below); however, the functions $\psi(p)$ have very different shapes, so the selection of the appropriate equation is much easier and more exact than to do it by other methods. The second advantage is that the function $\psi(p)$ immediately provides approximate information on the change in free energy of the adsorbent surface and on the physical characteristics of the adsorption process.

So let us see the practical conclusions as results in the calculation of the function $\psi(p)$.

The function ψ(p) *is constant*, i.e., it is similar to functions shown in the lower part of Fig. 6. In this case the appropriate isotherm equation is the *Freundlich equation*; however, it must be kept in mind that the F equation does not have a concrete and real physical meaning (infinitely great heterogeneity?) and, therefore, in most cases the constant value of function $\psi(p)$ relates to the fact that the number of measured data are not sufficient to determine the exact physical character of adsorption. So in this case it is recommended to extend the measurements towards lower or greater equilibrium pressures than the measured ones.

The function ψ(p) *is linear but the initial (to zero extrapolated) value of* $\psi(p)$ *is greater than unity*. This function $\psi(p)$ corresponds to the modified and generalized Freundlich (GF) equation describing isotherms of Type I shown in the lower part of Fig. 8. Since this equation is semiconsistent, i.e., the condition of thermodynamic consistence defined by Eq. (48) is not met, one can make a choice from two possibilities. First, to extend the measurements toward lower equilibrium pressures (lower coverages) than the measured ones in order to obtain information on the location of the Henry region (is it in a measurable range of the coverage?). If the aim of the isotherm determination was the calculation of the specific surface area, then the second possibility can be chosen. Since the second condition of consistence, Eq. (46), is fulfilled the modified GF equation is immediately applicable to describe the measured isotherm and to calculate the total monolayer capacity. Furthermore, we obtain information on the heterogeneity of the adsorbents. The greater the value of n the greater the heterogeneity of the adsorbent, especially in the range of low coverages.

The function ψ(p) *is linear and its initial (to zero extrapolated) value is equal to unity*. This function $\psi(p)$ represents the original and modified Langmuir (ML) equation, that is, the adsorption takes place on a homogeneous surface. The ML equation is directly applicable to calculate the total monolayer capacity (see Fig. 8 when $n = 1$).

The function ψ(p) *is concave from below but its initial (to zero extrapolated) value is greater than unity*. In this case the modified Sips equation (105) can be directly applied to calculation of the total monolayer capacity because one of the conditions of consistence, Eq. (46), is met. This function $\psi(p)$ represents adsorption on heterogeneous surfaces, and the interactions between molecules adsorbed may be so great that the isotherms transform into Type III or V (see Fig. 9). The same transformation can also be observed with the FG equation (see Fig. 8).

The function ψ(p) *is convex from below and its initial (to zero extrapolated) value is equal to unity*. This type of function $\psi(p)$ represents the modified Jovanovic equation and extensive interactions between molecules adsorbed. Since the equation fulfils both conditions of thermodynamic consistence it can be applied to both the determination of the Henry section the isotherm and to the calculation of the total monolayer capacity (see Fig. 10).

The function ψ(p) *is concave from below and its initial (to zero extrapolated) value is equal to unity*. As shown in Fig. 11 this type of function $\psi(p)$ corresponds to the Tóth (T) equation, which describes monolayer adsorption on heterogeneous surfaces. Furthermore, taking Figs. 12–15 into account, the T equation can be applied to the following types of adsorption on *heterogeneous* surfaces: (1) localized, with interactions (T instead of FGT), (2) mobile, without interactions (T instead of VT), and (3) mobile, with interactions (T instead of BHT). These three types of adsorption are reflected in the values of m present in the T equation applied instead of the FGT, VT, and BHT equations. Let us nominate the values of m in these four equations as m_{T}, m_{FGT}, m_{VT}, and m_{BHT}. According to the calculations shown in Figs. 12–14 it is always valid that

$$m_{\mathrm{T}} > m_{\mathrm{FGT}} \tag{152}$$

$$m_{\mathrm{T}} < m_{\mathrm{VT}} \tag{153}$$

and

$$m_{\mathrm{T}} = m_{\mathrm{BHT}} \tag{154}$$

It means that, if these three equations are applied for the same isotherm having the functions $\psi_{\mathrm{T}}(p)$, then Eqs. (152)–(154) are always valid.

The function ψ(p) *has an increasing part and after a maximum value it decreases.* This type of function $\psi(p)$ (see lower part of Figs 17 and 18) always indicates that the isotherm has two parts. In the increasing section of function $\psi(p)$ the monolayer, and in the decreasing section the multilayer, adsorption is dominant. This type of function $\psi(p)$ has great importance in the calculation of real values of specific surface area (see Sections V.B and V.C).

The function ψ(p) *has only a decreasing part.* This type indicates that, in the whole measured equilibrium pressure range, the adsorption has a multilayer character.

To obtain a better understanding of the practical applicability of the eight types of function $\psi(p)$, relating to isotherms of Types I and II, they are summarized in Fig. 19.

B. Application of the Selected Isotherm Equation and Calculation of Specific Surface Area

Gas adsorption measurements are widely used for calculating the surface area of a variety of different solid materials, such as industrial adsorbents, catalysts, pigments, ceramics, and building materials. The measurement of adsorption at the gas/solid interface also forms an essential part of many fundamental and applied investigations of the nature and behavior of solid surfaces. After the theoretical discussion of these possibilities some concrete examples are presented here.

The calculation of the specific surface area should be taken in three steps: (1) determination of the gas (vapor) adsorption isotherm on the solid materials to be investigated, (2) calculation of the function $\psi(p)$ from the measured isotherm, and (3) based on the function $\psi(p)$ selection of the appropriate isotherm equation and the calculation of the specific surface area from this equation which has to include the value of the total monolayer capacity ($n^{\mathrm{s}}_{\mathrm{m}}$).

The first step is not discussed here because there are excellent books dealing with the measurement technique for adsorption isotherms (for example Ref. 1 pp. 29–103); furthermore, there is up-to-date automatic equipment for this purpose. The second step needs an accurate determination of the adsorption isotherm; how-

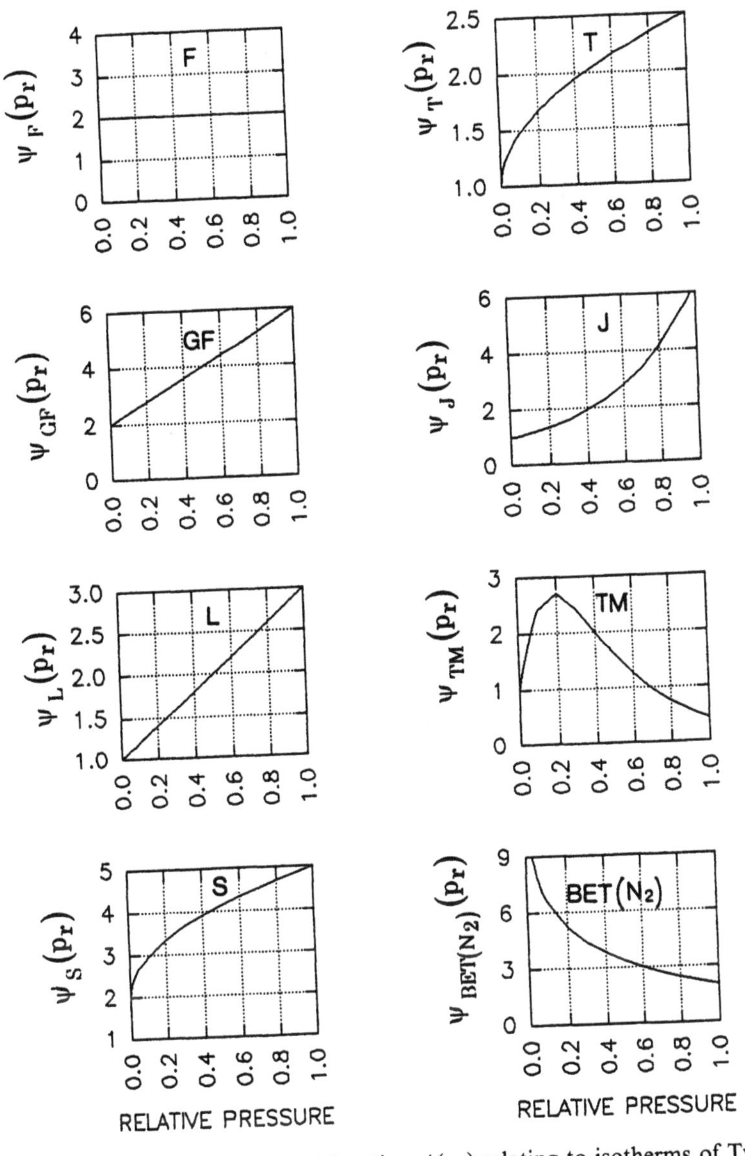

FIG. 19 The eight types of function $\psi(p_r)$ relating to isotherms of Types I and II.

ever, the accurate isotherms and the selected mathematical and thermodynamically consistent isotherm equations may provide more information on the adsorbents and on the physical character of adsorption than a simple nitrogen isotherm evaluated by the original BET equation.

The calculation of the specific surface area (a^s) from adsorption isotherms is based on the following relationship:

$$a^s = La_m n_m^s \tag{155}$$

where a_m is the molecular cross-sectional area occupied by the adsorbed molecule in the complete monolayer, and L is Avogadro's number. If n_m^s is expressed in mmol g^{-1} and a_m in nm^2, than we have from Eq. (155):

$$a^s(\text{m}^2\,\text{g}^{-1}) = 602.3 a_m n_m^s \tag{156}$$

In order to calculate the value of a_m it is assumed that the monolayer of molecules is a close-packed array and the molecules have spherical form. In this sense we have

$$a_m(\text{nm}^2) = 0.02\sqrt{3}\left(\frac{M\rho}{4\sqrt{2}L}\right)^{2/3} \tag{157}$$

where M is the molecular mass and ρ is the specific density of the *liquid* adsorbate expressed in g and $\text{cm}^3\,\text{g}^{-1}$, respectively. The values of a_m as a function of temperature (°C) calculated from Eq. (157) and relating to adsorptives, mentioned as examples in this chapter, are shown in Fig. 20.

In most cases the total monolayer capacity can be calculated from isotherms of Types I and II; however, in recent years a radical change has been taking place in the interpretation of the isotherms of Type I for porous solids. According to the classical Langmuir theory, the limiting adsorption (at the plateau) represents completion of the monolayer and may therefore be used for the calculation of the surface area. The alternative view, which is now widely accepted and also proven in this chapter, is that the initial (steep) part of the Type I isotherms represents monolayer adsorption or micropore filling and that the low slope of the plateau is due to multilayer adsorption on the small external area. In the following subsection it is shown how this multilayer section of Type I isotherms can be determined and taken into account by calculation of the specific surface area. Since the micropore filling does not represent exactly the surface coverage it is suggested that the term monolayer equivalent area should be applied to microporous solids.

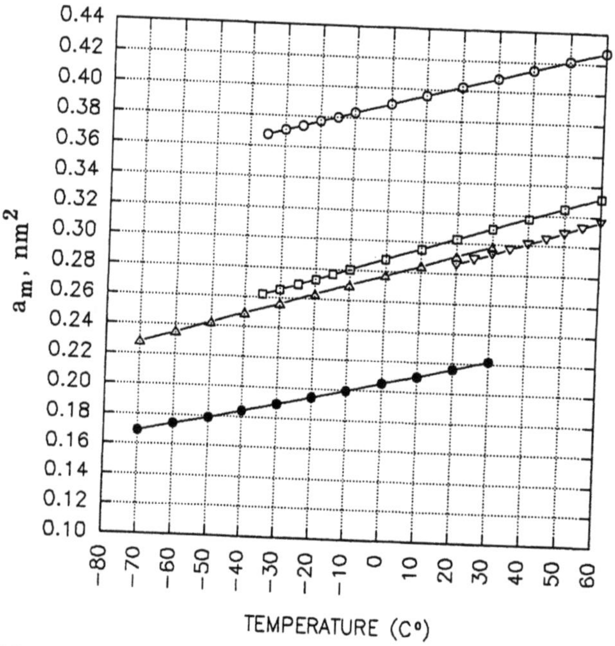

FIG. 20 Area occupied by one molecule (nm^2) as function of temperature calculated from Eq. (157). Carbon dioxide: ●; ethane: △; propylene: ▽; propane: □; *n*-butane: ○.

1. Calculation of Specific Surface Area from Isotherms of Type I with a Multilayer Plateau

In Section V.A it has been shown that a monolayer isotherm measured on a heterogeneous surface has always an increasing function $\psi(p_r)$ which, in most cases, can be described by the Tóth equation. It was also shown that in the case of multilayer adsorption the function $\psi(p_r)$ always has a decreasing domain which may begin at a relative pressure greater than 0.02. The consequence of the two above-mentioned facts is evident: if function $\psi(p_r)$ shows a decreasing part, then the adsorption has become a multilayer phenomenon. The transformation of a monolayer adsorption into a multilayer one can be seen in Figs 21 and 22.

In Fig. 21 the ethane isotherm measured on BPL-activated carbon ($a^s = 988\,\mathrm{m^2\,g^{-1}}$) at 28.2°C is represented. (For more details about BPL-activated carbon see Ref. 8 p. 78.) In the lower part of this figure the corresponding function $\psi(p_r)$ is shown, which has only an increasing character according to the T equation. (The accuracy in describing the measured data by the T equation can be seen in a handbook [8].) In Fig. 22 can be seen the ethane isotherm measured on the same adsorbent, but at much lower temperature (−60.5°C) [8]. The corresponding function $\psi(p_r)$ has a definite decreasing domain, indicating multilayer adsorption at the

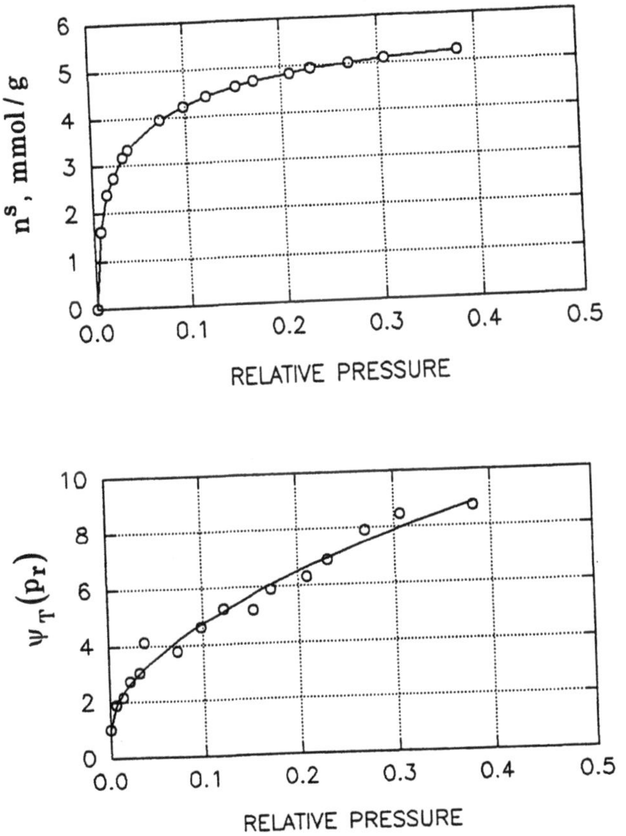

FIG. 21 Ethane isotherm measured on BPL-activated carbon at 28.2°C (upper diagram) and the corresponding function $\psi_T(p_r)$ (lower diagram). (From Ref. 8.)

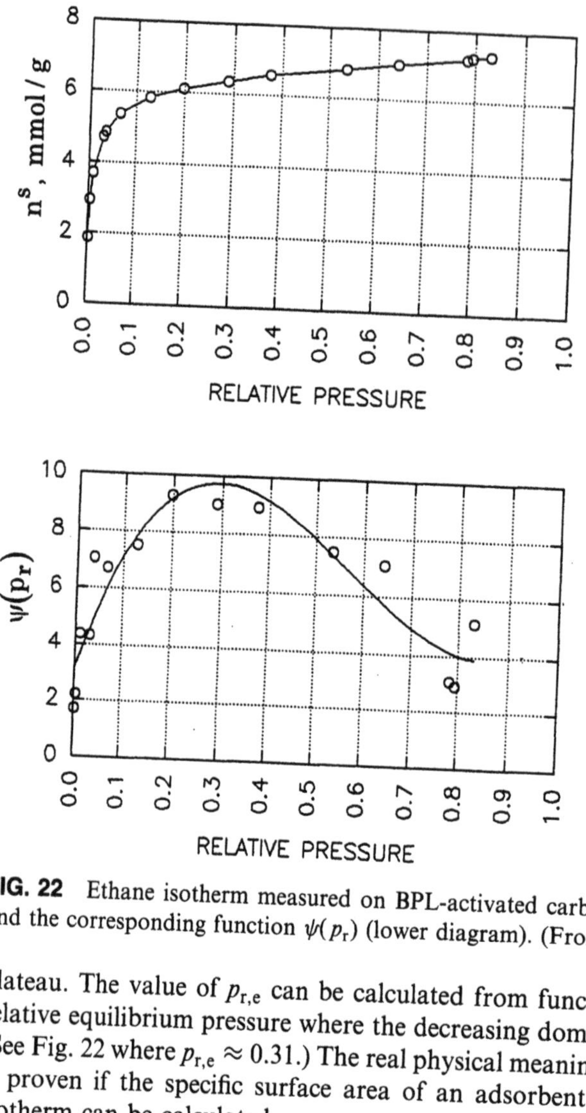

FIG. 22 Ethane isotherm measured on BPL-activated carbon at −60.5°C (upper diagram) and the corresponding function $\psi(p_r)$ (lower diagram). (From Ref. 8.)

plateau. The value of $p_{r,e}$ can be calculated from function $\psi(p_r)$. It is equal to the relative equilibrium pressure where the decreasing domain of function $\psi(p_r)$ begins. (See Fig. 22 where $p_{r,e} \approx 0.31$.) The real physical meaning of equilibrium pressure $p_{r,e}$ is proven if the specific surface area of an adsorbent from three domains of the isotherm can be calculated:

1. From the monolayer part of the isotherm, applying one of the consistent monolayer isotherm equations.
2. From the whole measured isotherm (including mono- and multi-layer parts), applying the TM equation (148).
3. From the multilayer part of the isotherm, applying the original or one of the modified BET equations.

All three values of a^s, within calculation errors, must be identical supposing that the selection of the applied isotherm equations has been made correctly.

The fulfilment of these conditions are shown in Table 2 where the applied equations, their parameters, the calculated values of a^s, and the average accuracy of application (δ) are shown. These last values have been calculated from the relationship:

TABLE 2 Parameters of the T, TM, *k*-BET, and $p_{r,e}$-*k*-BET Equations Applied to the Ethane Isotherm Measured on BPL-activated Carbon at 213 K.* Surface Area is also Calculated from the BET(N_2) Equation [$p_{r,e} = 0.31$; a_m(Ethane) $= 0.235\,nm^2$; a_m(Nitrogen) $= 0.162\,nm^2$]

Equation	p_r domain of application	n^s_m (mmol g^{-1})	$K_{T,r}$	m	χ_T	k	δ(%)	a^s ($m^2\,g^{-1}$) calc.	Deviation from a^s calculated from the BET(N_2) equation (%)
T	0.009–$p_{r,e}$ monolayer	7.0202	14.3675	0.3223	1.0696	0.0000	0.11	994	0.61
TM	0.009–0.83 all data, mono- and multilayers	0.6099	21.1019	0.4198	1.04746	0.1223	0.22	936	−5.3
			C						
k-BET	$p_{r,e}$–0.83 multilayer	6.3993	59.545			0.1671	0.14	906	−8.6
$p_{r,e}$-*k*-BET	$p_{r,e}$–0.83 multilayer	6.7253	62.572			0.1756	0.14	952	−3.6
BET(N_2)	According to the producers' measurements							988	0.0

**Source*: Ref. 8.

Applied equations:

T	$n^s_1(p_r) = n^s_m(\chi_T)^{1/m} p_r/(1/K_{T,r} + p_r^m)^{1/m}$	in Eq. (150)
TM	$n^s = n^s_m(\chi_T)^{1/m} p_r/\left\{(1/K_{T,r} + p_r^m)^{1/m}(1 - kp_r)\right\}$	Eq. (148)
k-BET	$n^s = n^s_m c p_r/\{[1 + (c - k)p_r][1 - kp_r]\}$	Eqs. (141–142)
$p_{r,e}$-k-BET	$n^s = n^s_m c p_r/\{[1 + p_{r,e} + (c - k)p_r][1 - k(p_r - p_{r,e})]\}, \quad p_r > p_{r,e}$	Eqs. (143–145)
BET(N_2)	$n^s = n^s_m c p_r/\{[1 + (c - 1)p_r][1 - p_r]\}$ only for N_2 isotherms at 77.3 K!	Eq. (135)

$$\delta = \left\{ \sum_{i'=1}^{N} \left| \left(\frac{n^s - n_c^s}{n_c^s} \right) 100 \right| \right\} \frac{1}{N} \tag{158}$$

where n^s is the measured, n_c^s is the calculated values of the adsorbed amounts, and N is the number of the measured data. From values of a^s shown in Table 2 the following conclusions can be drawn:

1. Application of the T equation to the monolayer domain of the ethane isotherm yields a value of a^s equal to that calculated by the equation $BET(N_2)$. The term $BET(N_2)$ indicates the original BET equation applied to the *nitrogen* isotherm measured at 77.3 K. Evidently, all other equations relate to the ethane isotherm in question.
2. The TM equation applied to the *whole* measured isotherm and the $p_{r,e}$-k-BET equation applied only to the *multilayer* section of the isotherm also yield approximately equal values of a^s, and the deviation of these from the a^s value calculated from the $BET(N_2)$ equation is less than 6.0%.
3. The k-BET equation applied to the multilayer domain yields the greatest a^s deviation from the $BET(N_2)$ value, but this deviation is also less than 10%.
4. It is evident that, based on Eq. (149), both the monolayer and the multilayer component isotherms can be calculated. The sum of these two component isotherms is equal to the measured original isotherm. These component isotherms can be seen in Fig. 23.

Since the value of $p_{r,e}$ can be determined from function $\psi(p_r)$, application of the isotherm equations presented in Table 2 needs three-parameter (n_m^s, c, k) fitting procedures, except the TM equation which needs a two-step four-parameter (n_m^s, $K_{T,r}$, m, k) fitting procedure.

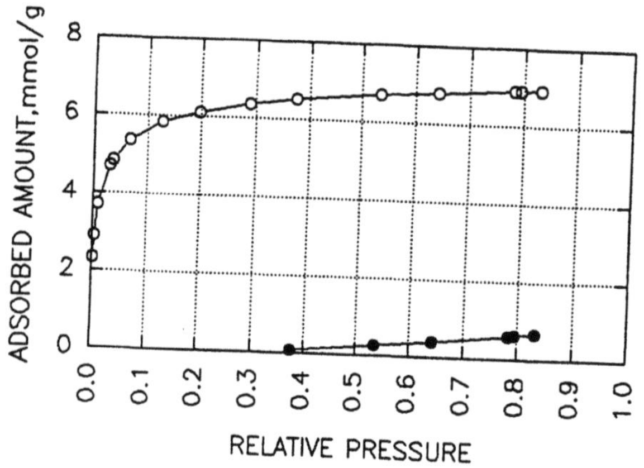

FIG. 23 Component isotherms of ethane isotherm measured on BPL-activated carbon at −60.5°C. The monolayer isotherm (○) is calculated from the T equation (116); the multilayer component (●) is calculated from Eq. (143) when $p_r > p_{r,e}$. (For parameters see Table 2.)

2. Calculation of Specific Surface Area from Isotherms of Type II Measured on Oxide Surfaces

The isotherms of type II can also be divided into two groups. To the first group belong isotherms having a definite monolayer domain, i.e., $p_{r,e}$ can be determined from the function $\psi(p_r)$, and in the second group are isotherms without monolayer parts, that is, the function $\psi(p_r)$ has only a decreasing character.

First, let us investigate the isotherms of Type II with a monolayer domain. For this aim it is very suitable to measure argon and oxygen isotherms on rutile at 85 K [15]. In order to follow and control the calculations discussed below the measured data for the argon isotherm are presented in Table 3. The first step is the calculation of the function $\psi(p_r)$, which is shown in Fig. 24.

From the enlarged part of function $\psi(p_r)$ (see the lower graph in Fig. 24) it can be seen that the monolayer domain is in the equilibrium relative pressure range 0–0.045, that is, $p_{r,e} = 0.045$, and in this domain the monolayer T equation can be applied. The equations in Table 4 are applied to the multilayer domain of the argon

TABLE 3 Measured Data for Argon Isotherms Determined on Rutile at 85 K

Equilibrium relative pressure	Adsorbed amount ($mmol\,g^{-1}$)
2.196×10^{-5}	0.0359
5.405×10^{-5}	0.0643
1.588×10^{-4}	0.1136
5.507×10^{-4}	0.1947
6.503×10^{-4}	0.2077
8.108×10^{-4}	0.2291
1.892×10^{-3}	0.3167
3.311×10^{-3}	0.3825
4.324×10^{-3}	0.4233
7.179×10^{-3}	0.5005
7.939×10^{-3}	0.5126
0.0136	0.6000
0.0182	0.6409
0.0332	0.7233
0.0437	0.7616
0.1137	0.9069
0.2127	1.0693
0.3129	1.2436
0.3725	1.3573
0.4838	1.5827
0.5789	1.8208
0.6515	2.0355
0.7008	2.2217
0.7389	2.4052
0.7703	2.6040
0.7965	2.7936

Source: Ref. 15.

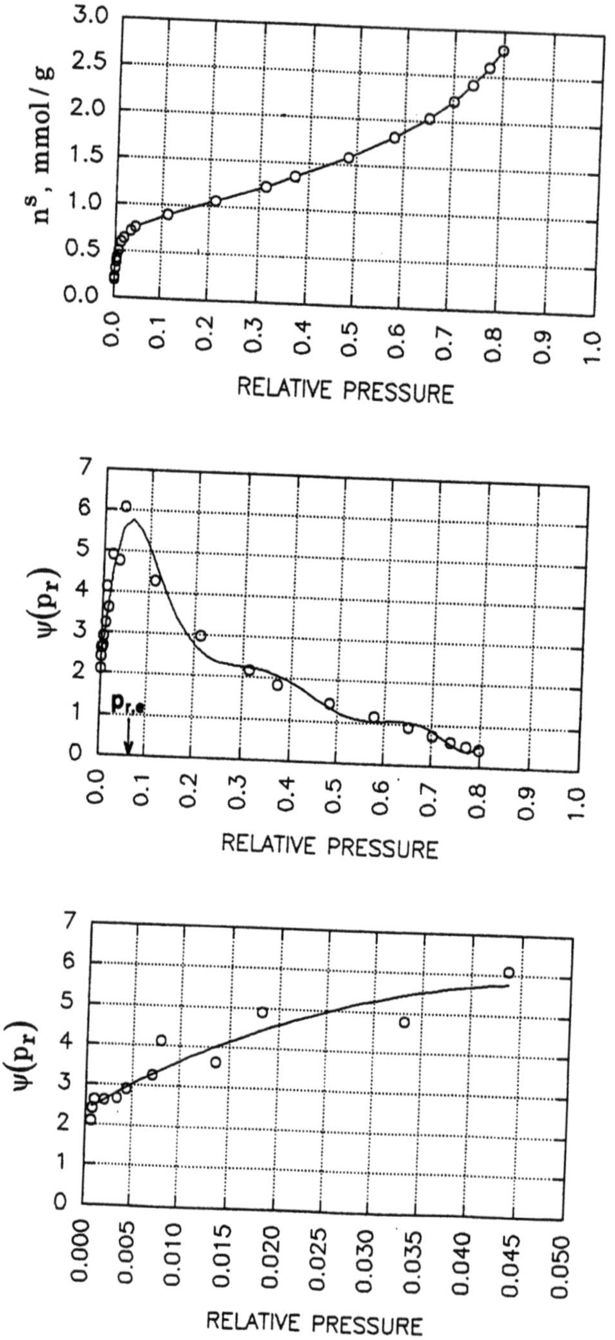

FIG. 24 Argon isotherm measured on rutile at 85 K (upper diagrams), the corresponding function $\psi(p_r)$, and its enlarged part in equilibrium relative pressure domain of 0–0.044 (lower diagrams).

TABLE 4 Parameters of the T, TM, k-BET, $p_{r,e}$-k-BET and BET(Ar) Equations Applied to the Argon Isotherm Measured on Rutile at 85 K.* Surface Area is also Calculated from the BET(N_2) Equation [$p_{r,e} = 0.045$; a_m(Argon) $= 0.142\ nm^2$; a_m(Nitrogen) $= 0.162\ nm^2$ (77.3 K)]

Equation	p_r domain of application	n^s_m (mmol g^{-1})	$K_{T,r}$	m	χ_T	k	δ(%)	a^s ($m^2\ g^{-1}$) calc.	Deviation from a^s calculated from the BET(N_2) equation (%)
T	0–0.045 monolayer	1.1579	10.7565	0.2983	1.0930	0.0000	0.98	99	12.5
TM	0.0007–0.80 mono- and multilayer	0.9942	18.1232	0.3927	1.0552	0.8091	0.89	85	−3.4
				C					
k-BET	$p_{r,e}$–0.80 multilayer	0.9710		44.541		0.8189	0.78	83	−5.7
$p_{r,e}$-k-BET	$p_{r,e}$–0.80 multilayer	1.0082		46.245		0.8502	0.78	86	−2.3
BET(Ar)	$p_{r,e}$–0.37 multilayer	0.8704		97.283		1.0000	0.37	74	−15.9
BET(N_2)	0.05–0.323 multilayer	0.9050		135.70		1.0000	1.10	88	0.0

**Source*: Ref. 15.

isotherm. The oxygen isotherm, also measured on rutile at 85 K [15], provides more information on the applicability of different equations to different (mono- and multi-layer) domains of Type II isotherms.

In Fig. 25 the oxygen isotherm and its function $\psi(p_r)$ are represented, and it is also shown that a monolayer domain exists and its upper equilibrium relative pressure limit is $p_{r,e} = 0.058$. According to this fact the applied equations, their parameters, and the calculated values of a^s are summarized in Table 5.

These data also prove that there are no essential differences between the values of a^s calculated from the mono- and multi-layer parts of the oxygen isotherm measured on rutile at 85 K. Exceptions are the k-BET and BET(O_2) relationships, but in these cases the deviation from the value a^s calculated by the BET(N_2) equation also does not exceed 15%.

Silica gel has an order of magnitude greater specific surface area than that of rutile; therefore, it is very instructive to evaluate the nitrogen, oxygen, and argon isotherms measured on this adsorbent at different temperatures [16]. The nitrogen

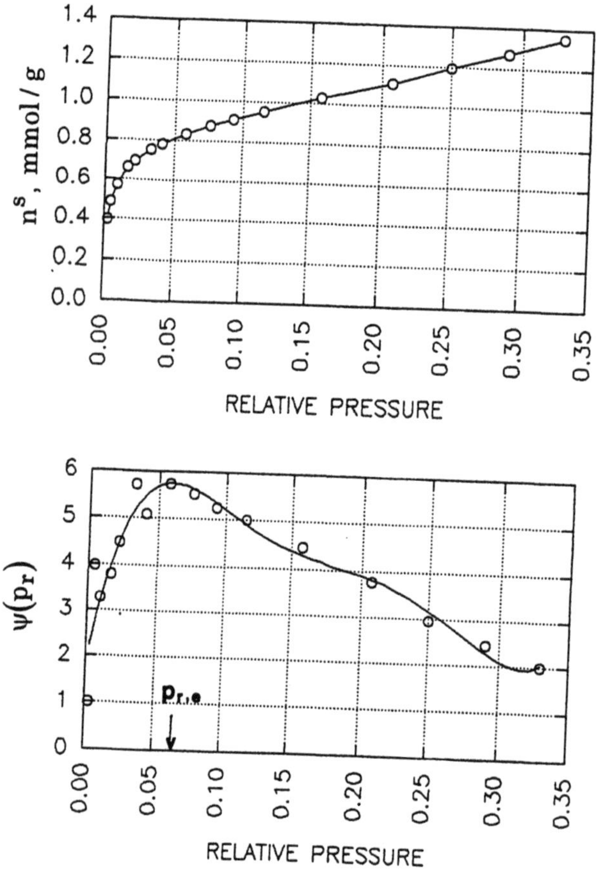

FIG. 25 Oxygen isotherm measured on rutile at 85 K (upper diagram) and the corresponding function $\psi(p_r)$ (loweer diagram). (From Ref. 15.)

TABLE 5 Parameters of the T, TM, k-BET, $p_{r,e}$-k-BET and BET(O_2) Equations Applied to the Oxygen Isotherm Measured on Rutile at 85 K.* Surface Area is also Calculated from the BET(N_2) Equation [$p_{r,e} = 0.058$; a_m(Oxygen) $= 0.139\,nm^2$; a_m(Nitrogen) $= 0.162\,nm^2$]

Equation	p_r domain of application	n_m^s (mmol g^{-1})	$K_{T,r}$	m	χ_T	k	δ(%)	a^s ($m^2\,g^{-1}$) calc.	Deviation from a^s calculated from the BET(N_2) equation (%)
T	0.0026–0.058 monolayer	1.1049	15.0161	0.3422	1.0666	0.0000	0.32	93	5.7
TM	0.0026–0.33 all data, mono- and multilayer	0.9560	27.3841	0.4536	1.0365	0.9938	0.20	80	−9.1
				C					
k-BET	$p_{r,e}$–0.33 multilayer	0.9002		103.53		1.0535	0.27	75	−14.8
$p_{r,e}$-k-BET	$p_{r,e}$–0.33 multilayer	0.9588		110.27		1.1221	0.27	80	−9.1
BET(O_2)	$p_{r,e}$–0.33 multilayer	0.9238		85.555		1.0000	0.46	77	−12.5
BET(N_2)	0.05–0.33 multilayer	0.9050		135.70		1.0000	1.10	88	0.0

**Source*: Ref. 15.

isotherm and its function $\psi(p_r)$ can be seen in Fig. 26, and the parameters of the applied equations are presented in Table 6.

The function $\psi(p_r)$ of the nitrogen isotherm proves that the isotherm does not have a monolayer domain (or a possible monolayer domain cannot be measured in the very low equilibrium relative pressure range). We have evaluated many nitrogen isotherms measured on very different solid adsorbents at 77.3 K. All these isotherms have functions $\psi(p_r)$ *without* any monolayer domain. In our opinion the value of $p_{r,e} \approx 0$ of nitrogen isotherms measured at 77.3 K, i.e., the fact that these nitrogen isotherms do not have a monolayer domain (or these domains are limited to very low equilibrium pressure range) is the main reason for the practical applicability of the BET(N_2) equation to the calculation of specific surface area.

This statement is supported by data for oxygen and argon isotherms measured on the same silica gel. As can be seen in Figs 27 and 28 these isotherms have measurable – identifiable from the functions $\psi(p_r)$ – monolayer domains, but there are not a sufficient number of data for a fitting procedure needed to calculate the parameters of a monolayer isotherm equation.

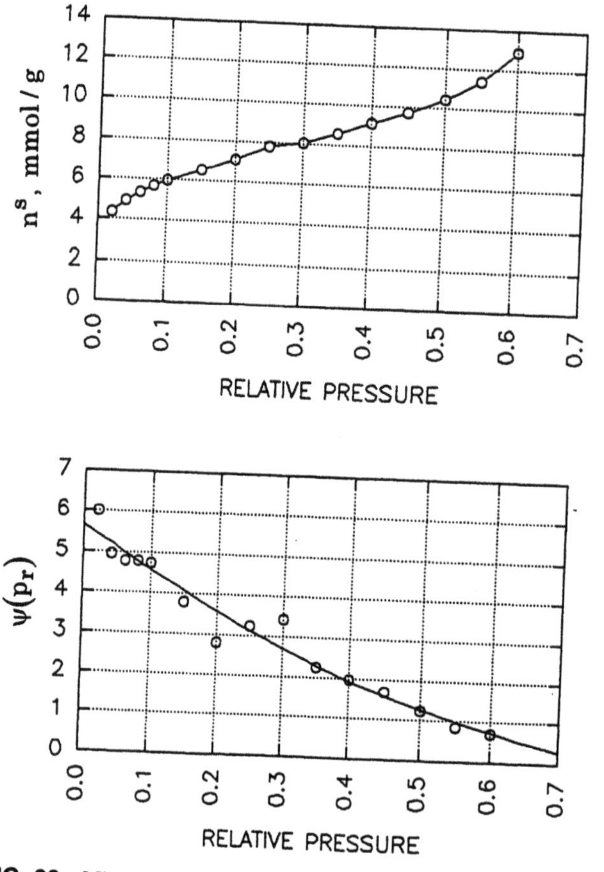

FIG. 26 Nitrogen isotherm measured on silica gel at 77.3 K and the corresponding function $\psi(p_r)$ (lower diagram). (From Ref. 16.)

TABLE 6 Parameters of the TM, k-BET, and BET (N_2) Equations Applied to the Nitrogen Isotherm Measured on Silica Gel at 77.3 K ($a_m = 0.162\,nm^2$; $p_{r,e} = 0.0$)

Equation	p_r domain of application	n_m^s ($mmol\,g^{-1}$)	$K_{T,r}$	m	χ_T	k	δ(%)	a^s ($m^2\,g^{-1}$) calc.	Deviation from a^s calculated from the BET(N_2) equation (%)
T	0.02–0.60 all data, multilayer	6.2260	37.7620	0.6337	1.0265	0.8567	1.37	607	5.6
				C					
k-BET	0.02–0.60 all data, multilayer	5.9825		106.453		0.8858	1.85	584	1.6
BET(N_2)	0.04–0.30 multilayer	5.8894		95.078		1.0000	0.87	575	0.0

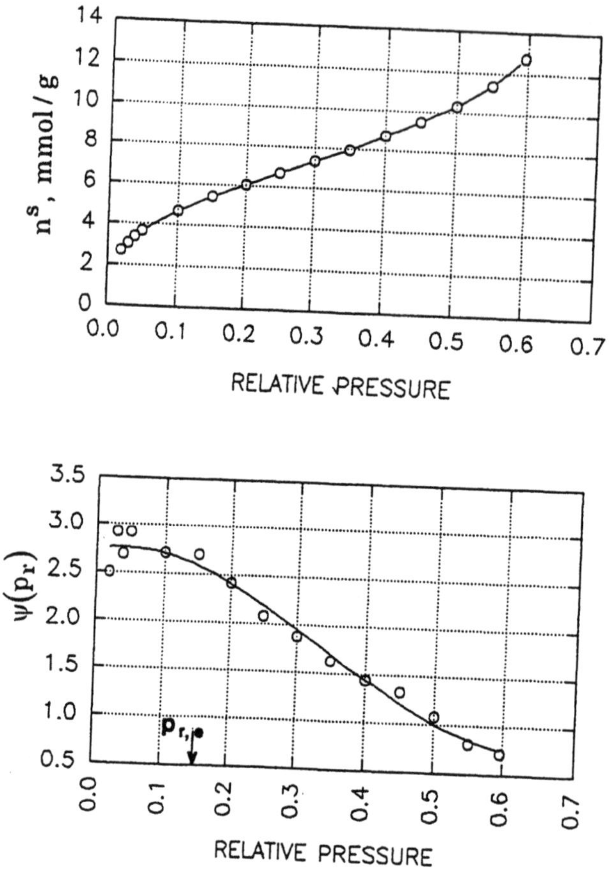

FIG. 27 Oxygen isotherm measured on silica gel at 90.2 K and the corresponding function $\psi(p_r)$ (lower diagram).

This is the reason why the parameters of the T equation and the value of a^s calculated from this relationship are not presented in Tables 7 and 8. However, the $p_{r,e}$ values have an important role in the application of the $p_{r,e}$-k-BET equation and they have the same role in the calculation of the a^s values approximately equal to those calculated from the BET(N_2) relationship.

In Fig. 29 are shown the deviations in the values of a^s calculated with the help of the isotherm equations presented in Tables 4–8 from the value of a^s calculated by the widely used and accepted BET(N_2) equation. In Table 9 are summarized the p_r domains of applicability of the different isotherm equations discussed in this section.

Taking Fig. 29 and Table 9 into account it can be stated that the T, TM, and $p_{r,e}$-k-BET equations are also applicable to the calculation of real specific surface areas; however, their p_r domain of applicability is much greater than that of the BET(N_2) equation.

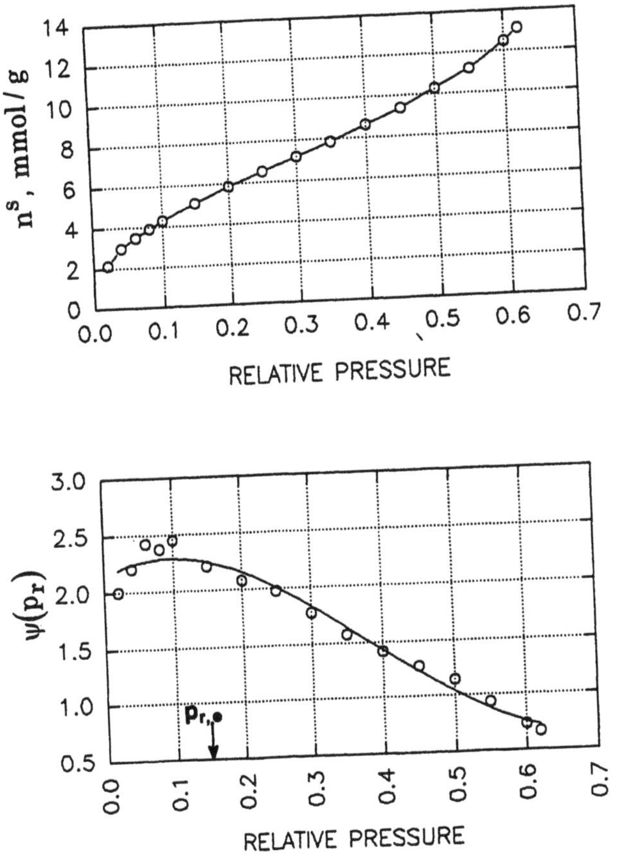

FIG. 28 Argon isotherm measured on silica gel at 88.2 K and the corresponding function $\psi(p_r)$ (lower diagram).

C. Summarizing Statements Referring to Calculations of Specific Surface Area

In Section V, concrete examples have demonstrated that the calculation of specific surface area includes some new or modified ideas and based on these the calculation methods for a^s from different isotherm equations can be widened. These ideas are summarized as follows:

1. In all evaluations concerning the measured adsorptive isotherms the first step is the calculation of function $\psi(p_r)$ with the help of a PC.
2. The fucntion $\psi(p_r)$ provides information on the mono- and multi-layer domains of the measured isotherm and also gives exact directions for choosing the physically and mathematically best applicable isotherm equations.
3. From function $\psi(p_r)$ can be determined the equilibrium relative pressure, $p_{r,e}$, where multilayer adsorption begins and takes place in parallel with monolayer adsorption.

TABLE 7 Parameters of the TM, k-BET, $p_{r,e}$-k-BET, and BET(O_2) Equations Applied to the Oxygen Isotherm Measured on Silica Gel at 90.2 K. Surface Area is also Calculated from the BET(N_2) Equation [$p_{r,e} = 0.15$; a_m(Oxygen) $= 0.141\ nm^2$; a_m(Nitrogen) $= 0.162\ nm^2$ (77.3 K)]

Equation	p_r domain of application	n_m^s (mmol g^{-1})	$K_{T,r}$	m	χ_T	k	δ(%)	a^s ($m^2\ g^{-1}$) calc.	Deviation from a^s calculated from the BET(N_2) equation (%)
TM	0.02–0.60 all data, mono- and multilayer	6.1733	9.3611	0.4686	1.10683	0.902	0.68	524	−8.9
				C					
BET(O_2)	$p_{r,e}$–0.30 multilayer	5.5658		27.015		1.000	0.15	473	−17.7
k-BET	$p_{r,e}$–0.60 multilayer	5.7706		24.070		0.9363	0.45	490	−14.8
$p_{r,e}$-k-BET	$p_{r,e}$–0.60 multilayer	6.7134		28.004		1.0893	0.45	570	−0.9
BET(N_2)	0.04–0.30 multilayer	5.8894		5.078		1.000	0.87	575	0.0

TABLE 8 Parameters of the TM, k-BET, $p_{r,e}$-k-BET, and BET(Ar) Equations Applied to the Argon Isotherm Measured on Silica Gel at 90.2 K. Surface Area is also Calculated from the BET(N_2) Equation [$p_{r,e} = 0.15$; a_m(Argon) $= 0.144\,nm^2$; a_m(Nitrogen) $= 0.162\,nm^2$ (77.3 K)]

Equation	p_r domain of application	n_m^s (mmol g^{-1})	$K_{T,r}$	m	χ_T	k	δ(%)	a^s ($m^2\,g^{-1}$) calc.	Deviation from a^s calculated from BET(N_2) equation (%)
TM	0.02–0.60 all data, mono-and multilayer	6.6224	5.9857	0.4424	1.1671	0.85002	0.51	574	0.2
				C					
BET(Ar)	$p_{r,e}$–0.30 multilayer	5.6498		19.150		1.000	0.24	490	−14.8
k-BET	$p_{r,e}$–0.62 multilayer	6.0917		15.490		0.8915	0.25	528	−8.2
$p_{r,e}$-k-BET	$p_{r,e}$–0.62 multilayer	7.0320		17.890		1.0291	0.25	610	6.1
BET(N_2)	0.04–0.30 multilayer	5.8894		95.078		1.000	0.87	575	0.0

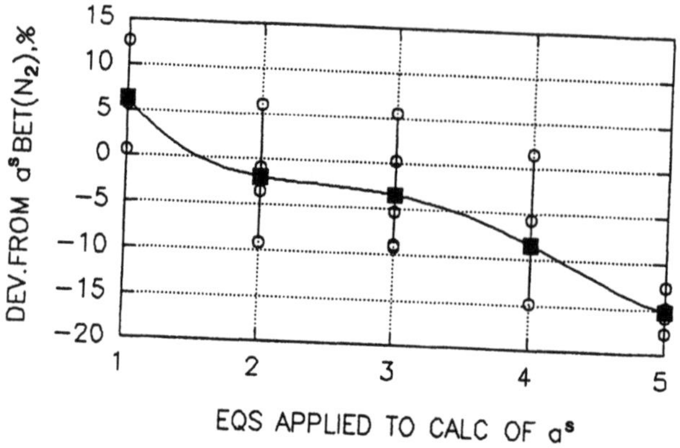

FIG. 29 Deviations of a^s values calculated by the BET(N_2) equation from a^s values calculated by other equations: 1: T; 2: $p_{r,e}$-k-BET; 3: TM; 4: k-BET; 5: BET.

4. From pure monolayer isotherms of Type I and from the monolayer domain of Type II isotherms the specific surface area can be calculated from a monolayer equation identified by the shape of the function $\psi(p_r)$.
5. From pure multilayer isotherms of Type II and from the multilayer domain of isotherms of Type I or II the specific surface area can be calculated from one of the modified BET equations.
6. The modified BET equations take into account that the adsorptive heat of the second and other layers is not equal to the condensation heat and that the equilibrium pressure $p_{r,e}$ exists.
7. The original BET equation can be applied for calculation of the specific surface area on conditions that:
   a. The adsorptive is nitrogen.
   b. The temperature is 77.3 K; therefore, the cross-sectional area occupied by one molecule is equal to 0.162 nm^2.
   c. The function $\psi(p_r)$ proves that multilayer adsorption takes place in the p_r domain 0.05–0.3, where the BET equation can be fitted to the measured data with an average accuracy of 1–2%.
8. The statements in points 1–6 make possible calculation of the specific surface area from suitable equations applied to different adsorptives at different temperatures. These calculated values of a^s are – within a deviation of ±5% – identical to those calculated from the BET(N_2) equation. It means that, in order to calculate the specific surface area, we are not obliged to measure only nitrogen isotherms at 77.3 K.
9. The statement in point 8 also means that we are not forced to apply arbitrary adjustment of the cross-sectional area of different molecules (CO_2, C_2H_6, Ar, O_2, etc.) in order to bring the surface areas into agreement with the BET(N_2) value. In our opinion this "adjustment" is the value of $p_{r,e}$ having a real and proved physical meaning.
10. The uniform interpretation of physical adsorption has a determining role in these statements. Namely, without integration of differential equations (52)–(53) with upper limit $p_m = p_0$ the monolayer equations discussed in this chapter do not include the values of χ which assure realistic values of n_m^s applicable to the calculation of realistic values of a^s.

TABLE 9 Equilibrium Relative Pressure Domains of Applicability of Different Isotherm Equations Applied to Isotherms of Types I and II and Average Deviations of a^s from a^s Calculated with the BET(N_2) Equation

Equation	p_r-domain of applicability	Average accuracy of application, δ (%)	Average deviation of a^s from a^s calculated from the BET(N_2) equation see (Fig. 29) (%)
T	$0 < p_r \leq 1$ for pure monolayer isotherms, $0 < p_r \leq p_{r,e}$ for monolayer domain of isotherms of Types I and II	< 1	6.3
TM	$0 < p_r < 0.8$–0.9 for mono + multilayer domains of isotherms	1–2	−3.5
k-BET	$p_{r,e} < p_r < 0.6-0.7$ for multilayer domain	1–2	−8.4
$p_{r,e}$-*k*-BET	$p_{r,e} < p_r < 0.7$–0.8 for multilayer domain	1–2	−2.0
BET	$p_{r,e} < p_r < 0.3$ for multilayer domain	1–2	−15.2
BET(N_2)	$0.05 < p_r < 0.3$ for multilayer domain and only for nitrogen isotherms measured at 77.3 K	1–2	0.0

REFERENCES

1. S Ross, JP Olivier. On Physical Adsorption. New York: Interscience, 1964, p 121.
2. JH deBoer. The Dynamical Character of Adsorption. Oxford: Clarendon Press, 1953, p 55.
3. PG Menon. J Am Chem Soc 87:3057, 1965.
4. PG Menon. Chem Rev 68:277, 1968.
5. Y Wakasugi, S Ozava, Y Ogino. J Colloid Interface Sci 79:399, 1981.
6. P Malbrunot, D Vidal, I Vermesse. Langmuir 8:577, 1992.
7. I Vernesse, D Levesque. J Chem Phys 101:9063, 1994.
8. DP Valenzuela, AL Myers. Adsorption Equilibrium Data Handbook. Englewood Cliffs, NJ: Prentice-Hall, 1989.
9. Z Király, I Dékány. Fundamentals of adsorption. Proceedings of the Third International Conference, United Engineering Trustees, New York, 1989, pp 425–435.
10. JR Sips. J Chem Phys 19:1024, 1950.
11. DS Jovanovic. Colloid Polym Sci 42:552, 1974.
12. J Tóth. Acta Chim Acad Sci Hung 30:415, 1962; 31:393, 1962; 32:39, 1962; 33:153, 1962; 69:311, 1971.
13. J Tóth. Acta Chim Acad Sci Hung 48:27, 1966.
14. RB Anderson. J Am Chem Soc 68:686, 1946.
15. LT Drain, JA Morrison. Trans Faraday Soc 48:840, 1952.
16. S Brunauer, PH Emmett. J Am Chem Soc 10:2682, 1937.

Index

Index

Index

Index